# Cubical Homotopy Theory

Graduate students and researchers alike will benefit from this treatment of various classical and modern topics in the homotopy theory of topological spaces, with an emphasis on cubical diagrams. The book contains more than 300 examples and provides detailed explanations of many fundamental results.

Part I focuses on foundational material on homotopy theory, viewed through the lens of cubical diagrams: fibrations and cofibrations, homotopy pullbacks and pushouts, and the Blakers–Massey Theorems. Part II includes a brief example-driven introduction to categories, limits, and colimits, and an accessible account of homotopy limits and colimits of diagrams of spaces. It also discusses cosimplicial spaces and relates this topic to the cubical theory of Part I, and provides computational tools via spectral sequences. The book finishes with applications to some exciting new topics that use cubical diagrams: an overview of two versions of calculus of functors and an account of recent developments in the study of the topology of spaces of knots.

BRIAN A. MUNSON is an Associate Professor of Mathematics at the US Naval Academy. He has held postdoctoral and visiting positions at Stanford University, Harvard University, and Wellesley College, Massachusetts. His research area is algebraic topology, and his work spans topics such as embedding theory, knot theory, and homotopy theory.

ISMAR VOLIĆ is an Associate Professor of Mathematics at Wellesley College, Massachusetts. He has held postdoctoral and visiting positions at the University of Virginia, Massachusetts Institute of Technology, and Louvain-la-Neuve University in Belgium. His research is in algebraic topology and his articles span a wide variety of subjects such as knot theory, homotopy theory, and category theory. He is an award-winning teacher whose research has been recognized by several grants from the National Science Foundation.

NEW MATHEMATICAL MONOGRAPHS

*Editorial Board*

All the titles listed below can be obtained from good booksellers or from Cambridge University Press. For a complete series listing visit www.cambridge.org/mathematics.

# Cubical Homotopy Theory

BRIAN A. MUNSON
*United States Naval Academy, Maryland*

ISMAR VOLIĆ
*Wellesley College, Massachusetts*

# CAMBRIDGE
## UNIVERSITY PRESS

University Printing House, Cambridge CB2 8BS, United Kingdom

Cambridge University Press is part of the University of Cambridge.

It furthers the University's mission by disseminating knowledge in the pursuit of education, learning and research at the highest international levels of excellence.

www.cambridge.org
Information on this title: www.cambridge.org/9781107030251

© Brian A. Munson and Ismar Volić 2015

First published 2015

Printed in the United Kingdom by Clays, St Ives plc

*A catalog record for this publication is available from the British Library*

*Library of Congress Cataloging in Publication Data*
Munson, Brian A., 1976–
Cubical homotopy theory / Brian A. Munson, United States Naval Academy, Maryland, Ismar Volić, Wellesley College, Massachusetts.
pages cm
Includes bibliographical references and index.
ISBN 978-1-107-03025-1 (alk. paper)
1. Homotopy theory.   2. Cube.   3. Algebraic topology.   I. Volić, Ismar, 1973–
II. Title.
QA612.7.M86   2015
514'.24–dc23
2015004584

ISBN 978-1-107-03025-1 Hardback

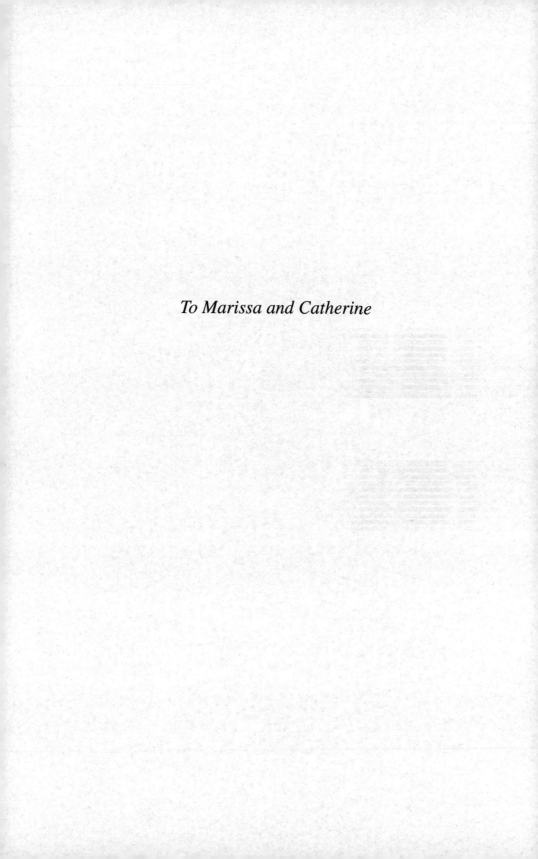

*To Marissa and Catherine*

# Contents

# Preface

**Purpose.** Cubical diagrams have become increasingly important over the last two decades, both as a powerful organizational tool and because of their many applications. They provide the language necessary for the Blakers–Massey Theorems, which unify many classical results; they lie at the heart of calculus of functors, which has many uses in algebraic and geometric topology; and they are intimately related to homotopy (co)limits of diagrams and (co)simplicial spaces. The growing importance of cubical diagrams demands an up-to-date, comprehensive introduction to this subject.

In addition, self-contained, expository accounts of homotopy (co)limits and (co)simplicial spaces do not appear to exist in the literature. Most standard references on these subjects adopt the language of model categories, thereby usually sacrificing concreteness for generality. One of the goals of this book is to provide an introductory treatment to the theory of homotopy (co)limits in the category of topological spaces.

This book makes the case for adding the homotopy limit and colimit of a punctured square (homotopy pullback and homotopy pushout) to the essential toolkit for a homotopy theorist. These elementary constructions unify many basic concepts and endow the category of topological spaces with a sophisticated way to "add" (pushout) and "multiply" (pullback) spaces, and so "do algebra". Homotopy pullbacks and pushouts lie at the core of much of what we do and they build a foundation for the homotopy theory of cubical diagrams, which in turn provides a concrete introduction to the theory of general homotopy (co)limits and (co)simplicial spaces.

**Features.** We develop the homotopy theory of cubical diagrams in a gradual way, starting with squares and working up to cubes and beyond. Along the way, we show the reader how to develop competence with these topics with over 300 worked examples. Fully worked proofs are provided for the most part, and the reader will be able to fill in those that are not provided or have only been sketched. Many results in this book are known, but their proofs do not appear to exist. If we were not able to find a proof in the literature, we have indicated that this is the case. The reader will also benefit from an abundance of suggestions for further reading.

Cubical diagrams are an essential concept for stating and understanding the generalized Blakers–Massey Theorems, fundamental results lying at the intersection of stable and unstable homotopy theory. Our proofs of these theorems are new, purely homotopy-theoretic in nature, and use only elementary methods. We show how many important results, such as the Whitehead, Hurewicz, and Freudenthal Suspension Theorems, follow from the Blakers–Massey Theorems. Another new feature is our brief but up-to-date discussion of quasifibrations and of the Dold–Thom Theorem from the perspective of homotopy pullbacks and pushouts. Lastly, most of the material on spectral sequences of cubical diagrams also does not seem to have appeared elsewhere.

Our expositional preference is for (homotopy) limits rather than (homotopy) colimits. This is partly due to a quirk of the authors, but also because the applications in this book use (homotopy) limits more than (homotopy) colimits. We have, however, at least stated all the results in the dual way, but have omitted many proofs for statements about (homotopy) colimits that are duals of those for (homotopy) limits. If a proof involving (homotopy) colimits had something new to offer, we have included it.

**Audience.** A wide variety of audiences can benefit from reading this book. A novice algebraic topology student can learn the basics of some standard constructions such as (co)fibrations, homotopy pullbacks and pushouts, and the classical Blakers–Massey Theorem. A person who would like to begin to study the calculus of functors or its recent applications can read about cubical diagrams, the generalized Blakers–Massey Theorem, and briefly about calculus of functors itself. An advanced reader who does not want to adopt the cubical point of view can delve deeper into general homotopy (co)limits (while staying rooted in topological spaces and not going through the model-theoretic machinery that other literature adopts), cosimplicial spaces, or Bousfield–Kan spectral sequences. In addition, geometrically-minded topologists will appreciate some of main examples involving configuration spaces, applications to knot and link theory (and more general embedding spaces), as well as the geometric proof of the Blakers–Massey Theorem.

**Organization.** The book is naturally divided into two parts. The first two chapters of Part I can be thought of as the necessary background and may be skimmed or even skipped and returned to when necessary. The book really begins in Chapter 3, where we study squares, homotopy pullbacks and pushouts, and develop their arithmetic and algebraic properties. We can deduce many standard results in classical homotopy theory using this language, and this also provides the foundation of the theory for higher-dimensional cubes. The payoff at the end of this introductory material is in Chapter 4, where we present the Blakers–Massey Theorems for squares, tools for comparing homotopy pushouts and pullbacks. These are central results in homotopy theory, and we give many applications. We then move to higher-dimensional cubes in Chapter 5, and are able to bootstrap the material on squares to give an accessible account which generalizes most of the material encountered thus far. This treatment also provides a concrete introduction to more general homotopy (co)limits. Again we end with the Blakers–Massey Theorems in Chapter 6, but this time for higher-dimensional cubes.

Part II of the book explores more general categories and the definitions and main properties of homotopy (co)limits. The hope is that the reader will have acquired enough intuition through studying Part I, which is very concrete, to be able to transition to some of the general abstract notions of Part II. We review some general category theory in Chapter 7 but in Chapter 8 return to the category of topological spaces in order to preserve concreteness and continue to supply ample and familiar examples. We then move on to cosimplicial spaces in Chapter 9, which are closely allied both with the general theory of homotopy limits and with the cubical theory developed in Part I of the book. We end in Chapter 10 with a sequence of applications representing brief forays into current research, including introductory material of both homotopy and manifold calculus of functors and some applications. All of this uses the material developed earlier – cubical and cosimplicial machinery as well as the Blakers–Massey Theorem – in an essential way. An appendix serves to illustrate or give background on some topics which are used throughout the text but are not central to its theme, such as simplicial sets, spectra, operads, and transversality.

We have included a flowchart at the end of this preface to indicate the interdependence of the chapters.

**Acknowledgments.** Above all, we thank Tom Goodwillie, who taught us most of what we know about cubical diagrams. This project grew out of the notes from a series of lectures he gave us while we were still graduate students. His influence and generosity with the mathematics he has passed on to us cannot be overstated.

We thank Phil Hirschhorn for his generous help, technical and mathematical. His careful reading and input into various parts of the book, especially Chapters 2, 8, and 9, was invaluable.

We thank Greg Arone for many helpful conversations and for his encouragement and enthusiasm for this project.

We also thank Emanuele Dotto, Dan Dugger, John Oprea, Ben Walter, and the anonymous reviewers for useful comments and suggestions.

Lastly, we thank the editors and the staff at Cambridge University Press for their support and patience during the writing of this book.

BRIAN A. MUNSON
ISMAR VOLIĆ

**Flowchart.** The dotted lines in the chart below indicate that a reader who has had some exposure to category theory can skip Chapter 7, which contains fairly standard material (there might need to be an occasional look back at this chapter for notation or statements of results). The left and the right columns of the chart are precisely Parts I and II.

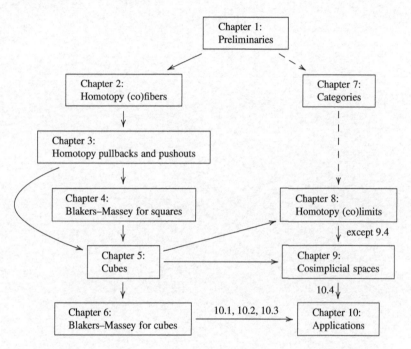

# PART I

Cubical diagrams

# 1

# Preliminaries

This chapter establishes some notational conventions and fundamental constructions. It is not comprehensive, nor is all of it even necessary (the unnecessary bits are meant to supply context), and we assume the reader is familiar with most of it already. Some of the material presented in this chapter is redundant in the sense that it will be revisited later. For instance, the cone, wedge, and suspension of a space will be discussed later in terms of colimits, a perspective more in line with the philosophy of this book. The essential topics presented here which are utilized elsewhere are topologies on spaces and spaces of maps, homotopy equivalences, weak equivalences, and a few properties of the class of CW complexes whose extra structure we will need from time to time. Some familiarity with homotopy groups (mostly their definition) will also be useful, and to a much lesser extent some exposure to homology. Many proofs are omitted, and references are given instead. We will clarify which is which along the way.

Most references given in this chapter are from Hatcher's *Algebraic topology* [Hat02]. There are a few other modern references which the authors have found useful, and which contain most, if not all, of these preliminary results as well, such as [AGP02, Gra75, May99, tD08] (we especially like [AGP02] since it seems to be the most elementary text which follows this book's philosophy; [Gra75] is neither modern nor in print, but still a unique and valuable resource). We owe all of these sources a debt, in this chapter and elsewhere.

## 1.1 Spaces and maps

A topological space is a pair $(X, \tau)$, where $\tau$ is a collection of subsets of $X$, the members of which are called *open* sets, which contains both the empty set and $X$, and which is closed under finite intersections and arbitrary unions. However, it is customary to suppress the topology from the notation, so we simply write $X$ in place of $(X, \tau)$, and typically denote generic topological spaces using capital Roman letters. A *subbase* for a topology $\tau$ on $X$ is a subset of $\tau$ for which every element of $\tau$ is a union of finite intersections of elements in the subset; it is a sort of generating set for $\tau$.

The complement of an open set $U \subset X$ is called *closed*, and to specify a topology on $X$ we may equivalently describe a system of closed sets which contains both the empty set and $X$, and which is closed under arbitrary intersections and finite unions. If $A \subset X$, we write $\mathring{A}$ for its *interior*; $\mathring{A}$ is the union of all open sets in $X$ which are contained in $A$, and as such is an open set. We will write $\overline{A}$ for the *closure* of $A$; by definition it is the intersection of all the closed subsets of $X$ which contain $A$, and is evidently closed.

By a *subspace* $A \subset X$, we mean the topology on $A$ whose open sets are of the form $A \cap U$ where $U$ is open in $X$. A collection of subspaces of a space $X$ is called a *cover* of $X$ if their union is $X$. For a nested sequence $X_1 \subset X_2 \subset \cdots$ of topological spaces, we endow the union $X = \bigcup_{i=1}^{\infty} X_i$ with the *weak topology*: a subset $C \subset X$ is closed if $C \cap X_i$ is closed in $X_i$ for each $i$. For a topological space $X$ with an equivalence relation $\sim$, we let $X/\sim$ denote the set of equivalence classes of $X$ under $\sim$, and endow this space with the *quotient topology*: a set of equivalence classes is called open if the union of those equivalence classes forms an open set in $X$. For $x$ in $X$, we denote by $[x]$ the corresponding point in $X/\sim$. For the quotient of $X$ by a subspace $A$, we write $X/A$ for this space and mean the quotient of $X$ by the equivalence relation $\sim$ on $X$ generated by $x \sim y$ if $x, y \in A$.

We will write $X \times Y$ for the product, $X \sqcup Y$ for the disjoint union, and when $X$ and $Y$ are subspaces of some larger space, $X \cup Y$ for the union along the intersection (which may be empty). Open sets of the product are generated by products of open sets, and open sets in the disjoint union are disjoint unions of open sets.

Several spaces are worthy of mention.

**Definition 1.1.1**

- The empty set $\emptyset$, that is, the space with no points.
- The one-point space $*$.
- The real numbers $\mathbb{R}$, topologized using the metric $d(x, y) = |x - y|$.
- The unit interval $I = \{t \in \mathbb{R} : 0 \leq t \leq 1\}$, topologized as a subspace of $\mathbb{R}$.
- Euclidean $n$-space $\mathbb{R}^n$, topologized using the usual metric $|-|$. By definition $\mathbb{R}^0 = \{0\}$.

- The $n$-dimensional cube $I^n = \{(t_1, \ldots, t_n) \in \mathbb{R}^n : 0 \leq t_i \leq 1 \text{ for all } i\}$, topologized as a subspace of $\mathbb{R}^n$.
- The unit $n$-disk $D^n = \{x \in \mathbb{R}^n : |x| \leq 1\}$ for $n \geq 0$, topologized as a subset of $\mathbb{R}^n$. We define $D^{-1} = \emptyset$.
- The unit $(n-1)$-sphere $S^{n-1} = \partial D^n = \{x \in \mathbb{R}^n : |x| = 1\}$ for $n \geq 0$, topologized as a subspace of $\mathbb{R}^n$. Note that $S^{-1} = \emptyset$; we define $S^{-2} = \emptyset$.
- The $n$-simplex $\Delta^n = \{(t_1, \ldots, t_n) : 0 \leq t_1 \leq \cdots \leq t_n \leq 1\}$. Alternatively, $\Delta^n = \{(x_0, \ldots, x_n) \in \mathbb{R}^{n+1} : 0 \leq x_i \leq 1 \text{ for all } i \text{ and } \sum_i x_i = 1\}$ (it is a standard exercise to prove the two descriptions are equivalent; when we have need of coordinates in the simplex we will utilize the latter description). A simplex is topologized as a subspace of the Euclidean space of which it is a subset. For $0 \leq k \leq n$, let $\partial_k \Delta^n \subset \Delta^n$ denote the subset of $\Delta^n$ consisting of those tuples $(x_0, \ldots, x_n)$ for which $x_k = 0$. This is called a *face* of $\Delta^n$, and it is itself a simplex of dimension $n-1$.

We assume our topological spaces to be compactly generated Hausdorff. To be Hausdorff means that any two distinct points are contained in disjoint open neighborhoods.

**Definition 1.1.2** A space $X$ is said to be *compactly generated* if it has the property that a subset $C$ of $X$ is closed if and only if the intersection $C \cap K$ with each compact subset $K$ of $X$ is also closed in $X$.

All of the spaces described in Definition 1.1.1 are compactly generated Hausdorff. We will denote by Top the category of compactly generated spaces. The definition of a category can be found in Definition 7.1.1 (and we will not use the language of categories in a serious way before Chapter 7). Any Hausdorff space $X$ can be made into a compactly generated space $kX$ (same point set, different topology: we take the smallest compactly generated topology which contains the given one). The identity function $kX \to X$ is continuous and a homeomorphism if and only if $X$ is compactly generated. Moreover, $kX$ and $X$ have the same compact subsets and the same homotopy groups (defined below). The product of two compactly generated spaces is given the topology of $k(X \times Y)$. Locally compact Hausdorff spaces, manifolds, metric spaces, and CW complexes (see below) are all compactly generated spaces. One benefit of working with compactly generated spaces is that this makes the duality between the notions of cofibration and fibration cleaner to state by eliminating the hypothesis of local compactness.

Maps between spaces will typically be denoted by a lower-case Latin letter such as $f$ or $g$; thus $f \colon X \to Y$ denotes a map between the topological spaces

$X$ and $Y$. In this case we say $X$ is the *domain* of $f$ and $Y$ the *codomain*. The term "map" means continuous map.

Several maps are worthy of mention.

## Definition 1.1.3

- The *identity map* from a space $X$ to itself will be denoted by $1_X$, defined by $1_X(x) = x$ for all $x$.
- The map from $X$ to $Y$ which has constant value $y \in Y$ will be denoted $c_y \colon X \to Y$, and is referred to as a *constant map*.
- For a subspace $A \subset X$, we usually write $\iota \colon A \to X$ for the inclusion map; occasionally we may use $i$ for this map. The subspace $A$ is a *retract* of $X$ if there exists a map $r \colon X \to A$, called a *retraction*, such that $r \circ \iota = 1_A$.
- Given a map $f \colon X \to Y$, we say a map $g \colon Y \to X$ is a *section* of $f$ if the composite $f \circ g = 1_Y$. Thus the inclusion map for a subspace which is a retract is a section of the retraction.
- For a space $X$ with equivalence relation $\sim$, there is a *quotient map* $q \colon X \to X/\!\sim$ which sends each point to its equivalence class.
- For a space $X$, we write $\Delta \colon X \to X \times X$ for the *diagonal map*, defined by $\Delta(x) = (x, x)$.
- For a space $X$, we write $\nabla \colon X \coprod X \to X$ for the *fold map*, defined to be the identity $1_X$ on each summand.
- Given a map $f \colon X \to Y$ and a subspace $A \subset X$, we let $f|_A \colon A \to Y$ denote the *restriction* of $f$ to $A$.
- A map $f \colon X \to Y$ is called a *homeomorphism* if there exists a continuous inverse $g \colon Y \to X$ for $f$; i.e. we have $g \circ f = 1_X$ and $f \circ g = 1_Y$. Spaces $X$ and $Y$ are then *homeomorphic* and we write $X \cong Y$. This is the most basic equivalence relation on the class of spaces we will consider.

Here is a useful result we will need later.

**Lemma 1.1.4** *If $X$ is a Hausdorff space and $A$ is a retract of $X$, then $A$ is closed in $X$.*

*Proof* Let $r \colon X \to A$ be the retraction and consider the map $X \to X \times X$ given by $x \mapsto (x, r(x))$. The preimage of the diagonal in $X \times X$ is the set of fixed points of $r$, which is $A$. But since $X$ is Hausdorff, the diagonal is closed, and hence $A$ is closed.                                    □

Returning again to a nested sequence $X_1 \subset X_2 \subset \cdots$ of topological spaces, we note that the weak topology on $X = \bigcup_{i=1}^{\infty} X_i$ has the property that, given a collection of maps $f_i : X_i \to Y$ such that $f_i|_{X_{i-1}} = f_{i-1}$ for all $i$, there is a unique map $f : X \to Y$ whose restriction to $X_i$ is equal to $f_i$ for all $i$.

Often we consider *diagrams* of spaces, which simply means families of spaces and maps between them. The language is chosen to suggest we are thinking of a sort of picture of these spaces and maps. We will usually deal with *commutative* diagrams, which means that all ways of getting from one space to another by following maps are the same. The two we will most frequently encounter are

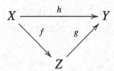

where commutativity means $g \circ f = h$ and

$$
\begin{array}{ccc}
W & \xrightarrow{f'} & Y \\
{\scriptstyle g'}\downarrow & & \downarrow{\scriptstyle g} \\
X & \xrightarrow{f} & Z
\end{array}
$$

where commutativity means $g \circ f' = f \circ g'$. We typically omit drawing the composed map $W \to Z$ in squares such as the above. We will later use the language of categories and functors to talk about diagrams (see Remark 7.1.16). If a diagram is not necessarily commutative, we will be explicit about this. One notable generally non-commutative diagram we will encounter first in Chapter 7 is

$$
X \mathrel{\substack{\xrightarrow{f} \\ \xrightarrow[g]{}}} Y
$$

where $f \neq g$.

The spaces in a diagram will often be parametrized by subsets of some finite set, and so we shall encounter spaces such as $X_U$ to denote which member of the family of spaces labeled "$X$" we mean. In the case that $U$ is a subset of a finite set, say $U = \{1, 2\} \subset \{1, 2, 3\}$, we will usually write $X_{12}$ in place of $X_{\{1,2\}}$ for cleaner presentation.

We will also often consider *pairs* of spaces $(X, A)$. where $X$ is a space and $A \subset X$ is a subspace. A *map of pairs* $f : (X, A) \to (Y, B)$ is a map $f : X \to Y$ such that $f(A) \subset B$. When $A = \{x_0\}$ is a single point, then $X$ will be called *based* or *pointed*, $x_0$ will be called the *basepoint*, and we will typically write $(X, x_0)$ in place of $(X, \{x_0\})$. A map $f : X \to Y$ of based spaces $X$ and $Y$ with basepoints $x_0$ and $y_0$ respectively is *based* if it is a map of pairs $(X, x_0)$ and $(Y, y_0)$. We let Top$_*$ denote the category of based spaces (meaning the objects of interest are based spaces and the maps of interest are based maps). A map of pairs $f : (X, A) \to (Y, B)$ is called a *homeomorphism of pairs* if it has a continuous inverse $g : (Y, B) \to (X, A)$. We can also consider based pairs $(X, A, x_0)$, where $x_0 \in A$ is the basepoint, and consider based maps $(X, A, x_0) \to (Y, B, y_0)$ of pairs, whose definition should be apparent.

The basepoint will often be suppressed from notation, but the reader should be warned that many constructions which require a choice of basepoint are not independent of that choice, such as the fundamental group. This should not cause confusion, as we will be clear about choosing basepoints when we are forced to do so.

**Definition 1.1.5**   A based space $(X, x_0)$ is called *well-pointed* if $X \times \{0\} \cup \{x_0\} \times I$ is a retract of $X \times I$. If $(X, x_0)$ is well-pointed, we call the basepoint $x_0$ *non-degenerate*. We assume all spaces to be well-pointed.

**Remark 1.1.6**   We prefer the equivalent definition that the inclusion of the basepoint $\{x_0\} \to X$ is a cofibration, but we do not have this language available yet. We will revisit this definition in Remark 2.3.20 once we have established the notion of a cofibration. The reason we assume our spaces to be well-pointed is to preserve homotopy invariance of various standard constructions, such as the suspension. □

CW complexes, mentioned above, play an important role at various points in this text and so it is worth recalling at least the idea of their construction. For instance, in Chapter 8 we will frequently deal with the realization of a simplicial complex, and it is easy to see how these can be considered as CW complexes. We refer the reader to [Hat02, Chapter 0, Appendix A] for more details on CW complexes. An *n-cell* $e^n$ is simply the *n*-disk $D^n$, and $\partial e^n = \partial D^n$ is its boundary, the $(n-1)$-sphere. A CW complex $X$ is a space built inductively starting with the empty set $X^{-1} = \emptyset$, with $X^n$ built from $X^{n-1}$ by attaching cells $e^n_\alpha$ to $X^{n-1}$ via maps $a_\alpha : \partial e^n \to X^{n-1}$. Here $\alpha$ ranges through a (possibly empty) indexing set $A_n$. Thus $X^0$ is a discrete set of points, and in general $X^n$ is a quotient space of $X^{n-1} \sqcup_{\alpha \in A_n} e^n_\alpha$. The space $X$ is then defined as $\cup_{n \geq 0} X^n$ and is given the weak topology: A subset $C \subset X$ is closed if $C \cap X^n$ is closed in $X^n$ for all $n$. We call $X^n$ the *n-skeleton* of $X$. A *subcomplex* of a CW complex $X$ is a subset $A$ which is a union of cells of $X$ such that the closure of each cell is contained in $A$.

A *relative CW complex* is a pair $(X, A)$ where $A$ is a topological space and $X$ a space which has been built from $A$ by attaching cells as above. That is, we use the same definition as above only with $X^{-1} = A$. The case where

$A = \emptyset$ specializes to an ordinary CW complex, so we may assume all CW complexes are relative CW complexes for the purposes of any statements about such spaces. The space $A$ need not have a cell structure itself. A relative CW complex $(X, A)$ has *dimension* $n$ if $X = X^n$ and $X \neq X^{n-1}$. We say it is *finite* if the number of its cells is finite; that is, each indexing set $A_n$ above is finite and there exists $N$ such that $A_n = \emptyset$ for $n \geq N$. Theorem 1.3.7 below says that any space can be approximated by a CW complex in a suitable sense.

The notion of CW complex furnishes an enormous number of examples of topological spaces built from disks. We now review some others basic constructions: the cone, suspension, join, wedge, and smash product of spaces. We will encounter all of these again in Chapters 2 and 3 as examples of homotopy (co)fibers and homotopy (co)limits, and will derive many results combining and comparing them there.

**Definition 1.1.7** For a topological space $X$, define $X_+$ to be $X$ with a disjoint basepoint.

The space $X_+$ is in fact the quotient of $X$ by the empty set. This is easiest to see diagrammatically by consideration of universal properties, as in Example 3.5.6.

**Definition 1.1.8** For a space $X$, the *cone* $CX$ on $X$ is the quotient space

$$CX = X \times I/(X \times \{1\}).$$

If $X = \emptyset$, then $CX$ is a point. If $X$ is based with basepoint $x_0$, the *reduced cone* on $X$, by abuse also called $CX$, is the quotient space

$$X \times I/(X \times \{1\} \cup \{x_0\} \times I).$$

Any map $f: X \rightarrow Y$ induces a map of cones,

$$Cf: CX \rightarrow CY, \qquad (1.1.1)$$

induced from the map $f \times 1_I: X \times I \rightarrow Y \times I$ by taking quotient spaces of the domain and codomain. See Example 3.6.8 for an alternative definition of the cone.

**Remark 1.1.9** The quotient map from the cone to the reduced cone is a homotopy equivalence because $(X, x_0)$ is well-pointed by assumption. See Remark 2.3.20 for a discussion. □

The cone on $X$ is so named as it is created from the cylinder $X \times I$ by collapsing one end. The space $X$ is naturally a subspace of $CX$ by the inclusion of $X \times \{0\}$ in $CX$.

**Definition 1.1.10** For a based space $X$ with basepoint $x_0$, the *reduced suspension* $\Sigma X$ of $X$ is the quotient space

$$\Sigma X = X \times I/(X \times \{0\} \cup \{x_0\} \times I \cup X \times \{1\}).$$

Its basepoint is the image of $X \times \{0\} \cup \{x_0\} \times I \cup X \times \{1\}$ by the quotient map. The *unreduced suspension* of a space $X$, based or not, is the quotient space

$$X \times I/\sim$$

where $\sim$ is the equivalence relation generated by $(x_1, 0) \sim (x_2, 0)$ and $(x_1, 1) \sim (x_2, 1)$. Inductively we define $\Sigma^n X = \Sigma\Sigma^{n-1} X$.

The suspension is the union of two copies of the reduced cone $CX$ glued together by the identity map $X \times \{0\} \to X \times \{0\}$. See Example 3.6.9 for an alternative definition of suspension.[1]

**Remark 1.1.11** We will also use $\Sigma X$ to denote the unreduced suspension, and we will refer to both versions simply as the *suspension*. It should always be clear to the reader which version we mean, usually depending on the existence of a basepoint, although we will try to be clear in instances where this may cause confusion. In any case, if $X$ is well-pointed (see Remark 1.1.6), then the quotient map from the unreduced suspension to the reduced suspension of $X$ is a homotopy equivalence, so this is usually not a concern.                    □

**Example 1.1.12** It is an easy exercise to see that there is a homeomorphism $\Sigma S^n \cong S^{n+1}$ for all $n \geq -1$. (The sphere $S^{-1}$ is empty, but its suspension is the quotient of the empty set by two points, and hence may be identified with $S^0$.)                    □

Any map $f \colon X \to Y$ induces a map of suspensions,

$$\Sigma f \colon \Sigma X \to \Sigma Y, \qquad\qquad (1.1.2)$$

induced from the map $f \times 1_I \colon X \times I \to Y \times I$.

---

[1] Using the language of Section 2.4, another way to think of this is as the cofiber of the cofibration $X \to CX$.

**Definition 1.1.13** Let $X$ and $Y$ be based spaces with basepoints $x_0$ and $y_0$ respectively. Define the *wedge* or *wedge sum* $X \vee Y$ by

$$X \vee Y = X \times \{y_0\} \cup_{(x_0 \times y_0)} \{x_0\} \times Y.$$

Alternatively, it is the quotient of $X \sqcup Y$ by the equivalence relation generated by $x_0 \sim y_0$. See Example 3.6.6 for an alternative definition of the wedge.

**Example 1.1.14**

- $S^1 \vee S^1 \cong$ "figure-8".
- $\Sigma(X \vee Y) \simeq \Sigma X \vee \Sigma Y$.
- Suppose $X$ is an $n$-dimensional CW complex, whose $n$-cells are indexed by the set $A_n$. Then the quotient space $X/X^{n-1} = \vee_{A_n} S^n$. If $A_n = \emptyset$ then this wedge sum is a single point. □

**Definition 1.1.15** Let $X$ and $Y$ be based spaces. Define the *smash product* $X \wedge Y$ by

$$X \wedge Y = X \times Y / X \vee Y.$$

See Example 3.6.10 for an alternative way to define the smash product.

Note that if $X$ is a based space with basepoint $x_0$, then we can still smash it with an unbased space $Y$ by adding a disjoint basepoint,

$$X \wedge Y_+ = X \times Y / \{x_0\} \times Y.$$

This is sometimes called the *half-smash product*.

**Example 1.1.16**

- Given based spheres $S^n$ and $S^m$, we have $S^n \wedge S^m \cong S^{n+m}$.
- More generally if $X$ is a based space, we have $S^n \wedge X \cong \Sigma^n X$.
- If $Y$ is another based space, then $\Sigma(X \wedge Y) \cong (\Sigma X) \wedge Y \cong X \wedge (\Sigma Y)$.
- If $X$ and $Y$ are unbased, then $X_+ \wedge Y_+ = X \times Y$. □

**Definition 1.1.17** Let $X$ and $Y$ be spaces. Define the *join* $X * Y$ by

$$X * Y = CX \times Y \cup_{X \times Y} X \times CY.$$

If $X$ and $Y$ are based spaces with basepoints $x_0$ and $y_0$ respectively, we define the *reduced join* of $X$ with $Y$ as the quotient space $X * Y / (X * \{y_0\} \cup \{x_0\} * Y)$. See Example 3.6.12 for an alternative way to define the join.

If $Y$ is a point, then the join of $X$ with $Y$ is the cone on $X$, and if $Y$ is two points it is the unreduced suspension of $X$ (two cones joined along their bases). In general the join can be thought of as a $Y$-parameter version of the cone. The reduced join is homotopy equivalent to the join $X * Y$ when $X$ and $Y$ are well-pointed spaces.

**Example 1.1.18**

- $X * \{*\} \cong CX$.
- $X * S^0 \cong \Sigma X$.
- $S^n * S^m \cong S^{n+m+1}$.                                      □

## 1.2 Spaces of maps

It is useful to think of the set of maps between topological spaces $X$ and $Y$ as a topological space itself. For instance, the set of all maps of the one-point space $*$ to a topological space $X$ is, as a point-set, isomorphic with $X$ itself, and so we would like to topologize this space to reflect this.

**Definition 1.2.1**  For spaces $X$ and $Y$, let $\mathrm{Map}(X, Y)$ denote the space of continuous maps from $X$ to $Y$. It is occasionally useful to write $Y^X$ in place of $\mathrm{Map}(X, Y)$. A subbase for its topology consists of the sets

$$U^K = \{f \in \mathrm{Map}(X, Y) \mid f(K) \subset U\},$$

where $K$ is a compact subset of $X$ and $U$ is an open subset of $Y$. This is called the *compact-open* topology.

Of course, when we write $\mathrm{Map}(X, Y)$, we really mean $k\,\mathrm{Map}(X, Y)$.

**Definition 1.2.2**  For subsets $A \subset X$ and $B \subset Y$, we let $\mathrm{Map}((X, A), (Y, B))$ denote the space of continuous maps of pairs. Here $f \in \mathrm{Map}((X, A), (Y, B))$ means that $f: X \to Y$ is continuous and $f(A) \subset B$. It is topologized as a subspace of $\mathrm{Map}(X, Y)$.

When $A = \{x_0\}$ and $B = \{y_0\}$ are one-point sets, i.e. when $X$ and $Y$ are based, we refer to this as the *based mapping space*. We write $\mathrm{Map}_*(X, Y)$ in place of $\mathrm{Map}((X, \{x_0\}), (Y, \{y_0\}))$ for brevity. This space is itself naturally a based space with basepoint the constant map $c_{y_0}: X \to Y$ which sends every point of $X$ to the basepoint of $Y$.

A few examples are worth pointing out.

**Example 1.2.3**   In the case where $A = \emptyset$, the space of maps of pairs $(X, A) \to$ $(Y, B)$ is just the space $\mathrm{Map}(X, Y)$. ☐

**Example 1.2.4** (Path space)   The space $X^I = \mathrm{Map}(I, X)$ is called the *space of paths in X*, or the *path space of X*. We will also denote it by $PX$. ☐

**Example 1.2.5** (Loop space)   If $X = I$, $A = \partial I$, and $B = \{y_0\}$ is a point in $Y$, it is customary to write $\Omega Y$ instead of $\mathrm{Map}((I, \partial I), (Y, y_0))$ (the notation $\Omega Y$ omits the basepoint $y_0$ for brevity). This is called the *loop space* of $Y$. Its basepoint is, as above, $c_{y_0}$, the constant map which sends $I$ to the basepoint $y_0$. Inductively we define $\Omega^n Y = \Omega \Omega^{n-1} Y$. Since $I/\partial I \cong S^1$, $\Omega Y \cong \mathrm{Map}_*(S^1, Y)$. Using Theorem 1.2.7 (see below) and the homeomorphism $I^n/\partial I^n \cong S^n$, it is not hard to see that $\Omega^n Y \cong \mathrm{Map}_*(S^n, Y)$.

A based map $f : X \to Y$ can also be "looped", by which we mean that $f$ induces a map

$$\Omega^n f : \Omega^n X \longrightarrow \Omega^n Y \tag{1.2.1}$$

given by composing a based map $S^n \to X$ with $f$. ☐

**Example 1.2.6** (Free loop space)   For $X = S^1$, the space $\mathrm{Map}(S^1, Y)$ is called the *free loop space* of $Y$ and is denoted by $LY$. ☐

There are two useful consequences of using the compact-open topology. The first is that the composition map

$$\mathrm{Map}(X, Y) \times \mathrm{Map}(Y, Z) \longrightarrow \mathrm{Map}(X, Z)$$

which sends $(f, g)$ to $g \circ f$ is continuous. The same is true for based mapping spaces. The other is known as the *exponential law*, most concisely expressed as

$$Y^{X \times Z} \cong (Y^X)^Z. \tag{1.2.2}$$

More precisely,

**Theorem 1.2.7** (Exponential law, unbased version)   *For spaces $X, Y, Z$, the map*

$$\mathrm{Map}(X \times Z, Y) \longrightarrow \mathrm{Map}(Z, \mathrm{Map}(X, Y))$$

*which associates $F : X \times Z \to Y$ with the map $z \mapsto (x \mapsto F(x, z))$ is a homeomorphism.*

It is important to note the standing assumptions here: the spaces $X, Y$, and $Z$ are compactly generated Hausdorff, and by $X \times Y$ we mean $k(X \times Y)$, and by

Map$(A, B)$ we mean $k\,\mathrm{Map}(A, B)$. This theorem is not true for arbitrary topological spaces. The proof of this theorem can be found, for example, in [Hat02, Proposition A.16]. The based version requires the smash product construction:

**Theorem 1.2.8** (Exponential law, based version)   *If X, Y, and Z are based, the map*

$$\mathrm{Map}_*(X \wedge Z, Y) \longrightarrow \mathrm{Map}_*(Z, \mathrm{Map}_*(X, Y))$$

*which associates* $F \colon X \times Z \to Y$ *with the map* $z \mapsto (x \mapsto F(x, z))$ *is a homeomorphism.*

One consequence of the above theorem is that the suspension operation $\Sigma$ is dual to[2] the loop space operation $\Omega = \mathrm{Map}_*(S^1, -)$ in the following sense.

**Theorem 1.2.9**   *Let Z and Y be based spaces. Then there is a homeomorphism*

$$\mathrm{Map}_*(\Sigma Z, Y) \cong \mathrm{Map}_*(Z, \Omega Y).$$

*Proof*   This follows from the based exponential law, Theorem 1.2.8, by setting $X = S^1$. Then by Example 1.1.16, $S^1 \wedge Z \cong \Sigma Z$ and $\mathrm{Map}_*(S^1, Y)$ is by definition $\Omega Y$.                                                                              □

## 1.3 Homotopy

The natural notion of isomorphism of topological spaces is that of homeomorphism. Determining whether two spaces are homeomorphic is in general a hopelessly difficult problem. The notion of homotopy, described below, gives rise to a weaker notion of equivalence called homotopy equivalence.

**Definition 1.3.1**

- Maps $f_0, f_1 \colon X \to Y$ are *homotopic* if there exists a continuous map $F \colon X \times I \to Y$ such that $F|_{X \times \{i\}} = f_i$ for $i = 0, 1$, and we write $f_0 \sim f_1$. We regard a homotopy $F$ as an element of $\mathrm{Map}(I, \mathrm{Map}(X, Y))$, and in the case where $X$ and $Y$ are based as an element of $\mathrm{Map}(I, \mathrm{Map}_*(X, Y))$ (thus a homotopy of based maps is simply a path in the based mapping space).
- A map $f \colon X \to Y$ is *null-homotopic* if it is homotopic to a constant map.
- A subspace $A \subset X$ with inclusion map $\iota \colon A \to X$ is a *deformation retraction* of $X$ if there exists a retraction $r \colon X \to A$ such that $\iota \circ r$ is homotopic to

---

[2]   Really, left adjoint to it; see Example 7.1.20.

the identity $1_X$ relative to $A$. That is, there exists $H: X \times I \to X$ such that $H(x, 0) = x$, $H(x, 1) \in A$, and $H(a, t) = a$ for all $t \in I$, and all $a \in A$.

- A map $f: X \to Y$ is a *homotopy equivalence* if there exists a map $g: Y \to X$ and homotopies from $f \circ g$ and $g \circ f$ to the identity functions $1_Y$ and $1_X$ on $Y$ and $X$ respectively. In this case $X$ and $Y$ are said to be *homotopy equivalent* and have the same *homotopy type*, and we write $X \simeq Y$. We call the map $g$ a *homotopy inverse* to $f$.
- A space $X$ is *contractible* if it has the homotopy type of a point. For instance, for any space $X$ the cone $CX$ is contractible, since the inclusion of the cone point (the equivalence class of $(x, 1)$ in $CX$) is a homotopy equivalence.

If $f_i: X_i \to Y_i$, $i \in I$, are a collection of homotopy equivalences, then the induced maps $\prod_i f_i: \prod_i X_i \to \prod_i Y_i$ and $\coprod_i f_i: \coprod_i X_i \to \coprod_i Y_i$ are homotopy equivalences.

Homotopy is an equivalence relation on the space of maps from $X$ to $Y$ (and more generally on the space of maps of pairs). For spaces $X$ and $Y$, let $[X, Y]$ denote the homotopy classes of maps $X \to Y$. For a map $f: X \to Y$ we let $[f]$ denote the corresponding element of $[X, Y]$. If $A \subset X$ and $B \subset Y$, we write $[(X, A), (Y, B)]$ for the homotopy classes of maps of pairs $(X, A) \to (Y, B)$; in this case the restriction of a homotopy of maps $(X, A) \to (Y, B)$ to $A$ is required to have image contained in $B$. In particular, when $A = \{x_0\}$ and $B = \{y_0\}$ are the basepoints of $X$ and $Y$, respectively, we get the homotopy classes of based maps $[(X, x_0), (Y, y_0)]$. When the basepoints do not need to be mentioned explicitly, we will write $[X, Y]_*$.

Homotopy theory studies those properties of topological spaces invariant under homotopy equivalence (and more generally weak equivalence, which we will describe below), and so we are largely interested in constructions which depend only on the homotopy type of spaces. We are also especially interested in "fixing" otherwise useful constructions which are not invariant under homotopy equivalence. For instance, homotopy and (co)homology groups (discussed in the next section) are invariant under homotopy equivalence. One way to characterize the homotopy type of a space $X$ is by the homotopy classes of maps of all spaces into or out of $X$.

**Proposition 1.3.2** *Let $f: X \to Y$ be a map. The following are equivalent:*

1. *$f$ is a homotopy equivalence.*
2. *The induced map*

$$f^*: \mathrm{Map}(Y, Z) \longrightarrow \mathrm{Map}(X, Z)$$
$$h \longmapsto h \circ f$$

*is a homotopy equivalence for all spaces $Z$.*

3. *The induced map*

$$f_* \colon \mathrm{Map}(Z, X) \longrightarrow \mathrm{Map}(Z, Y)$$
$$h \longmapsto f \circ h$$

*is a homotopy equivalence for all spaces Z.*
4. *The induced map* $f^* \colon [Y, Z] \to [X, Z]$ *is a bijection for all Z.*
5. *The induced map* $f_* \colon [Z, X] \to [Z, Y]$ *is a bijection for all Z.*

*In particular, X has the homotopy type of a point if and only if* $\mathrm{Map}(Z, X) \simeq *$
*if and only if* $\mathrm{Map}(X, Z) \simeq Z$ *for all spaces Z.*

*Proof*   It is clear from the definition that if $f$ is a homotopy equivalence then
the induced maps $\mathrm{Map}(Z, X) \to \mathrm{Map}(Z, Y)$ and $\mathrm{Map}(Y, Z) \to \mathrm{Map}(X, Z)$ are
homotopy equivalences for all spaces $Z$. If the induced map $\mathrm{Map}(Z, X) \to$
$\mathrm{Map}(Z, Y)$ is a homotopy equivalence for all $Z$, then setting $Z = *$ shows that $f$
is a homotopy equivalence. The rest of the proof follows from Proposition 1.3.5
below.                                                                        □

The last item above says that a map $f \colon X \to Y$ is a homotopy equivalence if
and only if the induced map $\mathrm{Map}(Z, X) \to \mathrm{Map}(Z, Y)$ is a bijection on the set
of path components. This leads to the notion of weak equivalence.

**Definition 1.3.3**   A map $f \colon X \to Y$ is a *weak equivalence* (or *weak homotopy
equivalence*) if $f$ induces a bijection $[K, X] \to [K, Y]$ for all CW complexes $K$.
If $X$ and $Y$ are weakly equivalent, we write $X \simeq Y$, and we say $X$ and $Y$ have
the same *weak homotopy type*.

**Remark 1.3.4**   We will use the same notation for weakly equivalent and
homotopy equivalent spaces, namely $X \simeq Y$, but will in each instance be clear
about which one we mean.                                                      □

A map $f \colon (X, A) \to (Y, B)$ of pairs is a *weak equivalence of pairs* if for all
relative CW pairs $(Z, W)$, where $W$ is a subcomplex of $Z$, the induced map
$[(Z, W), (X, A)] \to [(Z, W), (Y, B)]$ is a bijection.
    Note that we recover the definition of weak equivalence by taking $A = B =$
$\emptyset$; this forces $Z$ to be CW. Letting $W = \emptyset$, we see that a weak equivalence of
pairs $(X, A) \to (Y, B)$ implies that $X \to Y$ is a weak equivalence, and letting
$Z = W$ we see that $A \to B$ is a weak equivalence. It is for this reason we insist
$W$ is a subcomplex of $Z$ rather than letting $W$ be an arbitrary space; otherwise
this implies $A \to B$ is a homotopy equivalence by Proposition 1.3.2.

What follows is an analog of Proposition 1.3.2, the proof of which completes the proof of that proposition. Note, however, that this is a comparatively weaker statement.

**Proposition 1.3.5** *Let $f\colon X \to Y$ be a map. Suppose either of the induced maps*

$$\mathrm{Map}(Y, Z) \longrightarrow \mathrm{Map}(X, Z)$$
$$h \longmapsto h \circ f$$

*or*

$$\mathrm{Map}(Z, X) \longrightarrow \mathrm{Map}(Z, Y)$$
$$h \longmapsto f \circ h$$

*is a weak equivalence for all spaces $Z$. Then $f$ is a homotopy equivalence, and hence a weak equivalence.*

*Proof* If the induced map $\mathrm{Map}(Z, X) \to \mathrm{Map}(Z, Y)$ is a weak equivalence, then $f$ induces a bijection $[Z, X] \to [Z, Y]$, and letting $Z = Y$ we may choose $g \in \mathrm{Map}(Y, X)$ such that $[g] \in [Y, X]$ corresponds with $[1_Y]$ via $f$. This means $[f \circ g] = [1_Y]$. Now letting $Z = X$, note that since $f \circ g \sim 1_Y$ as above, this implies $[f \circ g \circ f] = [f]$, and hence both $[1_X]$ and $[g \circ f]$ map to $[f]$ via the induced map $[X, X] \to [X, Y]$, and since this map is assumed to be a bijection, $[1_X] = [g \circ f]$. So, in fact, $f$ is a homotopy equivalence.

If the induced map $\mathrm{Map}(Y, Z) \to \mathrm{Map}(X, Z)$ is a weak equivalence, then the same pattern of argument used above will show that $f$ is a weak equivalence. By choosing $g$ such that $[g] \in [Y, X]$ corresponds with $[1_X]$ via the bijection induced by $f$, we get $[g \circ f] = [1_X]$. Then noting that $[1_Y]$ and $[f \circ g]$ have equal image in $[X, Y]$ via the induced map since $g \circ f \sim 1_X$ shows that the classes must be equal, so $f \circ g \sim 1_Y$. $\square$

**Remark 1.3.6** If $f\colon X \to Y$ is a weak equivalence, it is not necessarily true that the induced maps $\mathrm{Map}(Z, X) \to \mathrm{Map}(Z, Y)$ and $\mathrm{Map}(Y, Z) \to \mathrm{Map}(X, Z)$ are weak equivalences. The following counterexample was communicated to us by Phil Hirschhorn. Let $X$ be the subspace of the plane that is the union of:

- for each positive integer $n$, the straight line joining $(0, 1)$ to $(1/n, 0)$;
- for each positive integer $n$, the straight line joining $(0, -1)$ to $(-1/n, 0)$;
- the straight line joining $(0, 1)$ and $(0, -1)$.

The map from $X$ to the one-point space $*$ is a weak equivalence, but $\mathrm{Map}(X, X)$ is not path-connected, since the identity map is not homotopic to a constant, and hence the induced map of mapping spaces $\mathrm{Map}(*, X) \to \mathrm{Map}(X, X)$ fails to be surjective on path components, since the domain is clearly path-connected (any two points can be joined by a path). Moreover, $X \to *$ cannot be a homotopy equivalence by Proposition 1.3.2.

However, if $Z$ is a CW complex and $f\colon X \to Y$ is a weak equivalence, then it is true that the induced map $\mathrm{Map}(Z, X) \to \mathrm{Map}(Z, Y)$ is also a weak equivalence. Similarly, if $X$ and $Y$ are CW complexes, then $\mathrm{Map}(Y, Z) \to \mathrm{Map}(X, Z)$ is a weak equivalence for all $Z$. The latter statement follows from Theorem 1.3.10 and Proposition 1.3.2.[3]                                                                    $\square$

One may rightly ask why in the definition of weak equivalence we favor maps from a CW complex rather than maps into a CW complex. The weak topology on a CW complex tells us that a map from a CW complex is (essentially) determined by its value on the cells, and as we will discuss below, it suffices to understand $[Z, X]$ when $Z$ is a single cell. In the next section we will discuss a local criterion, Proposition 1.4.4, which guarantees that a map is a weak equivalence.

Sometimes it is more convenient to work with CW complexes instead of general spaces owing to their extra structure. One such instance is in the proof of Theorem 4.2.1, the centerpiece of the first half of this book. Every space admits a weak equivalence from a CW complex. The proof of the following can be found in [Hat02, Proposition 4.13, Corollary 4.19] (see also [AGP02, Theorem 6.3.20]).

**Theorem 1.3.7** (CW Approximation Theorem)   *Any space $Y$ can be approximated by a CW complex $X$ in the sense that there exists a weak homotopy equivalence $f\colon Y \to X$. Moreover, any two such approximations have the same homotopy type.*

The relative version of the previous result is the following. For a proof, see [AGP02, Theorem 6.3.21].[4] A refinement appears later as Theorem 2.6.26.

---

[3]   In the more general settings of model categories, $\mathrm{Map}(-, -)$ preserves weak equivalences when the domains are cofibrant and the targets are fibrant (see, for example, [Hir03, Corollary 9.3.3]). In the setting of spaces, CW complexes are cofibrant and all spaces are fibrant, so that is how we get these two statements.

[4]   The statements of Theorem 1.3.7 and Theorem 1.3.8 in [AGP02] are slightly different than ours and require a connectivity hypothesis. However, it is straghtforward to reduce the general case presented here to the case with the connectivity hypothesis.

**Theorem 1.3.8** *Any pair* $(Y, B)$ *can be approximated by a CW pair* $(X, A)$ *in the sense that there exists a weak homotopy equivalence of pairs* $f: (X, A) \to (Y, B)$.

Furthermore, given a map $(Y, B) \to (Y', B')$, there is a map $(X, A) \to (X', A')$ of CW approximations where the obvious square commutes up to homotopy (meaning the two compositions are not equal but homotopic; see for example [Hat02, Proposition 4.18]). We can actually do better than this, and we thank Phil Hirschhorn for pointing out the following result, the detailed proof of which can be found in [Hir15a]. The idea is that, in the standard construction that yields commutativity up to homotopy, one chooses maps of spheres to represent elements of homotopy groups to be killed by attaching disks; in this one, no such choices are made, but instead one attaches cells using every possible map of a sphere. This construction thus attaches many more disks than are needed but produces commutativity on the nose.

**Theorem 1.3.9** *Suppose we have a commutative diagram*

$$
\begin{array}{ccc}
W & \xrightarrow{f} & X \\
\downarrow & & \downarrow \\
Y & \xrightarrow{g} & Z
\end{array}
$$

*Then there is a commutative diagram*

$$
\begin{array}{ccccc}
W & \xrightarrow{i} & X' & \xrightarrow{p} & X \\
\downarrow & & \downarrow & & \downarrow \\
Y & \xrightarrow{j} & Z' & \xrightarrow{q} & Z
\end{array}
$$

*such that* $i$ *and* $j$ *are inclusion maps of relative CW complexes,* $p$ *and* $q$ *are weak equivalences,* $p \circ i = f$, $q \circ j = g$, *and the vertical maps are the ones appearing in the first square.*

It is clear from the definitions that a homeomorphism is a homotopy equivalence and that a homotopy equivalence is a weak equivalence. However, for CW complexes weak equivalence and homotopy equivalence give the same equivalence classes of topological spaces.

**Theorem 1.3.10** (Homotopical Whitehead Theorem) *A map* $f: X \to Y$ *between CW complexes is a weak equivalence if and only if it is a homotopy equivalence.*

For a proof, see [Hat02, Theorem 4.5].

We express Theorems 1.3.7 and 1.3.10 by the following "equations":

$$\frac{\text{CW}}{\text{homotopy equivalence}} = \frac{\text{CW}}{\text{weak equivalence}} = \frac{\text{Spaces}}{\text{weak equivalence}}.$$

Many results in this book will be stated and proved (often in two different ways) for homotopy and weak equivalence. On occasion, we will state and prove an equivalence that may be weaker than the best one possible in order to present the cleanest proof or to stay true to the scope and focus of this book. We will indicate whenever this is the case to the best of our knowledge. That said, the preferred notion of equivalence in this book is weak equivalence.

## 1.4 Algebra: homotopy and homology

Here we give a brief overview of homotopy and homology groups. Details can be found in many sources, such as those referenced at the start of this chapter. Our presentation is aimed at comparing properties of homotopy and homology groups. Although these look vastly different from the definition, they share many important properties, but one important one enjoyed by homology but not by homotopy is "excision", which accounts for the difference in computational difficulty (homology groups being relatively easier to compute than homotopy groups). The major theorems of the first part of this book, Theorem 4.2.1 and Theorem 6.2.1, are concerned with the extent to which homotopy groups do satisfy excision. Facts in the first section on homotopy groups are essential to this text, whereas those in the section on homology are not.

### 1.4.1 Homotopy groups

In the special case of a map of pairs where $X = S^n$, $A = \{x_0\}$ is any point, and $B = \{y_0\}$, the common notation for $[(S^n, x_0), (Y, y_0)]$ is $\pi_n(Y, y_0)$. This is called the *nth homotopy group of Y*, denoted $\pi_n Y$ or $\pi_n(Y)$ when the basepoint is suppressed. In general $\pi_n(Y)$ is a pointed set for $n \geq 0$ (and it is the empty set for $Y$ the empty space), a group for $n \geq 1$, and an abelian group for $n \geq 2$. If $Y$ is empty, then only the set $\pi_0(Y) = \emptyset$ is defined. A map $f \colon (X, x_0) \to (Y, y_0)$ of based spaces induces a homomorphism of groups

$$f_* \colon \pi_n(X, x_0) \to \pi_n(Y, y_0) \tag{1.4.1}$$

for all $n$. The set $\pi_0$ is the set of *path components* of $Y$, and we say $Y$ is *path-connected* if this set consists of a single point. We say $Y$ is *simply-connected* if it is path-connected and $\pi_1(Y) = \{e\}$ is the trivial group. The sets/groups $\pi_n(Y, y_0)$ in general depend on $y_0$, but if $Y$ is a path-connected space they do not.

**Example 1.4.1** The homotopy groups of a one-point space are all trivial, i.e. $\pi_k(*) = \{e\}, k \geq 0$. □

**Proposition 1.4.2** *If based maps $f, g \colon (X, x_0) \to (Y, y_0)$ are homotopic, then the induced homomorphisms on homotopy groups are equal: $f_* = g_*$.*

We also consider homotopy classes of based maps of pairs,

$$[(D^n, \partial D^n, *), (Y, B, y_0)],$$

where $* \in \partial D^n$ and $y_0 \in B$ are the respective basepoints. In this case we write $\pi_n(Y, B, y_0)$ or simply $\pi_n(Y, B)$; these are called the *relative* homotopy groups. Clearly $\pi_n(Y, Y, y_0) = \{e\}$. They are only defined for $n > 0$, and are only groups for $n > 1$. None of them are defined if $B = \emptyset$. Equivalently, using the homeomorphism of pairs $(I^n, \partial I^n) \cong (D^n, \partial D^n)$, these could have been defined in terms of cubes; we will use this later in the text at various points, but for the most part it is convenient enough to work with disks. We refer the reader to [Hat02, Section 4.1] for a discussion of the group structure on these sets.

To the best of our knowledge, there does not exist a satisfactory definition of relative $\pi_0$. One could adopt one of two sensible approaches to defining a relative version of $\pi_0$. The first is to define it as homotopy classes of maps of pairs $(D^0, S^{-1}) \to (X, A)$, noting that this is the same as homotopy classes of maps $D^0 \to X$ since $S^{-1} = \emptyset$. This is undesirable since for $X = A \neq \emptyset$, $\pi_0(X, X)$ is not necessarily trivial; it is isomorphic to $\pi_0(X, x_0)$ (as unpointed sets). We will discuss the second approach after discussing the long exact sequence.

Associated to a pair $(X, A)$ is a long exact sequence of homotopy groups. For a space-level version, see Theorem 2.2.17. For a more standard (and detailed) treatment, see [Hat02, Theorem 4.3 and Theorem 2.20].

**Theorem 1.4.3** (Homotopy long exact sequence of a pair) *Let $(X, A)$ be a pair and $x_0 \in A$ be the basepoint. There is a long exact sequence*

$$\cdots \to \pi_n(A) \to \pi_n(X) \to \pi_n(X, A) \to \pi_{n-1}(A) \to \cdots \to \pi_0(A) \to \pi_0(X).$$

Exactness means the image of one map is equal to the kernel of the next map. This still makes sense near the end of the sequence where the group structures are not defined. The map $\pi_n(A) \to \pi_n(X)$ is the map induced by the inclusion, and the map $\pi_n(X) \to \pi_n(X, A)$ is similarly induced by the inclusion of maps of pairs $(D^n, \partial D^n, *) \to (X, A, x_0)$ which send $\partial D^n$ to the basepoint $x_0 \in A \subset X$. The *connecting map* or *boundary map* $\pi_n(X, A) \to \pi_{n-1}(A)$ is induced by the restriction of a map $(D^n, \partial D^n, *) \to (X, A, x_0)$ to $(\partial D^n, *) \to (A, x_0)$.

To finish the discussion about relative $\pi_0$ started above, the second idea for its definition, and in our opinion the most reasonable, is to declare $\pi_0(X, A, x_0) = \pi_0(X, x_0)/\pi_0(A, x_0)$. In this case the long exact sequence above can be extended to the right to $\pi_0(A) \to \pi_0(X) \to \pi_0(X, A) \to \{e\}$ provided $A \neq \emptyset$. The trouble occurs when $A = \emptyset$. In this case, the set is not even defined because of a lack of a basepoint. But we could then declare $\pi_0(X, \emptyset) = \pi_0(X)/\emptyset = \pi_0(X)_+$, since we have defined $\pi_0(\emptyset) = \emptyset$. We are therefore dealing with a sequence of sets $\pi_0(\emptyset) = \emptyset \to \pi_0(X) \to \pi_0(X)_+$, which cannot be sensibly called exact at $\pi_0(X)$. For one, this is not a sequence of pointed sets, so exactness is hard to pin down. Even still, if $X \neq \emptyset$, then the image of the empty set in $\pi_0(X)$ is of course the empty set, and this is not equal to the "kernel" of the inclusion $\pi_0(X) \to \pi_0(X)_+$ since the latter is clearly the component of $X$ containing the basepoint. One fix would be to declare the new basepoint of $\pi_0(X)_+$ to be the disjoint point added, but then $\pi_0(X) \to \pi_0(X)_+$ is no longer a map of pointed sets. If $X = \emptyset$, then defining relative $\pi_0$ to be empty is the only reasonable solution.

An important identity involving homotopy groups that we will have use for arises from the fact that the relationship from Theorem 1.2.9 passes to homotopy classes (see e.g. [Hat02, Section 4.3] for details). Namely, we have

$$\pi_{n+1}(X) \cong \pi_n(\Omega X). \tag{1.4.2}$$

For $f: X \to Y$ to be a weak equivalence, it is enough to verify the definition for $K = S^n$ for all $n \geq 0$. Thus for a map to be a weak equivalence it is enough to verify that it induces an isomorphism on path components and all homotopy groups, for every possible choice of compatible basepoints. The following is a consequence of our Proposition 2.6.17 where we use the case $(Z, A) = (D^j, \partial D^j)$ to show both injectivity and the surjectivity of the map $f_*$ and additionally setting $h$ to be the constant map to the basepoint for surjectivity.

**Proposition 1.4.4** *A map $f: X \to Y$ is a weak equivalence if and only if the induced map $f_*: \pi_0(X) \to \pi_0(Y)$ is an isomorphism and $f_*: \pi_k(X, x_0) \to \pi_k(Y, f(x_0))$ is an isomorphism for all $x_0 \in X$ and all $k \geq 1$.*

**Remarks 1.4.5**

- The characterization of a weak equivalence in terms of homotopy groups given above is sometimes taken as the definition of a weak equivalence. It is frequently incorrectly stated by saying $f$ is a weak equivalence if the induced map $f_*: \pi_k(X, x) \to \pi_k(Y, f(x))$ is an isomorphism for all $x \in X$ and for all $k \geq 0$. The problem with this is that if $X$ is empty, then the empty set is weakly equivalent to every space. This is why we separated the isomorphism on the path components from the isomorphisms on the higher homotopy groups.

- We have now two competing and non-equivalent notions of "equivalence up to homotopy" for spaces: homotopy equivalence and weak equivalence. The notion of homotopy equivalence is both natural and provides a strong duality between the notions of fibration and cofibration. Weak equivalence is the natural limiting notion of $k$-connected map, a concept which will be discussed Section 2.6 (also see below for a more elementary discussion). We will see as we progress that weak equivalence gives a weaker duality between fibrations and cofibrations, but as a limiting version of "$k$-connected map" it is more useful for approximations of the homotopy types of spaces.

□

The characterization of weak equivalences in Proposition 1.4.4 can shed some light on why the notion of weak equivalence is useful and natural. For a space $X$ a natural question is whether two points $x_0, x_1 \in X$ can be connected by a path $\gamma: I \to X$ (do $x_0$ and $x_1$ represent the same elements of $\pi_0(X)$?). If so, a next natural question to ask is whether any two such paths $\gamma, \gamma': I \to X$ are themselves homotopic relative to the boundary. Together $\gamma$ and $\gamma'$ determine a map $\Gamma: S^1 \to X$, and the existence of a homotopy between $\gamma$ and $\gamma'$ is a question about whether the class in $\pi_1(X)$ determined by $\Gamma$ is zero. If so, it determines a map $D^2 \to X$, and we can similarly ask whether any such two are homotopic relative to the boundary, and so on. Such questions reduce to questions about the homotopy groups $\pi_k(X)$.

## 1.4.2 Homology groups

While homotopy groups are clearly central to differentiating homotopy types, they are largely uncomputable. This stands in stark contrast to homology groups. The main difference in the two is "excision", which is the focus of most of our discussion here beyond the definitions. We focus on singular homology with integer coefficients, and give the standard treatment based on simplices. A treatment based on cubes would better parallel our later discussion of excision for homotopy groups (see Sections 4.4.2 and 6.3.2), but the technical details are too numerous to justify given our brevity with background on homology given here. None of the material in this section is crucial for this text and it is only meant to provide context for the story yet to be told.

For a space $X$ and integer $k \geq 0$, let $C_k(X)$ denote the free abelian group generated by maps $\sigma_k \colon \Delta^k \to X$, called *singular simplices*, and define $C_{-1}(X) = \{0\}$. We define a homomorphism $\partial_k \colon C_k(X) \to C_{k-1}(X)$, called a *differential*, by the unique linear extension of the map defined on a simplex $\sigma_k$ by

$$\partial_k(\sigma_k) = \sum_{i=0}^{k} (-1)^k \sigma_k|_{\partial_i \Delta^k}.$$

Then it is easy to check that $\partial_k \circ \partial_{k+1} = 0$ is the zero homomorphism (the alternating nature of the sum together with the combinatorics of the faces of a simplex make this work out), and so we have a *chain complex* of abelian groups

$$\cdots \xrightarrow{\partial_{k+2}} C_{k+1}(X) \xrightarrow{\partial_{k+1}} C_k(X) \xrightarrow{\partial_k} C_{k-1}(X) \xrightarrow{\partial_{k-1}} \cdots \to C_0(X) \to \{0\}.$$

Setting $Z_k(X) = \ker(\partial_k)$ (the $k$-cycles) and $B_k(X) = \mathrm{im}(\partial_{k+1})$ (the $k$-boundaries), since $\partial_k \circ \partial_{k+1} = 0$, $B_k(X) \subset Z_k(X)$ and we define the quotient group

$$H_k(X; \mathbb{Z}) = Z_k(X)/B_k(X),$$

and call it the *kth singular homology group*. For a pair $(X, A)$ we can define the *relative homology groups* by considering the quotient chain complex $C_k(X, A) = C_k(X)/C_k(A)$, which inherits a differential from $C_k(X)$, and defining $H_k(X, A; \mathbb{Z}) = Z_k(X, A)/B_k(X, A)$.

The group $H_0(X; \mathbb{Z})$ was defined by letting $C_{-1}(X) = \{0\}$. Alternatively, we can define the *reduced homology groups* $\widetilde{H}_k(X; \mathbb{Z})$ by instead using $C_{-1}(X) = \mathbb{Z}$, and letting $\partial_0 \colon C_0(X) \to \mathbb{Z}$ be defined by $\partial_0(\sum_i n_i \sigma_i) = \sum_i n_i$.

In parallel with Example 1.4.1, we have the following

**Example 1.4.6** The reduced homology groups of the one-point space are all trivial: $\widetilde{H}_k(*; \mathbb{Z}) = \{0\}$, $k \geq 0$. □

We also have the following parallel with (1.4.1).

**Proposition 1.4.7** *A continuous map* $f \colon X \to Y$ *induces homomorphisms* $f_* \colon \widetilde{H}_k(X; \mathbb{Z}) \to \widetilde{H}_k(Y; \mathbb{Z})$ *for all k.*

Moreover, associated with a pair $(X, A)$ we have a long exact sequence, much as with Theorem 1.4.3. For more details, see [Hat02, Theorem 2.16].

**Theorem 1.4.8** (Homology long exact sequence of a pair) *Let $(X, A)$ be a pair. There is a long exact sequence of abelian groups*

$$\cdots \to H_k(A) \to H_k(X) \to H_k(X, A) \to H_{k-1}(A) \to \cdots .$$

In contrast with homotopy groups, homology groups satisfy "excision."

**Theorem 1.4.9** (Homological excision) *For subspaces $A, B$ of a space $X$ whose interiors cover $X$, the inclusion of pairs $(B, A \cap B) \to (X, A)$ induces isomorphisms $H_k(B, A \cap B) \to H_k(X, A)$ for all $k$.*

Another formulation that better explains why Theorem 1.4.9 is referred to as excision is the following. Suppose $Z \subset A \subset X$ has the property that the closure of $Z$ is contained in the interior of $A$. Then the inclusion of pairs $(X - Z, A - Z) \to (X, A)$ induces isomorphisms $H_k(X - Z, A - Z) \to H_k(X, A)$ for all $k$. Putting $Z = X - B$ translates from the version above to this one. Thus excising the subspace $Z$ does not change the relative homology. This has many useful computational consequences that we will not explore here.

A different version of homology, called *cellular homology*, is usually better suited for computations. In this version, if $X$ is a CW complex (which we can always assume to be the case by Theorem 1.3.7), then one can construct the *cellular chain complex* which in $n$th degree has basis with one generator for each $n$-simplex of $X$. Denoting by $H_n^{CW}(X)$ the $n$th homology of this chain complex, we then have the following standard result the proof of which can be found, for example, in [Hat02, Theorem 2.35].

**Theorem 1.4.10** *For all $n \geq 0$, there are isomorphisms*

$$H_n^{CW}(X) \cong H_n(X).$$

Perhaps remarkably, the value of reduced homology groups on finite CW complexes is characterized by Example 1.4.6, Proposition 1.4.7, Theorem 1.4.8, and Theorem 1.4.9, in the sense that the value of $H_k(X)$ on a CW complex $X$ is entirely determined by the above data. From this perspective, the fact that homotopy groups satisfy analogs of all of these with the exception of excision means that homotopy and homology groups are more closely

related than is apparent from the definitions. We will explore an explicit relationship between homotopy and homology when we discuss the Hurewicz
Theorem, Theorem 4.3.2. We will also briefly note during our discussion of
the Dold–Thom Theorem, Theorem 2.7.23, that the reduced homology groups
as we have defined them here instead could instead have been defined as the
homotopy groups of the infinite symmetric product.

It is worth making some remarks about how one proves Theorem 1.4.9, since
a central theme of Part I this text is concerned with the extent to which homotopy groups satisfy excision. The following lemma, known as the "Lebesgue
Lemma" or the "Lebesgue Covering Lemma" is a key to excision arguments,
both the one used to prove Theorem 1.4.9 as well as those we will give later
verifying that excision for homotopy groups holds "in a range of dimensions".
For the proof, see for example [Mun75, Lemma 27.5].

**Lemma 1.4.11** (Lebesgue Lemma)    *Let $(X, d)$ be a compact metric space with
an open cover $\mathcal{U} = \{U_i\}_{i \in I}$. Then there exists a number $\delta > 0$ such that every
open set of diameter less than $\delta$ is contained in some $U_i$.*

There are a lot of technical details in the proof of Theorem 1.4.9, but the main
idea is to take a singular simplex $\sigma_k \colon \Delta^k \to X$ and barycentrically subdivide it
(roughly speaking, cut it into smaller simplices) so that the image of each of
these smaller simplices by the map $\sigma_k$ lies in the interiors of one of $A$ or $B$. This
can be achieved rather evidently by Lemma 1.4.11, since $\Delta^k$ inherits its metric
space structure by being a subspace of Euclidean space. This process leaves the
homology class $[\sigma_k] \in H_k(X; \mathbb{Z})$ unchanged, and in this sense the homology
classes on $A$ and $B$ determine (and are determined by) the classes on $X$.

Although homotopy groups do not in general satisfy excision (see the beginning of Section 4.1 for an example), a result related to Theorem 1.4.9 computes
the fundamental group $\pi_1(X)$ of a space covered by two open subsets. The
version of this theorem, known as the Seifert–van Kampen Theorem (sometimes just the "van Kampen Theorem"), presented below is a simplistic version
which serves a comparative purpose for us later on when we discuss excision
for homotopy groups and the Blakers–Massey Theorem. More general versions
are discussed both by Hatcher [Hat02, Section 1.2] and even more generally
by May [May99, Chapter 2.7].

For groups $G$ and $H$, we let $G * H$ denote their free product, and for homomorphisms $\phi_g \colon K \to G$ and $\phi_H \colon K \to H$ we let $G *_K H$ denote the quotient of
$G * H$ by the normal subgroup generated by elements of the form $\phi_G(k)\phi_H(k)^{-1}$
for $k \in K$.

**Theorem 1.4.12** (Seifert–van Kampen Theorem)  *Let $(X, x_0)$ be a pointed space with path-connected open subsets $A, B$ containing the basepoint $x_0$ such that $X = A \cup B$ and $A \cap B$ contains the basepoint and is also path-connected. Then*

$$\pi_1(X, x_0) \cong \pi_1(A, x_0) *_{\pi_1(A \cap B, x_0)} \pi_1(B, x_0).$$

# 2

# 1-cubes: Homotopy fibers and cofibers

Cubical diagrams are the central objects of study in this book. Some of these are quite familiar: a 0-cube of spaces is just a space $X$, and a 1-cube is a map of 0-cubes, or simply a map $f \colon X \to Y$ of spaces. Spaces were the subject of the previous chapter, and maps are the subject of this one. Maps of maps are squares, or 2-cubes, and these will be studied Chapter 3, although they will appear at times here (sometimes implicitly when we reference results from the next chapter).

A natural question in topology asks how well a map $f \colon X \to Y$ captures the difference between $X$ and $Y$. One might try to answer this by looking at its fibers or cofiber:

**Definition 2.1** For a map $f \colon X \to Y$ and for $y \in Y$, the space $F_y = f^{-1}(y)$ is called the *fiber of $f$ over $y$*. The space $Y/f(X)$ is called the *cofiber* of $f$.

If $f$ is a homeomorphism, then the fiber and cofiber of $f$ are both the one-point space.[1] The problem is that these spaces do not behave well under homotopy, as illustrated by the following examples.

**Example 2.2** (Fiber not homotopy invariant)  Consider the unit circle $S^1 \subset \mathbb{R}^2$, the projection map to the $x$-axis $f \colon S^1 \to \mathbb{R}$, and the unique map $f' \colon S^1 \to *$. While $\mathbb{R} \simeq *$, the fiber of $f$ is either empty, one point, or two points, the fiber of $f'$ is $S^1$. □

**Example 2.3** (Cofiber not homotopy invariant)  Let $f \colon S^1 \to D^2$ be the usual inclusion of $S^1$ as the boundary of $D^2$ and compare this map to $f' \colon S^1 \to *$.

---

[1] The fibers of the identity map $1_\emptyset \colon \emptyset \to \emptyset$ are empty but its cofiber is the one-point space.

Again the corresponding spaces are homotopy equivalent but the cofiber of $f$ is $S^2$ while the cofiber of $f'$ is the one-point space. □

However, if $f$ is a *fibration*, the fibers are preserved by homotopy equivalences (see Theorem 2.1.20), and if $f$ is a *cofibration*, the cofibers are preserved by homotopy equivalences (see Theorem 2.3.17). This is the reason fibrations and cofibrations are the most important kinds of maps in this book. In this chapter, we shall see how any map can be "replaced" by a fibration or a cofibration. Before studying these replacements, we review some of the most important properties of fibrations and cofibrations. More details can be found in many introductory algebraic topology books, including [AGP02, Hat02, May99].

## 2.1 Fibrations

**Definition 2.1.1** A map $p : X \to Y$ is a *fibration* if it satisfies the *homotopy lifting property*. That is, $p$ is a fibration if for all spaces $W$ and commutative diagrams

$$
\begin{array}{ccc}
W & \xrightarrow{\ g\ } & X \\
{\scriptstyle i_0}\downarrow & {\scriptstyle \widehat{h}}\nearrow & \downarrow{\scriptstyle p} \\
W \times I & \xrightarrow{\ h\ } & Y
\end{array}
\tag{2.1.1}
$$

the dotted arrow $\widehat{h}$ exists. Here $i_0$ is the inclusion of $W$ in $W \times I$ sending $w$ to $(w, 0)$.

For example, the unique map $X \to *$ to the one-point space is a fibration for all spaces $X$ (even $X = \emptyset$). Every homeomorphism is a fibration. Another example is a covering space. Fiber bundles and vector bundles are also important examples of fibrations.

**Remark 2.1.2** The above defines a *Hurewicz* fibration. If $p$ has the lifting property with respect to CW complexes $W$, then it is called a *Serre* fibration. In general one can define a $C$-fibration to be a map which has the homotopy lifting property with respect to a class of topological spaces $C$. □

Returning to Definition 2.1.1, we often refer to $X$ as the *total space* of the fibration $p$ and $Y$ as the *base space*, and all three $F_y \to X \to Y$ as a *fibration sequence* (where $F_y$ is as usual the fiber of $p$ over $y \in Y$). We will also say that $X$ *fibers over* $Y$.

Many authors require their fibrations to be surjective maps. We make no such assumption, but note the following.

**Proposition 2.1.3** (Fibrations are (mostly) surjective)   *Let $p\colon X \to Y$ be a fibration. Suppose $y \in Y$ is in the image of $p$, and $y' \in Y$ is in the same path component as $y$. Then $y'$ is also in the image of $p$.*

*Proof*   To see this, in the definition above take $W = *$, $g\colon * \to X$ the inclusion of $x$ in $X$ for $x$ a preimage of $y$, and $h\colon * \times I \to Y$ a path from $p(x)$ to a point $y'$. Then $\widehat{h}(*, 1)$ is a point of $X$ which maps to $y'$ via $p$.   □

Here are a few more useful examples and elementary results.

**Example 2.1.4** (Product of fibrations)   If $p_i\colon X_i \to Y_i$ are fibrations for $i = 1, 2$, then so is their product $p_1 \times p_2\colon X_1 \times X_2 \to Y_1 \times Y_2$. More generally, if $\mathcal{I}$ is any indexing set and $p_i\colon X_i \to Y_i$ is a fibration for each $i$, the map $\prod_i p_i\colon \prod_i X_i \to \prod_i Y_i$ is a fibration.   □

**Example 2.1.5** (Trivial fibration)   Let $X = Y \times F$, and let $p\colon X \to Y$ be the projection. This is a fibration, called the *trivial* fibration.   □

**Example 2.1.6** (Hopf fibration)   Identify $S^2 = \mathbb{C} \cup \infty$ and $S^3$ as the unit sphere $|z_0|^2 + |z_1|^2 = 1$ in $\mathbb{C}^2$. The *Hopf fibration* is the map $h\colon S^3 \to S^2$ given by $h(z_0, z_1) = z_1/z_0$. The fibers of this map are circles, which can be seen by writing the map $h$ in polar coordinates:

$$h(r_0 e^{i\theta_0}, r_1 e^{i\theta_1}) = (r_0/r_1)e^{i(\theta_0 - \theta_1)}.$$

Then for $re^{i\theta} \in S^2$, there is a unique $(r_0, r_1)$ such that $r = r_0/r_1$ and $r_0^2 + r_1^2 = 1$, and clearly then the fiber is parametrized by one of the angles $\theta_i$. We leave to the reader the verification that this map is a fibration. We thus obtain a fibration sequence $S^1 \to S^3 \to S^2$ which is called the *Hopf fibration*. For more on this and related fibration sequences involving spheres, see [Hat02, Example 4.45].   □

**Proposition 2.1.7** (Evaluation map)   *Let $X$ and $Y$ be spaces, and let $x_0 \in X$ be the basepoint. The evaluation map*

$$ev_{x_0}\colon \operatorname{Map}(X, Y) \longrightarrow Y$$

*given by $f \mapsto f(x_0)$ is a fibration.*

*Proof* We may replace $Y$ with $\mathrm{Map}(\{x_0\}, Y)$, and regard the evaluation map as the restriction map $\mathrm{Map}(X, Y) \to \mathrm{Map}(\{x_0\}, Y)$. We must show the dotted arrow exists in the following lifting problem:

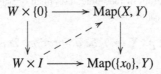

By Theorem 1.2.7, the top and bottom horizontal maps correspond respectively to maps $X \to \mathrm{Map}(W \times \{0\}, Y)$ and $\{x_0\} \to \mathrm{Map}(W \times I, Y)$. Moreover, they fit into a commutative diagram

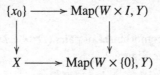

where the vertical maps are, from left to right, the inclusion of the basepoint in $X$ and the restriction to $W \times \{0\}$. Again using Theorem 1.2.7 on the horizontal maps, we have maps $\{x_0\} \times I \to \mathrm{Map}(W, Y)$ and $X \times \{0\} \to \mathrm{Map}(W, Y)$ which agree on $\{x_0\} \times \{0\} = \{x_0\} \times I \cap X \times \{0\}$ since the square above commutes. Hence we have a map

$$b : \{x_0\} \times I \cup X \times \{0\} \longrightarrow \mathrm{Map}(W, Y).$$

Since $X$ is well-pointed (we always assume this), there exists a retraction $r :$ $X \times I \to \{x_0\} \times I \cup X \times \{0\}$, and so we may define a map

$$b \circ r : X \times I \longrightarrow \mathrm{Map}(W, Y).$$

The map $b \circ r$ can be regarded using Theorem 1.2.7 as a map $W \times I \to$ $\mathrm{Map}(X, Y)$, and it is straightforward to check that this gives the desired lift. □

**Remark 2.1.8** This result follows immediately from Proposition 2.5.1, and we prefer the perspective of that result although we have not yet discussed cofibrations. Its proof follows the same general outline as the one above. The crucial point is the well-pointedness of the space $X$. □

At times we will want to solve a lifting problem when the space $W$ in Definition 2.1.1 is a CW complex. The following tells us such solutions can be built locally, cell by cell.

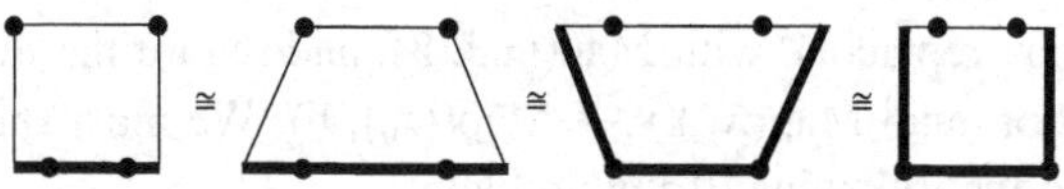

Figure 2.1  A picture proof of the homeomorphism of pairs $(D^k \times I, D^k \times \{0\}) \cong (D^k \times I, D^k \times \{0\} \cup_{\partial D^k \times \{0\}} \partial D^k \times I)$.

**Proposition 2.1.9**   *Let* $p \colon X \to Y$ *be a fibration. Then there exists a solution to the lifting problem*

$$
\begin{array}{ccc}
D^k \times \{0\} \cup_{\partial D^k \times \{0\}} \partial D^k \times I & \xrightarrow{\;g\;} & X \\
{\scriptstyle i}\downarrow & \nearrow & \downarrow{\scriptstyle p} \\
D^k \times I & \xrightarrow{\;h\;} & Y
\end{array}
$$

*Proof*   There is a homeomorphism of pairs $(D^k \times I, D^k \times \{0\}) \cong (D^k \times I, D^k \times \{0\} \cup_{\partial D^k \times \{0\}} \partial D^k \times I)$ (see Figure 2.1).

□

Solutions to the lifting problem posed in the definition of fibration are not unique, but any two lifts are homotopic.

**Proposition 2.1.10**   *Suppose* $p \colon X \to Y$ *is a fibration, and consider the lifting problem*

$$
\begin{array}{ccc}
W \times \{0\} & \xrightarrow{\;g\;} & X \\
{\scriptstyle i_0}\downarrow & \nearrow & \downarrow{\scriptstyle p} \\
W \times I & \xrightarrow{\;h\;} & Y
\end{array}
$$

*If* $\hat{h}_0, \hat{h}_1 \colon W \times I \to X$ *are solutions to this lifting problem, then there is a homotopy* $\hat{h}_0 \sim \hat{h}_1$ *relative to* $W \times \{0\}$.

*Proof*   Since $p \circ \hat{h}_0 = h = p \circ \hat{h}_1$, we let $H \colon W \times I \times I \to Y$ be the constant homotopy from $h$ to itself: $H(w, s, t) = h(w, t)$. Consider the lifting problem

$$
\begin{array}{ccc}
W \times (I \times \{0\} \cup \partial I \times I) & \xrightarrow{\;c\;} & X \\
{\scriptstyle i}\downarrow & \nearrow & \downarrow{\scriptstyle p} \\
W \times I \times I & \xrightarrow{\;H\;} & Y
\end{array}
$$

where $c \colon W \times (I \times \{0\} \cup \partial I \times I) \to X$ is defined as the composition $W \times I \times \{0\} \to W \to X$ of the natural projection with $g$, $\hat{h}_0$ on $W \times \{0\} \times I$, and $\hat{h}_1$ on $W \times \{1\} \times I$. The homeomorphism of pairs $(I \times I, I \times \{0\}) \cong (I \times I, I \times \{0\} \cup \partial I \times I)$ (as in Proposition 2.1.9) transforms the above diagram into one of the form appearing in Definition 2.1.1, with $W \times I$ in place of $W$. Hence a solution $\widehat{H} \colon W \times I \times I \to X$ exists. Moreover, $\widehat{H}_1 = \widehat{H}|_{W \times \{1\} \times I}$ has the property that $\widehat{H}_1(z, 0) = \hat{h}_0$ and $\widehat{H}_1(z, 1) = \hat{h}_1$, giving the desired homotopy. □

We then have the following useful result which tells us that maps into a fibration induce a fibration of mapping spaces.

**Proposition 2.1.11** (Maps preserve fibrations) *Suppose $p \colon X \to Y$ is a fibration. Then, for any space $Z$, the induced map*

$$p_* \colon \operatorname{Map}(Z, X) \longrightarrow \operatorname{Map}(Z, Y)$$

*is a fibration. If $F_y$ is the fiber of $p$ over $y \in Y$, then the fiber of $p_*$ over the constant map which sends $Z$ to $y$ is $\operatorname{Map}(Z, F_y)$. The same result is true if $X$ and $Y$ are based and $p$ is a based map.*

*Proof* In the lifting problem

$$
\begin{array}{ccc}
W \times \{0\} & \longrightarrow & \operatorname{Map}(Z, X) \\
\downarrow & \nearrow & \downarrow {\scriptstyle p_*} \\
W \times I & \longrightarrow & \operatorname{Map}(Z, Y)
\end{array}
$$

we can regard the horizontal arrows as elements of $\operatorname{Map}(W \times Z, X)$ and $\operatorname{Map}(W \times Z \times I, Y)$ respectively, which leads us to the lifting problem

$$
\begin{array}{ccc}
W \times Z \times \{0\} & \longrightarrow & X \\
\downarrow & \nearrow & \downarrow {\scriptstyle p} \\
W \times Z \times I & \longrightarrow & Y
\end{array}
$$

In this case the lift $W \times Z \times I \to X$ exists since $X \to Y$ is a fibration. Then, regarding this lift as a map $W \times I \to \operatorname{Map}(Z, X)$ using Theorem 1.2.7, it is straightforward to check that this is the desired lift in the original square.

The proof is the same in the based case, and in both cases the identification of the fibers is straightforward. □

Since the loop space is a mapping space, we immediately have the following

**Corollary 2.1.12** (Loops preserve fibration sequences) *If $F \to X \to Y$ is a based fibration sequence, then so is*

$$\Omega F \longrightarrow \Omega X \longrightarrow \Omega Y.$$

*Here the maps are the "loopings" of the maps in the original fibration sequence (see (1.2.1)), and the basepoint of $\Omega Y$ is the constant loop at the basepoint of $Y$.*

Fibrations satisfy a number of properties important to us. The first, and most basic, says that a fibration $p: X \to Y$ gives a measure of the difference in homotopy between $X$ and $Y$.

**Theorem 2.1.13** (Homotopy long exact sequence of a fibration)   *Let $p : X \to Y$ be a fibration of based spaces, where $x_0 \in X$ and $y_0 \in Y$ are the basepoints of $X$ and $Y$ respectively, and let $F = p^{-1}(y_0)$ be the fiber over $y_0$ with basepoint $x_0$. Then there is a long exact sequence of groups*

$$\cdots \to \pi_n(F) \to \pi_n(X) \to \pi_n(Y) \to \pi_{n-1}(F) \to \cdots$$

*(where we have omitted the basepoints in the notation for brevity). This sequence ends at $\pi_0(Y)$.*

For a proof of this theorem, see, for example, [Hat02, Theorem 4.41]. Note that implicit in our assumption above is that the fibers are non-empty; there is no long exact sequence for basepoints $y_0$ for which $F = \emptyset$.

**Remark 2.1.14**   For a trivial fibration $p: Y \times F \to Y$, the *connecting map* $\pi_n(Y) \to \pi_{n-1}(F)$ in the long exact sequence above is equal to zero, because any map $s: Y \to Y \times F$ which is the identity on the $y$-coordinate produces a map $\pi_n(Y) \to \pi_n(Y \times F)$ whose composition with the map to $\pi_n(Y)$ is the identity, and exactness then demands the map $\pi_n(Y) \to \pi_{n-1}(F)$ be zero. More generally if $p: X \to Y$ is a fibration and there exists a section $s: Y \to X$ of $p$, then the connecting map will be zero. In general, the connecting map of course need not be zero.                                                                    □

**Remark 2.1.15**   The long exact sequence in Theorem 2.1.13 is related to the long exact sequence of the pair $(X, F)$ (see Theorem 1.4.3) in that

$$\pi_n(X, F) \xrightarrow{\cong} \pi_n(Y),$$

where the isomorphism is induced by $p$. Noting that $p(F) = y_0$, we see that a map of pairs $(D^n, \partial D^n) \to (X, F)$ gives rise to a based map $D^n/\partial D^n \to Y$ upon composing the map $D^n \to X$ with $Y$.                                                            □

Clearly the composition of fibrations is a fibration, but there are other useful constructions which preserve this property. One is that the class of fibrations is stable under "pullback". For a fibration $p: X \to Y$ and a map $f: Z \to Y$, define the *pullback (of $Y$ by $f$)* as

$$f^*Y = X \times_Y Z = \{(x, z) \in X \times Z : f(x) = p(y)\}.$$

This is a space "over $Z$" in a sense that it is equipped with a map to $Z$, namely the projection. Pullbacks will be studied in detail in Section 3.1 (see Definition 3.1.1 in particular), though material from that section is not necessary for the following result.

**Proposition 2.1.16** (Fibrations are preserved by pullbacks)  *Suppose $p\colon X \to Y$ is a fibration, and $f\colon Z \to Y$ is a map. Then the map*

$$p^*\colon f^*Y \longrightarrow Z$$
$$(x,z) \longmapsto z$$

*is a fibration, and the fibers of $p^*$ over $z$ are canonically homeomorphic to the fibers of $p$ over $f(z)$.*

*Proof*   Consider the following commutative diagram

$$
\begin{array}{ccccc}
W & \longrightarrow & X \times_Y Z & \longrightarrow & X \\
{\scriptstyle i_0}\downarrow & \overset{\widehat{h}}{\nearrow} & \downarrow {\scriptstyle p^*} & & \downarrow {\scriptstyle p} \\
W \times I & \overset{h}{\longrightarrow} & Z & \overset{f}{\longrightarrow} & Y
\end{array}
$$

We are interested in whether the dotted arrow exists. Since $p$ is a fibration, there exists a lift $\widehat{f \circ h}\colon W \times I \to X$. Define $\widehat{h}\colon W \times I \to X \times_Y Z$ by

$$\widehat{h}(w,t) = (\widehat{f \circ h}(w,t), h(w,t)).$$

It is straightforward to check both the commutativity of the diagram and that the fibers of $p$ and $p^*$ are equal.   □

**Remark 2.1.17**   The above property also follows immediately from the universal property of the pullback discussed after Definition 3.1.1.   □

The following result is proved as part of Corollary 3.2.18 with tools developed in Chapter 3 and it is logically independent of those that follow. Alternatively, see, for example, [Hat02, Proposition 4.61].

**Proposition 2.1.18** (Fibrations have equivalent fibers)  *If $p\colon X \to Y$ is a fibration, then the fibers $F_y = p^{-1}(y)$ are homotopy equivalent for all $y$ in the same path component of $Y$.*

This result allows us to speak of *the* fiber of a fibration without too much ambiguity.

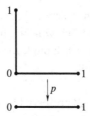

Figure 2.2 Map with fibers that are homotopy equivalent but which is not a fibration.

**Example 2.1.19** It is not enough for all the fibers of a map to be homotopy equivalent for that map to be a fibration. Let $I_1$ and $I_2$ be unit intervals with basepoint 0, let $L = I_1 \vee I_2$ be the wedge of two intervals, and define $p: L \to I$ by

$$p(t) = \begin{cases} 0, & \text{if } t \in I_1; \\ t, & \text{if } t \in I_2. \end{cases}$$

See Figure 2.2 for the picture of $p$.

The fibers of this map are either an interval $I$ or a point, all of which are clearly homotopy equivalent. This map is not a fibration. Let $h: I \to I$ be the identity map, and define $g: * \to L$ by $g(*) = 1 \in I_1$. Then in the commutative diagram

$$\begin{array}{ccc} * & \xrightarrow{\ g\ } & L \\ {\scriptstyle i_0}\Big\downarrow & \nearrow & \Big\downarrow{\scriptstyle p} \\ * \times I & \xrightarrow{\ f\ } & I \end{array}$$

the dotted arrow clearly cannot exist, so $p$ fails to have the homotopy lifting property. □

While this last example is not a fibration, it is the standard basic example of what is called a quasifibration, a useful and important notion we will discuss in Section 2.7.

The next result says that the homotopy type of the fibers is invariant under homotopy equivalences of the base and total space.

**Theorem 2.1.20** *Suppose in the commutative diagram*

$$\begin{array}{ccc} X_1 & \xrightarrow{\ f_X\ } & X_2 \\ {\scriptstyle p_1}\Big\downarrow & & \Big\downarrow{\scriptstyle p_2} \\ Y_1 & \xrightarrow{\ f_Y\ } & Y_2 \end{array}$$

*the vertical maps $p_1$ and $p_2$ are fibrations with fibers $F_1 = \text{fiber}_{x_1}(p_1)$ and $F_2 = \text{fiber}_{f_Y(x_1)}(p_2)$, respectively, and the maps $f_X$ and $f_Y$ are homotopy equivalences. Then the induced map of fibers $F_1 \to F_2$ is a homotopy equivalence.*

**Remark 2.1.21** If one only cares about the weak homotopy type then this follows immediately from Theorem 2.1.13 by comparing the long exact sequence of each fibration and using the five lemma. While we like this breezy explanation, we present a proof for homotopy equivalences that uses the homotopy lifting property and produces a homotopy, rather than weak, equivalence of the fibers. Our proof is based on [May99, Section 7.4]. □

*Proof of Theorem 2.1.20* Using Proposition 2.1.16, it is enough to prove this in the case where $Y_1 = Y_2$ and $f_Y = 1_Y$, as the fibration $p_2^*: X_2 \times_{Y_2} Y_1 \to Y_1$ has the same fibers as $p_2$ and the natural map $e: X_1 \to X_2 \times_{Y_2} Y_1$ has the property that $p_2^* \circ e = p_1$. Henceforth let us assume we have a diagram

where $f$ is a homotopy equivalence. Our goal is to show that the induced map of fibers $F_1 \to F_2$ over $y \in Y$ is a homotopy equivalence. Let $g': X_2 \to X_1$ be a homotopy inverse to $f$. Note that any map homotopic to $g'$ is also a homotopy inverse to $f$. The map $g'$ may not induce a map $F_2 \to F_1$, and the homotopies between $g' \circ f$ (resp. $f \circ g'$) and $1_{X_1}$ (resp. $1_{X_2}$) may not restrict to homotopies which remain within the fibers. The idea is to use the homotopy lifting property to make a homotopy of $g'$ and then a homotopy of those homotopies so that the above two conditions are true.

Since $f \circ g' \sim 1_{X_2}$ and $p_2 \circ f = p_1$, we have that $p_1 \circ g' \sim p_2$. Let $h': X_2 \times I \to Y$ be a homotopy from $p_1 \circ g'$ to $p_2$. Consider the lifting problem

$$
\begin{array}{ccc}
X_2 \times \{0\} & \overset{g'}{\longrightarrow} & X_1 \\
{\scriptstyle i_0}\big\downarrow & \nearrow & \big\downarrow{\scriptstyle p_1} \\
X_2 \times I & \underset{h'}{\longrightarrow} & Y
\end{array}
$$

Since $p_1$ is a fibration, the dotted arrow exists, and we see that this implies $g'$ is homotopic to a map $g$ such that $p_1 \circ g = p_2$. We claim that the restrictions of $f|_{F_1}: F_1 \to F_2$ and $g|_{F_2}: F_2 \to F_1$ to the fibers are homotopy inverses of one another.

To see this, let $h: X_1 \times I \to X_1$ be a homotopy from $g \circ f$ to $1_{X_1}$. Consider the lifting problem

$$
\begin{array}{ccc}
X_1 \times \{0\} & \xrightarrow{\ g \circ f\ } & X_1 \\
{\scriptstyle i_0} \downarrow & \nearrow & \downarrow {\scriptstyle p_1} \\
X_1 \times I & \xrightarrow{\ p_1 \circ h\ } & Y
\end{array}
$$

Since $p_1$ is a fibration, the dotted arrow exists, and it is a map $\widehat{p_1 \circ h}: X_1 \times I \to X_1$ with the properties that $\widehat{p_1 \circ h}(-, 0) = g \circ f$, $\widehat{p_1 \circ h}(-, 1) = 1_{X_1}$ and that the lower triangle commutes. This means that the family of maps $H_t: X_1 \to X_1$ defined by this homotopy has the property that

$$
\begin{array}{ccc}
X_1 & \xrightarrow{\ H_t\ } & X_1 \\
& {\scriptstyle p_1} \searrow & \downarrow {\scriptstyle p_1} \\
& & Y
\end{array}
$$

commutes for every $t$, which tells us that the restriction of $H_t$ is a map from $F_1$ to itself for all $t$. The homotopy from $f \circ g$ to $1_{X_2}$ is treated in a similar manner. $\qquad\square$

A direct consequence of the proof of Theorem 2.1.20, which we will need to establish the homotopy invariance of the homotopy pullback, Theorem 3.2.12, is the following.

**Corollary 2.1.22** *Let*

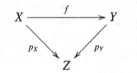

*be a commutative diagram where the maps to $Z$ are fibrations. If $f$ is a homotopy equivalence, then there exists a homotopy inverse $g: Y \to X$ such that*

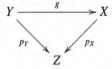

*commutes. In short, $f$ is a homotopy equivalence if and only if it is a homotopy equivalence of spaces which fiber over $Z$.*

A related useful result is that the pullback of a homotopy (resp. weak) equivalence by a fibration is a homotopy (resp. weak) equivalence.

**Proposition 2.1.23** *If $p\colon Y \to Z$ is a fibration and $f\colon X \to Z$ is a homotopy equivalence (resp. weak equivalence), then $X \times_Z Y \to Y$ is also a homotopy equivalence (resp. weak equivalence).*

The reader may wish to look back at the following proof after studying pull-backs, specifically Definition 3.1.1 and the discussion immediately following. In fact, this proposition will be restated in terms of pullbacks as Proposition 3.2.3. The reader should also consult Remark 3.2.2 for an explanation of why we state and prove this result for both weak and homotopy equivalences. The proof below is inspired by Theorem 13.1.2 of [Hir03], who in turn cites an unpublished manuscript of Reedy.

*Proof* Write $W = X \times_Z Y$ and let $g\colon W \to Y$ and $p'\colon W \to X$ be the projections. By Proposition 1.3.2 it is enough to show that the induced map $g_*\colon [V, W] \to [V, Y]$ is a bijection for all $V$. To show that $g_*$ is surjective, let $v\colon V \to Y$ be a map, and consider the composed map $p \circ v\colon V \to Z$. Since $f$ is a homotopy equivalence, it follows again from Proposition 1.3.2 that there exists a map $v'\colon V \to X$ such that $f \circ v' \sim p \circ v$. Let $H\colon V \times I \to Z$ be a homotopy from $p \circ v$ to $f \circ v'$. Consider the lifting problem

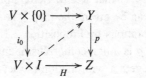

Since $p$ is a fibration there exists a solution $\hat{H}\colon V \times I \to Y$ to this lifting problem, and if we put $v'' = \hat{H}|_{V\times\{1\}}\colon V \to Y$, then since $\hat{H}|_{V\times\{0\}} = v$, $v'' \sim v$ and $p \circ v'' = H|_{V\times\{1\}} = f \circ v'$. Now define $u\colon V \to W$ by $u = (v', v'')$. The fact that $f \circ v' = p \circ v''$ ensures this defines a map to $W = X \times_Z Y$. Moreover, $g \circ u = v'' \sim v$, so $g_*$ is surjective.

The proof that $g_*$ is injective is similar. Let $v, v'\colon V \to W$ be maps such that $g \circ v \sim g \circ v'\colon V \to Y$. We will construct a homotopy $v \sim v'$. Let $H$ be a homotopy from $g \circ v$ to $g \circ v'$. This gives rise to a homotopy $p \circ H\colon V \times I \to Z$ from $p \circ g \circ v = f \circ p' \circ v$ to $p \circ g \circ v' = f \circ p' \circ v'$. Since $f$ is a homotopy equivalence, Proposition 1.3.2 applied to the bijection $f_*\colon [V \times I, X] \to [V \times I, Z]$ implies that $p' \circ v \sim p' \circ v'$. Choose a homotopy $G\colon V \times I \to Y$ from $p' \circ v$ to $p' \circ v'$; thus $f \circ G \sim p \circ H$ relative to $V \times \partial I$. Let $K$ be a homotopy from $f \circ G$ to $p \circ H$ relative to $V \times \partial I$. Let $c\colon V \times (I \times \{0\} \cup \partial I \times I) \to W$ denote the map whose restriction to $V \times I$ is $p \circ H$, $\{0\} \times I$ is $p \circ g \circ v$, and $\{1\} \times I$ is $p \circ g \circ v'$. Consider the lifting problem

$$V \times (I \times \{0\} \cup \partial I \times I) \xrightarrow{\ c\ } W$$

$$i_0 \downarrow \qquad\qquad \downarrow p'$$

$$V \times I \times I \xrightarrow{\ K\ } X$$

There is a homeomorphism of pairs $(I \times I, I \times \{0\}) \cong (I \times I, I \times \{0\} \cup \partial I \times I)$, and the map $p'$ is a fibration by Proposition 2.1.16, and so by Proposition 2.1.9 a solution $\hat{K} \colon V \times I \times I$ exists with the properties

- $\hat{K}|_{V \times I \times \{0\}} = p \circ H$;
- $\hat{K}|_{V \times \{0\} \times I} = p \circ g \circ v$;
- $\hat{K}|_{V \times \{1\} \times I} = p \circ g \circ v'$;
- $p' \circ \hat{K}|_{V \times I \times \{1\}} = f \circ G$.

Let $\hat{K}_1 = \hat{K}|_{V \times I \times \{1\}}$. Define $U \colon V \times I \to W$ by $u = (\hat{K}_1, G)$; we have already seen that $p' \circ \hat{K}_1 = f \circ G$ so this defines a map to $W$, and the properties above tell us it represents the desired homotopy.

The argument in the case of a weak equivalence is identical to the above, except we only need restrict attention to CW complexes $V$ and note that $V \times I$ can be given a CW structure (the reader may additionally note that all we need to know about $p$ is that it is a Serre fibration). But a simpler, more algebraic, argument can be given: the map $f$ is a weak equivalence and both maps $p'$ and $p$ are fibrations by Proposition 2.1.16. The map from the fibers of $p'$ to the fibers of $p$ is a homeomorphism, and hence a weak equivalence, and the five lemma together with the long exact sequence of a fibration from Theorem 2.1.13 finishes the argument. □

We finish with a statement about how fibrations are locally determined. A proof can be found in [May99, Chapter 7.4].

**Proposition 2.1.24**  *Suppose $p \colon X \to Y$ is a map and suppose there exists a countable open cover $\{U_i\}_{i \in I}$ of $Y$ such that*

$$p|_{U_i} \colon p^{-1}(U_i) \longrightarrow U_i$$

*is a fibration for all $i$. Then $p$ is a fibration.*

## 2.2  Homotopy fibers

In the sense of Theorem 2.1.13, the fibers of a map can give one measure of the difference between $X$ and $Y$. In general though, given a commutative diagram

$$X_1 \longrightarrow X_2$$

$$f_1 \Big\downarrow \qquad \Big\downarrow f_2$$

$$Y_1 \longrightarrow Y_2 \qquad\qquad (2.2.1)$$

where the horizontal arrows are homotopy equivalences, it is not necessarily true that the induced map from the fibers of $f_1$ to the fibers of $f_2$ is a homotopy equivalence. This was illustrated by Example 2.2. However, if $f_1$ and $f_2$ are fibrations, the induced map of fibers *is* a homotopy equivalence by Theorem 2.1.20. Fortunately, there is a way to turn any map of spaces $f: X \to Y$ into a fibration, but it comes at the cost of replacing $X$ with a more complicated but homotopy equivalent space. This is accomplished essentially by using the fact that the evaluation map $\mathrm{Map}(I, Y) \to Y$ is a fibration by Proposition 2.1.7 and that the space of maps of $I$ into $Y$ which are fixed at one end is contractible.

**Definition 2.2.1** Given a map $f: X \to Y$, define the *mapping path space of* $f$, denoted by $P_f$, to be the subspace of $X \times \mathrm{Map}(I, Y)$ given by

$$P_f = \{(x, \alpha) \ : \ x \in X, \alpha: I \to Y, \alpha(0) = f(x)\}.$$

For instance, the mapping path space of the identity map $1_X: X \to X$ is homeomorphic with the space of paths $X^I$. See Example 3.1.6 for an alternative description of the mapping path space in terms of limits.

Let $c_y: I \to Y$ be the constant map at $y$. Then it is clear that $f$ factors through $P_f$ as

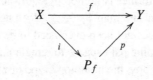

where

$$i(x) = (x, c_{f(x)}), \quad p(x, \alpha) = \alpha(1).$$

**Proposition 2.2.2** *The evaluation map $p$ is a fibration and the inclusion map $i$ is a homotopy equivalence; in fact $i(X)$ is a deformation retract of $P_f$.*

*Proof* A homotopy inverse to $i$ is the map $P_f \to X$ given by $(x, \alpha) \mapsto x$. The composition $X \to P_f \to X$ is equal to the identity, and the composition $P_f \to X \to P_f$ is homotopic to the identity by shrinking the paths $\alpha$; the map $H: P_f \times I \to P_f$ given by $H((x, \alpha), s) = (x, \alpha_s)$, where $\alpha_s(t) = \alpha(t(1 - s))$, is a homotopy from the identity to the composite $P_f \to X \to P_f$. It is fixed on $i(X)$.

To prove the first part, consider the lifting problem

$$
\begin{array}{ccc}
W \times \{0\} & \xrightarrow{\ g\ } & P_f \\
{\scriptstyle i_0}\downarrow & \nearrow & \downarrow{\scriptstyle p} \\
W \times I & \xrightarrow[\ h\ ]{} & Y
\end{array}
$$

Write $g(w) = (x_w, \gamma_w)$. Define $\widehat{h} \colon W \times I \to Y$ by

$$\widehat{h}(w, s) = (x_w, \alpha_{(w,s)})$$

where

$$
\alpha_{(w,s)}(t) = \begin{cases} \gamma_w(t + ts), & \text{if } 0 \le t \le 1/(1+s); \\ h(w, t + st - 1), & \text{if } 1/(1+s) \le t \le 1. \end{cases}
$$

Clearly $\alpha_{(w,0)} = \gamma_w$ and $p \circ \widehat{h}(w, s) = h(w, (t + st - 1))|_{t=1} = h(w, s)$.          □

The mapping path space $P_f$ is homotopy invariant by construction. Suppose the horizontal maps in the following commutative diagram are homotopy (or weak) equivalences:

$$
\begin{array}{ccc}
X_1 & \xrightarrow{\ \simeq\ } & X_2 \\
{\scriptstyle f_1}\downarrow & & \downarrow{\scriptstyle f_2} \\
Y_1 & \xrightarrow[\ \simeq\ ]{} & Y_2
\end{array}
$$

Then $P_{f_1} \simeq P_{f_2}$ since the former is homotopy equivalent to $X_1$ and the latter to $X_2$. For the same reason, the map $Y_1 \to Y_2$ does not affect the homotopy type of $P_{f_1}$ and $P_{f_2}$ (and does not need even need to be an equivalence for the homotopy type of the mapping path spaces to remain unchanged). In addition, if $f$ is changed by a homotopy, the homotopy type of $P_f$ remains unchanged. The proof of this, however, has to wait until the next chapter (Corollary 3.2.16).

**Definition 2.2.3**   Given a map $f \colon X \to Y$, the *homotopy fiber of $f$ over $y \in Y$*, denoted by $\mathrm{hofiber}_y(f)$, is the fiber over $y$ of the map $p \colon P_f \to Y$. Explicitly, $\mathrm{hofiber}_y(f)$ is the subspace of $X \times \mathrm{Map}(I, Y)$ given by

$$\mathrm{hofiber}_y(f) = \{(x, \alpha) \ : \ x \in X, \alpha \colon I \to Y, \alpha(0) = f(x), \alpha(1) = y\}.$$

**Example 2.2.4**   Let $X$ be a non-empty space. Then, for any $x_0 \in X$, $\mathrm{hofiber}_{x_0}(1_X)$ is contractible. The map $* \to \mathrm{hofiber}_{x_0}(1_X)$ sending $*$ to $(x_0, c_{x_0})$ is a homotopy equivalence, since the homotopy $(\alpha(s), (t \mapsto \alpha((1 - t)s + t)))$ is homotopic to the identity map of $\mathrm{hofiber}_{x_0}(1_X)$.          □

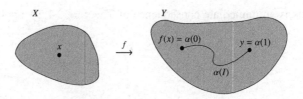

Figure 2.3  A point $(x, \alpha)$ in hofiber$_y(f)$.

A pictorial representation of the homotopy fiber hofiber$_y(f)$ is given in Figure 2.3.

Since $p$ is a fibration, we now by definition have a homotopy invariant fiber hofiber$_y(f)$ (by Theorem 2.1.20) as well as the homotopy long exact sequence described in Theorem 2.1.13 for the fibration sequence

$$\text{hofiber}_y(f) \longrightarrow P_f \longrightarrow Y.$$

We shall also see in Corollary 3.2.17 that the homotopy type of $P_f$ is unchanged if $f$ is changed by a homotopy.

**Remark 2.2.5**  The homotopy fiber of a map is one of the most basic examples of a *homotopy limit*, and will be considered again from this more general point of view in Example 3.2.8. □

As a result of Proposition 2.1.18, all homotopy fibers over a path-connected space are homotopy equivalent, so we will often omit $y$ from the notation, although clearly the space itself depends on $y$, and we will always be careful to name some choice of basepoint. Thus we often write hofiber$(f)$ instead of hofiber$_y(f)$. We will also often write hofiber$(X \xrightarrow{f} Y)$ when we want to remember the spaces or just hofiber$(X \to Y)$ when $f$ is clear from the context.

There is a canonical inclusion

$$\text{fiber}_y(f) \longrightarrow \text{hofiber}_y(f) \tag{2.2.2}$$
$$x \longmapsto (x, c_{f(x)}).$$

One indication that the way of replacing a map by a fibration described above is natural is that this map is a homotopy equivalence when the original map is itself a fibration:

**Proposition 2.2.6**  *If* $f : X \to Y$ *is a fibration, then, for any* $y \in Y$*, the inclusion of the fiber into the homotopy fiber from* (2.2.2) *is a homotopy equivalence.*

*Proof*  This is an immediate consequence of Theorem 2.1.20, applied to the diagram

$$
\begin{array}{ccc}
X & \longrightarrow & P_f \\
{\scriptstyle f}\downarrow & & \downarrow \\
Y & \underset{1_Y}{\longrightarrow} & Y
\end{array}
$$

where the top horizontal map is the canonical one from (2.2.2).            □

Another useful property of homotopy fibers is that their homotopy type is independent of the homotopy class of the map. If $\alpha, \beta \colon I \to X$ are paths in $X$ such that $\alpha(1) = \beta(0)$, we let $\alpha * \beta \colon I \to X$ be the path defined by $\alpha(2t)$ for $0 \le t \le 1/2$ and $\beta(2t-1)$ for $1/2 \le t \le 1$. We also let $\alpha^{-1} \colon I \to X$ be given by $t \mapsto \alpha(1-t)$.

**Proposition 2.2.7**  *Suppose $f_0, f_1 \colon X \to Y$ are homotopic. Then, for all $y \in Y$, there is a homotopy equivalence* $\mathrm{hofiber}_y(f_0) \simeq \mathrm{hofiber}_y(f_1)$ *induced by any homotopy $f_t$ from $f_0$ to $f_1$.*

*Proof*  Let $f_t$ be a homotopy from $f_0$ to $f_1$. Consider the map

$$
\mathrm{hofiber}_y(f_0) \longrightarrow \mathrm{hofiber}_y(f_1)
$$

given by $(x, \alpha) \mapsto (x, (f_t(x))^{-1} * \alpha)$, where $(f_t(x))^{-1}$ is the inverse of the path $f_t(x)$ from $f_0(x)$ to $f_1(x)$ given by $f$, and the superscript denotes this path is to run backwards. The path multiplication $\gamma * (f_t(x))^{-1}$ makes sense since $\alpha(0) = f_0(x)$, and the map has the correct codomain since $(f_t(x))^{-1} * \alpha$ is at $f_1(x)$ at time zero. The homotopy inverse of this map is the one which maps $(x, \beta)$ to $(x, f_t(x) * \beta)$, and it is straightforward to check these are homotopy inverses of one another.            □

**Remark 2.2.8**  Proposition 2.2.7 appears later as Corollary 3.2.17 with a different proof. It is also true that the mapping path spaces of homotopic maps are homotopy equivalent, which we will prove in Corollary 3.2.16, and that the homotopy fibers of a map over points in the same component are homotopy equivalent, which we will prove in Corollary 3.2.18; in particular, fibers of a fibration over points in the same component are homotopy equivalent.            □

**Example 2.2.9** (Loop space as homotopy fiber)  If $f \colon y_0 \to Y$ is the inclusion of a basepoint, then

$$
\mathrm{hofiber}_{y_0}(f) = \Omega Y,
$$

the based loop space of $Y$.            □

**Example 2.2.10** If $f: X \to Y$ is homotopic to a constant, then

$$\text{hofiber}(f) \simeq X \times \Omega Y.$$

To see this, choose a basepoint $y \in Y$ and consider $\text{Map}_b(I, Y)$, the space of maps where endpoints are sent to fixed points in the component of $y$. This space is homotopy equivalent to $\Omega Y$; this is easy to see by choosing paths from the images of the endpoints of $I$ to $y$ and gluing them to an element of $\text{Map}_b(I, Y)$ to produce an element of $\Omega Y$. The same can be done for the map back by gluing those same paths onto a loop. These are homotopy inverses via reparametrization. A sleeker way to see this, without writing down explicit homotopies, is to use Theorem 3.3.15 with $X_1 = X_2 = Y_1 = Y_2 = *$, $X_{12} = Z_{12} = Y_{12} = Y$, and $Z_1 = Z_2 = Y$.

Now, if $f$ is homotopic to a constant, it maps $X$ to some component of $Y$. Choose an arbitrary basepoint $y$ in that component of $Y$ so we can consider $\Omega Y$. By Proposition 2.2.7, we have that hofiber($f$) is homotopy equivalent to the homotopy fiber of a constant map from $X$ to $Y$. By definition, this homotopy fiber is the set of all $(x, \gamma)$ such that $\gamma(0) = f(x)$ and $\gamma(1) = y$. Since the map from $X$ to $Y$ is now constant, this is clearly $X \times \text{Map}_b(I, Y)$, and by the above, this is homotopy equivalent to $X \times \Omega Y$. □

**Proposition 2.2.11** *Suppose $X$, $Y$, and $Z$ are based spaces with base-points $x_0, y_0$, and $z_0$ respectively. Let $f: X \to Y$ be a based map and $f_*: \text{Map}_*(Z, X) \to \text{Map}_*(Z, Y)$ be the induced map. There is a natural homeomorphism*

$$\text{hofiber}_{c_{y_0}}(f_*) \cong \text{Map}_*(Z, \text{hofiber}_{y_0}(f)),$$

*where the basepoint of $\text{Map}_*(Z, Y)$ is the constant map $c_{y_0}: Z \to Y$.*

*Proof* Let $(g, \alpha) \in \text{Map}_*(Z, X) \times \text{Map}_*(Z, Y)^I$ be a point in the homotopy fiber of $f_*$, so that $f \circ g = \alpha(0)$ (here, for each $t$, $\alpha(t): Z \to Y$ is a based map). Using Theorem 1.2.7, we may regard this as the map $Z \to \text{hofiber}_{y_0}(f)$ which sends $z$ to $(g(z), \alpha_z)$, where $\alpha_z: I \to Y$ is given by $t \mapsto \alpha_z(t)$. This defines the desired homeomorphism. □

**Example 2.2.12** (Homotopy fiber and loops commute) An important special case of Proposition 2.2.11 is the case $Z = S^1$, in which case we get that there is a homeomorphism

$$\text{hofiber}(\Omega f) \cong \Omega \text{hofiber}(f) \tag{2.2.3}$$

(where by $\Omega f$ we mean the map induced by $f$ on loop spaces of $f$ as in (1.2.1)). □

**Remark 2.2.13**    As we shall see in Corollaries 3.3.16 and 8.5.8, the above is an instance of two homotopy limits commuting.    □

The above example plays a role in the following construction of a space-level version of the homotopy long exact sequence for a map $f: X \to Y$ of based spaces. The construction of the homotopy fiber can be iterated to define a sequence

$$\cdots \xrightarrow{p_3} F_2 \xrightarrow{p_2} F_1 \xrightarrow{p_1} F_0 \xrightarrow{p} X \xrightarrow{f} Y \qquad (2.2.4)$$

where $F_0 = \text{hofiber}(f)$, $F_1 = \text{hofiber}(p)$, and $F_i = \text{hofiber}(F_i \to F_{i-1})$ for $i \geq 2$. Exactness of this sequence is expressed by the following proposition and its corollary.

**Proposition 2.2.14**    *Suppose $f: X \to Y$ is a map of based spaces and let $F = \text{hofiber}(f)$. If $g: Z \to X$ is any based map, then $f \circ g$ is null-homotopic if and only if there exists a lift $\widehat{g}: Z \to F$.*

*Proof*    Suppose $g: Z \to X$ is map such that $f \circ g: Z \to Y$ is homotopic to $c_{y_0}$, where $y_0$ is the basepoint of $Y$. Let $H: Z \times I \to Y$ be a based null-homotopy from $f \circ g$ to $c_{y_0}$ (meaning if $z_0 \in Z$ is the basepoint, $H(\{z_0\} \times I) = y_0$). Regard $H$ as a map $\widetilde{H}: Z \to Y^I$ using Theorem 1.2.7. Define $\widehat{g}: Z \to F$ by $\widehat{g}(z) = (g(z), \widetilde{H}(z))$. To see that this really defines a map to the homotopy fiber $F$, note that, since $H(z,0) = (f \circ g)(z)$, $ev_0 \circ \widetilde{H}(z) = f \circ g(z)$, and since $H(z,1) = y_0$, $ev_1 \circ \widetilde{H}(z) = y_0$.

If $\widehat{g}: Z \to F$ exists, it is necessarily a map of the form $\widehat{g}(z) = (g(z), \gamma_z)$, where $\gamma_z \in Y^I$ satisfies $\gamma_z(0) = f \circ g(z)$ and $\gamma_z(1) = y_0$, and $\gamma_{z_0}(t) = c_{y_0}$. Regard $\gamma_z$ as a map $\Gamma: Z \times I \to Y$ and we see the above conditions imply $\Gamma(z,0) = f \circ g(z)$, $\Gamma(z,1) = y_0$, and $\Gamma(z_0,t) = c_{y_0}$, which is precisely a based null-homotopy of the composition $f \circ g$.    □

**Corollary 2.2.15**    *With $f, X, Y, F_0$ as above, for any based space $Z$ the sequence*

$$[Z, F_0]_* \longrightarrow [Z, X]_* \longrightarrow [Z, Y]_*$$

*is exact at $[Z, X]_*$, that is, the image of $[a] \in [Z, X]_*$ in $[Z, Y]_*$ is the homotopy class of the constant map if and only if there exists $[b] \in [Z, F_0]_*$ mapping to $[a]$.*

In fact, up to homotopy equivalences, the sequence appearing in Equation (2.2.4) is

$$\cdots \xrightarrow{\Omega^2 f} \Omega^2 Y \xrightarrow{-\Omega j} \Omega F_0 \xrightarrow{-\Omega i} \Omega X \xrightarrow{-\Omega f} \Omega Y \xrightarrow{j} F_0 \xrightarrow{i} X \xrightarrow{f} Y. \qquad (2.2.5)$$

This is called the *fiber sequence* or the *dual Barratt–Puppe sequence*. That is,

**Lemma 2.2.16**   *There are natural homotopy equivalences*

$$F_1 = \mathrm{hofiber}(F_0 \to X) \simeq \Omega Y$$

*and*

$$F_2 = \mathrm{hofiber}(F_1 \to F_0) \simeq \Omega X,$$

*and the map $F_2 \to F_1$ is $-\Omega f$.*

*Proof*   These follow from results which appear later in this book. The first equivalence is Example 3.3.13 and the second is Example 3.4.5. The statement that the map $F_2 \to F_1$ is the map $-\Omega f$ follows by inspection. (For the negative sign, see the proof of Lemma 2.4.19 for a geometric explanation of why this appears in the dual situation. Also see [tD08, Section 4.7].)   □

From Corollary 2.2.15 and Lemma 2.2.16, we thus have

**Theorem 2.2.17**   *The sequence (2.2.5) is homotopy exact, i.e. for $Z$ a based space, applying $[Z, -]_*$ to the sequence (2.2.5) produces an exact sequence (of pointed sets which are groups starting at $\Omega Y$ and abelian groups starting at $\Omega^2 Y$).*

Considering again the fibration sequence $\mathrm{hofiber}(f) \to P_f \to Y$ for a map $f : X \to Y$, we see that setting $Z = S^0$ in Theorem 2.2.17 and using Equation (1.4.2) gives precisely the homotopy long exact sequence for this fibration. This is the sense in which the Barratt–Puppe sequence generalizes the long exact sequence of homotopy groups and also realizes it on the level of spaces.

## 2.3  Cofibrations

Dual to Definition 2.1.1, we have

**Definition 2.3.1**   A map $f : X \to Y$ is a *cofibration* if it satisfies the *homotopy extension property*. That is, $f$ is a cofibration if for all spaces $Z$ and commutative diagrams

$$
\begin{array}{ccc}
X & \xrightarrow{h} & W^I \\
f \downarrow & \nearrow^{\widehat{h}} & \downarrow ev_0 \\
Y & \xrightarrow{g} & W
\end{array}
\qquad (2.3.1)
$$

the dotted arrow $\widehat{h}$ exists. Here $ev_0$ is evaluation at 0.

For instance, the inclusion of the empty set $\emptyset \to Y$ is a cofibration for every space $Y$. Every homeomorphism is a cofibration. In the definition above, it is common to regard the map $X \to W^I$ as a map $X \times I \to W$, and the "extension problem" in the definition is to extend this homotopy to a homotopy $Y \times I \to W$ with given initial value at $Y \times \{0\}$. This can be neatly organized in the diagram

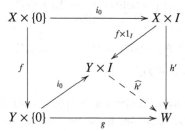

Here $h'$ and $\widehat{h'}$ correspond respectively to $h, \widehat{h}$ using the exponential law, Theorem 1.2.7. Note that if $W = X \times I \cup_f Y \times \{0\}$ (meaning $X \times I \sqcup Y \times \{0\}/\sim$, where $(x, 0) \sim (f(x), 0)$), and if $h'$ and $g$ are the natural inclusions, then the natural map $X \times I \cup Y \times \{0\} \to Y \times I$ is a section of the solution to the extension problem $\widehat{h'}$. We will in fact see in the proof of Proposition 2.3.2 that $X \times I \cup Y \times \{0\}$ is a retract of $Y \times I$ (the only observation that needs to be made to deduce this is that a cofibration is an inclusion map).

Note that by reversing all the arrows in the diagram (2.1.1) in the definition of a fibration and replacing $W \times I$ with its "dual", $\mathrm{Map}(I, W) = W^I$ (see (1.2.2)), we obtain the diagram in the definition of cofibration.

For a cofibration $f \colon X \to Y$, we will refer to the sequence $X \to Y \to Y/f(X)$ as a *cofibration sequence*.

Dual to Proposition 2.1.3 and an important early observation is the following.

**Proposition 2.3.2** (Cofibrations are injective with closed images) *Let $f \colon X \to Y$ be a cofibration. Then $f$ is injective, $X \times I \cup Y \times \{0\}$ is a retract of $Y \times I$, and $f(X)$ is closed in $Y$.*

*Proof* Consider the following extension problem

$$
\begin{array}{ccc}
X & \xrightarrow{\ h\ } & (X \times I \cup_f Y \times \{0\})^I \\
{\scriptstyle f}\big\downarrow & \nearrow & \big\downarrow{\scriptstyle ev_0} \\
Y & \xrightarrow{\ g\ } & X \times I \cup_f Y \times \{0\}
\end{array}
$$

where $g(y) = (y, 0)$ and $h(x)$ is the path which sends $t$ to $(x, 1 - t)$ for $t < 1$ and $(x, 0) \sim (f(x), 0)$ for $t = 1$. Since $f$ is a cofibration, there is a solution $\hat{h}$ to the extension problem. Let $x_1, x_2 \in X$ and suppose $f(x_1) = f(x_2) = y_0$. Since $\hat{h} \circ f = h$, $h(x_1) = h(x_2)$; in particular $ev_1 \circ h(x_1) = x_1 = x_2 = ev_1 \circ h(x_2)$.

To see that $f(X)$ is closed in $Y$, we already noted above that the natural map $X \times I \cup Y \times \{0\} \rightarrow Y \times I$ is a section of the map $Y \times I \rightarrow X \times I \cup Y \times \{0\}$. That is, since $f$ is injective and hence $X$ a subspace of $Y$, $f(X) \times I \cup Y \times \{0\}$ is a retract of $Y \times I$. Since $Y$ is Hausdorff, so is $Y \times I$, and so $f(X) \times I \cup Y \times \{0\}$ is closed in $Y \times I$ by Lemma 1.1.4 since it is a retract of $Y \times I$. Hence $f(X) \times \{1\}$ is closed in $Y \times \{1\}$ as well.[2]                                                                              □

In fact, the inclusion map $\iota \colon A \rightarrow X$ of a closed subspace is a cofibration if and only if $A \times I \cup X \times \{0\}$ is a retract of $X \times I$, which we shall see below in Proposition 2.3.10.

We leave the details omitted from the following examples to the reader.

**Example 2.3.3** (Coproduct of cofibrations)  If $f_1 \colon X_1 \rightarrow Y_1$ and $f_2 \colon X_2 \rightarrow Y_2$ are cofibrations, then so is their coproduct $f_1 \sqcup f_2 \colon X_1 \sqcup X_2 \rightarrow Y_1 \sqcup Y_2$. In the case where these are based maps of based spaces, the map $f_1 \vee f_2 \colon X_1 \vee X_2 \rightarrow Y_1 \vee Y_2$ is a cofibration.                                                                              □

**Example 2.3.4**  The inclusion $\{0\} \rightarrow I$ is a cofibration. It is not difficult to prove this directly. It is also a special case of Proposition 2.3.10.                         □

**Example 2.3.5**  For all $n \geq -1$, the inclusion of the boundary $S^n = \partial D^{n+1} \rightarrow D^{n+1}$ is a cofibration.

We will sketch the idea. By Proposition 2.3.10 below, it is sufficient to check that $D^{n+1} \times \{0\} \cup S^n \times I$ is a retract of $D^{n+1} \times I$. Let $S^n \subset \mathbb{R}^{n+1}$ be the natural inclusion of the unit sphere, and embed $S^n \times I \subset \mathbb{R}^{n+2} \cong \mathbb{R}^{n+1} \times \mathbb{R}$ in the evident way. Let $P = (0, 2) \in \mathbb{R}^{n+1} \times \mathbb{R}$ be a point just above the cylinder $D^{n+1} \times I$. Given a point $(x, t) \in D^{n+1} \times I$, there is a unique point $(y, s) \in D^{n+1} \times \{0\} \cup S^n \times I$ on the ray emanating from $P$ through $(x, t)$. Define $r \colon D^{n+1} \times I \rightarrow D^{n+1} \times \{0\} \cup S^n \times I$ by $r(x, t) = (y, s)$. It is not hard to check that this is continuous, and it is clearly equal to the identity on $D^{n+1} \times \{0\} \cup S^n \times I$, and hence is a retraction. This is also a special case of Example 2.3.11, also using Proposition 2.3.10.                      □

**Example 2.3.6**  Generalizing the previous example, the inclusion of a space $X \rightarrow CX$ into its cone is a cofibration.                                                         □

---

[2]  The limit of a convergent sequence of points in $f(X) \times \{1\}$ lies in $f(X) \times \{1\}$ because it lies in $f(X) \times I \cup Y \times \{0\}$.

Example 2.3.5 also implies that the inclusion of a subcomplex of a CW complex is a cofibration. More precisely, we have the following.

**Example 2.3.7** For any space $X$, the inclusion of $X$ in the space obtained by attaching an $n$-cell to $X$ is a cofibration. More generally, if $(X, A)$ is a relative CW complex, the inclusion $A \to X$ is a cofibration.

To see why the first statement is true, let $e^n$ be an $n$-cell and $f: \partial e^n \to X$ a map. The result now follows from Proposition 2.3.15 since, by Example 2.3.5, the inclusion $\partial e^n \to e^n$ is a cofibration, and hence so is the map $X \to X \cup_f e^n$. For finite-dimensional CW complexes, the proof is similar. Let $X^i$ denote the $i$-skeleton of $X$, and let $A_i$ denote the indexing set for the $i$-cells of $X$. For each $\alpha \in A_i$ let $f_\alpha: \partial e^i \to X^{i-1}$ be the attaching map. The inclusion $\sqcup_{A_i} \partial e^i \to \sqcup_{A_i} e^i$ is a cofibration by Example 2.3.3, and since $X^i = X^{i-1} \cup_{\sqcup_{\alpha \in A_i} f_\alpha} \sqcup_{A_i} e^i$, Proposition 2.3.15 tells us that the right vertical map in the diagram

$$\begin{array}{ccc} \sqcup_{A_i} \partial e^i & \longrightarrow & X^{i-1} \\ \downarrow & & \downarrow \\ \sqcup_{A_i} e^i & \longrightarrow & X^i \end{array}$$

is a cofibration. Since the composition of cofibrations is a cofibration, this completes the proof. The result is still true if $X$ has cells of arbitrarily high dimension; the above defines a lift on all the cells because the lifts agree where they have to and because $X$ is given the weak topology. See Remark 2.6.23 for an alternative argument. $\square$

Dual to Proposition 2.1.10, with dual proof, the extension problems posed in the definition of a cofibration are not unique, but any two such are homotopic.

**Proposition 2.3.8** *Suppose $f: X \to Y$ is a cofibration, and consider the extension problem*

$$\begin{array}{ccc} X & \xrightarrow{h} & W^I \\ f \downarrow & \nearrow^{\widehat{h}} & \downarrow ev_0 \\ Y & \longrightarrow & W \end{array}$$

*If $\widehat{h}_0, \widehat{h}_1: W \times I \to X$ are solutions to this extension problem, regarded as maps $\hat{H}_0, \hat{H}_1: X \times I \to W$ using Theorem 1.2.7, then there is a homotopy $\hat{H}_0 \sim \hat{H}_1$ relative to $X \times \{0\}$.*

Dual to Proposition 2.1.11, we have the following.

**Proposition 2.3.9** *Suppose $f: X \to Y$ is a cofibration. Then, for any space Z, the map*

$$f \times 1_Z: X \times Z \longrightarrow Y \times Z$$

*is also a cofibration (whose hocofiber will be described in Example 2.4.13).*

*Proof* We must show that the dotted arrow exists in the following square

By Theorem 1.2.7, both $g$ and $h$ correspond with maps $G: Y \to W^Z$ and $H: X \to (W^I)^Z$. Using Theorem 1.2.7 again to obtain $(W^I)^Z \cong (W^Z)^I$ (because $I \times Y \cong Y \times I$), we have a commutative diagram

$$
\begin{array}{ccc}
X & \xrightarrow{\ H\ } & (W^Z)^I \\
f \downarrow & \nearrow & \downarrow ev_0 \\
Y & \xrightarrow[\ G\ ]{} & W^Z
\end{array}
$$

Here the dotted arrow exists because $f$ is a cofibration, and it corresponds, via Theorem 1.2.7, to the desired map $Y \times Z \to W^I$. □

According to Proposition 2.3.2, a cofibration is an inclusion map with closed image. We give several useful equivalent formulations of the notion of a cofibration.

**Proposition 2.3.10** ([May99, Section 6.4]) *For A a closed subspace of X, the following are equivalent.*

1. *The inclusion $i: A \to X$ is a cofibration.*
2. *$A \times I \cup X \times \{0\}$ is a retract of $X \times I$.*
3. *There exists a function $u: X \to I$ such that $u^{-1}(0) = A$ and a homotopy $h: X \times I \to X$ such that $h(x, 0) = x$, $h(a, t) = a$ for all $a \in A$, and $h(x, 1) \in A$ if $u(x) < 1$.*

If $A \times I \cup X \times \{0\}$ is a retract of $X \times I$, then it is in fact a deformation retract of $X \times I$. Let $\iota: A \times I \cup X \times \{0\}$ denote the inclusion. If the retraction $r: X \times I \to A \times I \cup X \times \{0\}$ is given by $r(x, t) = (r_1(x, t), r_2(x, 2))$, then the homotopy $H: X \times I \times I \to X \times I$ given by $H(x, t, s) = (r_1(x, (1 - s)t), (1 - s)r_2(x, t) + st)$ is easily seen to give a homotopy $\iota \circ r \sim 1_{X \times I}$ relative to $A \times I$ (see the lemma in [Str66]).

*Proof of Proposition 2.3.10*   We will show that conditions 1 and 2 are equiv-alent and refer the reader to [May99] for the rest. In fact, we have already seen that 1 implies 2 in Proposition 2.3.2. For the other direction, consider the square

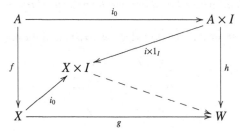

Suppose $i$ is a cofibration. Let $W = A \times I \cup X \times \{0\}$ (by Example 3.5.7, $A \times I \cup X \times \{0\} = \text{colim}(X \leftarrow A \rightarrow A \times I) \cong M_i$ is the mapping cylinder), and $g$ and $h$ be the natural inclusion maps. Let $u\colon X \times I \rightarrow A \times I \cup \times \{0\}$ be the dotted arrow. Then $u$ defines a retraction, as is easily seen by the commutativity of the diagram above.

Suppose now that $A \times I \cup X \times \{0\}$ is a retract of $X \times I$, and let $r\colon X \times I \rightarrow A \times I \cup X \times \{0\}$ be a retraction. In the extension problem above, a solution $e\colon X \times I \rightarrow W$ is given by the composition of $r$ with the given maps $g$ and $h$. The fact that $r$ is a retraction assures us that the diagram commutes.          □

The third condition above can be used to show that the inclusion of a closed submanifold is a cofibration:

**Example 2.3.11**   Let $N^n$ be a smooth manifold and $M \subset N$ be a smooth closed submanifold (meaning it is compact and without boundary). Then the inclusion $i\colon M \rightarrow N$ is a cofibration.

To see why, let $v(M, N)$ denote the normal bundle of $M$ in $N$. By the tubular neighborhood theorem, we may identify $v(M, N)$ with an open neighborhood of $M$ in $N$. In this identification $M$ is identified with the zero section of $v(M, N)$. Moreover, a choice of Riemannian metric on $N$ allows us to speak of the disk bundles $D_\rho v(M, N)$ of radius $\rho > 0$. Define $u\colon N \rightarrow I$ as follows. If $x \in N$ lies outside the unit disk bundle $D_1 v(M, N)$, we set $u(x) = 1$. If $x \in D_1 v(M, N)$, then there is a unique pair $(y, v) \in v(M, N)$ which represents $x$, where $y \in M$ and $v \in T_y N$. In this case we set $u(x) = |v|$. Then $u$ is continuous and $u^{-1}(0)$ is the zero section of the normal bundle, which we have identified with $M$. The idea for the homotopy $h\colon N \times I \rightarrow N$ is simple enough: fiberwise contract the unit disk bundle to the zero section, leaving everything outside $D_{1+\delta} v(M, N)$ fixed for some fixed $\delta > 0$. The details are slightly tedious to write down, but we include them in any case.

With $\delta > 0$ as above, let $\epsilon \colon \mathbb{R} \to \mathbb{R}$ be a map such that $\epsilon(y) = 1$ for $|y| \leq 1$ and $\epsilon(y) = 0$ for $|y| \geq 1 + \delta$. We will use polar coordinates on each fiber of $D_{1+\delta}v(M,N)$, so a point in the fiber over $m \in M$ is a pair $(v, r)$ where $v$ is a unit vector and $0 \leq r \leq 1 + \delta$. Define $h \colon N \times I \to N$ by $h(x,t) = x$ if $x$ lies outside $D_{1+\delta}v(M,N)$. Each $x \in D_{1+\delta}v(M,N)$ is uniquely represented as a pair $(m, (v, r))$ (it is understood that we identify $(m, v, 0)$ with $(m, v', 0)$). Then for $x = (m, (v, r))$ we let $h((m, (v, r)), t) = (m, (v, (1 - t)r + t\epsilon(r)))$. Then $h(x, 0) = x$ for all $x$ and $h((m, (v, 0)), t) = (m, v, 0)$. Moreover, $h(x, 1) \in M$ if $u(x) < 1$ since $u(x) < 1$ if and only if $n = (m, (v, r))$ with $r < 1$, so that $\epsilon(r) = 0$, and in this case $h((m, v, r), 1) = (m, v, 0)$. Thus $h$ satisfies condition 3 in Proposition 2.3.10. $\qquad\square$

**Remark 2.3.12** One can make all the relevant functions in the above proof smooth if so desired. The same proof as given above shows that the inclusion of $M$ in the Thom space of the normal bundle of $M$ is a cofibration. The example above is meant to illustrate the point that condition 3 in Proposition 2.3.10 is telling us that a subspace for which the inclusion is a cofibration has an open neighborhood of which it is a deformation retract. $\qquad\square$

Another useful property of cofibrations, which can be used to establish Proposition 2.3.10, is the following.

**Proposition 2.3.13** ([May99, Lemma in Section 6.4]) *If* $i \colon A \to X$ *and* $j \colon B \to Y$ *are cofibrations, then so is the induced map*

$$A \times Y \cup_{A \times B} X \times B \longrightarrow X \times Y.$$

As was the case with fibrations, clearly the composition of cofibrations is a cofibration. Once again we have three important properties: a long exact sequence, this time in (co)homology, stability under "pushout", and homotopy invariance of the cofibers.

**Theorem 2.3.14** (Homology long exact sequence of a cofibration) *Let* $f \colon X \to Y$ *be a cofibration, and let* $C = Y/f(X)$ *be the cofiber. Then there is a long exact sequence of groups*

$$\cdots \to \widetilde{H}_n(X) \to \widetilde{H}_n(Y) \to \widetilde{H}_n(C) \to \widetilde{H}_{n-1}(X) \to \cdots .$$

A similar statement is true for cohomology groups.

*Sketch of proof* It is a standard fact that the homology of a pair $(X, A)$ is isomorphic to the reduced homology of $X/A$ if the inclusion $A \to X$ is a cofibration (see e.g. [Bre93, Chapter VII, Corollary 1.7]). The result then follows from the homology long exact sequence of a pair (Theorem 1.4.8) and the fact that cofibrations are injective (Proposition 2.3.2).                    □

In Proposition 2.1.16, we saw that fibrations are preserved under pullbacks, and dually cofibrations are preserved under pushouts. Namely, given maps $f: X \to Y$ and $g: X \to Z$, one can consider the space $Z \cup_X Y$, called the *pushout of $Z$ with $Y$ along $X$*, defined to be the quotient of the disjoint union $Z \sqcup Y$ by the equivalence relation $\sim$ generated by $z \sim y$ if there exists $x \in X$ such that $f(x) = y$ and $g(x) = z$. Pushouts will be treated in detail in Section 3.5; see Definition 3.5.1 in particular.

**Proposition 2.3.15** (Cofibrations are preserved by pushouts) *Suppose $f: X \to Y$ is a cofibration and $g: X \to Z$ is a map. Then the map $f_*: Z \to Z \cup_X Y$ given by the inclusion of $Z$ is a cofibration.*

*Proof* Dual to proof of Proposition 2.1.16.                    □

**Remark 2.3.16** A remark similar to Remark 2.1.17 is in order: Proposition 2.3.15 follows immediately from the universal property of the pushout $Z \cup_X Y$; more will be said about this in Section 3.5.                    □

**Theorem 2.3.17** *Suppose in the commutative diagram*

$$\begin{array}{ccc} X_1 & \xrightarrow{\ f_X\ } & X_2 \\ {\scriptstyle g_1}\downarrow & & \downarrow{\scriptstyle g_2} \\ Y_1 & \xrightarrow{\ f_Y\ } & Y_2 \end{array}$$

*the vertical maps $g_1$ and $g_2$ are cofibrations with cofibers $C_1$ and $C_2$, respectively, and maps $f_X$ and $f_Y$ are homotopy equivalences. Then the induced map of cofibers $C_1 \to C_2$ is a homotopy equivalence.*

A proof dual to the proof of Theorem 2.1.20 is left to the reader.

Dually to Corollary 2.1.22, we have the following statement. Its proof is also a consequence of the proof of Theorem 2.3.17.

**Corollary 2.3.18** *Let*

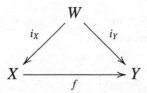

*be a commutative diagram where the maps from W are cofibrations. If f is a homotopy equivalence, then there exists a homotopy inverse $g: Y \to X$ such that*

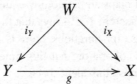

*commutes. In short, g is a homotopy equivalence if and only if it is a homotopy equivalence of spaces under W.*

We also have the analog of Proposition 2.1.23.

**Proposition 2.3.19** *If $p: X \to Y$ is a cofibration and $f: X \to W$ is a homotopy equivalence (resp. weak equivalence), then $Y \to W \cup_X Y$ is also a homotopy equivalence (resp. weak equivalence).*

*Proof* For the homotopy equivalence case, the proof is dual to that of Proposition 2.1.23 and we omit it. However, the proof for the case of weak equivalence does not dualize. (The short algebraic proof for the weak equivalence we give at the end of the proof of Proposition 2.1.23 does not dualize either.) We present the argument in this case in our restatement of this result later as Proposition 3.6.2. □

**Remark 2.3.20** With the notion of a cofibration under our belt, we can elaborate on the issue of basepoints a bit. As we mentioned in Remark 1.1.6, a based space $(X, x_0)$ is well-pointed if and only if the inclusion of the basepoint $\{x_0\} \to X$ is a cofibration. This is a technical condition that makes a variety of constructions easier, and in particular makes the theories of based and unbased spaces parallel. We focus only on the question of homotopy invariance. We would like constructions which depend on basepoints to be homotopy invariant to the extent that if we change basepoints within a path component, we get homotopy equivalent constructions. This may not be the case for spaces that are not well-pointed. □

## 2.4 Homotopy cofibers

As we saw in Example 2.3, the cofiber of a map also fails to be homotopy invariant. In the commutative square

$$\begin{array}{ccc} X_1 & \longrightarrow & X_2 \\ {\scriptstyle f_1}\downarrow & & \downarrow{\scriptstyle f_2} \\ Y_1 & \longrightarrow & Y_2 \end{array} \tag{2.4.1}$$

where the horizontal maps are homotopy equivalences, it does not follow that the cofibers of the vertical maps are homotopy equivalent. However, if $f_1$ and $f_2$ are cofibrations, then Theorem 2.3.14 says that these cofibers are homotopy equivalent. As was the case for fibrations, there is a way to turn any map of spaces $f \colon X \to Y$ into a cofibration, but this time it requires replacing the codomain $Y$ with a homotopy equivalent space.

**Definition 2.4.1** Given a map $f \colon X \to Y$, define the *mapping cylinder of $f$*, denoted by $M_f$, to be the quotient of $(X \times I) \sqcup Y$ given by

$$M_f = (X \times I) \sqcup Y/\sim,$$

where $\sim$ is the equivalence relation generated by $(x, 0) \sim f(x)$.

For instance, the mapping cylinder of the identity map $1_X \colon X \to X$ is the cylinder $X \times I$. We already encountered the mapping cylinder in Proposition 2.3.2.

**Remark 2.4.2** If $X$ is based with basepoint $\{x_0\}$, then one can define the *reduced* mapping cylinder by also identifying $\{x_0\} \times I$ to a point. The canonical map from unreduced to reduced mapping cylinder is an equivalence if the space is well-pointed. Since we always assume this is the case, we will not explicitly make a distinction between these two versions of the mapping cylinder. □

**Remark 2.4.3** Following up on Remark 2.3.20, one way to make a well-pointed space out of a based space $X$ is to replace it with a homotopy equivalent space which is the mapping cylinder of the map $* \to X$. □

It is again clear that $f$ factors through $M_f$ as

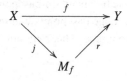

where

$$j(x) = (x, 1), \quad r(x, t) = f(x), \quad r(y) = y.$$

In analogy with Proposition 2.2.2, we have the following.

**Proposition 2.4.4** *The inclusion map $j$ is a cofibration and $r$ is a homotopy equivalence. In fact, $Y$ is a deformation retract of $M_f$.*

*Proof* That $r$ is a homotopy equivalence is easy to see by collapsing $X \times I$ to the image of $X \times \{0\}$ in $Y$. The inclusion $i \colon Y \to M_f$ clearly satisfies $r \circ i = 1_Y$, and the homotopy $H \colon M_f \times I \to M_f$ given by $H(x, t, s) = (x, t(1 - s))$ on $X \times I$ in $M_f$ and $H(y, s) = y$ on $i(Y)$ is a homotopy from $1_{M_f}$ to $i \circ r$ relative to $i(Y)$. The proof that $j$ is a cofibration is dual to the proof given in Proposition 2.2.2, and is left to the reader.                                                  □

Just as in the case of the mapping path space, the mapping cylinder is homotopy invariant. Replacing either of the spaces $X$ or $Y$ by a homotopy (or weakly) equivalent one does not change the (weak) homotopy type of $M_f$. This space is also unchanged if $f$ is changed by a homotopy, as we shall see in Corollary 3.6.20. Consequently, what is also unchanged by a homotopy (see Corollary 3.6.21) is the notion of the homotopy cofiber:

**Definition 2.4.5** The *homotopy cofiber of $f$*, denoted by hocofiber($f$), is the cofiber of the inclusion $j \colon X \to M_f$. That is,

$$\text{hocofiber}(f) = \frac{M_f}{X \times \{1\}}.$$

More explicitly, it is the quotient space of $(X \times I) \sqcup Y$ given by the equivalence relation generated by $(x, 1) \sim (x', 1)$ and $(x, 0) \sim f(x)$.

Figure 2.4 gives a picture of hocofiber($f$). When we want to emphasize the spaces, we will also write hocofiber($X \xrightarrow{f} Y$) or just hocofiber($X \to Y$) when $f$ is understood.

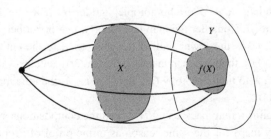

Figure 2.4  Homotopy cofiber hocofiber($f$).

**Remarks 2.4.6**

1. The homotopy cofiber of a map $f: X \to Y$ is one of the most basic examples of a *homotopy colimit*, and we shall see it again in Example 3.6.8.
2. Another name for hocofiber($f$) is the *mapping cone of $f$*, often denoted by $C_f$. We will use the notation and terminology interchangeably.
3. Looking back at Remark 2.4.2, Definition 2.4.5 has an obvious based analog. If $\{x_0\}$ is the basepoint of $X$, then the reduced mapping cone also has $\{x_0\} \times I$ identified to a point. We will not distinguish between the two versions unless it is necessary to do so.                                      □

Dual to (2.2.2), we have a canonical map

$$\text{hocofiber}(f) \longrightarrow \text{cofiber}(f) = Y/f(X) \qquad (2.4.2)$$
$$[x, t] \longmapsto [f(x)]$$
$$y \longmapsto [y]$$

and an analog of Proposition 2.2.6, the proof of which is dual and is left to the reader.

**Proposition 2.4.7**  *If $f: X \to Y$ is a cofibration, then the map from (2.4.2) is a homotopy equivalence.*

Dual to Proposition 2.2.7 we have following.

**Proposition 2.4.8**  *Suppose $f_0, f_1: X \to Y$ are homotopic. Then there is a homotopy equivalence $\text{hocofiber}(f_0) \simeq \text{hocofiber}(f_1)$, naturally induced by a choice of homotopy $f_t$ from $f_0$ to $f_1$.*

*Proof*  This proof was communicated to us by Phil Hirschhorn. Suppose $f_0: X \to Y$ and $f_1: X \to Y$ are homotopic, and let $H: X \times I \to Y$ be a homotopy from $f_0$ to $f_1$. Consider the map $\text{hocofiber}(f_0) \to \text{hocofiber}(f_1)$ that is the identity on $Y$, takes the lower half of the cone on $X$ to the entire cone on $X$ (i.e. if $0 \le t \le 1/2$, then the map takes $(x, t)$ to $(x, 2t)$), and takes the upper half of the cone on $X$ to the homotopy (i.e. if $1/2 \le t \le 1$, then the map takes $(x, t)$ to $H(x, 2t - 1)$).

One can define a map back using $H^{-1}$, and the compositions are homotopic to the identity maps for the same reason as in the proof of Proposition 2.2.7.                                      □

**Remark 2.4.9** The above result appears again later as Corollary 3.6.21 with a different proof. We will also prove in Corollary 3.6.20 that, for homotopic maps, the mapping cylinders are homotopy equivalent. □

**Example 2.4.10** We have

$$\text{hocofiber}(X \xrightarrow{1_X} X) \cong CX \simeq *.$$ □

**Example 2.4.11** (Suspension as homotopy cofiber) There is a natural homeomorphism

$$\text{hocofiber}(X \longrightarrow *) \cong \Sigma X,$$

where $\Sigma X$ is the suspension of $X$. □

**Example 2.4.12** Let $f: W \to X$ and $g: Y \to Z$ be maps of based spaces. Then there is a natural homeomorphism

$$\text{hocofiber}(f) \vee \text{hocofiber}(g) \cong \text{hocofiber}(f \vee g).$$

In particular, if $g = 1_Z: Z \to Z$ is the identity map, using Example 2.4.10 we see that

$$\text{hocofiber}(f \vee 1_Z) \cong CZ \vee \text{hocofiber}(f) \simeq \text{hocofiber}(f).$$

The verification of this is straightforward from the definition of homotopy cofiber. The evident map of quotient spaces

$$(W \vee Y) \times I \sqcup X \vee Z/ \sim \longrightarrow (W \times I \sqcup X/\sim) \vee (Y \times I \sqcup Z/\sim)$$

is a homeomorphism. We prefer to view the special case where $g = 1_Z$ through the lens of the last item in Corollary 3.7.19. □

The next two examples serve as the dual to Proposition 2.2.11.

**Example 2.4.13** Let $X, Y, Z$ be based spaces, and $f: X \to Y$ a map. Consider the map

$$f \times 1_Z: X \times Z \longrightarrow Y \times Z.$$

There is a natural homeomorphism

$$\text{hocofiber}(f \times 1_Z) \cong \text{hocofiber}(f) \wedge Z_+.$$

If $Y = *$, so that $\text{hocofiber}(f) \cong \Sigma X$ by Example 2.4.11, then for the projection $X \times Z \to Z$ we have

$$\text{hocofiber}(X \times Z \to Z) \cong \Sigma X \wedge Z_+.$$

This can also be verified from the definition, but it is somewhat cumbersome. We refer the reader instead to Example 3.7.20, where this claim is verified. □

Along the lines of the previous example, we also have the following.

**Example 2.4.14**   If $f: X \to Y$ is a map of based spaces and $Z$ is any based space, then for the map

$$f \wedge 1_Z : X \wedge Z \longrightarrow Y \wedge Z,$$

there is a homeomorphism

$$\text{hocofiber}(f \wedge 1_Z) \cong Z \wedge \text{hocofiber}(f).$$

The proof of this will be given as part of Corollary 3.7.19 (Examples 2.4.10, 2.4.12, and 2.4.13 all play a role in that proof).                    □

**Example 2.4.15** (Homotopy cofiber and suspension commute)   One important special case of Example 2.4.14 is $Z = S^1$, which is dual to Example 2.2.12. Recall that the smash product of a based space $X$ with $S^1$ is the suspension of $X$. We have

$$\text{hocofiber}(\Sigma f) \simeq \Sigma \text{hocofiber}(f),$$

where $f: X \to Y$ and $\Sigma f: \Sigma X \to \Sigma Y$ is the suspension of $f$ from (1.1.2). More succinctly (this is how this statement often appears in the literature),

$$C_{\Sigma f} \simeq \Sigma C_f.$$                    □

**Remark 2.4.16**   We will revisit this example in Corollaries 3.7.19, 5.8.10, and 8.5.8.                    □

In analogy with the sequence from (2.2.4), one can iterate the homotopy cofiber construction to get a sequence

$$X \xrightarrow{f} Y \xrightarrow{i} C_0 \xrightarrow{i_1} C_1 \xrightarrow{i_2} C_2 \xrightarrow{i_3} \cdots \qquad (2.4.3)$$

where $C_0 = \text{hocofiber}(f)$, $C_1 = \text{hocofiber}(i)$, and $C_i = \text{hocofiber}(C_{i-2} \to C_{i-1})$ for $i \geq 2$. Exactness is expressed by the following duals to Proposition 2.2.14 and Corollary 2.2.15.

**Proposition 2.4.17**   *Suppose $f: X \to Y$ is a map of based spaces and let $C = \text{hocofiber}(f)$ be the reduced homotopy cofiber. If $g: Y \to Z$ is any based map, then $f \circ g$ is null-homotopic if and only if there exists an extension $\widehat{g}: C \to Z$.*

*Proof*   Dual to the proof of Proposition 2.2.14.                    □

**Corollary 2.4.18** *With $f, X, Y, C_0$ as above, for any based space $Z$ the sequence*

$$[C_0, Z]_* \longrightarrow [Y, Z]_* \longrightarrow [X, Z]_*$$

*is exact at $[Y, Z]_*$.*

Up to homotopy equivalences, the sequence from (2.4.3) is what is known as the *cofiber sequence* or the *Barratt–Puppe sequence*

$$X \xrightarrow{f} Y \xrightarrow{i} C_0 \xrightarrow{q} \Sigma X \xrightarrow{\Sigma f} \Sigma Y \xrightarrow{\Sigma i} \Sigma C_0 \xrightarrow{\Sigma q} \Sigma^2 X \xrightarrow{\Sigma^2 f} \cdots. \qquad (2.4.4)$$

This follows from the dual of Lemma 2.2.16, which is the following.

**Lemma 2.4.19** *There are natural homotopy equivalences*

$$C_1 = \mathrm{hocofiber}(Y \to C_0) \simeq \Sigma X$$

*and*

$$C_2 = \mathrm{hocofiber}(C_0 \to C_1) \simeq \Sigma Y,$$

*and the map $C_1 \to C_2$ is $-\Sigma f$.*

*Proof* The first statement is Example 3.7.16 and the second is Example 3.8.4. That the map $C_1 \to C_2$ is $-\Sigma f$ follows by inspection. (As explained to us by Phil Hirschhorn, the negative sign can be seen from the pictures of the cones that are collapsed to form suspensions. Two of those cones meet at what is the 0 end for each of them. The map $\Sigma X \to \Sigma Y$ is induced by sliding the cone on $X$ into the one on $Y$, and it is the 0 end of the cone on $X$ that is moving toward the 1 end of the cone on $Y$. This is the reversal that produces the negative sign). □

From the above results, we thus have the following dual to Theorem 2.2.17.

**Theorem 2.4.20** *The sequence (2.4.4) is* homotopy coexact, *i.e. for $Z$ a based space, applying $[-, Z]_*$ to the sequence (2.4.4) produces an exact sequence (of pointed sets which are groups starting at $\Sigma Y$ and abelian groups starting at $\Sigma^2 Y$).*

Setting $Z = K(G, n)$, the Eilenberg–MacLane space that has an abelian group $G$ as its $n$th homotopy group and trivial homotopy groups otherwise in the above recovers the cohomology long exact sequence of the pair $(Y, X)$ (we have already encountered this in Theorem 2.3.14). This is because of the iso-morphism $[X, K(G, n)] \cong H^n(X, G)$ (see, for example, [Hat02, Theorem 4.57]). In fact, this relationship can be used as the definition of singular cohomology and further leads to the definition of a *spectrum* as a sequence of spaces satisfying certain properties that gives rise to a cohomology theory. For more details, see [Hat02, Section 4.3] and for more on spectra, see Section A.3.

## 2.5 Algebra of fibrations and cofibrations

In this section we state and prove results which involve both fibrations and cofibrations. With the exception of Proposition 2.5.1 and Theorem 2.5.4, which we will use in Chapter 3, we will not use these results until Chapter 8. Our first result is a counterpart to Proposition 2.1.11 and a special case of Proposition 3.9.1.

**Proposition 2.5.1** (Maps takes cofibrations to fibrations)  *If $f: X \to Y$ is a cofibration, then, for any space $Z$, the induced map $f^*: \mathrm{Map}(Y, Z) \to \mathrm{Map}(X, Z)$ is a fibration. Furthermore, the fiber of $f^*$ over the map which sends all of $X$ to $z \in Z$ is $\mathrm{Map}_*(Y/X, Z)$, where the basepoint of $Z$ is $z$ and the basepoint of $Y/X$ is the image of $X$ in the quotient space.*

*Proof*   Consider the lifting problem

$$
\begin{array}{ccc}
W & \longrightarrow & \mathrm{Map}(Y, Z) \\
\downarrow & \nearrow & \downarrow {\scriptstyle f^*} \\
W \times I & \longrightarrow & \mathrm{Map}(X, Z)
\end{array}
$$

Since $f$ is a cofibration, the dotted arrow exists in the diagram

$$
\begin{array}{ccc}
X & \longrightarrow & \mathrm{Map}(W, Z)^I \\
{\scriptstyle f}\downarrow & \nearrow & \downarrow {\scriptstyle ev_0} \\
Y & \longrightarrow & \mathrm{Map}(W, Z)
\end{array}
$$

Using Theorem 1.2.7, the above diagram becomes

$$
\begin{array}{ccc}
X & \longrightarrow & \mathrm{Map}(W \times I, Z) \\
{\scriptstyle f}\downarrow & \nearrow & \downarrow \\
Y & \longrightarrow & \mathrm{Map}(W, Z)
\end{array}
$$

Again applying Theorem 1.2.7, the top and bottom arrows correspond to maps $W \times I \to \mathrm{Map}(X, Z)$ and $W \to \mathrm{Map}(Y, Z)$, and the dotted arrow corresponds to a map $W \times I \to \mathrm{Map}(Y, Z)$, which is the desired lift. □

The next two results discuss (co)fibrations which are also homotopy equivalences, and are necessary for the proof of Theorem 2.5.4. Like Theorem 2.5.4, they are independently useful facts extracted from [Str68, Theorems 8 and 9].

**Proposition 2.5.2** *Suppose* $i: A \to B$ *is a cofibration and a homotopy equivalence. Then there exists a homotopy inverse* $j: B \to A$ *such that*

1. $j \circ i = 1_A$ *(that is, A is a retract of B);*
2. *there is a homotopy* $i \circ j \sim 1_B$ *relative to* $i(A)$*; that is, if* $i: A \to B$ *is a cofibration and a homotopy equivalence, then A is a deformation retract of B.*

*Proof* Since $i$ is a closed inclusion by Proposition 2.3.2, we will identify $A$ with $i(A)$, and omit $i$ whenever possible. Choose a homotopy inverse $k$ to $i$ and let $H$ be a homotopy from $k \circ i$ to $1_A$. Consider the extension problem

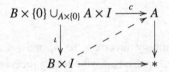

where the left vertical arrow is the inclusion and $c = k \cup H$ is the map which is equal to $k$ on $B \times \{0\}$ and $H$ on $A \times I$. To see that this defines a map on the union along $A \times \{0\}$, we only need to observe that $H(a, 0) = (k \circ i)(a) = k(a)$. Using Proposition 2.3.2, we have a retraction $r: B \times I \to B \times \{0\} \cup_{A \times \{0\}} A \times I$. The composite map $c \circ r: B \times I \to A$ defines the dotted arrow above, and the resulting diagram evidently commutes. Let $j = c \circ r|_{B \times \{1\}}$. Then $j: B \to A$ is homotopic to $k$, as the extension $c \circ r$ defines this homotopy, so $j$ is also a homotopy inverse to $i$. Note that, for $a \in A$, $c \circ r \circ \iota(a, 1) = c(a, 1) = 1_A(a) = a$, and so $j \circ i = 1_A$. This establishes the first claim.

For the second claim, let $G: B \times I \to B$ be a homotopy from $i \circ j$ to $1_B$. Consider the extension problem

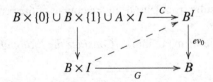

where the top horizontal map $C$ is given as follows: $C(b, 0) = (s \mapsto i \circ j(b))$ is a constant path, $C(b, 1) = (s \mapsto G(b, 1 - s))$, and $C(a, t) = (s \mapsto G(a, t(1 - s)))$. It is straightforward to check that this gives a well-defined map from the union to $B^I$, and it is also straightforward to check that the diagram above commutes.

By Proposition 2.3.13, the left vertical map in the above square is a cofibration, since both $A \to B$ and $\partial I \to I$ are cofibrations, the first by assumption, and the second by Example 2.3.5. Hence an extension $\widehat{G} \colon B \times I \to B^I$ exists. This corresponds by Theorem 1.2.7 to a map $\widetilde{G} \colon B \times I \times I \to B$, and we claim $\widetilde{G}|_{B \times I \times \{1\}} \colon B \times I \to B$ gives a homotopy between $i \circ j$ and $1_B$ relative to $A$. This is again straightforward, and follows from $ev_1 \circ C(a, t) = G(a, 0) = i \circ j(a) = a$ (identifying $a$ with $i(a)$), $ev_1 \circ C(b, 0) = i \circ j(b)$ and $ev_1 \circ C(b, 1) = G(b, 1) = 1_B(b)$. $\square$

The next result is a dual of Proposition 2.5.2.

**Proposition 2.5.3** *Suppose $p \colon X \to Y$ is a fibration and a homotopy equivalence. Then there exists a homotopy inverse $q \colon Y \to X$ such that*

1. *$p \circ q = 1_Y$ (i.e. $q$ is a section of $p$);*
2. *if $q \colon Y \to X$ is a cofibration, there is a homotopy $q \circ p \sim 1_X$ relative to $q(Y)$.*

*Proof* Choose a homotopy inverse $r$ to $p$, and let $H \colon Y \times I \to Y$ be a homotopy from $p \circ r$ to $1_Y$. Consider the lifting problem

$$
\begin{array}{ccc}
Y \times \{0\} & \xrightarrow{\ r\ } & X \\
{\scriptstyle i_0}\downarrow & \nearrow & \downarrow{\scriptstyle p} \\
Y \times I & \xrightarrow[\ H\ ]{} & Y
\end{array}
$$

Since $p$ is a fibration, the dotted arrow exists, and there is a map $\widehat{H} \colon Y \times I \to X$ such that $\widehat{H} \circ i_0 = r$ and $p \circ \widehat{H} = H$. Let $q = \widehat{H}|_{Y \times \{1\}}$. Then $q$ is homotopic to $r$ via $\widehat{H}$, and $p \circ q(y) = p \circ \widehat{H}(y, 1) = H(y, 1) = 1_Y(y)$.

The second statement follows from condition 2 of Proposition 2.5.2. $\square$

We now have the following result, which will be used in the proof of Theorem 8.3.2. The first statement is [Str68, Theorems 8 and 9]. We will also recreate various results from [Str66] in the course of the proof.

**Theorem 2.5.4** (The lifting axiom) *Consider the lifting problem*

$$
\begin{array}{ccc}
A & \xrightarrow{\ g\ } & X \\
{\scriptstyle i}\downarrow & \nearrow & \downarrow{\scriptstyle p} \\
B & \xrightarrow[\ h\ ]{} & Y
\end{array}
$$

*where $i$ is a cofibration and $p$ is a fibration.*

1. *If either i or p is a homotopy equivalence, then the dotted arrow exists and makes the diagram commute.*
2. *If $(B, A)$ is a relative CW complex, i an equivalence, and p a Serre fibration, then the dotted arrow exists and makes the diagram commute.*

**Remark 2.5.5** For the reader interested in model categories, the first part of the following theorem is the verification of the "lifting axiom" for the model category structure on Top in which the cofibrations and fibrations are as in our definitions here, and the weak equivalences are the homotopy equivalences. The second part is quite close to the same thing for the model structure in which the fibrations are the Serre fibrations and the weak equivalences are the weak homotopy equivalences, but in this model structure the class of cofibrations is much smaller than what we call a cofibration here. □

*Proof of Theorem 2.5.4* We prove the first statement. The proof of the second can be found in [Hir15d]. Assume $p$ is a homotopy equivalence. By Proposition 2.5.3, there exists a homotopy inverse $q\colon Y \to X$ such that $p \circ q = 1_Y$. Let $H\colon X \times I \to X$ be a homotopy from $q \circ p$ to $1_X$. We first show how to construct a solution to the lifting problem

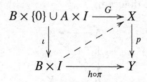

Here $\pi\colon B \times I \to B$ is the projection, and $G(b, 0) = q \circ h(b)$ and $G(a, t) = H(g(a), t)$. The formulas defining $G$ agree on $A \times \{0\}$ and so define a map from the union. It is straightforward to check that the square commutes. Let us see how this is related to the original question. Suppose $\widehat{H}\colon B \times I \to X$ is a solution to this lifting problem, and let $k = \widehat{H}|_{B \times \{1\}}$. Then $k\colon B \to X$ satisfies $p \circ k(b) = h(b)$ since $p \circ \widehat{H}(b, t) = h(b)$, and clearly $k \circ i(a) = G(a, 1) = H(g(a), 1) = g(a)$. Thus $k$ defines a solution to the original lifting problem.

Now we show how to construct $\widehat{H}$. We are now in a situation where $p$ is a fibration and a homotopy equivalence and $\iota$ is the inclusion of a subspace which is a retract of $B \times I$. Let $u\colon B \times I \to I$ be the function given by Proposition 2.3.10, so that $u^{-1}(0) = B \times \{0\} \cup A \times I$. Moreover, the remark following Proposition 2.3.10 tells us that, since $i$ is a cofibration, $\iota$ is inclusion of a subspace which is a deformation retract of $B \times I$. Let $r\colon B \times I \to B \times \{0\} \cup A \times I$ be a retraction, and $K\colon B \times I \times I \to B \times I$ a homotopy from $\iota \circ r$ to $1_{B \times I}$ relative to $B \times \{0\} \cup A \times I$. Define $\widetilde{K}\colon B \times I \times I \to B \times I$ by

$$\widetilde{K}(b,t,s) = \begin{cases} K(b,t,s/u(b,t)), & \text{if } s < u(b,t); \\ K(b,t,1), & \text{if } s \geq u(b,t). \end{cases}$$

This formula defines a continuous function. Consider now the lifting problem

$$
\begin{array}{ccc}
B \times I \times \{0\} & \xrightarrow{\;G \circ r\;} & X \\
{\scriptstyle i_0}\downarrow & \nearrow & \downarrow{\scriptstyle p} \\
B \times I \times I & \xrightarrow[\;h \circ \pi \circ \widetilde{K}\;]{} & Y
\end{array}
$$

This diagram commutes since $\widetilde{K}(b,t,0) = K(b,t,0) = \iota \circ r(b,t)$ and $p \circ G = h \circ \pi$. Since $p$ is a fibration, a solution $\widehat{K}\colon B \times I \times I \to X$ to this lifting problem exists. We have $\widehat{K}(b,t,0) = G \circ r(b,t)$ and $p \circ \widehat{K} = h \circ \pi \circ \widetilde{K}$. If we let $\widehat{H}(b,t) = \widehat{K}(b,t,u(b,t))\colon B \times I \to X$, then $p \circ \widehat{H}(b,t) = h \circ \pi \circ \widehat{K}(b,t,u(b,t)) = h \circ \pi \circ K(b,1) = h \circ \pi(b,1) = h \circ \pi(b,t)$. Moreover, $\widehat{H} \circ \iota = G$ since $K$ was a homotopy relative to $B \times \{0\} \cup A \times I$. This completes the construction of $\widehat{H}$ and this part of the proof.

Now assume $i$ is a homotopy equivalence. By Proposition 2.5.2, there exists a homotopy inverse $j\colon B \to A$ for $i$ such that $j \circ i = 1_A$, and $i \circ j$ is homotopic to $1_B$ relative to $A$.

Let $\tilde{h} = g \circ j$. Then the diagram

$$
\begin{array}{ccc}
A & \xrightarrow{\;g\;} & X \\
{\scriptstyle i}\downarrow & {\scriptstyle \tilde{h}}\nearrow & \downarrow{\scriptstyle p} \\
B & \xrightarrow[\;h \circ i \circ j\;]{} & Y
\end{array}
$$

commutes. Now we will deform this lift to fit in the original diagram, using the fact that $i \circ j \sim 1_B$ relative to $A$, which implies $h \circ i \circ j \sim h$ relative to $A$. Choose a homotopy $H\colon B \times I \to Y$ from $h \circ i \circ j$ to $h$ relative to $A$. Consider the lifting problem

$$
\begin{array}{ccc}
B \times \{0\} \cup A \times I & \xrightarrow{\;G\;} & X \\
{\scriptstyle \iota}\downarrow & \nearrow & \downarrow{\scriptstyle p} \\
B \times I & \xrightarrow[\;H\;]{} & Y
\end{array}
$$

where $G$ is the map $g \circ j$ on $B \times \{0\}$ and the composition of the projection $A \times I \to A$ with the map $g\colon A \to X$. This gives a well-defined map on the union since $j(a) = a$ for $a \in A$, and the square commutes. A solution $\widehat{H}$ to this lifting problem then provides the desired solution $\widehat{h}$ to the lifting problem in the statement of the theorem by letting $\widehat{h} = \widehat{H}|_{B \times \{1\}}$. But $\widehat{H}$ exists using precisely the same argument as above, as we are in the same situation: $\iota$ is the inclusion of a subspace which is a deformation retract of $B \times I$, and $p$ is a fibration. The same argument as above proves that the desired extension exists.    □

## 2.6  Connectivity of spaces and maps

The notion of connectivity of spaces and maps will be used heavily throughout this book, so we review it here. Our definitions are all on the level of spaces, for reasons we will elaborate on in Remark 2.6.8. Important algebraic consequences of connectivity will be discussed as well.

**Definition 2.6.1**  A space $X$ is *k-connected* if, for all $-2 \le i \le k$, every map $S^i \to X$ extends to a map $D^{i+1} \to X$. If this is true for all $k$, we say $X$ is ∞-*connected* or *weakly contractible*.

The above definition makes sense for $i = -1, -2$ since $S^i$ in both those cases is defined to be the empty set. If $X$ is $k$-connected then it is also $j$-connected for every $j \le k$. Every non-empty space is $(-1)$-connected, path-connected spaces are 0-connected, and simply-connected spaces are 1-connected. Every space, including the empty set, is $(-2)$-connected. The proof of the following can be found, for example, in [Hat02, p. 346].

**Proposition 2.6.2**  *The following are equivalent for an integer $k \ge 0$.*

1. *$X$ is $k$-connected.*
2. *For all $i \le k$, every map $S^i \to X$ is homotopic to a constant map.*
3. *For all $0 \le i \le k$, $\pi_i(X, x_0) = 0$ for all basepoints $x_0$.*

**Remark 2.6.3**  In line with the last part of the previous result, it is also true that the homology of a $k$-connected space $X$ vanishes up to dimension $k$. This is because, by Theorem 1.3.7, we can assume $X$ is a CW complex, and the $k$-connectedness implies by Theorem 2.6.26 that we can assume that it has a single 0-cell and no others cells of dimension $< k + 1$. Recalling the setup for cellular homology groups from the discussion preceding Theorem 1.4.10, this immediately implies that the cellular homology groups vanish up to dimension $k$, and hence, by Theorem 1.4.10, that the reduced singular homology groups also vanish up to dimension $k$.    □

We also have the notion of the connectivity of a pair $(X, A)$.

**Definition 2.6.4** A pair $(X, A)$ is *k-connected* if every map of pairs $(D^i, \partial D^i) \to (X, A)$ is homotopic relative to $\partial D^i$ to a map $D^i \to A$ for all $-1 \leq i \leq k$. We say the pair is *∞-connected* if this is true for all integers $k$.

If $X$ is non-empty and $k$-connected, and $*$ is any point in $X$, then the pair $(X, *)$ is $k$-connected. The pair $(X, \emptyset)$ is $(-1)$-connected if $X$ is non-empty and ∞-connected if $X$ is empty.

The above definition is equivalent to saying that the relative homotopy groups $\pi_i(X, A, x_0)$ are trivial for $0 < i \leq k$ and the map $\pi_0(A) \to \pi_0(X)$ is surjective.[3]

To relate the relative and absolute cases we have the following result, the verification of which we leave to the reader.

**Proposition 2.6.5** *A space $X$ is $k$-connected if and only if the pair $(CX, X)$ is $(k + 1)$-connected.*

**Definition 2.6.6** A map $f: X \to Y$ is *k-connected* if $\mathrm{hofiber}_y(f)$ is $(k - 1)$-connected for all $y \in Y$.

It follows that every map is $(-1)$-connected. As with the connectivity of spaces, if $f$ is $k$-connected, then it is $j$-connected for every $j \leq k$. If $k = \infty$, that is, $f$ is $k$-connected for all $k$, then $f$ is a weak equivalence. This follows from Proposition 2.6.17.

**Remark 2.6.7** We have mentioned already that there are two notions of equivalence we care about: homotopy equivalence and weak equivalence. Weak equivalence is the natural limiting notion of a $k$-connected map, and so we prefer this notion in settings where we pay attention to connectivites of maps. The most important instances are in Theorems 4.2.1 and 4.2.2 and their generalizations, Theorems 6.2.1 and 6.2.2. However, the notion of homotopy equivalence is useful because dualizing certain arguments is usually more straightforward or even trivial when using homotopy equivalences whereas this is not always the case with weak equivalences. A key instance of this, which we have already discussed, is Proposition 1.3.2, which stands in contrast with Proposition 1.3.5.                                                                          □

**Remark 2.6.8** Some authors define $k$-connectivity of a space in terms of the homotopy groups, but we prefer the space-level version given above, especially

---

[3] We have separated the case of $\pi_0$ since, as mentioned in Section 1.3, we do not believe there is a satisfactory definition of relative $\pi_0$.

since we give a space-level definition of $k$-connected map. The standard algebraic definition is to say that $f: X \to Y$ is $k$-connected if for all $x \in X$ the induced map $f_*: \pi_i(X, x) \to \pi_i(Y, f(x))$ is an isomorphism for all $0 \le i < k$ and onto when $i = k$. Unfortunately this is not quite correct if $X$ is empty. We have noted above that every pair $(Y, \emptyset)$ is $(-1)$-connected, and so the map $\emptyset \to Y$ should be a $(-1)$-connected map, but according to our algebraic definition the map $\emptyset \to Y$ is $\infty$-connected for logical reasons. □

**Proposition 2.6.9** *The following are equivalent for a map $f: X \to Y$ of spaces.*

1. *$f$ is $k$-connected.*
2. *The pair $(M_f, X)$, where $M_f$ is the mapping cylinder of $f$, is $k$-connected.*

*Either of these conditions implies the following one for any space $X$, and if $X$ is non-empty, the following condition is equivalent to the ones above.*

3. *For every $x \in X$ the induced map $f_*: \pi_i(X, x) \to \pi_i(Y, f(x))$ is an isomorphism for $0 \le i < k$ and a surjection for $i = k$.*

*In particular, if $Y = *$, then by Proposition 2.6.5 $f: X \to *$ is $k$-connected if and only if $X$ is $(k - 1)$-connected. If instead $X = *$ and $f: * \to Y$ is a cofibration, then $f$ is $k$-connected if and only if $Y$ is $k$-connected.*

**Remark 2.6.10** If $f: X \to Y$ is a cofibration, the equivalence of parts 1 and 2 says that the pair $(Y, X)$ is $k$-connected if and only if $f$ is $k$-connected. In particular, $(CX, X)$ is $(k + 1)$-connected if and only if $X \to CX$ is $(k + 1)$-connected if and only if $X$ is $k$-connected. □

*Sketch of proof of proposition 2.6.9* The equivalence of the first two items is trivial since the mapping cylinder is homotopy equivalent to $Y$ (the codomain of the map in question). The third item is obtained from the remarks preceding the statement of this result. The "in particular" parts follow from conditions 1 and 2 directly. We leave it to the reader to fill in the details. □

**Example 2.6.11** (Connectivity of a loop space) Let $X$ be a based space. If $X$ is $k$-connected, then $\Omega X$ is $(k - 1)$-connected. To prove this requires 2(a) of Proposition 3.3.11 applied to the square and using that the map $* \to X$ is $k$-connected by Proposition 2.6.9.          □

**Proposition 2.6.12** *Let* $f: X \to Y$ *be a map. If* $f$ *is* $k$-*connected, then* hocofiber($f$) *is* $k$-*connected.*

*Proof* The proof uses a result from the next chapter: apply 2(a) of Proposition 3.7.13 by considering the square

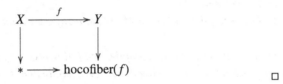

          □

**Example 2.6.13** (Connectivity of a suspension)    Suppose $X$ is $k$-connected. Then $\Sigma X$ is $(k + 1)$-connected. This is true for both based and unbased suspensions. In any case, $X \to *$ is $(k + 1)$-connected by Proposition 2.6.9, and by Example 2.4.11, hocofiber($X \to *$) $\cong \Sigma X$, so $\Sigma X$ is $(k + 1)$-connected by Proposition 2.6.12.          □

The converse to Proposition 2.6.12 is not true. For example, Remark 3.7.14 gives an example of a space $X$ for which the homotopy cofiber of $X \to *$, which is $\Sigma X$, is $\infty$-connected, yet the map $X \to *$ is only 0-connected. Nevertheless, there are instances when the connectivity of the homotopy cofiber does say something about the connectivity of the map. In particular, we have the following result, which will be proved later as Proposition 4.3.6.

**Proposition 2.6.14** *Suppose* $X$ *is simply-connected and* $f: X \to Y$ *is a map such that* hocofiber($f$) *is* $k$-*connected. Then* $f$ *is* $k$-*connected.*

The following basic result regarding compositions and connectivity will be used and generalized in Propositions 3.3.20 and 5.4.14. We omit the proof, as it amounts to unravelling the definitions.

**Proposition 2.6.15** *Given maps* $f: X \to Y$ *and* $g: Y \to Z$, *we have the following:*

1. *If $f, g$ are k-connected, then $g \circ f$ is k-connected.*
2. *If $f$ is $(k - 1)$-connected and $g \circ f$ is k-connected, then $g$ is k-connected.*
3. *If $g$ is $(k + 1)$-connected and $g \circ f$ k-connected, then $f$ is k-connected.*

Note that these results are also true when $k = \infty$.

**Proposition 2.6.16** *Suppose $p : X \to Y$ is a fibration, and let $F_y = p^{-1}(y)$ be a fiber. Then we have the following:*

1. *If $Y$ is k-connected and $F_y$ is k-connected for all $y$, then $X$ is k-connected.*
2. *If $X$ is k-connected and $F_y$ is $(k - 1)$-connected for all $y$, then $Y$ is k-connected.*
3. *If $X$ is k-connected and $Y$ is $(k + 1)$-connected, then $F_y$ is k-connected for all $y$.*

*Proof* For $k \geq 2$, the statements follow from the long exact sequence of homotopy groups associated to $p$. For $k < 2$, separate arguments are needed, but at most they involve simple uses of the homotopy lifting property of $p$. □

We will use the following characterization of weak equivalences in the proof of Lemma 3.6.14, which is an essential component of the theory of homotopy cocartesian squares.

**Proposition 2.6.17** *A map $f : X \to Y$ is k-connected if and only if, for all relative CW pairs $(Z, A)$ of dimension $\leq k$, the dotted arrow in the diagram*

$$
\begin{array}{ccc}
A & \xrightarrow{\ h\ } & X \\
{\scriptstyle i}\downarrow & {\scriptstyle G}\nearrow & \downarrow{\scriptstyle f} \\
Z & \xrightarrow[\ g\ ]{} & Y
\end{array}
$$

(2.6.1)

*exists and satisfies $G \circ i = h$ and $f \circ G \sim g$ relative to $A$. Futhermore, this is true if and only if it holds for all pairs $(D^j, \partial D^j)$ with $-1 \leq j \leq k$.*

*Moreover, $f$ is a weak equivalence if and only if this is true for all relative CW pairs $(Z, A)$, if and only if this is true for all pairs $(D^i, \partial D^j)$ for all $-1 \leq j$.*

As a reminder, recall that Proposition 1.4.4, an important result which characterizes weak equivalences in terms of homotopy groups, follows from the above.

Before we embark on the proof of this result, we will need to state some definitions and prove some lemmas that were given to us by Phil Hirschhorn; they will be used in one part of the proof.

For $k \geq 1$ we let $S_+^k$ and $S_-^k$ be the upper and the lower hemispheres of $S^k$, respectively, i.e. the sets of points $(x_1, \ldots, x_{k+1})$ on the sphere where $x_{k+1} \geq 0$ or $x_{k+1} \leq 0$. Also let $p_+ \colon S_+^k \to D^k$ and $p_- \colon S_-^k \to D^k$ be the homeomorphisms given by forgetting the last coordinate of the point on the hemisphere.

**Definition 2.6.18** If $X$ is a space and $\alpha, \beta \colon D^k \to X$ agree on $\partial D^k$, let $d(\alpha, \beta) \colon S^k \to X$ be the map that is $\alpha \circ p_+ \colon S_+^k \to X$ on the upper hemisphere of $S^k$ and $\beta \circ p_- \colon S_-^k \to X$ on the lower hemisphere of $S^k$. We call this the *difference map* of $\alpha$ and $\beta$.

**Lemma 2.6.19** *Let $X$ be a space and let $\alpha \colon D^k \to X$ be a map. For any $[g] \in \pi_k(X, \alpha(p_0))$ (where $p_0$ is the basepoint of $D^k$), there is a map $\beta \colon D^k \to X$ such that $\beta|_{\partial D^k} = \alpha|_{\partial D^k}$ and $[d(\alpha, \beta)] = [g]$ in $\pi_k(X, \alpha(p_0))$.*

*Proof* The basepoint of $D^k$ is a strong deformation retract of $D^k$, and so any two maps $D^k \to X$ are homotopic relative to the basepoint. Thus, the restriction of $g$ to $S_+^k$ is homotopic relative to the basepoint to $\alpha \circ p_+$. Since the inclusion $S_+^k \hookrightarrow S^k$ is a cofibration, there is a homotopy of $g$ to a map $h \colon S^k \to X$ such that $h|_{S_+^k} = \alpha \circ p_+$; we let $\beta = h \circ (p_-^{-1})$, and we have $h = d(\alpha, \beta)$.            □

**Lemma 2.6.20** (Additivity of difference elements) *If $X$ is a space and $\alpha, \beta, \gamma \colon D^k \to X$ are maps that agree on $\partial D^k$, then in $\pi_k(X, \alpha(p_0))$ (where $p_0$ is the basepoint of $D^k$) we have*

$$[d(\alpha, \beta)] + [d(\beta, \gamma)] = [d(\alpha, \gamma)].$$

*(If $k = 1$, addition should be replaced by multiplication.)*

*Proof* Let $T^k = S^k \cup D^k$, where we view $D^k$ as the subset of $\mathbb{R}^{k+1}$ given by

$$D^k = \{(x_1, x_2, \ldots, x_{k+1}) \colon x_1^2 + x_2^2 + \cdots x_k^2 \leq 1, x_{k+1} = 0\}.$$

Then $T^k$ is a CW complex that is the union of the $k$-cells $S_+^k$, $S_-^k$, and $D^k$, which all share a common boundary. We let $t(\alpha, \beta, \gamma) \colon T^k \to X$ be the map such that

$$t(\alpha, \beta, \gamma)|_{S_+^k} = \alpha \circ p_+,$$
$$t(\alpha, \beta, \gamma)|_{D^k} = \beta,$$
$$t(\alpha, \beta, \gamma)|_{S_-^k} = \gamma \circ p_-.$$

Thus,

- the composition $S^k \to T^k \xrightarrow{t(\alpha, \beta, \gamma)} X$ is $d(\alpha, \gamma)$;
- the composition $S^k \to S_+^k \cup D^k \subset T^k \xrightarrow{t(\alpha, \beta, \gamma)} X$ (where the first map is the identity on $S_+^k$ and is $p_-$ on $S_-^k$) is $d(\alpha, \beta)$;

- the composition $S^k \to D^k \cup S^k_- \subset T^k \xrightarrow{t(\alpha,\beta,\gamma)} X$ (where the first map is $p_+$ on $S^k_+$ and is the identity on $S^k_-$) is $d(\beta,\gamma)$.

The basepoint of $D^k$ is a strong deformation retract of $D^k$, and so the map $\beta \colon D^k \to X$ is homotopic relative to the basepoint to the constant map to $\alpha(p_0)$ (where $p_0$ is the common basepoint of $S^k$ and $D^k$). Since the inclusion $D^k \hookrightarrow T^k$ is a cofibration, there is a homotopy of $t(\alpha,\beta,\gamma)$ relative to the basepoint to a map $\hat{t}(\alpha,\beta,\gamma)$ that takes all of $D^k$ to the basepoint $\alpha(p_0)$. If $T^k \to S^k \vee S^k$ is the map that collapses $D^k$ to a point, then $\hat{t}(\alpha,\beta,\gamma)$ factors as $T^k \to S^k \vee S^k \xrightarrow{\alpha_\beta \vee \beta_\gamma} X$, where $\alpha_\beta \colon S^k \to X$ is homotopic to $d(\alpha,\beta)$ and $\beta_\gamma \colon S^k \to X$ is homotopic to $d(\beta,\gamma)$. Since the composition

$$S^k \hookrightarrow T^k \to S^k \vee S^k \xrightarrow{\alpha_\beta \vee \beta_\gamma} X$$

is homotopic relative to the basepoint to $d(\alpha,\gamma)$, we have $[d(\alpha,\gamma)] = [d(\alpha,\beta)] + [d(\beta,\gamma)]$ if $k > 1$ and $[d(\alpha,\gamma)] = [d(\alpha,\beta)] \cdot [d(\beta,\gamma)]$ if $k = 1$. $\qquad\square$

**Lemma 2.6.21** *If $X$ is a space, $\alpha,\beta \colon D^k \to X$ maps that agree on $\partial D^k$, and $[d(\alpha,\beta)]$ the identity element of $\pi_k(X)$, then $\alpha$ and $\beta$ are homotopic relative to $\partial D^k$.*

*Proof* Since $[d(\alpha,\beta)]$ is the identity element of $\pi_k(X)$, there is a map $h \colon D^{k+1} \to X$ whose restriction to $\partial D^{k+1}$ is $d(\alpha,\beta)$. View $D^k \times I$ as the cone on $\partial(D^k \times I) = (D^k \times \{0\}) \cup (S^{k-1} \times I) \cup (D^k \times \{1\})$ with vertex at the center of $D^k \times I$. Let $p \colon D^k \times I \to D^{k+1}$ be the map that

- on $D^k \times \{0\}$ is the composition $D^k \times \{0\} \xrightarrow{pr} D^k \xrightarrow{(p_+)^{-1}} S^k_+ \hookrightarrow D^{k+1}$;
- on $D^k \times \{1\}$ is the composition $D^k \times \{0\} \xrightarrow{pr} D^k \xrightarrow{(p_-)^{-1}} S^k_- \hookrightarrow D^{k+1}$;
- on $S^{k-1} \times I$ is the composition $S^{k-1} \times I \xrightarrow{pr} S^{k-1} \hookrightarrow S^k_+ \cap S^k_- \subset D^{k+1}$;
- takes the center point of $D^k \times I$ to the center point of $D^{k+1}$;
- is linear on each straight line connecting the center point of $D^k \times I$ to its boundary.

Here $pr$ is the projection map. The composition $D^k \times I \xrightarrow{p} D^{k+1} \xrightarrow{h} X$ is then a homotopy from $\alpha$ to $\beta$ relative to $\partial D^k$. $\qquad\square$

*Proof of Proposition 2.6.17* We first argue that showing the statement for all CW pairs $(Z,A)$ is equivalent to showing it for all pairs $(D^j, \partial D^j)$. Clearly $(D^j, \partial D^j)$ is an example of a CW pair so one direction is trivial.

For the other direction, let $Z^j$ denote the relative $j$-skeleton; recall that this is $A$ together with all the cells of $Z$ of dimension $\leq j$, and that $Z^{-1} = A$. Let $A_j$ be the indexing set for the $j$-cells of $Z^j$, and let $a_\alpha \colon \partial D^j \to Z^{j-1}$ be the attaching map for the cell labeled by $\alpha \in A_j$. To begin, we have the lifting problem

$$
\begin{array}{ccc}
Z^{-1} = A & \xrightarrow{\ h\ } & X \\
\downarrow & \nearrow & \downarrow{\scriptstyle f} \\
Z^0 & \xrightarrow[g_0|_{Z^0}]{} & Y
\end{array}
$$

where $g_0 = g$. For each $\alpha \in A_0$ we have an induced lifting problem

$$
\begin{array}{ccc}
\partial D^0 = \emptyset & \longrightarrow & X \\
\downarrow & \nearrow & \downarrow{\scriptstyle f} \\
D^0 & \xrightarrow[g_{0,\alpha}]{} & Y
\end{array}
$$

where $g_{0,\alpha}$ is the restriction of $g_0$ to the 0-cell $D^0$ labeled by $\alpha$. A solution $G_{0,\alpha}$ exists for all $\alpha$ since $-1 \leq j \leq k$ and $X \to Y$ is $k$-connected. We then define a solution to the lifting problem for $g_0|_{Z^0}$ above by letting $G_0 \colon Z^0 \to X$ be equal to $G_{0,\alpha}$ on the cell labeled by $\alpha$. This map is continuous because relative CW complexes have the weak topology. The weak topology is also the reason that the homotopies $f \circ G_{0,\alpha} \sim g_{o,\alpha}$ glue together to a homotopy (relative to $A$) $f \circ G_0 \sim g_0|_{Z^0}$.

The inclusion $Z^0 \to Z$ is a cofibration, and so the homotopy from $g_0|_{Z^0} \colon Z^0 \to Y$ to $f \circ G_0$ can be extended to a homotopy (relative to $A$) $H_0 \colon Z \times I \to Y$ from $g_0$ to a map $g_1 \colon Z \to Y$, and we then have the lifting problem

$$
\begin{array}{ccc}
Z^0 & \xrightarrow{\ G_0\ } & X \\
\downarrow & \nearrow & \downarrow{\scriptstyle f} \\
Z^1 & \xrightarrow[g_1|_{Z^1}]{} & Y
\end{array}
$$

(i.e. the solid arrow square commutes on the nose). For each $\alpha \in A_1$ we have an induced lifting problem

$$
\begin{array}{ccc}
\partial D^1 & \xrightarrow{\ G_0 \circ a_\alpha\ } & X \\
\downarrow & \nearrow & \downarrow{\scriptstyle f} \\
D^1 & \xrightarrow[g_{1,\alpha}]{} & Y
\end{array}
$$

where $g_{1,\alpha}$ is the restriction of $g_1$ to the 1-cell $D^1$ labeled by $\alpha$. A solution $G_{1,\alpha}$ exists for all $\alpha$ since $-1 \leq j \leq k$ and $X \to Y$ is $k$-connected. We then define a solution to the lifting problem for $g_1|_{Z^1}$ above by letting $G_1$ be equal to $G_{1,\alpha}$ on the cell labeled by $\alpha$. This map is continuous because relative CW complexes have the weak topology. The weak topology is also the reason that the homotopies $f \circ G_{1,\alpha} \sim g_{1,\alpha}$ glue together to a homotopy (relative to $Z^0$) $f \circ G_1 \sim g_1|_{Z^1}$.

The inclusion $Z^1 \to Z$ is a cofibration, and so the homotopy from $g_1|_{Z^1} : Z^1 \to Y$ to $f \circ G_1$ can be extended to a homotopy (relative to $Z^0$) $H_1 : Z \times I \to Y$ from $g_1$ to a map $g_2 : Z \to Y$, and we then have the lifting problem

For each $\alpha \in A_2$ we have an induced lifting problem

where $g_{2,\alpha}$ is the restriction of $g_2$ to the 2-cell $D^2$ labeled by $\alpha$. A solution $G_{2,\alpha}$ exists for all $\alpha$ since $-1 \leq j \leq k$ and $X \to Y$ is $k$-connected. We then define a solution to the lifting problem for $g_2|_{Z^2}$ above by letting $G_2$ be equal to $G_{2,\alpha}$ on the cell labeled by $\alpha$. This map is again continuous because relative CW complexes have the weak topology, and the weak topology is also again the reason that the homotopies $f \circ G_{2,\alpha} \sim g_{2,\alpha}$ glue together to a homotopy (relative to $Z^1$) $f \circ G_2 \sim g_2|_{Z^2}$. The inclusion $Z^2 \to Z$ is a cofibration, and so the homotopy from $g_2|_{Z^2} : Z^2 \to Y$ to $f \circ G_2$ can be extended to a homotopy (relative to $Z^1$) $H_2 : Z \times I \to Y$ from $g_2$ to a map $g_3 : Z \to Y$.

For the general step, we have defined the map $G_{j-1} : Z^{j-1} \to X$ and a map $g_j : Z \to Y$ such that we have the lifting problem

$$
\begin{array}{ccc}
Z^{j-1} & \xrightarrow{G_{j-1}} & X \\
\downarrow & \nearrow & \downarrow f \\
Z^j & \xrightarrow[g_j|_{Z^j}]{} & Y
\end{array}
$$

For each $\alpha \in A_j$ we have an induced lifting problem

$$
\begin{array}{ccc}
\partial D^j & \xrightarrow{G_{j-1} \circ a_\alpha} & X \\
\downarrow & \nearrow & \downarrow f \\
D^j & \xrightarrow[g_{j,\alpha}]{} & Y
\end{array}
$$

where $g_{j,\alpha}$ is the restriction of $g_j$ to the $j$-cell $D^j$ labeled by $\alpha$. A solution $G_{j,\alpha}$ exists for all $\alpha$ since $-1 \le j \le k$ and $X \to Y$ is $k$-connected. We then define a solution to the lifting problem for $g_j|_{Z^j}$ above by letting $G_j$ be equal to $G_{j,\alpha}$ on the cell labeled by $\alpha$. This map is continuous because relative CW complexes have the weak topology and, for the same reason, the homotopies $f \circ G_{j,\alpha} \sim g_{j,\alpha}$ glue together to a homotopy (relative to $Z^{j-1}$) $f \circ G_j \sim g_j|_{Z^j}$. The inclusion $Z^j \to Z$ is a cofibration, and so the homotopy from $g_j|_{Z^j} : Z^j \to Y$ to $f \circ G_j$ can be extended to a homotopy (relative to $Z^{j-1}$) $H_j : Z \times I \to Y$ from $g_j$ to a map $g_{j+1} : Z \to Y$.

Having completed those steps for all $j \ge 0$, the maps $G_j$ determine a map $G : Z \to X$ (the restriction of $G$ to $Z_j$ is $G_j$) and we have $G \circ i = h$. We define a homotopy $H : Z \times I \to Y$ from $f \circ G$ to $g$ by letting

$$H(z,t) = \begin{cases} H_0(z, 2t), & \text{if } 0 \le t \le 1/2; \\ H_1(z, 4t - 2), & \text{if } 1/2 \le t \le 3/4; \\ H_2(z, 8t - 6), & \text{if } 3/4 \le t \le 7/8; \\ H_3(z, 16t - 14), & \text{if } 7/8 \le t \le 15/16; \\ \vdots & \\ H_n(z, 2^{n+1}t + (2 - 2^{n+1})), & \text{if } 1 - \frac{1}{2^n} \le t \le 1 - \frac{1}{2^{n+1}}; \\ \vdots & \\ H_j(z, 1), & \text{if } t = 1 \text{ and } z \in Z^j - Z^{j-1}. \end{cases}$$

We must show that $H$ is continuous. Since $Z \times I$ has the weak topology with respect to the subspaces $Z^j \times I$, it is sufficient to show that the restriction of $H$ to each $Z^j \times I$ is continuous, and since each homotopy $H_j$ is a homotopy relative to $Z^{j-1}$, the restriction of $H$ to $Z^j \times I$ is the concatenation of only finitely many homotopies. That is, the restriction of $H$ to $Z^j \times I$ is continuous when restricted to each element of a finite cover of $Z^j \times I$ by closed sets, and is thus continuous.

So now we want to show that $f$ is $k$-connected if and only if, in the solid arrow diagram

$$\begin{array}{ccc} \partial D^j & \overset{h}{\longrightarrow} & X \\ {\scriptstyle i}\downarrow & {\scriptstyle G} \nearrow & \downarrow{\scriptstyle f} \\ D^j & \underset{g}{\longrightarrow} & Y \end{array} \qquad (2.6.2)$$

the dotted arrow exists for $-1 \le j \le k$ and makes the upper triangle commute on the nose and the lower triangle up to homotopy. First assume $f$ is

$k$-connected. We will first show that $G$ as in the above diagram exists for $j = k$ as follows:

The map $h$ defines an element $[h]$ of $\pi_{k-1}(X)$ (at some basepoint) such that $f_*([h]) = 0$ in $\pi_{k-1}(Y)$. Let $F$ be the homotopy fiber of $f$. Since $\pi_{k-1}(F) = 0$ (as $f$ is $k$-connected), the long exact homotopy sequence of a fibration implies that $[h] = 0$ in $\pi_{k-1}(X)$, and so there is a map $\gamma\colon D^k \to X$ such that $\gamma \circ i = h$.

The maps $f\gamma\colon D^k \to Y$ and $g\colon D^k \to Y$ agree on $\partial D^k$, and so there is a difference map $d(f\gamma, g)\colon S^k \to X$ (see Definition 2.6.18) that defines an element $\alpha$ of $\pi_k(Y)$. Since $\pi_{k-1}(F) = 0$, the long exact homotopy sequence implies that there is an element $\beta$ of $\pi_k(X)$ such that $f_*(\beta) = -\alpha$ (if $k > 1$) or $f_*(\beta) = \alpha^{-1}$ (if $k = 1$), and Lemma 2.6.19 implies that we can choose a map $G\colon D^k \to X$ that agrees with $\gamma\colon D^k \to X$ on $\partial D^k$ such that $[d(G, \gamma)] = \beta$ in $\pi_k(X)$. Thus, $Gi = h$, and since $[d(fG, f\gamma)] = [f \circ d(G, \gamma)] = f_*[d(G, \gamma)] = f_*(\beta) = -\alpha$ (if $k > 1$) or $\alpha^{-1}$ (if $k = 1$), Lemma 2.6.20 implies that $[d(fG, g)] = [d(fG, f\gamma)] + [d(f\gamma, g)] = -\alpha + \alpha = 0$ (with a similar statement if $k = 1$). Thus Lemma 2.6.21 implies that $fG$ is homotopic to $g$ relative to $\partial D^k$.

The above argument only used that $\pi_{k-1}(F) = 0$ and can be repeated for any $j \le k$ since $\pi_j(F) = 0$ as $f$ is $k$-connected.

Now assume $G$ as in (2.6.2) exists for $j \le k$ and it makes the upper triangle commute and the lower triangle commute up to homotopy. We want to show $f$ is $k$-connected. Let $g\colon S^{j-1} \to \text{hofiber}_y(f)$ be a (based) map. The goal is to prove that it extends to a map of all of $D^j$ provided that $j$ is at most $k$. let $F = \text{hofiber}_y(f)$, and let $i$ be the inclusion $F \to X$. The composition $f(i(g))\colon S^j \to Y$ maps all of $S^j$ to $y$, and so trivially extends to the disk. Then we have the diagram

where the top map is $i(g)$ and the bottom map the trivial extension of $f(i(g))$ to the disk $D^j$. By hypothesis a lift exists with the property that the upper triangle commutes on the nose. This means that $[i(g)]$ is equal to zero in $\pi_{j-1}(X)$, so $[g]$ is in the kernel of the map $\pi_{j-1}(F) \to \pi_{j-1}(X)$. Now we just need to argue that $\pi_{j-1}(X) \to pi_{j-1}(Y)$ is injective to prove this kernel is in fact equal to $\pi_{j-1}(F)$ by the long exact sequence of homotopy groups. But this is easy since the existence of the lifting solution above shows that the induced map on homotopy groups is injective (if $\partial D^j \to X$ has null-homotopic composite to $Y$, then it is also null-homotopic itself).

Now we wish to prove that $f$ is a weak equivalence if and only if, in the diagram (2.6.1) (or diagram (2.6.2)), the lift $G$ exists, making the top and bottom triangle commute and commute up to homotopy, respectively.

For one direction, suppose $G$ exists and set $A = \emptyset$, so $Z$ is a CW complex. Existence of the lift shows that $[Z, X] \to [Z, Y]$ is surjective. To get injectivity we proceed similarly, this time with $Z = Z \times I$ and $A = Z \times \partial I$ as the relative CW pair. Thus by definition $f$ is a weak equivalence.

For the other direction, consider the square

$$
\begin{array}{ccc}
\partial D^j & \xrightarrow{\ h\ } & X \\
{\scriptstyle i}\downarrow & & \downarrow{\scriptstyle f} \\
D^i & \xrightarrow[\ g\ ]{} & Y
\end{array}
$$

Since $f$ is a weak equivalence, there is a lift $H$ of $g$ but it need not agree with the given lift $h$ to $X$ on $\partial D^i$. To fix this, both $H \circ i$ and $h$ are maps $\partial D^i \to X$. They are equal in $[\partial D^i, Y]$ (i.e. $[f \circ H \circ i] = [f \circ h]$) since the bottom triangle commutes up to homotopy (but, again, not necessarily relative to $\partial D^j$). But $f$ is a weak equivalence so $[\partial D^i, X] = [\partial D^i, Y]$, and hence $H \circ i$ and $h$ are homotopic. Glue this homotopy to the lift $D^j \to X$ (essentially make a slightly larger disk) to produce a lift $H'$ which makes the upper triangle commute on the nose, while the lower triangle still commutes up to homotopy (essentially since the larger disk can be retracted to the smaller one). $\qquad\square$

A very closely related and useful result, called the Homotopy Extension Lifting Property (HELP), is the following. We omit the proof, but the reader can fill it in using the techniques from the last proof (or see [AGP02, Theorem 5.1.26]).

**Theorem 2.6.22** (Homotopy Extension Lifting Property (HELP)) *Suppose $f \colon X \to Y$ is $k$-connected, and $(Z, A)$ is a relative CW complex of dimension $\leq k$, with $\iota \colon A \to Z$ the inclusion. Let $G \colon Z \to Y$ and $g \colon A \to X$ be maps, and suppose there exists a homotopy $h \colon A \times I \to Y$ (represented by $h_t \colon A \to Y$ using Theorem 1.2.7) such that $h_0 = G \circ \iota$ and $h_1 = f \circ g$. That is, the diagram*

$$
\begin{array}{ccc}
A & \xrightarrow{\ g\ } & X \\
{\scriptstyle \iota}\downarrow & & \downarrow{\scriptstyle f} \\
Z & \xrightarrow[\ G\ ]{} & Y
\end{array}
$$

*commutes up to homotopy. Then there exists a map $\widetilde{g} \colon Z \to X$ and a homotopy $\widetilde{h} \colon Z \times I \to Y$ ($\widetilde{h}_t \colon Z \to Y$) such that $\widetilde{g} \circ \iota = g$, $\widetilde{h} \circ \iota = h_t$ for all $t$, $\widetilde{h}_0 = G$, and $\widetilde{h}_1 = f \circ \widetilde{g}$.*

**Remark 2.6.23** An immediate consequence of Theorem 2.6.22 is that the inclusion map $\iota\colon A \to Z$ for a relative CW complex $(Z, A)$ is a cofibration. To see this, let $f = 1_X\colon X \to X$ in the statement, and use the definition of cofibration. This gives an alternative argument for the result in Example 2.3.7. □

We end this section with a discussion (without proofs) of some results on cellular approximation. As we mentioned in Section 1.3, it is often useful to have a CW structure on a space, and in that case a connectivity assumption can be translated into an assumption about the dimension of various cells. We state results from [Hat02] without proof.

Let $f\colon (X, A) \to (Y, B)$ be a map of pairs of relative CW complexes. We say $f$ is *cellular* if $f(X^k) \subset Y^k$ for all $k$. The proof of the following can be found, for example, in [AGP02, Theorem 5.1.44].

**Theorem 2.6.24** *Every map $f\colon (X, A) \to (Y, B)$ of pairs of relative CW complexes is homotopic relative to A to a cellular map.*

The reader may also see [May99, Section 10.5] for a proof. The key ingredients are Theorem 2.6.22 and the following lemma, which we will prove using transversality (see Example A.2.16). We will utilize a related argument in one version of our proofs of the Blakers–Massey Theorems (4.2.1 and 6.2.1). For a proof using purely homotopy-theoretic techniques, see [Hat02, Lemma 4.10 and Corollary 4.12]. This is a non-trivial and delicate result which says that maps of CW complexes may be assumed to behave well with respect to dimension.

**Lemma 2.6.25** *The inclusion map $X \to X \cup e^k$ for attaching a k-cell to a space X is $(k-1)$-connected.*

Finally, we may translate a hypothesis about connectivity into one about dimensions for relative CW pairs. For the proof of the following, see for example [Hat02, Corollary 4.16].

**Theorem 2.6.26** *Suppose $(X, A)$ is a k-connected relative CW pair. Then there exists a relative CW pair $(Z, A)$ such that $Z - A$ contains only cells of dimension greater than k and a weak equivalence $f\colon (Z, A) \to (X, A)$ restricting to the identity on A.*

**Remark 2.6.27** Related to Remark 2.6.3, if $X$ is non-empty and $k$-connected, then $H_i(X) = 0$ for $i < k+1$. To see why, note that if $* \in X$ is any point, then

$(X, *)$ is $k$-connected, and Theorem 2.6.26 implies that there is a CW pair $(Z, *)$ weakly equivalent to $(X, *)$ such that $Z$ has cells in dimension $> k$. It follows immediately that $H_i(X) \cong H_i(Z) = 0$ for $i < k + 1$.                    □

## 2.7 Quasifibrations

Proposition 2.2.2 says that every map of spaces can be turned into a fibration. However, in many instances the fiber of the replacement, which is the homotopy fiber, is difficult to analyze. Quasifibrations are maps where the fiber and the homotopy fiber are weakly equivalent. Such a map need not satisfy the homotopy lifting property (see Example 2.1.19, to be discussed further below), but it still gives rise to a long exact sequence in homotopy. Moreover, unlike fibrations, quasifibrations are not stable under pullback. This will lead to two related notions – local quasifibration and universal quasifibration – which we will also explore here. Roughly speaking, a local quasifibration is one which is stable under restriction (pullback by an injective map), and a universal quasifibration is one which is stable under arbitrary pullbacks. Our modest contribution here is to introduce the notion of a universal quasifibration, due to Tom Goodwillie, and compare it with the notions of quasifibration and local quasifibration.

Quasifibrations were first defined by Dold and Thom [DT58] who used them in the proof of their famous theorem relating homotopy groups of the infinite symmetric product on a space $X$ and the homology of $X$; we state this result at the end of the section, and prove a mild generalization of it in Theorem 5.10.5 (the generalization is mild in the sense that it uses the classical Dold–Thom Theorem). Properties of quasifibrations were further studied and generalized in [Har70, May90, Sta68], among other places. They have been used in a variety of situations, ranging from the above-mentioned Dold–Thom Theorem and the classical Dold–Lashof construction [DL59] to a recent proof of the Bott periodicity [AP99, Beh02]. A more modern reference is Appendix A of [AGP02] where a detailed account of quasifibrations and a proof of the Dold–Thom Theorem is given. Another reference is the article [May90] by May.

We will find a use for quasifibrations in one version of the proof of Theorem 4.2.2, in one version of the proof of Theorem 5.10.8, as well as in the proof of Theorem 8.6.11 (Quillen's Theorem B).

**Definition 2.7.1**    A map $f: X \to Y$ is called a *quasifibration* if, for all $y \in Y$, the canonical map

$$\text{fiber}_y(f) \longrightarrow \text{hofiber}_y(f)$$

from (2.2.2) is a weak equivalence.

An immediate consequence is that a quasifibration $f: X \to Y$ is surjective onto those components in the image of the induced map $f_*: \pi_0(X) \to \pi_0(Y)$. Since both the fiber and homotopy fiber of any map $f: X \to Y$ are empty over components not in the image of $f$, we may as well assume our quasifibrations are surjective, but this is not necessary.

**Remark 2.7.2** For the reader who is familiar with homotopy pullback squares or who has already looked at Chapter 3, Definition 2.7.1 is equivalent to saying that $f$ is a quasifibration if and only if, for all $y \in Y$, the (pullback) square

$$
\begin{array}{ccc}
f^{-1}(y) & \longrightarrow & X \\
\downarrow & & \downarrow \\
\{y\} & \longrightarrow & Y
\end{array}
$$

is a homotopy pullback square.                                                             □

A fibration is a quasifibration. This is immediate from examining the long exact sequence (the same proof as Proposition 2.7.4), but the converse is not true since a quasifibration need not possess the homotopy lifting property, as illustrated in the following example.

**Example 2.7.3** (Quasifibration but not a fibration) Recall from Example 2.1.19 the space $L = I_1 \vee I_2$, the wedge of two intervals along 0, and the map $p: L \to I$ which maps $I_1$ to 0 and $I_2$ by the identity map. We have seen that it is not a fibration, but it is a quasifibration since its fibers and homotopy fibers are all contractible. Alternatively, we can use Proposition 2.7.4 below.        □

A quasifibration can also be thought of as a map which gives rise to a long exact sequence in homotopy, just like a fibration does (see Theorem 2.1.13).

**Proposition 2.7.4** *A map $f: X \to Y$ is a quasifibration if and only if, for all $y \in Y$ and $x \in f^{-1}(y)$, there is a long exact sequence*

$$
\cdots \longrightarrow \pi_i(f^{-1}(y), x) \longrightarrow \pi_i(X, x) \longrightarrow \pi_i(Y, y) \longrightarrow \cdots
$$

*Proof* Consider the diagram (where we omit the basepoints for simplicity)

$$
\begin{array}{ccccccccc}
\cdots \longrightarrow & \pi_i(\mathrm{hofiber}_y(f)) & \longrightarrow & \pi_i(P_f) & \longrightarrow & \pi_i(Y) & \longrightarrow & \pi_{i-1}(\mathrm{hofiber}_y(f)) & \longrightarrow \cdots \\
& \uparrow & & \uparrow & & \uparrow & & \uparrow & \\
\cdots \longrightarrow & \pi_i(\mathrm{fiber}_y(f)) & \longrightarrow & \pi_i(X) & \longrightarrow & \pi_i(Y) & \longrightarrow & \pi_{i-1}(\mathrm{hofiber}_y(f)) & \longrightarrow \cdots
\end{array}
$$

If $f$ is a quasifibration, the exactness of the bottom row follows immediately from Theorem 2.1.13, the commutativity of the diagram, and the fact that the vertical maps are all isomorphisms. If the bottom row is exact, then the isomorphism $\pi_i(\text{fiber}_y(f)) \to \pi_i(\text{hofiber}_y(f))$ follows from the five lemma (this is true even for $i = 0, 1$, where the five lemma does not apply), and this implies $f$ is a weak equivalence.                                                                    □

**Corollary 2.7.5**  *For $Y$ contractible, $f\colon X \to Y$ is a quasifibration if and only if the inclusion of the fiber into the total space $f^{-1}(y) \to X$ is a weak equivalence for all $y \in Y$.*

We can use this immediately to give an example of a map which is not a quasfibration, but whose cone is. In turn this gives an example of how quasifibrations are not stable under restriction, and therefore under pullback.

**Example 2.7.6**  (Restriction of a quasifibration need not be a quasifibration) We will utilize the topological sine curve to build various examples later, so we introduce it now as an example of a map which is not a quasifibration but whose cone is. This also gives a way to see the usefulness of Corollary 2.7.5.
    Let $X = \{(x, y) \in \mathbb{R}^2 : 0 \le x \le 1, y = \sin(1/x) \text{ if } x > 0, y \in [-1, 1] \text{ if } x = 0\}$ be the topological sine curve, topologized in the usual way as a subset of the plane. Let $p\colon X \to I$ be the projection map $p(x, y) = x$. This is not a quasifibration by Corollary 2.7.5 since the fibers of $p$ are either a point or an interval, and in either case contractible, while the target $I$ is contractible but $X$ is not (it is not path-connected).
    Let $Cp\colon CX \to CI$ denote the cone of the map $p\colon Cp(x, y, t) = (x, t)$. Then $Cp$ is a quasifibration by Corollary 2.7.5, as the fibers of $Cp$ are contractible since the fibers of $p$ are, and $CX$ is contractible because it is a cone.
    This shows that the restriction of a quasifibration need not be a quasifibration, for if we restrict the quasifibration $Cp$ to the base of the cone, we obtain the map $p\colon X \to I$ described above, which we have seen is not a quasifibration. So the restriction, and hence pullback, of a quasifibration need not be a quasifibration. We will build on this example below in Example 2.7.15.     □

We also have the following useful "patching criterion" for quasifibrations due to Dold and Thom [DT58, Section 2] which says that quasifibrations can be recognized locally. An exposition of the proof can be found in [Hat02, Lemma 4K.3] (that lemma actually lists two other equivalent conditions characterizing quasifibrations). Also see [May90, Corollary 2.3].

**Proposition 2.7.7** *A map $f: X \to Y$ is a quasifibration if and only if $Y$ is a union of open sets $U_1$ and $U_2$ such that the restrictions*

$$f^{-1}(U_1) \longrightarrow U_1,$$
$$f^{-1}(U_2) \longrightarrow U_2,$$
$$f^{-1}(U_1 \cap U_2) \longrightarrow U_1 \cap U_2$$

*are quasifibrations.*

**Definition 2.7.8** For $f: X \to Y$ and $U \subset Y$, we call $U$ *distinguished* with respect to $f$ if $f^{-1}(U) \to U$ is a quasifibration.

**Proposition 2.7.9** ([AGP02, Theorem A.1.17]) *Suppose $f: X \to Y$ is a map, and $Y = \bigcup_{i=1}^{\infty} Y_i$, $Y_i \subset Y_{i+1}$, is Hausdorff with the union topology. If each $Y_i$ is distinguished with respect to $f$, then $f$ is a quasifibration.*

*Proof* By Proposition 2.7.4 it is enough to show that the induced map

$$f_*: \pi_n(X, \text{fiber}_y(f)) \to \pi_n(Y, y)$$

is an isomorphism for all $n$. For each $i$, let $X_i = f^{-1}(Y_i)$ and let $f_i: X_i \to Y_i$ be the induced map. Our hypothesis says all maps $f_i$ are quasifibrations. Since $S^n$ is compact for all $n$, any map $g: S^n \to Y$ factors through some $Y_i$ for some $i$ (this uses the Hausdorff assumption). Using Proposition 2.7.4, we have isomorphisms

$$(f_i)_*: \pi_n(X_i, \text{fiber}_y(f_i)) \longrightarrow \pi_n(Y_i, y)$$

for all $i$, and our given element $[g] \in \pi_n(Y_i, y)$ therefore arises from an element of $\pi_n(X_i, \text{fiber}_y(f_i))$ under this isomorphism, and then gives rise to an element of $\pi_n(X, \text{fiber}_y(f))$ by the map of pairs $(X_i, \text{fiber}_y(f_i)) \to (X, \text{fiber}_y(f))$. This proves that $f_*$ is surjective. The proof that it is injective is similar: given $g_0, g_1: (D^n, \partial D^n) \to (X, \text{fiber}_y(f))$ such that $[f(g_0)]$ and $[f(g_1)]$ are equal in $\pi_n(Y, y)$, there is a (based) homotopy $S^n \times I \to Y$ from $f(g_0)$ to $f(g_1)$. Once again we use compactness to deduce that this map factors through some $Y_j$.  □

**Remark 2.7.10** For the argument above, it is enough if the inclusion $Y_i \to Y_{i+1}$ is relatively $T_1$, meaning that for any open set $U \subset Y_i$ and any point $y \in Y_{i+1} \setminus Y_i$ there exists an open set $W \subset Y_{i+1}$ such that $U \subset W$ and $y \notin W$. See [DI04, Appendix A] for more details.  □

The following technical lemma, the proof of which we omit, is due to Dold and Thom [DT58]. We give a different reference in the statement of the lemma since that one is in English. The result gives conditions under which we can "thicken" a distinguished set. We will use it in the proof of Theorem 8.6.11.

**Lemma 2.7.11** ([DL59, Lemma 1.3])   *Let $f: X \to Y$ be a map, let $Y_0 \subset Y$ be distinguished, and let $X_0 = f^{-1}(Y_0)$. Suppose there exist fiber-preserving deformations $H_t: X \to X$ and $h_t: Y \to Y$ such that the diagram*

$$
\begin{array}{ccc}
X & \xrightarrow{\ H_t\ } & X \\
{\scriptstyle f}\downarrow & & \downarrow{\scriptstyle f} \\
Y & \xrightarrow[\ h_t\ ]{} & Y
\end{array}
$$

*commutes and such that*

- $H_0 = 1_X$, $H_t(X_0) \subset X_0$, $H_1(X) \subset X_0$;
- $h_0 = 1_Y$, $h_t(Y_0) \subset Y_0$, $h_1(Y) \subset Y_0$.

*Assume furthermore that for all $y \in Y$ and all $i \geq 0$ the induced map*

$$
(H_1)_*: \pi_i(f^{-1}(y)) \to \pi_i(f^{-1}(h_1(y)))
$$

*is an isomorphism. Then $f$ is a quasifibration.*

Proposition 2.7.7 leads to the following definition, coined "local quasifibration" by Tom Goodwillie. This is actually the original definition of a quasifibration given by Dold and Thom in [DT56].

**Definition 2.7.12**   We say a map $f: X \to Y$ a *local quasifibration* if there exists a cover of distinguished open sets $\mathcal{U} = \{U_i\}_{i \in I}$ of $Y$ such that for all $y \in Y$ and all open neighborhoods $V$ of $y$ there exists some $U_i \in \mathcal{U}$ so $y \in U_i \subset V$.

A local quasifibration is one for which there exists sufficiently small open neighborhoods of each point over which $f$ is a quasifibration. That is, there exists a basis for the topology of $Y$ such that the restriction of $f$ to each open set in the basis is a quasifibration. Local quasifibrations behave well with respect to restriction to arbitrary open sets.

**Proposition 2.7.13**   *If $f: X \to Y$ is a local quasifibration and $W \subset Y$ is open, then the restriction $f^{-1}(W) \to W$ is a local quasifibration.*

*Proof*  With the cover $\mathcal{U} = \{U_i\}_{i \in I}$ from the definition of local quasifibration, take a subcover consisting of those $U_i$ such that $U_i \subset W$ for all $i$. It is clear that this covers $W$ since every open neighborhood of $y \in W$ contains some member of the cover, and such a cover contains sufficiently small open neighborhoods of each point by definition. □

For the proof of the following, see, for example, [AGP02, Theorem A.1.2].

**Theorem 2.7.14** (Local fibration is a quasifibration)  *If* $f: X \to Y$ *is a local quasifibration, then it is a quasifibration.*

The converse of this result is false:

**Example 2.7.15** (Quasifibration but not local quasifibration)  In Example 2.7.6, we showed that the map $Cp: CX \to CI$ is a quasifibration. However, it is not a local quasifibration, essentially for the same reasons given in that example. To see this, we simply need to fatten things up into an open set. Namely, let $0 < \epsilon < 1$ be given and let $U = I \times [0, \epsilon) \subset CI$. Then the natural map $Cp^{-1}(U) \to U$ cannot be a quasifibration by Corollary 2.7.5. The fibers of this map are still contractible, the base is still contractible, but the total space is homotopy equivalent to $X$, the topological sine curve, which is not path-connected. Hence by Proposition 2.7.13, $Cp$ cannot be a local quasifibration. □

Finally we have the notion of a universal quasifibration, which is a map whose pullback by an arbitrary map is always a quasifibration. Recall the notion of a pullback from the discussion prior to Proposition 2.1.16 (or see Definition 3.1.1).

**Definition 2.7.16**  A map $f: X \to Y$ is a *universal quasifibration* if, for every map $g: Z \to Y$ of any space $Z$ to $Y$, the pullback map

$$Z \times_Y X \longrightarrow Z$$

is a quasifibration.

Specializing to the case where $Z$ is an open subset of $Y$ and $g$ is the inclusion we see that a universal quasifibration is a local quasifibration. The case where $Z = Y$ and $g = 1_Y$ shows that a universal quasifibration is a quasifibration without reference to the fact that a local quasifibration is a quasifibration. We have seen that not every quasifibration is a local quasifibration. It is also true that not every local quasifibration is a universal quasifibration, since local quasifibrations are only stable under restriction to open sets, not arbitrary pullbacks.

**Example 2.7.17**  (Local quasifibration but not a universal quasifibration) Once again we will build our example from the topological sine curve. Let $X = \{(x, y, z) \in \mathbb{R}^3 : 0 \leq y \leq x \leq 1, z = \sin(1/x) \text{ if } x > y, -1 \leq z \leq 1 \text{ if } x = y\}$, and let $Y = \{(x, y) \in \mathbb{R}^2 : 0 \leq y \leq x \leq 1\}$. Topologize both $X$ and $Y$ as subspaces of $\mathbb{R}^3$ with the usual topology. Let $f : X \to Y$ be the projection $f(x, y, z) = (x, y)$. The map $f$ is a quasifibration by Corollary 2.7.5. First, it is easy to check that the fibers of $f$ are all contractible (they are either a single point or an interval), and the base $Y$ is clearly contractible, so it is enough to show that $X$ is contractible. To see this, let $X_\Delta \subset X$ be the subset consisting of those $(x, y, z)$ such that $x = y$. Then $X_\Delta$ is a deformation retract of $X$, since the family of maps $h_t : X \to X$ given by $h_t(x, y, z) = (x, xt + y(1 - t), z)$ gives the desired homotopy relative to $X_\Delta$ between the identity of $X$ and the retraction $X \to X_\Delta$ sending $(x, y, z)$ to $(x, x, z)$.

We claim that $f$ is a local quasifibration but not a universal quasifibration. The latter claim is easy: let $A = \{(x, 0) \in Y\}$, and $A \to Y$ be the inclusion. Then $A \times_Y X = f^{-1}(A) \to A$ is not a quasifibration, as we have already seen in Example 2.7.6.

Now we must show that $f$ is a local quasifibration. By definition it is enough to show that $f^{-1}(B) \to B$ is a quasifibration for all open sets $B$ which are the intersection of an open ball in $\mathbb{R}^2$ with $Y$. Clearly any such $B$ is contractible, and since the fibers of $f$ are either a single point or an interval (and in either case contractible), it is enough by Corollary 2.7.5 to show that $f^{-1}(B)$ is (weakly) contractible. There are two cases: $(0, 0) \notin B$ and $(0, 0) \in B$. In the first case it is easy to verify that $f^{-1}(B)$ is contractible (it is the union, along a contractible set, of the graph of a function over a contractible set with another contractible set). In the second case, this amounts to a microscopic version of checking that $f$ itself is a quasifibration. That is, the same proof as used above shows $f^{-1}(B)$ deformation retracts to a contractible subspace.            □

The following result says that to check if a map is a universal quasifibration it suffices to check the pullback over disks.

**Proposition 2.7.18**  *Suppose a map $f : X \to Y$ satisfies the property that, for every map $D \to Y$ of a closed disk of any dimension, the projection $D \times_Y X \to D$ is a quasifibration. That is, for every $* \in D$, the inclusion $* \times_Y X \to D \times_Y X$ is a weak equivalence. Then $f$ is a universal quasifibration.*

**Remark 2.7.19** A weaker version of this statement appears in [Goo92, p. 316], where the conclusion is that $f$ is a quasifibration. Since the proof of this result is omitted in [Goo92], we provide it here. □

*Proof of Proposition 2.7.18* Let $g\colon Z \to Y$ be a map. We are to show that the canonical map $p\colon Z \times_Y X \to Z$ is a quasifibration. We may assume $Z$ is connected since a map is a quasifibration if and only if it is a quasifibration over each path component of the target. Write $E = Z \times_Y X$, and let $F = \mathrm{fiber}_{z_0}(E \to Z)$, be the fiber. It is enough to show that the induced map $p_* \pi_n(E, F, x_0) \cong \pi_n(Z, z_0)$ for all basepoints $z_0$ and $x_0$ where $x_0 \in F$ and $p(x_0) = z_0$. It is most convenient here to speak of homotopy groups based on cubes and their boundaries rather than disks. For this map to be an isomorphism means that for all $n \geq 0$ and for all maps $a\colon (I^n, \partial I^n) \to (Z, z_0)$ there is a map (unique up to homotopy) $\widehat{a}\colon (I^n, \partial I^n) \to (E, F)$ such that $p \circ \widehat{a} = a$. Writing $I^n = I^{n-1} \times I$, we may choose this lift to be the constant map at $x_0$ on the contractible subspace $I^{n-1} \times \{0\} \cup \partial I^{n-1} \times I \subset I^{n-1} \times I$, and we are led to the lifting problem

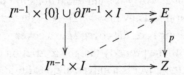

This lifting problem lives over $I^{n-1} \times I$ in the sense that if we can find the dotted arrow in the diagram

$$I^{n-1} \times \{0\} \cup \partial I^{n-1} \times I \longrightarrow (I^{n-1} \times I) \times_Z E \longrightarrow E$$

$$I^{n-1} \times I \xrightarrow{\quad = \quad} I^{n-1} \times I \longrightarrow Z$$

then we have our desired lift. But such a lift must exist because we assume that $f$ is a quasifibration when pulled back over disks, and $I^{n-1} \times I \cong D^n$. □

**Example 2.7.20** (Universal quasifibration but not fibration) Additionally, not every universal quasifibration is a fibration. In this case, we can use Example 2.7.3, which we claim is a universal quasifibration by Proposition 2.7.18. Let $p\colon L \to I$ be the map from Example 2.7.3, and let $g\colon D \to I$ be a map from a disk to the interval. It is enough to check that $D \times_I L$ is contractible. Let $A = g^{-1}(0)$, and assume that over each $a \in A$ the fibers of $D \times_I L \to D$ are intervals. Define a deformation retraction of $D \times_I L$ to $D$ by shrinking all of the intervals over $A$ simultaneously to zero. $\square$

Summarizing, we have strict containments

$$\{\text{fibrations}\} \subsetneq \{\text{universal quasifibrations}\} \subsetneq \{\text{local quasifibrations}\}$$
$$\subsetneq \{\text{quasifibrations}\}.$$

We also have a "patching" result for universal quasifibrations, due to Tom Goodwillie. This result shows that universal quasifibrations are local in a strong sense.

**Proposition 2.7.21** *Suppose $f \colon X \to Y$ is a map and there exists an open cover $\mathcal{U} = \{U_i\}_{i \in I}$ of $Y$ such that*

$$f|_{U_i} \colon f^{-1}(U_i) \longrightarrow U_i$$

*is a universal quasifibration for all $U_i$. Then $f$ is a universal quasifibration.*

*Proof* By Proposition 2.7.18, $f$ is a universal quasifibration if for every closed disk $D$ of any dimension and any map $g \colon D \to Y$, the projection $D \times_Y X \to D$ is a quasifibration. Pulling back the cover $\mathcal{U}$ to a cover $g^{-1}\mathcal{U} = \{g^{-1}(U_i)\}_{i \in I}$, we obtain a cover of $D$ with the property that the restriction of the map $D \times_Y X \to D$ to each member of the cover is a universal quasifibration. Thus we may also assume $Y = D$ is a disk. By compactness of $D$ we may also assume the open cover is finite.

Note that if $U, V$ are open sets in the cover, then since $f^{-1}(U) \to U$ is a universal quasifibration, $f^{-1}(U \cap V) \to U \cap V$ is a quasifibration (and evidently a universal quasifibration), and by Proposition 2.7.7, $f^{-1}(U \cup V) \to U \cup V$ is a quasifibration as well. In fact, $f^{-1}(U \cup V) \to U \cup V$ is a universal quasifibration, since every open $W \subset U \cup V$ can be written as a union $W = W_U \cup W_V$, where $W_U = W \cap U$ and $W \cap V = W_V$. Once again, since $f^{-1}(U) \to U$ and $f^{-1}(V) \to V$ are universal quasifibrations, $f^{-1}(W_U) \to W_U$, $f^{-1}(W_V) \to W_V$ and $f^{-1}(W_U \cap W_V) \to W_U \cap W_V$ are quasifibrations, and hence so is $f^{-1}(W) \to W$ by Proposition 2.7.7.

Now we induct on the size $k$ of the cover $\mathcal{U} = \{U_i\}_{i=1}^{k}$. The case $k = 1$ is trivial. The base case of the induction, $k = 2$, is Proposition 2.7.7 and the argument presented above. Assuming the result for all covers of size less than $k$, let $\mathcal{U} = \{U_i\}_{i=1}^{k}$ be given. Letting $\mathcal{U}' = \{U_i'\}_{i=1}^{k-1}$ be the open cover defined by $U_i' = U_i$ for $i < k - 1$ and $U_{k-1}' = U_{k-1} \cup U_k$, we are done by induction if $f^{-1}(U_{k-1} \cup U_k) \to U_{k-1} \cup U_k$ is a universal quasifibration. But this is precisely the argument presented above. $\square$

We close by briefly discussing the Dold–Thom Theorem. A detailed proof can be found, for example, in [AGP02, Appendix A]. A slight generalization (or reinterpretation) will be given in Theorem 5.10.5.

**Definition 2.7.22** For a space $X$, let $SP_n(X)$ be the quotient of $X^n$ by the action of $\Sigma_n$ which permutes the coordinates of $X^n$. This is called the *nth symmetric product of X*. Let $[x_1, \ldots, x_n]$ denote the equivalence class of a tuple $(x_1, \ldots, x_n) \in X^n$. Suppose now that $X$ is based with basepoint $*$. We have inclusion maps $SP_n(X) \rightarrow SP_{n+1}(X)$ given by sending $[x_1, \ldots, x_n]$ to $[x_1, \ldots, x_n, *]$. Then one can define the space

$$SP(X) = \bigcup_n SP_n(X)$$

called the *infinite symmetric product of X*.

Note that a map $f: X \rightarrow Y$ induces a map $SP(f): SP(X) \rightarrow SP(Y)$ (in the language of Chapter 7, the construction $SP(-)$ is functorial). Moreover, if $f$ is a weak equivalence, then $SP(f): SP(X) \rightarrow SP(Y)$ is also a weak equivalence.

One version of the Dold–Thom Theorem is as follows.

**Theorem 2.7.23** (Dold–Thom Theorem)    *Let $A \rightarrow X$ be a map, and let $C = $ hocofiber$(A \rightarrow X)$. The map $SP(X) \rightarrow SP(C)$ is a quasifibration with fiber $SP(A)$.*

A consequence of this theorem is that the homotopy groups of $SP(X)$ satisfy excision. In fact, there is an isomorphism

$$\pi_i(SP(X)) \cong \widetilde{H}_i(X; \mathbb{Z}) \tag{2.7.1}$$

for all $i$. Some authors call this statement the Dold–Thom Theorem; see for example [Hat02, Theorem 4K.6]. Moreover, one could define homology in this way, a point of view taken in [AGP02]. More about Dold–Thom and excision will be said in Theorem 5.10.5 and Example 10.1.7 in the context of homotopy calculus of functors.

We should remark that the isomorphism in (2.7.1) is not at all immediate from the statement of Theorem 2.7.23. We have not even defined a map between the homotopy groups of the infinite symmetric product and the singular homology of $X$. Ignoring this, we noted after Theorem 1.4.9 that the reduced homology groups of a finite CW complex are characterized by four properties, three of which homotopy groups share. The one they do not share is excision, and the Dold–Thom Theorem says that homotopy groups of infinite symmetric products do satisfy excision (see especially our reformulation,

Theorem 5.10.5). From this perspective it is enough to define a map between the homotopy groups of the infinite symmetric product and the reduced homology groups, and then prove this map is an isomorphism when $X$ is a point (which is easy to check).

# 3

# 2-cubes: Homotopy pullbacks and pushouts

In this chapter, we take a close look at square diagrams. In many ways, this is the most essential chapter in this book since most of the main ideas involved in understanding general cubical diagrams are built from and best illustrated by those for squares. In fact, we wish to emphasize that most of the results about cubes can, via combinatorics and bookkeeping, be reduced to results about squares. Moreover, many results which do not appear to be results about squares have proofs which are simplest with the language and techniques we develop here.

This chapter is where we will first encounter and use homotopy (co)limits, but the material presented here can be read independently of the material from Chapter 8.

We cannot hope to learn much about spaces (0-cubes) without knowing something about maps of spaces (1-cubes). This chapter makes the case for the next obvious statement: If we want to learn more about maps of spaces, we are going to need to know something about maps of maps of spaces. These are exactly the square diagrams

$$
\mathcal{S} = \begin{array}{ccc}
W & \longrightarrow & Y \\
\downarrow & & \downarrow \\
X & \longrightarrow & Z
\end{array}
$$

or 2-cubes $\mathcal{S}$, which have the additional convenient symmetry in that they can be regarded as a map of maps in two ways:

$$(W \to X) \longrightarrow (Y \to Z)$$

and

$$(W \to Y) \longrightarrow (X \to Z).$$

91

It will be natural even in this chapter to think of a "map of a map of maps" as a three-dimensional cube (both Proposition 3.3.24 and Proposition 3.7.29 use maps of squares), but such generalizations to higher-dimensional cubes will be explored in more detail in Chapter 5.

We saw in the previous chapter that one purpose of replacing an arbitrary map by a fibration or cofibration is to obtain a homotopy invariant fiber or cofiber (the replacement also yields lifting properties etc., but we are deliberately focusing just on homotopy invariance). We have also seen in Propositions 2.1.16 and 2.3.15 that fibrations are stable under pullback and cofibrations under pushout. That is, if $p\colon Y \to Z$ is a fibration and $f\colon X \to Z$ is a map, there is a naturally induced fibration over $X$, and dually if $i\colon W \to X$ is a cofibration and $g\colon W \to Y$ is a map, there is a naturally induced cofibration with domain $Y$.

The aforementioned properties of (co)fibrations are best organized by considering the "pullback" or "pushout" of a diagram consisting of three spaces and two maps. These are diagrams

$$S_0 = (X \longrightarrow Z \longleftarrow Y)$$

and

$$S_1 = (X \longleftarrow W \longrightarrow Y).$$

Each of these is a subdiagram of the square $S$ above. The subscripts 0 and 1 are meant to indicate that the "first" and "last" spaces in the square $S$ have been removed, respectively. We will thus also refer to $S_0$ and $S_1$ as *punctured squares*.

Other notation which we will employ here, and to a greater extent in Chapter 5, was already introduced in Section 1.1. We often index the spaces in the square by the subsets $S$ of $\{1, 2\}$ (without writing brackets $\{-\}$ or commas) and will thus also consider squares

$$\mathcal{X} = \begin{array}{ccc} X_\emptyset & \longrightarrow & X_2 \\ \downarrow & & \downarrow \\ X_1 & \longrightarrow & X_{12} \end{array}$$

and their relevant subdiagams

$$\mathcal{X}_0 = (X_1 \longrightarrow X_{12} \longleftarrow X_2)$$

and

$$\mathcal{X}_1 = (X_1 \longleftarrow X_\emptyset \longrightarrow X_2).$$

There are two natural situations in which square diagrams and the two relevant subdiagrams appear. For the first, suppose $p: Y \to Z$ is a fibration (or a fiber bundle, vector bundle, etc.), and let $f: X \to Z$ be a map, so that there is a diagram $X \xrightarrow{f} Z \xleftarrow{p} Y$. One often considers the pullback of the fibration $p$ by the map $f$; that is, one forms the space $p^*Y = \{(y, x) \in Y \times X : p(y) = f(x)\}$ which naturally fits into a commutative square diagram

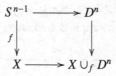

For the second, consider the situation of building a CW complex by attaching one cell at a time. At some intermediate stage, there is a space $X$ and a map $f: S^{n-1} \to X$ (the attaching map for that $n$-cell), and we form the space $X \cup_f D^n$. That is, we begin with a diagram $X \xleftarrow{f} S^{n-1} \to D^n$, where the right map is the inclusion of the boundary of the disk, and we form a commutative square diagram

$$
\begin{array}{ccc}
S^{n-1} & \longrightarrow & D^n \\
\downarrow{\scriptstyle f} & & \downarrow \\
X & \longrightarrow & X \cup_f D^n
\end{array}
$$

The spaces $p^*E$ and $X \cup_f D^n$ above are examples of pullbacks (limits) and pushouts (colimits) respectively.

We are thus led to study pullbacks and pushouts of two important subdiagrams $S_0$ and $S_1$ of a general commutative square $S$. For a generic square, neither the limit of $S_0$ nor the colimit of $S_1$ is a homotopy invariant of the spaces in the diagram (examples are given later) so the immediate goal in Sections 3.2 and 3.6 will be to replace them by something which is invariant under homotopy equivalences, namely the homotopy pullback (homotopy limit) and homotopy pushout (homotopy colimit), respectively. Of further interest is the comparison between $W$ and the homotopy pullback of $S_0$ and between $Z$ and the homotopy pushout of $S_1$. This is the subject of Sections 3.3 and 3.7.

Many examples and applications homotopy pullbacks and pushouts will be given in Sections 3.3 and 3.7. We will give further examples of their interaction in Section 3.9. The most interesting applications, however, will have to wait until Chapter 4 where we use the Blakers–Massey (or Triad Connectivity) Theorem for further comparisons between homotopy pushouts and pullbacks.

## 3.1 Pullbacks

The notion of a pullback was already discussed in the last chapter, immediately before Proposition 2.1.16. We give a formal definition and more careful study here.

**Definition 3.1.1**   The *pullback*, or *limit* of the diagram $S_0 = (X \xrightarrow{f} Z \xleftarrow{g} Y)$, denoted by $\lim(X \xrightarrow{f} Z \xleftarrow{g} Y)$ or $\lim(S_0)$, is the subspace of $X \times Y$ defined as

$$\lim(X \xrightarrow{f} Z \xleftarrow{g} Y) = \{(x, y) \colon f(x) = g(y)\}.$$

We may also write $X \times_Z Y$ for this space and call it the *fiber product* of $X$ and $Y$ over $Z$.

The terminology "limit" and the notation "lim" is used because the pullback is an example of a more general notion of limit, as will be shown in Example 7.3.13. Related to this is the fact that the pullback enjoys a "universal property": any map from a space $W$ into the diagram $S_0$ factors through the limit. That is, suppose we are given a space $W$ and maps $g' \colon W \to X$, $f' \colon W \to Y$, and $h' \colon W \to Z$ such that $h' = f' \circ g = g' \circ f$. The universal property means that there is a unique dotted arrow in the diagram

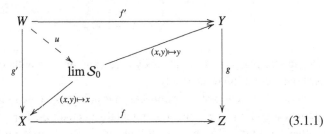

$$(3.1.1)$$

that makes the resulting diagram commute. It is easy to see that $u(w) = (g'(w), f'(w))$. As we mentioned in Remark 2.1.17, the fact that fibrations are preserved under pullback is an immediate consequence of this universal property of the limit, and as such it allows us to build new fibrations by mapping into the base of an existing one. We will study universal properties and limits in much more detail in Section 7.3; the pullback appears again in Example 7.3.13.

We now give a few examples of familiar constructions which are naturally described as pullbacks.

**Example 3.1.2** (Product as pullback)   Given spaces $X$ and $Y$, their product can be thought of as a pullback:

$$\lim(X \to * \leftarrow Y) \cong X \times Y$$

(we could write "=" instead of "≅" and take this as the definition of the product.) □

**Example 3.1.3** The pullback of the diagram $X \xrightarrow{1_X} X \xleftarrow{1_X} X$ is $X$. In fact, the canonical map

$$X \longrightarrow \lim(X \xrightarrow{1_X} X \xleftarrow{1_X} X)$$

given by $x \mapsto (x, x)$ is a homeomorphism. (This also follows from Proposition 3.1.9 below.) □

**Example 3.1.4** Let $f: X \to Y$ be a map. The pullback of the diagram $Y \xrightarrow{1_Y} Y \xleftarrow{f} X$ is $X$. The projection

$$\lim(Y \xrightarrow{1_Y} Y \xleftarrow{f} X) \to X$$

given by $(y, x) \mapsto x$ is a homeomorphism. □

**Example 3.1.5** (Fiber as pullback) Let $Y$ be a space, $y_0 \in Y$, and $f: X \to Y$ a map. The map

$$\mathrm{fiber}_{y_0}(f) \longrightarrow \lim(\{y_0\} \hookrightarrow Y \xleftarrow{f} X)$$
$$x \longmapsto (y_0, x)$$

is a homeomorphism. Indeed, by definition, $\mathrm{fiber}_{y_0}(f) = f^{-1}(y_0)$, while the limit is the set of all pairs $(y_0, x)$ such that $f(x) = y_0$. These are clearly the same set. □

**Example 3.1.6** (Mapping path space as pullback) Let $f: X \to Y$ be a map. Then

$$P_f = \lim(X \xrightarrow{f} Y \xleftarrow{ev_0} \mathrm{Map}(I, Y))$$

is precisely the mapping path space of $f$ from Definition 2.2.1. □

**Example 3.1.7** (Homotopy fiber as pullback) There is a homeomorphism

$$\mathrm{hofiber}_y(X \to Y) \cong \lim(P_f \to Y \leftarrow \{y_0\}),$$

where $P_f \to Y$ is the evaluation of the path at time $t = 1$. This follows directly from the definition of the homotopy fiber. □

**Example 3.1.8** (Pullback bundle)   If $p: X \to Y$ is a fibration (or a fiber bundle or a vector bundle), then for any map $f: Z \to Y$, the map $Z \times_Y X \to Z$ is a fibration (or a fiber bundle or a vector bundle). The fact for fibrations was established in Proposition 2.1.16. We leave the rest to the interested reader.   □

Proposition 2.1.16 says that fibrations are preserved by pullback. So are two other important classes of maps.

**Proposition 3.1.9** (Pullbacks preserve injections and homeomorphisms)   *Let* $p: X \to Y$ *be an injection (resp. homeomorphism). Then for any map* $f: Z \to Y$, *the induced map* $p^*: X \times_Y Z \to Z$ *is also an injection (resp. homeomorphism).*

*Proof*   If $p^*(x, z) = p^*(x', z')$ then $z = z'$ and $p(x) = f(z) = p(x')$ so $x = x'$ and thus injectivity is preserved under pullback. If $p: X \to Y$ is a homeomorphism with inverse $q: Y \to X$, the map $Z \to X \times_Y Z$ given by $z \mapsto ((q \circ f)(z), z)$ is clearly a homeomorphism inverse to the canonical map $X \times_Y Z \to Z$.   □

In the case where the map $f: Z \to Y$ is itself an injection, the first part of this result simply states the obvious fact that the restriction of an injective map is injective. Pullbacks do not, however, preserve homotopy equivalences. This is precisely the defect of limits/pullbacks that we aim to fix in the next section.

Sometimes a map of punctured squares

$$
\begin{array}{ccccc}
X_1 & \longrightarrow & X_{12} & \longleftarrow & X_2 \\
\downarrow & & \downarrow & & \downarrow \\
X_1' & \longrightarrow & X_{12}' & \longleftarrow & X_2'
\end{array}
$$

(meaning three vertical maps $X_S \to X_S'$ that make the above diagram commute) which is an objectwise fibration (meaning $X_S \to X_S'$ is a fibration for each $S$) induces a fibration of limits. This hypothesis is not quite strong enough in general,[1] but the hypotheses given below are enough.[2] We will require this result in the proof of Theorem 8.6.1; it is not used elsewhere in this section.

**Proposition 3.1.10**   *Let* $S_0 = (X_1 \to X_{12} \leftarrow X_2)$ *and* $S_0' = (X_1' \to X_{12}' \leftarrow X_2')$ *be punctured squares in which the maps* $X_2 \to X_{12}$ *and* $X_2' \to X_{12}'$ *are*

---

[1] Let $X_{12}' = *$ and let all other spaces be a space $X$, with all maps the identity or the unique map to a point. Then the map of limits is the diagonal inclusion of $X$ in $X \times X$, which is rarely a fibration.

[2] These hypotheses might seem strange, but they really just say that a certain diagram is *fibrant*. This will be discussed in Section 8.4.2.

*fibrations, and $S_0 \to S_0'$ a map of punctured squares such that $X_S \to X_S'$ are fibrations for all $S$. Further assume $X_2 \to \lim(X_{12} \to X_{12}' \leftarrow X_2')$ is a fibration.[3] Then the induced map $\lim S_0 \to \lim S_0'$ is a fibration.*

*Proof*   In the lifting problem

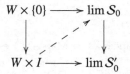

we use the canonical projections to obtain the lifting problem

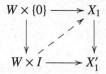

which can be solved since $X_1 \to X_1'$ is a fibration. We then use the maps $X_1 \to X_{12}$ and $X_1' \to X_{12}'$ to use this solution to give a compatible one for the induced lifting problem for $X_{12} \to X_{12}'$. Now consider the induced diagram and lifting problem

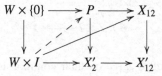

where $P = \lim(X_{12} \to X_{12}' \leftarrow X_2')$. By above, we already have a solution to the lifting problem for the outer square, as depicted by the solid diagonal arrow. Since the left square is a pullback, the dotted arrow exists by the universal property of $P$. Finally, as $X_2 \to P$ is a fibration, the evident lifting problem

$$
\begin{array}{ccc}
W \times \{0\} & \longrightarrow & X_2 \\
\downarrow & \nearrow & \downarrow \\
W \times I & \longrightarrow & P
\end{array}
$$

has a solution. All of this amounts to a solution to the original lifting problem. □

The following result will be used in Chapter 8 to prove that homotopy limits preserve fibrations of diagrams. Formally it is similar to Theorem 2.5.4. In fact, we will use the proof of Theorem 2.5.4 to prove it.

---

[3]  This hypothesis implies the map $X_2 \to X_2'$ is a fibration.

**Proposition 3.1.11** (Pullback corner map)    *Suppose $i\colon A \to B$ is a cofibration and $p\colon X \to Y$ is a fibration. Then the canonical map, the* pullback corner map,

$$\mathrm{Map}(B, X) \longrightarrow \lim \left(\mathrm{Map}(A, X) \to \mathrm{Map}(A, Y) \leftarrow \mathrm{Map}(B, Y)\right)$$

*is a fibration which is a homotopy equivalence if either $i$ or $p$ is a homotopy equivalence.*

*Proof*   Let $P = \lim \left(\mathrm{Map}(A, X) \to \mathrm{Map}(A, Y) \leftarrow \mathrm{Map}(B, Y)\right)$, and consider the lifting problem

$$
\begin{array}{ccc}
Z \times \{0\} & \xrightarrow{\ g\ } & \mathrm{Map}(B, X) \\
{\scriptstyle i_0}\big\downarrow & \nearrow & \big\downarrow {\scriptstyle p_*} \\
Z \times I & \xrightarrow[\ h\ ]{} & P
\end{array}
$$

Using Theorem 1.2.7, the map $g$ corresponds to a map $\tilde{g}\colon B \times Z \times \{0\} \to X$, and $h$ corresponds to maps $\tilde{h}_A\colon A \times Z \times I \to X$ and $\tilde{h}_B\colon B \times Z \times I \to Y$ such that the diagram

$$
\begin{array}{ccc}
A \times Z \times I & \xrightarrow{\ \tilde{h}_A\ } & X \\
{\scriptstyle i \times 1_{Z \times I}}\big\downarrow & & \big\downarrow {\scriptstyle p} \\
B \times Z \times I & \xrightarrow[\ \tilde{h}_B\ ]{} & Y
\end{array}
$$

commutes. Since $\tilde{h}_A$ and $\tilde{g}$ agree on their common intersection $A \times Z \times \{0\}$, together they define a map $c\colon A \times Z \times I \cup_{A \times Z \times \{0\}} B \times Z \times \{0\} \to X$, and we may then consider the lifting problem

$$
\begin{array}{ccc}
A \times Z \times I \cup_{A \times Z \times \{0\}} B \times Z \times \{0\} & \xrightarrow{\ c\ } & X \\
{\scriptstyle \iota}\big\downarrow & \nearrow & \big\downarrow {\scriptstyle p} \\
B \times Z \times I & \xrightarrow[\ \tilde{h}_B\ ]{} & Y
\end{array}
$$

A solution $\tilde{H}$ to this lifting problem gives rise to the desired solution $\hat{H}\colon Z \times I \to \mathrm{Map}(B, X)$ using Theorem 1.2.7, which solves the original lifting problem. To see why a solution exists, we observe that Proposition 2.3.9 tells us that $i \times 1_Z\colon A \times Z \to B \times Z$ is a cofibration since $i$ is, and so there exists a retraction $r\colon B \times Z \times I \to A \times Z \times I \cup B \times Z \times \{0\}$ by Proposition 2.3.10. We then proceed exactly as in the proof of Theorem 2.5.4 to show that a solution to this lifting problem exists (at the start of that proof, we found ourselves in precisely the situation we are in now).

Now consider the diagram

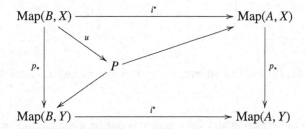

If $p$ is a homotopy equivalence, then so are the induced maps $p_*\colon \mathrm{Map}(B,X) \to \mathrm{Map}(B,Y)$, and $p_*\colon \mathrm{Map}(A,X) \to \mathrm{Map}(A,Y)$. By Proposition 2.1.23, the map $P \to \mathrm{Map}(B,Y)$ is a homotopy equivalence. It follows that $\mathrm{Map}(B,X) \to P$ is a homotopy equivalence. The argument assuming $i$ is a homotopy equivalence is analogous. □

**Remark 3.1.12** The above proof fails if we assume $i$ or $p$ to be only a weak equivalence, since if $i$ (resp. $p$) is a weak equivalence it does not necessarily follow that $i^*$ (resp. $p_*$) is a weak equivalence, by the remark following Proposition 1.3.5. □

In light of the diagram (3.1.1), for a given commutative square it makes sense to ask whether its "initial" space is in fact the pullback of the rest of the diagram. More formally, we have the following.

**Definition 3.1.13** We say a square diagram

$$\begin{array}{ccc} W & \xrightarrow{f'} & Y \\ {\scriptstyle g'}\downarrow & & \downarrow{\scriptstyle g} \\ X & \xrightarrow{f} & Z \end{array}$$

is *cartesian* (or *categorically cartesian*, or a *pullback*, or a *strict pullback*) if the canonical map

$$W \longrightarrow \lim(X \xrightarrow{f} Z \xleftarrow{g} Y)$$
$$w \longmapsto (f'(w), g'(w))$$

is a homeomorphism.

**Example 3.1.14** Examples 3.1.2, 3.1.3, 3.1.4, and 3.1.5 can all be recast in the language of cartesian squares. Respectively, they say that the squares

$$\begin{array}{ccc} X \times Y & \longrightarrow & Y \\ \downarrow & & \downarrow \\ X & \longrightarrow & * \end{array} \qquad \begin{array}{ccc} X & \xrightarrow{1_X} & X \\ {\scriptstyle 1_X}\downarrow & & \downarrow {\scriptstyle 1_X} \\ X & \xrightarrow[1_X]{} & X \end{array} \qquad \begin{array}{ccc} X & \xrightarrow{f} & Y \\ {\scriptstyle 1_X}\downarrow & & \downarrow {\scriptstyle 1_Y} \\ X & \xrightarrow[f]{} & Y \end{array} \qquad \begin{array}{ccc} \mathrm{fiber}_{y_0}(f) & \longrightarrow & X \\ \downarrow & & \downarrow {\scriptstyle g} \\ * & \longrightarrow & Y \end{array}$$

are cartesian. (In the last square, $* \to Y$ is the inclusion of the basepoint $y_0 \in Y$.)                                                                 □

The notion of "cartesianness" of a square is one of the central themes of this book.

## 3.2 Homotopy pullbacks

The fiber of a map is not homotopy invariant, and neither is the pullback of a diagram $S_0 = (X \to Z \leftarrow Y)$, since Example 3.1.5 says the fiber of a map is an example of a limit. Here is another more concrete example.

**Example 3.2.1** (Pullback not homotopy invariant)    Consider the pullback of the inclusion maps of the endpoints of the unit interval $[0, 1]$. The pullback of the diagram

$$\{0\} \longrightarrow [0, 1] \longleftarrow \{1\}$$

is empty. But $[0, 1] \simeq *$, and the pullback of

$$\{0\} \longrightarrow * \longleftarrow \{1\}$$

is a single point.

In fact, there is even a map of diagrams from the first to the second (by which we mean that there are maps of corresponding spaces that commute with the maps in the diagrams) which is a homotopy equivalence objectwise, but the map of pullbacks is not.                                                                 □

However, if our generic diagram $S_0 = X \to Z \leftarrow Y$ has the property that one of the maps is a fibration, then the pullback is homotopy invariant. We will establish this in Corollary 3.2.15. As a special case, consider $X = *$ and $Y \to Z$ a fibration – this is now simply Theorem 2.1.20; the fibers of a fibration are homotopy invariant. An essential result in this direction is that the property of being a homotopy equivalence is preserved by pullback of a fibration. This was the content of Proposition 2.1.23, which we restate below as Proposition 3.2.3 for convenience using the language introduced in this chapter.

**Remark 3.2.2** In the proof of Proposition 3.2.3/Proposition 2.1.23, we gave separate arguments for homotopy equivalence and weak equivalence. In light of Theorem 1.3.7, Theorem 1.3.10 and the remarks following, the reader may wonder why we worry about the distinction between the two at all. The reason is that everything in this and the previous section will be dualized in Sections 3.5 and 3.6 for pushouts and homotopy pushouts. Arguments such as those in the proof of Proposition 2.1.23 (3.2.3) above, which utilize the lifting property together with a homotopy inverse, dualize in a straightforward manner. The same cannot be said about the analog of the more algebraically flavored arguments which utilize the long exact sequence of a fibration, such as Proposition 3.3.18. The difficulty here is that the dual of the long exact sequence of a fibration is algebraically a homological, not a homotopical, object. Moreover, there is a stronger notion of duality between fibrations and cofibrations when the notion of equivalence is homotopy equivalence, and we exploit this at various points in the text. □

**Proposition 3.2.3** (Restatement of Proposition 2.1.23) *Suppose $p: Y \to Z$ is a fibration and $f: X \to Z$ is a homotopy (resp. weak) equivalence. Then the canonical projection*

$$\lim(X \xrightarrow{f} Z \xleftarrow{p} Y) \longrightarrow Y$$
$$(x, y) \longmapsto y$$

*is a homotopy (resp. weak) equivalence. That is, if in the cartesian square*

$$
\begin{array}{ccc}
\lim(X \to Z \leftarrow Y) & \longrightarrow & Y \\
\downarrow & & \downarrow p \\
X & \xrightarrow{\quad f \quad} & Z
\end{array}
$$

*$p$ is a fibration and $f$ is a homotopy (resp. weak) equivalence, then the canonical projection $\lim(X \xrightarrow{f} Z \xleftarrow{p} Y) \longrightarrow Y$ is a homotopy (resp. weak) equivalence.*

To make a homotopy invariant limit for an arbitrary diagram

$$S_0 = (X \xrightarrow{f} Z \xleftarrow{g} Y),$$

one idea is to change one of $f$ or $g$ into a fibration as discussed in Section 2.2 and then take the pullback. This works perfectly well, and is often how the homotopy pullback is defined in the literature. However, we will give a more symmetric and equivalent definition below which at least visually does not

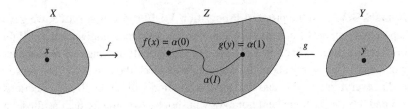

Figure 3.1  A picture of a point $(x, \alpha, y)$ in the homotopy pullback.

replace any of the spaces $X$, $Y$, or $Z$. We will discuss the equivalence of various models in Proposition 3.2.5 that this model is equivalent to replacing a map by a fibration and then taking the pullback.

**Definition 3.2.4**  The *homotopy pullback*, or *homotopy limit*, or *homotopy fiber product* of the diagram $\mathcal{S}_0 = (X \xrightarrow{f} Z \xleftarrow{g} Y)$, denoted by $\operatorname{holim} \mathcal{S}_0$ or $\operatorname{holim}(X \xrightarrow{f} Z \xleftarrow{g} Y)$, is the subspace of $X \times \operatorname{Map}(I, Z) \times Y$ consisting of

$$\{(x, \alpha, y) : \alpha(0) = f(x), \alpha(1) = g(y)\}.$$

A picture of the homotopy pullback is given in Figure 3.1.

The terminology "homotopy limit" and the notation "holim" is used because the homotopy pullback is an example of a homotopy limit, as we shall see in Example 8.2.8.

There are canonical maps

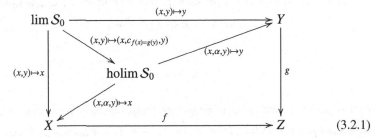

$$(3.2.1)$$

This may seem unusual, since every space which maps to $X$ and $Y$ in a way that is compatible with the maps $f$ and $g$ to $Z$ is supposed to factor through $\lim \mathcal{S}_0$. However, while the outer square and the two triangles commute, *the inner square does not commute* (but it does up to homotopy). Nevertheless this is something like the universal property that the limit enjoys. That said, we do *not* want to work with diagrams that commute up to homotopy.

The homotopy pullback of $X \to Z \leftarrow Y$ can be expressed in a number of ways as an ordinary pullback. The proof of the following is clear by inspection.

**Proposition 3.2.5** *The map*

$$\lim(P_f \xrightarrow{p} Z \xleftarrow{g} Y) \longrightarrow \text{holim}(X \xrightarrow{f} Z \xleftarrow{g} Y)$$

$$((x, \alpha), y) \longmapsto (x, \alpha, y)$$

*is a homeomorphism. In fact, the following five spaces are homeomorphic:*

- $\text{holim}(X \xrightarrow{f} Z \xleftarrow{g} Y)$;
- $\lim(P_f \to Z \xleftarrow{g} Y)$;
- $\lim(X \xrightarrow{f} Z \leftarrow P_g)$;
- $\lim(P_f \to Z \leftarrow P_g)$;
- $\lim\left(\lim(X \xrightarrow{f} Z \xleftarrow{ev_0} Z^I) \longrightarrow Z^I \longleftarrow \lim(Y \xrightarrow{g} Z \xleftarrow{ev_1} Z^I)\right).$

**Remark 3.2.6** We will not have much use for the last of these in this chapter, but it is the closest of all to how we will define general homotopy limits in Chapter 8. □

From now on, we will freely use any of the objects appearing in Proposition 3.2.5 as models for the homotopy pullback of the diagram $\mathcal{S}_0 = (X \xrightarrow{f} Z \xleftarrow{g} Y)$.

Here are some examples of homotopy pullbacks; we will encounter many more throughout this chapter.

**Example 3.2.7** We have that

$$\text{holim}(X \to * \leftarrow Y) \simeq X \times Y$$

and

$$\text{holim}(X \xrightarrow{1_X} X \xleftarrow{1_X} X) \simeq X.$$

Both are easy to argue from the definition; in the second square a homotopy inverse to the natural map $X \to \text{holim}(X \xrightarrow{1_X} X \xleftarrow{1_X} X)$ is given by $(x, \alpha, x') \mapsto \alpha(1/2)$. Alternatively, one could use Proposition 3.3.5 for both and refer to Examples 3.1.2 and 3.1.3. □

**Example 3.2.8** (Homotopy fiber as homotopy pullback) For a map $f : X \to Y$ and $\{y_0\} \to Y$ the inclusion of a basepoint, we have a homeomorphism

$$\text{holim}(\{y_0\} \to Y \xleftarrow{f} X) \cong \text{hofiber}_{y_0}(f).$$

This is immediate from the definitions. Thus Examples 2.2.9 and 2.2.10 could be viewed as examples of homotopy pullbacks. □

**Example 3.2.9** (Mapping path space as homotopy pullback)   Let $f: X \to Y$ be a map. We have a natural homeomorphism

$$\mathrm{holim}(X \xrightarrow{f} Y \xleftarrow{1_Y} Y) \cong P_f$$

given by sending $(x, \gamma, y)$ to $(x, \gamma)$. Since $X$ is a deformation retract of $P_f$, we then have a homotopy equivalence

$$\mathrm{holim}(X \xrightarrow{f} Y \xleftarrow{1_Y} Y) \simeq X.$$      $\square$

**Example 3.2.10** (Loop space as homotopy pullback)   As a special case of the last example, in conjunction with Example 2.2.9, we have for a space $X$ with basepoint $x_0$ a homeomorphism

$$\mathrm{holim}(\{x_0\} \longrightarrow X \longleftarrow \{x_0\}) \cong \Omega X.$$

Explicitly, a point in the homotopy pullback is a triple $(x_0, \alpha, x_0)$ where $\alpha: I \to X$ satisfies $\alpha(0) = \alpha(1) = x_0$. This defines a based quotient map $\alpha': S^1 = I/\partial I \to X$.      $\square$

**Example 3.2.11** (Free loop space as homotopy pullback)   If $\Delta: X \to X \times X$ is the diagonal map $\Delta(x) = (x, x)$, then we have a homeomorphism

$$\mathrm{holim}(X \xrightarrow{\Delta} X \times X \xleftarrow{\Delta} X) \cong LX,$$

where $LX = \mathrm{Map}(S^1, X)$ is the free loop space on $X$ (see Example 1.2.6). To see why, we note that a point in the homotopy limit is a triple $(x, \gamma, x')$ where $\gamma = (\gamma_1, \gamma_2): I \to X \times X$ satisfies $\gamma(0) = (x, x)$ and $\gamma(1) = (x', x')$. We associate to this data the map $S^1 \to X$ given by $\gamma_1 * \gamma_2^{-1}$, where $*$ denotes path multiplication, and $S^1$ is identified with the quotient space $I \amalg I / \sim$, where the copies of $0$ (resp. $1$) are identified. This defines the required homeomorphism.      $\square$

Now we establish the homotopy invariance of the homotopy pullback. It is the dual of what is known as the "gluing lemma" (see Theorem 3.6.13). Our proof dualizes the proof of Theorem 3.6.13 we learned from Rognes [Rog]. The remainder of the section is devoted to deducing various consequences of this result.

**Theorem 3.2.12** (The matching lemma)   *Suppose we have a commutative diagram*

$$X \xrightarrow{\ f\ } Z \xleftarrow{\ g\ } Y$$

with vertical maps $e_X$, $e_Z$, $e_Y$ to

$$X' \xrightarrow{\ f'\ } Z' \xleftarrow{\ g'\ } Y'$$

where the vertical arrows are homotopy (resp. weak) equivalences. Then the induced map

$$\mathrm{holim}(X \xrightarrow{f} Z \xleftarrow{g} Y) \longrightarrow \mathrm{holim}(X' \xrightarrow{f'} Z' \xleftarrow{g'} Y')$$

$$(x, \gamma, y) \longmapsto (e_X(x), e_Z \circ \gamma, e_Y(y))$$

is a homotopy (resp. weak) equivalence.

*Proof* We deal with the case of homotopy equivalence first. The main idea is to factor the map to create two subproblems. Consider the related diagram

$$X \xrightarrow{\ f\ } Z \xleftarrow{\ g\ } Y$$
with maps $1_X$, $e_Z$, $1_Y$ to
$$X \xrightarrow{\ e_Z \circ f\ } Z' \xleftarrow{\ e_Z \circ g\ } Y$$
with maps $e_X$, $1_{Z'}$, $e_Y$ to
$$X' \xrightarrow{\ f'\ } Z' \xleftarrow{\ g'\ } Y'$$

It is enough to show the homotopy limit of the first row is homotopy equivalent to the homotopy limit of the second, and the homotopy limit of the second is homotopy equivalent to the homotopy limit of the third.

Let us deal with the first two rows first. Our task is to find a homotopy inverse to the induced map

$$\mathrm{holim}(X \to Z \leftarrow Y) \longrightarrow \mathrm{holim}(X \to Z' \leftarrow Y).$$

Choose a homotopy inverse $e_Z'$ for $e_Z$ and homotopies $h_t \colon Z \to Z$ with $h_0 = 1_Z$ and $h_1 = e_Z' \circ e_Z$, and $h_t' \colon Z' \to Z'$ with $h_0' = 1_{Z'}$ and $h_1' = e_Z \circ e_Z'$. Define a map

$$\mathrm{holim}(X \to Z' \leftarrow Y) \longrightarrow \mathrm{holim}(X \to Z \leftarrow Y)$$

by

$$(x, \gamma', y) \longmapsto (x, (h_t \circ f(x)) * (e_Z' \circ \gamma'(t)) * (h_t \circ g(y))^{-1}, y),$$

where $*$ denotes path multiplication and the inverse superscript denotes that the path is to run backwards. Note first that $h_1 \circ f(x) = e_Z' \circ \gamma'(0)$ and $e_Z' \circ \gamma'(1) = (h_1 \circ g)(y)$, so the expression above defines a path as required. Also note that at $t = 0$ we have $h_0 \circ f(x) = f(x)$ and at $t = 1$ we have $h_0 \circ g(y) = g(y)$,

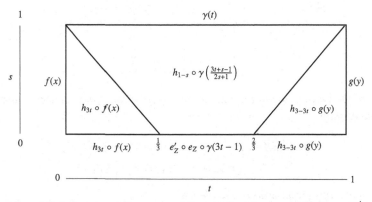

Figure 3.2 A homotopy between $\gamma$ and $(h_t \circ f(x)) * (e'_Z \circ e_Z \circ \gamma(t)) * (h_t \circ g)^{-1}$. In the figure all paths and homotopies have been reparametrized, whereas in the proof they are not.

so the formula above defines a map to the homotopy pullback. We claim it is homotopy inverse to the canonical map

$$\mathrm{holim}(X \to Z \leftarrow Y) \longrightarrow \mathrm{holim}(X \to Z' \leftarrow Y)$$

which sends $(x, \gamma, y)$ to $(x, e_Z \circ \gamma, y)$. Consider the composed map

$$\mathrm{holim}(X \to Z \leftarrow Y) \longrightarrow \mathrm{holim}(X \to Z' \leftarrow Y) \longrightarrow \mathrm{holim}(X \to Z \leftarrow Y)$$

which is given by

$$(x, \gamma, y) \mapsto (x, (h_t \circ f(x)) * (e'_Z \circ e_Z \circ \gamma(t)) * (h_t \circ g)^{-1}(y), y).$$

Figure 3.2 indicates why $\gamma$ is homotopic relative to its endpoints to $(h_t \circ f(x)) * (e'_Z \circ e_Z \circ \gamma(t)) * (h_t \circ g(y))^{-1}$ using the homotopy $h_t$, and hence why this composed map is homotopic to the identity. The other composed map is similar, except that here the natural thing is to construct the homotopy between $\gamma'$ and the resulting path in two steps. First we show as above that the composed path is homotopic to $e \circ e' \circ \gamma'$, and next we use the homotopy $h'_t$ to make a homotopy of $e \circ e'$ to the identity on $Z$.

Next we deal with the second and third rows and seek a homotopy inverse to the induced map

$$\mathrm{holim}(X \to Z' \leftarrow Y) \longrightarrow \mathrm{holim}(X' \to Z' \leftarrow Y').$$

First we call upon Proposition 3.2.5 and use $\lim(P_{e_Z \circ f} \to Z' \leftarrow P_{e_Z \circ g})$ and $\lim(P_{f'} \to Z' \leftarrow P_{g'})$ as models for $\mathrm{holim}(X \to Z' \leftarrow Y)$ and $\mathrm{holim}(X' \to Z' \leftarrow Y')$ respectively. It is enough to show that these are homotopy equivalent. We have a commutative diagram

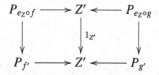

and Corollary 2.1.22 tells us the outer two vertical maps have homotopy inverses over $Z'$. It is clear by inspection that this is homotopy inverse to the map in question.

The argument for weak equivalences is considerably easier. Consider the diagram

$$\text{holim}(X \to Z \leftarrow Y) \longrightarrow \lim(P_f \to Z \leftarrow Y)$$
$$\downarrow \qquad\qquad\qquad\qquad \downarrow$$
$$\text{holim}(X' \to Z' \leftarrow Y') \longrightarrow \lim(P_{f'} \to Z' \leftarrow Y')$$

The horizontal arrows are the homeomorphisms described in Proposition 3.2.5. We will then argue that the right vertical map above is a weak equivalence. First consider the square

$$P_f \longrightarrow P_{f'}$$
$$\downarrow \qquad\quad \downarrow$$
$$Z \longrightarrow Z'$$

The map of fibers of the vertical maps is a weak equivalence by comparing the long exact sequences of the fibrations $P_f \to Z$ and $P_{f'} \to Z'$ together with the fact that the horizontal maps are weak equivalences (and using the five lemma).

Consider the square

$$\lim(P_f \to Z \leftarrow Y) \longrightarrow \lim(P_{f'} \to Z' \leftarrow Y')$$
$$\downarrow \qquad\qquad\qquad\qquad\qquad \downarrow$$
$$Y \longrightarrow Y'$$

The map from the fibers of the left vertical map to the fibers of the right vertical map is a weak equivalence since the fibers of the left (resp. right) vertical map are weakly equivalent to the fibers of $P_f \to Z$ (resp. $P_{f'} \to Z'$). Since the map $Y \to Y'$ is a weak equivalence and the vertical maps are fibrations by Proposition 2.1.16, once again comparing the long exact sequences we see that the top horizontal map above is a weak equivalence. □

It is useful to know when we can work with the ordinary limit rather than the homotopy limit, as the space of paths is often unwieldy.

**Proposition 3.2.13**  *Consider the diagram*

$$X \xrightarrow{f} Z \xleftarrow{g} Y.$$

*If f or g is a fibration, then the canonical map (see (3.2.1))*

$$\lim(X \xrightarrow{f} Z \xleftarrow{g} Y) \longrightarrow \operatorname{holim}(X \xrightarrow{f} Z \xleftarrow{g} Y)$$

*is a homotopy equivalence.*

*Proof*  Without loss of generality assume $f$ is a fibration. Consider the diagram

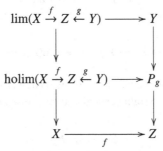

The bottom square is a pullback by Proposition 3.2.5, and by Proposition 2.1.16 the middle horizontal arrow is a fibration since $f$ is. Since $Y \to P_g$ is a homotopy equivalence, it follows from Proposition 3.2.3 that the top left vertical arrow is a homotopy equivalence if we can show the top square is a strict pullback. In this case, we are already implicitly using $\lim(X \to Z \leftarrow P_g)$ in place of $\operatorname{holim}(X \to Z \leftarrow Y)$. It is straightforward to check that

$$\lim\Big( \lim(X \to Z \leftarrow P_g) \to P_g \leftarrow Y \Big) \cong \lim(X \to Z \leftarrow Y),$$

as the former consists of those $(x, y, \gamma, y')$ such that $\gamma \colon I \to Z$ satisfies $\gamma(0) = f(x)$ and $\gamma(1) = g(y)$, and furthermore $\gamma = c_{g(y')}$ is a constant path, so that the projection map associating $(x, y, \gamma, y')$ with $(x, y)$ defines a homeomorphism from the space above to $\lim(X \to Z \leftarrow Y)$.  □

**Remark 3.2.14**  To verify that the upper square in the diagram in the above example is a pullback easily follows from the analog of 1(a) of Proposition 3.3.20 for strict pullback squares. We leave it to the reader to fill in the details.  □

Thus limits of diagrams $X \to Z \leftarrow Y$ in which at least one map is a fibration preserve homotopy equivalences.

**Corollary 3.2.15** *Consider the commutative diagram*

$$
\begin{array}{ccccc}
X & \xrightarrow{f} & Z & \xleftarrow{g} & Y \\
\simeq\downarrow & & \simeq\downarrow & & \downarrow\simeq \\
X' & \xrightarrow{f'} & Z' & \xleftarrow{g'} & Y'
\end{array}
$$

*and suppose that the vertical maps are homotopy (resp. weak) equivalences and at least one of $f, g$ and at least one of $f', g'$ are fibrations. Then the map*

$$
\lim(X \xrightarrow{f} Z \xleftarrow{g} Y) \longrightarrow \lim(X' \xrightarrow{f'} Z' \xleftarrow{g'} Y')
$$

*is a homotopy (resp. weak) equivalence.*

*Proof*   Consider the diagram

$$
\begin{array}{ccc}
\lim(X \to Z \leftarrow Y) & \longrightarrow & \operatorname{holim}(X \to Z \leftarrow Y) \\
\downarrow & & \downarrow \\
\lim(X' \to Z' \leftarrow Y') & \longrightarrow & \operatorname{holim}(X' \to Z' \leftarrow Y')
\end{array}
$$

The horizontal arrows are homotopy (resp. weak) equivalences by Proposition 3.2.13, and the right vertical arrow is a homotopy (resp. weak) equivalence by Theorem 3.2.12, and so the left vertical arrow is a homotopy (resp. weak) equivalence.                                                                    □

Another consequence of Theorem 3.2.12 is the homotopy invariance of the mapping path space $P_f$ (recall Definition 2.2.1) of a map $f\colon X \to Y$ under homotopy of $f$. Thematically this belongs in the previous chapter, and although we could have proved it there, we much prefer conceptually the proof we present here.

**Corollary 3.2.16**   *If $f_0, f_1\colon X \to Y$ are homotopic, then there is a homotopy equivalence*

$$
P_{f_0} \simeq P_{f_1}.
$$

*Proof*   Suppose $F\colon X \times I \to Y$ is a homotopy from $f_0$ to $f_1$. Consider the commutative diagram

$$
\begin{array}{ccccc}
X \times \{0\} & \xrightarrow{f_0} & Y & \xleftarrow{ev_0} & Y^I \\
\iota_0\downarrow & & \downarrow 1_Y & & \downarrow 1_{Y^I} \\
X \times I & \xrightarrow{F} & Y & \xleftarrow{ev_0} & Y^I \\
\iota_1\uparrow & & \uparrow 1_Y & & \uparrow 1_{Y^I} \\
X \times \{1\} & \xrightarrow{f_1} & Y & \xleftarrow{ev_0} & Y^I
\end{array}\;,
$$

where $\iota_0$ and $\iota_1$ are the obvious inclusions. The limits of the rows are, from top to bottom, $P_{f_0}$, $P_F$, and $P_{f_1}$. Since the evaluation map is a fibration by Proposition 2.1.7, these limits are equivalent to the homotopy limits by Proposition 3.2.13. Then because all of the vertical maps are homotopy equivalences, by Theorem 3.2.12 the induced maps in the diagram $P_{f_0} \to P_F \leftarrow P_{f_1}$ are homotopy equivalences. □

Recalling Definition 2.2.3, the definition of the homotopy fiber, we now revisit Proposition 2.2.7 and give a proof using techniques of this chapter.

**Corollary 3.2.17**   *If $f_0, f_1 : X \to Y$ are homotopic, then, for any $y \in Y$,*

$$\text{hofiber}_y(f_0) \simeq \text{hofiber}_y(f_1).$$

*Proof*   Again suppose $F : X \times I \to Y$ is a homotopy from $f_0$ to $f_1$ and consider the diagram

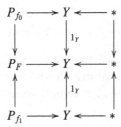

The maps $* \to Y$ are the inclusion of $y$ in $Y$, and the other maps are all of the evident ones. The limit of each row, from top to bottom, is $\text{hofiber}_y(f_0)$, $\text{hofiber}_y(F)$, and $\text{hofiber}_y(f_1)$. The limit of each row is homotopy equivalent to its homotopy limit by Proposition 3.2.13 because the left horizontal maps are fibrations, and all of the vertical maps are homotopy equivalences using Corollary 3.2.16. Thus Theorem 3.2.12 implies that the homotopy limits of the rows are all homotopy equivalent, and hence the limits of the rows are all homotopy equivalent as well. □

**Corollary 3.2.18**   *For a map $f : X \to Y$ and any $y_1$, $y_2$ in the same path component of $Y$,*

$$\text{hofiber}_{y_1}(f) \simeq \text{hofiber}_{y_2}(f).$$

*If $f$ is a fibration, then it follows that its fibers are homotopy equivalent.*

*Proof*   The proof is essentially the same as in the previous corollary: Let $H : I \to Y$ be a homotopy between two maps $*_1 : * \to Y$ and $*_2 : * \to Y$ with images $y_1$ and $y_2$, respectively. Let $P_i$ be the mapping path space of $*_i : * \to Y$, so we have a commutative diagram

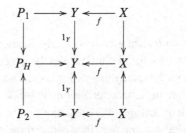

As in the last proof, the limits of all the rows are homotopy equivalent. But the limit of the top row is $\text{hofiber}_{y_1}(f)$ and the limit of the bottom row is $\text{hofiber}_{y_2}(f)$.

The last statement now follows from Proposition 2.2.6. □

## 3.3 Arithmetic of homotopy cartesian squares

As mentioned in the introduction to this chapter, we will also be interested in comparing the initial space in a square to the homotopy limit of the rest of it. This gives us a way to measure how far a square

$$S = \begin{array}{ccc} W & \longrightarrow & Y \\ \downarrow & & \downarrow \\ X & \longrightarrow & Z \end{array}$$

is from being homotopy cartesian. The natural way to measure this is in the difference of the homotopy types of $W$ and $\text{holim}(X \to Z \leftarrow Y)$. This is simply the square diagram analog of asking how highly connected a map of spaces is. To formalize this, recall that there is a canonical map

$$a(S) \colon W \longrightarrow \text{holim}(X \to Z \leftarrow Y) = \text{holim}(S_0) \qquad (3.3.1)$$

given by composing the canonical maps $W \to \lim(S_0)$ (see (3.1.1) and the discussion around it) and $\lim(S_0) \to \text{holim}(S_0)$ (see discussion following Definition 3.2.4). We then have

**Definition 3.3.1** A square diagram

$$S = \begin{array}{ccc} W & \longrightarrow & Y \\ \downarrow & & \downarrow \\ X & \longrightarrow & Z \end{array}$$

is

- *homotopy cartesian* (or ∞-*cartesian*, or a *homotopy pullback*) if $a(S)$ is a weak equivalence;
- *k-cartesian* if the map $a(S)$ is $k$-connected.

**Remarks 3.3.2**

1. If a square is $k$-cartesian, then it is clearly $j$-cartesian for all $j \leq k$.
2. The second part of the definition clearly generalizes the first since $\infty$-connected means the same thing as a weak equivalence by Proposition 1.4.4 (which relies on the characterization of weak equivalences in Proposition 2.6.17). It is natural to define homotopy cartesian and $k$-cartesian squares alongside one another since many of the properties we will develop in this section hold equally well for both. However, while we will see an abundance of interesting examples of homotopy cartesian squares in this chapter, the main examples of $k$-cartesian squares will have to wait until Chapter 4, since the Blakers–Massey Theorems will provide us with the tools to construct them.
3. We remark for the final time on homotopy versus weak equivalence. In addition to it being the limiting notion of $k$-connected map, we have chosen weak equivalence in the definition above because we wish to freely pass between spaces and CW complexes, the latter being a more convenient category of spaces to work with at times because we can exploit the cell structure.                                                                                  □

**Example 3.3.3**   Examples 3.2.7–3.2.11 can be recast as statements about homotopy cartesian squares. Namely, they simply say that the squares

$$
\begin{array}{ccc}
X \times Y & \longrightarrow & Y \\
\downarrow & & \downarrow \\
X & \longrightarrow & *
\end{array}
\qquad
\begin{array}{ccc}
X & \xrightarrow{1_X} & X \\
{\scriptstyle 1_X}\downarrow & & \downarrow{\scriptstyle 1_X} \\
X & \xrightarrow{1_X} & X
\end{array}
\qquad
\begin{array}{ccc}
P_f & \longrightarrow & X \\
\downarrow & & \downarrow{\scriptstyle f} \\
Y & \xrightarrow{1_Y} & Y
\end{array}
$$

$$
\begin{array}{ccc}
\mathrm{hofiber}_{y_0}(f) & \longrightarrow & X \\
\downarrow & & \downarrow{\scriptstyle f} \\
* & \longrightarrow & Y
\end{array}
\qquad
\begin{array}{ccc}
\Omega X & \longrightarrow & * \\
\downarrow & & \downarrow \\
* & \longrightarrow & X
\end{array}
\qquad
\begin{array}{ccc}
LX & \longrightarrow & X \\
\downarrow & & \downarrow{\scriptstyle \Delta} \\
X & \xrightarrow{\Delta} & X \times X
\end{array}
$$

are homotopy cartesian. Most of these examples will be generalized to cubes in Section 5.4.                                                                                  □

**Remarks 3.3.4**

1. Several of the squares above are not commutative, for example the square

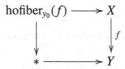

What we really mean here is to replace $X$ by $P_f$, and $f$ by the canonical map $P_f \to Y$ to make an honest commutative square.

2. The squares in the previous example whose homotopy pullbacks are $\Omega X$ and $LX$ are homotopy cartesian but are *not* cartesian. For example, the pull-back of $* \to X \leftarrow *$, where both maps include the basepoint of $X$ is a point (it is empty if the two maps are distinct). To obtain an equivalent square which is both cartesian and homotopy cartesian one needs to replace one of the maps $* \to X$ with the mapping path space construction on this inclusion. More generally we can obtain the space of paths between any two points in $X$ by a similar construction. □

The following is immediate from Proposition 3.2.13, and is a useful recognition principle for homotopy cartesian squares.

**Proposition 3.3.5** *Suppose*

$$
\begin{array}{ccc}
W & \longrightarrow & Y \\
\downarrow & & \downarrow \\
X & \longrightarrow & Z
\end{array}
$$

*is cartesian and $X \to Z$ or $Y \to Z$ is a fibration. Then the square is homotopy cartesian.*

We now turn to studying the "arithmetic" of homotopy cartesian and $k$-cartesian squares, that is, how they interact with themselves and various kinds of maps. Some consequences and examples will also be given throughout.

To start, the pointwise product of homotopy cartesian squares is homotopy cartesian. More precisely, we have the following

**Proposition 3.3.6** *Suppose*

$$
S_1 = \begin{array}{ccc}
W_1 & \longrightarrow & Y_1 \\
\downarrow & & \downarrow \\
X_1 & \longrightarrow & Z_1
\end{array}
\quad and \quad
S_2 = \begin{array}{ccc}
W_2 & \longrightarrow & Y_2 \\
\downarrow & & \downarrow \\
X_2 & \longrightarrow & Z_2
\end{array}
$$

*are homotopy cartesian. Then the square*

$$
\begin{array}{ccc}
W_1 \times W_2 & \longrightarrow & Y_1 \times Y_2 \\
\downarrow & & \downarrow \\
X_1 \times X_2 & \longrightarrow & Z_1 \times Z_2
\end{array}
$$

*is homotopy cartesian. If $S_1$ and $S_2$ are $k_1$-cartesian and $k_2$-cartesian, respectively, then the square of products is $\min\{k_1, k_2\}$-cartesian.*

*Proof*  We prove the homotopy cartesian statement; the $k$-cartesian version is essentially the same.

It is clear that $W_1 \times W_2$ is homotopy equivalent to the product

$$\operatorname{holim}(X_1 \to Z_1 \leftarrow Y_1) \times \operatorname{holim}(X_2 \to Z_2 \leftarrow Y_2).$$

Since maps of any space to a product are expressed as a pair of maps,

$$\operatorname{holim}(X_1 \to Z_1 \leftarrow Y_1) \times \operatorname{holim}(X_2 \to Z_2 \leftarrow Y_2)$$

is homeomorphic to

$$\operatorname{holim}(X_1 \times X_2 \to Z_1 \times Z_2 \leftarrow Y_1 \times Y_2).$$        □

**Remark 3.3.7**  The homotopy cartesian part of Proposition 3.3.6 can also be proved using Theorem 3.3.15, essentially because products can be thought of as homotopy pullbacks.        □

In particular, if $W_1 = X_1 = Y_1 = Z_1$ and the maps in the first square above are all the identity, then we have the following

**Corollary 3.3.8** (Products commute with homotopy pullbacks)  *If*

$$S = \begin{array}{ccc} W & \longrightarrow & Y \\ \downarrow & & \downarrow \\ X & \longrightarrow & Z \end{array}$$

*is a homotopy cartesian diagram and $V$ is any space, then*

$$S \times V = \begin{array}{ccc} W \times V & \longrightarrow & Y \times V \\ \downarrow & & \downarrow \\ X \times V & \longrightarrow & Z \times V \end{array}$$

*where all maps are induced from the previous diagram using the identity on $V$, is homotopy cartesian. Similarly, if $S$ is $k$-cartesian, so is $S \times V$.*

The cubical version is Corollary 5.4.6.

The following result says that the exponential of some homotopy cartesian squares is homotopy cartesian. (The reader should compare this with Proposition 3.9.1, and eventually with Proposition 5.4.7.) Recall the notation $A^B = \operatorname{Map}(B, A)$.

**Proposition 3.3.9** (Maps and homotopy cartesian squares)  *Suppose in the square*

$$S = \begin{array}{ccc} W & \longrightarrow & Y \\ \downarrow & & \downarrow \\ X & \longrightarrow & Z \end{array}$$

*the map $W \to \mathrm{holim}(X \to Z \leftarrow Y)$ is a homotopy equivalence (so in particular S is homotopy cartesian). Then, for any space V, the square*

$$S^V = \mathrm{Map}(V, S) = \begin{array}{ccc} W^V & \longrightarrow & Y^V \\ \downarrow & & \downarrow \\ X^V & \longrightarrow & Z^V \end{array}$$

*is homotopy cartesian. If S is k-cartesian and V has the homotopy type of a CW complex of dimension d, then $S^V$ is $(k - d)$-cartesian.*

**Remark 3.3.10** The first statement of this result is not true if $S$ is just required to be homotopy cartesian. For example, let $W$ be the space from Remark 1.3.6. Then the square

is homotopy cartesian, but letting $V = W$ and taking the square of mapping spaces as in the statement of the result, the three corners other than the upper left are a point, but the upper left is not path-connected.

However, the first statement could have been given for homotopy cartesian squares if the $V$ was assumed to be a CW complex; see the end of Remark 1.3.6. □

*Proof of Proposition 3.3.9* For the first statement, it suffices to prove that there is a homotopy equivalence

$$\mathrm{Map}\big(V, \mathrm{holim}(X \to Z \leftarrow Y)\big)$$
$$\simeq \mathrm{holim}\big(\mathrm{Map}(V, X) \to \mathrm{Map}(V, Z) \leftarrow \mathrm{Map}(V, Y)\big).$$

First consider the case where $Y \to Z$ is a fibration. By Proposition 3.2.13,

$$\mathrm{holim}(X \to Z \leftarrow Y) \simeq \lim(X \to Z \leftarrow Y).$$

Moreover, by Proposition 2.1.11 and Proposition 3.2.13 again, we also have a homotopy equivalence

$$\mathrm{holim}\big(\mathrm{Map}(V, X) \to \mathrm{Map}(V, Z) \leftarrow \mathrm{Map}(V, Y)\big)$$
$$\simeq \lim\big(\mathrm{Map}(V, X) \to \mathrm{Map}(V, Z) \leftarrow \mathrm{Map}(V, Y)\big).$$

It is clear that the map

$$\mathrm{Map}\left(V, \lim(X \to Z \gets Y)\right)$$
$$\longrightarrow \lim\left(\mathrm{Map}(V, X) \to \mathrm{Map}(V, Z) \gets \mathrm{Map}(V, Y)\right)$$

which sends $h$ to $(h_X, h_Y)$ (the induced maps from $W$ to $X$ and $Y$ respecitvely), is a homeomorphism. If $Y \to Z$ is not a fibration, we replace $g \colon Y \to Z$ by the fibration $P_g \to Z$ and note that $\mathrm{Map}(V, Y) \simeq \mathrm{Map}(V, P_g)$ since $\mathrm{Map}(V, -)$ preserves homotopy equivalences by Proposition 1.3.2. We then again appeal to Proposition 2.1.11 to see that $\mathrm{Map}(V, P_g) \to \mathrm{Map}(V, Z)$ is a fibration. The proof then follows from the previous argument, using the homotopy invariance of the homotopy pullback, Theorem 3.2.12.

For the latter statement, the key observation is that, if $V$ is of dimension $d$ and $f \colon X \to Y$ is a $k$-connected map, then $\mathrm{Map}(V, X) \to \mathrm{Map}(V, Y)$ is $(k - d)$-connected.[4]                                                                 □

The next result says that the homotopy pullback of a weak equivalence is a weak equivalence and that any square with parallel maps which are weak equivalences is a homotopy pullback.

**Proposition 3.3.11**    *Consider the square diagram*

$$
\begin{array}{ccc}
W & \longrightarrow & Y \\
\downarrow & & \downarrow \\
X & \longrightarrow & Z
\end{array}
$$

1. (a) *If the square is homotopy cartesian and the map $Y \to Z$ is a homotopy (resp. weak) equivalence, then the map $W \to X$ is a homotopy (resp. weak) equivalence;*
   (b) *If both $W \to X$ and $Y \to Z$ are homotopy (resp. weak) equivalences, then the square is homotopy cartesian.*
2. (a) *If the square is $k$-cartesian and the map $Y \to Z$ is $k$-connected, then the map $W \to X$ is $k$-connected.*
   (b) *If $W \to X$ is $k$-connected and $Y \to Z$ is $(k + 1)$-connected, then the square is $k$-cartesian.*

**Remarks 3.3.12**

1. The third obvious statement, which tries to deduce the connectivity of $Y \to Z$ from the square being $k$-cartesian and the map $W \to X$ being a $k$-connected is not true. For example, if $W = X = \emptyset$, then the square is trivially homotopy cartesian for any map $Y \to Z$.

---

[4] This comes down to the fact that $\Omega^d X$ is $(k - d)$-connected if $X$ is $k$-connected, which follows from Example 2.6.11. We leave the details to the reader.

2. Statements 1(a) and 1(b) imply each other (as do 2(a) and 2(b)). We will use 1(a) to obtain 1(b), and we leave it as an exercise to the reader to prove the second independently and use it to prove the first statement.
3. One way to read this result is that the homotopy pullback of a weak equivalence is a weak equivalence and that a homotopy cartesian square is a "map of weak equivalences that is a weak equivalence". That is, what we mean by a weak equivalence of maps (1-cubes) is a homotopy cartesian square (homotopy cartesian 2-cube). □

*Proof* For simplicity we work with weak equivalences, though the proof is the same for homotopy equivalences. For 1(a) it suffices to prove that the map

$$\text{holim}(X \to Z \leftarrow Y) \longrightarrow X$$

is a weak equivalence, since the map

$$W \longrightarrow \text{holim}(X \to Z \leftarrow Y)$$

is assumed to be a weak equivalence. Let $g$ denote the map from $Y$ to $Z$. Consider the pullback square

$$
\begin{array}{ccc}
\text{holim}(X \to Z \leftarrow Y) & \longrightarrow & P_g \\
\downarrow & & \downarrow \\
X & \longrightarrow & Z
\end{array}
$$

By Proposition 3.2.3, the map $\text{holim}(X \to Z \leftarrow Y) \to X$ is a weak equivalence. For 1(b), consider the diagram

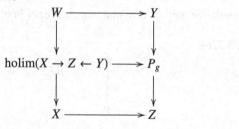

The map $P_g \to Z$ is a weak equivalence since $Y \to P_g$ and $Y \to Z$ are. Since the lower square is homotopy cartesian, by 1(a) the map $\text{holim}(X \to Z \leftarrow Y) \to X$ is a weak equivalence, and since by hypothesis $W \to X$ is a weak equivalence, so is $W \to \text{holim}(X \to Z \leftarrow Y)$.

The proofs of 2(a) and 2(b) are immediate from the proofs of 1(a) and 1(b) and an application of Proposition 2.6.15. □

We encountered the following example in the discussion of the dual Barratt–Puppe sequence (2.2.5) (see Lemma 2.2.16). We are now ready to give its proof as a consequence of Proposition 3.3.11. We could also deduce this immediately from Proposition 3.3.18.

**Example 3.3.13**   If $F \to X \to Y$ is a fibration sequence, then

$$\text{hofiber}(F \to X) \simeq \Omega Y.$$

To see this, we have a homotopy cartesian square

That is, $F \to \text{holim}(X \to Y \leftarrow *)$ is a weak equivalence by Proposition 2.2.6. Consider the square

$$
\begin{array}{ccc}
F & \longrightarrow & \text{holim}(X \to Y \leftarrow *) \\
\downarrow & & \downarrow \\
X & \longrightarrow & X
\end{array}
$$

where $X \to X$ is the identity, $F \to X$ the inclusion, and

$$\text{holim}(X \to Y \leftarrow *) \longrightarrow X$$

is projection onto $X$. This is also homotopy cartesian by Proposition 3.3.11, because the two horizontal arrows are weak equivalences. Since the homotopy fiber of $\text{holim}(X \to Y \leftarrow *) \to X$ is $\Omega Y$ (this can be seen directly from the definitions or by using Theorem 3.3.15 below), we have a weak equivalence

$$\text{hofiber}(F \to X) \longrightarrow \Omega Y. \qquad \square$$

We can generalize the last example to an arbitrary homotopy cartesian square.

**Example 3.3.14**   Suppose

$$
\begin{array}{ccc}
W & \longrightarrow & X \\
\downarrow & & \downarrow \\
Y & \longrightarrow & Z
\end{array}
$$

is homotopy cartesian. Then there is a weak equivalence

$$\text{hofiber}(W \to X \times Y) \simeq \Omega Z.$$

The argument is the same as above. The square

$$
\begin{array}{ccc}
W & \longrightarrow & \text{holim}(Y \to Z \leftarrow X) \\
\downarrow & & \downarrow \\
X \times Y & \longrightarrow & X \times Y
\end{array}
$$

is homotopy cartesian since both horizontal arrows are weak equivalences (the reader should think about what all the maps in this diagram are). Taking vertical homotopy fibers gives the desired weak equivalence since

$$\text{hofiber}\big(\text{holim}(X \to X \leftarrow Y) \to X \times Y\big) \simeq \Omega Z.$$

Again we may use Theorem 3.3.15 to see this last weak equivalence. □

We next establish a useful fact about iterated homotopy pullbacks. It is a special case of a more general result (Proposition 8.5.5) which says that homotopy limits commute.

**Theorem 3.3.15** *Suppose we have a commutative diagram*

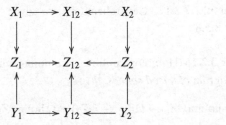

*Denote by* $\text{holim}(\mathcal{X}), \text{holim}(\mathcal{Z})$, *and* $\text{holim}(\mathcal{Y})$ *the homotopy pullbacks of the first, second, and third rows of this diagram respectively, and by* $\text{holim}(C_1)$, $\text{holim}(C_{12})$, *and* $\text{holim}(C_2)$ *the homotopy pullbacks of the first, second, and third columns respectively. Then there is a natural homeomorphism*

$$\text{holim}(\text{holim}(\mathcal{X}) \to \text{holim}(\mathcal{Z}) \leftarrow \text{holim}(\mathcal{Y}))$$

$$\downarrow \cong$$

$$\text{holim}(\text{holim}(C_1) \to \text{holim}(C_{12}) \leftarrow \text{holim}(C_2))$$

Taking $X_1 = X$, $X_2 = Y$, $Y_2 = Z$ and the rest of the spaces as points, this says $(X \times Y) \times Z \cong X \times (Y \times Z)$, and we think of this result as a sort of associative law for multiplication of punctured squares.

*Proof* To avoid cumbersome notation, we assume all the maps in the diagram above are inclusions (i.e. we suppress the functions). The proof in the general case is the same, but with more notation. A point in

$$\text{holim}\big(\text{holim}(\mathcal{X}) \to \text{holim}(\mathcal{Z}) \leftarrow \text{holim}(\mathcal{Y})\big)$$

is a tuple

$$(x_1, \alpha_{12}, x_2, \gamma_1, \Gamma_{12}, \gamma_2, y_1, \beta_{12}, y_2)$$

where $\alpha_{12}: I \to X_{12}, \beta_{12}: I \to Y_{12}$ satisfy the obvious properties, and $\gamma_i: I \to Z_i$ for $i = 1, 2$ satisfy $\gamma_i(0) = x_i$ and $\gamma_i(1) = y_i$ for $i = 1, 2$. We may regard $\Gamma_{12}$ as a map $\Gamma_{12}: I \times I \to Z_{12}$ such that $\Gamma_{12}(\{0\} \times I) = \gamma_1$ and $\Gamma_{12}(\{1\} \times I) = \gamma_2$. From this we produce a point in

$$\mathrm{holim}\Big(\mathrm{holim}(C_1) \to \mathrm{holim}(C_{12}) \leftarrow \mathrm{holim}(C_2)\Big)$$

by sending the above tuple to

$$(x_1, \gamma_1, y_1, \Gamma_{12}|_{I \times \{0\}}, \Gamma_{12}, \Gamma_{12}|_{I \times \{1\}}, x_2, \gamma_2, y_2).$$

This map is clearly a homeomorphism.                                          □

We then have the following consequence. The reader should compare this with Example 2.2.12. Generalizations are given in Corollary 5.4.10 and Corollary 8.5.8.

**Corollary 3.3.16** (Homotopy pullbacks commute with loops)   *If $Z_1 \to Z_{12} \leftarrow Z_2$ is a diagram of based spaces, then*

$$\mathrm{holim}(\Omega Z_1 \to \Omega Z_{12} \leftarrow \Omega Z_2) \cong \Omega\,\mathrm{holim}(Z_1 \to Z_{12} \leftarrow Z_2).$$

*In particular, if we let $Z_2 = *$ and $f_1$ denote the map $Z_1 \to Z_{12}$, then* $\mathrm{hofiber}(\Omega f_1) \cong \Omega\,\mathrm{hofiber}(f_1)$.

*Proof*   In the diagram from Theorem 3.3.15, set all the $X_i$ and $Y_i$ to be one-point spaces, with vertical maps inclusions of basepoints. Using Example 3.2.10, which describes the loop space as a homotopy limit, the result then immediately follows from Theorem 3.3.15.                                          □

Even more generally, the square of homotopy fibers of a map of homotopy cartesian squares is itself homotopy cartesian. Another way to say this is as follows.

**Corollary 3.3.17**   *There is a natural homeomorphism*

$$\mathrm{holim}\Big(\mathrm{hofiber}(X_1 \to Z_1) \to \mathrm{hofiber}(X_{12} \to Z_{12}) \leftarrow \mathrm{hofiber}(X_2 \to Z_2)\Big)$$
$$\cong \mathrm{hofiber}\Big(\mathrm{holim}(X_1 \to X_{12} \leftarrow X_2) \to \mathrm{holim}(Z_1 \to Z_{12} \leftarrow Z_2)\Big).$$

*Proof*   In the statement of Theorem 3.3.15, take $Y_S = *$.                                          □

The following will be generalized later as Proposition 5.4.12.

**Proposition 3.3.18** *A square*

$$
\begin{array}{ccc}
W & \longrightarrow & Y \\
\downarrow & & \downarrow \\
X & \xrightarrow{\ f\ } & Z
\end{array}
$$

*is homotopy cartesian (k-cartesian) if and only if for all $x \in X$, the map*

$$\mathrm{hofiber}_x(W \to X) \longrightarrow \mathrm{hofiber}_{f(x)}(Y \to Z)$$

*is a weak equivalence (k-connected).*

**Remark 3.3.19** When the above square is a square of inclusions, Proposition 3.3.18 says that homotopy cartesian means that the map

$$\pi_i(X, W) \longrightarrow \pi_i(Z, Y)$$

is an isomorphism for all $i$. For the $k$-cartesian version, the map is an isomorphism for $i \le k$ and onto for $i = k + 1$. □

*Proof of Proposition 3.3.18* We will prove the homotopy cartesian version of the statement; the $k$-cartesian variant follows the same pattern.

Replacing the map $Y \to Z$ by a fibration, we have that the homotopy pullback is the pullback and the fiber of the pullback is homeomorphic to the homotopy fiber of the map $Y \to Z$. So we want to prove that the map from $W$ to the pullback is a weak equivalence if and only if for every point $x \in X$ the homotopy fiber $\mathrm{hofiber}(W \to X)$ is weakly equivalent to the actual fiber (which is weakly equivalent to the homotopy fiber) of the pullback. But this follows from the long exact sequence.[5] □

The following is a useful result on "factorizations" of squares that is analogous to Proposition 3.3.11 (see Remark 3.3.22 for more on what we mean by this). We will make use of such results frequently, especially in the proof of Theorem 4.2.1. The more general version for cubes of arbitrary dimension is Proposition 5.4.14.

**Proposition 3.3.20** *In the diagram*

$$
\begin{array}{ccc}
X_\emptyset \longrightarrow X_1 \longrightarrow X_{13} \\
\downarrow \qquad\quad \downarrow \qquad\quad \downarrow \\
X_2 \longrightarrow X_{12} \longrightarrow X_{123}
\end{array}
$$

---

[5] Note the use of the long exact sequence in this proof, as this result does not dualize. It does dualize if we replace "weak equivalence" by "homology equivalence", in which case the dual proof works as well.

*let the left square be denoted by* $\boxed{1}$ *, the right by* $\boxed{2}$ *, and the outer square by* $\boxed{1\,2}$ *. Then*

1. (a) *If* $\boxed{1}$ *and* $\boxed{2}$ *are homotopy cartesian, then* $\boxed{1\,2}$ *is homotopy cartesian.*
   (b) *If* $\boxed{2}$ *and* $\boxed{1\,2}$ *are homotopy cartesian, then* $\boxed{1}$ *is homotopy cartesian.*
2. (a) *If* $\boxed{1}$ *and* $\boxed{2}$ *are k-cartesian, then so is* $\boxed{1\,2}$ *.*
   (b) *If* $\boxed{2}$ *is* (k + 1)-*cartesian and* $\boxed{1\,2}$ *is k-cartesian, then* $\boxed{1}$ *is k-cartesian.*

**Remark 3.3.21** The third obvious statement is false in general. In the diagram

$$\begin{array}{ccccc} \emptyset & \longrightarrow & X_1 & \longrightarrow & X_{13} \\ \downarrow & & \downarrow & & \downarrow \\ \emptyset & \longrightarrow & X_{12} & \longrightarrow & X_{123} \end{array}$$

$\boxed{1}$ and $\boxed{1\,2}$ are trivially homotopy cartesian, but $\boxed{2}$ can be an arbitrary square. Note that this does not contradict Proposition 3.3.18 in either way one may view the homotopy fibers of the left square above. □

*Proof of Proposition 3.3.20* We prove statements 1(a) and 1(b). The proof of 2(a) and 2(b) is immediate from the proof of 1(a) and 1(b) together with Proposition 2.6.15.

Consider the diagram

$$\begin{array}{ccc} X_\emptyset \longrightarrow \mathrm{holim}(X_2 \to X_{12} \leftarrow X_1) & \longrightarrow & \mathrm{holim}(X_2 \to X_{123} \leftarrow X_{13}) \\ \downarrow & & \downarrow \\ X_1 \longrightarrow & & \mathrm{holim}(X_{12} \to X_{123} \leftarrow X_{13}) \end{array}$$

We will show in a moment that the square appearing in this diagram is homotopy cartesian. The square above is *not* strictly commutative; what we are claiming though is that the homotopy limit of the evident punctured square is weakly equivalent to $\mathrm{holim}(X_2 \to X_{12} \leftarrow X_1)$. Assuming this is so, note that both sets of hypotheses in statement 1 assume $\boxed{2}$ is homotopy cartesian, which means that the map

$$X_1 \longrightarrow \mathrm{holim}(X_{12} \to X_{123} \leftarrow X_{13})$$

is a weak equivalence. By Proposition 3.3.11, the map

$$\mathrm{holim}(X_2 \to X_{12} \leftarrow X_1) \longrightarrow \mathrm{holim}(X_2 \to X_{123} \leftarrow X_{13})$$

is a weak equivalence. Both results now immediately follow. Now we establish that the square

$$\text{holim}(X_2 \to X_{12} \leftarrow X_1) \longrightarrow \text{holim}(X_2 \to X_{123} \leftarrow X_{13})$$

$$X_1 \longrightarrow \text{holim}(X_{12} \to X_{123} \leftarrow X_{13})$$

is homotopy cartesian. Consider the commutative diagram

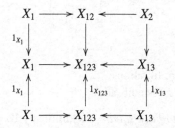

Since the homotopy pullback of the first column is homotopy equivalent to $X_1$, it is evident that the homotopy pullback of the homotopy pullback of the columns is homotopy equivalent to the homotopy pullback of the square in question. By Theorem 3.3.15, the homotopy pullback in question is homeomorphic to the homotopy pullback of the homotopy pullback of the rows. Let $\text{holim}(X)$ denote the homotopy pullback of the last two (identical) rows. Then the homotopy pullback of the homotopy pullback of the rows is

$$\text{holim}\Big(\text{holim}(X_1 \to X_{12} \leftarrow X_2) \longrightarrow \text{holim}(X) \xleftarrow{=} \text{holim}(X)\Big).$$

This is homotopy equivalent to $\text{holim}(X_1 \to X_{12} \leftarrow X_2)$ by 1(a) of Proposition 3.3.11. □

**Remark 3.3.22** We will see in Proposition 5.4.13 that Proposition 3.3.20 is equivalent to a result which says that a map of homotopy cartesian squares (a 3-cube) is homotopy cartesian, and the homotopy pullback of a homotopy cartesian square is homotopy cartesian. This makes it more apparent how this result is a generalization of Proposition 3.3.11. □

One might say we have taken the perspective that the limit operation is defective and so requires a replacement which satisfies the homotopy invariance property. Another perspective takes the stance that the spaces and maps in the diagram are defective, and so we should replace them with equivalent spaces so that the limit does have the homotopy invariance property, as we explain below. The reason we are discussing such replacements in this "arithmetic" section is that Proposition 3.3.24 can be thought of as saying that any square can be replaced by a convenient representative; i.e. the equivalence classes of squares have suitable representatives. We will generalize what follows from squares to cubes in Definition 5.4.22 and Theorem 5.4.23.

**Definition 3.3.23**  We say that a square

$$\mathcal{S} = \begin{array}{ccc} W & \longrightarrow & X \\ \downarrow & & \downarrow \\ Y & \longrightarrow & Z \end{array}$$

is *fibrant* if

- $X \to Z$ and $Y \to Z$ are fibrations;
- $W \to \lim(Y \to Z \leftarrow X)$ is a fibration.

We say $\mathcal{S}$ is a *fibrant pullback square* if additionally $W = \lim(Y \to Z \leftarrow X)$.

Note that, by Proposition 3.2.13, for a fibrant square we have that

$$\lim(Y \to Z \leftarrow X) \simeq \operatorname{holim}(Y \to Z \leftarrow X). \tag{3.3.2}$$

This is the observation we will seek to generalize in Proposition 5.4.26.

The next result says that every square $\mathcal{S}$ is weakly equivalent to a fibrant square, called a *fibrant replacement of* $\mathcal{S}$, and hence, by the observation above, every associated punctured square (the diagram $Y \to Z \leftarrow X$ obtained by removing the initial space $W$) has the property that its limit has the homotopy invariance property.

By a *homotopy (resp. weak) equivalence of squares* $\mathcal{S} \to \mathcal{S}'$, we mean a commutative diagram

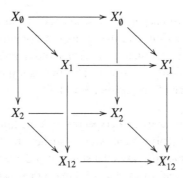

such that the maps $X_S \to X'_S$ are homotopy (resp. weak) equivalences for each $S$.

**Proposition 3.3.24** *Every square admits a homotopy (and hence weak) equivalence to a fibrant square. That is, given a square*

$$S = \begin{array}{ccc} W & \longrightarrow & X \\ \downarrow & & \downarrow \\ Y & \longrightarrow & Z \end{array}$$

*there is a fibrant square*

$$S' = \begin{array}{ccc} W' & \longrightarrow & X' \\ \downarrow & & \downarrow \\ Y' & \longrightarrow & Z' \end{array}$$

*and a map of squares (a commutative diagram that looks like a 3-cube)*

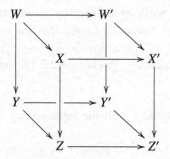

*such that $W \to W'$, $X \to X'$, $Y \to Y'$, and $Z \to Z'$ are homotopy equivalences. Every homotopy cartesian square admits an equivalence to a fibrant pullback square.*

*Proof* Define $Z' = Z$. Let $f: X \to Z$ and $g: Y \to Z$ be the maps in the square. Define $X' = P_f$ and $Y' = P_g$. Let $h: W \to \mathrm{holim}(X \xrightarrow{f} Z \xleftarrow{g} Y)$ be the canonical map, and let $W' = P_h$. Then

$$\mathcal{X}' = \begin{array}{ccc} W' & \longrightarrow & X' \\ \downarrow & & \downarrow \\ Y' & \longrightarrow & Z' \end{array}$$

where the maps from $W' \to X'$, $Y'$ are the canonical projections and $X' \to Z'$ and $Y' \to Z'$ are evaluation of the path at time $t = 1$, is a fibrant square. This is because Proposition 3.2.5 says that $\mathrm{holim}(X \xrightarrow{f} Z \xleftarrow{g} Y)$ is homeomorphic to $\lim(P_f \to Z \leftarrow P_g)$, so that the map $P_h \to \lim(P_f \to Z \leftarrow P_g)$ is a fibration (this is the composition of the fibration $P_h \to \mathrm{holim}(X \xrightarrow{f} Z \xleftarrow{g} Y)$ with the homeomorphism to $\lim(P_f \to Z \leftarrow P_g)$).

The maps $W \to W'$, $X \to X'$, and $Y \to Y'$ are all homotopy equivalences as in Proposition 2.2.2, and it is straightforward to check that the diagram (three-dimensional cube) commutes.

In the case where $S$ is homotopy cartesian, we define $W' = \lim(X' \to Z' \leftarrow Y')$, and let $W \to W'$ be the canonical map given by the universal property of $W'$. Then in the square

$$
\begin{array}{ccc}
W & \longrightarrow & \mathrm{holim}(X \to Z \leftarrow Y) \\
\downarrow & & \downarrow \\
W' & \longrightarrow & \mathrm{holim}(P_f \to Z \leftarrow P_g)
\end{array}
$$

the right vertical map is a homotopy equivalence by Theorem 3.2.12, the top vertical arrow is a weak equivalence by assumption, and the lower horizontal arrow is a weak equivalence since the first map in the composition $W' \to \lim(P_f \to Z \leftarrow P_g) \to \mathrm{holim}(P_f \to Z \leftarrow P_g)$ is an equality and the second is a weak equivalence by Proposition 3.2.13. Hence $W \to W'$ is a weak equivalence.                                             □

**Remark 3.3.25** Continuing with the same spaces as above, consider the square

$$
\begin{array}{ccc}
W & \longrightarrow & \mathrm{holim}(Y \to Z \leftarrow X) \\
\downarrow & & \downarrow \\
W' & \longrightarrow & \mathrm{holim}(Y' \to Z' \leftarrow X')
\end{array}
$$

By homotopy invariance of the homotopy pullback, Theorem 3.2.12, the right vertical map is a weak equivalence (it is a pointwise weak equivalence), and so is the left vertical map. Thus the square is homotopy cartesian by 1(b) of Proposition 3.3.11. If the lower horizontal map is $k$-connected, so is the top horizontal map by 2(a) of Proposition 3.3.11 (see also Remark 3.3.12). So to prove that $S$ is $k$-cartesian, it suffices to prove $S'$ is. This goes the other way as well; since both vertical maps are equivalences, if $S'$ is $k$-cartesian, then so is $S$ by 2(a) of Proposition 3.7.13.                                             □

We will encounter fibrant replacements of cubical diagrams in Theorem 5.4.23 (and of more general diagrams in Section 8.4.2). The following result will be useful there.

**Lemma 3.3.26** *Suppose*

$$\mathcal{X} = \begin{array}{ccc} X_\emptyset & \longrightarrow & X_1 \\ \downarrow & & \downarrow \\ X_2 & \longrightarrow & X_{12} \end{array}$$

*is fibrant. Then every map $X_S \to X_T$ for $S \subset T$ is a fibration.*

*Proof* By definition $X_i \to X_{12}$ are fibrations, and so by symmetry it suffices to prove that $X_\emptyset \to X_2$ is a fibration since compositions of fibrations are fibrations. The map $X_0 \to X_2$ is the composition

$$X_0 \longrightarrow \lim(X_2 \to X_{12} \leftarrow X_1) \longrightarrow X_2.$$

By definition, the first of these is a fibration and the second is a pullback of the fibration $X_1 \to X_{12}$. Since a composition of fibrations is a fibration, we get the desired result. □

## 3.4 Total homotopy fibers

For the square

$$\mathcal{S} = \begin{array}{ccc} W & \longrightarrow & Y \\ \downarrow & & \downarrow \\ X & \longrightarrow & Z \end{array}$$

Definition 3.3.1 can be interpreted as saying that the homotopy fiber of

$$a(\mathcal{S}): W \longrightarrow \mathrm{holim}(X \to Z \leftarrow Y)$$

measures the extent to which $\mathcal{S}$ is a homotopy cartesian square. We explore this further in this section.

Note that, if $\mathcal{S}$ is a square of based spaces, then $\mathrm{holim}(X \to Z \leftarrow Y)$ has a natural basepoint consisting of the tuple $(x_0, \gamma, y_0)$, where $x_0$ and $y_0$ are the basepoints of $X$ and $Y$ respectively and $\gamma$ is the constant path at their images in $Z$ (the basepoint of $Z$).

**Definition 3.4.1** For $\mathcal{S}$ a square of based spaces, define its *total (homotopy) fiber* to be

$$\mathrm{tfiber}(\mathcal{S}) = \mathrm{hofiber}(a(\mathcal{S})),$$

where the homotopy fiber is taken over the natural basepoint of $\mathrm{holim}(X \to Z \leftarrow Y)$.

**Remark 3.4.2** The total fiber can be defined for a square of unbased spaces by a choice of basepoint $(x, \gamma, y) \in \text{holim}(X \to Z \leftarrow Y)$. Dealing with only the based case, as we do here, simplifies the presentation of certain facts.     □

Proposition 3.3.18 says that the square $S$ is $k$-cartesian if and only if the maps $\text{hofiber}(W \to X) \to \text{hofiber}(Y \to Z)$ are $k$-connected. Thus the connectivity of the homotopy fiber of the map of homotopy fibers also measures the extent to which $S$ is homotopy cartesian. This hints at an alternative description of the total homotopy fiber as an iterated homotopy fiber.

**Proposition 3.4.3** (Total homotopy fiber as iterated homotopy fiber) *Let $S$ be the commutative square diagram of based spaces*

$$
\begin{array}{ccc}
W & \longrightarrow & Y \\
\downarrow & & \downarrow \\
X & \longrightarrow & Z
\end{array}
\tag{3.4.1}
$$

*and let the basepoint of* $\text{holim}(X \to Z \leftarrow Y)$ *be given by the basepoints of $X$ and $Y$ and the constant path in $Z$ between their images. Then*

$$\text{tfiber}(S) \simeq \text{hofiber}\big(\text{hofiber}(W \to X) \longrightarrow \text{hofiber}(Y \to Z)\big),$$

*where the basepoint of* $\text{hofiber}(Y \to Z)$ *is the pair consisting of the basepoint of $Y$ and the constant path at the basepoint of $Z$. Moreover,*

$$\text{tfiber}(S) \simeq \text{hofiber}\big(\text{hofiber}(W \to Y) \longrightarrow \text{hofiber}(X \to Z)\big).$$

**Remark 3.4.4**

1. When we take (homotopy) fibers of (homotopy) fibers, we will refer to the result as the "iterated (homotopy) fiber".
2. In light of Proposition 3.3.18, a square is homotopy cartesian if and only if its total homotopy fiber is weakly contractible.     □

*Proof of Proposition 3.4.3* Consider the diagram

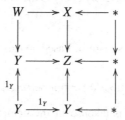

The maps are all the evident ones, with the maps from the one-point space the inclusion of the various basepoints. The homotopy pullbacks of the rows, from top to bottom, are hofiber($W \to X$), hofiber($Y \to Z$), and $*$, using Example 3.2.8 for the first two and Example 3.2.9 for the last. The homotopy limit of the resulting diagram is evidently the iterated homotopy fiber, hofiber(hofiber($W \to X$) $\to$ hofiber($Y \to Z$)). The homotopy pullbacks of the columns, from left to right, are $W$, holim($X \to Z \leftarrow Y$) and $*$, the first once again using Example 3.2.9. The homotopy limit of this is evidently the total homotopy fiber of the square by definition. The two are homeomorphic by Theorem 3.3.15. Note, however, that we get only a weak equivalence between the total fiber and the iterated homotopy fiber because, for instance, the homotopy pullback of the first column is really the mapping path space of the map $W \to Y$, which is homotopy equivalent to $W$, but not equal to it.

It is also easy to see that the two iterated homotopy fiber descriptions of the total homotopy fiber from Proposition 3.4.3 are homeomorphic directly by applying Theorem 3.3.15 to the diagram

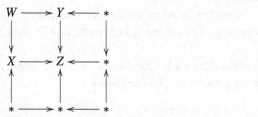

We leave the details to the reader. □

The next result is one of the inputs in the proof of Lemma 2.2.16 used in the construction of the dual Barratt–Puppe sequence.

**Example 3.4.5** The homotopy fiber of the map hofiber($F \to X$) $\longrightarrow F$ (i.e. what can be thought of as the map $\Omega X \to F$ by Example 3.3.13) is weakly equivalent to $\Omega Y$. To see this, consider the square

where the left vertical map is the identity. Computing its total homotopy fiber by taking horizontal homotopy fibers first gives (by definition) hofiber(hofiber($F \to X$) $\longrightarrow F$). Taking vertical homotopy fibers first gives hofiber($* \to X$) $\simeq \Omega X$. The two are weakly equivalent by Proposition 3.4.3 (again, the two iterated homotopy fibers are homeomorphic, but the above identifications are up to weak equivalence). □

**Example 3.4.6**    For any map $f\colon X \to Y$, and any choice of point $y_0 \in Y$, there is a homotopy equivalence

$$\text{hofiber}(X \xrightarrow{(1_X,f)} X \times Y) \simeq \Omega Y,$$

where $\Omega Y$ is the space of loops based at $y_0$. This is because the homotopy fiber $\text{hofiber}(X \xrightarrow{(1_X,f)} X \times Y)$ may be thought of as the total fiber of the diagram

$$
\begin{array}{ccc}
X & \xrightarrow{\;1_X\;} & X \\
\downarrow{\scriptstyle f} & & \downarrow \\
Y & \longrightarrow & *
\end{array}
$$

If we think of this total fiber as an iterated homotopy fiber using Proposition 3.4.3, then taking horizontal homotopy fibers yields $* \to Y$, whose homotopy fiber is $\Omega Y$.

Note that by doing the iterated fibers in the other direction, we see that the homotopy fiber of the natural projection $\text{hofiber}(f) \to X$ is $\Omega Y$.     $\square$

**Proposition 3.4.7**    *Suppose $X \to Y \to Z$ are maps of based spaces. Then we have a homotopy fiber sequence*

$$\text{hofiber}(X \to Y) \to \text{hofiber}(X \to Z) \to \text{hofiber}(Y \to Z).$$

*That is,* $\text{hofiber}(X \to Y)$ *is weakly equivalent to the homotopy fiber of the map*

$$\text{hofiber}(X \to Z) \to \text{hofiber}(Y \to Z),$$

*where the basepoint in* $\text{hofiber}(Y \to Z)$ *is the constant loop at the basepoint.*

*Proof*    Consider the square

$$
\mathcal{S} = \quad
\begin{array}{ccc}
X & \longrightarrow & Y \\
\downarrow & & \downarrow \\
Z & \xrightarrow{\;1_Z\;} & Z
\end{array}
$$

We may identify the total fiber of this square by taking iterated fibers vertically or horizontally using Proposition 3.4.3. This yields weak equivalences

$$\text{hofiber}(X \to Y) \simeq \text{tfiber}(\mathcal{S})$$

$$\simeq \text{hofiber}\big(\text{hofiber}(X \to Z) \to \text{hofiber}(Y \to Z)\big).\quad \square$$

We next have two results relating the homotopy fibers of a retraction to the homotopy fibers of the inclusion.

**Proposition 3.4.8** *Suppose $f\colon X \to Y$ and $g\colon Y \to X$ are maps of based such that $g \circ f$ is the identity (i.e. assume $X$ is a retract of $Y$). Then there is a weak equivalence*

$$\mathrm{hofiber}(f) \simeq \Omega\,\mathrm{hofiber}(g).$$

*Proof* Compare the vertical, horizontal, and total fibers of the square

$$
\begin{array}{ccc}
X & \xrightarrow{\ f\ } & Y \\
{\scriptstyle 1_X}\downarrow & & \downarrow{\scriptstyle g} \\
X & \xrightarrow{\ 1_X\ } & X
\end{array}
$$

using Proposition 3.4.3. □

As a generalization of Proposition 3.4.8, we have the following.

**Proposition 3.4.9** *Let $X$ be the diagram of based spaces*

$$
\begin{array}{ccc}
X_0 & \longrightarrow & X_1 \\
\downarrow & & \downarrow \\
X_2 & \longrightarrow & X_{12}
\end{array}
$$

*and suppose there are maps $X_{S \cup i} \to X_S$ for $S \subset \{1,2\}$ and $i \notin S$ such that the composition $X_S \to X_{S \cup i} \to X_S$ is the identity. Then*

$$\mathrm{tfiber}(X) \simeq \Omega^2\,\mathrm{tfiber}(X'),$$

*where $X'$ is the square*

$$
\begin{array}{ccc}
X_{12} & \longrightarrow & X_2 \\
\downarrow & & \downarrow \\
X_1 & \longrightarrow & X_0
\end{array}
$$

*Proof* This follows from repeated application of Proposition 3.4.8. Consider the diagram

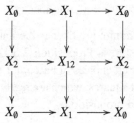

By Proposition 3.4.3,

$$\text{tfiber}(\mathcal{X}) \simeq \text{hofiber}\Big(\text{hofiber}(X_\emptyset \to X_1) \longrightarrow \text{hofiber}(X_2 \to X_{12})\Big),$$

and using Proposition 3.4.8 on each of the homotopy fibers above, we have

$$\text{tfiber}(\mathcal{X}) \simeq \text{hofiber}\Big(\Omega\,\text{hofiber}(X_1 \to X_\emptyset) \longrightarrow \Omega\,\text{hofiber}(X_{12} \to X_2)\Big).$$

The induced map of loop spaces above is the loop of a map of homotopy fibers, since it is induced by the map of total fibers of the left to the right square in the 3-cube

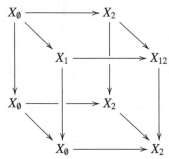

where we view the total fibers of the left and right squares iteratively (first vertical, then horizontal fibers) using Proposition 3.4.3. By Example 2.2.12, we have

$$\text{tfiber}(\mathcal{X}) \simeq \Omega\,\text{hofiber}\Big(\text{hofiber}(X_1 \to X_\emptyset) \longrightarrow \text{hofiber}(X_{12} \to X_2)\Big),$$

and the iterated homotopy fiber on the right side is precisely the total fiber of the upper-right square in the diagram at the start of this proof. The same argument shows that the total fiber of the upper-right square is weakly equivalent to the loop space of the total fiber of the lower-right square, which is tfiber($\mathcal{X}'$).                                                                 □

We can also use the iterated fiber description of the total fiber given in Proposition 3.4.3 to come to a reasonable definition of the homotopy groups of a square diagram. For a map of based spaces $f\colon X \to Y$ (a 1-cube), with $y \in Y$ the basepoint, we may define the homotopy groups of the map $f$ as

$$\pi_k(f) = \pi_{k-1}\,\text{hofiber}_y(f). \tag{3.4.2}$$

Here the basepoint of the homotopy fiber is the natural one, and $k \geq 1$. The reason for the dimension shift is Proposition 2.6.12, which says that $f$ is $k$-connected if and only if for all choices of basepoint $\in Y$, hofiber$_y(f)$ is $(k-1)$-connected. Iterating this for squares, which are really maps of 1-cubes, we have the following definition and proposition.

Let $S$ be the square

$$
\begin{array}{ccc}
W & \longrightarrow & Y \\
f_1 \downarrow & & \downarrow f_2 \\
X & \longrightarrow & Z
\end{array}
$$

of based spaces.

**Definition 3.4.10** For $k \geq 2$, define

$$\pi_k(S) = \pi_{k-2}(\text{tfiber } S).$$

It is easy to see, using the long exact sequence of homotopy groups together with Proposition 3.4.3, that we have the following generalization of Proposition 3.4.7 (obtained by setting $W = Y$ in the above square). A generalization for cubes is given as Proposition 5.5.11.

**Proposition 3.4.11** (Homotopy long exact sequence of a square) *There is a long exact sequence of homotopy groups*

$$\cdots \to \pi_i(f_1) \to \pi_i(f_2) \to \pi_i(\text{tfiber } S) \to \pi_{i-1}(f_1) \to \cdots.$$

*The sequence ends at* $\pi_1(f_2)$.

# 3.5 Pushouts

We now begin what is essentially a dualization of Sections 3.1–3.4. Nearly everything from the previous section has a dual, with a few exceptions which we point out along the way. The central object now will be the diagram

$$S_1 = (X \xleftarrow{f} W \xrightarrow{g} Y).$$

**Definition 3.5.1** The *pushout*, or *colimit*, of the diagram $S_1 = (X \xleftarrow{f} W \xrightarrow{g} Y)$, denoted by $\mathrm{colim}(X \xleftarrow{f} W \xrightarrow{g} Y)$, or $\mathrm{colim}(S_1)$, or $X \cup_W Y$, is the quotient space of $X \sqcup Y$ defined as

$$\mathrm{colim}(X \xleftarrow{f} W \xrightarrow{g} Y) = X \sqcup Y/\sim,$$

where $\sim$ is the equivalence relation on $X \sqcup Y$ generated by $f(w) \sim g(w)$ for $w \in W$.

Again, the terminology "colimit" and the notation "colim" is supposed to indicate that there is something more general at play; indeed, we will show in Example 7.3.32 that the pushout is an example of a more general notion of colimit. As was the case with the pullback, the pushout enjoys a "universal property" in that every map out of the diagram $S_1$ factors through the colimit. That is, in any commutative solid arrow diagram

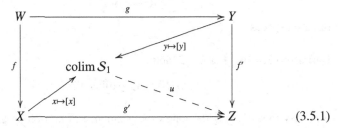

$$(3.5.1)$$

there is a unique dotted arrow $u$ that makes the resulting diagram commute. It is again easy to see that the formulas $u([x]) = g'(x)$ and $u([y]) = f'(y)$ determine a well-defined map to $Z$. That cofibrations are preserved under pushout, the content of Proposition 2.3.15, is immediate from the universal property. We will revisit the pushout from the point of view of universal properties in Example 7.3.32.

Here are some examples of familiar spaces and constructions which are naturally described in the language of pushouts.

**Example 3.5.2** (Disjoint union as pushout)   A pushout along the empty set is the disjoint union:

$$\mathrm{colim}(X \leftarrow \emptyset \rightarrow Y) \cong X \sqcup Y.$$

In this case the equivalence relation on $X \sqcup Y$ identifies no points.   □

**Example 3.5.3** (Wedge sum as pushout)   For based spaces the above pushout becomes the wedge sum.

$$\mathrm{colim}(X \leftarrow * \rightarrow Y) \cong X \vee Y.$$   □

**Example 3.5.4**   The pushout of the diagram $X \xleftarrow{1_X} X \xrightarrow{1_X} X$ is homeomorphic to $X$.   □

**Example 3.5.5**   Let $f\colon X \to Y$ be a map. The pushout of the diagram $X \xleftarrow{1_X} X \xrightarrow{f} Y$ is $Y$; the inclusion $Y \to \mathrm{colim}(X \xleftarrow{1_X} X \xrightarrow{f} Y)$ is a homeomorphism.   □

**Example 3.5.6** (Cofiber as pushout)   For a map $f\colon X \to Y$ we have

$$\mathrm{cofiber}(f) = Y/f(X) \cong \mathrm{colim}(Y \xleftarrow{f} X \to *).$$

This follows from the definition of the quotient space. One important special case is when $X = \emptyset$, in which case we obtain $Y_+$, the space $Y$ together with a disjoint point. Thus the quotient of a space $Y$ by the empty set is $Y_+$. □

**Example 3.5.7** (Mapping cylinder as pushout)  Let $f\colon X \to Y$ be a map. Let $i\colon X \to X \times I$ be the map that sends $x$ to $(x, 0)$. Then

$$\mathrm{colim}(Y \xleftarrow{f} X \xrightarrow{i} X \times I) \cong M_f,$$

the mapping cylinder of $f$. The colimit in question is precisely the space given in Definition 2.4.1. We leave the slightly different case of the reduced mapping cylinder to the reader; it is a quotient space of the colimit above. □

**Example 3.5.8** (Homotopy cofiber as pushout)  For a map $f\colon X \to Y$, there is a homeomorphism

$$\mathrm{hocofiber}(f) \cong \mathrm{colim}(M_f \leftarrow X \to *).$$

This is immediate from the definition of the homotopy cofiber. □

**Example 3.5.9** (Attaching a cell as pushout)  The procedure of attaching a cell is a pushout:

$$\mathrm{colim}(X \xleftarrow{f} \partial e^n \to e^n) = X \cup_f e^n.$$ □

We now have a dual to Proposition 3.1.9.

**Proposition 3.5.10** (Pushouts preserve surjections and homeomorphisms) *Suppose $i\colon X \to Y$ is a surjection (resp. homeomorphism), then for any map $f\colon X \to Z$, the induced map $i_*\colon Z \to Y \cup_X Z$ is also a surjection (resp. homeomorphism).*

*Proof*  Let $w \in Y \cup_X Z$ and consider the quotient map $Y \sqcup Z \to Y \cup_X Z$. Either $w$ is the image of some point in $Y$ or $Z$. If $w$ is the image of some $y \in Y$, then $i(x) = y$ for some $x \in X$ since $i$ is surjective and since $i(x) \sim f(x) \in Z$, we see $w$ is the image of $f(x) \in Z$. The other case is obvious.

If $i$ is a homeomorphism with inverse $j\colon Y \to X$, then the induced map $Z \to Y \cup_X Z$ has inverse $Y \cup_X Z \to Z$. This inverse is induced by the map $Y \sqcup Z \to Z$ defined by $w \mapsto (f \circ j)(w)$ for $w \in Y$ and $1_Z(w) = w$ for $w \in Z$. □

As with pullbacks, pushouts do not preserve homotopy equivalences, and we deal with this issue in the next section. Next we have a dual to

Proposition 3.1.10, proof is dual and which we omit. We will not need this result until Chapter 8.[6]

**Proposition 3.5.11**  *Let $S_1 = (X_1 \leftarrow X_0 \rightarrow X_2)$ and $S'_1 = (X'_1 \leftarrow X'_0 \rightarrow X'_2)$ be punctured squares in which the maps $X_0 \rightarrow X_2$ and $X'_0 \rightarrow X'_2$ are cofibrations, and $S_1 \rightarrow S'_1$ a map of punctured squares such that $X_S \rightarrow X'_S$ are cofibrations for all $S$. Further assume the canonical map $\mathrm{colim}(X'_0 \leftarrow X_0 \rightarrow X_2) \rightarrow X'_2$ is a cofibration. Then the induced map $\mathrm{colim}\, S_1 \rightarrow \mathrm{colim}\, S'_1$ is a cofibration.*

Given a commutative square, looking back at the diagram (3.5.1), it again makes sense to ask whether its "final" space is the pushout of the rest of the diagram.

**Definition 3.5.12**  A square diagram

$$
\begin{array}{ccc}
W & \xrightarrow{\ g\ } & Y \\
{\scriptstyle f}\downarrow & & \downarrow \\
X & \longrightarrow & Z
\end{array}
$$

is called *cocartesian* (or *categorically cocartesian*, or a *pushout*, or a *strict pushout*) if the canonical map

$$
\mathrm{colim}(X \xleftarrow{\ f\ } W \xrightarrow{\ g\ } Y) \longrightarrow Z
$$

from (3.5.1) is a homeomorphism.

**Example 3.5.13**  Examples 3.5.2–3.5.9 can all be recast in the language of cocartesian squares, that is, they all say that certain squares are cocartesian. For instance, Example 3.5.2 says that the square

$$
\begin{array}{ccc}
\emptyset & \longrightarrow & X \\
\downarrow & & \downarrow \\
Y & \longrightarrow & X \sqcup Y
\end{array}
$$

is cocartesian. We leave it to the reader to revisit the other examples and write down the obvious squares.                                    □

We finish with a result about the interaction of pullbacks and pushouts. They are both analogs of the familiar exponential laws from algebra – the first is the analog of $(a^b)^c = a^{bc}$, and the second the analog of $a^{b+c} = a^b a^c$. This will

---

[6]  As was the case with Proposition 3.1.10, the hypotheses below might seem unmotivated, but they mean that a certain diagram is *cofibrant*; more details can be found in Section 8.4.2.

be generalized in Proposition 7.5.1 (see also Proposition 3.9.1). We leave the straightforward proof to the reader.

**Proposition 3.5.14** *There are natural homeomorphisms*

$$\text{Map}\big(Z, \lim(X \leftarrow W \rightarrow Y)\big) \xrightarrow{\cong} \lim\big(\text{Map}(Z, X)$$
$$\rightarrow \text{Map}(Z, W) \leftarrow \text{Map}(Z, Y)\big),$$
$$\text{Map}\big(\text{colim}(X \leftarrow W \rightarrow Y), Z\big) \xrightarrow{\cong} \lim\big(\text{Map}(X, Z)$$
$$\rightarrow \text{Map}(W, Z) \leftarrow \text{Map}(Y, Z)\big).$$

## 3.6 Homotopy pushouts

Just as in the case of pullbacks, pushouts are not homotopy invariant. This is evident in the the following

**Example 3.6.1** (Pushout not homotopy invariant)  Consider the diagram

$$\text{colim}(D^1 \longleftarrow S^0 = \partial D^1 \longrightarrow D^1) \simeq S^1$$

$$\text{colim}(* \longleftarrow S^0 \longrightarrow *) \simeq *$$

where the two top maps are the standard inclusions. The pushouts clearly have different homotopy types.  □

However, if one of the maps in the diagram $\mathcal{S}_1 = (X \leftarrow W \rightarrow Y)$ is a cofibration, then the pushout is homotopy invariant. We will prove this in Corollary 3.6.19, but we first need the following dual to Proposition 3.2.3. This was already encountered as Proposition 2.3.19, but we now restate it and give a proof for the weak equivalence case in a way that uses the language of this chapter.

**Proposition 3.6.2** *Suppose $i\colon W \rightarrow Y$ is a cofibration and $f\colon W \rightarrow X$ is a homotopy (resp. weak) equivalence. Then the inclusion*

$$Y \longrightarrow \text{colim}(X \xleftarrow{f} W \xrightarrow{i} Y)$$

*is a homotopy (resp. weak) equivalence. That is, if in the cocartesian square*

$$
\begin{array}{ccc}
W & \xrightarrow{\ i\ } & Y \\
{\scriptstyle f}\downarrow & & \downarrow \\
X & \longrightarrow & \text{colim}(X \leftarrow W \rightarrow Y)
\end{array}
$$

*i* is a cofibration and *f* is a homotopy (resp. weak equivalence), then $Y \to$ colim$(X \leftarrow W \to Y)$ is a homotopy (resp. weak) equivalence.

*Proof* As previously promised, we give a proof of the statement for weak equivalences. It uses the homotopy equivalence version of this result, stated originally as Proposition 2.3.19. The proof for homotopy eqiuvalences was omitted there, but it is dual to the proof of Proposition 2.1.23 for homotopy equivalences.

By Theorem 1.3.9 we have a commutative diagram

$$
\begin{array}{ccccc}
X' & \xleftarrow{f'} & W' & \xrightarrow{i'} & Y \\
\downarrow & & \downarrow & & \downarrow \\
X & \xleftarrow{f} & W & \xrightarrow{i} & Y
\end{array}
$$

in which $X', W'$, and $Y'$ are all CW complexes and the vertical maps are all weak equivalences. Furthermore, using the same result we may assume that $i'$ is a cofibration. Since $f$ and the vertical maps are weak equivalences, it follows from Proposition 2.6.15 (with $k = \infty$) that $f'$ is also a weak equivalence. Since the domain and codomain of $f'$ are CW complexes, $f'$ is in fact a homotopy equivalence by Theorem 1.3.10. Now consider the diagram

$$
\begin{array}{ccc}
Y' & \longrightarrow & \text{colim}(X' \leftarrow W' \to Y') \\
\downarrow & & \downarrow \\
Y & \longrightarrow & \text{colim}(X \leftarrow W \to Y)
\end{array}
$$

By Proposition 3.6.17 and Theorem 3.6.13, the right vertical map is a weak equivalence, and by Proposition 2.3.19 (this is the statement of the result we are proving in the case of homotopy equivalences), the top horizontal map is a homotopy equivalence. The map $Y' \to Y$ is a weak equivalence by construction. Once again using Proposition 2.6.15, again with $k = \infty$, we can deduce that the lower horizontal map is a weak equivalence.    □

To make a homotopy invariant colimit for an arbitrary diagram

$$
\mathcal{S}_1 = (X \leftarrow W \to Y),
$$

one approach is therefore to change one of the maps into a cofibration using the mapping cylinder construction and then take the pushout. As we did for limits, we will give a more symmetric and equivalent definition, and then in Proposition 3.6.5 argue that this model is the same as the usual one.

**Definition 3.6.3** The *homotopy pushout*, or *homotopy colimit*, of the diagram $\mathcal{S}_1 = (X \xleftarrow{f} W \xrightarrow{g} Y)$, denoted by hocolim $\mathcal{S}_1$ or hocolim$(X \xleftarrow{f} W \xrightarrow{g} Y)$, is

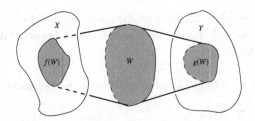

Figure 3.3 A picture of the homotopy pushout.

the quotient space of $X \sqcup (W \times I) \sqcup Y$ under the equivalence relation generated by $f(w) \sim (w, 0)$ and $g(w) \sim (w, 1)$ for $w \in W$. If $W$ is a based space with basepoint $w_0$, we add the relation $(w_0, t) \sim (w_0, s)$ for all $s, t \in I$.

### Remarks 3.6.4

1. The based homotopy colimit is a quotient of the unbased one where we collapse the copy of an interval $I$ associated with the basepoint. Since our spaces are well-pointed, the two versions are homotopy equivalent. Whenever we discuss statements for homotopy pushouts involving based spaces, we always mean the based homotopy colimit as above, with the natural basepoint.

2. Note that we can identify $\mathrm{hocolim}(X \xleftarrow{f} W \xrightarrow{g} Y)$ with the union of $M_f$ with $M_g$ along $W \times \{1\} \subset M_f, M_g$. Sometimes the homotopy pushout as we define it here is known as the *double mapping cylinder*. □

A picture of the homotopy pushout is given in Figure 3.3.

As before, we use the terminology "homotopy colimit" and the notation "hocolim" with an eye toward general homotopy colimits; we shall see in Example 8.2.21 that the homotopy pushout is an example of a homotopy colimit.

There are canonical maps

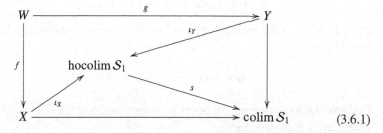

(3.6.1)

The maps $\iota_X$ and $\iota_Y$ are inclusions and $s: \mathrm{hocolim}\, X \to \mathrm{colim}\, X$ projects away from the interval coordinate. The inner square does not commute, since the images of $w \in W$ via $\iota_X \circ f$ and $\iota_Y \circ g$ lie at opposite ends of the cylinder $X \times I$.

Dual to Proposition 3.2.5, we have the following models for the homotopy pushout, which we will use interchangeably. Again the proof just consists of unraveling the definitions.

**Proposition 3.6.5**  *The following five spaces are homeomorphic:*

- $\operatorname{hocolim}(X \xleftarrow{f} W \xrightarrow{g} Y)$;
- $\operatorname{colim}(M_f \leftarrow W \rightarrow Y)$;
- $\operatorname{colim}(X \leftarrow W \rightarrow M_g)$;
- $\operatorname{colim}(M_f \leftarrow W \rightarrow M_g)$;
- $\operatorname{colim}(\operatorname{colim}(X \xleftarrow{f} W \xrightarrow{i_0} W \times I) \longleftarrow W \times I \longrightarrow \operatorname{colim}(Y \xleftarrow{g} W \xrightarrow{i_1} W \times I))$.

Here are some examples of homotopy pushouts.

**Example 3.6.6** (Disjoint union and wedge as homotopy pushouts)   We have

$$\operatorname{hocolim}(X \leftarrow \emptyset \rightarrow Y) \cong X \sqcup Y.$$

For based spaces, the corresponding statement is

$$\operatorname{hocolim}(X \leftarrow * \rightarrow Y) \simeq X \vee Y.$$

The former is clear from the definition and the latter is straightforward to see by shrinking the interval $* \times I$ to a point. Also clear is the case

$$\operatorname{hocolim}(X \xleftarrow{1_X} X \xrightarrow{1_X} X) \simeq X$$

because the homotopy pushout is $X \times I$, which is clearly homotopy equivalent to $X$. □

**Example 3.6.7** (Mapping cylinder as homotopy pushout)   For a map $f \colon X \rightarrow Y$, we have a homeomorphism

$$\operatorname{hocolim}(X \xleftarrow{1_X} X \xrightarrow{f} Y) \cong M_f.$$

Since $Y \rightarrow M_f$ is the inclusion of a deformation retract, we have a homotopy equivalence

$$\operatorname{hocolim}(X \xleftarrow{1_X} X \xrightarrow{f} Y) \simeq Y.$$ □

**Example 3.6.8** (Homotopy cofiber as homotopy pushout)   For a map $f \colon X \rightarrow Y$, we have a homeomorphism

$$\operatorname{hocolim}(* \leftarrow X \xrightarrow{f} Y) \cong \operatorname{hocofiber}(f).$$

This is true in both the based and unbased settings. □

**Example 3.6.9** (Suspension as homotopy pushout)  There is a homeomorphism

$$\text{hocolim}(* \leftarrow X \rightarrow *) \cong \Sigma X.$$

This is a homeomorphism both in the unbased and based settings, and the proof is immediate upon inspection. □

**Example 3.6.10** (Smash product as homotopy pushout)  For based spaces $X$ and $Y$, there is a weak equivalence

$$\text{hocolim}(* \leftarrow X \vee Y \rightarrow X \times Y) \simeq X \wedge Y.$$

To see why this is true, compare with Definition 1.1.15. There we defined

$$X \wedge Y = \text{colim}(* \leftarrow X \vee Y \rightarrow X \times Y).$$

Since our spaces are well-pointed, Proposition 2.3.13 implies the inclusion $X \vee Y \rightarrow X \times Y$ is a cofibration, then by Proposition 3.6.17, the canonical map from the homotopy colimit to the colimit is a weak equivalence. □

Here is a consequence of the previous example.

**Example 3.6.11**  If $X$ and $Y$ are based connected spaces, there is a homotopy equivalence

$$\Sigma(X \times Y) \simeq \Sigma(X \vee Y) \vee \Sigma(X \wedge Y)$$

induced by a section of the inclusion $\Sigma(X \vee Y) \rightarrow \Sigma(X \times Y)$.

To see this, by Example 3.6.10, we have a cofibration sequence

$$\Sigma(X \vee Y) \rightarrow \Sigma(X \times Y) \rightarrow \Sigma(X \wedge Y)$$

by applying $\Sigma$ everywhere. Define a map $s \colon \Sigma(X \times Y) \rightarrow \Sigma(X \vee Y)$ by

$$s(t \wedge (x,y)) = \begin{cases} 2t \wedge (x,*), & 0 \le t \le 1/2; \\ (2t-1) \wedge (*,y), & 1/2 \le t \le 1. \end{cases}$$

This map is continuous at $t = 1/2$ since it is the basepoint in both formulas. The composed map $\Sigma(X \vee Y) \rightarrow \Sigma(X \times Y) \rightarrow \Sigma(X \vee Y)$ is homotopic to the identity, and the result follows in a manner dual to that described at the end of Proposition 3.9.6. Alternatively, [Hat02, Proposition 4.I.1] gives a simple geometric argument for proving that this splitting exists. □

**Example 3.6.12** (Join as homotopy pushout)  There is a homeomorphism

$$\text{hocolim}(Y \longleftarrow X \times Y \longrightarrow X) \cong X * Y.$$

To see this, recall from Definition 1.1.17 that the join $X * Y = CX \times Y \cup_{X \times Y} X \times CY$. Denote a point in this space as a pair $([x, t], [y, s])$, and recall that in the cone $CZ$ on a space $Z$, $[z, 1] = [z', 1]$ for all $z, z' \in Z$, and that we may consider $Z$ as the subspace of all pairs $[z, 0]$ in $CZ$. By definition, $\text{hocolim}(Y \longleftarrow X \times Y \longrightarrow X)$ is the quotient space $X \sqcup X \times Y \times I \sqcup Y / \sim$, where $x \sim (x, y, 0)$ and $(x, y, 1) \sim y$. Define a map

$$h \colon \text{hocolim}(Y \longleftarrow X \times Y \longrightarrow X) \longrightarrow X * Y$$

by the formulas $h(x) = ([x, 0], [y, 1])$, $h(x, y, t) = ([x, t], [y, t])$, and $h(y) = ([x, 1], [y, 0])$. It is easy to see that $h$ is a homeomorphism.                               □

What follows is the result that establishes homotopy invariance of the homotopy pushout. The first proof of of this appears to be [Bro68, 7.5.7]. We give an outline, which is a dualization of the proof of Theorem 3.2.12; for details, see Rognes [Rog].

**Theorem 3.6.13** (The gluing lemma)    *Consider the commutative diagram*

$$
\begin{array}{ccccc}
X & \xleftarrow{\ f\ } & W & \xrightarrow{\ g\ } & Y \\
{\scriptstyle e_X}\downarrow & & {\scriptstyle e_W}\downarrow & & \downarrow{\scriptstyle e_Y} \\
X' & \xleftarrow{\ f'\ } & W' & \xrightarrow{\ g'\ } & Y'
\end{array}
$$

*where the vertical arrows are homotopy (resp. weak) equivalences. Then the induced map*

$$\text{hocolim}(X \xleftarrow{f} W \xrightarrow{g} Y) \longrightarrow \text{hocolim}(X' \xleftarrow{f'} W' \xrightarrow{g'} Y')$$

$$\left( [x] \cup [w, t] \cup [y] \right) \longmapsto \left( [e_X(x)] \cup [e_W(w), t] \cup [e_Y(y)] \right)$$

*is a homotopy (resp. weak) equivalence.*

*Proof of Theorem 3.6.13 in case of homotopy equivalences*    The    proof    for homotopy equivalences is dual to Theorem 3.2.12. In this case one factors the map of diagrams in the statement as

$$
\begin{array}{ccccc}
X & \xleftarrow{\ f\ } & W & \xrightarrow{\ g\ } & Y \\
{\scriptstyle e_X}\downarrow & & {\scriptstyle e_W}\downarrow & & \downarrow{\scriptstyle e_Y} \\
X' & \xleftarrow{\ f' \circ e_W\ } & W & \xrightarrow{\ g' \circ e_W\ } & Y' \\
{\scriptstyle 1_{X'}}\downarrow & & {\scriptstyle e_W}\downarrow & & \downarrow{\scriptstyle 1_{X'}} \\
X' & \xleftarrow{\ f'\ } & W' & \xrightarrow{\ g'\ } & Y'
\end{array}
$$

To see that the homotopy pushout of the first row is homotopy equivalent to the homotopy pushout of the second row, use Corollary 2.3.18 and colim($M_f \leftarrow W \rightarrow M_g$) from Proposition 3.6.5 for the homotopy pushouts of the first two rows. To show that the homotopy pushouts of the second and third rows are equivalent is to dualize the first part of the proof of Theorem 3.2.12. We will treat the case of weak equivalences separately below, as it is more subtle. $\square$

The proof of Theorem 3.6.13 in the case where the maps are assumed to be weak equivalences is more delicate than the argument given in Theorem 3.2.12. It follows from the following lemma due to Gray [Gra75], the proof of which we will present here since the text in question is out of print. Another reason that merits its inclusion is that the argument is also closely related to part of the argument (appearing in Section 4.4.2) we will present to prove one of the main theorems of this text: the Blakers–Massey Theorem, here Theorem 4.2.1. Both cut a cube up into little pieces and make small deformations, much the way one does in proving excision holds for homology groups, as we have breezily discussed in Section 1.4.2.

**Lemma 3.6.14** ([Gra75, Lemma 16.24]) *Let $X$ and $Y$ be spaces with sub-spaces $X_1, X_2 \subset X$ and $Y_1, Y_2 \subset Y$ such that the union of the interiors of the $X_i$ cover $X$ and the union of the interiors of the $Y_i$ cover $Y$. Suppose $f : X \rightarrow Y$ is a map such that, for $i = 1, 2$, $f(X_i) \subset Y_i$, and the restrictions $f : X_i \rightarrow Y_i$ and $f : X_1 \cap X_2 \rightarrow Y_1 \cap Y_2$ are weak equivalences. Then $f$ is a weak equivalence.*

We will present the proof shortly, but first we give some useful corollaries of this, beginning with the proof of Theorem 3.6.13 where we assume the vertical maps are weak equivalences.

*Proof of Theorem 3.6.13 in case of weak equivalences* By definition,

$$Z = \text{hocolim}(X \leftarrow W \rightarrow Y)$$

is a quotient of $X \sqcup W \times I \sqcup Y$, and similarly for

$$Z' = \text{hocolim}(X' \leftarrow W' \rightarrow Y').$$

Let $P_X, P_Y \subset Z$ be the evident quotients of $X \sqcup W \times [0, 2/3)$ and $W \times (1/3, 1] \sqcup Y$ respectively, and similarly define $P_{X'}, P_{Y'} \subset Z'$. Then $\{P_W, P_Y\}$ and $\{P_{X'}, P_{Y'}\}$ are open covers for $Z, Z'$ respectively. The canonical map $e : Z \rightarrow Z'$ has the property that $e(P_V) \subset P_{V'}$ for $V = X, Y$. The canonical projection $P_V \rightarrow V$ is a homotopy (and hence weak) equivalence, whose homotopy inverse is the inclusion of $V$ in $P_V$ (see Proposition 2.4.4; although $P_V$ is not exactly the mapping cylinder of $W \rightarrow V$, the same proof shows that $P_V \rightarrow V$ is a homotopy equivalence), also for $V = X, Y$. Hence the restriction of $e$ to $P_X$, $P_Y$, and $P_X \cap P_Y = W \times (1/3, 2/3) \simeq W$ are all weak equivalences. It follows from Lemma 3.6.14 that $e$ is a weak equivalence.                                    □

Two more corollaries of Lemma 3.6.14 are worth noting. The first is really a useful restatement, which we learned from [DI04].

**Corollary 3.6.15**   *Suppose $f: X \to Y$ is a map of spaces and $\{U, V\}$ is an open cover of $Y$. If $f^{-1}(U) \to U$, $f^{-1}(V) \to V$, and $f^{-1}(U \cap V) \to U \cap V$ are weak equivalences, then so is $f$.*

A useful corollary, due to May [May90], the proof of which we also learned from [DI04], is as follows.

**Corollary 3.6.16** ([May90, Corollary 1.4])   *Suppose $f: X \to Y$ is a map, and $\{U_i\}_{i \in I}$ is an open cover of $Y$. For a subset $S \subset I$, write $U_S = \cap_{i \in S} U_i$. If $f^{-1}(U_S) \to U_S$ is a weak equivalence for every finite set $S \subset I$, then $f$ is a weak equivalence.*

*Sketch of proof.*   Consider the set of all open sets $W \subset X$ such that $f^{-1}(W \cap U_S) \to W \cap U_S$ is a weak equivalence for all finite sets $S$ (including the empty set). Zorn's lemma says that this set has a maximal element, and by Lemma 3.6.14 it must be $X$ itself.                                    □

We now prove Lemma 3.6.14. We follow Gray's presentation with a few stylistic changes.

*Proof of Lemma 3.6.14*   Factor the map $f: X \to Y$ as $X \to M_f \to Y$ as in Proposition 2.4.4. Since $M_f \to Y$ is a homotopy equivalence, it is enough to prove $X \to M_f$ is a weak equivalence. Hence, without loss of generality, we may assume $X \to Y$ is a cofibration and hence an inclusion by Proposition 2.3.2. By Proposition 2.6.9, it is enough to prove that the pair $(Y, X)$ is $\infty$-connected, which means (using a cubical variant of Definition 2.6.4) that we must prove that every map of pairs $(I^n, \partial I^n) \to (Y, X)$ is homotopic relative to $\partial I^n$ to a map $I^n \to X$ for all $n$.

Suppose then that we are given a map of pairs $g: (I^n, \partial I^n) \to (Y, X)$, and write $g_\partial: \partial I^n \to X$ for the restriction of $g$ to the boundary. We will construct a map $\widehat{g}: I^n \to X$ whose restriction to $\partial I^n$ is $g_\partial$ and such that the composition $f \circ \widehat{g} \sim g$ relative to $\partial I^n$. Recalling that $\mathring{Z}$ stands for the interior of $Z$, for $i = 1, 2$ define

$$A_i = g^{-1}(Y_i - \mathring{Y}_i) \cup g_\partial^{-1}(X_i - \mathring{X}_i).$$

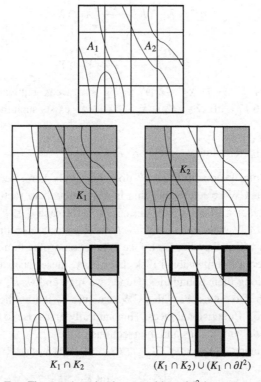

Figure 3.4 Top: The sets $A_i$ and a decomposition of $I^2$ into cubes. Middle: The cubes $K_i$ are those which do not meet $A_i$. Bottom: The sets over which we begin to construct the desired deformation.

Clearly the $A_i$ are closed, and they are disjoint since for $Z = X, Y$ the union of the interiors of the $Z_i$ cover $Z$. Let $U_i = I^n - A_i$. Since the $A_i$ are closed and disjoint, $\{U_1, U_2\}$ is an open cover of $I^n$, and by the Lebesgue covering lemma (here Lemma 1.4.11), we may subdivide $I^n$ into subcubes $W$ (subsets of $I^n = \prod_{i=1}^n [0, 1]$ of the form $\prod_{i=1}^n [a_i, a_i + \epsilon]$, $\epsilon > 0$) such that each $W$ is contained in some $U_i$ (i.e. no $W$ intersects both $A_1$ and $A_2$). See the top picture in Figure 3.4 for a schematic in the case $n = 2$. This means no such $W$ intersects both $A_1$ and $A_2$.

For $i = 1, 2$, define $K_i$ to be the union of those cubes $W$ such that $W \cap A_i = \emptyset$, captured schematically by the middle two pictures in Figure 3.4. Then $I^n = K_1 \cup K_2$. Note also that $g$ carries $K_i$ to $Y_i$. We will construct the desired lift by first constructing it on $K_1 \cap K_2$, and then make compatible lifts on $K_1$ and $K_2$. Restricting $g$ and $f$ we have then a commutative diagram and a lifting problem

$$\begin{array}{ccc}
\partial I^n \cap K_1 \cap K_2 & \xrightarrow{\ g_\partial\ } & X_1 \cap X_2 \\
\downarrow & \nearrow & \downarrow f \\
K_1 \cap K_2 & \xrightarrow[g]{} & Y_1 \cap Y_2
\end{array}$$

By hypothesis $f : X_1 \cap X_2 \rightarrow Y_1 \cap Y_2$ is a weak equivalence, and by Proposition 2.6.17 a lift $G_{12} : K_1 \cap K_2 \rightarrow X_1 \cap X_2$ exists such that

- $G_{12}|_{\partial I^n \cap K_1 \cap K_2} = g_\partial$;
- $f \circ G_{12} \sim g|_{K_1 \cap K_2}$ relative to $\partial I^n \cap K_1 \cap K_2$.

Strictly speaking, Proposition 2.6.17 does not apply here, because we have not specified a CW structure, but this is easy enough to rectify. Each cube $W = \prod_{i=1}^{n} [a_i, a_i + \epsilon]$ has faces of dimension $k$ for each $0 \leq k \leq n$ specified by subspaces of $W$ where precisely $n - k$ coordinates are on the boundary of the interval in which they lie. Each face is homeomorphic to a $k$-dimensional cube and hence a $k$-dimensional disk. Moreover, these faces are attached to one another along their boundaries. Hence we may endow $I^n$, with the given decomposition into cubes $W$, with a CW structure by letting the $k$-cells be the union of all the $k$-dimensional faces. This naturally gives rise to CW structures on the $K_i$ and $\partial I^n$ and unions and intersections of these.

Extend the map $G_{12}$ to a map $G'_{12 \rightarrow 1} : (K_1 \cap K_2) \cup (K_1 \cap \partial I^n) \rightarrow X_1$ by

$$G'_{12 \rightarrow 1} = \begin{cases} G_{12}, & \text{on } K_1 \cap K_2; \\ g, & \text{on } K_1 \cap \partial I^n. \end{cases}$$

Clearly this map is well-defined and continuous since by construction $G_{12}$ is equal to $g$ on $K_1 \cap K_2 \cap \partial I^n$. Moreover, the homotopy given by the solution to the lifting problem above extends to a homotopy $f \circ G'_{12 \rightarrow 1} \sim g|_{(K_1 \cap K_2) \cup (K_1 \cap \partial I^n)}$ relative to $K_1 \cap \partial I^n$; we simply extend on $K_1 \cap \partial I^n$ using the constant homotopy. Consider the extension problem

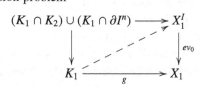

where the top horizontal map is the homotopy (relative to $K_1 \cap \partial I^n$) from $g|_{(K_1 \cap K_2) \cup (K_1 \cap \partial I^n)}$ to $f \circ G'_{12 \rightarrow 1}$. Since the left vertical map is a cofibration by Example 2.3.7 (using the CW structure we described above), a solution $H_1 : K_1 \rightarrow X_1^I$ to this extension problem exists. This amounts to a homotopy from $g|_{K_1}$ to a map $g_1 : K_1 \rightarrow X_1$ such that $g_1|_{(K_1 \cap K_2) \cup (K_1 \cap \partial I^n)} = f \circ G'_{12 \rightarrow 1}$, and this homotopy is constant on $K_1 \cap \partial I^n$. Now consider the diagram

$$(K_1 \cap K_2) \cup (K_1 \cap \partial I^n) \xrightarrow{G'_{12 \to 1}} X_1$$

Since $f|_{X_1} : X_1 \to Y_1$ is a weak equivalence, again using Proposition 2.6.17, a lift $G_1 : K_1 \to X_1$ exists, and satisfies

- $G_1|_{(K_1 \cap K_2) \cup (K_1 \cap \partial I^n)} = G'_{12 \to 1}$;
- $f \circ G_1 \sim g_1$ relative to $(K_1 \cap K_2) \cup (K_1 \cap \partial I^n)$.

Since $g \sim g_1$ relative to $K_1 \cap \partial I^n$, it follows that $f \circ G_1 \sim g$ relative to $K_1 \cap \partial I^n$. In a similar manner we construct $G_2 : K_2 \to X_2$ with the properties

- $G_2|_{(K_1 \cap K_2) \cup (K_2 \cap \partial I^n)} = G'_{12 \to 2}$;
- $f \circ G_2 \sim g_2$ relative to $(K_1 \cap K_2) \cup (K_2 \cap \partial I^n)$.

Similarly $f \circ G_2 \sim g$ relative to $K_2 \cap \partial I^n$.

Since $G'_{12 \to 1}$ and $G'_{12 \to 2}$ agree on $K_1 \cap K_2$, the restrictions of $G_1$ and $G_2$ to $K_1 \cap K_2$ are equal, and hence together define a map

$$\widehat{G} : I^n = K_1 \cup K_2 \longrightarrow X$$

such that $\widehat{G}|_{K_i} = G_i$, and hence by construction of the $G_i$, $\widehat{G}|_{\partial I^n} = g_\partial$.

All that is left to prove is that $f \circ \widehat{G} \sim g$ relative to $\partial I^n$. To see why, note that the homotopies $f \circ G_1 \sim g$ and $f \circ G_2 \sim g$ agree on $(K_1 \cap K_2) \times I$, and hence form a homotopy $f \circ \widehat{G} \sim g$. This homotopy is fixed on $\partial I^n = (K_1 \cap \partial I^n) \cup (K_2 \cap \partial I^n)$, as we have observed above that $f \circ G_1 \sim g$ relative to $K_1 \cap \partial I^n$ and $f \circ G_2 \sim g$ relative to $K_2 \cap \partial I^n$. □

Dual to Proposition 3.2.13 we have the following.

**Proposition 3.6.17** *Consider the diagram*

$$X \xleftarrow{f} W \xrightarrow{g} Y.$$

*If $f$ or $g$ is a cofibration, then the canonical map (see (3.6.1))*

$$\mathrm{hocolim}(X \xleftarrow{f} W \xrightarrow{g} Y) \longrightarrow \mathrm{colim}(X \xleftarrow{f} W \xrightarrow{g} Y)$$

*is a homotopy equivalence.*

*Proof* Assume $f$ is a cofibration. The proof is dual to the proof of Proposition 3.2.13 upon considering the diagram

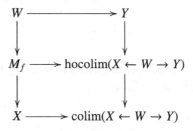

and using Propositions 3.6.5 and 3.6.2.                                              □

**Remark 3.6.18** In the previous proof, we only need to use the statement of Proposition 3.6.2 in the case of homotopy equivalences. We point this out since we use Proposition 3.6.17 in the proof of Proposition 3.6.2 for weak equivalences.                                              □

**Corollary 3.6.19** *Given a commutative diagram*

$$X \xleftarrow{\ f\ } W \xrightarrow{\ g\ } Y$$
$$\downarrow{\simeq} \qquad \downarrow{\simeq} \qquad \downarrow{\simeq}$$
$$X' \xleftarrow{\ f'\ } W' \xrightarrow{\ g'\ } Y'$$

*suppose that the vertical maps are weak equivalences and one of $f, g$ and one of $f', g'$ are cofibrations. Then the map*

$$\mathrm{colim}(X \xleftarrow{f} W \xrightarrow{g} Y) \longrightarrow \mathrm{colim}(X' \xleftarrow{f'} W' \xrightarrow{g'} Y')$$

*is a weak equivalence.*

*Proof* Dual to the proof of Corollary 3.2.15 using Proposition 3.6.17 and Theorem 3.6.13.                                              □

We now have the dual of Corollary 3.2.16. As was the case with that result, the statement that follows fits better thematically in the previous chapter, but the proof given here is much sleeker than the one that would only use the tools from the previous chapter. Recall the definition of the mapping cylinder $M_f$ from Definition 2.4.1.

**Corollary 3.6.20** *If $f_0, f_1 \colon X \to Y$ are homotopic, then there is a homotopy equivalence*

$$M_{f_0} \simeq M_{f_1}.$$

*Proof* Suppose $F: X \times I \to Y$ is a homotopy from $f_0$ to $f_1$. Consider the diagram

$$
\begin{array}{ccccc}
X \times I \times \{0\} & \longleftarrow & X \times \{0\} \times \{0\} & \xrightarrow{\;f_0\;} & Y \\
\downarrow & & \downarrow & & \downarrow{\scriptstyle 1_Y} \\
X \times I \times I & \longleftarrow & X \times I \times \{0\} & \xrightarrow{\;F\;} & Y \\
\uparrow & & \uparrow & & \uparrow{\scriptstyle 1_Y} \\
X \times I \times \{0\} & \longleftarrow & X \times \{1\} \times \{0\} & \xrightarrow{\;f_1\;} & Y
\end{array}
$$

where the unlabeled maps are all inclusions. The colimits of the rows are, from top to bottom, $M_{f_0}$, $M_F$, and $M_{f_1}$. These are equivalent to the homotopy pushouts since, in each row, the left horizontal arrow is a cofibration by Example 2.3.4 and Proposition 2.3.9. Since all of the vertical maps are homotopy equivalences, Theorem 3.6.13 says that the maps in the diagram $M_{f_0} \to M_F \leftarrow M_{f_1}$ are homotopy equivalences. $\qquad\square$

Recalling the definition of the homotopy cofiber, Definition 2.4.5, we immediately have the following, the proof of which is dual to that of Corollary 3.2.17.

**Corollary 3.6.21** *If $f_0, f_1 : X \to Y$ are homotopic, then*

$$
\mathrm{hocofiber}(f_0) \simeq \mathrm{hocofiber}_y(f_1).
$$

## 3.7 Arithmetic of homotopy cocartesian squares

We now turn to square diagrams and a comparison of the homotopy colimit of the subdiagram $\mathcal{S}_1$ to the last space of the square. That is, for any square

$$
\mathcal{S} = \quad
\begin{array}{ccc}
W & \longrightarrow & Y \\
\downarrow & & \downarrow \\
X & \longrightarrow & Z
\end{array}
$$

we have a canonical map

$$
b(\mathcal{S}): \mathrm{hocolim}(\mathcal{S}_1) = \mathrm{hocolim}(X \leftarrow W \to Y) \longrightarrow Z \tag{3.7.1}
$$

given by the composition of the canonical maps $\mathrm{hocolim}(\mathcal{S}_1) \to \mathrm{colim}(\mathcal{S}_1)$ (see the discussion following (3.6.1)) and $\mathrm{colim}(\mathcal{S}_1) \to Z$ (see (3.5.1) and the discussion around it). Analogously to Definition 3.3.1, we then have the following.

**Definition 3.7.1**    A square diagram

$$\mathcal{S} = \begin{array}{ccc} W & \longrightarrow & Y \\ \downarrow & & \downarrow \\ X & \longrightarrow & Z \end{array}$$

is

- *homotopy cocartesian* (or $\infty$-*cocartesian*, or a *homotopy pushout*) if $b(\mathcal{S})$ is a weak equivalence;
- *k-cocartesian*  if the map $b(\mathcal{S})$ is $k$-connected.

If a square is $k$-cocartesian, then it is clearly $j$-cocartesian for all $j \leq k$.

**Example 3.7.2**    Various examples from Section 3.6 can be restated in terms of homotopy cocartesian squares. For instance, Examples 3.6.6, 3.6.7, 3.6.8, 3.6.9, 3.6.10, and 3.6.12 say that the squares

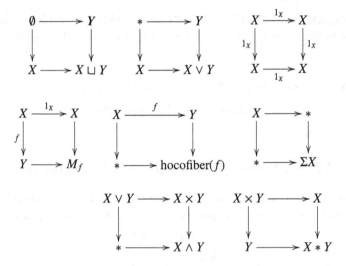

are homotopy cocartesian. Most of these results will be generalized to cubes in Section 5.8.                                                                                 □

We remark that several of the squares above are not commutative, and what we mean by them being homotopy cocartesian is that the homotopy colimit of the evident punctured square is homotopy equivalent to the "last" space in the square.

A common and useful square diagram is one that is given by a pair of open sets which cover a given space $Z$, so we treat this separately.

**Definition 3.7.3** We call a square

$$\mathcal{X} = \begin{array}{ccc} W & \longrightarrow & X \\ \downarrow & & \downarrow \\ Y & \longrightarrow & Z \end{array}$$

an *open pushout square* or the triple $(Z; X, Y)$ an *open triad* if $Z$ is the union of open sets $X$ and $Y$ along $W$.

**Proposition 3.7.4** *Every homotopy cocartesian square admits an equivalence from an open triad.*

*Proof* Given

$$\mathcal{S} = \begin{array}{ccc} W & \longrightarrow & X \\ \downarrow & & \downarrow \\ Y & \longrightarrow & Z \end{array}$$

write $\mathrm{hocolim}(Y \leftarrow W \rightarrow X) = Y \sqcup W \times I \sqcup X / \sim$, where $\sim$ is the evident equivalence relation. By assumption the canonical map $\mathrm{hocolim}(Y \leftarrow W \rightarrow X) \rightarrow Z$ is a weak equivalence. We may also assume $Z = \mathrm{hocolim}(Y \leftarrow W \rightarrow X)$. Let $U_Y = Y \sqcup W \times [0, 2/3) / \sim$ and $U_X = X \sqcup W \times (1/3, 1] / \sim$ be the evident open subspaces of $Z$. Then the obvious projections $U_X \cap U_Y = W \times (1/3, 2/3) \rightarrow W$, $U_Y \rightarrow Y$, $U_X \rightarrow X$ are homotopy equivalences, and we have a map of squares (a commutative diagram)

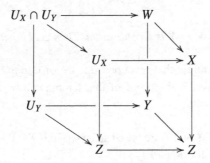

such that the horizontal maps are all homotopy equivalences.     □

Every open triad is itself homotopy cocartesian.

**Example 3.7.5** Suppose $U, V \subset X$ are open subsets such that $X = U \cup V$. The square of inclusions

$$
\begin{array}{ccc}
U \cap V & \longrightarrow & U \\
\downarrow & & \downarrow \\
V & \longrightarrow & X
\end{array}
$$

is homotopy cocartesian.

To see why this is true, let $X' = \mathrm{hocolim}(V \leftarrow U \cap V \to U)$ and let $P_U, P_V \subset X'$ be as in the proof of Theorem 3.6.13. Then $\{P_U, P_V\}$ forms an open cover for $X'$, and we have a map of diagrams

$$
\begin{array}{ccccc}
P_V & \longleftarrow & P_U \cap P_V & \longrightarrow & P_U \\
\downarrow & & \downarrow & & \downarrow \\
V & \longleftarrow & U \cap V & \longrightarrow & U
\end{array}
$$

where the vertical maps are given by projections, all of which are homotopy (and hence weak) equivalences. By Lemma 3.6.14, the induced map of colimits $X' = \mathrm{colim}(P_V \leftarrow P_U \cap P_V \to P_U) \to \mathrm{colim}(V \leftarrow U \cap V \to V) = X$ is a weak equivalence, and thus by definition the square in question is homotopy cocartesian. □

The following dual to Proposition 3.3.5 follows immediately from Proposition 3.6.17.

**Proposition 3.7.6**  *Suppose*

$$
S = \begin{array}{ccc}
W & \longrightarrow & Y \\
\downarrow & & \downarrow \\
X & \longrightarrow & Z
\end{array}
$$

*is cocartesian and $W \to Y$ is a cofibration. Then $S$ is homotopy cocartesian.*

Here is another example that is a consequence of Lemma 3.6.14. In fact, the same argument given in the proof of that lemma can be adapted to give the following useful result.

**Example 3.7.7**  If $X$ and $Y$ are based spaces, and if $X \vee Y \to X$ and $X \vee Y \to Y$ are the projections, the square

$$
\begin{array}{ccc}
X \vee Y & \longrightarrow & Y \\
\downarrow & & \downarrow \\
X & \longrightarrow & *
\end{array}
$$

is homotopy cocartesian.

To see this, note that Theorem 3.6.13 implies that the map induced by projection

$$\text{hocolim}(X \vee CY \leftarrow X \vee Y \rightarrow CX \vee Y) \longrightarrow \text{hocolim}(X \leftarrow X \vee Y \rightarrow Y)$$

is a weak equivalence. The inclusion $X \vee Y \rightarrow X \vee CY$ is a cofibration since $Y \rightarrow CY$ is. Now, by Proposition 3.6.17,

$$\text{hocolim}(X \vee CY \leftarrow X \vee Y \rightarrow CX \vee Y) \simeq \text{colim}(X \vee CY \leftarrow X \vee Y \rightarrow CX \vee Y),$$

and the latter space is evidently contractible since it is equal to $CX \vee CY$.

A generalization of this example to cubical diagrams can be found in Example 5.8.21. □

Although we are restricting most of our attention to pushouts and pullbacks of topological spaces, it is worth revisiting the Seifert–van Kampen Theorem, here Theorem 1.4.12, as it has a succinct statement in terms of pushouts. Recall that the notation $G *_K H$ stands for the amalgamated sum (see the discussion before Theorem 1.4.12).

**Theorem 3.7.8** (Seifert–van Kampen Theorem) *Suppose we have a homotopy cocartesian square of based spaces*

$$\begin{array}{ccc} W & \longrightarrow & A \\ \downarrow & & \downarrow \\ B & \longrightarrow & X \end{array}$$

*such that W, A, and B are path-connected. Then the square*

$$\begin{array}{ccc} \pi_1(W) & \longrightarrow & \pi_1(A) \\ \downarrow & & \downarrow \\ \pi_1(B) & \longrightarrow & \pi_1(X) \end{array}$$

*is a pushout of groups. That is, the canonical map*

$$\pi_1(A) *_{\pi_1(W)} \pi_1(B) \longrightarrow \pi_1(X)$$

*is an isomorphism.*

We now look at some "arithmetic" properties of homotopy pushouts. First, dual to Proposition 3.3.6, the pointwise union (coproduct) of homotopy cocartesian squares is homotopy cocartesian. The proof is dual, but the homotopy cocartesian part can also be proved using Theorem 3.7.18.

**Proposition 3.7.9**    *Suppose*

$$S_1 = \begin{array}{ccc} W_1 & \longrightarrow & Y_1 \\ \downarrow & & \downarrow \\ X_1 & \longrightarrow & Z_1 \end{array} \quad \text{and} \quad S_2 = \begin{array}{ccc} W_2 & \longrightarrow & Y_2 \\ \downarrow & & \downarrow \\ X_2 & \longrightarrow & Z_2 \end{array}$$

*are homotopy cocartesian. Then the square*

$$\begin{array}{ccc} W_1 \sqcup W_2 & \longrightarrow & Y_1 \sqcup Y_2 \\ \downarrow & & \downarrow \\ X_1 \sqcup X_2 & \longrightarrow & Z_1 \sqcup Z_2 \end{array}$$

*is homotopy cocartesian. For based spaces, if we replace $\sqcup$ with $\vee$, the result remains true. If $S_1$ and $S_2$ are $k_1$-cocartesian and $k_2$-cocartesian, respectively, then the squares of products and wedges are $\min\{k_1, k_2\}$-cocartesian.*

**Corollary 3.7.10** (Wedge commutes with homotopy pushout)    *If*

$$S = \begin{array}{ccc} W & \longrightarrow & Y \\ \downarrow & & \downarrow \\ X & \longrightarrow & Z \end{array}$$

*is a homotopy cocartesian square of based spaces and $V$ is any based space, then*

$$S \vee V = \begin{array}{ccc} W \vee V & \longrightarrow & Y \vee V \\ \downarrow & & \downarrow \\ X \vee V & \longrightarrow & Z \vee V \end{array}$$

*is homotopy cocartesian. Analogously, if $S$ is $k$-cocartesian, so is $S \vee V$.*

*Proof*    Dual to the proof of Proposition 3.3.6.                        □

One can think of the following dual of Corollary 3.3.8 as the distributive law of multiplication over addition.

**Proposition 3.7.11** (Products commute with homotopy pushouts)    *Suppose*

$$S = \begin{array}{ccc} W & \longrightarrow & Y \\ \downarrow & & \downarrow \\ X & \longrightarrow & Z \end{array}$$

*is homotopy cocartesian. Then for any space V, the square*

$$S \times V = \quad \begin{array}{ccc} W \times V & \longrightarrow & Y \times V \\ \downarrow & & \downarrow \\ X \times V & \longrightarrow & Z \times V \end{array}$$

*where all maps are induced from the previous diagram using the identity on the V coordinate, is homotopy cocartesian. If S is k-cocartesian, $S \times V$ is also k-cocartesian.*

*Proof*  This follows from the fact that there is a natural homeomorphism

$$\mathrm{hocolim}(X \times V \leftarrow W \times V \to Y \times V) \longrightarrow \big(\mathrm{hocolim}(X \leftarrow W \to Y)\big) \times V.$$

The k-cocartesian version is also straightforward. □

**Remark 3.7.12**  More generally, Proposition 3.7.11 is true for fiber products over a space. This will appear as Proposition 3.9.3. □

The next result, which is dual to Proposition 3.3.11, says that the homotopy pushout of a weak equivalence is a weak equivalence, and that any square with parallel maps which are weak equivalences is a homotopy pushout. Similarly for the k-cocartesian version. More precisely, we have the following.

**Proposition 3.7.13**  *Consider the square diagram*

$$\begin{array}{ccc} W & \longrightarrow & Y \\ \downarrow & & \downarrow \\ X & \longrightarrow & Z \end{array}$$

1. (a) *If the square is homotopy cocartesian and the map $W \to X$ is a homotopy (resp. weak) equivalence, then $Y \to Z$ is a homotopy (resp. weak) equivalence.*
   (b) *If both $W \to X$ and $Y \to Z$ are homotopy (resp. weak) equivalences, then the square is homotopy cocartesian.*
2. (a) *If $W \to X$ is k-connected and the square is k-cocartesian, then $Y \to Z$ is k-connected.*
   (b) *If $W \to X$ is $(k-1)$-connected and $Y \to Z$ is k-connected, then the square is k-cocartesian.*

*Proof*  Dual to the proof of Proposition 3.3.11, using Proposition 3.6.2. □

**Remark 3.7.14**   Analogously to the first part of Remark 3.3.12, if $Y \to Z$ is a weak equivalence and the square is homotopy cocartesian, then nothing in general can be said about the map $W \to X$. This happens due to the existence of "acyclic" spaces $X$ whose reduced homology groups vanish but which have non-trivial fundamental group. We borrow example 2.38 from [Hat02]. Let $X$ be the space obtained from $S^1 \vee S^1$ by attaching a pair of 2-cells via the words $a^5 b^{-3}$ and $b^3 (ab)^{-2}$ (i.e. these words give a formula for maps $S^1 \to S^1 \vee S^1$). The reduced homology groups of $X$ vanish, $\pi_1(X)$ is a non-trivial perfect group (a group whose abelianization is trivial), so $X$ is not weakly contractible. However, its suspension $\Sigma X$ is 1-connected by Example 2.6.13, and since $H_i(\Sigma X) \cong H_{i-1}(X)^7$ for $i \geq 1$ and $X$ is connected, the homology groups of $\Sigma X$ vanish. By Theorem 4.3.5 the unique map $\Sigma X \to *$ is a weak equivalence, so $\Sigma X$ is weakly contractible. Hence in the homotopy cocartesian square

the right-most map is a weak equivalence, whereas the map $X \to *$ is only 0-connected. For more on acyclic spaces, see [Dro72].

   The other parts of Remark 3.3.12 have valid analogs here, and we invite the reader to formulate them.                                                                □

An easy application of Proposition 3.7.13 is to show that the quotient of a space by a contractible subspace is equivalent to the original space.

**Corollary 3.7.15**   *Suppose $f: X \to Y$ is a map with $X$ contractible. Then the natural map $Y \to$ hocofiber$(f)$ is a weak equivalence. In particular, the quotient of a space $Y$ by a contractible subspace $X$ for which the inclusion $X \to Y$ is a cofibration is weakly equivalent to $Y$.*

*Proof*   This is not hard to see by considering the homotopy pushout square

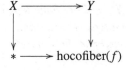

Since $X$ is contractible, the map $X \to *$ is a weak equivalence, by statement 1 of Proposition 3.7.13 the induced map $Y \to$ hocofiber$(f)$ is a weak equivalence.                                                                □

---

[7] This follows from Theorem 2.3.14 for the homotopy cofibration sequence $X \to * \to \Sigma X$ using that the reduced homology groups of a point are all zero.

The following is dual to Example 3.3.13. We already saw it in Lemma 2.4.19, giving rise to the Barratt–Puppe sequence (2.4.4).

**Example 3.7.16** Suppose $X \to Y \to C$ is a cofibration sequence. Then there is a weak equivalence

$$\Sigma X \simeq \text{hocofiber}(Y \to C).$$

To see this, we have a homotopy cocartesian square

$$
\begin{array}{ccc}
X & \longrightarrow & Y \\
\downarrow & & \downarrow \\
* & \longrightarrow & C
\end{array}
$$

which says that $\text{hocolim}(* \leftarrow X \to Y) \simeq C$. Consider the square

$$
\begin{array}{ccc}
Y & \xrightarrow{\ 1_Y\ } & Y \\
\downarrow & & \downarrow \\
\text{hocolim}(* \leftarrow X \to Y) & \longrightarrow & C
\end{array}
$$

where $Y \to \text{hocolim}(* \leftarrow X \to Y)$ is the inclusion of $Y$. This square is homotopy cocartesian by 1(b) of Proposition 3.7.13, since both horizontal arrows are weak equivalences. Since $\text{hocofiber}(Y \to \text{hocolim}(* \leftarrow X \to Y)) \simeq \Sigma X$ (easy to see from the definitions or from Theorem 3.7.18), we have a weak equivalence of vertical homotopy cofibers $\Sigma X \to \text{hocofiber}(Y \to C)$. □

The next example generalizes the previous one and is a dual to Example 3.3.14.

**Example 3.7.17** Suppose

$$
\begin{array}{ccc}
X & \longrightarrow & Y \\
\downarrow & & \downarrow \\
Z & \longrightarrow & W
\end{array}
$$

is a homotopy cocartesian square of based spaces. Then

$$\Sigma X \simeq \text{hocofiber}(Y \vee Z \to W).$$

The reasoning for this is the same as the above. The square

$$
\begin{array}{ccc}
Y \vee Z & \longrightarrow & Y \vee Z \\
\downarrow & & \downarrow \\
\text{hocolim}(Y \leftarrow X \to Z) & \longrightarrow & W
\end{array}
$$

(this is the homotopy pushout in based spaces), where the top map is the identity and the left vertical map is the wedge of the inclusion maps, is homotopy cocartesian since both horizontal arrows are equivalences. Taking vertical cofibers gives the desired weak equivalence since

$$\text{hocofiber}(Y \vee Z \to \text{hocolim}(Y \leftarrow X \to Z)) \simeq \Sigma X,$$

again from Theorem 3.7.18.                                        □

The following dual to Theorem 3.3.15 is a manifestation of the more general fact that homotopy colimits commute (see Proposition 8.5.5). It can be thought of as associativity of addition, dual to our remark following Theorem 3.3.15. Since the proof is exactly dual to that of Theorem 3.3.15, we leave it to the reader.

**Theorem 3.7.18**    *Suppose we have a commutative diagram*

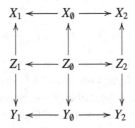

*Denote by* $\text{hocolim}(\mathcal{X})$, $\text{hocolim}(\mathcal{Z})$, *and* $\text{hocolim}(\mathcal{Y})$ *the homotopy pushouts of the first, second, and third rows of this diagram respectively, and by* $\text{hocolim}(C_1)$, $\text{hocolim}(C_0)$, *and* $\text{hocolim}(C_2)$ *the homotopy pushouts of the first, second, and third columns respectively. Then there is a natural homeomorphism*

$$\text{hocolim}(\text{hocolim}(\mathcal{X}) \leftarrow \text{hocolim}(\mathcal{Z}) \to \text{hocolim}(\mathcal{Y}))$$

$$\Big\downarrow \cong$$

$$\text{hocolim}(\text{hocolim}(C_1) \leftarrow \text{hocolim}(C_0) \to \text{hocolim}(C_2))$$

We now have the following corollary, the first part of which is dual to Corollary 3.3.16. The reader should compare this with Example 2.4.15. Related results also appeared as Examples 3.6.6, 3.6.9, 3.6.10, and 3.6.12. A generalization to cubical diagrams is given in Corollary 5.8.10 and for more general diagrams in Corollary 8.5.8.

**Corollary 3.7.19** *If* $Z_1 \leftarrow Z_0 \rightarrow Z_2$ *is a diagram of (based) spaces and V is a based space, we then have*

$$\text{hocolim}(V \vee Z_1 \leftarrow V \vee Z_0 \rightarrow V \vee Z_2) \cong V \vee \text{hocolim}(Z_1 \leftarrow Z_0 \rightarrow Z_2)$$
$$\text{hocolim}(V \times Z_1 \leftarrow V \times Z_0 \rightarrow V \times Z_2) \cong V \times \text{hocolim}(Z_1 \leftarrow Z_0 \rightarrow Z_2)$$
$$\text{hocolim}(V \wedge Z_1 \leftarrow V \wedge Z_0 \rightarrow V \wedge Z_2) \cong V \wedge \text{hocolim}(Z_1 \leftarrow Z_0 \rightarrow Z_2)$$
$$\text{hocolim}(V * Z_1 \leftarrow V * Z_0 \rightarrow V * Z_2) \cong V * \text{hocolim}(Z_1 \leftarrow Z_0 \rightarrow Z_2)$$

*In particular,*

$$\text{hocolim}(\Sigma Z_1 \leftarrow \Sigma Z_0 \rightarrow \Sigma Z_2) \cong \Sigma \text{hocolim}(Z_1 \leftarrow Z_0 \rightarrow Z_2).$$

*If* $Z_2 = *$ *is a point so that* $\text{hocolim}(Z_1 \leftarrow Z_0 \rightarrow *) \cong \text{hocofiber}(Z_0 \rightarrow Z_1)$, *then*

$$\text{hocofiber}(V \vee Z_0 \rightarrow V \vee Z_1) \cong CV \vee \text{hocofiber}(Z_0 \rightarrow Z_1)$$

*and*

$$\text{hocofiber}(V \wedge Z_0 \rightarrow V \wedge Z_1) \cong V \wedge \text{hocofiber}(Z_0 \rightarrow Z_1).$$

*Proof* The first item is straightforward. The second item is Proposition 3.7.11. The third uses the first two, Theorem 3.7.18 applied to the diagram

$$
\begin{array}{ccccc}
V \times Z_1 & \longleftarrow & V \times Z_0 & \longrightarrow & V \times Z_2 \\
\uparrow & & \uparrow & & \uparrow \\
V \vee Z_1 & \longleftarrow & V \vee Z_0 & \longrightarrow & V \vee Z_2 \\
\downarrow & & \downarrow & & \downarrow \\
* & \longleftarrow & * & \longrightarrow & *
\end{array}
$$

and Example 3.6.10. The fourth item uses the second, Example 3.6.6, and an application of Theorem 3.7.18, all applied to the diagram

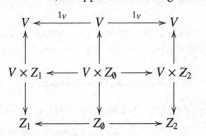

$$
\begin{array}{ccccc}
V & \xleftarrow{1_V} & V & \xrightarrow{1_V} & V \\
\uparrow & & \uparrow & & \uparrow \\
V \times Z_1 & \longleftarrow & V \times Z_0 & \longrightarrow & V \times Z_2 \\
\downarrow & & \downarrow & & \downarrow \\
Z_1 & \longleftarrow & Z_0 & \longrightarrow & Z_2
\end{array}
$$

The fifth item is a special case of the third, since suspension is the same as smash product with the circle. The statements about homotopy cofibers seem to require slightly different proofs. The first of them has already been established and is a matter of combining Example 2.4.10 with Example 2.4.12. For the second, we use Example 2.4.13 and apply all three of the aforementioned examples to the diagram

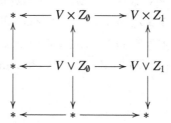

As promised in Example 2.4.13, we have the following result.

**Example 3.7.20** Let $X, Y, Z$ be based spaces, and $f: X \to Y$ a map. Theorem 3.7.18 gives a homeomorphism

$$\text{hocofiber}(f \times 1_Z) \cong \text{hocofiber}(f) \wedge Z_+.$$

To see why, consider the diagram

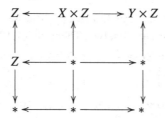

Taking homotopy colimits first along the rows from top to bottom yields $Z \times \text{hocofiber}(X \to Y)$ using the third item in Corollary 3.7.19, then $Z$, and finally $*$. Then taking the homotopy colimit of the result yields $\text{hocofiber}(X \to Y) \wedge Z_+$. Instead taking homotopy colimits along the columns first and then the homotopy colimit of the result yields $\text{hocofiber}(f \times 1_Z)$. □

Dual to Corollary 3.3.17, the square of homotopy cofibers of a map of homotopy cocartesian squares is itself a homotopy cocartesian square.

**Corollary 3.7.21** *There is a natural homeomorphism*

$$\text{hocolim}\left(\text{hocofiber}(Z_1 \to X_1) \leftarrow \text{hocofiber}(Z_0 \to X_0)\right.$$
$$\to \text{hocofiber}(Z_2 \to X_2)\big)$$
$$\cong \text{hocofiber}\left(\text{hocolim}(Z_1 \leftarrow Z_0 \to Z_2)\right.$$
$$\to \text{hocolim}(X_1 \leftarrow X_0 \to X_2)\big).$$

*Proof* Let $Y_S = *$ for all $S$ in the statement of Theorem 3.7.18. □

Theorem 3.7.18 relates the join and smash products in the following example.

**Proposition 3.7.22** *If $X$ and $Y$ are based spaces, then there is a weak equivalence*

$$X * Y \simeq \Sigma(X \wedge Y).$$

*Proof* Consider the diagram

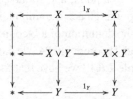

where the map $X \vee Y \to X \times Y$ is the inclusion of the wedge into the product, and the other maps are the obvious projections. The homotopy pushouts of the columns, from left to right, are weakly equivalent to $*$, $*$, and $X * Y$. The first of these is obvious, the second is Example 3.7.7, and the third is Example 3.6.12. Hence the homotopy pushout of the homotopy pushouts of the columns is homotopy equivalent to $X * Y$ by Example 3.6.7. By Theorem 3.7.18, this iterated homotopy pushout is homeomorphic to the homotopy pushout of the homotopy pushouts of the rows. From top to bottom, these are weakly equivalent to $*, X \wedge Y$, and $*$. The first and last statements follow from Example 3.6.7 and the second is Example 3.6.10. Thus the homotopy pushout of the homotopy pushouts of the rows is homotopy equivalent to $\text{hocolim}(* \leftarrow X \wedge Y \to *)$, which is weakly equivalent to $\Sigma(X \wedge Y)$ by Example 3.6.9. □

We will give an alternative proof of this equivalence in Example 5.8.17. Here is a consequence that we will find useful later.

**Proposition 3.7.23** (Connectivity of the join) *If $X$ is p-connected and $Y$ is q-connected, then $X * Y$ is $(p + q + 2)$-connected. Moreover, $X \wedge Y$ is $(p + q + 1)$-connected.*

*Proof* If one of $X$ or $Y$ is empty, then the join is the other space, and the result is still true; if they are both empty, the result is still true since it gives the connectivity of the empty set, which is $-2$. So we can assume $X$ or $Y$ is non-empty and can choose basepoints in each. By Proposition 3.7.22,

$$X * Y \simeq \Sigma(X \wedge Y).$$

Without loss of generality, using Theorem 1.3.8, we may assume $X$ and $Y$ are CW complexes, and by Theorem 2.6.26 we may additionally assume

$$X = * \cup \{\text{cells of dimension} \geq p + 1\},$$
$$Y = * \cup \{\text{cells of dimension} \geq q + 1\}.$$

Since $X$ and $Y$ are compactly generated, we may give a CW structure to the product $X \times Y$ using the products of the cells of $X$ and $Y$ (see [Hat02, Theorem A.6] for more on this point-set matter). To create the smash product from the product, we identify $X \vee Y$ to a point, and so the cells of the form $* \times e^n$ and $e^m \times *$ are identified to a point. Hence the minimal dimension of the cells in $X \wedge Y$ is $p + q + 2$. Using the definition of a $k$-connected pair, Definition 2.6.4 (with the pair $(X \wedge Y, *)$) and Theorem 2.6.24, we have that $X \wedge Y$ is $(p+q+1)$-connected. Then by Example 2.6.13 we have that $\Sigma(X \wedge Y) \simeq X * Y$ is $(p + q + 2)$-connected. □

We now have a partial dual of Proposition 3.3.18 that will be generalized later as Proposition 5.8.12.

**Proposition 3.7.24**   *Suppose the square*

$$\begin{array}{ccc} W & \longrightarrow & Y \\ \downarrow & & \downarrow \\ X & \longrightarrow & Z \end{array}$$

*is homotopy cocartesian (k-cocartesian). Then the map*

$$\text{hocofiber}(W \to X) \longrightarrow \text{hocofiber}(Y \to Z)$$

*is a weak equivalence (k-connected).*

**Remark 3.7.25**   The converse is not true in general. For instance, let $X$ be the space from Remark 3.7.14. The suspension $\Sigma X$ is weakly contractible, whereas $X$ is not. Consider the square

Then the map of vertical homotopy cofibers is a weak equivalence since $\Sigma X$ is weakly contractible, but the square is not homotopy cocartesian because $\text{hocolim}(* \leftarrow * \to X)$ has the homotopy type of $X$, which is not weakly contractible. □

*Proof* We prove the homotopy cocartesian version and leave it to the reader to provide the $k$-cocartesian version (by dualizing the $k$-cartesian part of Proposition 3.3.18).

Consider the square

$$\begin{array}{ccc} \operatorname{hocolim}(W \xleftarrow{1_W} W \to Y) & \longrightarrow & Y \\ \downarrow & & \downarrow \\ \operatorname{hocolim}(X \leftarrow W \to Y) & \longrightarrow & Z \end{array}$$

The top horizontal arrow (defined by the square below) is a weak equivalence by 1(b) of Proposition 3.7.13 applied to the square

$$\begin{array}{ccc} W & \xrightarrow{1_W} & W \\ \downarrow & & \downarrow \\ Y & \xrightarrow{1_Y} & Y \end{array}$$

By assumption the lower horizontal arrow is a homotopy equivalence. It suffices to show that the homotopy cofiber $C$ of the map

$$\operatorname{hocolim}(W \xleftarrow{1_W} W \to Y) \longrightarrow \operatorname{hocolim}(X \leftarrow W \to Y)$$

is homotopy equivalent to $\operatorname{hocofiber}(W \to X)$. To see this, consider the diagram

$$\begin{array}{ccccc} X & \longleftarrow & W & \longrightarrow & Y \\ \uparrow & & \uparrow{\scriptstyle 1_W} & & \uparrow \\ W & \xleftarrow{1_W} & W & \longrightarrow & Y \\ \downarrow & & \downarrow & & \downarrow \\ * & \longleftarrow & * & \longrightarrow & * \end{array}$$

The homotopy pushout of the homotopy pushout of the rows is the homotopy cofiber $C$ described above, using Example 3.6.7 on the middle row. By Theorem 3.7.18, it is homeomorphic to the homotopy pushout of the homotopy pushouts of the columns; the middle and right-most columns are weakly contractible by Example 3.6.7. But this latter iterated homotopy pushout is equivalent to $\operatorname{hocolim}(X \leftarrow W \to *)$, which is the homotopy cofiber of $W \to X$. $\qquad\square$

Here is a dual to Proposition 3.3.20. Its generalization to higher-dimensional cubes appears later as Proposition 5.8.14.

**Proposition 3.7.26** *In the diagram*

$$X_\emptyset \longrightarrow X_1 \longrightarrow X_{13}$$
$$\downarrow \qquad\qquad \downarrow \qquad\qquad \downarrow$$
$$X_2 \longrightarrow X_{12} \longrightarrow X_{123}$$

*let the left square be denoted by* $\boxed{1}$ *, the right by* $\boxed{2}$ *, and the outer square by* $\boxed{1\,2}$ *. Then*

1.  (a) *If* $\boxed{1}$ *and* $\boxed{2}$ *are homotopy cocartesian, then* $\boxed{1\,2}$ *is homotopy cocartesian.*
    (b) *If* $\boxed{1}$ *and* $\boxed{1\,2}$ *are homotopy cocartesian, then* $\boxed{2}$ *is homotopy cocartesian.*
2.  (a) *If* $\boxed{1}$ *and* $\boxed{2}$ *are k-cocartesian, then so is* $\boxed{1\,2}$ *.*
    (b) *If* $\boxed{1}$ *is* $(k-1)$*-cocartesian and* $\boxed{1\,2}$ *is k-cocartesian, then* $\boxed{2}$ *is k-cocartesian.*

**Remark 3.7.27** The third obvious statement is false in general. Let $X$ be the space from Remark 3.7.14, and consider the diagram

$$* \longrightarrow X \longrightarrow *$$
$$\downarrow \qquad\quad \downarrow \qquad\quad \downarrow$$
$$* \longrightarrow * \longrightarrow \Sigma X$$

The outer and right-most squares are homotopy cocartesian, but the left-most square is not, because $\mathrm{hocolim}(* \leftarrow * \to X) \simeq X$ is not contractible. $\qquad\square$

*Proof of Proposition 3.7.26* The proofs of parts 1(a) and 1(b) are dual to the proof of Proposition 3.3.20, using Proposition 3.7.13 on the top horizontal arrow in the diagram

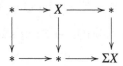

The observation that the square in this diagram is homotopy cocartesian follows from an application of Theorem 3.7.18, and then Proposition 3.7.13 again, to the diagram

$$
\begin{array}{ccccc}
X_{13} & \longleftarrow & X_1 & \longrightarrow & X_{12} \\
\uparrow & & \uparrow & & \uparrow {\scriptstyle 1_{x_{12}}} \\
X_\emptyset & \overset{1_{x_\emptyset}}{\longleftarrow} & X_\emptyset & \longrightarrow & X_{12} \\
\downarrow & & \downarrow & & \downarrow {\scriptstyle 1_{x_{12}}} \\
X_2 & \overset{1_{x_2}}{\longleftarrow} & X_2 & \longrightarrow & X_{12}
\end{array}
$$

The proofs of the other two statements now follow from the above along with Proposition 2.6.15. □

We close with the discussion dual to that of fibrant squares. Mirroring Definition 3.3.23, we have the following.

**Definition 3.7.28** We say a square

$$
\mathcal{S} = \begin{array}{ccc}
W & \longrightarrow & X \\
\downarrow & & \downarrow \\
Y & \longrightarrow & Z
\end{array}
$$

is *cofibrant* if

- $W \to X$ and $W \to Y$ are cofibrations;
- $\operatorname{colim}(Y \leftarrow W \to X) \to Z$ is a cofibration.

If in addition $Z = \operatorname{colim}(Y \leftarrow W \to X)$, we say $\mathcal{S}$ is a *cofibrant pushout square*.

We observe that, in particular, a cofibrant square satisfies

$$
\operatorname{hocolim}(Y \leftarrow W \to X) \simeq \operatorname{colim}(Y \leftarrow W \to X)
$$

by Proposition 3.6.17.

The proposition that follows is dual to Proposition 3.3.24, with a dual proof using the mapping cylinders from Definition 2.4.1.

**Proposition 3.7.29** *Every square admits a homotopy (and hence weak) equivalence from a cofibrant square. That is, given a square*

$$
\mathcal{S} = \begin{array}{ccc}
W & \longrightarrow & X \\
\downarrow & & \downarrow \\
Y & \longrightarrow & Z
\end{array}
$$

*there is a cofibrant square*

$$S' = \begin{array}{ccc} W' & \longrightarrow & X' \\ \downarrow & & \downarrow \\ Y' & \longrightarrow & Z' \end{array}$$

*and a map of squares (a commutative diagram in the shape of a 3-cube)*

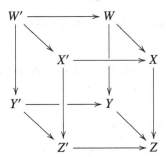

*such that* $W' \to W$, $X' \to X$, $Y' \to Y$, *and* $Z' \to Z$ *are a homotopy equivalences. Every homotopy cocartesian square admits an equivalence from a cofibrant pushout square.*

A square like $S'$ in the above result is called a *cofibrant replacement* of $S$.

*Proof*  We sketch the proof, which is dual to the proof of Proposition 3.3.24. Let $f: W \to X$ and $g: W \to Y$ be the maps in the diagram. Let $W' = W$, $X' = M_f$, and $Y' = M_y$. Let $h$: hocolim$(X \overset{f}{\leftarrow} W \overset{g}{\to} Y) \to Z$ be the canonical map and let $Z' = M_h$. The resulting square is cofibrant by a dual argument to that in the proof of Proposition 3.3.24. For $V = W, X, Y, Z$ the maps $V' \to V$ are the evident ones, as discussed after Definition 2.4.1 and in Proposition 2.4.4. If $S$ is homotopy cocartesian, we let $Z' = \text{colim}(M_f \leftarrow W \to M_g)$.        □

**Remark 3.7.30**  In analogy with Remark 3.3.25, $X$ is $k$-cocartesian if and only if $X'$ is. We leave it to the reader to fill in the details.        □

We will encounter cofibrant replacements in Theorem 5.8.23 and in Definition 8.4.12.

Let $X_1, X_2$ be based spaces. We have seen in Example 3.7.7 that the square

$$X = \begin{array}{ccc} X_1 \vee X_2 & \longrightarrow & X_1 \\ \downarrow & & \downarrow \\ X_2 & \longrightarrow & * \end{array}$$

is homotopy cocartesian. It turns out that a great many homotopy cocartesian squares are of this form. We thank John Klein for explaining to us the following result, due to Goodwillie, which appears in [KW12, Lemma 3.5]. The cubical generalization is Proposition 5.8.29.

**Proposition 3.7.31**    *Let*

$$\mathcal{X} = \begin{array}{ccc} X_{12} & \longrightarrow & X_1 \\ \downarrow & & \downarrow \\ X_2 & \longrightarrow & * \end{array}$$

*be a homotopy cocartesian square, and let $\Sigma\mathcal{X}$ be the square of suspensions. Then $\Sigma\mathcal{X}$ is weakly equivalent to*

$$\mathcal{X}' = \begin{array}{ccc} \Sigma X_1 \vee \Sigma X_2 & \longrightarrow & \Sigma X_1 \\ \downarrow & & \downarrow \\ \Sigma X_2 & \longrightarrow & * \end{array}$$

*Proof*   This proof is adapted from [KW12, Lemma 3.5]. Let $X_\emptyset = *$. For each $S \subset \{1, 2\}$, define a map $\Sigma X_S \to \bigvee_{i \in S} \Sigma X_i$ as follows. First use the pinch map $\Sigma X_S \to \Sigma X_S \vee \Sigma X_S$ (collapse the middle of the suspension to a point), and then map the $i$th copy of $\Sigma X_S$ to $\Sigma X_i$ if $i \in S$ by the given maps, or to a point if $i \notin S$. This gives a map of squares from $\Sigma\mathcal{X}$ to $\mathcal{X}'$.

Clearly the maps $\Sigma X_S \to \bigvee_{i \in S} \Sigma X_i$ are homotopic to the identity if $|S| \leq 1$. We are then left to show that the map

$$a \colon \Sigma X_{12} \to \Sigma X_1 \vee \Sigma X_2$$

is a weak equivalence. By Proposition 3.7.29 we may also assume $X_{12} \to X_1$ and $X_{12} \to X_2$ are cofibrations. Since $\mathcal{X}$ is homotopy cocartesian, we have a weak equivalence $X_1/X_{12} \simeq \text{hocofiber}(X_{12} \to X_1) \to \text{hocofiber}(X_2 \to *) \simeq \Sigma X_2$ by Proposition 3.7.24, and similarly a weak equivalence $X_2/X_{12} \to \Sigma X_1$.

By the Barratt–Puppe sequence (2.4.4), we have maps $X_i/X_{12} \to \Sigma X_{12}$ for $i = 1, 2$, and so have an induced map

$$b \colon X_2/X_{12} \vee X_1/X_{12} \longrightarrow \Sigma X_{12} \vee \Sigma X_{12} \longrightarrow \Sigma X_{12},$$

where the last map is the fold map. The compositions $a \circ b$ and $b \circ a$ are homotopic to the identities, and the result follows.     □

# 3.8 Total homotopy cofibers

For

$$S = \begin{array}{ccc} W & \longrightarrow & Y \\ \downarrow & & \downarrow \\ X & \longrightarrow & Z \end{array}$$

the connectivity of the map

$$b(S)\colon \mathrm{hocolim}(X \longleftarrow W \to Y) \to Z$$

measures the extent to which $S$ is a homotopy cocartesian square. Hence it is most natural to ask about the connectivity of the homotopy fiber of this map when asking how cocartesian a square is. However, it is usually easier to compute the homotopy cofiber of $b(S)$, which provides partial information on the connectivity of $b(S)$. In analogy with Definition 3.4.1, we thus have the following.

**Definition 3.8.1**    Define the *total (homotopy) cofiber* of the square $S$ as

$$\mathrm{tcofiber}(S) = \mathrm{hocofiber}(b(S)).$$

Proposition 3.7.13 says that if the square is homotopy cocartesian, then $Y \to Z$ has the same connectivity as the map $W \to X$; that is, it says that connectivity is preserved under pushouts. In fact, the map

$$\mathrm{hocofiber}(W \to X) \to \mathrm{hocofiber}(Y \to Z)$$

is a weak equivalence. To see this, we will interpret total homotopy cofiber as an *iterated homotopy cofiber*, just as we did with total homotopy fibers.

**Proposition 3.8.2** (Total homotopy cofiber as iterated homotopy cofiber)    *Let $S$ be the square*

$$\begin{array}{ccc} W & \longrightarrow & Y \\ \downarrow & & \downarrow \\ X & \longrightarrow & Z \end{array} \tag{3.8.1}$$

*Then*

$$\mathrm{tcofiber}(S) \simeq \mathrm{hocofiber}\Big(\mathrm{hocofiber}(W \to X) \to \mathrm{hocofiber}(Y \to Z)\Big).$$

*Moreover,*

$$\mathrm{tcofiber}(S) \simeq \Big(\mathrm{hocofiber}(W \to Y) \to \mathrm{hocofiber}(X \to Z)\Big).$$

*Proof* Dual to the proof of Proposition 3.4.3. □

**Remarks 3.8.3**

1. In analogy with Remark 3.4.4, when we take (homotopy) cofibers of (homotopy) cofibers, the result will be called the "iterated (homotopy) cofiber".
   It is not hard to see that the two iterated homotopy cofibers in the above result are in fact homeomorphic.
2. From the above result, along with Proposition 3.7.24, a square is homotopy cocartesian if and only if its total homotopy cofiber is weakly contractible. □

The following is dual to Example 3.4.5 and was already encountered in the discussion of the Barratt–Puppe sequence (Lemma 2.4.19).

**Example 3.8.4** Suppose $X \to Y \to C$ is a homotopy cofiber sequence. The homotopy cofiber of the canonical map $C \longrightarrow \text{hocofiber}(Y \to C)$ (i.e. what can be thought of as the map $C \to \Sigma X$ by Example 3.7.16) is weakly equivalent to $\Sigma Y$. To see this, consider the square

$$
\begin{array}{ccc}
* & \longrightarrow & Y \\
\downarrow & & \downarrow \\
C & \xrightarrow{1_C} & C
\end{array}
$$

Computing its total homotopy cofiber in two ways using Proposition 3.8.2 gives the desired result. □

The following is dual to Proposition 3.4.7.

**Proposition 3.8.5** *For a sequence* $X \to Y \to Z$ *of maps, we have a cofibration sequence*

$$\text{hocofiber}(X \to Y) \longrightarrow \text{hocofiber}(X \to Z) \longrightarrow \text{hocofiber}(Y \to Z).$$

*Proof* Consider the square

$$
\mathcal{X} = \quad
\begin{array}{ccc}
X & \xrightarrow{1_X} & X \\
\downarrow & & \downarrow \\
Y & \longrightarrow & Z
\end{array}
$$

where the right map is the composite $X \to Y \to Z$. Comparing the total cofiber by computing it iteratively using Proposition 3.8.2 in the horizontal and vertical directions, we get

$$\mathrm{tcofiber}(\mathcal{X}) \simeq \mathrm{hocofiber}(Y \to Z)$$

and

$$\mathrm{tcofiber}(\mathcal{X}) \simeq \mathrm{hocofiber}\left(\mathrm{hocofiber}(X \to Y) \longrightarrow \mathrm{hocofiber}(X \to Z)\right). \qquad \square$$

The following are generalizations of Proposition 3.4.8 and Proposition 3.4.9.

**Proposition 3.8.6** *Suppose $f\colon X \to Y$ and $g\colon Y \to X$ are maps such that $g \circ f$ is the identity. Then*

$$\Sigma \, \mathrm{hocofiber}(f) \simeq \mathrm{hocofiber}(g).$$

*Proof* Compare the vertical, horizontal, and total cofibers of the square

$$\begin{array}{ccc} X & \xrightarrow{\ f\ } & Y \\ {\scriptstyle 1_X}\downarrow & & \downarrow{\scriptstyle g} \\ X & \xrightarrow{\ 1_X\ } & X \end{array}$$

using Proposition 3.8.2. $\qquad \square$

Generalizing this further, we have the following.

**Proposition 3.8.7** *Let $\mathcal{X}$ be the diagram*

$$\begin{array}{ccc} X_\emptyset & \longrightarrow & X_1 \\ \downarrow & & \downarrow \\ X_2 & \longrightarrow & X_{12} \end{array}$$

*and suppose there are maps $X_{S \cup i} \to X_S$ for $S \subset \{1,2\}$ and $i \notin S$ such that the composition $X_S \to X_{S \cup i} \to X_S$ is the identity. Then*

$$\Sigma^2 \, \mathrm{tcofiber}(\mathcal{X}) \simeq \mathrm{tcofiber}(\mathcal{X}'),$$

*where $\mathcal{X}'$ is the square*

$$\begin{array}{ccc} X_{12} & \longrightarrow & X_2 \\ \downarrow & & \downarrow \\ X_1 & \longrightarrow & X_\emptyset \end{array}$$

*Proof* Dual to the proof of Proposition 3.4.9. $\qquad \square$

Just as we defined the homotopy groups of a map and of a square diagram, we can also define the homology groups of a map and of a square diagram in terms of the total homotopy cofiber. Here a dimension shift is not required, as it was in Definition 3.4.10, since a $k$-connected map has $k$-connected cofiber (even if its domain is empty).

To start, for a map $f\colon X \to Y$, we can define

$$H_k(f) = H_k(M_f, j(X)), \tag{3.8.2}$$

where $j$ is the map from Proposition 2.4.4. Now let $S$ be the square

$$
\begin{array}{ccc}
W & \longrightarrow & Y \\
{\scriptstyle f_1}\big\downarrow & & \big\downarrow{\scriptstyle f_2} \\
X & \longrightarrow & Z
\end{array}
$$

**Definition 3.8.8**  For $i \geq 0$, define

$$H_i(S) = H_i(\text{tcofiber}\, S).$$

Analogous to Proposition 3.4.11, we have the following.

**Proposition 3.8.9** (Homology long exact sequence of a square)  *There is a long exact sequence of homology groups*

$$\cdots \to H_i(f_1) \to H_i(f_2) \to H_i(\text{tcofiber}\, S) \to H_{i-1}(f_1) \to \cdots.$$

## 3.9 Algebra of homotopy cartesian and cocartesian squares

We now present several "algebraic" examples in which various kinds of squares interact. The homotopy pullback of $X \to Z \leftarrow Y$ is the homotopy invariant product of $X$ with $Y$ over $Z$, and the homotopy pushout of $X \leftarrow W \to Y$ is the homotopy invariant coproduct (sum) of $X$ with $Y$ under $W$, so combining them is something like "doing algebra". The first result is related to Proposition 3.3.9 and says that exponentiating a certain kind of homotopy pushout results in a homotopy pullback (i.e. exponentiation takes sums to products). This will be generalized to cubical diagrams in Proposition 5.10.1, and to arbitrary diagrams in Proposition 8.5.4.

**Proposition 3.9.1**  *Suppose in the square*

$$
S = \quad
\begin{array}{ccc}
W & \longrightarrow & Y \\
\big\downarrow & & \big\downarrow \\
X & \longrightarrow & Z
\end{array}
$$

*the map* $\mathrm{hocolim}(X \leftarrow W \rightarrow Y) \rightarrow Z$ *is a homotopy equivalence (so in particular $\mathcal{S}$ is homotopy cocartesian). Then, for any space $V$, the square*

$$V^{\mathcal{S}} = \mathrm{Map}(\mathcal{S}, V) = \begin{array}{ccc} V^Z & \longrightarrow & V^Y \\ \downarrow & & \downarrow \\ V^X & \longrightarrow & V^W \end{array}$$

*is homotopy cartesian.*

**Remark 3.9.2** As in Proposition 3.3.9, this result is not true if $\mathcal{S}$ is just required to be homotopy cocartesian. Again let $W$ be the "double comb space" from Remark 1.3.6. Then the square

is homotopy cocartesian, but letting $V = W$ and taking the square of mapping spaces as in the statement of the result, the three corners other than the upper-left are homeomorphic to $W$, but the upper-left is not path-connected.

A way to state the result in terms of homotopy cartesian squares (i.e. in terms of weak equivalences) would have been to require that $\mathrm{hocolim}(X \leftarrow W \rightarrow Y)$ and $Z$ be CW complexes (see the end of Remark 1.3.6). □

*Proof of Proposition 3.9.1* If $A \rightarrow B$ is a homotopy equivalence, then the induced map $V^B \rightarrow V^A$ is a homotopy equivalence by Proposition 1.3.2, so by Proposition 3.7.29 we may also assume $\mathcal{S}$ is of the form

$$\begin{array}{ccc} X_\emptyset & \longrightarrow & X_2 \\ \downarrow & & \downarrow \\ X_1 & \longrightarrow & X_1 \cup_{X_\emptyset} X_2 \end{array}$$

where $X_\emptyset \rightarrow X_i$ are cofibrations for $i = 1, 2$. By inspection, the square

$$\begin{array}{ccc} \mathrm{Map}(X_1 \cup_{X_\emptyset} X_2, V) & \longrightarrow & \mathrm{Map}(X_2, V) \\ \downarrow & & \downarrow \\ \mathrm{Map}(X_1, V) & \longrightarrow & \mathrm{Map}(X_\emptyset, V) \end{array}$$

is cartesian (maps defined on a union of two sets are precisely those pairs of maps, one for each of the sets, which agree on the intersection). By Proposition 2.5.1, the map $\mathrm{Map}(X_2, V) \rightarrow \mathrm{Map}(X_\emptyset, V)$ is a fibration, and hence by Proposition 3.3.5 the square $\mathrm{Map}(\mathcal{S}, V)$ is homotopy cartesian. □

The following result follows directly from Proposition 3.5.14 since homotopy pullback and homotopy pushout can be thought of as ordinary pullback and pushout (Propostions 3.2.5 and 3.6.5). The result says that exponentiation takes (homotopy amalgamated) sums to (homotopy fiber) products. Generalizations of this result can be found in Propositions 5.10.1 and 8.5.4.

**Proposition 3.9.3** *Let*

$$S = \begin{array}{ccc} X_\emptyset & \longrightarrow & X_1 \\ \downarrow & & \downarrow \\ X_2 & \longrightarrow & X_{12} \end{array}$$

*Let $Y, Z$ be spaces, $Z \to Y$ a map, and suppose there is a map $X_S \to Y$ for each $S$ such that the resulting diagram commutes. Then we have a natural homeomorphism*

$$\Phi \colon \operatorname{hocolim}(X_1 \times_Y Z \leftarrow X_\emptyset \times_Y Z \to X_2 \times_Y Z)$$
$$\longrightarrow \operatorname{hocolim}(X_1 \leftarrow X_\emptyset \to X_2) \times_Y Z.$$

*Hence the square*

$$\begin{array}{ccc} X_\emptyset \times_Y Z & \longrightarrow & X_1 \times_Y Z \\ \downarrow & & \downarrow \\ X_2 \times_Y Z & \longrightarrow & X_{12} \times_Y Z \end{array}$$

*is homotopy cocartesian if $S$ is homotopy cocartesian, and is k-cocartesian if $S$ is k-cocartesian.*

*Proof* Define

$$\phi \colon X_1 \times_Y Z \sqcup X_\emptyset \times_Y Z \times I \sqcup X_2 \times_Y Z \longrightarrow (X_1 \sqcup X_\emptyset \times I \sqcup X_2) \times_Y Z$$

by $\phi(x_i, z) = (x_i, z)$ for $x_i \in X_i$, $i = 1, 2$, and $\phi(((x_0, z), t)) = ((x_0, t), z)$ for $x_0 \in X_\emptyset$. It is straightforward to check that this gives rise to a map of the quotient spaces which define the domain and codomain of $\Phi$ in the statement and that this map has a continuous inverse. $\qquad\square$

More generally, the above result is true if we replace $X_S \times_Y Z$ with $\operatorname{holim}(X_S \to Y \leftarrow Z)$ for each $S$. See the remarks following Theorem 5.10.10.

If in the statement above we take $Y = *$, then we note that we obtain Proposition 3.7.11 as a corollary. If instead we take $Z = *$ we also have the following corollary, which will be generalized in Proposition 5.10.2 for cubical diagrams and in Proposition 8.5.9 for arbitrary ones.

**Corollary 3.9.4**    *Suppose the square*

$$
\begin{array}{ccc}
X_\emptyset & \longrightarrow & X_1 \\
\downarrow & & \downarrow \\
X_2 & \longrightarrow & X_{12}
\end{array}
$$

*is homotopy cocartesian and there is a map $X_S \to Y$ for each $S$ such that the resulting diagram commutes. Then, for any $y \in Y$, the square of homotopy fibers*

$$
\begin{array}{ccc}
\mathrm{hofiber}_y(X_\emptyset \to Y) & \longrightarrow & \mathrm{hofiber}_y(X_1 \to Y) \\
\downarrow & & \downarrow \\
\mathrm{hofiber}_y(X_2 \to Y) & \longrightarrow & \mathrm{hofiber}_y(X_{12} \to Y)
\end{array}
$$

*is homotopy cocartesian. That is, there is a weak equivalence*

$$
\mathrm{holim}\left(\mathrm{hocolim}(X_1 \leftarrow X_0 \to X_2) \longrightarrow Y \longleftarrow *\right)
$$
$$
\simeq \mathrm{hocolim}\left(\mathrm{holim}(X_1 \to Y \leftarrow *) \longleftarrow \mathrm{holim}(X_0 \to Y \leftarrow *)\right.
$$
$$
\left. \longrightarrow \mathrm{holim}(X_2 \to Y \leftarrow *)\right)
$$

*Moreover, if we replace "homotopy pushout" with "k-cocartesian" in the original square, the square of homotopy fibers is then k-cocartesian.*

This result says that "factoring out" the space $Y$ preserves the quality of being homotopy cocartesian, which we think of as the distributive law of multiplication (homotopy pullback) over addition (homotopy pushout). As we mentioned above, both this and Proposition 3.7.11 are special cases of Proposition 3.9.3, but we can also see Proposition 3.7.11 as a special case of Corollary 3.9.4 if we let $X_S = Z_S \times Y$ for all $S$ and the maps in the diagram $S \mapsto X_S$ are the identity on the $Y$-coordinate.

*Proof*    For the homotopy cocartesian version, let $y \in Y$ denote the basepoint, and let $f_i \colon X_\emptyset \to X_i$ be the evident maps. By definition,

$$
\mathrm{hocolim}\left(\mathrm{hofiber}_y(X_2 \to Y) \leftarrow \mathrm{hofiber}_y(X_\emptyset \to Y) \to \mathrm{hofiber}_y(X_1 \to Y)\right)
$$

is equal to the quotient space

$$
\mathrm{hofiber}_y(X_2 \to Y) \cup \mathrm{hofiber}_y(X_\emptyset \to Y) \times I \cup \mathrm{hofiber}_y(X_1 \to Y)/\sim
$$

where $(x_0, \gamma, 0) \sim (f_2(x_0), \gamma)$ and $(x_0, \gamma, 1) \sim (f_1(x_0), \gamma)$. Clearly we have a natural homeomorphism $\text{hofiber}_y(X_0 \to Y) \times I \cong \text{hofiber}_y(X_0 \times I \to Y)$ where the map $X_0 \times I \to Y$ is the projection to $X_0$ followed by the map $X_0 \to Y$. Using this it is clear by inspection that the quotient space in question is, by definition, the homotopy fiber over $y$ of the map

$$\text{hocolim}(X_2 \leftarrow X_0 \to X_1) \longrightarrow Y,$$

which is homotopy equivalent to $\text{hofiber}_y(X_{12} \to Y)$ since the square was assumed to be homotopy cocartesian.

For the $k$-cocartesian statement, let $P = \text{hocolim}(X_2 \leftarrow X_0 \to X_1)$. Then by the above, the square

$$
\begin{array}{ccc}
\text{hofiber}_y(X_0 \to Y) & \longrightarrow & \text{hofiber}_y(X_1 \to Y) \\
\downarrow & & \downarrow \\
\text{hofiber}_y(X_2 \to Y) & \longrightarrow & \text{hofiber}_y(P \to Y)
\end{array}
$$

is a homotopy pushout. By assumption, the canonical map $P \to X_{12}$ is $k$-connected, and it follows from the long exact sequence of a fibration that the induced map $\text{hofiber}_y(P \to Y) \to \text{hofiber}_y(X_{12} \to Y)$ is also $k$-connected, hence the square in question is $k$-cocartesian. $\qquad\square$

The next few examples are applications of Corollary 3.9.4.

**Example 3.9.5**  For based spaces $X$ and $Y$ and for

$$f \colon X \vee Y \longrightarrow X \times Y$$

the inclusion of the wedge into the product, we have

$$\text{hofiber}(f) \simeq \Omega X * \Omega Y.$$

This is because the square

$$
\begin{array}{ccc}
* & \longrightarrow & X \\
\downarrow & & \downarrow \\
Y & \longrightarrow & X \vee Y
\end{array}
$$

is a homotopy pushout as seen in Example 3.6.6 (it is a pushout and either of the maps $* \to X$ and $* \to Y$ is a cofibration, as we assume all spaces are well-pointed). We then apply Corollary 3.9.4 by taking fibers everywhere over $X \times Y$, and obtain a homotopy pushout

$$\begin{array}{ccc} \Omega(X \times Y) & \longrightarrow & \Omega Y \\ \downarrow & & \downarrow \\ \Omega X & \longrightarrow & \text{hofiber}(X \vee Y \to X \times Y) \end{array}$$

That hofiber$(X \to X \times Y) \simeq \Omega Y$ (and hofiber$(Y \to X \times Y) \simeq \Omega X$) is Example 3.4.6.                                                                                            □

We can say a bit more about Example 3.9.5 upon taking loop spaces.

**Proposition 3.9.6**   *There is a homotopy equivalence*

$$\Omega(X \vee Y) \simeq \Omega(X \times Y) \times \Omega(\Omega X * \Omega Y)$$

*induced by a section of the inclusion map*

$$\Omega(X \vee Y) \longrightarrow \Omega(X \times Y).$$

*Proof*   Example 3.9.5 establishes a fibration sequence

$$\Omega(\Omega X * \Omega Y) \to \Omega(X \vee Y) \to \Omega(X \times Y)$$

after applying $\Omega$ (and using Corollary 2.1.12). Denote a point in $\Omega(X \times Y)$ by a pair $(\gamma_X, \gamma_Y)$, and define a map $s \colon \Omega(X \times Y) \to \Omega(X \vee Y)$ by

$$(\gamma_X, \gamma_Y) \longmapsto \gamma_X * \gamma_Y,$$

where

$$\gamma_X * \gamma_Y(t) = \begin{cases} \gamma_X(2t), & \text{if } 0 \le t \le 1/2; \\ \gamma_Y(2t-1), & \text{if } 1/2 \le t \le 1. \end{cases}$$

The composite map

$$\Omega(X \times Y) \longrightarrow \Omega(X \vee Y) \to \Omega(X \times Y)$$

is the identity, and hence $s$ is a section of the inclusion $\Omega(X \vee Y) \to \Omega(X \times Y)$. The fibration $\Omega(\Omega X * \Omega Y) \to \Omega(X \vee Y) \to \Omega(X \times Y)$ is a principal fibration with a section, and hence we have the desired splitting

$$\Omega(X \vee Y) \simeq \Omega(X \times Y) \times \Omega(\Omega X * \Omega Y).$$

See [Hat02, p. 412] for the definition of principal fibration and [Hat02, Exercise 22, Chapter 4.3] for the fact that principal fibrations with a section give rise to the splitting we describe here.                                                              □

**Theorem 3.9.7** (Ganea's Theorem)   *If $F \to X \to Y$ is a fibration sequence, then*

$$\mathrm{hofiber}(\mathrm{hocofiber}(F \to X) \to Y) \simeq F * \Omega Y.$$

*Proof*   Consider the homotopy cocartesian square

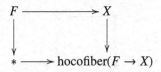

Now take homotopy fibers everywhere over $Y$ to obtain by Corollary 3.9.4 another homotopy cocartesian square

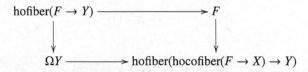

The composed map $F \to Y$ is null-homotopic (it is constant) and by Example 2.2.10 it is homotopy equivalent (in fact equal to in this case) $F \times \Omega Y$. Thus we have a homotopy cocartesian square

$$
\begin{array}{ccc}
F \times \Omega Y & \longrightarrow & F \\
\downarrow & & \downarrow \\
\Omega Y & \longrightarrow & \mathrm{hofiber}(\mathrm{hocofiber}(F \to X) \to Y)
\end{array}
$$

It is not hard to check that the maps from $F \times \Omega Y$ to $F$ and $\Omega Y$ are the projections, and by Example 3.6.12 $\mathrm{hofiber}(\mathrm{hocofiber}(F \to X) \to Y)$ admits a weak equivalence from the join $F * \Omega Y$.     □

**Example 3.9.8**   If $f\colon X \vee X \to X$ is the fold map which sends each copy of $X$ in the wedge to $X$ by the identity, then

$$\mathrm{hofiber}(f) \simeq \Sigma \Omega X.$$

This follows from Corollary 3.9.4 by fibering the homotopy cocartesian square

$$
\begin{array}{ccc}
* & \longrightarrow & X \\
\downarrow & & \downarrow \\
X & \longrightarrow & X \vee X
\end{array}
$$

over $X$ and using Example 3.6.9.     □

**Example 3.9.9**   If $f: X \vee Y \to Y$ is the map that sends $X$ to the wedge point and is the identity on $Y$, then

$$\mathrm{hofiber}(f) \simeq X \wedge (\Omega Y)_+.$$

To see this, consider the homotopy cocartesian square

Fibering over $Y$ everywhere, using both $f$ and the constant map at the basepoint $X \to Y$, an application of Corollary 3.9.4 gives a homotopy cocartesian square

$$
\begin{array}{ccc}
\Omega Y & \longrightarrow & X \times \Omega Y \\
\downarrow & & \downarrow \\
* & \longrightarrow & \mathrm{hofiber}(X \vee Y \to Y)
\end{array}
$$

and hence a weak equivalence

$$X \wedge (\Omega Y)_+ = (\Omega Y \times X)/(\Omega Y \times *) \simeq \mathrm{hofiber}(X \vee Y \to Y). \qquad \square$$

One application of Proposition 3.4.8 is

**Example 3.9.10**   There is a weak equivalence

$$\mathrm{hofiber}(Y \to X \vee Y) \simeq \Omega(X \wedge \Omega Y_+).$$

This follows from Proposition 3.4.8 and Example 3.9.9. $\qquad \square$

We can also use Corollary 3.9.4 to analyze the map in Lemma 2.6.25.

**Example 3.9.11**   There is a weak equivalence, induced by a map of homotopy cofibers,

$$S^n \wedge \Omega(X \cup e^n)_+ \xrightarrow{\simeq} \Sigma\,\mathrm{hofiber}(X \to X \cup e^n).$$

Alternatively, suppose $Y$ is a CW complex and $y \in Y$ is in the interior of an $n$-cell. Then there is a weak equivalence

$$S^n \wedge \Omega Y_+ \xrightarrow{\simeq} \Sigma\,\mathrm{hofiber}(Y - \{y\} \to Y).$$

This is clear by considering the homotopy pushout

$$
\begin{array}{ccc}
\partial e^n & \longrightarrow & e^n \\
\downarrow & & \downarrow \\
X & \longrightarrow & X \cup e^n
\end{array}
$$

Then taking fibers over $X \cup e^n$ and using Example 2.2.10 yields a homotopy pushout

$$
\begin{array}{ccc}
\partial e^n \times \Omega(X \cup e^n) & \longrightarrow & e^n \times \Omega(X \cup e^n) \\
\downarrow & & \downarrow \\
\mathrm{hofiber}(X \to X \cup e^n) & \longrightarrow & *
\end{array}
$$

which, upon taking horizontal homotopy cofibers, produces a weak equivalence

$$
S^n \wedge \Omega(X \cup e^n)_+ \longrightarrow \Sigma\,\mathrm{hofiber}(X \to X \cup e^n)
$$

by Proposition 3.7.13. $\qquad\Box$

The following consequence of Corollary 3.9.4 will be generalized to cubical diagrams in Proposition 5.10.4.

**Proposition 3.9.12** *Suppose*

$$
\mathcal{X} = \begin{array}{ccc}
X_\emptyset & \longrightarrow & X_1 \\
\downarrow & & \downarrow \\
X_2 & \longrightarrow & X_{12}
\end{array}
$$

*is a homotopy pullback, and let* $D = \mathrm{hocolim}(X_1 \leftarrow X_\emptyset \to X_2)$. *Then the homotopy fiber of the canonical map* $b(\mathcal{X}) \colon D \to X_{12}$ *is*

$$
\mathrm{hofiber}(X_1 \to X_{12}) * \mathrm{hofiber}(X_2 \to X_{12}).
$$

*Proof* Fiber the homotopy pushout square

$$
\begin{array}{ccc}
X_\emptyset & \longrightarrow & X_1 \\
\downarrow & & \downarrow \\
X_2 & \longrightarrow & D
\end{array}
$$

everywhere over $X_{12}$. By Corollary 3.9.4, the square of fibers

$$\begin{array}{ccc}
\text{hofiber}(X_\emptyset \to X_{12}) & \longrightarrow & \text{hofiber}(X_1 \to X_{12}) \\
\downarrow & & \downarrow \\
\text{hofiber}(X_2 \to X_{12}) & \longrightarrow & \text{hofiber}(D \to X_{12})
\end{array}$$

is also a homotopy pushout. Since $X_\emptyset \simeq \text{holim}(X_1 \to X_{12} \leftarrow X_2)$, fibering over $X_{12}$ yields an equivalence

$$\text{hofiber}(X_\emptyset \to X_{12}) \simeq \text{hofiber}(X_1 \to X_{12}) \times \text{hofiber}(X_2 \to X_{12}).$$

(This step requires an application of Theorem 3.3.15 which we leave to the reader.) That is, we have a homotopy pushout

$$\begin{array}{ccc}
\text{hofiber}(X_1 \to X_{12}) \times \text{hofiber}(X_2 \to X_{12}) & \longrightarrow & \text{hofiber}(X_1 \to X_{12}) \\
\downarrow & & \downarrow \\
\text{hofiber}(X_2 \to X_{12}) & \longrightarrow & \text{hofiber}(D \to X_{12})
\end{array}$$

and, by Example 3.6.12, a weak equivalence

$$\text{hofiber}(D \to X_{12}) \simeq \text{hofiber}(X_1 \to X_{12}) * \text{hofiber}(X_2 \to X_{12}). \qquad \square$$

**Example 3.9.13** (Fiberwise join)   Let $X \to Y$ be a map. We define the *fiberwise join* $X *_Y U$ of $X$ with $U$ over $Y$ to be

$$X *_Y U = \text{hocolim}(X \leftarrow X \times U \to Y \times U).$$

To justify its name, we apply Corollary 3.9.4 to to show that

$$\text{hofiber}(X *_Y U \to Y) \simeq \text{hofiber}(X \to Y) * U.$$

Consider the homotopy pushout square

$$\begin{array}{ccc}
X \times U & \longrightarrow & Y \times U \\
\downarrow & & \downarrow \\
X & \longrightarrow & X *_Y U
\end{array}$$

and fiber this over the square of identity maps

$$\begin{array}{ccc}
Y & \longrightarrow & Y \\
\downarrow & & \downarrow \\
Y & \longrightarrow & Y
\end{array}$$

in the obvious way. Corollary 3.9.4 gives us another homotopy pushout square

$$\begin{array}{ccc}
\text{hofiber}(X \to Y) \times U & \longrightarrow & \text{hofiber}(Y \to Y) \times U \\
\downarrow & & \downarrow \\
\text{hofiber}(X \to Y) & \longrightarrow & \text{hofiber}(X *_Y U \to Y)
\end{array}$$

which proves that the canonical map

$$\text{hofiber}(X \to Y) * U \to \text{hofiber}(X *_Y U \to Y)$$

is a weak equivalence.

If $U$ is empty then the fiberwise join is simply the space $X$. Two non-trivial interesting cases of note are when $U$ is a finite set of size one or two. These give, respectively, the *fiberwise cone* $C_Y X$ of $X$ over $Y$ and the *fiberwise suspension* $\Sigma_Y X$ of $X$ over $Y$. The fiberwise cone $C_Y X$ is homeomorphic with the mapping cylinder of the map $X \to Y$. More generally, if $U$ is a finite set, the fibers (over $Y$) of the fiberwise join $X *_Y U$ are the union along hofiber$(X \to Y)$ of $|U| - 1$ copies of $\Sigma_Y$ hofiber$(X \to Y)$. Continuing to assume $U$ is a finite set, if the map $X \to Y$ is $k$-connected, then so long as $U$ is non-empty, the map $X *_Y U \to Y$ is $(k + 1)$-connected, which can be seen upon consideration of the fibers over $Y$.

There is a natural homeomorphism $(X *_Y U) *_Y V \cong X *_Y (U * V)$ as spaces over $Y$ (where a space over $Y$ is a space that comes with a map to $Y$ and maps of spaces over $Y$ make the obvious triangle commute), the details of which we leave to the reader. The homotopy fibers of $(X *_Y U) *_Y V \to Y$ are easily seen to be hofiber$(X \to Y) * U * V$, which are the same as the homotopy fibers of $X *_Y (U * V) \to Y$. As a special case, consider when $V$ is a finite set of size two. Then $\Sigma_Y(X *_Y U) \cong (\Sigma_Y X) *_Y U$ as spaces over $Y$. □

**Example 3.9.14** If $U$ is a finite set, then $X * U \cong$ hocofiber$(\bigvee_U X \to X)$, where $\bigvee_U X \to X$ is the "fold map". To prove this we induct on $|U|$. When $|U| = 1$ both spaces are clearly $CX$.[8] For the inductive step, let $u \in U$ and write $\bigvee_U X = X \vee \bigvee_{U-\{u\}} X$. Consider the diagram

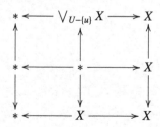

---

[8] When $U = \emptyset$, both sides result in $X$, as the wedge over no copies of $X$ is a point.

Here all maps are either the identity, the unique map to a point, or the fold map. We use Theorem 3.7.18. First taking homotopy colimits along the rows and using Example 3.6.7, we obtain the diagram

$$\text{hocofiber}\left(\bigvee_{U-\{u\}} X \to X\right) \leftarrow X \to *.$$

By induction, $\text{hocofiber}(\bigvee_{U-\{u\}} X \to X) \cong X * (U - \{u\})$, and so the homotopy colimit of the above diagram is diagram is

$$\text{hocofiber}\left(X \to X * (U - \{u\})\right),$$

where $X$ maps to the common "base" of the cones comprising the join. This is evidently a cofibration, and so the homotopy cofiber above is homotopy equivalent to the ordinary cofiber, which is easily seen to be $X * U$.

If instead we first take homotopy colimits along the columns we obtain the diagram

$$* \leftarrow \bigvee_{U} X \to X$$

whose homotopy colimit is $\text{hocofiber}(\bigvee_U X \to X)$. By Theorem 3.7.18, the two descriptions are homeomorphic. □

# 4

# 2-cubes: The Blakers–Massey Theorems

The last chapter explored two different definitions of "equivalence" of 1-cubes, or maps of spaces. These are the notions of a homotopy cartesian and homotopy cocartesian square. In Section 3.9 we were essentially combining the two notions. The Blakers–Massey Theorem (also known as the Triad Connectivity Theorem) and its dual are theorems which directly compare these, giving an idea to what degree a homotopy cocartesian square is homotopy cartesian and vice versa.

The Blakers–Massey Theorem has far-reaching consequences in homotopy theory and many classical results follow from it. For example, we will in this chapter use the Blakers–Massey Theorem to deduce the Freudenthal Suspension Theorem, Serre's Theorem, the (Relative) Hurewicz Theorem, and the Homological Whitehead Theorem. It should also be noted that the proof of the Blakers–Massey Theorem we give here is different than any other one we know of in the literature, although it follows the ideas of Goodwillie [Goo92] closely.

The general Blakers–Massey Theorem will be proved in Chapter 6.

## 4.1 Historical remarks

The original motivation for the work Blakers and Massey did in [BM49, BM51, BM52, BM53a, BM53b] is the simple observation that homology groups satisfy excision while homotopy groups do not. Namely, if a space $X$ is a union of two subspaces $A$ and $B$, then, under some mild hypotheses, the inclusions

$$(A, A \cap B) \longrightarrow (X, B)$$
$$(B, A \cap B) \longrightarrow (X, A)$$

induce isomorphisms on relative homology groups. However, this is not true for homotopy groups (take e.g. $X = S^2$ with $A$ and $B$ two hemispheres intersecting in a circle). Nevertheless, the Blakers–Massey Theorem states that excision for homotopy groups is satisfied through a range that depends on the connectivities of the pairs of spaces. The early version of this theorem is given below as Theorem 4.1.1. We include this as a convenient starting point for a brief discussion on the history of this theorem, but our preferred statement is Theorem 4.2.1, which is also more general.

Recall from Definition 2.6.4 the notion of the connectivity of a pair $(X, A)$. We then have the following.

**Theorem 4.1.1** (Blakers–Massey or homotopy excision)  *Suppose $X$ is a union of two open subspaces $A$ and $B$ such that $A$, $B$, and $A \cap B$ are path-connected. Suppose $(A, A \cap B)$ is $k_1$-connected and $(B, A \cap B)$ is $k_2$-connected, with $k_1, k_2 \geq 1$. Then the map*

$$\pi_i(A, A \cap B) \longrightarrow \pi_i(X, B)$$

*induced by inclusion is a bijection for $1 \leq i < k_1 + k_2$ and a surjection for $i = k_1 + k_2$. In other words, the inclusion $(A, A \cap B) \to (X, B)$ is $(k_1 + k_2)$-connected.*

The original statement of this theorem appears in [BM52, Theorem I] (with stronger hypotheses than stated above, but those are only necessary for studying the first non-vanishing homotopy group; see Theorem 4.1.2). Of interest is also [Whi78, Section VII.7], which gives various ways of altering the hypotheses to obtain a slightly weaker version of the theorem. Some of the alternative proofs and improvements of the Blakers–Massey Theorem are due to Araki [Ara53], tom Dieck *et al.* [tDKP70], Moore [Moo53], Namioka [Nam62], Spanier [Spa67], and Witbooi [Wit95], among others.

More modern expositions of the proof of Theorem 4.1.1 can also be found in [Hat02, Theorem 4.23] and [tD08, Theorem 6.4.1]. A proof that uses triad homotopy groups (see below for what these are) can be found in [May99, Chapter 11, Section 3]. The reader should be aware that, in some versions of the Blakers–Massey Theorem, such as in [Hat02], the authors will be explicit about requiring $(X, A)$ and $(X, B)$ to be CW pairs, but in light of Theorem 1.3.7, which says that any space can be approximated by a CW complex, this is not necessary (in fact, a proof that any excisive triad – see below for what this means – can be replaced by a CW triad can be found in [May99, Chapter 10, Section 7]). In addition, the definition of the homotopy groups of a pair $(X, A)$ in the literature sometimes includes $\pi_0$ (perhaps defined as

$\pi_0(X, A) = \pi_0(X)/\pi_0(A)$ because then one gets an exact sequence of homotopy groups that makes sense at the tail; see e.g. [Hat02, p. 476]).

The Blakers–Massey Theorem states a range in which the *triad homotopy groups*, are trivial, and hence it is sometimes referred to as the *Triad Connectivity Theorem*. Namely, a space $X$ with subspaces $A$ and $B$ is called a *triad* and denoted by $(X; A, B)$ (not to be confused with a *triple* $(X, A, B)$ where we require $B \subset A$). If, in addition, $X = \mathring{A} \cup \mathring{B}$ and $A \cap B \neq \emptyset$, then the triad is said to be *excisive*; this is of course the usual condition when discussing homological excision and the Mayer–Vietoris sequence. One can define homotopy groups of a triad, $\pi_k(X; A, B)$, for $k \geq 2$, as homotopy classes of maps

$$\alpha \colon I^k \to X$$

satisfying

$$\alpha(t_1, \ldots, t_k) \begin{cases} \in A, & t_{k-1} = 0; \\ \in B, & t_k = 0; \\ = *, & (t_1, \ldots, t_k) \in \partial I^k \text{ and } t_{k-1} \neq 0 \text{ or } t_k \neq 0. \end{cases}$$

For $k = 2$ this is a based set and for $k = 3$, this may not be abelian (it is for $k > 3$); for details, see [BM51, Part 2]. If $B \subset A$ and $k > 2$, this reduces to the usual definition of a relative homotopy group $\pi_k(X, A)$ from Section 1.3 [BM51, Theorem 3.2.1]. One then also has the *long exact homotopy sequence of a triad* [BM51, (3.5)] which generalizes the usual homotopy long exact sequence of a pair from Theorem 1.4.3:

$$\cdots \longrightarrow \pi_{k+1}(X; A, B) \longrightarrow \pi_k(A, A \cap B) \longrightarrow \pi_k(X, B)$$
$$\longrightarrow \pi_k(X; A, B) \longrightarrow \cdots$$
$$\cdots \longrightarrow \pi_2(X; A, B) \longrightarrow \pi_1(A, A \cap B) \longrightarrow \pi_1(X, B).$$

It follows from this exact sequence that relative homotopy groups will satisfy excision if the triad homotopy groups vanish. The original Blakers–Massey Theorem says precisely this: under the connectivity hypotheses from Theorem 4.1.1, the excisive triad $(X; A, B)$ is $(k_1 + k_2)$-connected (where the connectivity of the triad is defined in an analogous way to that of a pair; see [BM51, (4.1)]).

To complete the picture, many authors, including Blakers and Massey, also study the first non-vanishing triad homotopy group. We have the following.

**Theorem 4.1.2** *With hypotheses as in Theorem 4.1.1, the map*

$$\pi_{k_1+1}(A, A \cap B) \otimes \pi_{k_2+1}(B, A \cap B) \longrightarrow \pi_{k_1+k_2+1}(X; A, B) \qquad (4.1.1)$$

*given by the generalized Whitehead product, is an isomorphism.*

The original statement of this theorem, with more restrictive hypotheses, can be found in [BM53a, Theorem I]. Blakers and Massey put some conditions on the relative homotopy groups of $(B, A \cap B)$ and $(B, A \cap B)$ if one or both of $k_1$ and $k_2$ are equal to 1. Alternatively, one could require that $A \cap B$ be simply-connected. The improvement stated here is due to Brown and Loday [BL84], [BL87a, Theorem 4.2]. In particular, in low dimensions a non-abelian tensor product defined by those authors has to be used to describe the map in (4.1.1). The proof of Theorem 4.1.2 is algebraic, and we will give a space-level explanation for a special case of it in Section 4.6, where we will also discuss a space-level version of the generalized Whitehead products.

Theorem 4.1.1 can be recast as a statement about square diagrams. The hypothesis that $X$ is covered by subsets $A$ and $B$ (or that the triad $(X; A, B)$ is excisive) implies that the square

is homotopy cocartesian (this was Example 3.7.5). Recall from Remark 2.6.10 that an inclusion map $X \to Y$ is $k$-connected if and only if the pair $(Y, X)$ is $k$-connected, and recall also that $\pi_k \text{hofiber}(X \to Y) \cong \pi_{k+1}(Y, X)$. Then we can reinterpret Theorem 4.1.1 as saying that the map

$$\text{hofiber}(A \cap B \to B) \longrightarrow \text{hofiber}(A \to X)$$

is $(k_1 + k_2 - 1)$-connected for any choice of basepoint in $B$, and hence by Proposition 3.3.18 that the square

$$
\begin{array}{ccc}
A \cap B & \longrightarrow & A \\
\downarrow & & \downarrow \\
B & \longrightarrow & X
\end{array}
$$

is $(k_1 + k_2 - 1)$-cartesian. We can deduce the classical result, Theorem 4.1.1, from Theorem 4.2.1 directly once we know the square above is homotopy cocartesian; this is the content of Example 3.7.5. On the other hand, Example 3.7.5 relies on the Homological Whitehead Theorem (Theorem 4.3.5) and the Seifert–vanKampen Theorem (Theorem 1.4.12). The Homological Whitehead Theorem, in turn, itself depends on Theorem 4.2.1. This may seem like a convoluted path to follow, but it is the perspective of this book to emphasize the importance of the more modern statement of Theorem 4.1.1, appearing here as Theorem 4.2.1. Moreover, Theorem 4.2.1 is stronger in the sense that it has weaker hypotheses: not only is that a statement about any homotopy cocartesian square but we also eliminate the hypothesis that the connectivity of the pairs $(A, A \cap B)$ and $(B, A \cap B)$ be at least 1.

The Blakers–Massey Theorem can be regarded as a statement about how far a homotopy pushout square is from being a homotopy pullback square. There are several proofs in the literature that use this point of view. For example, Mather [Mat73] proves a weaker version where $k_1, k_2 \geq 2$ using the Second Cube Theorem (see Theorem 5.10.8). Another proof that uses the language of pushouts and pullbacks in combination with cellular approximations is due to Chachólski [Cha97].

As we have discussed above, Theorem 4.1.1 can be thought of as a special case of Goodwillie's general Blakers–Massey Theorem [Goo92], which appears in this text as Theorem 6.2.1. It is worth emphasizing again that Goodwillie improves the statement to eliminate the hypothesis on the connectivity of the pairs. In addition, his techniques apply to squares which are not homotopy cocartesian, but instead $k$-cocartesian for some $k$. These improvements are due in part to working purely on the space level and reducing to a special case, and in part due to the machinery of cubical diagrams. Goodwillie uses transversality arguments to argue a special case; we present this proof as well as a new, purely homotopy-theoretic, version (see Sections 4.4.1 and 4.4.2). The first time transversality techniques were used to prove the Blakers–Massey Theorem appears to be [Gra75, Chapters 13 and 16]. We will present a different proof that is related to Goodwillie's approach in Section 4.4.

Goodwillie's generalization of the Blakers–Massey Theorem uses cubical diagrams, and this is the proof that is most relevant to us. Early versions of the cubical approach are due to Barratt–Whitehead [BW56], Brown-Loday [BL87a], and Ellis-Steiner [ES87]. These will be discussed in Section 6.1 before we provide our own proof of the generalized Blakers–Massey Theorem in Section 6.3.

Finally, we should mention that the Blakers–Massey Theorem belongs in a circle of seminal results in homotopy theory that uses the interplay between homotopy and homology. The other such classical results from around the same time are the Serre, the (Relative) Hurewicz, and the Homotopical and Homological Whitehead Theorems. There are various ways to deduce these important theorems from each other – many of the proofs of the original Blakers–Massey Theorem mentioned above use the Hurewicz or Serre Theorems, Serre's Theorem can be deduced from the Hurewicz Theorem, etc. – but what we will demonstrate in Section 4.3 is that the Blakers–Massey Theorem (and its dual) can stand above all since the Serre, (Relative) Hurewicz, and Homological Whitehead Theorems can all be deduced from it. For further information about how these classical theorems are related, see [Ark11, Section 6.4] or [Sel97, Section 7.4].

## 4.2 Statements and applications

We first state the Blakers–Massey Theorem, its dual, and the generalizations of those two statements. Note that now there are no restrictions on the $k_i$. It is even possible for one or both to be negative (though in the latter case the statement of the theorem is vacuously true).

**Theorem 4.2.1** (Blakers–Massey Theorem for squares)　*Suppose that*

$$\mathcal{X} = \begin{array}{ccc} X_\emptyset & \longrightarrow & X_1 \\ \downarrow & & \downarrow \\ X_2 & \longrightarrow & X_{12} \end{array}$$

*is homotopy cocartesian and that the maps $X_\emptyset \to X_i$ are $k_i$-connected for $i = 1, 2$. Then $\mathcal{X}$ is $(k_1 + k_2 - 1)$-cartesian.*

In light of Proposition 3.3.18, we may also read Theorem 4.2.1 as saying that

$$\text{hofiber}(X_\emptyset \to X_2) \longrightarrow \text{hofiber}(X_1 \to X_{12})$$

is $(k_1 + k_2 - 1)$-connected for every choice of basepoint $x \in X_2$ (which automatically determines a basepoint in $X_{12}$). Or, in the case where the square in question is a square of inclusions, that the map of pairs $(X_2, X_\emptyset) \to (X_{12}, X_1)$ is $(k_1 + k_2)$-connected (recalling that $\pi_k(X, A) \cong \pi_{k-1}(\text{hofiber}(A \to X))$ and with the usual caveat about relative $\pi_0$). Here is the dual to Theorem 4.2.1.

**Theorem 4.2.2** (Dual Blakers–Massey Theorem for squares)　*Suppose that*

$$\mathcal{X} = \begin{array}{ccc} X_\emptyset & \longrightarrow & X_1 \\ \downarrow & & \downarrow \\ X_2 & \longrightarrow & X_{12} \end{array}$$

*is homotopy cartesian and that the maps $X_i \to X_{12}$ are $k_i$-connected for $i = 1, 2$. Then $\mathcal{X}$ is $(k_1 + k_2 + 1)$-cocartesian.*

This theorem follows immediately from Proposition 3.9.12 together with Proposition 3.7.23, but we will give a different easy proof in Section 4.5 that uses quasifibrations.

　　Just as in the above, Theorem 4.2.2 can be read as a statement about a map of homotopy cofibers, this time using Proposition 3.7.24 (which is related to Proposition 2.6.12), although we must be careful to say that Theorem 4.2.2 implies that the map

$$\text{hocofiber}(X_{\emptyset} \to X_2) \longrightarrow \text{hocofiber}(X_1 \to X_{12})$$

is $(k_1 + k_2 + 1)$-connected, but the connectivity of the map of homotopy cofibers does not in general imply that the square is highly cocartesian. When the square in question is a square of inclusion maps, this tells us that the map $H_i(X_2, X_{\emptyset}) \to H_i(X_{12}, X_1)$ is an isomorphism for $i < k_1 + k_2 + 1$ and onto for $i = k_1 + k_2 + 1$; see Remark 2.6.3. The reason for this is the isomorphism $\widetilde{H}_*(X, A) \cong \widetilde{H}_*(X/A)$ and, if $X$ is a $k$-connected space, then $\widetilde{H}_i(X) = 0$ for $i \leq k$.

We will also prove the following generalizations of Theorems 4.2.1 and 4.2.2.

**Theorem 4.2.3** *Suppose that*

$$\mathcal{X} = \begin{array}{ccc} X_{\emptyset} & \longrightarrow & X_1 \\ \downarrow & & \downarrow \\ X_2 & \longrightarrow & X_{12} \end{array}$$

*is $j$-cocartesian and that the maps $X_{\emptyset} \to X_i$ are $k_i$-connected for $i = 1, 2$. Then $\mathcal{X}$ is $\min\{k_1 + k_2 - 1, j - 1\}$-cartesian.*

The dual version is the following.

**Theorem 4.2.4** *Suppose that*

$$\mathcal{X} = \begin{array}{ccc} X_{\emptyset} & \longrightarrow & X_1 \\ \downarrow & & \downarrow \\ X_2 & \longrightarrow & X_{12} \end{array}$$

*is $j$-cartesian and that the maps $X_i \to X_{12}$ are $k_i$-connected for $i = 1, 2$. Then $\mathcal{X}$ is $\min\{k_1 + k_2 + 1, j + 1\}$-cocartesian.*

Before proving these theorems, we present a few consequences and examples of their uses. More are given in Section 4.3. The first few examples and results are consequences of Theorem 4.2.1.

**Example 4.2.5** If based spaces $X_i$ are $k_i$-connected for $i = 1, 2$, then the inclusion

$$X_1 \vee X_2 \longrightarrow X_1 \times X_2$$

is $(k_1 + k_2 + 1)$-connected. This is because

$$\mathcal{X} = \begin{array}{ccc} X_1 \vee X_2 & \longrightarrow & X_1 \\ \downarrow & & \downarrow \\ X_2 & \longrightarrow & * \end{array}$$

admits a homotopy equivalence from the square

$$
\begin{array}{ccc}
X_1 \vee X_2 & \longrightarrow & X_1 \vee CX_2 \\
\downarrow & & \downarrow \\
CX_1 \vee X_2 & \longrightarrow & CX_1 \vee CX_2
\end{array}
$$

the latter of which is evidently homotopy cocartesian. Since the map from a space $X$ to its cone $CX$ has connectivity equal to the connectivity of $X$, the square $\mathcal{X}$ is $(k_1 + k_2 + 1)$-cartesian by Theorem 4.2.1.          □

**Remark 4.2.6**   The above example is related to the Hilton–Milnor Theorem, here Theorem 6.5.9, which helps describe the homotopy type of a wedge of suspensions. This example is also a first step in determining the second derivative of the identity functor in homotopy calculus of functors (more on functor calculus can be found in Section 10.1).          □

**Example 4.2.7**   For $i = 1, 2$, let $X_i$ be $k_i$-connected spaces. There is a $(k_1 + k_2 + \min\{k_1, k_2\} + 1)$-connected map

$$
\Sigma(\Omega X_1 \wedge \Omega X_2) \longrightarrow \Omega(X_1 \wedge X_2),
$$

induced by a map of homotopy fibers. To see this consider the homotopy cocartesian square

$$
\begin{array}{ccc}
X_1 \vee X_2 & \longrightarrow & X_1 \times X_2 \\
\downarrow & & \downarrow \\
* & \longrightarrow & X_1 \wedge X_2
\end{array}
$$

which is $K = (k_1 + k_2 + \min\{k_1, k_2\} + 1)$-cartesian by Example 4.2.5 and Theorem 4.2.1. By Proposition 3.3.18 this implies that the map of horizontal homotopy fibers is $K$-connected. Now use Example 3.9.5 and Proposition 3.7.22 to rewrite the domain of this map of homotopy fibers to give the desired result.          □

Now recall from Theorem 1.2.9 that there is a natural homeomorphism

$$
\mathrm{Map}_*(X, \Omega Y) \cong \mathrm{Map}_*(\Sigma X, Y). \tag{4.2.1}
$$

The following is a foundational result in stable homotopy theory.

**Theorem 4.2.8** (Freudenthal Suspension Theorem)   *The map*

$$
S^n \longrightarrow \Omega \Sigma S^n,
$$

*which corresponds to the identity map $\Sigma S^n \to \Sigma S^n$ via (4.2.1) (setting $X = S^n$, $Y = \Sigma S^n$), is $(2n - 1)$-connected.*

*Proof* Writing $\Sigma S^n \simeq S^{n+1} = CS^n \cup_{S^n} CS^n$ as the union of hemispheres, we have a homotopy cocartesian square

$$
\begin{array}{ccc}
S^n & \longrightarrow & CS^n \\
\downarrow & & \downarrow \\
CS^n & \longrightarrow & S^{n+1}
\end{array}
$$

where the maps $S^n \to CS^n \simeq *$ are $n$-connected since $S^n$ is $(n-1)$-connected. It follows from Theorem 4.2.1 that the square is $(2n-1)$-cartesian, which implies that the map

$$S^n \longrightarrow \mathrm{holim}(CS^n \to \Sigma S^n \leftarrow CS^n) \simeq \Omega\Sigma S^n$$

is $(2n - 1)$-connected. $\qquad\square$

**Remark 4.2.9** The Blakers–Massey Theorem 4.2.1 gives the optimal connectivity estimate in the sense that it predicts the best possible connectivity in at least one case. For instance, Theorem 4.2.8 says that the map $S^0 \to \Omega\Sigma S^0$ is $(-1)$-connected, and it is in fact not $0$-connected since the codomain $\Omega\Sigma S^0 \simeq \Omega S^1$ has countably many path components, whereas the domain has only two. Of course, Theorem 4.2.1 will not give the best estimate in all cases. For an easy example, let $X$ be a well-pointed $k$-connected space, and let $* \to X$ and $* \to S^0$ be the inclusions of the basepoints. The cocartesian square

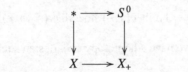

is homotopy cocartesian since $* \to X$ is a cofibration (using Proposition 3.7.6) but, according to Theorem 4.2.1, it is only $(k - 2)$-cartesian. To see that it is homotopy cartesian, note that $\mathrm{holim}(X \to X_+ \leftarrow S^0)$ is the space of all $(x, \gamma, y)$ such that $\gamma$ joins $x$ with the image of $y$ in $X_+$. The set of all tuples with $y$ not the basepoint of $S^0$ is empty, and hence the homotopy limit is the space of paths with one endpoint the basepoint of $X$, which is contractible. $\qquad\square$

An easy generalization of Theorem 4.2.8 replaces the sphere with an arbitrary highly connected space $X$.

**Theorem 4.2.10** *If $X$ is an $n$-connected space, then the map*

$$X \longrightarrow \Omega\Sigma X,$$

*which corresponds to the identity map* $\Sigma X \to \Sigma X$ *via* (4.2.1), *is* $(2n + 1)$-*connected.*

**Example 4.2.11**  Suppose $X$ and $Y$ are based spaces with $X$ $n$-connected, and let $f \colon X \to \Omega Y$ and $g \colon \Sigma X \to Y$ correspond via (4.2.1).

1. If $f$ is $k$-connected, then $g$ is $(\min\{2n + 2, k\} + 1)$-connected.
2. If $g$ is $(k + 1)$-connected, then $f$ is $\min\{2n + 1, k\}$-connected.

To see this, consider the commutative diagram

The top horizontal arrow is $(2n + 1)$-connected by Theorem 4.2.10. The result now follows from Proposition 2.6.15, noting that $\Omega g$ is $k$-connected if $g$ is $(k + 1)$-connected.                                   □

**Example 4.2.12**  Suppose $X \to Y$ is $k$-connected, and that $X$ itself is $l$-connected. Then there is a naturally induced $(k + l)$-connected map

$$\mathrm{hofiber}(X \to Y) \longrightarrow \Omega\,\mathrm{hocofiber}(X \to Y).$$

Equivalently, there is a $(k + l)$-connected map

$$X \longrightarrow \mathrm{hofiber}\big(Y \to \mathrm{hocofiber}(X \to Y)\big).$$

To see this, note that we have a homotopy cocartesian square

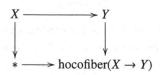

Since $X$ is $l$-connected, the map $X \to *$ is $(l + 1)$-connected and by hypothesis $X \to Y$ is $k$-connected. Then Theorem 4.2.1 implies that this square is $(k + l)$-cartesian. Using Proposition 3.3.18 in two different ways (on the horizontal and vertical homotopy fibers respectively) gives the two equivalent results.   □

We now give some applications of Theorem 4.2.2, the dual Blakers–Massey Theorem.

**Example 4.2.13** Let $X$ be a $k$-connected based space, and $* \to X$ the inclusion of the basepoint. The square

is a homotopy pullback, and the maps $* \to X$ are $k$-connected, so Theorem 4.2.2 implies that the square is $(2k + 1)$-cocartesian. This means that the map

$$\text{hocolim}(* \leftarrow \Omega X \to *) \simeq \Sigma \Omega X \to X$$

is $(2k + 1)$-connected. $\qquad\qquad\qquad\qquad\qquad\qquad\qquad\qquad\square$

It is easy to compute the connectivity of the join $X_1 * X_2$ (given in Proposition 3.7.23) using Theorem 4.2.2. However, it is this very connectivity estimate that is used in the proof of Theorem 4.2.2, so it would be dishonest to attempt to derive it in this way. It is still, however, a good exercise for the novice to verify the result in Proposition 3.7.23 using Theorem 4.2.2.

Here is an application of Propositions 3.7.23 and 3.9.12, due to Ganea, involving the so-called "fiber–cofiber" construction. Before we state it, note that if $F \to X \to Y$ is a (homotopy) fibration sequence, then there is a natural map $\text{hocofiber}(F \to X) \to Y$. Since $F \to X \to Y$ is null-homotopic, this null homotopy can be used to build a map $\text{hocofiber}(F \to X) = X \cup CF \to Y$ by using $X \to Y$ on $X$ and the null-homotopy to define a compatible map on the cone $CF$.

**Proposition 4.2.14** (Ganea's fiber–cofiber construction, [Gan65]) *Let $X$ be a based space. Consider the diagram*

$$
\begin{array}{ccccccc}
\Omega X & \longrightarrow & F_1 & \longrightarrow & F_2 & \longrightarrow & \cdots \\
\downarrow & & \downarrow & & \downarrow & & \\
PX & \longrightarrow & G_1 & \longrightarrow & G_2 & \longrightarrow & \cdots \\
\downarrow & & \downarrow & & \downarrow & & \\
X & \xrightarrow{1_X} & X & \xrightarrow{1_X} & X & \xrightarrow{1_X} & \cdots
\end{array}
$$

*where*

$$F_j = \text{hofiber}(G_j \to X),$$
$$G_1 = \text{hocofiber}(\Omega X \to PX) \simeq \Sigma \Omega X,$$
$$G_{j+1} = \text{hocofiber}(F_j \to G_j).$$

*Then $G^{j+1} \to X$ is $((j+2)(k+1)-1)$-connected, and hence $F_{j+1}$ is $((j+2)(k+1)-2)$-connected. Moreover,*

$$F_{j+1} \simeq \underset{j+2}{*} \, \Omega X.$$

*Proof*   The reason the diagram commutes is that the composition $F_i \to G_i \to X$ is null-homotopic and so $G_i \to X$ factors through $G_{i+1} = \text{hocofiber}(F_i \to G_i)$. The result follows by induction on $j$ using Proposition 3.9.12. We have a homotopy pushout square

$$
\begin{array}{ccc}
F_j & \longrightarrow & G_j \\
\downarrow & & \downarrow \\
* & \longrightarrow & G_{j+1}
\end{array}
$$

and a homotopy pullback square

$$
\begin{array}{ccc}
F_j & \longrightarrow & G_j \\
\downarrow & & \downarrow \\
* & \longrightarrow & X
\end{array}
$$

By Proposition 3.9.12,

$$F_{j+1} = \text{hofiber}(G_{j+1} \to X) \simeq \text{hofiber}(* \to X) * \text{hofiber}(G_j \to X),$$

which implies

$$F_{j+1} \simeq \Omega X * F_j.$$

By induction, $F_j \simeq \underset{j+1}{*} \, \Omega X$. As for the connectivity of $F_{j+1}$, since $F_1 \simeq \Omega X * \Omega X$ is $2k$-connected by Proposition 3.7.23 and $F_{j+1} \simeq \Omega X * F_j$, again using Proposition 3.7.23 and induction we compute that $F_{j+1}$ is $((j+2)(k+1)-2)$-connected.     □

## Remarks 4.2.15

1. Proposition 4.2.14 gives a sequence of spaces

$$PX \longrightarrow \Sigma\Omega X \simeq G_1 \longrightarrow G_2 \longrightarrow \cdots$$

admitting compatible maps to $X$ which increase in connectivity as one moves up the tower, provided that $X$ is at least connected. We also obtain an explicit description of the difference between $X$ and the approximating space $G_j$ as $F_j$ is the join of a number of copies of $\Omega X$.

2. If we were interested only in the connectivity of $F_{j+1}$ in Proposition 4.2.14, we could have used induction on $j$ together with Theorem 4.2.2 applied to the homotopy pullback square

$$
\begin{array}{ccc}
F_j & \longrightarrow & G_j \\
\downarrow & & \downarrow \\
* & \longrightarrow & X
\end{array}
$$

If $X$ is $k$-connected, then $* \to X$ is $k$-connected, and by induction $G_j \to X$ is $((j+1)(k+1)-1)$-connected, so by Theorem 4.2.2, the square is $((j+2)(k+1)-1)$-cocartesian. This means $G_{j+1} \to X$ is $((j+2)(k+1)-1)$-connected, and hence $F_{j+1} = \text{hofiber}(G_{j+1} \to X)$ is $((j+2)(k+1)-2)$-connected. $\square$

Here is a dual to Proposition 4.2.14, the proof of which we leave to the reader. The maps $X \to G^i$ are induced in a dual way to the maps $G_i \to X$ described before the statement of Proposition 4.2.14.

**Proposition 4.2.16** *Let $X$ be a space. Consider the diagram*

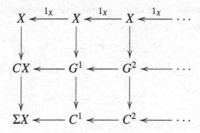

*where*

$$
\begin{aligned}
C^j &= \text{hocofiber}(X \to G^j), \\
G^1 &= \text{hofiber}(CX \to \Sigma X) \simeq \Omega \Sigma X, \\
G^{j+1} &= \text{hofiber}(G^j \to C^j).
\end{aligned}
$$

*Then $X \to G^{j+1}$ is $((j + 1)k + 1)$-connected, and hence $C^{j+1}$ is $((j + 1)k + 1)$-connected.*

**Remarks 4.2.17** In contrast to Proposition 4.2.14, we do not know of a good explicit description of the spaces $C^j$. We have a tower of spaces

$$
CX \longleftarrow \Omega \Sigma X \simeq G^1 \longleftarrow G^2 \longleftarrow \cdots
$$

admitting compatible maps from $X$ which increase in connectivity as one moves up the tower assuming that $X$ is at least 1-connected. Both of the last two results are related at least in spirit to Examples 6.2.6 and 6.2.7 which appear later.                                                                          □

**Remark 4.2.18**   Propositions 4.2.14 and Proposition 4.2.16 are some of the most important tools in the subject of *Lusternik–Schnirelmann category*. For more details, see [CLOT03].                                                          □

The following example is a multi-relative generalization of Example A.2.21. In Section A.2 we present the necessary machinery to discuss transversality, which is utilized in the proof below. The reader unfamiliar with this topic may still be able to do without the appendix and instead use the following hazy statements. If $P$ and $Q$ are smooth $p$- and $q$-dimensional submanifolds of a smooth $n$-dimensional manifold $N$ respectively, they should "generically" intersect in a manifold of dimension $n - p - q$ (or not at all). In particular the intersection is generically empty if $p + q < n$. Transversality theory says that one can always arrange for the intersections to be generic.

**Example 4.2.19**   Let $N$ be a smooth manifold and $P_1, P_2$ smooth closed submanifolds of dimensions $p_1$ and $p_2$, respectively, whose intersection in $N$ is transverse. The square of inclusions

$$C = \begin{array}{ccc} N - (P_1 \cup P_2) & \longrightarrow & N - P_1 \\ \downarrow & & \downarrow \\ N - P_2 & \longrightarrow & N \end{array}$$

is $(2n - p_1 - p_2 - 3)$-cartesian. The square

$$\begin{array}{ccc} N - (P_1 \cup P_2) & \longrightarrow & N - P_1 \\ \downarrow & & \downarrow \\ N - P_2 & \longrightarrow & N - (P_2 \cap P_2) \end{array}$$

is an open pushout square (see Definition 3.7.3), and by Example 3.7.5 is homotopy cocartesian. Since $P_1$ and $P_2$ intersect transversely, $P_1 \cap P_2$ has dimension $p_1 + p_2 - n$. It follows that the natural map $N - (P_1 \cap P_2) \to N$ is $(2n - p_1 - p_2 - 1)$-connected by Example A.2.21. Again by Example A.2.21, the maps $N - (P_1 \cup P_2) \to N - P_i$ are $(n - p_i - 1)$-connected for $i = 1, 2$, and hence by Theorem 4.2.3, since $\min\{2n - p_1 - p_2 - 3, 2n - p_1 - p_2 - 2\} = 2n - p_1 - p_2 - 3$, the square $C$ is $(2n - p_1 - p_2 - 3)$-cartesian. An important special case occurs when the intersection of $P_1$ with $P_2$ is empty. In that case the square above is homotopy cocartesian and hence $(2n - p_1 - p_2 - 3)$-cartesian by Theorem 4.2.1.                                                              □

**Example 4.2.20** The utility of Theorem 4.2.3 can also be seen through considering the following *disjunction problem*. In the present context, this means the following. Suppose we have a smooth map from a smooth manifold $P$ to a smooth manifold $N$ which is homotopic to two different maps, one whose image is disjoint from a submanifold $Q_1 \subset N$ and another whose image is disjoint from a submanifold $Q_2 \subset N$. When is this map homotopic to a smooth map which is disjoint from both submanifolds? We can actually give an answer for families of such maps. Here is the setup. Recall from Definition A.2.7 the space of smooth maps $C^\infty(P, N)$ from $P$ to $N$.

Suppose $P$ is a smooth closed manifold of dimension $p$, and let $N$ be a smooth manifold of dimension $n$ with smooth closed disjoint submanifolds $Q_1, Q_2$ of dimensions $q_1, q_2$ respectively. Consider the square

$$\mathcal{M} = \begin{array}{ccc} C^\infty(P, N - (Q_1 \cup Q_2)) & \longrightarrow & C^\infty(P, N - Q_1) \\ \downarrow & & \downarrow \\ C^\infty(P, N - Q_2) & \longrightarrow & C^\infty(P, N) \end{array}$$

We are going to use Theorem 4.2.3 to prove that $\mathcal{M}$ is $(2n - 2p - q_1 - q_2 - 3)$-cartesian. We first claim that $\mathcal{M}$ is $(2n - 2p - q_1 - q_2 - 1)$-cocartesian. First note that if $Q \subset N$ is a closed subset, then the inclusion $C^\infty(P, N - Q) \subset C^\infty(P, N)$ is open. To see this, consider the evaluation map $ev \colon P \times C^\infty(P, N) \to N$, where $ev(x, f) = f(x)$. Since $Q \subset N$ is closed, $ev^{-1}(Q) \subset P \times C^\infty(P, N)$ is closed, and since $P$ is compact, the image of $ev^{-1}(Q)$ in $C^\infty(P, N)$ is closed. By construction, this is the subset of all $f \colon P \to N$ such that $f(P) \cap Q \neq \emptyset$. In particular, its complement, $C^\infty(P, N - Q)$, is open. Let $C = \mathrm{colim}(C^\infty(P, N - Q_1) \leftarrow C^\infty(P, N - (Q_1 \cup Q_2)) \to C^\infty(P, N - Q_2))$. If we replace $C^\infty(P, N)$ with $C$ in $\mathcal{M}$ above we obtain a homotopy cocartesian square by Example 3.7.5, and so to determine how cocartesian $\mathcal{M}$ is, we only need to compute the connectivity of the inclusion map $C \to C^\infty(P, N)$. Let $g \colon (D^i, \partial D^i) \to (C^\infty(P, N), C)$ be a map of pairs. This corresponds via Theorem 1.2.7 to a map $\widetilde{g} \colon D^i \times P \to N$ such that

- for each $t \in \partial D^i$, there exists $j = 1, 2$ so that $\widetilde{g}(t, -) \colon P \to N - Q_j$.

We are to deform $\widetilde{g}$ relative to $\partial D^i \times M$ to a map satisfying this condition for all $t \in D^i$, provided $i < 2n - 2p - q_1 - q_2$. Since the $Q_j$ are closed in $N$ and $P$ is compact, there exists a neighborhood $N$ of the boundary $\partial D^i$ such that the above condition holds for all $t \in N$. As in the proof of Theorem A.2.12, we can use Theorem A.2.10 to make a homotopy of $\widetilde{g}$ relative to $\partial D^i \times P$ such that $\widetilde{g}$ is smooth outside of $N \times P$.

Now we show that there is a dense set of maps $g$ whose associated maps $\widetilde{g}$ satisfy the bulleted condition above for all $t \in D^i$ provided $i < 2n-2p-q_1-q_2$. To do this we invoke Theorem A.2.23 as follows. Consider the multijet space $J_0^{(2)}(D^i \times P, N)$. We could have $(t, x_1, x_2)$ where $t \in D^i$, $x_1, x_2 \in P$ and $x_1 \neq x_2$ such that $\widetilde{g}(t, x_1) \in Q_1$ and $\widetilde{g}(t, x_2) \in Q_2$. Then the pair $((t, x_1), (t, x_2)) \in (D^i \times P)^{(2)}$ maps to a submanifold $B$ of $J_0^{(2)}(D^i \times P, N)$ of codimension $i+n-q_1+n-q_2$ ($i$ to make the two points of $D^i$ equal, and $n-q_j$ to account for $\widetilde{g}$ mapping $(t, x_j)$ to $Q_j$ for $j = 1, 2$). By Theorem A.2.23, a small homotopy of $\widetilde{g}$ (relative to $\partial D^i \times P$) makes it transverse to $B$, and transverse will mean empty intersection if $\mathrm{codim}(B) > \dim(D^i \times P)^{(2)} = 2i + 2p$. Hence $\widetilde{g}$ has a homotopy to a map $\widetilde{g}'$ enjoying the bulleted property for all $t \in D^i$ if $i + n - q_1 + n - q_2 > 2i + 2p$, or $i < 2n-2p-q_1-q_2$. It follows that there is $\widetilde{g}' : D^i \times P \to N$ which is associated with a map $g' : D^i \to C^\infty(P, N)$ via Theorem 1.2.7 having the desired property when $i < 2n - 2p - q_1 - q_2$, and so $M$ is $(2n - 2p - q_1 - q_2 - 1)$-cocartesian.

Finally we claim that, for $j = 1, 2$, the maps

$$\mathrm{Map}(P, N - (Q_1 \cup Q_2)) \longrightarrow \mathrm{Map}(P, N - Q_j)$$

are $(n - p - q_j - 1)$-connected. We can argue by a simpler version of the arguments above, and for this we leave the details to the reader. Alternatively, we can use Example A.2.21 together with the fact that since $P$ has dimension $p$, if $X \to Y$ is $k$-connected then the induced map $\mathrm{Map}(P, X) \to \mathrm{Map}(P, Y)$ is $(k - p)$-connected (see Proposition 3.3.9; it is fine to invoke that result since all the spaces here are manifolds and hence have the homotopy type of CW complexes). In any case, by Theorem 4.2.3 the square is $(2n - 2p - q_1 - q_2 - 3)$-cartesian.                              □

## 4.3 Hurewicz, Whitehead, and Serre Theorems

As promised earlier, in this section we show how the classical theorems of Hurewicz, Whitehead, and Serre follow from the Blakers–Massey Theorem and its dual.

First we turn our attention to the Hurewicz Theorem, which asserts a relationship between the first non-trivial homotopy and homology groups of a space. For the absolute version, we use Theorem 4.2.2 (dual Blakers–Massey), whereas for the proof of the relative version, we use the absolute version and Theorem 4.2.1 (Blakers–Massey).

To start, we need to discuss the *Hurewicz map* $H_n$, defined as follows: a based map $f : S^n \to X$ determines an element of $\pi_n(X, x_0)$, where $x_0$ is the basepoint of $X$, but it also defines an element of $H_n(X)$ – choose a generator $a$ of $H_n(S^n) \cong \mathbb{Z}$, and let $f_* : H_n(S^n) \to H_n(X)$ be the induced map. Then define

$$h_n : \pi_n(X, x_0) \longrightarrow H_n(X) \qquad (4.3.1)$$
$$[f] \longmapsto f_*(a).$$

See [Hat02, Proposition 4.26] for why this map is a homomorphism of groups.

**Remark 4.3.1** The Hurewicz map is not obviously induced by a map of spaces. In the spirit of our preference for space-level constructions, we make a note about a space-level version of the Hurewicz map which uses constructions from the Dold–Thom Theorem (Theorem 2.7.23). Namely, $h_n$ can be realized using infinite symmetric products: recall from Definition 2.7.22 the construction of the infinite symmetric product $SP(X) = \cup_n(SP_n(X))$ of a based space $X$. The inclusion $X \to SP(X)$ induces a map

$$\pi_i(X) \longrightarrow \pi_i(SP(X)) \cong \widetilde{H}_i(X)$$

with the latter being the isomorphism obtained using the consequence of Theorem 2.7.23 mentioned following its statement. It can be shown that this map coincides with the algebraically defined Hurewicz map $h_n$ in any degree $i$. For details, see for example the end of Section 4.K in [Hat02]. □

**Theorem 4.3.2** ((Absolute) Hurewicz Theorem) *Suppose $X$ is an $(n-1)$-connected space, with $n \geq 2$. Then $\widetilde{H}_i(X) = 0$ for all $i < n$ and the map*

$$h_n : \pi_n(X) \longrightarrow H_n(X)$$

*is an isomorphism.*

Our proof follows [Ark11, Theorem 6.4.8].

*Proof* Recall from Example 1.2.4 the definition of the path space $PX$ of $X$ (space of maps of $I$ in $X$), and consider the homotopy cartesian square

On the one hand, since this square is homotopy cartesian we have isomorphisms $\pi_i(PX, \Omega X) \to \pi_i(X, *)$ for all $i$; see Remark 3.3.19. On the other hand, as $X$ is $(n-1)$-connected, the maps $* \to X$ and $* \simeq PX \to X$ are both $(n-1)$-connected, and by Theorem 4.2.2, the square is $(2n-1)$-cocartesian. In particular, this means the induced map $H_i(PX, \Omega X) \to H_i(X, *) \cong H_i(X)$

is an isomorphism for $i < 2n - 1$. To justify this, we use the fact that a $k$-connected map has $k$-connected homotopy cofiber (Proposition 2.6.12), the long exact sequence of a cofibration (Theorem 2.3.14), and the vanishing homology of a highly connected space (see Remark 2.6.3). Finally, from the long exact sequence of a pair, we have isomorphisms $\pi_i(PX, \Omega X) \to \pi_{i-1}(\Omega X)$ and $H_i(PX, \Omega X) \to H_{i-1}(\Omega X)$ for all $i$, since $PX$ is contractible. We can summarize this in the commutative diagram

$$
\begin{array}{ccccc}
\pi_i(X) & \longleftarrow & \pi_i(PX, \Omega X) & \longrightarrow & \pi_{i-1}(\Omega X) \\
\downarrow & & \downarrow & & \downarrow \\
H_i(X) & \longleftarrow & H_i(PX, \Omega X) & \longrightarrow & H_{i-1}(\Omega X)
\end{array}
$$

By the above, the right horizontal arrows and the top left horizontal arrow are isomorphisms for all $i$, and the bottom left horizontal arrow is an isomorphism if $i < 2n - 1$. To prove $\pi_n(X) \to H_n(X)$ is an isomorphism, it suffices to prove that $\pi_{n-1}(\Omega X) \to H_{n-1}(\Omega X)$ is an isomorphism. This follows by induction: since $X$ is $(n - 1)$-connected, $\Omega X$ is $(n - 2)$-connected by Example 2.6.11, and so $\pi_{i-1}(\Omega X) \to H_{i-1}(\Omega X)$ is an isomorphism for all $i - 1 \leq n - 1$. We have to be a little careful when $i = 2$. In that case we must show that $\pi_1(\Omega X) \cong H_1(\Omega X)$. This follows from the fact that the abelianized fundamental group is the first homology group, and that the fundamental group of a loop space is abelian and hence equal to its own commutator subgroup.                          □

We can also define a relative version of the Hurewicz map,

$$
h_n : \pi_n(X, A, x_0) \longrightarrow H_n(X, A),
$$

using $(X, A)$ in place of $X$ and $(D^n, \partial D^n)$ in place of $S^n$. Here $a \in H_n(D^n, \partial D^n)$ is a fixed generator of this infinite cyclic group, and again $h_n[f] = f_*(a)$, where $f_* : H_n(D^n, \partial D^n) \to H_n(X, A)$ is the map induced by $f$.

   As a corollary of Theorem 4.3.2 and Theorem 4.2.1 we then obtain a relative version of the Hurewicz Theorem.

**Theorem 4.3.3** (Relative Hurewicz Theorem)  *Suppose $(X, A)$ is an $(n - 1)$-connected based pair with $n \geq 2$ and basepoint $x_0 \in A$, and suppose $A$ is 1-connected. Then $\widetilde{H}_i(X, A) = 0$ for $i < n$ and $\pi_n(X, A, x_0) \cong H_n(X, A)$.*

*Proof*  Using homotopy invariance of homotopy groups we may also assume the inclusion $A \to X$ is a cofibration by replacing $X$ with the mapping cylinder of the inclusion map. Let $C = \mathrm{hocofiber}(A \to X)$ and $F = \mathrm{hofiber}(A \to X)$. Consider the homotopy cocartesian square

Since $A$ is 1-connected, the map $A \to *$ is 2-connected. Since $(X, A)$ is $(n - 1)$-connected, it follows that $A \to X$ is $(n - 1)$-connected. By Theorem 4.2.1, the square is $n$-cartesian, which implies that the induced map

$$F = \text{hofiber}(A \to X) \longrightarrow \text{hofiber}(* \to C) \simeq \Omega C$$

is $n$-connected. That is,

$$\pi_i(F) \longrightarrow \pi_i(\Omega C)$$

is an isomorphism for $i < n$ and onto for $i = n$. But $\pi_i(\Omega Y) \cong \pi_{i+1}(Y)$ and $\pi_i(\text{hofiber}(Y \to Z)) \cong \pi_{i+1}(Z, Y)$ (comparing the long exact sequences of a fibration and a pair). Thus

$$\pi_j(X, A) \longrightarrow \pi_j(C)$$

is an isomorphism for $j < n + 1$ and onto for $j = n + 1$. Note that $H_j(X, A) \cong \widetilde{H}_j(C)$ for all $j$ (this time comparing the long exact sequence of a cofibration with the long exact sequence of a pair). Consider the commutative diagram

$$\begin{array}{ccc} \pi_n(X, A) & \longrightarrow & \pi_n(C) \\ \downarrow & & \downarrow \\ H_n(X, A) & \longrightarrow & \widetilde{H}_n(C) \end{array}$$

The horizontal maps are isomorphisms as described above. It suffices to show that the right vertical map is an isomorphism. Since $(X, A)$ is $(n-1)$-connected, by Theorem 1.3.8 we may assume $X - A$ contains only cells of dimension greater than $n - 1$, so that $C = X/A$ is clearly an $(n - 1)$-connected space. Thus the right vertical map is an isomorphism by Theorem 4.3.2. □

**Remark 4.3.4** One can eliminate the hypothesis that $A$ is 1-connected but the statement requires that the quotient of the relative homotopy groups by the action of $\pi_1(A)$ be taken into account. See, for example, [Hat02, Theorem 4.37]. □

We now show how the Homological Whitehead Theorem follows from the Relative Hurewicz Theorem.

**Theorem 4.3.5** (Homological Whitehead Theorem)   *If $f: X \to Y$ is a map of 1-connected spaces which induces an isomorphism in homology in all dimensions, then $f$ is a weak equivalence.*

*Proof*   Since both $X$ and $Y$ are 1-connected, $\pi_1(Y, X) = 0$ and the first non-trivial homotopy group of the pair $(Y, X)$ is isomorphic to the first non-trivial homology group by Theorem 4.3.3. But these are all zero if $f$ induces isomorphisms in homology.                                                         □

We can use Theorem 4.2.1 to deduce Proposition 2.6.14, which gives conditions under which we can deduce the connectivity of a map from the connectivity of its homotopy cofibers (recalling that the connectivity of a map was defined in terms of the connectivity of its homotopy fibers). We restate it here for convenience and give its proof.

**Proposition 4.3.6** (Restatement of Proposition 2.6.14)   *Suppose $X$ is simply-connected and $f: X \to Y$ is a map such that* hocofiber$(f)$ *is $k$-connected. Then $f$ is $k$-connected.*

*Proof*   If $k = -1$ the statement is vacuously true. It is also easy to verify that if $C$ is 0-connected, then $f$ is 0-connected. We only need to see that each homotopy fiber of $f$ is non-empty. Let $y \in Y$. Since $C$ is connected, $y$ must have a path to $f(X)$ (see Definition 2.4.5). So assume $k \geq 1$, let $C = $ hocofiber$(f)$, and consider the homotopy cocartesian square

Since $X$ is simply-connected, the map $X \to *$ is 2-connected, and if $C$ is $k$-connected with $k \geq 1$, we have seen above that $f$ is 0-connected. By Theorem 4.2.1, the square above is 1-cartesian, and by Proposition 3.3.18 the map hofiber$(f) \to$ hofiber$(* \to C) \simeq \Omega C$ is 1-connected for any choice of basepoint in $Y$. In particular, it is an isomorphism on path components, which means that $\pi_0$ hofiber$(f) \cong \pi_0 \Omega C \cong \pi_1 C = \{e\}$, using Equation (1.4.2). Using the long exact sequence of a fibration (Theorem 2.1.13) and Proposition 2.6.9, this implies that $f$ is in fact 1-connected. But this implies, using Theorem 4.2.1 once again, that the square above is in fact 3-cartesian, and we can repeat this argument above to show that $f$ is 2-connected provided $k \geq 2$. Continuing in this manner shows that $f$ is $k$-connected whenever $C$ is $k$-connected.                □

Lastly, we give the proof of the Serre Theorem.

**Theorem 4.3.7** (Serre Theorem)  *Suppose $F \to X \to Y$ is a fibration sequence, $X \to Y$ is $p$-connected, and $Y$ itself is $q$-connected. Then there is a $(p+q+1)$-connected map, naturally induced as a map of homotopy cofibers,*

$$\text{hocofiber}(F \to X) \longrightarrow Y.$$

*Proof*  We have a homotopy cartesian square

The map $* \to Y$ is $q$-connected and the map $X \to Y$ is $p$-connected, which implies, by Theorem 4.2.2, that there is a $(p + q + 1)$-connected map

$$\text{hocofiber}(F \to X) \longrightarrow \text{hocofiber}(* \to Y) \simeq Y. \qquad \square$$

**Remark 4.3.8**  Theorem 4.3.7 is in fact easy to read off from Theorem 3.9.7, so no reference to Theorem 4.2.2 is actually necessary. We just have to note that looping decreases connectivity by 1 (see Example 2.6.11), suspending increases it by 1 (see Example 2.6.13), and smashing sums the connectivities and adds 1 (see Proposition 3.7.23). $\qquad \square$

## 4.4 Proofs of the Blakers–Massey Theorems for squares

Before we embark on the proofs of the Blakers–Massey Theorems – Theorem 4.2.1 and its generalization Theorem 4.2.3 – we prove a key lemma which is the heart of the proof of both. We do this in two ways – once in Section 4.4.1 using transversality and once in Section 4.4.2 using purely homotopy-theoretic methods. The proof using transversality is easier in the sense that it invokes general position results that even the reader not familiar with this theory should be able to believe. We will also follow that thread by applying these theorems to situations involving smooth manifolds (see, e.g., Example 4.2.19). Since the transversality techniques needed for the proof of the key lemma are not needed anywhere else in the book, the required background has been relegated to Section A.2. The more homotopy-theoretic proof is included because it fits the general point of view of this book better. Moreover, both styles of proof will be generalized to prove a key lemma related to Theorem 6.2.1. The proofs of the two Blakers–Massey theorems, which will

follow by formal facts about squares developed in Chapter 3, will then be given
in Section 4.4.3.

### 4.4.1 The geometric step using transversality

The key lemma we need is concerned with the situation of attaching two cells
to a space $X$. It is this special case of Theorem 4.2.1 which implies the general
theorem.

**Lemma 4.4.1**  *The square*

$$\begin{array}{ccc} X & \longrightarrow & X \cup e^{d_1} \\ \downarrow & & \downarrow \\ X \cup e^{d_2} & \longrightarrow & X \cup e^{d_1} \cup e^{d_2} \end{array}$$

*where $e^{d_j}$ is a cell of dimension $d_j$ for $j = 1, 2$ and the maps are inclusions, is*
$(d_1 + d_2 - 3)$-*cartesian.*

*Proof*  For $j = 1, 2$, let $p_j \in e^{d_j}$ be points in the interiors of the cells. Note
that $p_j$ has a neighborhood homeomorphic to $\mathbb{R}^{d_j}$, so that $Y = X \cup e^{d_1} \cup e^{d_2}$ is
a manifold near the $p_j$. It suffices to show that

$$\begin{array}{ccc} Y - \{p_1, p_2\} & \longrightarrow & Y - \{p_2\} \\ \downarrow & & \downarrow \\ Y - \{p_1\} & \longrightarrow & Y \end{array}$$

is $(d_1 + d_2 - 3)$-cartesian, or, equivalently, that the map

$$Y - \{p_1, p_2\} \longrightarrow \mathrm{holim}\left((Y - \{p_1\}) \to Y \leftarrow Y - \{p_2\}\right)$$

is $(d_1 + d_2 - 3)$-connected.

By Example 3.3.3, the natural map $Y - \{p_1, p_2\} \to \mathrm{holim}(Y - \{p_1, p_2\} \to$
$Y - \{p_1, p_2\} \leftarrow Y - \{p_1, p_2\})$ is a homotopy equivalence, and hence it suffices
to prove the inclusion

$$\mathrm{holim}\left(Y - \{p_1, p_2\} \to Y - \{p_1, p_2\} \leftarrow Y - \{p_1, p_2\}\right)$$

$$\longrightarrow \mathrm{holim}\left((Y - \{p_1\}) \to Y \leftarrow Y - \{p_2\}\right)$$

is $(d_1 + d_2 - 3)$-connected. Let $Y_{1 \cap 2} = \mathrm{holim}(Y - \{p_1, p_2\} \to Y - \{p_1, p_2\} \leftarrow Y -$
$\{p_1, p_2\})$ and $Y_{1 \cup 2} = \mathrm{holim}((Y - \{p_1\}) \to Y \leftarrow Y - \{p_2\})$. The advantage $Y_{1 \cap 2}$ has
over $Y - \{p_1, p_2\}$ is that the inclusion $Y_{1 \cap 2} \to Y_{1 \cup 2}$ is open.[1] Consider a map of

---

[1] Let $I \times Y_{1 \cup 2} \to Y$ send $(t, (y_1, \gamma, y_2))$ to $\gamma(t)$. The inverse image of $\{p_1\} \cup \{p_2\} \subset Y$ is closed,
and the projection of its complement to $Y_{1 \cup 2}$ is the set $Y_{1 \cap 2}$, which is therefore open.

pairs $g\colon (D^i, \partial D^i) \to (Y_{1\cup 2}, Y_{1\cap 2})$. Since the $p_i$ are closed and $\partial D^i$ is compact, there is a neighborhood $T$ of $\partial D^i$ such that $g(T) \subset Y_{1\cap 2}$. Let $\widetilde{g}\colon D^i \times I \to Y$ be the map corresponding to $g$ via Theorem 1.2.7, so that $\widetilde{g}(T \times I) \subset Y - \{p_1, p_2\}$ by our previous observations.

Our first goal is to find a smooth approximation for $\widetilde{g}$. For $i = 1, 2$ let $B_i \subset \overset{\circ}{e}{}^{d_i}$ be open neighborhoods of $p_i$, chosen to lie in the interior of the cells so they can be given the structure of a smooth manifold. We can choose the $B_i$ small enough so that $\widetilde{g}^{-1}(B_i)$ has empty intersection with $T \times I$ for each $i$. Let $B = \widetilde{g}^{-1}(B_1) \cup \widetilde{g}^{-1}(B_2)$. Then Theorem A.2.10 clearly applies to $\widetilde{g}|_B\colon B \to \overset{\circ}{e}{}^{d_1} \cup \overset{\circ}{e}{}^{d_2}$, and we may assume that the homotopy used to smooth $\widetilde{g}|_B$ is constant outside of a small open neighborhood of $\widetilde{g}^{-1}(p_1) \cup \widetilde{g}^{-1}(p_2)$. Let $B' \subset B \times B$ be the set of all $((s, t_1), (s, t_2))$ such that $(s, t_i) \in B$ and $t_1 \neq t_2$. Another small homotopy makes the map

$$G\colon B' \longrightarrow (\overset{\circ}{e}{}^{d_1} \cup \overset{\circ}{e}{}^{d_2})^2$$

given by $G(s, t_1, t_2) = (\widetilde{g}(s, t_1), \widetilde{g}(s, t_1))$ transverse to $\{p_1\} \times \{p_2\}$ by Theorem A.2.23 (the argument is essentially the same as that given in Example 4.2.20). In this case transverse means empty if $i + 2 < d_1 + d_2$, or $i < d_1 + d_2 - 2$. Now the map $\widetilde{g}$ is homotopic to a map $\widetilde{g}'\colon D^i \times I \to Y$ which misses $\{p_1, p_2\}$, and the associated map $g'\colon D^i \to Y_{1\cap 2}$ is the desired lift. Hence $Y_{1\cap 2} \to Y_{1\cup 2}$ is $(d_1 + d_2 - 3)$-connected. $\qquad\square$

**Remark 4.4.2** As we mentioned in the course of the proof, the transversality argument used in this proof of Lemma 4.4.1 is in the same spirit as the one used in Example 4.2.20. $\qquad\square$

## 4.4.2 The geometric step using homotopy theory

In this section, we offer an alternative proof of Lemma 4.4.1 which does not rely on transversality arguments, and is purely homotopy-theoretic in nature. The proof given in the previous section is certainly shorter than the one below, but there we made use of significant machinery which is only sketched in the appendix, whereas here we rely on only elementary homotopy-theoretic constructions. Thematically the proof given in this section is similar to those for excision of ordinary homology in that it involves subdivision and small deformations. It is based on a proof we learned from tom Dieck's book [tD08], who in turn cites Puppe [tDKP70] for the original idea. The presentation here follows that of tom Dieck closely through Theorem 4.4.6, and then we follow Goodwillie's proof [Goo92] from there, replacing his transversality argument (not exactly replicated above) with a "coordinate counting" argument. This entire chain of reasoning is generalized later to give an alternative proof Lemma 6.3.1 (a crucial ingredient in the proof of Theorem 6.2.1).

We begin with a definition followed by a technical lemma.

**Definition 4.4.3**  Let $a = (a_1, \ldots, a_n) \in \mathbb{R}^n$, $\delta > 0$, and $L \subset \{1, \ldots, n\}$ (possibly empty). A *cube* $W$ in $\mathbb{R}^n$ is a set of the form

$$W = W(a, \delta, L) = \{x \in \mathbb{R}^n : a_i \leq x_i \leq a_i + \delta \text{ for } i \in L, x_i = a_i \text{ for } i \notin L\}.$$

Define $\dim(W) = |L|$. The *boundary* $\partial W$ of $W$ is the set of all $x$ in $W$ such that $x_i = a_i$ or $x_i = a_i + \delta$ for at least one value of $i \in L$. The boundary $\partial W$ is a union of *faces*. A face of a cube is also a cube.

**Definition 4.4.4**   With $W$ as above and for $j = 1, 2$ and $p \geq 1$, define

$$K_p^j(W)$$
$$= \left\{ x \in W : \frac{\delta(j-1)}{2} + a_i < x_i < \frac{\delta j}{2} + a_i \text{ for at least } p \text{ values of } i \in L \right\}.$$

Note that if $p > \dim(W)$, then $K_p^j(W) = \emptyset$. If $p \leq q$, then $K_q^j(W) \subset K_p^j(W)$. The following lemma gives the basic technical deformation result; it appears as Lemma 6.9.1 of [tD08].

**Lemma 4.4.5**   *Let $Y$ be a space with a subspace $A \subset Y$, $W$ a cube, $j \in \{1, 2\}$, and $f \colon W \to Y$ a map. For a given $p \leq \dim(W)$ suppose that*

$$f^{-1}(A) \cap W' \subset K_p^j(W')$$

*for all cubes $W' \subset \partial W$. Then there exists a map $g \colon W \to Y$ homotopic to $f$ relative to $\partial W$ such that*

$$g^{-1}(A) \subset K_p^j(W).$$

*Proof*  Without loss of generality $W = I^n$, $n \geq 1$. We will construct a map $h \colon I^n \to I^n$ homotopic to the identity and define $g$ to be the composition of $f$ with $h$. Let $x = \left( \frac{2j-1}{4}, \ldots, \frac{2j-1}{4} \right)$ be the center of the cube $\left[ \frac{j-1}{2}, \frac{j}{2} \right]^n$. For a ray $y$ emanating from $x$, let $P(y)$ be its intersection with $\partial \left[ \frac{j-1}{2}, \frac{j}{2} \right]^n$ and $Q(y)$ its intersection with $\partial I^n$. Let $h$ map the segment from $P(y)$ to $Q(y)$ onto the point $Q(y)$ and the segment from $x$ to $P(y)$ affinely onto the segment from $x$ to $Q(y)$. Clearly $h$ is homotopic to the identity of $I^n$ relative to $\partial I^n$, and so $g = f \circ h$ is homotopic to $f$ relative to $\partial I^n$. It remains to check that $g$ satisfies the property in the conclusion of the theorem.

Suppose $z \in I^n$ and $g(z) \in A$. Write $z = (z_1, \ldots, z_n)$. If $z \in \left(\frac{i-1}{2}, \frac{i}{2}\right)$, then $z \in K_n^j(W) \subset K_p^j(W)$ and we are done. Suppose then that there exists $i$ so that either $z_i \geq \frac{i}{2}$ or $z_j \leq \frac{i-1}{2}$. Then, by definition of $h$, we have $h(z) \in \partial I^n$, so $h(z) \in W'$ for some face $W'$ of dimension $n - 1$. Since $g(z) = f(h(z)) \in A$, $h(z) \in f^{-1}(A)$, then by assumption $h(z) \in K_p^j(W')$. Thus for at least $p$ values of $i$, we have $\frac{i-1}{2} < h(z)_i < \frac{i}{2}$, where $h(z)_i$ denotes the $i$th coordinate of $h(z)$. By definition of $h$,

$$h(z)_i = \frac{2j-1}{4} + t\left(z_i - \frac{2j-1}{4}\right) \qquad \text{for } t \geq 1.$$

Inserting this expression into the previous inequalities and solving for $z_i$ yields

$$-\frac{1}{4t} + \frac{2j-1}{4} < z_i < \frac{1}{4t} + \frac{2j-1}{4}.$$

Since the lower bound increases with $t$ and the upper bound decreases with $t$, substituting $t = 1$ into each gives

$$\frac{j-1}{2} < z_i < \frac{j}{2}$$

so that $z \in K_p^j(W)$. □

Suppose $Y$ is a space with open subsets $Y_\emptyset, Y_1, Y_2$ such that $Y$ is the union of $Y_1$ with $Y_2$ along $Y_\emptyset$. Let $f: I^n \to Y$ be a map. By the Lebesgue covering lemma (Lemma 1.4.11) we can decompose $I^n$ into cubes $W$ such that $f(W) \subset Y_j$ for some $j \in \{\emptyset, 1, 2\}$ depending on $W$. The following appears as Theorem 6.9.2 in [tD08]. The proof is nearly identical, and we include it for completeness and because we will generalize all of this in one of our proofs of Theorem 6.2.1.

**Theorem 4.4.6** *With the $Y_j$ and $f$ as above, assume that for each $j$, $(Y_j, Y_\emptyset)$ is $k_j$-connected, with $k_j \geq 0$ (i.e. the inclusion $Y_\emptyset \to Y_j$ is $k_j$-connected). Then there is a homotopy $f_t$ of $f$ with $f_0 = f$ such that*

1. *if $f(W) \subset Y_j$, then $f_t(W) \subset Y_j$ for all $t$;*
2. *if $f(W) \subset Y_\emptyset$, then $f_t(W) = f(W)$ for all $t$;*
3. *if $f(W) \subset Y_j$, $f_1^{-1}(Y_j \setminus Y_\emptyset) \cap W \subset K_{k_j+1}^j(W)$.*

*Proof* Let $C^l$ be the union of cubes $W$ with $\dim(W) \leq l$. The homotopy $f_t$ is constructed inductively over $C^l \times I$. If $\dim(W) = 0$, then if $f(W) \subset Y_\emptyset$, we let $f_t = f$, which achieves the second condition. If $f(W) \subset Y_j$ and $f(W) \not\subset Y_i$ for $i \neq j$, then since $(Y_j, Y_\emptyset)$ is $k_j$-connected and $k_j \geq 0$, we may choose a path from $f(W)$ to some point in $Y_\emptyset$ and use this as the homotopy, so that $f_1(W) \subset Y_\emptyset$.

Then clearly the first condition holds and so does the third. This proves the base case.

Since the inclusion $\partial W \subset W$ is a cofibration for any cube $W$ by Example 2.3.5, we may extend over all cubes $W$ so that the first and second conditions hold. By induction suppose that $f$ has been changed by a homotopy satisfying all three conditions for cubes of dimension less than $l$, and let $W$ be a cube with $\dim(W) = l$. If $f(W) \subset Y_\emptyset$, we let $f_t = f$ as usual. If $f(W) \subset Y_j$ and $f(W) \not\subset Y_i$ for $i \neq j$, then we have the following.

- If $\dim(W) = l \leq k_j$, since $(Y_j, Y_\emptyset)$ is $k_j$-connected by definition there is a homotopy $f_t$ of $f$ relative to $\partial W$ such that $f_1(W) \subset Y_\emptyset$, and clearly the first and third conditions hold.
- If $\dim(W) = l > k_j$, we use Lemma 4.4.5. Let $A = Y_j \backslash Y_\emptyset \subset Y_j$. By induction, for all $W' \subset \partial W$

$$f^{-1}(Y_j \backslash Y_\emptyset) \cap W' \subset K_l^j(W') \subset K_{k_j+1}^j(W'),$$

and by Lemma 4.4.5, there is a homotopy $f_t$ of $f$ relative to $\partial W$ such that $f_1^{-1}(Y_j \backslash Y_\emptyset) \cap W \subset K_{k_j+1}^j(W)$.      □

Before we use these results to give a proof of Lemma 4.4.1, we need to do two things:

1. Reduce to the case where the connectivities of the maps $X \to X \cup e^{d_j}$ for $j = 1, 2$ are at least zero (i.e. we want $d_j \geq 1$ for $j = 1, 2$).
2. Convert the square in the statement of Lemma 4.4.1 into one where the maps are inclusions of open sets in order to apply the previous results.

For the first item, note that it is enough by Proposition 3.3.18 to prove that, for all choices of basepoint $x \in X \cup e^{d_2}$, the induced map

$$\mathrm{hofiber}_x(X \to X \cup e^{d_2}) \to \mathrm{hofiber}_x(X \cup e^{d_1} \to X \cup e^{d_1} \cup e^{d_2})$$

has the desired connectivity. If $d_2 = 0$ and the basepoint is chosen as the single point $e^{d_2} = *$ disjoint from $X$, then the result is trivially true, since the map of homotopy fibers is the identity map of the empty set to itself. If not, then the cell $e^{d_2} = *$ plays no role in either of the homotopy fibers above, and the map is a weak equivalence. Hence we may assume that $d_2 \geq 1$, and in this case the basepoint may be chosen to lie in $X$ itself by homotopy invariance of homotopy fibers over path components (Corollary 3.2.18). Similarly we may assume $d_1 \geq 1$.

For the second item, as in the proof of Lemma 4.4.1 presented above for $j = 1, 2$ let $p_j \in e^{d_j}$ be interior points, and let $Y = X \cup e^{d_1} \cup e^{d_2}$. Then the square

$$Y - \{p_1, p_2\} \longrightarrow Y - \{p_2\}$$
$$\downarrow \qquad\qquad\qquad \downarrow$$
$$Y - \{p_1\} \longrightarrow Y$$

admits a weak equivalence from the square

$$X \longrightarrow X \cup e^{d_1}$$
$$\downarrow \qquad\qquad\qquad \downarrow$$
$$X \cup e^{d_2} \longrightarrow X \cup e^{d_1} \cup e^{d_2}$$

by the natural inclusions, and if we put $Y = Y_{12} = X \cup e^{d_1} \cup e^{d_2}$, $Y_\emptyset = Y - \{p_1, p_2\}$, $Y_1 = Y - \{p_2\}$, and $Y_2 = Y - \{p_1\}$, we are in the situation of Theorem 4.4.6, where $k_j = d_j - 1$ for $j = 1, 2$.

We now have everything set up to give an alternative proof of Lemma 4.4.1. Our presentation now follows that of Goodwillie [Goo92] more closely, although, as mentioned earlier, we have replaced his transversality argument with a "coordinate counting" one using the tools developed above.

*Alternative proof of Lemma 4.4.1* With $Y_S$ as above, choose a basepoint $y \in Y_\emptyset = Y - \{p_1, p_2\}$. We are to show that the map

$$\text{hofiber}_y(Y - \{p_1, p_2\} \to Y - \{p_1\}) \longrightarrow \text{hofiber}_y(Y - \{p_2\} \to Y)$$

is $(d_1 + d_2 - 3)$-connected. Let $C$ be the contractible space

$$C = \text{hofiber}_y(Y - \{p_2\} \longrightarrow Y - \{p_2\}) \simeq *.$$

Both

$$C \cap \text{hofiber}_y(Y - \{p_1, p_2\} \to Y - \{p_1\}) = \text{hofiber}_y(Y - \{p_1, p_2\} \to Y - \{p_1, p_2\})$$

and

$$C \cap \text{hofiber}_y(Y - \{p_2\} \to Y) = \text{hofiber}_y(Y - \{p_2\} \to Y - \{p_2\})$$

are contractible. The square

$$C \cap Z \longrightarrow Z$$
$$\downarrow \qquad\qquad\qquad \downarrow$$
$$C \longrightarrow C \cup Z$$

is homotopy cocartesian by Example 3.7.5.[2] It follows from statement 1 of Proposition 3.7.13 that the inclusions

$$\text{hofiber}_y(Y - \{p_1, p_2\} \to Y - \{p_1\}) \longrightarrow C \cup \text{hofiber}_y(Y - \{p_1, p_2\} \to Y - \{p_1\})$$

and

$$\text{hofiber}_y(Y - \{p_2\} \to Y) = C \cup \text{hofiber}_y(Y - \{p_2\} \to Y)$$

are weak equivalences. Hence it suffices to show that

$$C \cup \text{hofiber}_y(Y - \{p_1, p_2\} \to Y - \{p_1\}) \to C \cup \text{hofiber}_y(Y - \{p_2\} \to Y)$$

is $(d_1 + d_2 - 3)$-connected.

Let

$$\phi \colon (I^n, \partial I^n) \longrightarrow \Big( C \cup \text{hofiber}_y(Y - \{p_2\} \to Y),$$
$$C \cup \text{hofiber}_y(Y - \{p_1, p_2\} \to Y - \{p_1\}) \Big)$$

be a map of pairs. The map $\phi$ is associated via Theorem 1.2.7 to a map

$$\Phi \colon I^n \times I \to Y$$

with boundary conditions

(B0) $\Phi(z, 0) = y \in Y - \{p_1, p_2\}$ is the basepoint for all $z \in I^n$;

(B1) $\Phi(z, 1) \in Y - \{p_2\}$ for all $z \in I^n$;

(B2) For each $z \in \partial I^n$ there exists $i \in \{1, 2\}$ so that $\Phi(z, t) \in Y_i$ for all $t \in I$.

We will make a homotopy of $\Phi$ preserving (B0)–(B2) such that (B2) holds for all $z \in I^n$ when $n \leq d_1 + d_2 - 3$.

It is now convenient to change notation and recall that $Y = Y_{12} = X \cup e^{d_1} \cup e^{d_2}$, $Y_0 = Y - \{p_1, p_2\}$, $Y_1 = Y - \{p_2\}$, and $Y_2 = Y - \{p_1\}$. We apply Theorem 4.4.6 to $\Phi \colon I^n \times I \to Y$ and obtain a decomposition of $I^n \times I$ into cubes $W$ such that, for each $W$, there is some $j$ so $\Phi(W) \subset Y_j$, and a homotopy $\Phi_r$ for $0 \leq r \leq 1$ of $\Phi = \Phi_0$ such that

1. $\Phi(W) \subset Y_j$ implies $\Phi_r(W) \subset Y_j$ for all $r$;
2. $\Phi(W) \subset Y_0$ implies $\Phi_r(W) = \Phi(W)$ for all $r$;
3. $\Phi(W) \subset Y_j$ implies $\Phi_1^{-1}(Y_j \setminus Y_0) \cap W \subset K_{d_j}^{j,k}(W)$.

First we prove that $\Phi_r$ satisfies (B0)–(B2) for all $r$.

(B0)   Since $\Phi(z, 0) = y \in Y_0$ is the basepoint for all $z \in I^n$, we have for all cubes $W \subset I^n \times \{0\}$ that $\Phi(W) = y$, and the second condition above implies $\Phi_r(W) = \Phi(W)$ for all $r$, so that $\Phi_r(z, 0) = y$ for all $r$.

---

[2]  It is not difficult to verify that these are all open subsets of $\text{hofiber}_y(Y \to Y)$. We needed a similar result in our first proof of Lemma 4.4.1.

(B1) Since $\Phi(z, 1) \in Y_1$ for all $z \in I^n$, then for all cubes $W \subset I^n \times \{1\}$, $\Phi(W) \subset Y_1$ by the first condition above, and so $\Phi_r(z, q) \in Y_1$ for all $z \in I^n$ and for all $r$.

(B2) For each $z \in \partial I^n$ there exists $j(z) \in \{1, 2\}$ such that $\Phi(\{z\} \times I) \subset Y_{j(z)}$. Let $W_1, \ldots, W_h$ be cubes in $I^n \times I$ such that $\{z\} \times I \subset W_1 \cup \cdots \cup W_h$ and such that each $W_a$ contains a point of the form $(z, t)$ for some $t$. Since $\Phi(\{z\} \times I) \subset Y_{j(z)}$, for each $a = 1$ to $h$ we must have $\Phi(W_a) \subset Y_{j(z)}$, which implies $\Phi_r(W_1 \cup \cdots \cup W_h) \subset Y_{j(z)}$ for all $r$ by the first condition above.

Now we show that $\Phi_1$ actually satisfies the stronger condition that for each $z \in I^n$ there exists $j(z) \in \{1, 2\}$ so that $\Phi_1(z, t) \in Y_{j(z)}$ for all $t \in I$ when $n \leq d_1 + d_2 - 3$. Let $\pi \colon I^n \times I \to I^n$ be the projection. We claim that

$$\pi\left(\Phi_1^{-1}(Y_1 \setminus Y_\emptyset)\right) \cap \pi\left(\Phi_1^{-1}(Y_2 \setminus Y_\emptyset)\right) = \emptyset$$

if $n \leq d_1 + d_2 - 3$. Let $y$ be a point in the intersection above. Thus there exists $t_1, t_2$ such that $(y, t_j) \in \Phi_1^{-1}(Y_j \setminus Y_\emptyset)$, and of course $y = \pi(y, t_j)$ for $j = 1, 2$, and $(y, t_j) \in W_j$ for some cube $W_j \subset I^n \times I$. Hence $w(j) = (y, t_j) \in W_j \cap \Phi_1^{-1}(Y_j \setminus Y_\emptyset) \subset K_{d_j}^j(W_j)$ for $j = 1, 2$ by item 3 of Theorem 4.4.6. Thus $w(j)$ has at least $d_j$ coordinates $w(j)_i$ such that $a_i + \delta(j - 1)/2 < w(j)_i < a_i + \delta j/2$, where $W_j = W(a, \delta, L)$ and $a = (a_1, \ldots, a_{n+1})$. Hence, for each $j$, $y$ has at least $d_j$ coordinates $y_i$ satisfying the same bounds. For $j = 1, 2$ the projection $\pi(W_j)$ is a cube containing $y$, and we may assume $\pi(W_1) = \pi(W_2)$ by subdividing further if necessary, thus making the bounds above the same bounds for each $j$. Therefore $y$ has at least $d_j$ coordinates satisfying the above bounds for all $j$, which is impossible if $n \leq d_1 + d_2 - 3$, so that the intersection above is indeed empty. Hence there is some $i(z) \in \{1, 2\}$ such that $y \notin \pi\left(\Phi_1^{-1}(Y_{i(z)} \setminus Y_\emptyset)\right)$; that is, for all $t$, $z = (y, t) \notin \Phi_1^{-1}(Y_{i(z)} \setminus Y_\emptyset)$. When $n = 0$, to show the map of path components is surjective we require $d_1, d_2 \geq 1$. We already noted just before the start of this proof how the statement of the theorem was true if some $d_j = 0$. □

### 4.4.3 The formal step

Here we finally prove the Blakers–Massey Theorems for squares.

*Proof of Theorem 4.2.1* Recall that we are trying to show that the homotopy cocartesian square

$$\begin{array}{ccc} X_\emptyset & \longrightarrow & X_1 \\ \downarrow & & \downarrow \\ X_2 & \longrightarrow & X_{12} \end{array}$$

is $(k_1 + k_2 - 1)$-cartesian when the maps $X_\emptyset \to X_i$ are $k_i$-connected for $i = 1, 2$. Our goal is to reduce to the special case considered in Lemma 4.4.1. For simplicity, write $K = k_1 + k_2 - 1$.

By Proposition 3.7.29 we may also assume the square above is of the form

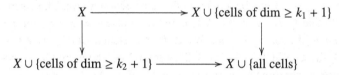

where $X_\emptyset \to X_1$ and $X_\emptyset \to X_2$ are cofibrations. Furthermore, by Theorem 1.3.8 and Theorem 2.6.26 we may also assume for each $i = 1, 2$ that $(X_i, X_\emptyset)$ is a relative CW complex with $X_i = X_\emptyset \cup \{\text{cells of dim} \geq k_i + 1\}$. Letting $X = X_\emptyset$, it suffices to prove that the square

$$
\begin{array}{ccc}
X & \longrightarrow & X \cup \{\text{cells of dim} \geq k_1 + 1\} \\
\downarrow & & \downarrow \\
X \cup \{\text{cells of dim} \geq k_2 + 1\} & \longrightarrow & X \cup \{\text{all cells}\}
\end{array}
$$

is $(k_1 + k_2 + 1)$-cartesian. Moreover, it is enough to prove this in the case where the number of attached cells is finite. Indeed, suppose $A$ is a space and $(B, A)$, $(C, A)$, and $(D, A)$ are relative CW complexes that fit into a square of inclusions

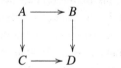

This square is $K$-cartesian if and only if hofiber$(A \to C) \to$ hofiber$(B \to D)$ is $K$-connected for all choice of basepoints in $C$ by Proposition 3.3.18, which amounts to proving every map of pairs $(D^i, \partial D^i) \to$ (hofiber$(B \to D)$, hofiber$(A \to C)$) is homotopic relative to $\partial D^i$ to a map $D^i \to$ hofiber$(A \to C)$ for $i \leq K$. But $D^i \to$ hofiber$(B \to D)$ corresponds via Theorem 1.2.7 to a map $D^i \times I \to D$, and by compactness of the domain and the topology on $D$, it can map to only finitely many of the cells which build $D$ from $A$.

To reduce further, consider the following diagram, where $e_1$ and $e_2$ are two of the cells.

$$
\begin{array}{ccccc}
X & \longrightarrow & X \cup e_1 & \longrightarrow & X \cup e_1 \cup e_2 \\
\downarrow & & \downarrow & & \downarrow \\
X \cup \{\text{cells}\} & \longrightarrow & X \cup e_1 \cup \{\text{cells}\} & \longrightarrow & X \cup e_1 \cup e_2 \cup \{\text{cells}\}
\end{array}
$$

If we can prove that the left square and the right square are $K$-cartesian, then so is

$$
\begin{array}{ccc}
X & \longrightarrow & X \cup e_1 \cup e_2 \\
\downarrow & & \downarrow \\
X \cup \{\text{cells}\} & \longrightarrow & X \cup e_1 \cup e_2 \cup \{\text{cells}\}
\end{array}
$$

by 1(a) of Proposition 3.3.20. One reduces to a single cell in the vertical direction in precisely the same manner, and now we may apply Lemma 4.4.1 to complete the proof. □

We can now prove Theorem 4.2.3 using Theorem 4.2.1.

*Proof of Theorem 4.2.3* Recall that we are to show that if the square

$$
\begin{array}{ccc}
X_\emptyset & \longrightarrow & X_1 \\
\downarrow & & \downarrow \\
X_2 & \longrightarrow & X_{12}
\end{array}
$$

is $j$-cocartesian and $X_\emptyset \to X_i$ is $k_i$-connected, then the square is $\min\{k_1 + k_2 - 1, j - 1\}$-cartesian. Let $D = \mathrm{hocolim}(X_2 \leftarrow X_\emptyset \to X_1)$. Consider the diagram

$$
\begin{array}{ccccc}
X_\emptyset & \longrightarrow & X_1' & \longrightarrow & X_1 \\
\downarrow & & \downarrow & & \downarrow \\
X_2 & \longrightarrow & D & \longrightarrow & X_{12}
\end{array}
$$

Here $X_1'$ is the mapping cylinder of the map $X_\emptyset \to X_1$, so the map $X_1 \to X_1'$ is a homotopy equivalence. The left square is homotopy cocartesian by Proposition 3.6.17 and hence $(k_1 + k_2 - 1)$-cartesian by Theorem 4.2.1. The right square is $(j - 1)$-cartesian by Proposition 3.3.11. Then 2(a) of Proposition 3.3.20 implies the outer square is $\min\{j - 1, k_1 + k + 2 - 1\}$-cartesian. □

## 4.5 Proofs of the dual Blakers–Massey Theorems for squares

The proof of Theorem 4.2.2, the dual Blakers–Massey Theorem for squares, is considerably easier than the proof of Theorem 4.2.1 because it does not require anything like Lemma 4.4.1. We already mentioned after its statement that it follows immediately from Proposition 3.9.12 together with Proposition 3.7.23; in this case the homotopy fiber of the canonical map from the homotopy pushout to the last space can be analyzed directly. Here we supply another proof using quasifibrations (the reader should review the definitions from Section 2.7) and the result about the connectivity of the join of two spaces, the aforementioned Proposition 3.7.23. Note that the proof below mimics the proof of Theorem 4.2.1 in that it seeks to reduce the theorem to a special case. For Theorem 4.2.1, this meant replacing the maps $X_0 \to X_i$ with cofibrations, and in this case we will replace maps $X_i \to X_{12}$ with fibrations. We begin with a lemma about quasifibrations.

**Lemma 4.5.1**   *Suppose*

*is a square in which all maps are fibrations. Then the canonical map*

$$\mathrm{hocolim}(X_2 \leftarrow X_0 \to X_1) \longrightarrow X_{12}$$

*is a quasifibration.*

*Proof*   We will in fact show that this map is a universal quasifibration (see Definition 2.7.16) using Proposition 2.7.18, and hence that it is a quasifibration (see remarks following Definition 2.7.16). Let $D$ be a disk and $D \to X_{12}$ a map. Write $\mathcal{X}_1 = X_2 \leftarrow X_0 \to X_1$ for short. By Remark 3.7.12, there is a natural homeomorphism

$$\mathrm{hocolim}(D \times_{X_{12}} \mathcal{X}_1) \longrightarrow D \times_{X_{12}} \mathrm{hocolim}(\mathcal{X}_1),$$

where the diagram $D \times_{X_{12}} \mathcal{X}_1$ is the entry-wise fiber products. Let $*$ be any point in $D$, and consider the square

$$\mathrm{hocolim}(* \times_{X_{12}} \mathcal{X}_1) \longrightarrow \mathrm{hocolim}(D \times_{X_{12}} \mathcal{X}_1)$$

$$\downarrow \qquad\qquad\qquad\qquad \downarrow$$

$$* \times_{X_{12}} \mathrm{hocolim}(\mathcal{X}_1) \longrightarrow D \times_{X_{12}} \mathrm{hocolim}(\mathcal{X}_1)$$

We have just discussed why the vertical arrows are homeomorphisms, and by Theorem 3.6.13, the top vertical map is a weak equivalence since $X_S \to X_{12}$ is a fibration, so the maps $* \times_{X_{12}} X_S = \mathrm{lim}(* \to X_{12} \leftarrow X_S) \to \mathrm{lim}(D \to X_{12} \leftarrow X_S) = D \times_{X_{12}} X_S$ are weak equivalences for each $S$ using Proposition 3.2.13. It follows that the bottom horizontal arrow is a weak equivalence.   □

We remark that by Proposition 3.3.24 that every square admits a homotopy equivalence to a square which satisfies the hypotheses of this lemma. We are now ready to proceed with a proof of Theorem 4.2.2.

*Proof of Theorem 4.2.2* We wish to show that the homotopy cartesian diagram

$$
\begin{array}{ccc}
X_0 & \longrightarrow & X_1 \\
\downarrow & & \downarrow \\
X_2 & \longrightarrow & X_{12}
\end{array}
$$

is $(k_1 + k_2 + 1)$-cocartesian when the maps $X_i \to X_{12}$ are $k_i$-connected for $i = 1, 2$. Again for simplicity write $K = k_1 + k_2 + 1$. By Proposition 3.3.24 we may also assume the square in question is of the form

$$
\begin{array}{ccc}
X_1 \times_{X_{12}} X_2 & \longrightarrow & X_1 \\
\downarrow & & \downarrow \\
X_2 & \longrightarrow & X_{12}
\end{array}
$$

where $X_1 \to X_{12}$ and $X_2 \to X_{12}$ are fibrations. Let $F_i = \text{fiber}(X_i \to X_{12})$ for $i = 1, 2$ (note that these are the strict fibers). By Lemma 4.5.1 the map

$$ b \colon \text{hocolim}(X_1 \leftarrow X_0 \to X_2) \longrightarrow X_{12} $$

is a quasifibration. By Corollary 3.9.4, the strict fiber of $b$ is

$$ \text{hocolim}(F_1 \leftarrow F_1 \times F_2 \to F_2), $$

which is homotopy equivalent to $F_1 * F_2$ by Example 3.6.12. Since the maps $X_i \to X_{12}$ are $k_i$-connected, $F_i$ is $(k_i - 1)$-connected, and by Proposition 3.7.23, $F_1 * F_2$ is $(k_1 + k_2)$-connected. It follows that the square is $(k_1 + k_2 + 1)$-cocartesian. □

We can now use Theorem 4.2.2 to prove Theorem 4.2.4, the more general version of the dual of the Blakers–Massey Theorem. The proof is dual to the proof of Theorem 4.2.3.

*Proof of Theorem 4.2.4* We are to show that if the square

$$
\begin{array}{ccc}
X_0 & \longrightarrow & X_1 \\
\downarrow & & \downarrow \\
X_2 & \longrightarrow & X_{12}
\end{array}
$$

is $j$-cartesian and $X_i \to X_{12}$ is $k_i$-connected, then the square is $\min\{k_1 + k_2 + 1, j + 1\}$-cocartesian.

Let $P = \text{holim}(X_2 \to X_{12} \leftarrow X_1)$. Consider the diagram

$$
\begin{array}{ccccc}
X_\emptyset & \longrightarrow & P & \longrightarrow & X_1 \\
\downarrow & & \downarrow & & \downarrow \\
X_2 & \longrightarrow & X_2' & \longrightarrow & X_{12}
\end{array}
$$

Here $X_2'$ is the path space construction for the map $X_2 \to X_{12}$, so that $X_2 \to X_2'$ is an equivalence. The right square is homotopy cartesian and hence $(k_1 + k_2 + 1)$-cocartesian by Theorem 4.2.2. The left square is $(j + 1)$-cocartesian by (2) of Proposition 3.7.13. Then Proposition 3.7.26 implies the outer square is $\min\{j + 1, k_1 + k_2 + 1\}$-cocartesian. $\qquad\qquad\square$

## 4.6 Homotopy groups of squares

The Blakers–Massey Theorem, Theorem 4.2.1, tells us a range in which homotopy groups satisfy excision. Given a homotopy cocartesian square

$$
X = \quad
\begin{array}{ccc}
X_\emptyset & \longrightarrow & X_1 \\
\downarrow & & \downarrow \\
X_2 & \longrightarrow & X_{12}
\end{array}
$$

in which the maps $X_\emptyset \to X_1$ and $X_\emptyset \to X_2$ are $k_1$- and $k_2$-connected respectively, the square is $(k_1 + k_2 - 1)$-cartesian, which means that the space $\text{tfiber}(X)$ is $(k_1 + k_2 - 2)$-connected. (According to Definition 3.4.10, this can also be thought of as $X$ being $(k_1 + k_2)$-connected.) Thus we may think of the first non-trivial homotopy group of $\text{tfiber}(X)$ as a measure of the failure of the homotopy groups of the square $X$ to satisfy excision.

We will begin by computing the first non-trivial homotopy group of the simplest kind of homotopy cocartesian square. We will then see in Proposition 6.5.1 that we can reduce the case of a generic homotopy cocartesian square to the special case considered below.

**Proposition 4.6.1**   *For $i = 1, 2$, suppose $X_i$ is a $k_i$-connected based space with $k_i \geq 1$. The square*

$$
X = \quad
\begin{array}{ccc}
* & \longrightarrow & X_1 \\
\downarrow & & \downarrow \\
X_2 & \longrightarrow & X_1 \vee X_2
\end{array}
$$

*is $(k_1 + k_2 - 1)$-cartesian, and*

$$
\pi_{k_1 + k_2 - 1}(\text{tfiber}(X)) \cong \pi_{k_1 + 1}(X_1) \otimes \pi_{k_2 + 1}(X_2).
$$

**Remark 4.6.2** The homotopy pullback $\mathrm{holim}(X_1 \to X_1 \vee X_2 \leftarrow X_2)$ is sometimes called the *cojoin of $X_1$ and $X_2$* and denoted by $X_1 \tilde{*} X_2$. The above results thus says that, if $X_i$ is $k_i$-connected, the cojoin is $(k_1 + k_2 - 1)$-connected. We will use this fact in the proof of the Berstein–Hilton Theorem, Theorem 6.2.15. $\square$

*Proof of proposition 4.6.1* The square is homotopy cocartesian and, since by assumption $X_i$ is $k_i$-connected, the maps $* \to k_i$ are $k_i$-connected. Thus, by Theorem 4.2.1, the square is $(k_1 + k_2 - 1)$-cartesian.

For the second part, by Proposition 3.4.9, $\mathrm{tfiber}(X)$ is homotopy equivalent to $\Omega^2 \, \mathrm{tfiber}(X')$, where

$$
X' = \quad
\begin{array}{ccc}
X_1 \vee X_2 & \longrightarrow & X_1 \\
\downarrow & & \downarrow \\
X_1 & \longrightarrow & *
\end{array}
$$

Since $\pi_k(\Omega X) \cong \pi_{k+1}(X)$, the homotopy group in question is therefore

$$\pi_{k_1+k_2+1} \, \mathrm{hofiber}(X_1 \vee X_2 \to X_1 \times X_2).$$

Consider the homotopy cocartesian square

$$
\begin{array}{ccc}
X_1 \vee X_2 & \longrightarrow & X_1 \times X_2 \\
\downarrow & & \downarrow \\
* & \longrightarrow & X_1 \wedge X_2
\end{array}
$$

This square is $(k_1 + k_2 + \min\{k_1, k_2\} + 1)$-cartesian by Example 4.2.7. In particular, we have isomorphisms

$$\pi_{k_1+k_2+1}\big(\mathrm{hofiber}(X_1 \vee X_2 \to X_1 \times X_2)\big) \cong \pi_{k_1+k_2+1}\big(\Omega(X_1 \wedge X_2)\big)$$

$$\cong \pi_{k_1+k_2+2}(X_1 \wedge X_2).$$

Since the $X_i$ are $k_i$-connected, $X_1 \wedge X_2$ is $(k_1 + k_2 + 1)$-connected by the proof of Proposition 3.7.23. Then, by the Hurewicz Theorem (Theorem 4.3.2),

$$\pi_{k_1+k_2+2}(X_1 \wedge X_2) \cong H_{k_1+k_2+2}(X_1 \wedge X_2).$$

By the Künneth formula and using the connectivities of $X_1$ and $X_2$,

$$H_{k_1+k_2+2}(X_1 \wedge X_2) \cong \widetilde{H}_{k_1+1}(X_1) \otimes \widetilde{H}_{k_2+1}(X_2).$$

This is isomorphic to $\pi_{k_1+1}(X_1) \otimes \pi_{k_2+2}(X_2)$ using the Hurewicz Theorem again.

$\square$

When the spaces $X_1, X_2$ above are both suspensions, we can not only identify this first non-trivial group but also give a stable-range space-level description of the homotopy groups of $\mathrm{tfiber}(X')$, and hence $\mathrm{tfiber}(X)$. The discussion below will continue in Section 6.5 after we have stated and proved the generalized Blakers–Massey Theorem, Theorem 6.2.1.

**Theorem 4.6.3**  *Suppose $X_1$ and $X_2$ are based spaces, $k_1$- and $k_2$-connected respectively. Consider the homotopy cocartesian square*

$$\mathcal{X} = \quad \begin{array}{ccc} \Sigma(X_1 \vee X_2) & \longrightarrow & \Sigma X_1 \\ \downarrow & & \downarrow \\ \Sigma X_2 & \longrightarrow & \Sigma * \end{array}$$

*Then there exists a $(k_1 + k_2 + \min\{k_1, k_2\} + 3)$-connected map*

$$P \colon \Sigma(X_1 \wedge X_2) \longrightarrow \text{tfiber}(\mathcal{X}).$$

Note that the square $\mathcal{X}$ is $(k_1 + k_2 + 3)$-cartesian by Theorem 4.2.1, and hence tfiber($\mathcal{X}$) is $(k_1 + k_2 + 2)$-connected, so that the map $P$ may be thought of as giving a "stable-range" description of the homotopy groups of tfiber($\mathcal{X}$).

The map $P$ above is the *generalized Whitehead product*, to be defined below (see [Ark62, Definition 2.2]; also see [Spe71, Definition 6.2] for a relative version).

In the situation described in Theorem 4.6.3, the Whitehead product map $P$ is induced by a map $W \colon \Sigma(X_1 \wedge X_2) \to \Sigma(X_1 \vee X_2)$ which is built from the commutator, defined below.

**Definition 4.6.4**  Suppose $X_1$ and $X_2$ are based spaces, and define the *commutator* map

$$C \colon \Omega X_1 \times \Omega X_2 \longrightarrow \Omega(X_1 \vee X_2)$$

by $C(\alpha, \beta) = \alpha * \beta * \alpha^{-1} * \beta^{-1}$, where $*$ stands for path multiplication.

Let $\ell \colon X \to \Omega\Sigma X$ be the map associated with $1_{\Sigma X} \colon \Sigma X \to \Sigma X$ via Theorem 1.2.7; this is the map which sends $x$ to the loop $(t \mapsto t \wedge x) = \ell(x)$ (the map which naturally occurs in the Freudenthal Suspension Theorem, Theorem 4.2.8). Let $\widetilde{C} \colon X_1 \times X_2 \to \Omega\Sigma(X_1 \vee X_2)$ be given by $\widetilde{C}(x_1, x_2) = C(\ell(x_1), \ell(x_2))$, and let $\widetilde{W} \colon \Sigma(X_1 \times X_2) \to \Sigma(X_1 \vee X_2)$ be the associated map via Theorem 1.2.7. We need to describe a few properties of $\widetilde{W}$, so we explicitly write its formula below to make filling in the necessary details easier for the reader. We have

$$\widetilde{W}(t \wedge (x_1, x_2)) = \begin{cases} 4t \wedge (x_1, *), & \text{if } 0 \leq t \leq 1/4; \\ (4t - 1) \wedge (*, x_2), & \text{if } 1/4 \leq t \leq 1/2; \\ (3 - 4t) \wedge (x_1, *), & \text{if } 1/2 \leq t \leq 3/4; \\ (4 - 4t) \wedge (*, x_2), & \text{if } 3/4 \leq t \leq 1. \end{cases}$$

It is not hard to see that $\widetilde{W}(t \wedge (x_1, *))$ and $\widetilde{W}(t \wedge (*, x_2))$ are null-homotopic. We leave the details to the reader, though this should be readily believable since the idea is that the expression $\alpha * \beta * \alpha^{-1} * \beta^{-1}$ is null-homotopic if one of $\alpha$ or $\beta$ is null-homotopic.[3] Hence the restriction of $\widetilde{W}$ to $\Sigma(X_1 \vee X_2) \subset \Sigma(X_1 \times X_2)$ is null-homotopic. Choose a null-homotopy $H_t \colon \Sigma(X_1 \vee X_2) \to \Sigma(X_1 \vee X_2)$ where $H_0 = \widetilde{W}$, and let $H_1 = \widetilde{W}_1$. Since the inclusion $\Sigma(X_1 \vee X_2) \to \Sigma(X_1 \times X_2)$ is a cofibration we may extend $H_t$ to all of $\Sigma(X_1 \times X_2)$. Thus we have a commutative diagram

$$
\begin{array}{ccc}
\Sigma(X_1 \vee X_2) & \xrightarrow{\ \iota\ } & \Sigma(X_1 \times X_2) \\
\downarrow & & \downarrow{\scriptstyle \widetilde{W}_1} \\
* & \longrightarrow & \Sigma(X_1 \vee X_2)
\end{array}
$$

**Definition 4.6.5**   Define

$$W \colon \Sigma(X_1 \wedge X_2) \longrightarrow \Sigma(X_1 \vee X_2)$$

to be the induced map of horizontal homotopy cofibers above. This is called the *generalized Whitehead product* of the inclusion maps $\Sigma X_i \to \Sigma(X_1 \vee X_2)$.

The homotopy class of $W$ is independent of the choice of homotopy $H_t$ used to define it. For the same reasons that $\widetilde{W}$ is null-homotopic when restricted to $\Sigma(X_1 \vee X_2)$, the composed maps $p_i \circ W \colon \Sigma(X_1 \wedge X_2) \to \Sigma X_i$, where $p_i \colon \Sigma(X_1 \vee X_2) \to \Sigma X_i$ is the projection, are null-homotopic for $i = 1, 2$. Hence $W$ induces a map

$$P \colon \Sigma(X_1 \wedge X_2) \to \mathrm{tfiber}(\mathcal{X}) \simeq \mathrm{hofiber}\big(\Sigma(X_1 \vee X_2) \to \Sigma X_1 \times \Sigma X_2\big).$$

We will sketch a proof that this map has a high connectivity indirectly, appealing to Theorem 4.2.1 and its special case, the Freudenthal Suspension Theorem.

**Theorem 4.6.6**   *There is a diagram*

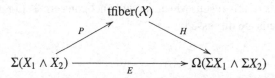

*which commutes up to homotopy, in which the map $E$ is the canonical map*

$$\Sigma(X_1 \wedge X_2) \longrightarrow \Omega\Sigma\Sigma(X_1 \wedge X_2) \cong \Omega(\Sigma X_1 \wedge \Sigma X_2),$$

---

[3] Analogously, the commutator of the identity element of a group with any other element is the identity.

*given by Theorem 4.2.10, and where $H$ is induced by a map of homotopy fibers in a homotopy cocartesian square. The map $E$ is $(2(k_1 + k_2 + 2) + 1)$-connected, $H$ is $(k_1 + k_2 + \min\{k_1, k_2\} + 4)$-connected, and hence $P$ is $(k + 1 + k_2 + \min\{k_1, k_2\} + 3)$-connected.*

*Proof* We will sketch the argument. Consider the homotopy cocartesian square (noting the canonical homeomorphism $\Sigma(X_1 \vee X_2) \cong \Sigma X_1 \vee \Sigma X_2$)

$$
\begin{array}{ccc}
\Sigma(X_1 \vee X_2) & \longrightarrow & \Sigma X_1 \times \Sigma X_2 \\
\downarrow & & \downarrow \\
* & \longrightarrow & \Sigma X_1 \wedge \Sigma X_2
\end{array}
$$

Using the fact that the square in Theorem 4.6.3 is evidently $(k_1 + k_2 + 3)$-cartesian, the top horizontal map in the above is thus $(k_1 + k_2 + 3)$-connected. The left vertical map is clearly $(\min\{k_1, k_2\} + 2)$-connected, and hence the square is $K = (k_1 + k_2 + \min\{k_1, k_2\} + 4)$-cartesian by Theorem 4.2.1. Thus the induced map

$$H : \operatorname{tfiber}(X) \simeq \operatorname{hofiber}\big(\Sigma(X_1 \vee X_2) \to \Sigma X_1 \times \Sigma X_2\big) \longrightarrow \Omega(\Sigma X_1 \wedge \Sigma X_2)$$

of horizontal homotopy fibers is $K$-connected. Now $\Sigma(X_1 \wedge X_2)$ is $(k_1 + k_2 + 2)$-connected, and the canonical map $\ell : \Sigma(X_1 \wedge X_2) \to \Omega\Sigma\Sigma(X_1 \wedge X_2) \cong \Omega(\Sigma X_1 \wedge \Sigma X_2)$ is $(2(k_1 + k_2 + 2) + 1)$-connected by the Freudenthal Suspension Theorem. It follows from Proposition 2.6.15 that $W : \Sigma(X_1 \wedge X_2) \to \operatorname{tfiber}(X)$ is $(K - 1)$-connected provided that the diagram

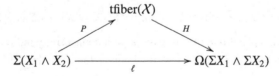

commutes up to homotopy. For the details on this, we suggest [Joh95, Proposition 6.8] and [Mun11, Theorem 4.3]. Both of these references are about generalizations of this result, which we will encounter in Section 6.5, where a few more details are discussed. $\qquad\square$

# 5

# $n$-cubes: Generalized homotopy pullbacks and pushouts

In Chapter 3, we were interested in comparing the initial (final) space of a square to the homotopy limit (colimit) of the rest of the square. This led to the notion of a homotopy (co)cartesian and $k$-(co)cartesian squares and we established many interesting properties of such squares. This chapter is the cubical version of this story. In particular, we will define homotopy (co)cartesian cubes and their $k$-(co)cartesian counterparts and then discuss their various features. The narrative here will very closely parallel that of Sections 3.3, 3.4, 3.7, 3.8, and 3.9.

## 5.1 Cubical and punctured cubical diagrams

Here we set some notation and terminology. Let $n \geq 0$ be an integer. Informally, a $n$-cube or $n$-cubical diagram of spaces is a commutative diagram in the shape of a $n$-cube. We have already studied 0-cubes (spaces), 1-cubes (maps of spaces), and 2-cubes (square diagrams), and even encountered the stray 3-cube. More formally, consider the poset $\mathcal{P}(n)$ of subsets of $\underline{n} = \{1, \ldots, n\}$. The poset structure is given by containment, so $S \leq T$ means $S \subset T$, and any poset may be considered as a *category*. In later chapters, we will refer to $\mathcal{P}(\underline{n})$ as a *cubical indexing category*. An $n$-cube of spaces $\mathcal{X}$ is a functor $\mathcal{X} \colon \mathcal{P}(\underline{n}) \to \mathrm{Top}$ (or to $\mathrm{Top}_*$ for a cube of based spaces). The language of categories is discussed in Chapter 7. In particular the reader can consult Definition 7.1.1 and Definition 7.1.15 for the formal definitions of category and functor. However, we can easily describe what this means without the need for the language of functors and categories. An $n$-cube $\mathcal{X}$ is:

- a space $\mathcal{X}(S) = X_S$ for each $S \subset \underline{n}$;
- a map $\mathcal{X}(S \subset T) = f_{S \subset T} \colon X_S \to X_T$ for each $S \subset T$ such that

221

- $f_{S \subset S} = 1_{X_S}$,
- for all $R \subset S \subset T$, the diagram

commutes.

As we already mentioned, a 0-cube is a space $X_\emptyset$, a 1-cube is a map of spaces $X_\emptyset \to X_1$, and a 2-cube is a commutative diagram

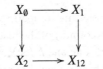

As in Chapter 3, we have written $X_1$ and $X_{12}$ in place of $X_{\{1\}}$ and $X_{\{1,2\}}$ above, and we will continue to do so. Moreover, we only draw those arrows $f_{S \subset T}$ for $|T - S| = 1$, as all other arrows are compositions of these. In addition, we do not draw in the identity maps. Thus we depict a typical 3-cubical diagram of spaces as

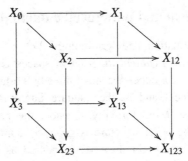

To create a cube of based spaces from an unbased one, we may choose a basepoint in $X_\emptyset$ (provided of course $X_\emptyset \neq \emptyset$) to create compatible basepoints in the other spaces using the unique map $X_\emptyset \to X_S$.

Two important subposets of $\mathcal{P}(\underline{n})$ are $\mathcal{P}_0(\underline{n})$, the subposet of non-empty subsets of $\underline{n}$, and $\mathcal{P}_1(\underline{n})$, the subposet of proper subsets of $\underline{n}$. Diagrams of spaces that are functors from these categories/posets are referred to as *punctured n-cubes*. We have already encountered punctured 2-cubes in Chapter 3. Here are pictures of punctured 3-cubes of spaces, the first indexed by $\mathcal{P}_0(\underline{3})$ and the second indexed by $\mathcal{P}_1(\underline{3})$:

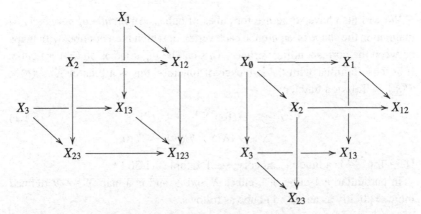

It is not hard to see that a punctured cube can be drawn in the shape of a barycentrically subdivided simplex of dimension 1 less than the dimension of the underlying cube. So, for example, we will often redraw the above punctured cubes as

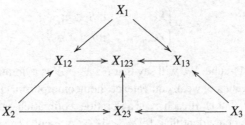

and

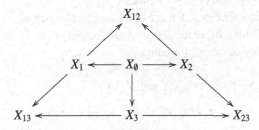

For brevity, we will say that a cube or a punctured cube of spaces is *indexed by* $\mathcal{P}(\underline{n})$ (or by $\mathcal{P}_0(\underline{n})$ or $\mathcal{P}_1(\underline{n})$ as appropriate). We may also speak of $S$-cubes for a finite set $S$. These are diagrams indexed by $\mathcal{P}(S)$, the poset of subsets of $S$.

**Definition 5.1.1** Let $\mathcal{X}$ be an $n$-cube of spaces. For subsets $U \subset T \subset S$, the $(|T| - |U|)$-cube $\partial_U^T \mathcal{X}$ is defined by $V \mapsto \mathcal{X}(V \cup U)$ for $V \subset T - U$. Such cubes are called the *faces* of $\mathcal{X}$.

It is common to let $\partial^T \mathcal{X}$ stand for $\partial_\emptyset^T \mathcal{X}$ and $\partial_U \mathcal{X}$ stand for $\partial_U^n \mathcal{X}$, although we will not use this notation extensively.

We will also have some use for cubes of cubes. An *n-cube of m-cubes* is a diagram in the shape of an *n*-cube each vertex of which is an *m*-cube, with maps between the corresponding vertices. This can be regarded as an $(n + m)$-cube. (For those familiar with the language of functors, this is a functor $X: \mathcal{P}(\underline{n}) \times \mathcal{P}(\underline{m}) \to$ Top or a functor

$$X: \mathcal{P}(\underline{n}) \longrightarrow \text{Top}^{\mathcal{P}(\underline{m})} \tag{5.1.1}$$
$$S \longmapsto (X(S): \mathcal{P}(\underline{m}) \to \text{Top}).$$

Here $\text{Top}^{\mathcal{P}(\underline{m})}$ is a functor category; see Example 7.1.30.)

In particular, a 1-cube of *n*-cubes $X$ and $\mathcal{Y}$ and is a map $X \to \mathcal{Y}$ defined more explicitly as an $(n + 1)$-cube as follows.

**Definition 5.1.2** Let $X, \mathcal{Y}: \mathcal{P}(\underline{n}) \to$ Top be *n*-cubes. A *map of n-cubes* or a *natural transformation of n-cubes* $F: X \to \mathcal{Y}$ is a map $f_S: X_S \to Y_S$ for all $S$ such that the assignment $Z: \mathcal{P}(\underline{n+1}) \to$ Top, $S \mapsto Z_S$ given by

$$Z_S = \begin{cases} X_S, & \text{if } S \subset \underline{n}; \\ Y_{S-\{n+1\}}, & \text{if } n + 1 \in S, \end{cases}$$

defines an $(n + 1)$-cube. We will say that $F: X \to \mathcal{Y}$ is a fibration, cofibration, homotopy equivalance, weak equivalence, homeomorphism, etc., if it is such a map objectwise, that is, if each $f_S$ is a fibration, cofibration, etc.

It should also be evident that, for $n \geq 1$, every *n*-cube can be viewed (in *n* distinct ways) as a map of $(n - 1)$-cubes.

For topological spaces $X, Y$ there is a topological space of maps $\text{Map}(X, Y)$, which we may think of as the space of all 1-cubes $X \to Y$. So too is there a topological space of maps of *n*-cubes.

**Definition 5.1.3** For *n*-cubes $X$ and $\mathcal{Y}$, let

$$\text{Nat}(X, \mathcal{Y}) \subset \prod_{S \in \mathcal{P}(\underline{n})} \text{Map}(X_S, Y_S)$$

denote the subspace consisting of all collections of maps $(f_S)_{S \in \mathcal{P}(\underline{n})}$ in the product of the mapping spaces $\text{Map}(X_S, Y_S)$ such that each collection determines a map $F: X \to \mathcal{Y}$ of *n*-cubes as in Definition 5.1.2. It is topologized as a subspace of this product of mapping spaces. The space $\text{Nat}(X, \mathcal{Y})$ is called the space of *natural transformations* from $X$ to $\mathcal{Y}$. Other notation we may use for this space is $\text{Nat}_{\mathcal{P}(\underline{n})}(X, \mathcal{Y})$ when we want to emphasize the indexing category, or $\text{Nat}_{\mathcal{P}(\underline{n})}(X(\bullet), \mathcal{Y}(\bullet))$ when we wish to emphasize the argument of the cubes, or some combination of the above.

We may also speak of the space of natural transformations between punctured $n$-cubes in the obvious way.

We close this section with two simple examples. The first will be used in the definition of the (co)limit of a punctured cube and the second in the definition of the homotopy (co)limits of the punctured cube.

**Example 5.1.4** (Constant punctured cubical diagram)  If $X_S = X$ for all $S$ in $\mathcal{P}_0(\underline{n})$ or in $\mathcal{P}_1(\underline{n})$ and if all the maps are the identity, then we will refer to such a diagram as a *constant punctured cube at $X$* and denote it by $C_X$. Of most interest is the case when $X = *$, a one-point space, in which case the punctured cube will be denoted by $C_*$.  □

**Example 5.1.5** (Punctured cubical simplices)  We will utilize two punctured cubical diagrams repeatedly, one for homotopy limits of punctured cubes and the other for homotopy colimits of punctured cubes.

Define $\Delta(\bullet)\colon \mathcal{P}_0(\underline{n}) \to \mathrm{Top}$ by

$$\Delta(S) = \left\{ (t_1, \ldots, t_n)\colon 0 \le t_i \le 1, \sum_i t_i = 1, t_i = 0 \text{ for } i \notin S \right\},$$

and for $S \subset T$ let $d^{S \subset T} = \Delta(S \subset T)\colon \Delta(S) \to \Delta(T)$ be the evident inclusion. Note that $\Delta(\underline{n}) \cong \Delta^{n-1}$, and in general $\Delta(S)$ is homeomorphic to a simplex of dimension $|S| - 1$. We will denote the value of $\Delta$ on $S$ by $\Delta(S)$ and, to be consistent with the notation above, will drop braces and commas from our notation for the set $S$ (as in the figure below).

When $n = 3$, we have the punctured cubical diagram depicted in Figure 5.1. For instance, $\Delta(12)$ is the face described in coordinates by tuples $(t_1, t_2, 0)$ such that $t_1 + t_2 = 1$.

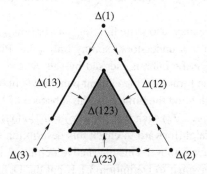

Figure 5.1  Punctured 3-cube of simplices.

Dually, we can consider

$$\Delta(\underline{n} - \bullet) \colon \mathcal{P}_1(\underline{n}) \longrightarrow \mathrm{Top},$$

where $S \subset T$ gives rise to a natural inclusion $d_{S \subset T}\Delta(\underline{n} - T) \to \Delta(\underline{n} - S)$ as above. Note that $\Delta(\underline{n}) = \emptyset$. A picture for $\Delta(\underline{3} - \bullet)$ would be the same as the one above.

Since our notation for a functor does not usually include the argument, when there is no danger of confusion we will use simply $\Delta$ for $\Delta(\bullet)$ or for $\Delta(\underline{n} - \bullet)$.

Both these punctured cubical simplex diagrams extend to the same cubical simplex functor $\mathcal{P}(\underline{n}) \to \mathrm{Top}$ in the obvious way. $\square$

## 5.2 Limits of punctured cubes

Let $X \colon \mathcal{P}_0(\underline{n}) \to \mathrm{Top}$ be a punctured $n$-cube, and as usual write $X(S) = X_S$ and $X(S \subset T) = f_{S \subset T} \colon X_S \to X_T$. Recall the constant punctured cube $C_*$ from Example 5.1.4.

**Definition 5.2.1** The *pullback* or *limit* of the punctured $n$-cube $X$, denoted $\lim_{\mathcal{P}_0(\underline{n})} X$, is the space

$$\lim_{\mathcal{P}_0(\underline{n})} X = \operatorname*{Nat}_{S \in \mathcal{P}_0(\underline{n})} (C_*(S), X(S)).$$

That is (using that a map from a point to a space is the same as choosing a point in the space), the limit of $X$ is the subspace of $\prod_{S \in \mathcal{P}_0(\underline{n})} X_S$ consisting of all tuples $(x_S)_{S \in \mathcal{P}_0(\underline{n})}$ such that $f_{S \subset T}(x_S) = f_{R \subset T}(x_R)$ for all $T$ and for all $R, S \subset T$.

**Remarks 5.2.2**

1. Note that when $n = 2$ this is slightly different than but equivalent to Definition 3.1.1.
2. Alternatively, we may also write $\lim_{S \in \mathcal{P}_0(\underline{n})} X_S$ or $\lim_{\emptyset \neq S \subset \underline{n}} X_S$ in place of the above, or, when $n$ is understood, simply $\lim_{S \neq \emptyset} X_S$. Fix finite sets $R \subset T$. In the case where the cube is indexed by sets $R \subset S \subset T$, we will write $\lim_{S \supsetneq R} X_S$ for the limit over the evident punctured cube.
3. We could have defined the limit to be the subspace of $\prod_{i \in \underline{n}} X_i$ consisting of those tuples $(x_1, \ldots, x_n)$ such that $f_{\{i\} \subset S}(x_i) = f_{\{j\} \subset S}(x_j)$ for all $S$ and for all $i, j \in S$. It is straightforward to check this description is homeomorphic to the definition given above. This alternative definition is simpler to visualize and closer in spirit to Definition 3.1.1, but the Definition 5.2.1 is more directly comparable to the homotopy limit, and more in line with a model for limits for general diagrams of spaces from Proposition 7.4.16. $\square$

**Remark 5.2.3** (Limit of a cube)   One can also define the limit of a cubical diagram $X \colon \mathcal{P}(\underline{n}) \to$ Top as the space

$$\lim_{\mathcal{P}(\underline{n})} X = \operatorname*{Nat}_{S \in \mathcal{P}(\underline{n})} (C_*(S), X(S)).$$

But it is not hard to see that this is homeomorphic to $X_\emptyset$. For a more general result, see Example 7.3.9.   □

**Example 5.2.4**   The limit, in the case of a punctured 3-cube $X \colon \mathcal{P}_0(\underline{3}) \to$ Top, can be visualized as

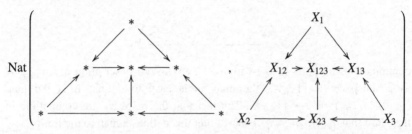

The reader should compare this picture with the one in Example 5.3.3.   □

The limit $\lim_{\mathcal{P}_0(\underline{n})} X$ enjoys a universal property analogous to the one enjoyed by the pullback (see discussion following Definition 3.1.1). There are canonical projection maps $p_S \colon \lim_{\mathcal{P}_0(\underline{n})} X \to X_S$ for each $S$ such that, for all $S \subset T$, the diagram

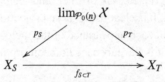

commutes. Thus the assignment

$$S \longmapsto \begin{cases} \lim_{\mathcal{P}_0(\underline{n})} X, & \text{if } S = \emptyset; \\ X_S, & \text{if } S \neq \emptyset \end{cases}$$

defines an $n$-cube. Now suppose $W$ is a space with maps $q_S \colon W \to X_S$ for each $S \in \mathcal{P}_0(\underline{n})$ so that the diagram

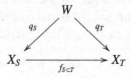

commutes for all pairs $(S, T)$ with $S \subset T$ (i.e. so the $X_S$ with $S \neq \emptyset$ together with $W$ as the space indexed by $\emptyset$ define an $n$-cube). There is a unique map

$$u_0 : W \longrightarrow \lim_{\mathcal{P}_0(\underline{n})} X$$

such that the diagram

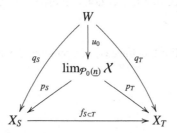

commutes for all pairs $(S, T)$ with $S \subset T$. With our description of $\lim_{\mathcal{P}_0(\underline{n})} X$ as a subspace of $\prod_{i \in \underline{n}} X_i$, the map $u_0$ is easy to define: It is the map $(q_1, q_2, \ldots, q_n) : W \to \prod_{i \in \underline{n}} X_i$. The fact that the maps $q_S$ are compatible in the sense that $f_{S \subset T} \circ q_S = q_T$ ensures that this defines a map to the limit.

Here are a few examples of limits of punctured cubes, the first two parallel to Examples 3.1.2 and 3.1.3.

**Example 5.2.5** Let $X_1, \ldots, X_n$ be spaces, and define $X : \mathcal{P}_0(\underline{n}) \to$ Top by $X(S) = \prod_{i \notin S} X_i$, with the maps $X(S \subset T) : X(S) \to X(T)$ the obvious projections. Then $\lim_{\mathcal{P}_0(\underline{n})} X \cong \prod_{i \in \underline{n}} X_i$. To see why, we note that the limit is the subspace of $\prod_{i \in \underline{n}} \prod_{i \neq j \in \underline{n}} X_j$. Fix $i, k$ distinct, and let $p_{ik} : \prod_{i \neq j \in \underline{n}} X_j \to X_k$ be the canonical projection. It is clear using these projections and the definition of the limit that all points in each appearance of the factor of $X_k$ in the limit must be equal. Hence the limit is the subspace homeomorphic to $\prod_{i \in \underline{n}} X_i$.  □

**Example 5.2.6** There is another obvious way to write the product of the spaces $X_1, \ldots, X_n$ as a pullback. Let $X : \mathcal{P}_0(\underline{n}) \to$ Top be given by $X(\{i\}) = X_i$ and $X(S) = *$ for $|S| \geq 2$. Then $\lim_{\mathcal{P}_0(\underline{n})} X \cong \prod_i X_i$.  □

**Example 5.2.7** Let $X : \mathcal{P}_0(\underline{n}) \to$ Top be the constant punctured cube $C_X$ from Example 5.1.4. Then $\lim_{\mathcal{P}_0(\underline{n})} X \cong X$ is the diagonal subspace of $X^n$.  □

The following useful result says that we may think of limits of punctured cubes in an iterative way in terms of limits of punctured squares. The corresponding fact for homotopy limits is Lemma 5.3.6, and is central to importing the theory of squares into the proofs of the corresponding results about cubes.

**Lemma 5.2.8** (Iterated pullback) *Given a punctured $n$-cube $X : \mathcal{P}_0(\underline{n}) \to$ Top, $n \geq 2$, there is a homeomorphism*

$$\lim_{S \in \mathcal{P}_0(\underline{n})} X_S \cong \lim\left(X_n \to \lim_{R \in \mathcal{P}_0(\underline{n-1})} X_{R \cup \{n\}} \leftarrow \lim_{R \in \mathcal{P}_0(\underline{n-1})} X_R\right).$$

*Proof* Both sides of the isomorphism are clearly subspaces of $\prod_{S \in \mathcal{P}(\underline{n})} X_S$. It is straightforward to check that they are contained in one another using Definition 5.2.1 (and the definition of the pullback, Definition 3.1.1). □

Lemma 5.2.8 can sometimes provide insight when used in multiple ways to view a punctured cube, as illustrated in the following.

**Example 5.2.9** Suppose $f: X \to Y$ and $g: Y \to Z$ are based maps of based spaces, with basepoints $x_0, y_0, z_0$. Let $h = g \circ f$ be the composed map. We are going to compare the fibers of $f, g$, and $h$. Consider the punctured 3-cube $X$ given by the diagram

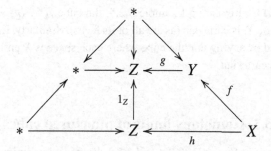

where the maps $* \to Y, Z$ are the inclusion of the basepoint. Using Lemma 5.2.8, considering the "rows" above, we have

$$\lim_{\mathcal{P}_0(\underline{3})} X \cong \lim\left(* \to \lim(* \to Z \leftarrow Y) \leftarrow \lim(* \to Z \leftarrow X)\right).$$

By Example 3.1.5, we have $\lim(* \to Z \leftarrow Y) = \text{fiber}_{z_0}(g)$, and similarly $\lim(* \to Z \leftarrow X) = \text{fiber}_{z_0}(h)$, so that, again applying Example 3.1.5,

$$\lim_{\mathcal{P}_0(\underline{3})} X \cong \text{fiber}_{y_0}\left(\text{fiber}_{z_0}(h) \to \text{fiber}_{z_0}(g)\right).$$

But we can also use Lemma 5.2.8 to see that

$$\lim_{\mathcal{P}_0(\underline{3})} X \cong \lim\left(* \to \lim(* \to Z \overset{1_Z}{\leftarrow} Z) \leftarrow \lim(* \to Y \overset{f}{\leftarrow} X)\right).$$

By Example 3.1.4, $\lim(* \to Z \overset{1_Z}{\leftarrow} Z) \cong *$, and clearly $\lim(* \to * \leftarrow *) = *$, so that

$$\lim_{\mathcal{P}_0(\underline{3})} X \cong \lim(* \to Y \leftarrow X) = \text{fiber}_{y_0}(f).$$

Hence

$$\text{fiber}_{y_0}(f) \cong \text{fiber}_{y_0}(\text{fiber}_{z_0}(h) \to \text{fiber}_{z_0}(g)).$$

The reader may wish to compare this with Proposition 3.4.7.      □

We now have a generalization of Definition 3.1.13.

**Definition 5.2.10**  We say an $n$-cube $X \colon \mathcal{P}(\underline{n}) \to$ Top is *cartesian* (or *categorically cartesian*, or a *(strict) pullback*) if the canonical map

$$X_{\emptyset} \longrightarrow \lim_{\mathcal{P}_0(\underline{n})} X$$

is a homeomorphism.

**Example 5.2.11**  Revisiting Example 5.2.5, the cube $X \colon \mathcal{P}(\underline{n}) \to$ Top given by $X(S) = \prod_{i \notin S} X_i$ is cartesian (as are all of its faces). Similarly, Example 5.2.7 can be restated by saying that the cube where each space is $X$ and all the maps are identity is cartesian.      □

## 5.3 Homotopy limits of punctured cubes

As we have seen in Examples 3.2.1 and 3.6.1, limits and colimits of punctured squares are not homotopy invariant. To fix this, we introduced *homotopy* limits and colimits of punctured cubical squares, and we now generalize this to punctured cubes. In Section 5.7, we will tell the dual story of homotopy colimits of punctured cubes. Heuristically, homotopy limits and colimits are "fattened up" limits and colimits, and it is this added "thickness" that endows them with homotopy invariance.

The motivation for our definition of the homotopy limit of a punctured cube is given by Lemma 5.3.6, which says that we may think of these homotopy limits as iterated homotopy limits of punctured squares. From a pedagogical viewpoint it may be better to use Lemma 5.2.8 as a motivation for defining the homotopy limit of a punctured cube (define homotopy limits inductively by replacing lim with holim), but we prefer to give a definition which is both more symmetric and follows the conventions for more general homotopy limits discussed in Chapter 8.

Recall the definition of a punctured cubical simplex $\Delta(\bullet) \colon \mathcal{P}_0(\underline{n}) \to$ Top from Example 5.1.5.

**Definition 5.3.1**   For $X \colon \mathcal{P}_0(\underline{n}) \to$ Top (or Top$_*$) a punctured cube, define the *homotopy limit of X* as

$$\operatorname*{holim}_{\mathcal{P}_0(\underline{n})} X = \operatorname*{Nat}_{S \in \mathcal{P}_0(\underline{n})} (\Delta(S), X(S)).$$

**Remark 5.3.2** (Homotopy limit of a cube)   One could define the homotopy limit of a cube (not punctured) as the space of natural transformations much as in the above definition; see Proposition 8.2.2. However, as we shall see in Proposition 9.3.4, the homotopy limit in this case is not so interesting since it is homotopy equivalent to $X_\emptyset$.   □

The homotopy limit of $X$ is thus a subspace of $\prod_S \operatorname{Map}(\Delta^{|S|-1}, X_S)$. In the case $n = 2$, this is precisely Definition 3.2.4. The statement that generalizes Definition 5.3.1 to arbitrary diagrams of spaces is Proposition 8.2.2.

**Example 5.3.3**   We can visualize the homotopy limit in the case of a punctured 3-cube $X \colon \mathcal{P}_0(\underline{3}) \to$ Top as follows:

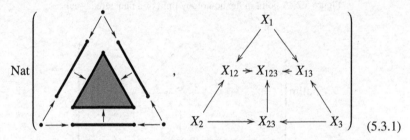

$$(5.3.1)$$

Comparing this picture above with the one in Example 5.2.4 illustrates the heuristic that homotopy limit is a "fattened up" limit.

Unravelling this gives that a point in $\operatorname{holim}_{\mathcal{P}_0(\underline{3})} X$ is a compatible list of points $x_i$ in the spaces $X_i$, homotopies $\alpha_{ij}$ in the spaces $X_{ij}$, and a 2-parameter homotopy $\alpha_{123}$ in $X_{123}$; one such point is depicted in Figure 5.2 below.   □

**Example 5.3.4**   For the punctured cube indexed on $\mathcal{P}_0(\underline{n})$ where all the spaces are $X$ and the maps are all identity, the homotopy limit is $\operatorname{Map}(\Delta^2, X) \simeq X$.   □

**Example 5.3.5** (Loop spaces as holims of punctured cubes)   Let $X$ be a based space, and let $* \to X$ be the inclusion of the basepoint. We have already seen in Example 3.2.10 that $\operatorname{holim}(* \to X \leftarrow *) \simeq \Omega X$ is the loop space of $X$. Consider now:

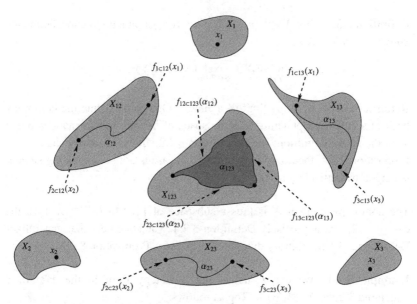

Figure 5.2  A point in the homotopy limit of a punctured 3-cube.

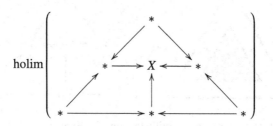

By definition a point in this homotopy limit is a map $\Delta^2 \to X$ such that the boundary $\partial\Delta^2$ maps to the basepoint of $X$. Thus this homotopy limit is homotopy equivalent to $\Omega^2 X$. We could have also used Lemma 5.3.6 to see this.

Similarly, let $\mathcal{X}\colon \mathcal{P}_0(\underline{n}) \to \mathrm{Top}_*$ be a punctured $n$-cubical diagram given by $S \mapsto *$ if $S \subsetneq \underline{n}$ and $\underline{n} \mapsto X$, where the maps in the diagram are either the identity or inclusions of the basepoint. Then by induction using Lemma 5.3.6 below, $\mathrm{holim}\,\mathcal{X} \simeq \Omega^{n-1} X$. $\qquad\square$

The following analog of Lemma 5.2.8 exhibits the homotopy limit of a punctured cube as an iterated homotopy pullback.

**Lemma 5.3.6** (Iterated homotopy pullback)  *Given a punctured $n$-cube $\mathcal{X}\colon \mathcal{P}_0(\underline{n}) \to \mathrm{Top}$, $n \geq 2$, there is a homeomorphism*

$$\operatorname*{holim}_{S \in \mathcal{P}_0(\underline{n})} X_S \cong \operatorname{holim}\left(X_n \longrightarrow \operatorname*{holim}_{R \in \mathcal{P}_0(\underline{n-1})} X_{R \cup \{n\}} \longleftarrow \operatorname*{holim}_{R \in \mathcal{P}_0(\underline{n-1})} X_R\right).$$

*Proof* Write $C\Delta^{n-1} = \Delta^{n-1} \times I/\Delta^{n-1} \times \{1\}$. Define $h \colon C\Delta^{n-1} \to \Delta^n$ by

$$h\big([(t_1, \ldots, t_{n-1}), t]\big) = \big(t_1(1-t), \ldots, t_{n-1}(1-t), t\big).$$

It is easy to see that $h$ is a homeomorphism. A point in $\operatorname{holim}_{S \in \mathcal{P}_0(\underline{n})} X_S$ is a collection of maps $f_S \colon \Delta(S) \to X_S$ for $\emptyset \neq S \subset \underline{n}$ satisfying certain compatibility properties. The homeomorphism in question is given by the map

$$(f_S)_{S \in \mathcal{P}_0(\underline{n})} \longmapsto \big(f_n(1), \Gamma, \{f_R\}_{R \subset \mathcal{P}_0(\underline{n-1})}\big)$$

where

$$\Gamma \colon I \longrightarrow \operatorname*{holim}_{R \in \mathcal{P}_0(\underline{n-1})} X_{R \cup \{n\}}$$

is the parametrized family of maps $(g_R(t))_{R \in \mathcal{P}_0(\underline{n-1})}$, $t \in I$, given by

$$g_R(t)(t_1, \ldots, t_{n-1}) = f_{R \cup \{n\}}\big(h([(t_1, \ldots, t_{n-1}), t])\big). \qquad \square$$

**Remark 5.3.7** In the above statement, we could replace $X_n$ by any of the $X_i$ but the way we have presented it is the cleanest to state and prove. $\qquad \square$

**Example 5.3.8** It is worth illustrating the previous proposition in the case $n = 3$, since we will use it repeatedly later. In this case, we simply have the homeomorphism

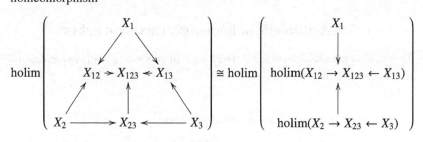

$\qquad \square$

The following result establishes the homotopy invariance of the homotopy limit, generalizing Theorem 3.2.12. The proof here is much easier because it boils down to Theorem 3.2.12.

Observe that, if $\mathcal{X}$ and $\mathcal{Y}$ are punctured cubes indexed on $\mathcal{P}_0(\underline{n})$ and if $N \colon \mathcal{X} \to \mathcal{Y}$ is a map of cubes, then there is an induced map $\operatorname{holim}_{\mathcal{P}_0(\underline{n})} \mathcal{X} \longrightarrow \operatorname{holim}_{\mathcal{P}_0(\underline{n})} \mathcal{Y}$ given by composing a map in $\operatorname{Nat}_{\mathcal{P}_0(\underline{n})}(\Delta, \mathcal{X})$ with $N$.

**Theorem 5.3.9** (Homotopy invariance of homotopy limits, cubes case)   *Suppose* $X, \mathcal{Y} \colon \mathcal{P}_0(\underline{n}) \to$ Top *are punctured n-cubes, and* $X \to \mathcal{Y}$ *is a map of punctured n-cubes which is a homotopy (resp. weak) equivalence (so* $X(S) \to \mathcal{Y}(S)$ *is a homotopy (resp. weak) equivalence for all S ). Then the induced map*

$$\operatorname*{holim}_{\mathcal{P}_0(\underline{n})} X \longrightarrow \operatorname*{holim}_{\mathcal{P}_0(\underline{n})} \mathcal{Y}$$

*is a homotopy (resp. weak) equivalence.*

*Proof*   These statements are vacuously true when $n = 0, 1$. For $n \geq 2$, these follow by induction on $n$ using Theorem 3.2.12 and Lemma 5.3.6 as follows:

The base case $n = 2$ is Theorem 3.2.12. Consider the diagram

$$
\begin{array}{ccccc}
X_n & \longrightarrow & \displaystyle\operatorname*{holim}_{S \in \mathcal{P}_0(\underline{n-1})} X_{S \cup \{n\}} & \longleftarrow & \displaystyle\operatorname*{holim}_{S \in \mathcal{P}_0(\underline{n-1})} X_S \\
\downarrow & & \downarrow & & \downarrow \\
Y_n & \longrightarrow & \displaystyle\operatorname*{holim}_{S \in \mathcal{P}_0(\underline{n-1})} Y_{S \cup \{n\}} & \longleftarrow & \displaystyle\operatorname*{holim}_{S \in \mathcal{P}_0(\underline{n-1})} X_S
\end{array}
$$

By hypothesis the left vertical map is a homotopy (resp. weak) equivalence, and by induction the middle and right vertical maps are homotopy (resp. weak) equivalences, and hence, by Theorem 3.2.12, the induced map of homotopy limits of the rows is a homotopy (resp. weak) equivalence. By Lemma 5.3.6 the homotopy limits of these rows are $\operatorname{holim}_{\mathcal{P}_0(\underline{n})} X$ and $\operatorname{holim}_{\mathcal{P}_0(\underline{n})} \mathcal{Y}$ respectively.   $\square$

## 5.4  Arithmetic of homotopy cartesian cubes

This section parallels Section 3.3. Here we will generalize many examples and results encountered there.

Let

$$X \colon \mathcal{P}(\underline{n}) \longrightarrow \text{Top}$$
$$S \longmapsto X_S$$

be an $n$-cube of spaces (target could be Top$_*$ as well). and recall that $\mathcal{P}_0(\underline{n})$ indexes the punctured cube missing the initial space $X_\emptyset$. We have canonical maps

$$X_\emptyset \longrightarrow \operatorname*{lim}_{\mathcal{P}_0(\underline{n})} X \quad \text{and} \quad \operatorname*{lim}_{\mathcal{P}_0(\underline{n})} X \longrightarrow \operatorname*{holim}_{\mathcal{P}_0(\underline{n})} X.$$

The map from $X_\emptyset$ to the limit was described immediately before Definition 5.2.10, where we defined a cube to be cartesian if this map is a homeomorphism. The map from the limit to the homotopy limit is given as follows. Recall the punctured cubical simplex $\Delta(\bullet)$: $\mathcal{P}_0(\underline{n}) \to$ Top from Example 5.1.5. There is a map of punctured $n$-cubes $\Delta(\bullet) \to C_*$, where $C_*$ is the constant punctured cube at the one-point space (see Example 5.1.4). This induces a map

$$\lim_{\mathcal{P}_0(\underline{n})} X = \operatorname*{Nat}_{\mathcal{P}_0(\underline{n})}(C_*, X) \longrightarrow \operatorname*{Nat}_{\mathcal{P}_0(\underline{n})}(\Delta(\bullet), X) = \operatorname*{holim}_{\mathcal{P}_0(\underline{n})} X.$$

As in (3.3.1), we will denote the composition of the above two maps by

$$a(X) \colon X_\emptyset \longrightarrow \operatorname*{holim}_{\mathcal{P}_0(\underline{n})} X. \qquad (5.4.1)$$

**Definition 5.4.1**  We say the $n$-cube $X \colon \mathcal{P}(\underline{n}) \to$ Top is

- *homotopy cartesian* (or $\infty$-*cartesian*) if the map $a(X)$ is a weak equivalence;
- $k$-cartesian if the map $a(X)$ is $k$-connected.

**Remark 5.4.2**  As the above definition generalizes that for a square, Definition 3.3.1, the remarks following that definition apply here. In particular, if a cube is $k$-cartesian, then it is clearly $j$-cartesian for $j \leq k$ (we will use this at various times throughout). □

**Example 5.4.3**  Examples 5.3.4 and 5.3.5 can be restated in terms of cartesian cubes. For example, Example 5.3.5 says that the $n$-cube with initial space $\Omega^{n-1}X$, final space $X$, and the rest one-point spaces, is homotopy cartesian. □

**Example 5.4.4**  Let $X$ be an $n$-cube, and define an $(n+1)$-cube $\widetilde{X} = X \to X$ as the identity map of $n$-cubes. Then $\widetilde{X}$ is homotopy cartesian. To see this, by Lemma 5.3.6 write

$$\operatorname*{holim}_{\mathcal{P}_0(\underline{n+1})} \widetilde{X} \cong \operatorname*{holim}\left( X_\emptyset \to \operatorname*{holim}_{\mathcal{P}_0(\underline{n})} X \leftarrow \operatorname*{holim}_{\mathcal{P}_0(\underline{n})} X \right),$$

where the right horizontal map is the identity map, so Example 3.2.9 implies the result. □

Here is a generalization of Proposition 3.3.6.

**Proposition 5.4.5**  *If $X$ and $\mathcal{Y}$ are homotopy cartesian $n$-cubes of spaces, then the $n$-cube $X \times \mathcal{Y}$ given by $(X \times \mathcal{Y})(S) = X_S \times Y_S$ (with the obvious maps) is also homotopy cartesian. The same is true if "homotopy cartesian" is replaced by "$k$-cartesian".*

*Proof*   The homeomorphism $\text{Map}(X, Y \times Z) \cong \text{Map}(X, Y) \times \text{Map}(X, Z)$ induces a homeomorphism $\text{Nat}(\Delta(\bullet), \mathcal{X} \times \mathcal{Y}) \cong \text{Nat}(\Delta(\bullet), \mathcal{X}) \times \text{Nat}(\Delta(\bullet), \mathcal{Y})$.          □

The following generalization of Corollary 3.3.8 is then immediate.

**Corollary 5.4.6** (Products commute with homotopy limits, cubes case)   *If $\mathcal{X}$ is a homotopy cartesian n-cube of spaces and V any space, then the n-cube $\mathcal{X} \times V$ (with the obvious maps) is also homotopy cartesian. The same is true if "homotopy cartesian" is replaced by "k-cartesian".*

The following generalizes Proposition 3.3.9. A related statement is the first part of Proposition 5.10.1 and, more generally, Proposition 8.5.4.

**Proposition 5.4.7** (Maps and homotopy cartesian cubes)   *Suppose $\mathcal{X}$ is an n-cube where the map $a(\mathcal{X}): X_\emptyset \to \text{holim}_{\mathcal{P}_0(\underline{n})} \mathcal{X}$ is a homotopy equivalence (so in particular $\mathcal{X}$ is homotopy cartesian). Then, for any space Z, the cube $\mathcal{X}^Z = \text{Map}(Z, \mathcal{X})$ is homotopy cartesian. If $\mathcal{X}$ is k-cartesian and Z has the homotopy type of a CW complex of dimension d, then $\mathcal{X}^Z$ is $(k - d)$-cartesian.*

**Remark 5.4.8**   As mentioned in Remark 3.3.10, the first result could have been stated in terms of homotopy cartesian cubes if Z was required to be a CW complex.          □

*Proof of Proposition 5.4.7*   The exponential law, Theorem 1.2.7, induces a homeomorphism

$$\text{Map}(Z, \text{Nat}(\Delta(\bullet), \mathcal{X})) \cong \text{Nat}(\Delta(\bullet), \mathcal{X}^Z).$$

The result now follows since $\text{Map}(Z, -)$ preserves homotopy equivalences by Proposition 1.3.2. For the other statement, see our remarks surrounding Proposition 3.3.9.          □

We next have the following analog of Theorem 3.3.15, and in fact this analog is a consequence of that result by induction. A generalization can be found in Proposition 8.5.5.

**Theorem 5.4.9** (Homotopy limits commute, cubes case)   *Suppose $\mathcal{X}: \mathcal{P}_0(\underline{n}) \times \mathcal{P}_0(\underline{m}) \to \text{Top}$ is a diagram $(S, T) \to X(S, T)$ (an n-punctured cube of m-punctured cubes, or an m-punctured cube of n-punctured cubes). Then there is a homeomorphism*

$$\underset{\mathcal{P}_0(\underline{n})}{\text{holim}} \underset{\mathcal{P}_0(\underline{m})}{\text{holim}} \mathcal{X} \cong \underset{\mathcal{P}_0(\underline{m})}{\text{holim}} \underset{\mathcal{P}_0(\underline{n})}{\text{holim}} \mathcal{X}.$$

*Proof* By Theorem 3.3.15, the result is true for $n = m = 2$. By induction suppose the result is true for $(n, m) = (2, m - 1)$. Then for $X \colon \mathcal{P}_0(\underline{2}) \times \mathcal{P}_0(\underline{m}) \to$ Top, consider the following sequence of homeomorphisms

$$\operatorname*{holim}_{\mathcal{P}_0(\underline{2})} \operatorname*{holim}_{\mathcal{P}_0(\underline{m})} X$$

$$\cong \operatorname*{holim}_{S \in \mathcal{P}_0(\underline{2})} \operatorname*{holim}\left( X(S, \underline{m}) \to \operatorname*{holim}_{R \in \mathcal{P}_0(\underline{m-1})} X(S, R) \right.$$

$$\left. \leftarrow \operatorname*{holim}_{R \in \mathcal{P}_0(\underline{m-1})} X(S, R \cup \{m\}) \right)$$

$$\cong \operatorname*{holim}\left( \operatorname*{holim}_{S \in \mathcal{P}_0(\underline{2})} X(S, \underline{m}) \to \operatorname*{holim}_{S \in \mathcal{P}_0(\underline{2})} \operatorname*{holim}_{R \in \mathcal{P}_0(\underline{m-1})} X(S, R) \right.$$

$$\left. \leftarrow \operatorname*{holim}_{S \in \mathcal{P}_0(\underline{2})} \operatorname*{holim}_{R \in \mathcal{P}_0(\underline{m-1})} X(S, R \cup \{m\}) \right)$$

$$\cong \operatorname*{holim}\left( \operatorname*{holim}_{S \in \mathcal{P}_0(\underline{2})} X(S, \underline{m}) \to \operatorname*{holim}_{R \in \mathcal{P}_0(\underline{m-1})} \operatorname*{holim}_{S \in \mathcal{P}_0(\underline{2})} X(S, R) \right.$$

$$\left. \leftarrow \operatorname*{holim}_{R \in \mathcal{P}_0(\underline{m-1})} \operatorname*{holim}_{S \in \mathcal{P}_0(\underline{2})} X(S, R \cup \{m\}) \right)$$

$$\cong \operatorname*{holim}_{\mathcal{P}_0(\underline{m})} \operatorname*{holim}_{\mathcal{P}_0(\underline{2})} X.$$

The first homeomorphism uses Lemma 5.3.6, the second follows from Theorem 3.3.15, the third by induction, and the last uses Lemma 5.3.6 again. A similar argument shows that the $(2, m)$ case implies the $(n, m)$ case for $n > 2$. ☐

Using the description of the loop space as a homotopy limit (Example 3.2.10), from the previous result we have the following consequence. This generalizes Corollary 3.3.16.

**Corollary 5.4.10** (Loops and homotopy limits commute, cubes case) *If $X$ is a punctured cube of based spaces, then*

$$\operatorname*{holim}_{\mathcal{P}_0(\underline{n})} \Omega X \cong \Omega \operatorname*{holim}_{\mathcal{P}_0(\underline{n})} X.$$

The commutativity of loops with the homotopy limit will be encountered again in Corollary 8.5.8.

Here is a generalization of Corollary 3.3.17.

**Corollary 5.4.11** *Suppose $X \to Y$ is a map of punctured n-cubes and suppose $Y$ is based. Consider the punctured n-cube $S \mapsto F_S = \text{hofiber}(X_S \to Y_S)$. Then there is a homeomorphism*

$$\underset{S \in \mathcal{P}_0(\underline{n})}{\operatorname{holim}} F_S \cong \operatorname{hofiber}\left(\underset{S \in \mathcal{P}_0(\underline{n})}{\operatorname{holim}} X_S \to \underset{S \in \mathcal{P}_0(\underline{n})}{\operatorname{holim}} Y_S\right).$$

*Proof* We simply note that hofiber$(X \to Y) \cong \operatorname{holim}(X \to Y \leftarrow *)$ is itself a homotopy limit (see Example 3.2.8) and then apply Theorem 5.4.9. □

Note that we could have deduced Corollary 5.4.10 from this result by recalling from Example 2.2.9 that $\Omega Y = \operatorname{hofiber}(* \to Y)$ and taking $X$ to be a cube of one-point spaces.

The following is a generalization of Proposition 3.3.18.

**Proposition 5.4.12** *A map $X \to \mathcal{Y}$ of n-cubes is a k-cartesian $(n + 1)$-cube if and only if, for all $y \in Y_\emptyset$, the n-cube $\mathcal{F}_y = \operatorname{hofiber}(X \to \mathcal{Y})$ is k-cartesian.*

*Proof* Let $\mathcal{Z} = (X \to \mathcal{Y})$ be the $(n+1)$-cube representing the map of $n$-cubes, so that $\mathcal{Z}$ is the cube

$$S \mapsto \begin{cases} X_S, & S \subset \underline{n}; \\ Y_{S-\{n+1\}}, & n+1 \in S. \end{cases}$$

Let $\widetilde{\mathcal{Y}} = \mathcal{Y} \to \mathcal{Y}$ be the $(n + 1)$-cube obtained by mapping $\mathcal{Y}$ to itself by the identity. Consider the square

$$\begin{array}{ccc} X_\emptyset & \longrightarrow & \underset{\mathcal{P}_0(\underline{n+1})}{\operatorname{holim}} \mathcal{Z} \\ \downarrow & & \downarrow \\ Y_\emptyset & \longrightarrow & \underset{\mathcal{P}_0(\underline{n+1})}{\operatorname{holim}} \widetilde{\mathcal{Y}} \end{array}$$

By Example 5.4.4, the lower horizontal arrow is a homotopy equivalence. Using the long exact sequence of a fibration, the top horizontal arrow has the same connectivity as the map

$$F_\emptyset = \operatorname{hofiber}(X_\emptyset \to Y_\emptyset) \to \operatorname{hofiber}\left(\underset{S \in \mathcal{P}_0(\underline{n+1})}{\operatorname{holim}} \mathcal{Z} \to \underset{S \in \mathcal{P}_0(\underline{n+1})}{\operatorname{holim}} \widetilde{\mathcal{Y}}\right).$$

By Corollary 5.4.11, we have a homeomorphism

$$\operatorname{hofiber}\left(\underset{S \in \mathcal{P}_0(\underline{n+1})}{\operatorname{holim}} \mathcal{Z} \to \underset{S \in \mathcal{P}_0(\underline{n+1})}{\operatorname{holim}} \widetilde{\mathcal{Y}}\right) \cong \underset{S \in \mathcal{P}_0(\underline{n+1})}{\operatorname{holim}} \operatorname{hofiber}(\mathcal{Z} \to \widetilde{\mathcal{Y}}).$$

Regarding $\mathcal{Z}$ and $\widetilde{\mathcal{Y}}$ as maps of $n$-cubes, hofiber($\mathcal{Z} \to \widetilde{\mathcal{Y}}$) is homotopy equivalent to the total homotopy fiber of the square

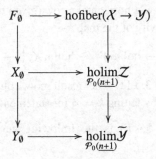

of $n$-cubes, which is evidently homotopy equivalent to the $n$-cube hofiber($\mathcal{X} \to \mathcal{Y}$). Thus we have a (horizontal) map of fibration sequences

$$
\begin{array}{ccc}
F_\emptyset & \longrightarrow & \text{hofiber}(\mathcal{X} \to \mathcal{Y}) \\
\downarrow & & \downarrow \\
X_\emptyset & \longrightarrow & \underset{\mathcal{P}_0(n+1)}{\text{holim}}\,\mathcal{Z} \\
\downarrow & & \downarrow \\
Y_\emptyset & \longrightarrow & \underset{\mathcal{P}_0(n+1)}{\text{holim}}\,\widetilde{\mathcal{Y}}
\end{array}
$$

and the result now follows.                                                  □

Now we have a generalization of Proposition 3.3.11. Recall that a map of $n$-cubes is an $(n + 1)$-cube (and that any $n$-cube can be regarded as a map of $(n - 1)$-cubes in various ways).

**Proposition 5.4.13** *Suppose $X$ and $\mathcal{Y}$ are n-cubes, and that $(X \to \mathcal{Y})$ is a map of n-cubes, considered as an $(n + 1)$-cube.*

1. (a) *If $(X \to \mathcal{Y})$ and $\mathcal{Y}$ are homotopy cartesian, then $X$ is homotopy cartesian.*
   (b) *If $X$ and $\mathcal{Y}$ are homotopy cartesian, then $(X \to \mathcal{Y})$ is homotopy cartesian.*
2. (a) *If $(X \to \mathcal{Y})$ and $\mathcal{Y}$ are k-cartesian, then $X$ is k-cartesian.*
   (b) *If $X$ is k-cartesian and $\mathcal{Y}$ is $(k + 1)$-cartesian, then $(X \to \mathcal{Y})$ is k-cartesian.*

All the remarks as in Remark 3.3.12 apply here as well.

*Proof*   Consider the diagram

$$X_\emptyset \longrightarrow \operatorname*{holim}\left(Y_\emptyset \to \operatorname*{holim}_{S \neq \emptyset} Y_S \leftarrow \operatorname*{holim}_{S \neq \emptyset} X_S\right) \longrightarrow \operatorname*{holim}_{S \neq \emptyset} X_S$$

$$P_{a(\mathcal{Y})} \longrightarrow \operatorname*{holim}_{S \neq \emptyset} Y_S$$

where $P_{a(\mathcal{Y})} \simeq Y_\emptyset$ is the mapping path space of the canonical map $a(\mathcal{Y})\colon Y_\emptyset \to \operatorname{holim}_{S \neq \emptyset} Y_S$, and the model for the homotopy limit in the upper-left corner of the square that we use is $\lim(P_{a(\mathcal{Y})} \to \operatorname{holim}_{S \neq \emptyset} Y_S \leftarrow \operatorname{holim}_{S \neq \emptyset} X_S)$. The square above is evidently homotopy cartesian by Proposition 3.2.13, and hence the connectivity of the lower horizontal arrow (given by how cartesian $\mathcal{Y}$ is) is the same as the connectivity of the map

$$\operatorname*{holim}\left(Y_\emptyset \to \operatorname*{holim}_{S \neq \emptyset} Y_S \leftarrow \operatorname*{holim}_{S \neq \emptyset} X_S\right) \longrightarrow \operatorname*{holim}_{S \neq \emptyset} X_S$$

by 2(a) of Proposition 3.3.11. The result now follows immediately from Proposition 2.6.15 (and by letting $k = \infty$ for statements 1(a) and 1(b)).    □

Related to Proposition 5.4.13 is the following generalization of Proposition 3.3.20.

**Proposition 5.4.14**   *Suppose $X = (\mathcal{U} \to \mathcal{V})$ and $\mathcal{Y} = (\mathcal{V} \to \mathcal{W})$ are n-cubes, and let $Z = (\mathcal{U} \to \mathcal{W})$, all considered as maps of $(n - 1)$-cubes. Then*

1. (a) *If $X$ and $\mathcal{Y}$ are homotopy cartesian, then $Z$ is homotopy cartesian.*
   (b) *If $\mathcal{Y}$ and $Z$ are homotopy cartesian, then $X$ is homotopy cartesian.*
2. (a) *If $X$ and $\mathcal{Y}$ are k-cartesian, then $Z$ is k-cartesian.*
   (b) *If $\mathcal{Y}$ is $(k + 1)$-cartesian and $Z$ is k-cartesian, then $X$ is k-cartesian.*

See Remark 3.3.21 for an example where the third obvious statement fails.

*Proof*   We will prove the $k$-cartesian version, which also proves the homotopy cartesian one by setting $k = \infty$. Consider the following diagram of $(n-1)$-cubes

$$\mathcal{U} \longrightarrow \mathcal{U} \longrightarrow \mathcal{V}$$
$$\mathcal{V} \longrightarrow \mathcal{W} \longrightarrow \mathcal{W}$$

To prove 2(a), view the right-most square as a map of $n$-cubes $X = (\mathcal{U} \to \mathcal{V}) \to (\mathcal{W} \to \mathcal{W})$. Since $(\mathcal{W} \to \mathcal{W})$ is homotopy cartesian and $X$ is $k$-cartesian, the right-most square, as an $(n + 1)$-cube, is $k$-cartesian by 2(b) of

Proposition 5.4.13. We may view this $(n + 1)$-cube as a map $\mathcal{Z} = (\mathcal{U} \to \mathcal{W}) \to (\mathcal{V} \to \mathcal{W}) = \mathcal{Y}$. Since $\mathcal{Y} = (\mathcal{V} \to \mathcal{W})$ is $k$-cartesian, 2(a) of Proposition 5.4.13 implies that $\mathcal{Z} = (\mathcal{U} \to \mathcal{W})$ is $k$-cartesian.

To prove 2(b), if $\mathcal{Y} = (\mathcal{V} \to \mathcal{W})$ is $(k + 1)$-cartesian, then since $\mathcal{U} \to \mathcal{U}$ is homotopy cartesian, by 2(b) of Proposition 5.4.13, the left-most square above is a $k$-cartesian $(n + 1)$-cube. Viewing it as a map $\mathcal{X} = (\mathcal{U} \to \mathcal{V}) \to (\mathcal{U} \to \mathcal{W}) = \mathcal{Z}$, since $\mathcal{Z}$ is $k$-cartesian, 2(a) of Proposition 5.4.13 implies $\mathcal{X} = (\mathcal{U} \to \mathcal{V})$ is $k$-cartesian as well. □

**Remarks 5.4.15** When $\mathcal{X} = \mathcal{U} \to \mathcal{V}$, $\mathcal{Y} = \mathcal{V} \to \mathcal{W}$, and $\mathcal{Z} = \mathcal{U} \to \mathcal{W}$, we will write $\mathcal{Z} = \mathcal{X}\mathcal{Y}$ and think of this as a factorization of the cube $\mathcal{Z}$. Such factorizations will be used extensively in the proof of Theorem 6.2.3.

Propositions 5.4.13 and 5.4.14 are in fact equivalent (one can use Proposition 5.4.14 to prove Proposition 5.4.13). After all, they generalize Propositions 3.3.11 and 3.3.20 but the latter is itself a generalization of the former. We will, however, need both statements in this and the next chapter, and this is why we stated them both. □

What follows are a few results about cubes of cubes. These will be used in Section 10.1. Recall from Section 5.1 that an *n-cube of m-cubes* is an $n$-cube whose each vertex is an $m$-cube and with maps between the corresponding vertices of of $m$-cubes (this can be viewed as an $(n + m)$-cube). An $(n + 1)$-cube, regarded as a 1-cube of $n$-cubes, that is, a map of $n$-cubes, is an example of this.

**Proposition 5.4.16** *Let $\mathcal{X}$ be an n-cube of m-cubes. Assume that $\mathcal{X}$, viewed as a $(n + m)$-cube, is k-cartesian, and that for each $S \subset \underline{n}$, $S \neq \emptyset$, the m-cube $\mathcal{X}(S)$ is $(k + |S| - 1)$-cartesian. Then the m-cube $\mathcal{X}(\emptyset)$ is k-cartesian.*

*Proof* We induct on $n$. If $n = 0$ there is nothing to prove. For $n = 1$, we have a map $\mathcal{X}(\emptyset) \to \mathcal{X}(1)$ of $m$-cubes. By hypothesis, considered as an $(m + 1)$-cube it is $k$-cartesian, and since we also assume $\mathcal{X}(1)$ is $k$-cartesian, 2(a) of Proposition 5.4.13 implies that $\mathcal{X}(\emptyset)$ is $k$-cartesian. Now assume the result is true for cubes of dimension $n$, and let $\mathcal{X}$ be an $(n + 1)$-cube of $m$-cubes. Write $\mathcal{X}$ as a map of $n$-cubes $(R \mapsto \mathcal{X}(R)) \to (R \mapsto \mathcal{X}(R \cup \{n + 1\}))$ for $R \in \underline{n}$.

We claim that, by induction, the $n$-cube of $m$-cubes $R \mapsto \mathcal{X}(R \cup \{n + 1\})$ is $k$-cartesian, and since we assume $\mathcal{X}$, considered as an $(n + 1)$-cube of $m$ cubes, to be $k$-cartesian, 2(a) of Proposition 5.4.13 implies that the $p$-cube $R \mapsto \mathcal{X}(R)$ is also $k$-cartesian. It follows that $\mathcal{X}(\emptyset)$ is $k$-cartesian.

To see why the $n$-cube of $m$-cubes $R \mapsto X(R \cup \{n + 1\})$ is $k$-cartesian, we need to use 2(b) of Proposition 5.4.13. Again we apply induction. When $n = 1$ we may write the above as $X(1) \to X(12)$, and by hypothesis $X(12)$ is $(k + 1)$-cartesian and $X(1)$ is $k$-cartesian, so the 1-cube of $m$-cubes $(X(1) \to X(12))$ is $k$-cartesian by 2(b) of Proposition 5.4.13. For $n \geq 2$, write $R \mapsto X(R \cup \{n + 1\})$ as $(U \mapsto X(U \cup \{n\})) \to (U \mapsto X(U \cup \{n, n + 1\}))$ for $U \in \underline{n-1}$. By induction $U \mapsto X(U \cup \{n\})$ is $k$-cartesian, and $U \mapsto X(U \cup \{n, n + 1\})$ is $(k + 1)$-cartesian, so by 2(b) of Proposition 5.4.13, $R \mapsto X(R \cup \{n + 1\})$ is $k$-cartesian.          $\square$

**Proposition 5.4.17**  *Let $X$ be a functor from $\mathcal{P}_0(n)$ to $m$-cubes of spaces (so we think of $X$ as an $n$-cube of $m$-cubes) and write $X(S, T) = (X(S))(T)$. If, for all $S \neq \emptyset$, $X(S)$ is a $k_S$-cartesian $m$-cube, then the $m$-cube $T \mapsto \mathrm{holim}_{S \neq \emptyset} X(S, T)$ is $\min_S\{1 - n + k_S\}$-cartesian.*

*Proof*  Define an $n$-cube of $m$-cubes $\mathcal{Y}$ by setting $\mathcal{Y}(S, T) = X(S, T)$ for $S \neq \emptyset$ and $\mathcal{Y}(\emptyset, T) = \mathrm{holim}_{S \neq \emptyset} \mathcal{Y}(U, V)$. As an $(n+m)$-cube, $\mathcal{Y}$ is homotopy cartesian. Proposition 5.4.16 now implies the result.          $\square$

We next discuss strongly cartesian cubes. Recall the notion of a face of a cube from Definition 5.1.1.

**Definition 5.4.18**  We call an $n$-cube $X$ *strongly homotopy cartesian* (resp. *strongly cartesian*) if each of its faces of dimension $\geq 2$ is homotopy cartesian (resp. cartesian).

By 1(b) of Proposition 5.4.13, it is enough to require that all the square faces be (homotopy) cartesian in order for the cube to be strongly cartesian. The same is true for strongly cartesian cubes by replacing "holim" by "lim" everywhere in Proposition 5.4.13 and using Proposition 3.1.9. In fact, by induction and using those results, we have the following.

**Proposition 5.4.19**  *A strongly (homotopy) cartesian cube is (homotopy) cartesian.*

**Example 5.4.20**  If $C_X$ is the constant $n$-cube with all spaces $X$ and all maps the identity, then $X$ is strongly (homotopy) cartesian. This follows from Example 3.3.3.          $\square$

**Example 5.4.21**  Suppose $X_1, \ldots, X_n$ are spaces. Generalizing Example 3.2.7, the cube

$$X: \mathcal{P}(\underline{n}) \longrightarrow \text{Top}$$

$$S \longmapsto \begin{cases} \prod_{i \notin S} X_i, & S \neq \underline{n}; \\ *, & S = \underline{n}, \end{cases}$$

where an inclusion $T \subset S$ induces a projection $X(T) \to X(S)$ away from those factors indexed by elements of $S - T$, is strongly (homotopy) cartesian and hence (homotopy) cartesian. This is true because each square face of the cube is of the form

$$S = \begin{array}{ccc} (\prod_{i \in T} X_i) \times X_j \times X_k & \longrightarrow & (\prod_{i \in T} X_i) \times X_j \\ \downarrow & & \downarrow \\ (\prod_{i \in T} X_i) \times X_k & \longrightarrow & \prod_{i \in T} X_i \end{array}$$

for some $T \subset \underline{n}$ and $j, k \notin T$. This square is (homotopy) cartesian by Example 3.3.3 and Corollary 3.3.8. □

Next we generalize the discussion of fibrant squares. To simplify the notation in what follows, we set

$$\lim_{T \supsetneq S} X_S = \lim_{R \subset \mathcal{P}_0(\underline{n} - S)} X_{R \cup S}$$

(and similarly for $T \supset S$). We have the following analog to Definition 3.3.23.

**Definition 5.4.22** We call a (punctured) $n$-cube $X$ *fibrant* if for all $S \subset \underline{n}$ ($\emptyset \neq S \subset \underline{n}$) the map

$$X_S = \lim_{T \supset S} X_T \longrightarrow \lim_{T \supsetneq S} X_T$$

is a fibration. An $n$-cube is called a *fibrant pullback cube* if, in addition, $X_S = \lim_{T \supsetneq S} X_T$ for all $|S| \geq 2$.

A straightforward induction on $|S|$ shows that a fibrant pullback cube has $X_S$ equal to the fiber product over $X_{\underline{n}}$ of $X_{\underline{n}-\{i\}}$ for $i \in \underline{n} - S$ (and the fiber product itself may be defined iteratively in the evident way).

The following is a generalization of Proposition 3.3.24 and says that one can always choose a fibrant cube as a "representative" for a given cube. Such a representative is called a *fibrant replacement* of the original cube. The second statement below is [Goo92, Proposition 2.2], but our terminology differs.

**Theorem 5.4.23** *Every (punctured) cube admits a homotopy equivalence to a (punctured) fibrant cube. Every strongly homotopy cartesian cube admits a homotopy equivalence to a fibrant pullback cube.*

*Proof* We give a proof for a cubical diagram; the proof is essentially the same for punctured cubes. Let $\mathcal{X} = (S \mapsto X_S) \colon \mathcal{P}(\underline{n}) \to$ Top be an *n*-cube. Define a cubical diagram $\mathcal{Y}$ by

$$\mathcal{Y}(S) = Y_S = \operatorname*{holim}_{T \supset S} X_T,$$

where this homotopy limit can be thought of as the mapping path space $P_{a(\partial_S^n \mathcal{X})}$ of the canonical map

$$a(\partial_S^n \mathcal{X}) \colon X_S \longrightarrow \operatorname*{holim}_{T \supsetneq S} X_T,$$

except in the case where $S = \underline{n}$, in which case we define $Y_{\underline{n}} = X_{\underline{n}}$. Note that, in light of Remark 5.3.2, it makes sense that the homotopy limit of the cube $\mathcal{Y}(\emptyset) = \operatorname{holim}_{T \supsetneq \emptyset} X_T$ would be homotopy equivalent to $X_\emptyset$; here we are essentially taking this as the definition of the homotopy limit of a cube, but the reader can verify that this is homeomorphic to the homotopy limit of the cube obtained from Definition 8.2.1 (or Proposition 8.2.2).

Evidently there is a map $X_S \to Y_S$ which is a homotopy equivalence for each $S$ by Proposition 2.2.2, and we claim this induces a map of cubes $\mathcal{X} \to \mathcal{Y}$. This will be clear once we establish that $\mathcal{Y}$ is itself a cube (i.e. the diagram commutes), which amounts to showing that if $R \subset S \subset T$, we have a commutative diagram

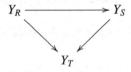

But this is straightforward to check from the definitions.

Now we must show that $\mathcal{Y}$ is fibrant. The case $n = 1$ is amounts to Proposition 2.2.2. The case $n = 2$ is Proposition 3.3.24. It is enough by induction to verify that the map

$$Y_\emptyset \longrightarrow \operatorname*{lim}_{S \in \mathcal{P}_0(\underline{n})} Y_S \tag{5.4.2}$$

is a fibration. That is, we may assume that for every $T \in \mathcal{P}_0(\underline{n})$, $Y_T \to \operatorname{lim}_{S \supsetneq T} X_S$ is a fibration. We claim this map factors through a homeomorphism $\operatorname{lim}_{S \in \mathcal{P}_0(\underline{n})} Y_S \cong \operatorname{holim}_{S \in \mathcal{P}_0(\underline{n})} X_S$, and that there is a commutative diagram

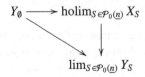

where the horizontal maps are the evident ones and the vertical map is the homeomorphism we will describe. This will complete the proof since the top horizontal map is by definition of $Y_S$ a fibration.

Now we prove $\mathrm{holim}_{S \in \mathcal{P}_0(\underline{n})} X_S \cong \lim_{S \in \mathcal{P}_0(\underline{n})} Y_S$. As in Remark 5.2.2 we will think of $\lim_{S \in \mathcal{P}_0(\underline{n})} Y_S$ as a subspace of $\prod_{i \in \underline{n}} Y_i$. Using Definition 5.3.1, a point in $\mathrm{holim}_{S \in \mathcal{P}_0(\underline{n})} X_S$ is a collection of maps $f_S : \Delta(S) \to X_S$, and for each $i$ the subcollection of those $f_S$ where $i \in S$ determines a point in $Y_i$, and also a point in $\lim_{S \in \mathcal{P}_0(\underline{n})} Y_S$. We leave it to the reader to verify that this map is a homoeomorphism.

Now suppose $X$ is strongly homotopy cartesian. We can define its replacement $\mathcal{Y}$ in a slightly different way to get a more concise formula for it. Let $Y_{\underline{n}} = X_{\underline{n}}$, and for each $i \in \underline{n}$ let $Y_{\underline{n}-\{i\}}$ be the mapping path space of the map $X_{\underline{n}-\{i\}} \to X_{\underline{n}}$. Then for each $|S| \geq 2$ let $Y_{\underline{n}-S}$ be the fiber product of $Y_{\underline{n}-\{i\}}$ for $i \in S$. This can be defined inductively as an iterated limit as follows. Let $s \in S$, and define

$$Y_{\underline{n}-S} = \lim\left( Y_{\underline{n}-(S-\{s\})} \to \prod_{S-\{s\}} Y_{\underline{n}} \leftarrow Y_{\underline{n}-\{s\}} \right),$$

where the left horizontal map is induced by the product over $j \in S - \{s\}$ of the maps $Y_{\underline{n}-\{j\}} \to Y_{\underline{n}}$, and the right horizontal map is $|S - \{s\}|$ copies of the map $Y_{\underline{n}-\{s\}} \to Y_{\underline{n}}$.

Since $X$ is strongly homotopy cartesian it follows by induction on $|\underline{n}-S|$ that the natural map $X \to \mathcal{Y}$ is a homotopy equivalence as follows. For $|\underline{n} - S| \geq 2$, consider the commutative diagram

$$
\begin{array}{ccc}
X_{\underline{n}-S} & \longrightarrow & Y_{\underline{n}-S} \\
\downarrow & & \downarrow \\
\mathrm{holim}_{T \supsetneq \underline{n}-S} X_T & \longrightarrow & \mathrm{holim}_{T \supsetneq \underline{n}-S} Y_T
\end{array}
$$

By hypothesis, the left vertical map is a weak equivalence. Since $X_T \to Y_T$ is a weak equivalence for all $T$, Theorem 5.3.9 implies that the lower horizontal arrow is a weak equivalence. The right vertical arrow is a weak equivalence because we have a factorization

$$Y_{\underline{n}-S} \longrightarrow \lim_{T \supsetneq \underline{n}-S} Y_T \longrightarrow \mathrm{holim}_{T \supsetneq \underline{n}-S} Y_T.$$

Since $\mathcal{Y}$ is fibrant (not hard to see), the first map is a weak equivalence, and by Proposition 5.4.26 the second map is a weak equivalence as well. It now follows that the top horizontal arrow in the square above is a weak equivalence, so that $X \to \mathcal{Y}$ is an equivalence of cubes. $\qquad\square$

**Remark 5.4.24** With $X$ an $n$-cube and $\mathcal{Y}$ its fibrant replacement as in the above, consider the square

$$
\begin{array}{ccc}
X_\emptyset & \longrightarrow & \underset{S \neq 0}{\operatorname{holim}} X_S \\
\downarrow & & \downarrow \\
Y_\emptyset & \longrightarrow & \underset{S \neq 0}{\operatorname{holim}} Y_S
\end{array}
$$

Both vertical maps are homotopy equivalences, and so exactly the same argument as in Remark 3.3.25 shows that $X$ is $k$-cartesian if and only if $\mathcal{Y}$ is. □

The following will be needed in the proof of Proposition 5.4.26.

**Lemma 5.4.25** *Suppose $X = T \mapsto X_T$ is a fibrant $n$-cube. Then every face $\partial_R^S X$, $R \subset S \subset \underline{n}$, is a fibrant $(|S| - |R|)$-cube.*

*Proof* It is enough to prove by induction that every face of dimension $n - 1$ is fibrant. The case $n = 1$ is trivial and the case $n = 2$ is the content of Lemma 3.3.26. By hypothesis, for each $i \in \underline{n}$, the face $\partial_{\{i\}}^n X$ is fibrant. Then by symmetry it is enough to prove that a face of the form $\partial_\emptyset^{n-\{i\}} X$ is fibrant for some $i$. Without loss of generality assume $i = n$. Consider the square

$$
\begin{array}{ccc}
\underset{S \in \mathcal{P}_0(\underline{n})}{\lim} X_S & \longrightarrow & \underset{R \in \mathcal{P}_0(\underline{n-1})}{\lim} X_R \\
\downarrow & & \downarrow \\
X_n & \longrightarrow & \underset{R \in \mathcal{P}_0(\underline{n-1})}{\lim} X_{R \cup \{n\}}
\end{array}
$$

By Lemma 5.2.8, this is a strict pullback square. Since the cube is fibrant, the lower horizontal arrow, is a fibration. Hence so is the upper horizontal arrow, by Proposition 2.1.16. We also have by hypothesis a fibration $X_\emptyset \to \lim_{S \in \mathcal{P}_0(\underline{n})} X_S$. Now consider the composite

$$
X_\emptyset \longrightarrow \underset{S \in \mathcal{P}_0(\underline{n})}{\lim} X_S \longrightarrow \underset{R \in \mathcal{P}_0(\underline{n-1})}{\lim} X_R.
$$

We have shown both maps are fibrations, and hence so is their composite. □

The first part of the following result will be extended to more general diagrams in Theorem 8.4.9 and will in fact be restated as the first part of Corollary 8.4.10.

**Proposition 5.4.26** *Let $X$ be a fibrant (punctured) $n$-cube. Then the canonical map*

$$\lim_{S \in \mathcal{P}_0(\underline{n})} X_S \longrightarrow \operatorname{holim}_{S \in \mathcal{P}_0(\underline{n})} X_S$$

is a homotopy equivalence. Hence a fibrant cube $X$ is $k$-cartesian if and only if

$$X_\emptyset \longrightarrow \lim_{S \in \mathcal{P}_0(\underline{n})} X_S$$

is $k$-connected.

*Proof* The case $n = 2$ is Proposition 3.2.13. By induction, assume the result is true for all fibrant cubes of dimension less than $n$. For an $n$-cube $X$, using Lemma 5.3.6 write

$$\operatorname{holim}_{\mathcal{P}_0(\underline{n})} X \cong \operatorname{holim}\left(X_n \to \operatorname{holim}_{\mathcal{P}_0(\underline{n-1})} X_{S \cup n} \leftarrow \operatorname{holim}_{\mathcal{P}_0(\underline{n-1})} X_S\right).$$

If we could rewrite all the homotopy limits as limits on the right ride of this expression, we would be done.

Since all the faces of $X$ are fibrant by Lemma 5.4.25, we have

$$\operatorname{holim}_{\mathcal{P}_0(\underline{n-1})} X_{S \cup n} \simeq \lim_{\mathcal{P}_0(\underline{n-1})} X_{S \cup n},$$

$$\operatorname{holim}_{\mathcal{P}_0(\underline{n-1})} X_S \simeq \lim_{\mathcal{P}_0(\underline{n-1})} X_S.$$

So now we have

$$\operatorname{holim}_{\mathcal{P}_0(\underline{n})} X \simeq \operatorname{holim}\left(X_n \to \lim_{\mathcal{P}_0(\underline{n-1})} X_{S \cup n} \leftarrow \lim_{\mathcal{P}_0(\underline{n-1})} X_S\right). \tag{5.4.3}$$

But since $X$ is fibrant, the left arrow in the diagram above is a fibration. Hence the outermost homotopy limit may also be replaced with the limit by Proposition 3.2.13. □

If a punctured square $X_1 \to X_{12} \leftarrow X_2$ is fibrant, then in particular by Proposition 3.2.13 its limit is weakly equivalent to its homotopy limit, although that result says that it suffices that one of the maps be a fibration. A related result is Proposition 3.3.5. Here is an analogous result for 3-cubes. We state it simply because the question of when the limit of a punctured cube is the same as its homotopy limit is a natural one to ask. The proof is motivated by the iterative description of homotopy limits of punctured cubes, Lemma 5.3.6.

**Proposition 5.4.27** *Let* $X : \mathcal{P}(\underline{3}) \to$ Top *be a 3-cube. If* $X$ *is cartesian and*

1. *either of* $X_3 \to \lim(X_{13} \to X_{123} \leftarrow X_{23})$ *or* $\lim(X_1 \to X_{12} \leftarrow X_2) \to$ $\lim(X_{13} \to X_{123} \leftarrow X_{23})$ *is a fibration,*
2. *either of* $X_{13} \to X_{123}$ *or* $X_{23} \to X_{123}$ *is a fibration,*
3. *either of* $X_1 \to X_{12}$ *or* $X_2 \to X_{12}$ *is a fibration.*

*then* $X$ *is homotopy cartesian.*

**Remark 5.4.28** Permutations of $\{1, 2, 3\}$ will generate other lists of sufficient statements. This can be generalized to higher-dimensional cubes, though it would require more bookkeeping and a better organization than we have attempted here. We will not pursue such generalizations because, including this very proposition, we really have no use for such statements. □

*Proof of Proposition 5.4.27* We provide a sketch which the reader can fill in. Consider the punctured 3-cube

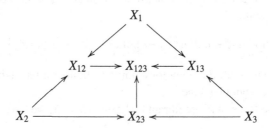

Now apply Lemma 5.2.8 and use Proposition 3.2.13 on each of the three evident punctured squares, the first two being $X_1 \to X_{12} \leftarrow X_2$ and $X_{13} \to X_{123} \leftarrow X_{23}$. If we let $Y_{12}$ and $Y_{123}$ denote the limits of these punctured squares respectively, the third is $X_3 \to Y_{123} \leftarrow Y_{12}$. □

We finish with a result that will be generalized in Theorem 8.6.1. Suppose $\mathcal{X} = S \mapsto X_S$ is an $n$-cube. The inclusion $S \subset \underline{n}$ of a non-empty subset $S$ gives rise to a projection map

$$\operatorname*{holim}_{T \in \mathcal{P}_0(\underline{n})} X_T = \operatorname{Nat}_{\mathcal{P}_0(\underline{n})}(\Delta(\bullet), \mathcal{X}) \longrightarrow \operatorname{Nat}_{\mathcal{P}_0(S)}(\Delta(\bullet), \mathcal{X}) = \operatorname*{holim}_{R \in \mathcal{P}_0(S)} X_R. \qquad (5.4.4)$$

**Proposition 5.4.29** *The map in (5.4.4) is a fibration for every non-empty $S$.*

*Proof* Since the composition of fibrations is a fibration, it suffices to prove this for $S \subset \underline{n-1}$ with $|S| = n-1$. By symmetry we may also assume $S = \underline{n-1}$. By Lemma 5.3.6 we have a homeomorphism

$$\operatorname*{holim}_{T \in \mathcal{P}_0(\underline{n})} X_T \cong \operatorname{holim}\left( X_n \to \operatorname*{holim}_{R \in \mathcal{P}_0(\underline{n-1})} X_{R \cup \{n\}} \leftarrow \operatorname*{holim}_{R \in \mathcal{P}_0(\underline{n-1})} X_R \right).$$

By Proposition 3.2.5, we then have a homeomorphism

$$\operatorname*{holim}_{T \in \mathcal{P}_0(\underline{n})} X_T \cong \lim\left( P_n \to \operatorname*{holim}_{R \in \mathcal{P}_0(\underline{n-1})} X_{R \cup \{n\}} \leftarrow \operatorname*{holim}_{R \in \mathcal{P}_0(\underline{n-1})} X_R \right),$$

where $P_n$ is the mapping path space of the map $X_n \to \operatorname{holim}_{R \in \mathcal{P}_0(\underline{n-1})} X_{R \cup \{n\}}$. That is, we have a (strict) pullback square

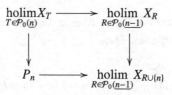

Since the lower horizontal arrow is a fibration, so is the upper horizontal arrow by Proposition 2.1.16. □

## 5.5 Total homotopy fibers

We now turn to the generalization of the definition of the total homotopy fiber of a square (Definition 3.4.1).

For an $n$-cube $X$ of spaces, recall from (5.4.1) the canonical map

$$a(X): X_\emptyset \longrightarrow \operatorname*{holim}_{\mathcal{P}_0(\underline{n})} X. \tag{5.5.1}$$

If $X$ is an $n$-cube in $\mathrm{Top}_*$ (a cube of based spaces), then $\operatorname{holim}_{\mathcal{P}_0(\underline{n})} X$ has a natural basepoint given by the element of $\mathrm{Nat}(\Delta(\bullet), X)$ which sends every face $\Delta(S)$ of $\Delta(\underline{n}) \cong \Delta^{n-1}$ to the basepoint of the corresponding space $X_S$.

**Definition 5.5.1** Let $X$ be an $n$-cube of based spaces. Define the *total homotopy fiber of $X$* to be

$$\mathrm{tfiber}(X) = \mathrm{hofiber}(a(X)),$$

where the homotopy fiber is taken over the natural basepoint of $\operatorname{holim}_{\mathcal{P}_0(\underline{n})} X$.

Combining the above definition with Definition 5.4.1, we have that a cube is $k$-cartesian if the total fiber is $(k-1)$-connected.

**Remark 5.5.2** The homotopy limit $\operatorname{holim}_{\mathcal{P}_0(\underline{n})} X$ did not need to be based in a natural way that comes from the basepoint of $X_\emptyset$. We could have used the homotopy fiber over any basepoint in this homotopy limit, but our choice will simplify matters later. □

**Example 5.5.3** The total homotopy fiber of a 1-cube of based spaces $X_\emptyset \to X_1$ is simply the space $\mathrm{hofiber}(X_\emptyset \to X_1)$. For a square, this is the same as Definition 3.4.1. □

In analogy with Proposition 3.4.3, we can think of the total homotopy fiber inductively as an iterated fiber.

**Proposition 5.5.4** (Total homotopy fiber as iterated homotopy fiber) *Let $X$ be an n-cube of based spaces. Then*

- *if $\underline{n} = \emptyset$, tfiber($X$) = $X_\emptyset$;*
- *for $\underline{n} \neq \emptyset$, view $X = \mathcal{Y} \to \mathcal{Z}$ as a map of $(n-1)$-cubes, and we have*

$$\text{tfiber}(X) \simeq \text{hofiber}(\text{tfiber}(\mathcal{Y}) \to \text{tfiber}(\mathcal{Z})),$$

*where the basepoint in* tfiber($\mathcal{Z}$) *is the natural one.*

It is not *a priori* obvious that the above iterative description is independent of the way we view the cube as a map of cubes of lower dimension, but we will clear this up shortly. As with squares, when we repeatedly take (homotopy) fibers, we will refer to the resulting space as the "iterated (homotopy) fiber".

*Proof* Recall the notation for faces of a cube from Definition 5.1.1. It should be evident from the proof of Lemma 5.3.6 that it is enough to prove that if we write $X = \partial^{n-1}X \to \partial^n_{\{n\}}X$ as a map of $(n-1)$-cubes then

$$\text{tfiber}(X) \simeq \text{hofiber}\left(\text{tfiber}(\partial^{n-1}X) \to \text{tfiber}(\partial^n_{\{n\}}X)\right).$$

By definition,

$$\text{tfiber}(\partial^{n-1}X) = \text{hofiber}\left(X_\emptyset \to \underset{R \in \mathcal{P}_0(\underline{n-1})}{\text{holim}} X_S\right),$$

and

$$\text{tfiber}(\partial^n_{\{n\}}X) = \text{hofiber}\left(X_n \to \underset{R \in \mathcal{P}_0(\underline{n-1})}{\text{holim}} X_{R\cup\{n\}}\right).$$

Hence

$$\text{hofiber}\left(\text{tfiber}(\partial^{n-1}X) \to \text{tfiber}(\partial^n_{\{n\}}X)\right)$$

is homeomorphic to the total homotopy fiber of the square

$$
\begin{array}{ccc}
X_\emptyset & \longrightarrow & \underset{R \in \mathcal{P}_0(\underline{n-1})}{\text{holim}} X_R \\
\downarrow & & \downarrow \\
X_n & \longrightarrow & \underset{R \in \mathcal{P}_0(\underline{n-1})}{\text{holim}} X_{R\cup\{n\}}
\end{array}
$$

by Proposition 3.4.3. But by definition the total homotopy fiber of this square is the homotopy fiber of the map

$$X_\emptyset \longrightarrow \operatorname{holim}\left(X_n \to \operatorname*{holim}_{R \in \mathcal{P}_0(\underline{n-1})} X_R \leftarrow \operatorname*{holim}_{R \in \mathcal{P}_0(\underline{n-1})} X_{R \cup \{n\}}\right).$$

The codomain of this map is homeomorphic to $\operatorname{holim}_{\mathcal{P}_0(\underline{n})} X$ by Lemma 5.3.6, and the result follows. It is now clear that this proof is independent of the manner in which we write $X$ as a map of $(n-1)$-cubes, and hence the iterative description of the total homotopy fiber is also independent of the choices necessary to define it. □

**Example 5.5.5** For each $S \subset \underline{n}$, Let $X_S$ be a based space. Define an $n$-cube $\mathcal{Y} = S \mapsto Y_S$ by $Y_S = \prod_{R \subset \underline{n} - S} X_R$, where the inclusion $S \subset T$ gives rise to a natural projection map $Y_S \to Y_T$ away from those factors $X_R$ for which $R$ is not a subset of $\underline{n} - T$. Then $\operatorname{tfiber}(\mathcal{Y}) \simeq X_{\underline{n}}$, as an easy induction will show.

The result is clear in the case $n = 1$. For $n \geq 2$, write $\mathcal{Y} = \mathcal{W} \to \mathcal{Z}$ as a map of $(n-1)$-cubes, where

$$\mathcal{Z}(S) = \prod_{R \subset \underline{n-1} - S} X_R$$

and

$$\mathcal{W}(S) = \prod_{R \subset \underline{n-1}} X_{R \cup \{n\}} \times \prod_{R \subset \underline{n-1} - S} X_R.$$

By Proposition 5.5.4, $\operatorname{tfiber}(\mathcal{Y}) \simeq \operatorname{hofiber}(\operatorname{tfiber}(\mathcal{W}) \to \operatorname{tfiber}(\mathcal{Z}))$. By induction, $\operatorname{tfiber}(\mathcal{Z}) \simeq X_{\underline{n-1}}$, and $\operatorname{tfiber}(\mathcal{W}) \simeq X_{\underline{n}} \times X_{\underline{n-1}}$. It is evident that the map

$$\operatorname{tfiber}(\mathcal{W}) \simeq X_{\underline{n}} \times X_{\underline{n-1}} \longrightarrow X_{\underline{n-1}} \simeq \operatorname{tfiber}(\mathcal{Z})$$

is the projection map, whose homotopy fiber is clearly equivalent to $X_{\underline{n}}$. □

**Example 5.5.6** (Cube of configuration spaces) For a space $X$, define the *configuration space of $n$ points in $X$* to be the space

$$\operatorname{Conf}(n, X) = \{(x_1, x_2, \ldots, x_n) \in X^n \mid x_i \neq x_j \text{ for } i \neq j\}.$$

This is therefore the space of distinct $n$-tuples of points in $X$, that is $X^n$ with the "fat" diagonal removed. We may also view $\operatorname{Conf}(n, X)$ as the subspace of $\operatorname{Map}(\underline{n}, X) = X^{\underline{n}}$ consisting of all maps $e \colon \underline{n} \to X$ such that $e(i) \neq e(j)$ for all $i \neq j$. Thus $\operatorname{Conf}(0, X) \cong *$ and $\operatorname{Conf}(1, X) \cong X$. It also easy to see that, for example, $\operatorname{Conf}(2, \mathbb{R}^m) \simeq S^{m-1}$ (the mapping $\mathbb{R}^m \times \mathbb{R}^m \setminus \Delta_{\mathbb{R}^m} \to S^{m-1}$ given by $(x, y) \mapsto (x - y)/|x - y|$ is a homotopy equivalence).

We have a cubical diagram of configuration spaces

$$\mathrm{Conf}(\underline{n} - \bullet, X) \colon \mathcal{P}(\underline{n}) \longrightarrow \mathrm{Top}$$
$$S \longmapsto \mathrm{Conf}(\underline{n} - S, X).$$

The inclusion $S \subset T$ gives rise to a restriction $\mathrm{Conf}(\underline{n} - S) \to \mathrm{Conf}(\underline{n} - T)$; these are maps that project a tuple to another tuple by forgetting the points not indexed by $\underline{n} - T$. To simplify notation, we are going to identify $\mathrm{Conf}(S, X)$ with $\mathrm{Conf}(|S|, X)$ when this will not cause confusion.

The total homotopy fiber of this diagram is important in calculus of functors and we will explore it further in Examples 6.2.9 and 10.2.25. The maps in the cube are generated by projections $\mathrm{Conf}(n, X) \to \mathrm{Conf}(n - 1, X)$ which are in fact fibrations [FN62]. When $X = \mathbb{R}^m$, the fiber is $\mathbb{R}^m$ with $n - 1$ points removed, which is homotopy equivalent to a wedge of $n - 1$ spheres of dimension $m - 1$. We thus have a fibration sequence

$$\bigvee_{n-1} S^{m-1} \longrightarrow \mathrm{Conf}(n, \mathbb{R}^m) \longrightarrow \mathrm{Conf}(n - 1, \mathbb{R}^m).$$

From this and from Example 3.9.5, we can compute that the total homotopy fiber of the 3-cube of configurations arising from three points in $\mathbb{R}^m$ is $\Omega S^{m-1} *$ $\Omega S^{m-1}$:

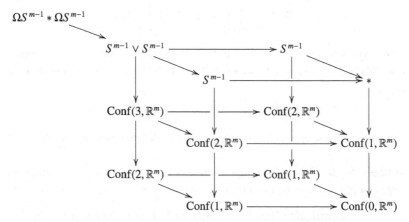

The top square is homotopy equivalent to the square of fibers of the vertical maps in the cube, and the upper-left space is the total homotopy fiber of the top square. □

The following generalizes Proposition 3.4.9.

**Proposition 5.5.7** *Let X be a n-cube of based spaces, and suppose there are maps $X_{S \cup \{i\}} \to X_S$ for $S \subset \underline{n}$ and $i \notin S$ such that the composition $X_S \to X_{S \cup \{i\}} \to X_S$ is the identity. Then*

$$\text{tfiber}(X) \simeq \Omega^n \, \text{tfiber}(X'),$$

*where $X'$ is the n-cube defined by $S \mapsto X_{\underline{n}-S}$.*

*Proof* This follows from induction on $n$ using a repeated application of Proposition 3.4.8. In constructing the proof the reader may wish to first take a look at the proof of Proposition 3.4.9. □

We next give a few alternative descriptions of the total homotopy fiber.

**Proposition 5.5.8** (Definition 1.1 [Goo92]) *Let X be an n-cube of based spaces. Then tfiber(X) is the subspace of $\prod_{S \subset \underline{n}} \text{Map}(I^S, X_S)$ consisting of those maps $\Phi_S : I^S \to X_S$ for which the following hold:*

1. *The diagram*

$$
\begin{array}{ccc}
I^T & \longrightarrow & I^S \\
\Phi_T \downarrow & & \downarrow \Phi_S \\
X_T & \longrightarrow & X_S
\end{array}
$$

   *commutes for all $T \subset S \subset \underline{n}$, where the upper horizontal arrow takes a function $T \to I$ and extends it by zero on $S - T$ to a function $S \to I$.*
2. *For each $S$, $\Phi_S((I^S)_1)$ maps to the basepoint, where $(I^S)_1 = \{u \in I^S : u_s = 1 \text{ for some } s \in S\}$.*

*Proof* We use induction on $n$. The result is evidently true for $n = 0$, and the case $n = 1$ is clear by inspection. Assume $X$ is an $n$-cube, $n \geq 2$, and write $X = \partial^{n-1}X \to \partial^n_{\{n\}}X$ as a map of $(n-1)$-cubes. By Proposition 5.5.4,

$$\text{tfiber}(X) = \text{hofiber}\left(\text{tfiber}(\partial^{n-1}X) \longrightarrow \text{tfiber}(\partial^n_{\{n\}}X)\right),$$

where the basepoint of $\text{tfiber}(\partial^n_{\{n\}}X)$ is the pair $(x_n, c)$, with $x_n \in X_{\{n\}}$ the basepoint and $c : I \to \text{holim}_{\mathcal{P}_0(R \in \underline{n-1})} X_{R \cup \{n\}}$ the constant path at the natural basepoint of $\text{holim}_{\mathcal{P}_0(R \in \underline{n-1})} X_{R \cup \{n\}}$ (the element of $\text{Nat}_{\mathcal{P}_0(\underline{n-1})}(\Delta(\bullet), X(\bullet \cup \{n\}))$ which sends $\Delta(R)$ to the basepoint of $X_{R \cup \{n\}}$ for all $R$).

By induction, the homotopy fiber in question is a subspace of

$$\prod_{R \subset \underline{n-1}} \text{Map}(I^R, X_R) \times \prod_{R \subset \underline{n-1}} \text{Map}(I^{R \cup \{n\}}, X_{R \cup \{n\}})$$

consisting of those maps satisfying conditions 1 and 2 for all $T \subset S \subset \underline{n-1}$ and for all $n \in T \subset S \subset \underline{n}$. For $T \subset S$ with $n \in S - T$, it is straightforward to check that conditions 1 and 2 are also satisfied.  □

We can mix the iterative description of the total homotopy fiber with the definition of total homotopy fiber of a square to obtain the following useful characterization of the total homotopy fiber.

**Proposition 5.5.9**  *Let $X$ be an n-cube of based spaces, and let $i \in \underline{n}$. Then* tfiber($X$) *is equivalent to the total homotopy fiber of the square*

$$
\begin{array}{ccc}
X_\emptyset & \longrightarrow & X_i \\
\downarrow & & \downarrow \\
\underset{S \in \mathcal{P}_0(\underline{n}-\{i\})}{\operatorname{holim}} X_S & \longrightarrow & \underset{S \in \mathcal{P}_0(\underline{n}-\{i\})}{\operatorname{holim}} X_{S \cup \{i\}}
\end{array}
$$

This is useful if $X = \mathcal{Y} \to \mathcal{Z}$ is a map of cubes.

*Proof*  This follows immediately from the proof of Proposition 5.5.4.  □

In analogy with the definition of homotopy groups of a square, Definition 3.4.10, we have the following.

**Definition 5.5.10**  For an $n$-cube $X \colon \mathcal{P}(\underline{n}) \to$ Top and for $i \geq n$, define

$$
\pi_i(X) = \pi_{i-n}(\text{tfiber } X).
$$

One can also get a long exact sequence of homotopy groups in a couple of different ways from the total homotopy fiber. There is of course the homotopy fiber sequence

$$
\text{tfiber}(X) \longrightarrow X_\emptyset \longrightarrow \underset{T \in \mathcal{P}_0(S)}{\operatorname{holim}} X_T,
$$

which gives rise to the usual long exact sequence in homotopy. But if we view $X = \mathcal{Y} \to \mathcal{Z}$ as a map of $(n-1)$-cubes, then we also have the following result, which is really a restatement of Proposition 5.5.4.

**Proposition 5.5.11**  *For $X = \mathcal{Y} \to \mathcal{Z}$ a map of cubes, there is a homotopy fiber sequence*

$$
\text{tfiber}(X) \longrightarrow \text{tfiber}(\mathcal{Y}) \longrightarrow \text{tfiber}(\mathcal{Z}).
$$

This fibration sequence then gives rise to a long exact sequence in homotopy. This reduces precisely to the long exact sequence from Proposition 3.4.11 since the total homotopy fiber of a 1-cube is just an ordinary homotopy fiber.

## 5.6 Colimits of punctured cubes

We now begin the dual discussion for pushouts of punctured cubes. Let $X\colon \mathcal{P}_1(\underline{n}) \to \mathrm{Top}$ be a punctured $n$-cube, and again write $X(S) = X_S$ and $X(S \subset T) = f_{S \subset T}\colon X_S \to X_T$.

**Definition 5.6.1** The *pushout* or *colimit* of the punctured $n$-cube $X$, denoted $\mathrm{colim}_{\mathcal{P}_1(\underline{n})} X$, is the quotient space

$$\operatorname*{colim}_{\mathcal{P}_1(\underline{n})} X = \coprod_{S \in \mathcal{P}_1(\underline{n})} X_S / \sim$$

where the equivalence relation $\sim$ is generated by

$$f_{S \subset T}(x_S) \sim f_{S \subset T'}(x_S)$$

for all $T, T'$ and all $S \subset T, T'$.

**Remarks 5.6.2**

1. In the case $n = 2$, the above definition reduces to Definition 3.5.1.
2. Alternatively, we may also write $\mathrm{colim}_{S \in \mathcal{P}_1(\underline{n})} X_S$ or $\mathrm{colim}_{S \neq \underline{n}} X_S$ in place of the above. Fix finite sets $R \subset T$. In the case where the cube is indexed by sets which contain $R$ and are contained in $T$, we will write $\mathrm{colim}_{S \subsetneq T-R} X_S$ for the limit over the evident punctured cube, and omit $R$ from this notation when it is empty.
3. The colimit could have been defined as the subspace of $\coprod_{i \in \underline{n}} X_{\underline{n}-\{i\}}$ modulo the equivalence relation generated by $f_{S \subset \underline{n}-\{i\}}(x_S) \sim f_{S \subset \underline{n}-\{j\}}(x_S)$ where $i \neq j$ ranges over all $i, j \in \underline{n}$ and $S$ ranges over all subsets of $\underline{n} - \{i, j\}$. It is straightforward to check that this description and the one in Definition 5.6.1 are homeomorphic. This alternative description is simpler to visualize than what is given in Definition 5.6.1 and natural from the perspective of Definition 3.5.1, but Definition 5.6.1 is closer to the general model for colimits of diagrams of spaces in Proposition 7.4.18. □

**Remark 5.6.3** (Colimit of a cube) In analogy with Remark 5.2.3, the colimit of a cube can be defined as in Definition 5.6.1 but over the category $\mathcal{P}(\underline{n})$. However, this colimit is simply $X_{\underline{n}}$. A more general result is Example 7.3.27. □

As was the case for the limit, the colimit of a punctured cube enjoys a universal property. There are canonical inclusions $i_S\colon X_S \to \mathrm{colim}_{\mathcal{P}_1(\underline{n})} X$ such that, for all $S \subset T$, the diagram

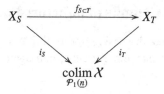

commutes. Thus the assignment

$$S \longmapsto \begin{cases} X_S, & \text{if } S \subsetneq \underline{n}; \\ \operatorname*{colim}_{\mathcal{P}_1(\underline{n})} X_S, & \text{if } S = \underline{n} \end{cases}$$

defines an *n*-cube. If $Z$ is a space with maps $j_S : X_S \to Z$ for all $S$ such that

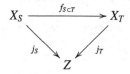

commutes, then there is a unique map

$$u_1 : \operatorname*{colim}_{\mathcal{P}_1(\underline{n})} X \longrightarrow Z$$

such that the diagram

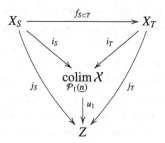

commutes.

Here are a few examples. Generalizing Examples 3.5.2 and 3.5.3, we have the following.

**Example 5.6.4** Let $X_1, \ldots, X_n$ be spaces. Define $X \colon \mathcal{P}_1(\underline{n}) \to \text{Top}$ by $X_S = \coprod_{i \in S} X_i$. Then it is straightforward to see that $\operatorname{colim}_{\mathcal{P}_1(\underline{n})} X = \coprod_{i \in \underline{n}} X_i$. In the category of based spaces, we replace $\emptyset$ with the one-point space $*$ and the maps $* \to X_i$ are the inclusion of the basepoint. In this case the colimit is the wedge sum $\bigvee_{i \in \underline{n}} X_i$. $\quad\square$

**Example 5.6.5** An alternative description of the disjoint union of $X_1, \ldots, X_n$ is to define $X \colon \mathcal{P}_1(\underline{n}) \to \text{Top}$ by

$$X(S) = \begin{cases} \emptyset, & \text{if } |S| < n - 1; \\ X_i, & \text{if } S = \underline{n} - \{i\}. \end{cases}$$

In this case it is not hard to see that $\operatorname{colim}_{\mathcal{P}_1(\underline{n})} X \cong \coprod_{i \in \underline{n}} X_i$. For based spaces, the evident changes give the wedge sum $\bigvee_{i \in \underline{n}} X_i$ as the colimit. □

The next example is a natural generalization of Example 3.5.4.

**Example 5.6.6** Let $X \colon \mathcal{P}_1(\underline{n}) \to$ Top be the constant punctured cube $C_X$. Then $\operatorname{colim}_{\mathcal{P}_1(\underline{n})} X \cong X$, as all copies of $X$ are identified via the identity map. □

We now have the following iterative description of the colimit of a punctured cube, dual to Lemma 5.2.8 with dual proof which we omit.

**Lemma 5.6.7** (Iterated pushout) *Given a punctured n-cube $X \colon \mathcal{P}_1(\underline{n}) \to$ Top, $n \geq 2$, there is a homeomorphism*

$$\operatorname*{colim}_{S \in \mathcal{P}_1(\underline{n})} X_S \cong \operatorname{colim}\left( X_{n-1} \leftarrow \operatorname*{colim}_{R \in \mathcal{P}_1(\underline{n-1})} X_R \to \operatorname*{colim}_{R \in \mathcal{P}_1(\underline{n-1})} X_{R \cup \{n\}} \right).$$

Dual to Example 5.2.9, we have the following application of Lemma 5.6.7.

**Example 5.6.8** Suppose $f \colon X \to Y$ and $g \colon Y \to Z$ are maps, and let $h = g \circ f$ be the composite. Consider the punctured 3-cube $X$ given by the diagram

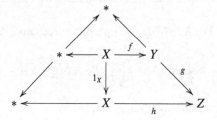

By Lemma 5.6.7 we have

$$\operatorname*{colim}_{\mathcal{P}_1(\underline{3})} X \cong \operatorname{colim}\left( Z \leftarrow \operatorname{colim}(Y \leftarrow X \xrightarrow{1_X} X) \to \operatorname{colim}(* \leftarrow * \to *) \right).$$

Clearly $\operatorname{colim}(* \leftarrow * \to *) = *$, and by Proposition 3.5.10, $\operatorname{colim}(Y \leftarrow X \xrightarrow{1_X} X) \cong Y$, so the colimit of the punctured square describing $\operatorname{colim}_{\mathcal{P}_1(\underline{3})} X$ above is isomorphic to $\operatorname{colim}(Z \leftarrow Y \to *)$. But this is precisely $\operatorname{cofiber}(g)$ by Example 3.5.6. We can also use Lemma 5.6.7 to write

$$\operatorname*{colim}_{\mathcal{P}_1(\underline{3})} X \cong \left( * \leftarrow \operatorname{colim}(* \leftarrow X \to Y) \to \operatorname{colim}(* \leftarrow X \to Z) \right)$$

or, again using Example 3.5.6,

$$\text{colim}\left(* \leftarrow \text{cofiber}(f) \rightarrow \text{cofiber}(h)\right),$$

so that

$$\text{cofiber}(g) \cong \text{cofiber}\left(\text{cofiber}(f) \rightarrow \text{cofiber}(h)\right). \qquad \square$$

We now generalize of Definition 3.5.12. Again first note that if $X$ is a cubical diagram, then $X_{\underline{n}}$ admits a canonical map from the colimit of the remaining punctured cube since it admits a map from each $X_S$.

**Definition 5.6.9**   We say an $n$-cube $X \colon \mathcal{P}(\underline{n}) \rightarrow$ Top is *cocartesian* (or *categorically cocartesian*, or a *(strict) pushout*) if the canonical map

$$\underset{\mathcal{P}_1(\underline{n})}{\text{colim}} X_S \longrightarrow X_{\underline{n}}$$

is a homeomorphism.

**Example 5.6.10**   From Examples 5.6.4 and 5.6.6, we have that the cubes $S \mapsto \coprod_{i \in S} X_S$ and $S \mapsto X$ are cocartesian.          $\square$

## 5.7  Homotopy colimits of punctured cubes

Recall Example 5.1.5 for the definition of the punctured cubical simplex $\Delta(\underline{n} - \bullet) \colon \mathcal{P}_1(\underline{n}) \rightarrow$ Top. Dual to Definition 5.3.1 we have the following.

**Definition 5.7.1**   For $X \colon \mathcal{P}_1(\underline{n}) \rightarrow$ Top a punctured cube, define the *homotopy colimit of $X$* as

$$\underset{\mathcal{P}_1(\underline{n})}{\text{hocolim}} X = \coprod_{S \in \mathcal{P}_1(\underline{n})} X_S \times \Delta(\underline{n} - S)/ \sim$$

where $\sim$ is the equivalence relation generated by

$$(x_S, d_{S \subset T}(t)) \sim (f_{S \subset T}(x_S), t)$$

for all $S \subset T \subsetneq \underline{n}$.

**Remark 5.7.2** (Pointed homotopy colimit)   The above definition is also valid for punctured cubes in $\text{Top}_*$, but more care has to be taken for general diagrams; see Remark 8.2.14.          $\square$

**Remark 5.7.3** (Homotopy colimit of a cube)   Dually to Remark 5.3.2, one could also define the homotopy colimit of a cube. The definition would resemble the one above; see Remark 8.2.13. However, as we shall see in Proposition A.1.16, this homotopy colimit would simply be a space homotopy equivalent to $X_{\underline{n}}$. □

When $n = 2$, the above definition gives Definition 3.6.3. In the category Top$_*$ of based spaces, the homotopy colimit is a quotient of the above formula by the subspace hocolim$_{S \in \mathcal{P}_1(\underline{n})} *_S$, where $*_S \in X_S$ is the basepoint. We will, however, use the same notation for both. For the extension of the above definition to general diagrams of spaces, the reader should jump ahead to Definition 8.2.12 and Remark 8.2.13.

**Example 5.7.4**  For the punctured cube indexed on $\mathcal{P}_1(\underline{n})$ where all the spaces are $X$ and the maps are all identity, the homotopy colimit is equal to $X \times \Delta^{n-1}$, and is clearly homotopy equivalent to $X$. □

**Example 5.7.5** (Suspensions as hocolims of punctured cubes)  Let $X$ be a space. By Example 3.6.9, hocolim($* \leftarrow X \rightarrow *$) $\simeq \Sigma X$ is the unreduced suspension (in Top$_*$, it is the reduced suspension). Similarly, by definition

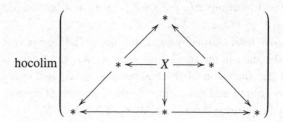

is the quotient of $X \times \Delta^2$ by $X \times \partial \Delta^2$, giving the unreduced two-fold suspension $\Sigma^2 X$. To get the reduced suspension, if $X$ is based then we quotient the above by the homotopy colimit of the diagram which has the basepoint of $X$ replacing $X$ above. It is also easy to use Lemma 5.7.6 to see this.

More generally, if $X \colon \mathcal{P}_1(\underline{n}) \rightarrow$ Top is defined by $X(\emptyset) = X$ and $X(S) = *$ for all $\emptyset \neq S \subsetneq \underline{n}$, then using induction and Lemma 5.7.6 we see hocolim$_{\mathcal{P}_1(\underline{n})} X \simeq \Sigma^{n-1} X$ is the $(n - 1)$-fold unreduced suspension of $X$. □

Dual to Lemma 5.3.6, with a dual proof which we leave to the reader, we have the following.

**Lemma 5.7.6** (Iterated homotopy pushout)  *Given a punctured n-cube* $X \colon \mathcal{P}_1(\underline{n}) \rightarrow$ Top, $n \geq 2$, *there is a homeomorphism*

$$\text{hocolim}_{\mathcal{P}_1(\underline{n})} \mathcal{X} \cong \text{hocolim}\left(X_{\underline{n-1}} \leftarrow \underset{R \in \mathcal{P}_1(\underline{n-1})}{\text{hocolim}} X_R \rightarrow \underset{R \in \mathcal{P}_1(\underline{n-1})}{\text{hocolim}} X_{R \cup \{n\}}\right).$$

A remark analogous to Remark 5.3.7 applies equally well here. Here is the first of many applications of Lemma 5.7.6 that we shall see.

**Example 5.7.7** Let $X_1, \ldots, X_n$ be based spaces, and consider the punctured $n$-cube $\mathcal{X} \colon \mathcal{P}_1(\underline{n}) \rightarrow \text{Top}$ given by

$$\mathcal{X}(S) = X_S = \left\{(x_1, \ldots, x_n) \in \prod_{i=1}^{n} X_i \mid x_j = *_i \text{ for all } j \notin S\right\},$$

where $*_i$ denotes the basepoint of $X_i$. The maps $X_S \rightarrow X_T$ for $S \subset T$ are the evident inclusions. Note that $X_S \cong \prod_{i \in S} X_i$.

We claim that $\text{hocolim}_{\mathcal{P}_1(\underline{n})} \mathcal{X}$ is homotopy equivalent to the space $W(X_1, \ldots, X_n)$ of all $(x_1, \ldots, x_n)$ such that $x_i = *_i$ for at least one $i$. We call $W(X_1, \ldots, X_n)$ the *fat wedge* of the spaces $X_1, \ldots, X_n$.

Moreover, we claim that the canonical map

$$\underset{\mathcal{P}_1(\underline{n})}{\text{hocolim}} \mathcal{X} \longrightarrow \underset{\mathcal{P}_1(\underline{n})}{\text{colim}} \mathcal{X}$$

induced by the projections $\Delta(\underline{n} - S) \times X_S \rightarrow X_S$ is a weak equivalence, and that $\text{colim}_{\mathcal{P}_1(\underline{n})} \mathcal{X} \cong W(X_1, \ldots, X_n)$.

We will prove both claims by induction on $n$. The case $n = 2$ is obvious; the punctured square in question is $X_1 \leftarrow * \rightarrow X_2$, and since our spaces are well-pointed, the inclusion of the basepoint $* \rightarrow X_i$ is a cofibration. Thus the homotopy colimit and the colimit of these punctured squares are weakly equivalent by Proposition 3.6.17, and $W(X_1, X_2) = X_1 \vee X_2$.

Assume the result is true for cubes of dimension less than $n$, and consider the $n$-cube at the beginning of this example. Using Lemma 5.7.6 we have a homeomorphism

$$\underset{\mathcal{P}_1(\underline{n})}{\text{hocolim}} \mathcal{X} \cong \text{hocolim}\left(\prod_{i \in \underline{n-1}} X_i \leftarrow \underset{\mathcal{P}_1(\underline{n-1})}{\text{hocolim}} \mathcal{X} \rightarrow \underset{\mathcal{P}_1(\underline{n-1})}{\text{hocolim}} (\mathcal{X} \times X_n)\right).$$

By Proposition 5.8.7, we have a weak equivalence of the above with

$$\text{hocolim}\left(\prod_{i \in \underline{n-1}} X_i \leftarrow \underset{\mathcal{P}_1(\underline{n-1})}{\text{hocolim}} \mathcal{X} \rightarrow X_n \times \underset{\mathcal{P}_1(\underline{n-1})}{\text{hocolim}} \mathcal{X}\right)$$

where the right-most map is the evident inclusion of the homotopy colimit using the basepoint of $X_n$. By induction, the canonical map $\text{hocolim}_{\mathcal{P}_1(\underline{n-1})} \mathcal{X} \rightarrow \text{colim}_{\mathcal{P}_1(\underline{n-1})} \mathcal{X}$ is a weak equivalence, and hence the homotopy colimit above admits a weak equivalence to

$$\text{hocolim}\left(\prod_{i \in \underline{n-1}} X_i \leftarrow \underset{\mathcal{P}_1(\underline{n-1})}{\text{colim}}\, \mathcal{X} \to X_n \times \underset{\mathcal{P}_1(\underline{n-1})}{\text{colim}}\, \mathcal{X}\right),$$

which, again by induction, is equal to

$$\text{hocolim}\left(\prod_{i \in \underline{n-1}} X_i \leftarrow W(X_1, \ldots, X_{n-1}) \to X_n \times W(X_1, \ldots, X_{n-1})\right).$$

Since the map $W(X_1, \ldots, X_{n-1}) \to \prod_{i \in \underline{n-1}} X_i$ is a cofibration (by induction and Proposition 2.3.13), using Proposition 3.6.17 the homotopy colimit above admits a weak equivalence to

$$\text{colim}\left(\prod_{i \in \underline{n-1}} X_i \leftarrow W(X_1, \ldots, X_{n-1}) \to X_n \times W(X_1, \ldots, X_{n-1})\right).$$

By inspection, this colimit is equal to $W(X_1, \ldots, X_n)$ by definition. □

We now have the dual of Theorem 5.3.9, which establishes the homotopy invariance of the homotopy colimit of a punctured cube.

**Theorem 5.7.8** (Homotopy invariance of homotopy colimits, cubes case) *Suppose* $\mathcal{X}, \mathcal{Y} \colon \mathcal{P}_1(\underline{n}) \to \text{Top}$ *are punctured n-cubes and* $\mathcal{X} \to \mathcal{Y}$ *is a homotopy (resp. weak) equivalence. Then the induced map*

$$\underset{\mathcal{P}_1(\underline{n})}{\text{hocolim}}\, \mathcal{X} \longrightarrow \underset{\mathcal{P}_1(\underline{n})}{\text{hocolim}}\, \mathcal{Y}$$

*is a homotopy (resp. weak) equivalence.*

*Proof* Dual to Theorem 5.3.9 using induction, Theorem 3.6.13, and Lemma 5.7.6. □

## 5.8 Arithmetic of homotopy cocartesian cubes

This section is dual to Section 5.4 and corresponds to Section 3.7. The reader should recall the definitions of homotopy cocartesian and $k$-cocartesian squares (Definition 3.7.1).

Let as before $\mathcal{X} \colon \mathcal{P}(\underline{n}) \to \text{Top}$, $S \to X_S$, be an $n$-cube of spaces (or based spaces), and recall that restricting $\mathcal{X}$ to $\mathcal{P}_1(\underline{n})$ gives the punctured cube missing the final space $X_{\underline{n}}$. Then we have canonical maps

$$\underset{\mathcal{P}_1(\underline{n})}{\text{colim}}\, \mathcal{X} \longrightarrow X_{\underline{n}} \quad \text{and} \quad \underset{\mathcal{P}_1(\underline{n})}{\text{hocolim}}\, \mathcal{X} \longrightarrow \underset{\mathcal{P}_1(\underline{n})}{\text{colim}}\, \mathcal{X}.$$

The map from the colimit to $X_{\underline{n}}$ was discussed prior to Definition 5.6.9. In that definition, we said $X$ is cocartesian if this map is weak equivalence. The map from the homotopy colimit to the colimit is the one induced by the projection map $\Delta(\underline{n} - S) \times X_S \to X_S$ for each $S$, which induces a map of the appropriate quotient spaces.

In analogy with (3.7.1), we have the composed map

$$b(X): \operatorname*{hocolim}_{\mathcal{P}_1(\underline{n})} X \longrightarrow X_{\underline{n}}. \tag{5.8.1}$$

**Definition 5.8.1**   We say the $n$-cube $X: \mathcal{P}(\underline{n}) \to$ Top is

- *homotopy cocartesian* (or $\infty$-*cartesian*)  if the map $b(X)$ is a weak equivalence;
- $k$-*cocartesian* if the map $b(X)$ is $k$-connected.

The comments made in Remark 5.4.2 apply here (in a dual sense). In particular, a $k$-cocartesian cube is also $j$-cocartesian for all $j \leq k$.

**Example 5.8.2**   Examples 5.3.4 and Example 5.7.5 can be restated in terms of cocartesian terminology; for example, the latter says that the $n$-cube with initial space a based space $X$, final space $\Sigma^{n-1} X$, and the rest one-point spaces, is homotopy cocartesian.                                               □

**Example 5.8.3**   Generalizing Example 3.7.5, suppose $U_1, \ldots, U_n$ are open subset of $X$ and $U_1 \cup \cdots \cup U_n = X$. Then the $n$-cube

$$X: \mathcal{P}(\underline{n}) \longrightarrow \text{Top}$$

$$S \longmapsto \begin{cases} \bigcap_{i \notin S} U_i, & S \neq \underline{n}; \\ U_1 \cup \cdots \cup U_n = X, & S = \underline{n}, \end{cases}$$

is homotopy cocartesian. This can be seen by employing the description of the homotopy pushout of the punctured cube as an iterated homotopy pushout by Lemma 5.7.6 and using Lemma 3.6.14. When $n = 3$, the punctured cube indexed on $\mathcal{P}_1(\underline{3})$ is

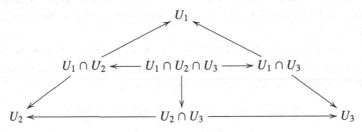

The homotopy pushouts of the rows, from the top, are $U_1$, $(U_1 \cap U_2) \cup (U_1 \cap U_3)$, and $U_2 \cup U_3$, where we have used Example 3.7.5 for the latter two (replacing the homotopy colimit with the ordinary colimit). The homotopy pushout of the punctured cube is then

$$\text{hocolim}(U_3 \leftarrow (U_1 \cap U_2) \cup (U_1 \cap U_3) \rightarrow U_2 \cup U_3).$$

But this is $U_1 \cup U_2 \cup U_3 = X$, again by Example 3.7.5. The details of the general case are left to the reader. □

**Example 5.8.4** Dual to Example 5.4.4, if $\mathcal{X}$ is an $n$-cube and $\widetilde{\mathcal{X}} = \mathcal{X} \rightarrow \mathcal{X}$ is an $(n + 1)$-cube obtained from the identity map of $\mathcal{X}$, then $\widetilde{\mathcal{X}}$ is homotopy cocartesian. We leave the details to the reader. □

Here is a generalization of Proposition 3.7.9, dual to Proposition 5.4.5 with dual proof. Parts 1 and 3 of this result, as well as the corollary following it, will be generalized in Proposition 8.5.7.

**Proposition 5.8.5**

1. *If $\mathcal{X}$ and $\mathcal{Y}$ are homotopy cocartesian n-cubes of spaces, then the n-cube $\mathcal{X} \sqcup \mathcal{Y}$ given by $S \mapsto X_S \sqcup Y_S$ (with the obvious maps) is also homotopy cocartesian.*
2. *If $\mathcal{X}$ and $\mathcal{Y}$ and $k_1$-cocartesian and $k_2$-cocartesian, respectively, then $\mathcal{X} \sqcup \mathcal{Y}$ is $\min\{k_1, k_2\}$-cocartesian.*
3. *If $\mathcal{X}$ and $\mathcal{Y}$ are cubes of based spaces, both statements above hold for the cube $\mathcal{X} \vee \mathcal{Y}$ given by $S \mapsto X_S \vee Y_S$.*

As a consequence, we have a generalization of Corollary 3.7.10.

**Corollary 5.8.6** (Wedge commutes with homotopy colimit, cubes case) *If $\mathcal{X}$ is a homotopy cocartesian cube of based spaces, so is the cube $\mathcal{X} \vee V$, for any based space $V$. If $\mathcal{X}$ is $k$-cartesian, so is $\mathcal{X} \vee V$.*

Dually to Proposition 5.4.7, we also have a generalization of Proposition 3.7.11. A related general statement is Corollary 8.5.6.

**Proposition 5.8.7** (Products commute with homotopy colimits, cubes case) *If $\mathcal{X}$ is a homotopy cocartesian cube, so is the cube $\mathcal{X} \times V$, for any space $V$, where the maps $X_S \times V \rightarrow X_T \times V$ for $S \subset T$ are $f_{S \subset T} \times 1_V$. If $\mathcal{X}$ is $k$-cocartesian, so is $\mathcal{X} \times V$.*

*Proof* The space $\text{hocolim}_{\mathcal{P}_1(\underline{n})}\, X \times V$ is a quotient of $\coprod_{S \in \mathcal{P}_1(\underline{n})} X_S \times V$, and the evident homeomorphism $\coprod_{S \in \mathcal{P}_1(\underline{n})} X_S \times V \cong V \times \coprod_{S \in \mathcal{P}_1(\underline{n})} X_S$ induces a homeomorphism of quotient spaces $\text{hocolim}_{\mathcal{P}_1(\underline{n})}\, X \times V \cong V \times \text{hocolim}_{\mathcal{P}_1(\underline{n})}\, X$. $\qquad\square$

That products of cubes with fixed spaces commute with homotopy colimits gives the following generalization of Example 3.6.12. It is an application of Proposition 5.8.7.

**Example 5.8.8**  Let $X_1, \ldots, X_n$ be spaces and consider the punctured $n$-cube

$$X \colon \mathcal{P}_1(\underline{n}) \longrightarrow \text{Top}$$
$$S \longmapsto \prod_{i \notin S} X_i,$$

where all the maps are projections; an inclusion $T \subset S$ induces a projection $X(T) \to X(S)$ away from those factors indexed by elements of $S - T$. Then

$$\text{hocolim}_{\mathcal{P}_1(\underline{n})} X \simeq X_1 * \cdots * X_n.$$

To illustrate, consider the case $n = 3$. The punctured 3-cube in question is

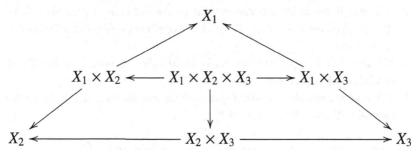

Using Lemma 5.7.6, we can take homotopy colimits along the rows to see that the homotopy colimit of the above diagram is homeomorphic to

$$\text{hocolim}\left( \begin{array}{c} X_1 \\ \uparrow \\ \text{hocolim}(X_1 \times X_2 \leftarrow X_1 \times X_2 \times X_3 \to X_1 \times X_3) \\ \downarrow \\ \text{hocolim}(X_2 \leftarrow X_2 \times X_3 \to X_3) \end{array} \right)$$

Recalling the definition of the join and using Proposition 5.8.7, we see the above is weakly equivalent to

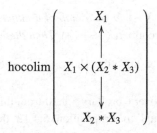

$$\mathrm{hocolim}\left(\begin{array}{c} X_1 \\ \uparrow \\ X_1 \times (X_2 * X_3) \\ \downarrow \\ X_2 * X_3 \end{array}\right)$$

Again using the definition of the join, the above is weakly equivalent to $X_1 * (X_2 * X_3) \cong X_1 * X_2 * X_3$. Note that the associativity of the join is easily proved using these methods, since we already know that the various ways of viewing homotopy colimits iteratively are equivalent. The general proof follows by straightforward induction. $\qquad\square$

We now state the dual to Theorem 5.4.9, with dual proof using Lemma 5.7.6 and Theorem 3.7.18, which we omit. A generalization can be found in the second part of Proposition 8.5.5.

**Theorem 5.8.9** (Homotopy colimits commute, cubes case) *Suppose* $X \colon \mathcal{P}_1(\underline{n}) \times \mathcal{P}_1(\underline{m}) \to \mathrm{Top}$ *is a diagram* $(S, T) \to X(S, T)$ *(an n-punctured cube of m-punctured cubes, or an m-punctured cube of n-punctured cubes). Then there is a homeomorphism*

$$\mathrm{hocolim}_{\mathcal{P}_1(\underline{n})} \mathrm{hocolim}_{\mathcal{P}_1(\underline{m})} X \cong \mathrm{hocolim}_{\mathcal{P}_1(\underline{m})} \mathrm{hocolim}_{\mathcal{P}_1(\underline{n})} X.$$

In particular we have the following result that generalizes Corollary 3.7.19. The proof is identical to that special case of squares. We shall see this result again in Corollary 8.5.8. Notice that the second equivalence essentially restates Corollary 5.8.6.

**Corollary 5.8.10** *For a punctured cube* $X$ *(in* $\mathrm{Top}$ *or* $\mathrm{Top}_*$*) and* $V$ *a (based) space, there are weak equivalences*

$$\mathrm{hocolim}_{\mathcal{P}_1(\underline{n})} \Sigma X \simeq \Sigma \, \mathrm{hocolim}_{\mathcal{P}_1(\underline{n})} X,$$

$$\mathrm{hocolim}_{\mathcal{P}_1(\underline{n})} (V \vee X) \simeq V \vee \mathrm{hocolim}_{\mathcal{P}_1(\underline{n})} X,$$

$$\mathrm{hocolim}_{\mathcal{P}_1(\underline{n})} (V \wedge X) \simeq V \wedge \mathrm{hocolim}_{\mathcal{P}_1(\underline{n})} X,$$

$$\mathrm{hocolim}_{\mathcal{P}_1(\underline{n})} (V * X) \simeq V * \mathrm{hocolim}_{\mathcal{P}_1(\underline{n})} X.$$

Dual to Corollary 5.4.11, we have the following.

**Corollary 5.8.11** *Let $X \to Y$ be a map of n-cubes, and define $Z \colon \mathcal{P}(\underline{n}) \to$ Top by $C(S) = C_S = \mathrm{hocofiber}(X_S \to Y_S)$. Then there is a homeomorphism*

$$\mathrm{hocolim}_{S \in \mathcal{P}_1(\underline{n})} C_S \cong \mathrm{hocofiber}\left(\mathrm{hocolim}_{S \in \mathcal{P}_1(\underline{n})} X_S \to \mathrm{hocolim}_{S \in \mathcal{P}_1(\underline{n})} Y_S\right).$$

Note that one could recover Corollary 5.8.10 from this result as well.

Here is a partial dual to Proposition 5.4.12 that generalizes Proposition 3.7.24.

**Proposition 5.8.12** *Suppose a map $X \to Y$ of n-cubes is a k-cocartesian $(n + 1)$-cube. Then the n-cube $\mathrm{hocofiber}(X \to Y)$ is k-cocartesian.*

The converse is not true, as was already discussed along with Proposition 3.7.24.

*Proof* The proof is essentially dual to the proof of Proposition 5.4.12. The only difference is that we need to apply Proposition 3.7.13 to the evident square that arises by taking cofibers. □

The following two results generalize Propositions 3.7.13 and 3.7.26. They are dual to Propositions 5.4.13 and 5.4.14.

**Proposition 5.8.13** *Suppose $X$ and $Y$ are n-cubes, and that $(X \to Y)$ is a map of n-cubes, considered as an $(n + 1)$-cube.*

1. (a) *If $(X \to Y)$ and $X$ are homotopy cocartesian, then $Y$ is homotopy cocartesian.*
   (b) *If $X$ and $Y$ are homotopy cocartesian, then $(X \to Y)$ is homotopy cocartesian.*
2. (a) *If $X$ and $(X \to Y)$ are k-cocartesian, then $Y$ is k-cocartesian.*
   (b) *If $X$ is $(k - 1)$-cocartesian and $Y$ is k-cocartesian, then $(X \to Y)$ is k-cocartesian.*

*Proof* Consider the diagram

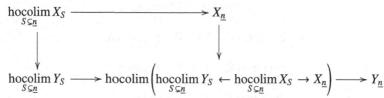

The square above is homotopy cocartesian by inspection, hence the connectivity of the upper horizontal arrow (given by how cocartesian $X$ is) is the same as the connectivity of the map

$$\text{hocolim}\left(\underset{S \subsetneq n}{\text{hocolim}}\, Y_S \leftarrow \underset{S \subsetneq n}{\text{hocolim}}\, X_S \to X_{\underline{n}}\right) \longrightarrow Y_{\underline{n}}.$$

The result now follows immediately from Proposition 2.6.15 (the homotopy cartesian version is obtained by setting $k = \infty$). □

**Proposition 5.8.14** *Suppose* $X = (\mathcal{U} \to \mathcal{V})$ *and* $\mathcal{Y} = (\mathcal{V} \to \mathcal{W})$ *are n-cubes, and let* $\mathcal{Z} = (\mathcal{U} \to \mathcal{W})$, *all considered as maps of* $(n-1)$*-cubes. Then*

1. (a) *If* $X$ *and* $\mathcal{Y}$ *are homotopy cocartesian, then* $\mathcal{Z}$ *is homotopy cocartesian.*
   (b) *If* $X$ *and* $\mathcal{Z}$ *are homotopy cocartesian, then* $\mathcal{Y}$ *is homotopy cocartesian.*
2. (a) *If* $X$ *is k-cocartesian and* $\mathcal{Y}$ *is k-cocartesian, then* $\mathcal{Z}$ *is k-cocartesian.*
   (b) *If* $X$ *is* $(k-1)$*-cocartesian and* $\mathcal{Z}$ *is k-cocartesian, then* $\mathcal{Y}$ *is k-cocartesian.*

*Proof* Consider the following diagram of $(n-1)$-cubes:

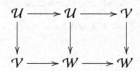

To prove 2(a), view the left-most square as a map $(\mathcal{U} \to \mathcal{U}) \to (\mathcal{V} \to \mathcal{W}) = \mathcal{Y}$. Since $\mathcal{Y}$ is $k$-cocartesian and $\mathcal{U} \to \mathcal{U}$ is homotopy cocartesian, 2(b) of Proposition 5.8.13 implies the left-most square is a $k$-cocartesian $(n+1)$-cube. Viewing this $(n+1)$-cube as a map $X = (\mathcal{U} \to \mathcal{V}) \to (\mathcal{U} \to \mathcal{W}) = \mathcal{Z}$ and using the hypothesis that $X$ is $k$-cocartesian, 2(a) of Proposition 5.8.13 now implies $\mathcal{Z}$ is $k$-cocartesian.

To prove 2(b), view the right-most square as a map $X = (\mathcal{U} \to \mathcal{V}) \to (\mathcal{W} \to \mathcal{W})$ of $n$-cubes. Since $X$ is $(k-1)$-cocartesian and $\mathcal{W} \to \mathcal{W}$ is homotopy cocartesian, 2(b) of Proposition 5.8.13 implies the right square is a $k$-cocartesian $(n+1)$-cube. Viewing this $(n+1)$-cube as a map $\mathcal{Z} = (\mathcal{U} \to \mathcal{W}) \to (\mathcal{V} \to \mathcal{W}) = \mathcal{Y}$ and using the hypothesis that $\mathcal{Z}$ is $k$-cocartesian, 2(a) of Proposition 5.8.13 implies that $\mathcal{Y}$ is $k$-cocartesian. □

The next two propositions are duals of Proposition 5.4.16 and Proposition 5.4.17. Recall the notation and terminology for "cubes of cubes" from there.

**Proposition 5.8.15** *Let* $X$ *be an n-cube of m-cubes. Assume that* $X$, *viewed as a* $(m+n)$*-cube, is k-cocartesian, and that, for each* $S \subsetneq \underline{n}$, *the m-cube* $X(S)$ *is* $(k + |S| - 1 + n)$*-cocartesian. Then the m-cube* $X(\underline{n})$ *is k-cocartesian.*

*Proof* Dual to the proof of Proposition 5.4.16. □

**Proposition 5.8.16**   *Let $X$ be a functor from $\mathcal{P}_1(n)$ to m-cubes of spaces and write $X(S, T) = (X(S))(T)$. If, for all $S \subsetneq \underline{n}$, $X(S)$ is a $k_S$-cocartesian m-cube, then the m-cube $T \mapsto \operatorname{hocolim}_{U \neq \emptyset} X(S, T)$ is $\min_S\{|T| - |S| - 1 + k_S\}$-cocartesian.*

*Proof*   Dual to the proof of Proposition 5.4.17.                                                    □

**Example 5.8.17**   Here is another proof, first given in Proposition 3.7.22, of the equivalence $X * Y \simeq \Sigma(X \wedge Y)$ for based spaces $X$ and $Y$. It uses Proposition 5.8.16. Consider the following diagram of squares.

$$
\begin{array}{ccc}
X \vee CY \longrightarrow X & \quad X \vee Y \longrightarrow X \times Y & \quad CX \vee Y \longrightarrow Y \\
\downarrow \qquad\quad \downarrow & \quad \downarrow \qquad\quad \downarrow & \quad \downarrow \qquad\quad \downarrow \\
CY \longrightarrow * & \quad * \longrightarrow X \wedge Y & \quad CX \longrightarrow *
\end{array}
$$

Each of the three square diagrams depicted are homotopy cocartesian. We may view the above diagram as a square of punctured squares

$$
\begin{array}{ccc}
(X \vee CY \leftarrow X \vee Y \to CX \vee Y) & \longrightarrow & (X \leftarrow X \times Y \to Y) \\
\downarrow & & \downarrow \\
(CY \leftarrow * \to CX) & \longrightarrow & (* \leftarrow X \wedge Y \to *)
\end{array}
$$

By Proposition 5.8.16, the square of homotopy colimits of the diagram associated with each vertex in the cube is homotopy cocartesian. This is weakly equivalent to the square

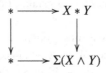

$$
\begin{array}{ccc}
* & \longrightarrow & X * Y \\
\downarrow & & \downarrow \\
* & \longrightarrow & \Sigma(X \wedge Y)
\end{array}
$$

Thus $X * Y \to \Sigma(X \wedge Y)$ is a weak equivalence by 1(a) of Proposition 3.7.13.   □

We now discuss strongly (homotopy) cocartesian cubes. Recall the notation for faces of the cube from Definition 5.1.1.

**Definition 5.8.18**   We call an $n$-cube $X$ *strongly homotopy cocartesian* (resp. *strongly cocartesian*) if each face of dimension $\geq 2$ is homotopy cocartesian (resp. cocartesian).

By 1(b) of Proposition 5.8.13, the definition could have just required two-dimensional faces to be homotopy cocartesian. Similarly for the cocartesian case, again by using Proposition 5.8.13 with "colim" in place of "hocolim" everywhere, and Proposition 3.5.10. This implies the following.

**Proposition 5.8.19** *A strongly (homotopy) cocartesian cube is (homotopy) cocartesian.*

**Example 5.8.20** Let $X \to Y$ be a map of spaces. Recall the notion of the fiberwise join $X *_Y U$ from Example 3.9.13. Then, for $U \subset \underline{n}$, the $n$-cube $X = U \mapsto X *_Y U$ is strongly homotopy cocartesian. This follows from homotopy pushouts commuting with joins; see Corollary 3.7.19.                    □

**Example 5.8.21** We can now generalize Examples 3.6.6 and 3.7.7. Consider the $n$-cubes

$$X \colon \mathcal{P}(\underline{n}) \longrightarrow \mathrm{Top}$$

$$S \longmapsto \begin{cases} \bigvee_{i \in S} X_i, & S \neq \emptyset; \\ *, & S = \emptyset, \end{cases}$$

and

$$\mathcal{Y} \colon \mathcal{P}(\underline{n}) \longrightarrow \mathrm{Top}$$

$$S \longmapsto \begin{cases} \bigvee_{i \notin S} X_i, & S \neq \underline{n}; \\ *, & S = \underline{n}. \end{cases}$$

The maps in the first cube are inclusions and those in the second projections. These cubes are strongly homotopy cocartesian and hence homotopy cocartesian.

To see this, for $X$, each square is of the form

$$\begin{array}{ccc} \bigvee_{i \in T} X_i & \longrightarrow & (\bigvee_{i \in T} X_i) \vee X_j \\ \downarrow & & \downarrow \\ (\bigvee_{i \in T} X_i) \vee X_k & \longrightarrow & (\bigvee_{i \in T} X_i) \vee X_j \vee X_k \end{array}$$

for some $T \subset \underline{n}$ and $j, k \notin T$. These squares are homotopy cocartesian by Corollary 3.7.10 and Example 3.7.2.                    □

Dually to Definition 5.4.22, we have the following generalization of Definition 3.7.28.

**Definition 5.8.22** We call a (punctured) $n$-cube $X$ a *cofibrant cube* if, for all $S \subset \underline{n}$ ($S \subsetneq \underline{n}$), the map

$$\operatorname*{colim}_{T \subsetneq S} X_T \longrightarrow \operatorname*{colim}_{T \subset S} X_T = X_S$$

is a cofibration. An $n$-cube $X$ is called a *cofibrant pushout cube* if, in addition, $X_S = \operatorname{colim}_{T \subsetneq S} X_T$ for all $|S| \geq 2$.

It is easy to show that a cofibration pushout cube $X$ has $X_S$ equal to the union over $X_\emptyset$ of the $X_i$ for $i \in S$.

The following generalizes Proposition 3.7.29 and dualizes Theorem 5.4.23. The cofibrant cube that is equivlent to the given one is called a *cofibrant replacement* of the original cube.

**Theorem 5.8.23** *Every (punctured) cube admits a homotopy equivalence from a (punctured) cofibrant cube. Every strongly homotopy cocartesian n-cube admits a homotopy equivalence from a cofibrant pushout cube.*

*Proof*  Dual to the proof of Theorem 5.4.23, but replace initial maps in the cube by cofibrations.          □

**Lemma 5.8.24** *Suppose $X = T \mapsto X_T$ is a cofibrant n-cube. Then every face $\partial_R^S X$, $R \subset S \subset \underline{n}$, is a cofibrant $(|S| - |R|)$-cube.*

*Proof*  Dual to the proof of Lemma 5.4.25.          □

As in Proposition 5.4.26, the proof of the following is by induction and we leave it to the reader. We shall see a restatement of the first part in Corollary 8.4.10 (and a generalization in Theorem 8.4.9).

**Proposition 5.8.25** *Let $X$ be a cofibrant (punctured) n-cube. Then the canonical map*

$$\operatorname*{hocolim}_{S \in \mathcal{P}_1(\underline{n})} X_S \longrightarrow \operatorname*{colim}_{S \in \mathcal{P}_1(\underline{n})} X_S$$

*is a homotopy equivalence. Hence a cofibrant cube $X$ is k-cocartesian if and only if*

$$\operatorname*{colim}_{S \in \mathcal{P}_1(n)} X_S \to X_{\underline{n}}$$

*is k-connected.*

**Proposition 5.8.26** *Suppose $X$ is an n-cube such that, for each $S \subset \underline{n}$, the map $X_S \to X_{\underline{n}}$ is the inclusion of an open subset of $X_{\underline{n}}$ and $X_{S \cap T} = X_S \cap X_T$ for all $S, T \subset \underline{n}$. Then the canonical map*

$$\operatorname*{hocolim}_{T \in \mathcal{P}_1(\underline{n})} X_T \longrightarrow \operatorname*{colim}_{T \in \mathcal{P}_1(\underline{n})} X_T$$

*is a weak equivalence.*

*Proof* This follows by induction on $n$ using Lemma 5.6.7, Lemma 5.7.6, and Example 3.7.5. □

This last proposition covers many "standard" cases of cubes, such as the square that arises when a space $X$ is covered by two open subsets $X_1, X_2 \subset X$.

We invite the reader to state and prove a dual to Proposition 5.4.27; we have no use for such statements so we omit it.

**Remark 5.8.27** In analogy with Remark 5.4.24 it is easy to see that a cube is $k$-cocartesian if and only if its cofibrant replacement is. We leave the details to the reader. □

We now have the dual of Proposition 5.4.29, which will also be generalized in Theorem 8.6.1. For any proper subset $S \subsetneq \underline{n}$, we have a natural inclusion map

$$\operatorname*{hocolim}_{\mathcal{P}_1(S)} \mathcal{X} \longrightarrow \operatorname*{hocolim}_{\mathcal{P}_1(\underline{n})} \mathcal{X}. \tag{5.8.2}$$

**Proposition 5.8.28** *The map (5.8.2) is a cofibration.*

*Proof* Dual to the proof of Proposition 5.4.29. □

To close, here is the generalization of Proposition 3.7.31 the proof of which also follows immediately from that result.

**Proposition 5.8.29** ([KW12, Lemma 3.5.12]) *Let $\mathcal{X} = S \mapsto X_S$ be a strongly homotopy cocartesian $n$-cube of based spaces with $X_{\underline{n}} = *$, and let $\Sigma \mathcal{X}$ denote the cube of suspensions. Then $\Sigma \mathcal{X}$ is weakly equivalent to an $n$-cube $\mathcal{Y} = S \mapsto Y_S$ with $Y_{\underline{n}-\{i\}} = \Sigma X_{\underline{n}-\{i\}}$ for all $i$ and $Y_S = \vee_{i \notin S} \Sigma X_{\underline{n}-\{i\}}$ for $|S| \geq 2$.*

## 5.9 Total homotopy cofibers

This section is essentially the dualization of Section 5.5.

Given an $n$-cube $\mathcal{X}$ of spaces, we have from (5.8.1) the usual canonical map

$$b(\mathcal{X}): \operatorname*{hocolim}_{\mathcal{P}_1(\underline{n})} \mathcal{X} \longrightarrow X_{\underline{n}}. \tag{5.9.1}$$

**Definition 5.9.1**   Let $X$ be an $n$-cube of spaces. Define the *total homotopy cofiber of* $X$ to be

$$\text{tcofiber}(X) = \text{hocofiber}(b(X)).$$

In the case of a cube of based spaces, we make the evident adjustments (using the based homotopy colimit and the reduced mapping cone for the homotopy cofiber).

**Example 5.9.2**   The total homotopy cofiber of a 1-cube of spaces $X_0 \to X_1$ is just the homotopy cofiber and, for a square, this reduces to Definition 3.8.1.                                                                                                              □

Dual to Proposition 5.5.4 and generalizing Proposition 3.8.2, we have the following.

**Proposition 5.9.3** (Total homotopy cofiber as iterated homotopy cofiber)   *Let $X$ be an $n$-cube of spaces.*

- *The total homotopy cofiber of $X$ is equivalent to $X_0$ if $n = 0$.*
- *For $n > 0$, view $X = \mathcal{Y} \to \mathcal{Z}$ as a map of $(n - 1)$-cubes and define* $\text{tcofiber}(X) = \text{hocofiber}(\text{tcofiber}(\mathcal{Y}) \to \text{tcofiber}(\mathcal{Z}))$.

*Proof*   Dual to the proof of Proposition 5.5.4.                                                      □

**Example 5.9.4**   Let $X_1, \ldots, X_n$ be based spaces. Recall the punctured $n$-cube $X$ from Example 5.7.7 given by

$$X(S) = X_S = \left\{ (x_1, \ldots, x_n) \in \prod_{i=1}^{n} X_i : x_j = *_i \text{ for all } j \notin S \right\},$$

where $*_i$ denotes the basepoint of $X_i$, and the maps $X_S \to X_T$ for $S \subset T$ are the evident inclusions. This can be extended to a cubical diagram in the obvious way with $X(\underline{n})$ the product of the $X_i$. By Example 5.7.7, there is a homotopy equivalence

$$\operatorname*{hocolim}_{\mathcal{P}_1(\underline{n})} X \simeq W(X_1, \ldots, X_n),$$

where $W(X_1, \ldots, X_n)$ is the fat wedge, that is, the space of all tuples $(x_1, \ldots, x_n)$ where $x_i = *_i$ for at least one $i$. Then $\text{tcofiber}(X) \simeq X_1 \wedge \ldots \wedge X_n$, since the inclusion of $W(X_1, \ldots, X_n)$ in $X_1 \times \cdots \times X_n$ is a cofibration (use Proposition 2.3.13 and induction) and so the homotopy cofiber of this map is equivalent to the cofiber.

We claim that $X_1 \wedge \cdots \wedge X_n \cong (X_1 \times \cdots \times X_n)/W(X_1, \ldots, X_n)$. This follows by induction on $n$, with the base case $n = 2$ using the definition of the smash product. Consider the diagram

$$
\begin{array}{ccccc}
\prod_{i \in \underline{n-1}} X_i & \xleftarrow{\ 1_{\prod_{i \in \underline{n-1}} X_i}\ } & \prod_{i \in \underline{n-1}} X_i & \xrightarrow{\hspace{3cm}} & \prod_{i \in \underline{n}} X_i \\
{\scriptstyle 1_{\prod_{i \in \underline{n-1}} X_i}} \big\uparrow & & \big\uparrow & & \big\uparrow \\
\prod_{i \in \underline{n-1}} X_i & \longleftarrow & W(X_1, \ldots, X_{n-1}) & \longrightarrow & X_n \times W(X_1, \ldots, X_{n-1}) \\
\big\downarrow & & \big\downarrow & & \big\downarrow \\
* & \longleftarrow & * & \longrightarrow & *
\end{array}
$$

The homotopy colimits of the columns are homotopy equivalent to the colimits of the columns by Proposition 3.6.17 since the top vertical map in each column is a cofibration. These are, from left to right, $*$, $X_1 \wedge \cdots \wedge X_{n-1}$ by induction, and $(X_n)_+ \wedge (X_1 \wedge \cdots \wedge X_{n-1})$ by induction and Example 2.4.13. The homotopy colimits of the rows are also homotopy equivalent to the colimits of the rows since all of the left-most horizontal arrows are cofibrations. These are, from top to bottom, $\prod_{i \in \underline{n}} X_i$ by Example 3.5.5, $W(X_1, \ldots, X_n)$ by Example 5.7.7, and $*$. By Theorem 3.7.18, we obtain a homeomorphism

$$
\mathrm{hocolim}\, (* \leftarrow X_1 \wedge \cdots \wedge X_{n-1} \to (X_n)_+ \wedge (X_1 \wedge \cdots \wedge X_{n-1}))
$$
$$
\cong \mathrm{hocolim}\, \Big( \prod_{i \in \underline{n}} X_i \leftarrow W(X_1, \ldots, X_n) \to * \Big).
$$

The first homotopy colimit here is, by inspection, $X_1 \wedge \cdots \wedge X_n$. The homotopy colimits just appearing are colimits, once again due to the evident cofibrations, and so the homotopy equivalence we have just derived here gives a homeomorphism

$$
X_1 \wedge \cdots \wedge X_n \cong X_1 \times \cdots \times X_n / W(X_1, \ldots, X_n). \qquad \square
$$

We now have a generalization of Proposition 3.8.7. This is dual to Proposition 5.5.7. The proof is by induction on $n$ using a repeated application of Proposition 3.8.7.

**Proposition 5.9.5** *Let $X$ be an $n$-cube of spaces, and suppose there are maps $X_{S \cup \{i\}} \to X_S$ for $S \subset \underline{n}$ and $i \notin S$ such that the composition $X_S \to X_{S \cup \{i\}} \to X_S$ is the identity. Then*

$$
\Sigma^n \mathrm{tcofiber}(X) \simeq \mathrm{tcofiber}(X'),
$$

*where $X'$ is the $n$-cube defined by $S \mapsto X_{\underline{n}-S}$.*

**Example 5.9.6** For based spaces $X_1, \ldots, X_n$, the $n$-cube $\mathcal{X}$ from Example 5.9.4 satisfies the hypotheses of Proposition 5.9.5. In this case the cube $\mathcal{X}'$ in the statement of Proposition 5.9.5 is the one considered in Example 5.8.8. Thus, combining Example 5.9.4 with Example 5.8.8 using Proposition 5.9.5, we have a weak equivalence

$$\Sigma^n(X_1 \wedge \ldots \wedge X_n) \longrightarrow \Sigma(X_1 * \cdots * X_n).$$

We could get a "desuspension" of this result using only Proposition 3.7.22, which says that for based spaces $X$ and $Y$ there is a weak equivalence $\Sigma(X \wedge Y) \to X * Y$. If $X_1, \ldots, X_n$ are based spaces, then repeated application of Proposition 3.7.22 together with the homeomorphism $\Sigma(X \wedge Y) \cong \Sigma X \wedge Y$ gives us a weak equivalence

$$\Sigma^{n-1}(X_1 \wedge \ldots \wedge X_n) \longrightarrow X_1 * \cdots * X_n. \qquad \square$$

We invite the reader to formulate a dual to Proposition 5.5.8. This is straightforward and comes down to bookkeeping, but we will not have any use for this result and therefore omit it.

The following is dual to Proposition 5.5.9, with a dual proof.

**Proposition 5.9.7** *Let $\mathcal{X}$ be an $n$-cube of spaces, and let $i \in \underline{n}$. Then* tcofiber($\mathcal{X}$) *is homotopy equivalent to the total homotopy cofiber of the square*

$$
\begin{array}{ccc}
\underset{S \in \mathcal{P}_1(\underline{n}-\{i\})}{\mathrm{hocolim}\, X_T} & \longrightarrow & \underset{S \in \mathcal{P}_1(\underline{n}-\{i\})}{\mathrm{hocolim}\, X_{S \cup \{i\}}} \\
\downarrow & & \downarrow \\
X_{S-\{i\}} & \longrightarrow & X_{\underline{n}}
\end{array}
$$

We can also now generalize the notion of the homology groups of a square from Definition 3.8.8.

**Definition 5.9.8** For an $n$-cube $\mathcal{X} \colon \mathcal{P}(\underline{n}) \to \mathrm{Top}$ and for $i \geq 0$, define

$$H_i(\mathcal{X}) = H_i(\text{tcofiber}\,\mathcal{X}).$$

In analogy with Proposition 3.8.9, we can now view $\mathcal{X} = \mathcal{Y} \to \mathcal{Z}$ as a map of $(n-1)$-cubes. Then we have

**Proposition 5.9.9** *For $\mathcal{X} = \mathcal{Y} \to \mathcal{Z}$ a map of cubes, there is a cofibration sequence*

$$\text{tcofiber}(\mathcal{Y}) \longrightarrow \text{tcofiber}(\mathcal{Z}) \longrightarrow \text{tcofiber}(\mathcal{X}).$$

Using Theorem 2.3.14, this cofibration gives rise to a long exact sequence in homology. Because the total homotopy cofiber of a map of spaces (1-cube) is simply the homotopy cofiber, this reduces to the long exact sequence from Proposition 3.8.9.

## 5.10 Algebra of homotopy cartesian and cocartesian cubes

We now examine how homotopy cartesian and cocartesian cubes interact with each other and various constructions. Several results here are generalizations of the statements appearing in Section 3.9 and will in turn be generalized in Section 8.5.

Here is a generalization of Proposition 3.9.1 (see also Proposition 8.5.4).

**Proposition 5.10.1** (Maps and homotopy (co)limits (cubes case)) *For $X$ an n-cube of spaces, there are homeomorphisms*

$$\mathrm{Map}\left(Z, \operatorname*{holim}_{S \in \mathcal{P}_0(\underline{n})} X_S\right) \xrightarrow{\cong} \operatorname*{holim}_{S \in \mathcal{P}_0(\underline{n})} \mathrm{Map}(Z, X_S),$$

$$\mathrm{Map}\left(\operatorname*{hocolim}_{S \in \mathcal{P}_1(\underline{n})} X_S, Z\right) \xrightarrow{\cong} \operatorname*{holim}_{R \in \mathcal{P}_0(\underline{n})} \mathrm{Map}(X_{\underline{n}-R}, Z).$$

*In particular, if $X$ is k-cartesian and $Z$ has the homotopy type of a CW complex of dimension d, then $\mathrm{Map}(Z, X)$ is $(k - d)$-cartesian. If $X$ is homotopy cocartesian then $\mathrm{Map}(X, Z)$ is homotopy cartesian.*

*Proof* We have a homeomorphism $\mathrm{Map}(Z, \mathrm{Nat}_{\mathcal{P}_0(\underline{n})}(\Delta, X)) \cong \mathrm{Nat}_{\mathcal{P}_0(\underline{n})}(\Delta \times Z, X)$, where $\Delta \times Z \colon \mathcal{P}_0(\underline{n}) \to \mathrm{Top}$ is given by $\Delta \times Z(S) = \Delta(S) \times Z$ and, for $S \subset T$, $\Delta \times Z(S) \to \Delta \times Z(T)$ is $d_{S \subset T} \times 1_Z$ (maps $d_{S \subset T}$ are described in Example 5.1.5). Thus the first statement follows. The connectivity statement follows from Proposition 3.3.9.

For the second homeomorphism, recall by definition that $\operatorname{hocolim}_{\mathcal{P}_1(\underline{n})} X$ is equal to

$$\coprod_{S \in \mathcal{P}_1(\underline{n})} \Delta(\underline{n} - S) \times X_S / \sim$$

where $\sim$ is the equivalence relation generated by

$$(d^{S \subset T}(t), x_S) \sim (t, f_{S \subset T}(x_S)).$$

Thus a map $g \in \mathrm{Map}(\operatorname{hocolim}_{\mathcal{P}_1(\underline{n})} X, Z)$ consists of a collection of maps

$$g_S : \Delta(\underline{n} - S) \times X_S \longrightarrow Z$$

such that, for all $S \subset T$, $g_S(d^{S \subset T}(t), x_S) = g_T(t, f_{S \subset T}(x_S))$ for all $t \in \Delta(\underline{n} - T)$ and $x_S \in X_S$. By Theorem 1.2.7 the $g_S$ determine maps $\widetilde{g}_S : \Delta(\underline{n} - S) \to X_S \times Z$, and they satisfy

$$\widetilde{g}_S(d^{S \subset T}(t)) = \widetilde{g}_T(t)$$

for all $S \subset T$, and for all $t \in \Delta(\underline{n} - T)$. The assignment $g \mapsto \{\widetilde{g}_R\}_{R \in \mathcal{P}_0(\underline{n} - R)}$ then determines the desired homeomorphism.                                                      □

We also have a generalization of Corollary 3.9.4 and a special case of Proposition 8.5.9.

**Proposition 5.10.2**  *Suppose $X$ is a strongly homotopy cocartesian $n$-cube and suppose that, for some space $Y$, there is a map $X_S \to Y$ for each $S$ such that*

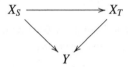

*commutes for all $S \subset T$ (equivalently, that there exists a natural transformation from $X$ to a cube $\mathcal{Y}$ such that $\mathcal{Y}(S) = Y$ and $\mathcal{Y}(S \subset T) = 1_Y$ for all $S \subset T$). Then, for any $y \in Y$, the $n$-cube of homotopy fibers $\mathrm{hofiber}_y(X \to Y)$ is strongly homotopy cocartesian.*

*Proof*  This is immediate from the $n = 2$ case, covered in Corollary 3.9.4 since all of the square faces are homotopy cocartesian.                                      □

**Remark 5.10.3**  More generally, if the original cube is $k$-cocartesian, then so is the cube of fibers. We leave the details to the reader.                                      □

The following is used in the proof of Theorem 6.2.2, the dual of the generalized Blakers–Massey Theorem. It is a consequence of Proposition 5.10.2 and generalizes Proposition 3.9.12 (using the latter result in the proof).

**Proposition 5.10.4**  *Suppose $X$ is a strongly homotopy cartesian $n$-cube of spaces. Recall the canonical map $b(X)$: $\mathrm{hocolim}_{\mathcal{P}_1(\underline{n})} \to X_{\underline{n}}$. Then, for any basepoint $x \in X_{\underline{n}}$,*

$$\mathrm{hofiber}\, b(X) \simeq \underset{i \in \underline{n}}{*}\, \mathrm{hofiber}(X_{\underline{n} - \{i\}} \to X_{\underline{n}}).$$

*Proof*  The proof proceeds by induction on $n$. We first give the proof of the case $n = 3$ as an illustration ($n = 2$ is Proposition 3.9.12). Let $D = \mathrm{hocolim}_{S \subsetneq \underline{3}} X_S$. The claim is that for the strongly homotopy cocartesian cube

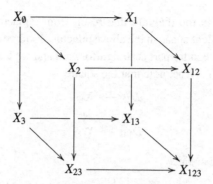

we have hofiber$(D \to X_{123})$ = hofiber$(X_{12} \to X_{123})$ * hofiber$(X_{13} \to X_{123})$ * hofiber$(X_{23} \to X_{123})$. Let $D_{12}$ = hocolim$(X_1 \leftarrow X_0 \to X_2)$ and $D_{123}$ = hocolim$(X_{13} \leftarrow X_3 \to X_{23})$. By definition of $D$, $D_{12}$, $D_{123}$ and using the iterated homotopy colimit description of a punctured cube, Lemma 5.7.6, we have

$$D = \mathrm{hocolim}\left(\begin{array}{ccc} & X_{12} & \\ & \nearrow \quad \nwarrow & \\ X_1 \leftarrow & X_0 \longrightarrow & X_2 \\ \nearrow & \downarrow & \searrow \\ X_{13} \leftarrow & X_3 \longrightarrow & X_{23} \end{array}\right) \simeq \mathrm{hocolim}\left(\begin{array}{c} X_{12} \\ \uparrow \\ D_{12} \\ \downarrow \\ D_{123} \end{array}\right)$$

By Corollary 3.9.4, we have

$$\mathrm{hofiber}(D \to X_{123}) = \mathrm{hocolim}\left(\begin{array}{c} \mathrm{hofiber}(X_{12} \to X_{123}) \\ \uparrow \\ \mathrm{hofiber}(D_{12} \to X_{123}) \\ \downarrow \\ \mathrm{hofiber}(D_{123} \to X_{123}) \end{array}\right)$$

Since the bottom face is homotopy cartesian, Proposition 3.9.12 implies that

$$\mathrm{hofiber}(D_{123} \to X_{123}) \simeq \mathrm{hofiber}(X_{13} \to X_{123}) * \mathrm{hofiber}(X_{23} \to X_{123}).$$

It suffices to prove that

$$\mathrm{hofiber}(D_{12} \to X_{123}) \simeq \qquad\qquad (5.10.1)$$
$$\mathrm{hofiber}(X_{12} \to X_{123}) \times \Big(\mathrm{hofiber}(X_{13} \to X_{123}) * \mathrm{hofiber}(X_{23} \to X_{123})\Big)$$

and that the maps in the diagram that result from replacing hofiber($D_{12} \to X_{123}$) by this equivalent space in the above punctured square are the projections onto each factor. The latter part is straightforward and we leave it to the reader. To see why (5.10.1) is true, note that the square

$$\begin{array}{ccc} D_{12} & \longrightarrow & X_{12} \\ \downarrow & & \downarrow \\ D_{123} & \to & X_{123} \end{array}$$

is homotopy cartesian since

$$\text{hofiber}(D_{12} \to X_{12}) \simeq \text{hofiber}(X_1 \to X_{12}) * \text{hofiber}(X_2 \to X_{12})$$
$$\simeq \text{hofiber}(X_{13} \to X_{123}) * \text{hofiber}(X_{23} \to X_{123})$$
$$\simeq \text{hofiber}(D_{123} \to X_{123})$$

The first and third equivalences use Proposition 3.9.12, and the second uses the fact that the original cubical diagram is strongly homotopy cartesian. Therefore the map

$$D_{12} \to \text{holim}(X_{12} \to X_{123} \leftarrow D_{123})$$

is a weak equivalence, which gives a weak equivalence

$$\text{hofiber}(D_{12} \to X_{123}) \simeq \text{holim}\Big(\text{hofiber}(X_{12} \to X_{123}) \to$$
$$* \leftarrow \text{hofiber}(D_{123} \to X_{123})\Big)$$
$$\simeq \text{hofiber}(X_{12} \to X_{123}) \times \text{hofiber}(D_{123} \to X_{123})$$

Combining this with

$$\text{hofiber}(D_{123} \to X_{123}) \simeq \text{hofiber}(X_{13} \to X_{123}) * \text{hofiber}(X_{23} \to X_{123})$$

proved above, this establishes the result for the case $n = 3$.

For the general case, let $D = \text{hocolim}_{S \subsetneq n} X_S$, $D_{n-1} = \text{hocolim}_{S \subsetneq n-1} X_S$, and $D_n = \text{hocolim}_{S \subsetneq n-1} X_{S \cup \{n\}}$. We have $D \simeq \text{hocolim}(X_{n-1} \leftarrow D_{n-1} \to D_n)$. By Corollary 3.9.4, we have

$$\text{hofiber}(D \to X_n) \simeq \text{hocolim} \left( \begin{array}{c} \text{hofiber}(X_{n-1} \to X_n) \\ \uparrow \\ \text{hofiber}(D_{n-1} \to X_n) \\ \downarrow \\ \text{hofiber}(D_n \to X_n) \end{array} \right)$$

By induction, $\mathrm{hofiber}(D_{\underline{n}} \to X_{\underline{n}}) \simeq *_{i \in \underline{n}-1} \mathrm{hofiber}(X_{\underline{n}-\{i\}} \to X_{\underline{n}})$. It suffices to prove

$$\mathrm{hofiber}(D_{n-1} \to X_{\underline{n}}) \simeq \mathrm{hofiber}(X_{\underline{n}-1} \to X_{\underline{n}}) \times \underset{i \in \underline{n}-1}{*} \mathrm{hofiber}(X_{\underline{n}-\{i\}} \to X_{\underline{n}}).$$

The square

$$\begin{array}{ccc} D_{n-1} & \longrightarrow & X_{n-1} \\ \downarrow & & \downarrow \\ D_{\underline{n}} & \longrightarrow & X_{\underline{n}} \end{array}$$

is homotopy cartesian by comparing the vertical fibers using Proposition 3.9.12 and the fact that $\mathcal{X}$ is strongly homotopy cartesian. Hence the canonical map

$$D_{n-1} \to \mathrm{holim}\left(X_{n-1} \to X_{\underline{n}} \leftarrow D_{\underline{n}}\right)$$

is a weak equivalence, which gives rise to a weak equivalence

$$\mathrm{hofiber}(D_{n-1} \to X_{\underline{n}}) \simeq \mathrm{holim}\left(\mathrm{hofiber}(X_{n-1} \to X_{\underline{n}}) \to \right.$$
$$\left. * \leftarrow \mathrm{hofiber}(D_{\underline{n}} \to X_{\underline{n}})\right)$$
$$\simeq \mathrm{hofiber}(X_{n-1} \to X_{\underline{n}}) \times \mathrm{hofiber}(D_{\underline{n}} \to X_{\underline{n}}). \qquad \square$$

The following reconception of the Dold–Thom Theorem uses Proposition 3.7.29. Recall the definition of the symmetric product from Definition 2.7.22. First we restate the Dold–Thom Theorem, Theorem 2.7.23, in language that was not available to us before. The theorem states that if we have a homotopy cocartesian square

$$\begin{array}{ccc} A & \longrightarrow & X \\ \downarrow & & \downarrow \\ * & \longrightarrow & C \end{array}$$

then $SP(X) \to SP(C)$ is a quasifibration with fiber $SP(A)$. By definition, this means the natural map $SP(A) \to \mathrm{hofiber}(SP(X) \to SP(A))$ is a weak equivalence. In other words, the square

$$\begin{array}{ccc} SP(A) & \longrightarrow & SP(X) \\ \downarrow & & \downarrow \\ * & \longrightarrow & SP(C) \end{array}$$

is homotopy cartesian.

**Theorem 5.10.5** (Dold–Thom Theorem for squares)   *Suppose*

$$X = \begin{array}{ccc} X_\emptyset & \longrightarrow & X_1 \\ \downarrow & & \downarrow \\ X_2 & \longrightarrow & X_{12} \end{array}$$

*is homotopy cocartesian. Then*

$$SP(X) = \begin{array}{ccc} SP(X_\emptyset) & \longrightarrow & SP(X_1) \\ \downarrow & & \downarrow \\ SP(X_2) & \longrightarrow & SP(X_{12}) \end{array}$$

*is homotopy cartesian.*

*Proof*   By Proposition 3.7.29, we may assume $X$ is a cofibrant pushout square, so that in particular both $X_\emptyset \to X_2$ and $X_1 \to X_{12}$ are cofibrations. Consider the 3-cube

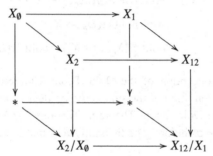

Applying SP everywhere we obtain the 3-cube

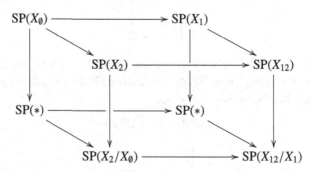

By the Dold–Thom Theorem (Theorem 2.7.23), the left and right squares are homotopy pullbacks, and hence the 3-cube is homotopy cartesian by Proposition 5.4.13. Since SP takes weak equivalences to weak equivalences

and the original square was a homotopy pushout, we have a weak equivalence $X_2/X_\emptyset \to X_{12}/X_1$ which induces a weak equivalence $\mathrm{SP}(X_2/X_\emptyset) \to \mathrm{SP}(X_{12}/X_1)$. Thus the bottom square is homotopy cartesian. By 1(a) of Proposition 5.4.13, the top square is homotopy cartesian as well.  □

Next we discuss Mather's First and Second Cube Theorems. The proofs of the Cube Theorems do not require the language of (co)cartesian cubical diagrams, but are thematically similar to the results we have encountered in this chapter. In fact, the case $n = 3$ of the proof of Proposition 5.10.4 above shares some common ideas with the proof of the First Cube Theorem. Mather's Cube Theorems were originally stated and proved for *homotopy* commutative cubes, but we will as usual always assume our diagrams are strictly commutative.

We first need the following lemma. Recall the definition of a quasifibration, Definition 2.7.1.

**Lemma 5.10.6** *Suppose*

$$
\begin{array}{ccc}
X_\emptyset & \xrightarrow{\ f\ } & X_1 \\
\downarrow & & \downarrow{\scriptstyle p} \\
X_2 & \xrightarrow{\ g\ } & X_{12}
\end{array}
$$

*is a strict pullback square and $p$ is a fibration. Then the induced map of mapping cylinders $M_f \to M_g$ is a quasifibration.*

*Proof* Let $\widetilde{p} \colon M_f \to M_g$ be the induced map of mapping cylinders. We will show that, for all $x \in M_g$, the inclusion $\mathrm{fiber}_x(\widetilde{p}) \to \mathrm{hofiber}_x(\widetilde{p})$ is a weak equivalence. First note that, depending on $x$, the strict fiber of $\widetilde{p}$ is either $\mathrm{fiber}_x(X_\emptyset \to X_2)$ or $\mathrm{fiber}_x(X_1 \to X_{12})$.

Next consider the diagram

$$
\begin{array}{ccccc}
X_\emptyset & \longrightarrow & M_f & \xrightarrow{\ r_f\ } & X_1 \\
\downarrow & & \downarrow & & \downarrow{\scriptstyle p} \\
X_2 & \longrightarrow & M_g & \xrightarrow[\ r_g\ ]{} & X_{12}
\end{array}
$$

where the original horizontal arrows have been factored through the mapping cylinders as in Proposition 2.4.4. The outer square is homotopy cartesian by Proposition 3.3.5 since it is a strict pullback and $p$ is a fibration. The right square is homotopy cartesian by 1(b) of Proposition 3.3.11, and so the left square is homotopy cartesian by 1(b) of Proposition 3.3.20. From this we conclude that the homotopy fibers of all of the vertical maps are weakly equivalent by Proposition 3.3.18, and since $X_1 \to X_{12}$ and $X_\emptyset \to X_2$ are fibrations (the latter by Proposition 2.1.16), the homotopy fibers of the outer two vertical maps are weakly equivalent to their homotopy fibers. But we have already observed these to be the strict fibers of $M_f \to M_g$.        □

**Theorem 5.10.7** (First Cube Theorem [Mat76])    *In the diagram*

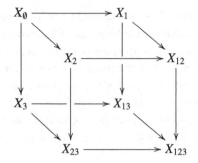

*suppose the left and back faces are homotopy pullbacks, and the top and bottom faces are homotopy pushouts. Then the front and right faces are homotopy pullbacks.*

*Proof*    It suffices to prove the front face is a homotopy pullback; the argument for the right face is identical. We have to show that the square

$$
\begin{array}{ccc}
X_2 & \longrightarrow & X_{12} \\
\downarrow & & \downarrow \\
X_{23} & \longrightarrow & X_{123}
\end{array}
$$

is a homotopy pullback. Since the top and bottom faces are homotopy pushouts, we have therefore to prove that

$$
\begin{array}{ccc}
X_2 & \longrightarrow & \mathrm{hocolim}(X_2 \leftarrow X_\emptyset \to X_1) \\
\downarrow & & \downarrow{\scriptstyle q} \\
X_{23} & \longrightarrow & \mathrm{hocolim}(X_{23} \leftarrow X_3 \to X_{13})
\end{array}
$$

is homotopy cartesian, where $q$ stands for the naturally induced map. We may assume by Proposition 3.3.24 that the maps $X_S \to X_{S \cup \{3\}}$ are fibrations for $S = \emptyset, \{1\}, \{2\}$. We claim that the map $\mathrm{hocolim}(X_2 \leftarrow X_\emptyset \to X_1) \to \mathrm{hocolim}(X_{23} \leftarrow X_3 \to X_{13})$ is a quasifibration by Proposition 2.7.7. To see this, write

$$
\mathrm{hocolim}(X_{23} \leftarrow X_3 \to X_{13}) = X_{23} \sqcup X_3 \times I \sqcup X_{13}/\!\sim
$$

where $\sim$ is the evident equivalence relation. Let $U_1, U_2 \subset \mathrm{hocolim}(X_{23} \leftarrow X_3 \to X_{13})$ be the evident quotients of $X_{13} \sqcup X_3 \times (1/3, 1]$ and $X_{23} \sqcup X_3 \times [0, 2/3)$

(a similar construction was considered in the proof of Theorem 3.6.13 in the case of weak equivalences, which appears following the statement of Lemma 3.6.14). Then $\{U_1, U_2\}$ forms an open cover for $\mathrm{hocolim}(X_{23} \leftarrow X_3 \rightarrow X_{13})$, and $q^{-1}(U_1) \rightarrow U_1$, $q^{-1}(U_2)$, and $q^{-1}(U_1 \cap U_2) \rightarrow U_1 \cap U_2$ are quasifibrations. The first two statements follow essentially from Lemma 5.10.6 (our statement is for the mapping cylinder, and here we consider a slightly elongated version which is open at one end, but the technique of proof still applies). The statement about the intersection follows since $q^{-1}(U_1 \cap U_2) \rightarrow U_1 \cap U_2$ is a fibration, as it is the product of a fibration with an identity map. Hence, by Proposition 2.7.7, $q$ is a quasifibration. The strict fiber of $q$ can be identified with the fiber($X_\emptyset \rightarrow X_3$). But since the left face is a homotopy pullback and $X_2 \rightarrow X_{23}$ is a fibration, this is weakly equivalent to the fiber of $X_2 \rightarrow X_{23}$. □

Mather's Second Cube Theorem can be proven using similar techniques.

**Theorem 5.10.8** (Second Cube Theorem [Mat76]) *In the diagram*

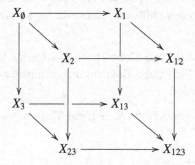

*suppose all the vertical faces are homotopy pullbacks and the bottom face is a homotopy pushout. Then the top face is a homotopy pushout.*

*Proof* We have to show that the map $\mathrm{hocolim}(X_2 \leftarrow X_\emptyset \rightarrow X_1) \rightarrow X_{12}$ is a weak equivalence. Consider the diagram

$$\begin{array}{ccc} \mathrm{hocolim}(X_2 \leftarrow X_\emptyset \rightarrow X_1) & \longrightarrow & X_{12} \\ \downarrow & & \downarrow \\ \mathrm{hocolim}(X_{23} \leftarrow X_3 \rightarrow X_{13}) & \longrightarrow & X_{123} \end{array}$$

By hypothesis the bottom arrow is a weak equivalence. By 1(a) of Proposition 3.3.11, it suffices to show that the square is homotopy cartesian. To see this, we once again compare the map of fibers. Without loss of generality assume $X_S \rightarrow X_{S \cup \{3\}}$ is a fibration for $S = \emptyset, \{1\}, \{2\}$. Then the left vertical map above is a quasifibration as in the proof of Theorem 5.10.7, since the left and back faces are homotopy pullbacks. Moreover the fibers of this map are weakly equivalent to fiber$(X_0 \to X_3)$. But since the front face is also a homotopy pullback, these are the same as the homotopy fibers of the map $X_{12} \to X_{123}$. Hence the square is homotopy cartesian, and hocolim$(X_2 \leftarrow X_0 \to X_1) \to X_{12}$ is a weak equivalence, and so the top face of the original cube is a homotopy pushout.                                                                                □

Here is an alternative proof, which does not use quasifibrations, due to Tom Goodwillie.

*Alternative proof of Theorem 5.10.8*    We may assume by Proposition 3.7.4 that $X_{123}$ is the union of the open sets $X_{13}$ and $X_{23}$ along the open set $X_3$, so that the bottom face is an open pushout square (see Definition 3.7.3). We may also assume that $X_{12} \to X_{123}$ is a fibration and that all of the vertical squares are strict pullbacks by replacing $X_{12} \to X_{123}$ by its mapping path space. Pulling back along $X_{12} \to X_{123}$ yields a new square which is also an open pushout square and whose map to the top square is a weak equivalence because the vertical squares are homotopy pullbacks, and so the top square is also a homotopy pushout.                                                                          □

Our first proof of the Second Cube Theorem suggests that it may be more closely related to the First Cube Theorem than is initially obvious. This is the content of our third proof.

*Third proof of Theorem 5.10.8: First Cube Theorem implies Second Cube Theorem*
Let
$$D_{12} = \mathrm{hocolim}(X_1 \leftarrow X_0 \to X_2).$$
Then the hypotheses of Theorem 5.10.7 are satisfied for the diagram

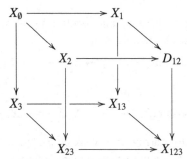

and hence the front and right faces are homotopy pullbacks. This immediately implies that the canonical map $D_{12} \to X_{12}$ is a weak equivalence, so the top square is a homotopy pushout.                                                                        □

Here is a result whose statement is similar to the Cube Theorems and will be used in the proof of Theorem 5.10.10 below. It is an immediate consequence of Proposition 3.3.20 and Proposition 3.7.26, and we sketch the proof with a picture and let the reader fill in the details.

**Proposition 5.10.9** *Consider the 3-cube*

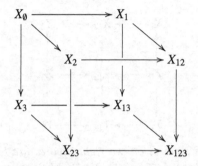

1. *If the front face of the cube is $(k + 1)$-cartesian and if the right and back faces are $k$-cartesian, then the left face is $k$-cartesian.*
2. *If the front and left faces of the cube are $k$-cocartesian and if the back face is $(k - 1)$-cocartesian, then the right face is $k$-cocartesian.*

Of course, when $k = \infty$, we obtain statements in terms of homotopy pullbacks and pushouts.

*Proof* Consider the diagram obtained from our cube by adding two diagonals:

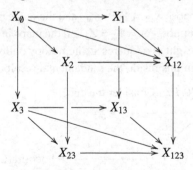

Now apply Proposition 3.3.20 and Proposition 3.7.26 as appropriate. □

The following is an application of the Cube Theorems, due to Mather and Walker [MW80]. It says that, under some circumstances, homotopy pullbacks and pushouts commute. Just as for the Cube Theorems, the original statement of this result is for homotopy commutative diagrams. The reader should also look back at Theorems 3.3.15 and 3.7.18; those results say that homotopy pullbacks, as well as homotopy pushouts, commute with themselves, so the result below establishes the third remaining possibility, of course under much stricter hypotheses.

**Theorem 5.10.10** ([MW80, Theorem 1])  *Suppose we have a commutative diagram*

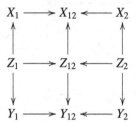

*and suppose two left squares or two right squares are homotopy pullbacks. Denote by* $\mathrm{holim}(\mathcal{X})$, $\mathrm{holim}(\mathcal{Z})$, *and* $\mathrm{holim}(\mathcal{Y})$ *the homotopy pullbacks of the first, second, and third rows of this diagram respectively, and by* $\mathrm{hocolim}(C_1)$, $\mathrm{hocolim}(C_{12})$, *and* $\mathrm{hocolim}(C_2)$ *the homotopy pushouts of the first, second, and third columns respectively. Then there is a homotopy equivalence*

$$\mathrm{hocolim}\big(\mathrm{holim}(\mathcal{X}) \leftarrow \mathrm{holim}(\mathcal{Z}) \rightarrow \mathrm{holim}(\mathcal{Y})\big),$$

$$\Big\downarrow \simeq$$

$$\mathrm{holim}\big(\mathrm{hocolim}(C_1) \rightarrow \mathrm{hocolim}(C_{12}) \leftarrow \mathrm{hocolim}(C_2)\big).$$

As a special case, suppose $X_{12} = Y_{12} = Z_{12}$ and the maps in the middle column are the identity, and suppose $X_2 = Y_2 = Z_2$ and the maps in the right column are the identity. Then the right two squares are homotopy pullbacks and we recover Proposition 3.9.3 from this result, the fundamental result of Section 3.9.

*Proof of Theorem 5.10.10*    Consider the cube

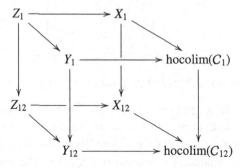

where the top and bottom faces are homotopy pushouts and the left and back faces are homotopy pullbacks. Thus by the First Cube Theorem, Theorem 5.10.7, the front and right faces are also homotopy pullbacks.

Now consider the cube

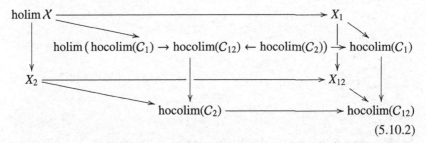

$$(5.10.2)$$

The front, right, and back faces of this cube are homotopy pullbacks, and hence by statement 1 of Proposition 5.10.9 so is the left one.

By considering analogous cubes, we can show that the remaining vertical faces of the cube

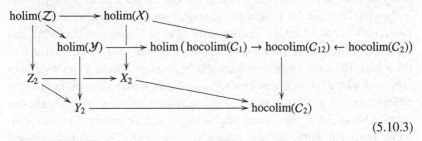

$$(5.10.3)$$

are all homotopy pullbacks. Since the bottom face of this cube is a homotopy pushout, it follows by the Second Cube Theorem, Theorem 5.10.8, that the top face is also a homotopy pushout. In other words, the canonical map

$$\text{hocolim}\big(\text{holim}(\mathcal{X}) \leftarrow \text{holim}(\mathcal{Z}) \rightarrow \text{holim}(\mathcal{Y})\big)$$

$$\downarrow$$

$$\text{holim}\big(\text{hocolim}(C_1) \rightarrow \text{hocolim}(C_{12}) \leftarrow \text{hocolim}(C_2)\big)$$

is a homotopy equivalence. □

# 6

# The Blakers–Massey Theorems for *n*-cubes

In this chapter we prove the Blakers–Massey Theorem and its dual for *n*-dimensional cubes, which encompass and generalize Theorems 4.2.1 and 4.2.2. We will focus on the three-dimensional case before presenting the complete *n*-dimensional result. This three-dimensional case, which is much easier to digest than the general case, contains most of the main ideas of the proofs.

As with Theorem 4.2.1, there is a geometric step for which we will provide two proofs – one based on transversality (due to Goodwillie [Goo92]), and the other using purely homotopy-theoretic methods (due to the first author [Mun14]). The latter version replaces the "dimension counting" transversality argument with a "coordinate counting" one. There is also a second formal step which relies on general facts about connectivities of maps and (co)cartesian cubes. Our proof of the formal step is organized differently than in Goodwillie's original work, but it is otherwise very close. The spirit of this part of the proof is best captured by the three-dimensional case.

## 6.1 Historical remarks

Before looking at this section, the reader might want to glance at Section 4.1 for the history of the original Blakers–Massey Theorem for triads and the variant in terms of squares.

The general Blakers–Massey Theorem, also known as the *Blakers–Massey Theorem for n-cubes* or *(n + 1)-ad Connectivity Theorem* is essentially a statement about higher-order excision for homotopy groups. Some of its far-reaching consequences will be recounted in Section 6.2 and in Chapter 10.

In its early form, the statement concerns the homotopy groups and the connectivity of an $(n + 1)$-ad $(X; X_1, \ldots, X_n)$ which are defined analogously to that of a triad (see Section 4.1). For details, see [BW56, Section 2].

**Theorem 6.1.1** ((*n* + 1)-ad Connectivity Theorem)   *Suppose* $(X; X_1, \ldots, X_n)$ *is an* (*n* + 1)-*ad where* $\{X_i\}_{1 \leq i \leq n}$ *is an open cover of X. Suppose* $\bigcap_{i=1}^{n} X_i$ *is path-connected. Let*

$$X_{\{i\}} = \bigcap_{\substack{1 \leq j \leq n \\ j \neq i}} X_j$$

*and suppose*

$$X = \bigcup_{i=1}^{n} X_{\{i\}}.$$

*Finally suppose the pairs* $(X_{\{i\}}, \bigcap_{i=1}^{n} X_i)$ *are* $k_i$-*connected,* $k_i \geq 1$. *Then*

$$\pi_k(X; X_1, \ldots, X_n) = 0 \text{ for } k \leq k_1 + \cdots + k_n.$$

The first proof of this theorem is due to Barratt and Whitehead [BW56] although the hypotheses there require $\bigcap_{i=1}^{n} X_i$ to be simply-connected and $k_i \geq 2$. The improvement stated above is due to Ellis and Steiner [ES87] who generalize techniques of Brown and Loday [BL87a].

As in the classical Blakers–Massey case, of special importance is also the first non-vanishing homotopy group of the (*n* + 1)-ad, which Ellis and Steiner [ES87], using work of Brown and Loday [BL87b], show to be given by an isomorphism

$$\bigoplus_{(n-1)!} \bigotimes_{i=1}^{n} \pi_{k_i+1}(X_{\{i\}}, \bigcap_{i=1}^{n} X_i) \cong \pi_{k_1 + \cdots + k_n + 1}(X; X_1, \ldots, X_n).$$

The group on the right is not necessarily abelian, and some abelian cases were initially considered by Barratt and Whitehead. This is of course a generalization of Theorem 4.1.2.

A simpler proof of Theorem 6.1.1 involving only space-level constructions was given by Goodwillie [Goo92]. In that setup, Theorem 6.1.1 is translated into the statement about cubical diagrams, just as in the classical case described in Section 4.1. From the data in the hypotheses of Theorem 6.1.1, one can form a cube of intersections and inclusions, as in Example 5.8.3, with $\bigcap_{i=1}^{n} X_i$ as the initial space. By Example 3.7.5, this is a strongly homotopy cocartesian cube. The connectivities of the pairs $(X_{\{i\}}, \bigcap_{i=1}^{n} X_i)$ translate into the connectivities of the initial maps of the cube. Then Theorem 6.1.1 can be interepreted as saying that this cube is $(1 - n + k_1 + \cdots + k_n)$-cartesian. This is because, the homotopy groups of an (*n* + 1)-ad can be related to the homotopy groups of the total fiber of this cube. This was explained in details for squares in Section 4.1 and we leave the details to the reader.

The cubical version of Theorem 6.1.1, which is a statement about any strongly homotopy cocartesian cube and thus covers the situation above, is stated as Theorem 6.2.1 below. Note that the hypotheses there are even more general than in Theorem 6.1.1. This is essentially since Goodwillie [Goo92] did not use homology in his proof but rather general position arguments. We will provide a slightly different proof from Goodwillie's in Section 6.3.

## 6.2 Statements and applications

**Theorem 6.2.1** (Generalized Blakers–Massey Theorem)   *Let* $X = (T \mapsto X_T)$ *be a strongly homotopy cocartesian $S$-cube with $|S| = n \geq 1$. Suppose that, for each $i \in S$, the map $X_\emptyset \to X_i$ is $k_i$-connected. Then $X$ is $(1 - n + \sum_{i \in S} k_i)$-cartesian.*

The dual version is the following.

**Theorem 6.2.2** (Dual of the generalized Blakers–Massey Theorem)   *Suppose $X = (T \mapsto X_T)$ is a strongly homotopy cartesian $S$-cube with $|S| = n \geq 1$. Suppose that, for each $i \in S$, the map $X_{S-\{i\}} \to X_{\underline{n}}$ is $k_i$-connected. Then $X$ is $(-1 + n + \sum_{i \in S} k_i)$-cocartesian.*

Both of these theorems have generalizations analogous to Theorems 4.2.3 and 4.2.4. We need some extra language regarding partitions of sets to state them. If $S$ is a finite set, a *partition* of $S$ is a set $\{T_\alpha\}_{\alpha \in A}$ of pairwise disjoint nonempty subsets of $S$ which cover $S$. Each member $T_\alpha$ of a partition is called a *block*.

**Theorem 6.2.3**   *Let $X = (T \mapsto X_T)$ be an $S$-cube with $|S| = n \geq 1$. Suppose that*

1. *for each $\emptyset \neq U \subset S$, the $U$-cube $\partial^U X$ is $k_U$-cocartesian;*
2. *for $U \subset V$, $k_U \leq k_V$.*

*Then $X$ is $(1 - n + \min\{\sum_\alpha k_{T_\alpha} : \{T_\alpha\}$ is a partition of $S\})$-cartesian.*

Dually, we have the following.

**Theorem 6.2.4**   *Let $X = (T \mapsto X_T)$ be an $S$-cube with $|S| = n \geq 1$. Suppose that*

1. *for each* $\emptyset \neq U \subset S$, *the* $U$-*cube* $\partial_{S-U}X$ *is* $k_U$-*cartesian;*
2. *for* $U \subset V$, $k_U \leq k_V$.

*Then* $X$ *is* $(-1 + n + \min\{\sum_\alpha k_{T_\alpha} : \{T_\alpha\}$ *is a partition of* $S\})$-*cocartesian.*

**Remark 6.2.5** The hypotheses $k_U \leq k_V$ for $U \subset V$ in the statements of Theorem 6.2.3 and Theorem 6.2.4 can be weakened, although we will not pursue this here. See Remark 6.3.5 for a comment on this in the case $n = 3$. □

First we present a few examples and applications.

**Example 6.2.6** Let $X$ be a $k$-connected space, and consider the $n$-cube $S \mapsto X_S$ where $X_\emptyset = X$, $X_i = CX$, the maps $X_\emptyset \to X_i$ are the inclusion of $X$ in the cone, and $X_S = \cup_{i \in S} X_i$ is the union along $X$ of $X_i$ for $i \in S$ (when $|S| = 2$, $X_S = \Sigma X$ is the suspension). This is precisely the fiberwise join cube $X_S = X * S$ studied in Example 3.9.13 and Example 5.8.20 (taking $Y$ to be the one-point space). From the latter example, we also know that this cube is by construction strongly homotopy cocartesian (Example 5.8.20), and since $X$ is $k$-connected, the maps $X_\emptyset \to X_i$ are $(k+1)$-connected. Hence the $n$-cube is $(nk+1)$-cartesian by Theorem 6.2.1. That is, the map

$$X \longrightarrow \underset{S \neq \emptyset}{\mathrm{holim}}\, X_S$$

is $(nk + 1)$-connected. Thus we have a space, built from contractible spaces by homotopy pullbacks, which approximates the homotopy type of $X$ in an increasingly large range if $X$ is simply-connected.

Fiberwise join will play a key role in Section 10.1. □

**Example 6.2.7** Dual to the previous example, consider a $k$-connected based space $(X, x_0)$, and let $PX = \{\gamma : I \to X \mid \gamma(0) = x_0\}$ be the path space. Define an $n$-cube as follows: $X_{\underline{n}} = X$, $X_{\underline{n}-i} = PX$, and $X_{\underline{n}-S}$ is equal to the fiber product over $X$ of the $X_{\underline{n}-\{i\}}$ where $i$ ranges over $S$. The $n$-cube $S \mapsto X_S$ is strongly cartesian by construction (it is a fibrant pullback cube), and the canonical maps $PX \to X$ are $k$-connected. By Theorem 6.2.2, the cube is $(n(k+1)-1)$-cocartesian. That is, the map

$$\underset{S \neq \underline{n}}{\mathrm{hocolim}}\, X_S \longrightarrow X$$

is $(n(k+1)-1)$-connected. Thus we have built a space from contractible spaces by homotopy pushouts which approximates the homotopy type in an increasing range depending on $n$ if $X$ is at least connected.

It is not hard to show that $X_S \simeq (\Omega X)^{n-|S|-1}$ when $|\underline{n} - S| \geq 2$ using induction on $n$. To see this, it suffices to show $X_\emptyset \simeq (\Omega X)^{n-1}$ for $n \geq 2$. When $n = 2$, $X_\emptyset = \Omega X$. For an $n$-cube $S \mapsto X_S$, $n > 2$, we have by Lemma 5.3.6

$$X_\emptyset \simeq \operatorname{holim}\left(X_n \to \operatorname*{holim}_{\emptyset \neq R \subset \underline{n-1}} X_{R \cup \{n\}} \leftarrow \operatorname*{holim}_{\emptyset \neq R \subset \underline{n-1}} X_R\right).$$

By definition, the map

$$X_n \longrightarrow \operatorname*{holim}_{\emptyset \neq R \subset \underline{n-1}} X_{R \cup \{n\}}$$

is a weak equivalence, hence

$$X_\emptyset \simeq \operatorname*{holim}_{\emptyset \neq R \subset \underline{n-1}} X_R.$$

By induction, it follows that $X_\emptyset \simeq (\Omega X)^{n-1}$.     □

**Example 6.2.8**  Let $S^{p_1}, \ldots, S^{p_n}$ be spheres of dimensions $p_1, \ldots, p_n$ considered as based spaces, and consider the $n$-cube

$$\mathcal{W} = (S \longmapsto \vee_{i \in S} S^{p_i}).$$

Then $\mathcal{W}$ is $(1 - n + \sum_{i=1}^{n} p_i)$-cartesian.

To see why, note that there is a natural trasformation of cubes $\widetilde{\mathcal{W}} \to \mathcal{W}$, where

$$\widetilde{\mathcal{W}} = \left(S \longmapsto \vee_{i \in S} S^{p_i} \vee_{j \notin S} D^{p_j+1}\right)$$

is given by projecting away from the disks $D^{p_j+1}$. The cube $\widetilde{\mathcal{W}}$ is strongly cocartesian since every square face is a pushout square of cofibrations, and hence homotopy cocartesian by Proposition 3.7.6. Moreover, the natural inclusion $S^{p_i} \to D^{p_i+1}$ is $p_i$-connected, and hence the maps $\vee_{i \in \underline{n}} S^{p_i} \to \vee_{i \in \underline{n}-\{j\}} S^{p_i} \vee D^{p_j+1}$ are $p_j$-connected for all $j \in \underline{n}$. It follows from Theorem 6.2.1 that the $n$-cube in question is $(1 - n + \sum_{i=1}^{n} p_i)$-cartesian.     □

Before our next example, which is important in calculus of functors, recall from Example 5.5.6 that the configuration space of $n$ points in $X$ is denoted by $\operatorname{Conf}(n, X)$ and that there is a natural $n$-cube associated to this space, namely the one given by $\operatorname{Conf}(\underline{n} - \bullet, X) \colon \mathcal{P}(\underline{n}) \to \mathrm{Top}$, where an inclusion $S \subset T$ gives rise to a restriction map $\operatorname{Conf}(\underline{n} - S, X) \to \operatorname{Conf}(\underline{n} - T, X)$.

**Example 6.2.9**  Let $M$ be a smooth $m$-dimensional manifold. Consider the $n$-cube $C = (S \mapsto \operatorname{Conf}(\underline{n} - S, M))$. The cube $C$ is $((n-1)(m-2)+1)$-cartesian.

To see this, regard the $m$-cube in question as a map of $(m-1)$-cubes

$$\left(R \mapsto \mathrm{Conf}(\{n\} \cup \underline{n-1} - R, M)\right) \longrightarrow \left(R \mapsto \mathrm{Conf}(\underline{n-1} - R, M)\right),$$

where $R \subset \underline{n-1}$. Choose a basepoint $(x_1, \dots, x_{n-1}) \in \mathrm{Conf}(\underline{n-1}, M)$. This gives rise to a basepoint in $\mathrm{Conf}(\underline{n-1} - R, M)$ for all $R \subset \underline{n-1}$. By Proposition 5.4.12, the map of $(n-1)$-cubes, as an $n$-cube, is as cartesian as the $(n-1)$-cube of (homotopy) fibers. For each $R$, the restriction map $\mathrm{Conf}(\{n\} \cup \underline{n-1} - R, M)) \to \mathrm{Conf}(\underline{n-1} - R, M))$ above is a fibration whose fiber is $M \setminus \{x_i : i \in \underline{n-1} - R\}$ (see Example 5.5.6), which we write as $M - (\underline{n-1} - R)$ for short. The $(m-1)$-cube of fibers is

$$R \longmapsto M - (\underline{n-1} - R),$$

with $M - (\underline{n-1} - R) \to M - (\underline{n-1} - S)$ the evident inclusion for $R \subset S$. Further, $M - (\underline{n-1} - R)$ is an open subset of $M - (\underline{n-1} - S)$ since it is the complement of the closed subset $S - R$. Each square face of this $(n-1)$-cube is of the form

$$
\begin{array}{ccc}
M - (\underline{n-1} - R) & \longrightarrow & M - \left(\underline{n-1} - (R \cup \{i\})\right) \\
\downarrow & & \downarrow \\
M - \left(\underline{n-1} - (R \cup \{j\})\right) & \longrightarrow & M - \left(\underline{n-1} - (R \cup \{i,j\})\right)
\end{array}
$$

for $R \subset \underline{n-1}$ $i, j \notin R$, and $i \neq j$. It is clear by inspection that

$$M - (\underline{n-1} - R) = M - \left(\underline{n-1} - (R \cup \{i\})\right) \cap M - \left(\underline{n-1} - (R \cup \{j\})\right)$$

and

$$M - \left(\underline{n-1} - (R \cup \{i\})\right) \cup M - \left(\underline{n-1} - (R \cup \{j\})\right) = M - \left(\underline{n-1} - (R \cup \{i,j\})\right).$$

Hence, by Example 3.7.5, this square is homotopy cocartesian, and so the $(n-1)$-cube in question is strongly homotopy cocartesian. The maps

$$M - (\underline{n-1}) \longrightarrow M - (\underline{n-1} - \{i\})$$

are $(m-1)$-connected for each $i \in \underline{n-1}$ by Example A.2.16. By Theorem 6.2.1, this $(n-1)$-cube, and hence the original $n$-cube, is $((n-1)(m-2)+1)$-cartesian.

□

**Remark 6.2.10** As mentioned in Example 5.5.6, when $M = \mathbb{R}^m$, the space $\mathbb{R}^m - (\underline{n-1} - R)$ from the previous example is homotopy equivalent to a wedge of $n - 1 - |R|$ spheres of dimension $m - 1$. Moreover, the cube

$$R \longmapsto \mathbb{R}^m - (\underline{n-1} - R)$$

is weakly equivalent to the cube

$$R \longmapsto \bigvee_{\underline{n-1}-R} S^{m-1},$$

where the maps are projections away from wedge summands. This can be related to Example 6.2.8 using Proposition 5.5.7: the total homotopy fiber of the cube $R \mapsto \bigvee_R S^{m-1}$ is weakly equivalent to the $(n-1)$-fold loop space of the total homotopy fiber of the cube $R \mapsto \bigvee_{\underline{n-1}-R} S^{m-1}$ by Proposition 5.5.7, and so we can compute how cartesian the cube in question is by using Example 6.2.8.                                                                         □

In the previous example, we encountered the complement of a finite set of points. More generally, we can consider a cube of pairwise disjoint submanifolds.

**Example 6.2.11**   Let $M$ be a smooth manifold of dimension $m$ and $P_1, \ldots, P_n$ pairwise disjoint closed submanifolds of dimensions $p_1, \ldots, p_n$. Consider the $n$-cube $S \mapsto M - \bigcup_{i \in \underline{n}-S} P_i$. Clearly this cube is strongly homotopy cocartesian; every square face is of the form

$$
\begin{array}{ccc}
A \cap B & \longrightarrow & A \\
\downarrow & & \downarrow \\
B & \longrightarrow & A \cup B
\end{array}
$$

for some open sets $A$ and $B$, and such squares are homotopy cocartesian by Example 3.7.5. For each $j \in \underline{n}$, the inclusion map

$$M - \bigcup_{i \in \underline{n}} P_i \longrightarrow M - \bigcup_{i \in \underline{n}-\{j\}} P_i$$

is $(m - p_j - 1)$-connected by Example A.2.21, and hence the cube itself is $(nm + 1 - \sum_{i \in \underline{n}}(p_i - 2))$-cartesian by Theorem 6.2.1.                   □

Even more generally we can consider the complement of submanifolds which may intersect, but only in a "nice" way. See Section A.2.3 for any unfamiliar terms.

**Example 6.2.12**   Let $M$ be a smooth manifold and $P_1, \ldots, P_n$ smooth closed submanifolds of dimensions $p_1, \ldots, p_n$ respectively whose intersections are all transverse, in the sense that, for every $S \subset \underline{n}$, the intersection $\bigcap_{i \in S} P_i$ is a smooth closed submanifold of $M$ of codimension $|S|m - \sum_{i \in S} p_i$, or is empty. We claim that, for $S \subset \underline{n}$, the $n$-cube of inclusions

$$C = S \longmapsto N - \bigcup_{i \notin S} P_i$$

is $(1 + n(m-2) - \sum_i p_i)$-cartesian.

This will follow from Theorem 6.2.3 once we establish that the faces $\partial^S C$ are $(-1 + |S|m - \sum_{j \in S} p_j)$-cocartesian, since the numbers $k_U = (-1 + |S|m - \sum_{j \in S} p_j)$ satisfy $k_U \leq k_V$ when $U \subset V$, as $p_i \leq m$ for all $i$. Moreover, the sum over partitions is minimized when the partition of $\underline{n}$ consists of all singletons because the sum of the numbers $|S|m - \sum_{j \in S} p_j$, where $S$ ranges over the blocks of a given partition, is in fact independent of the chosen partition.

Fix $S \subset \underline{n}$. For $i \in S$, let $U_i = M - P_i$, and note that for a proper subset $R \subset S$, $\partial^S C(R) = M - \bigcup_{j \notin R} P_j = \bigcap_{j \notin R} U_j$. By Example 5.8.3, the $|S|$-cube

$$S \longmapsto \begin{cases} \bigcap_{j \notin R} U_j, & R \neq S; \\ \bigcup_{j \in S} U_j = (M - \bigcup_{i \notin S} P_i) - \bigcap_{j \in S} P_j, & R = S, \end{cases}$$

is homotopy cocartesian. By assumption $\bigcap_{i \in S} P_i$ is a submanifold of $M$ of codimension $|S|m - \sum_{i \in S} p_i$ (or is empty), and by Example A.2.21, the inclusion $(M - \bigcup_{i \notin S} P_i) - \bigcap_{j \in S} P_j \to M - \bigcup_{i \notin S} P_i = \partial^S C(S)$ is therefore $(-1 + |S|m - \sum_{i \in S} p_i)$-connected, so that $\partial^S C$ is $(-1 + |S|m - \sum_{i \in S} p_i)$-cocartesian. $\quad\square$

**Remark 6.2.13** It should be apparent that the only assumption we really need about the intersections $\bigcap_{i \in S} P_i$ is that they are submanifolds. The assumption on their codimension is a simplification which makes the computations above easier to follow, but is unnecessary. $\quad\square$

Let $P$ be a smooth closed manifold and $N$ a smooth manifold. An *embedding* $P \to N$ is a smooth map which is a homeomorphism onto its image and for which the derivative map $TP \to TN$ is a fiberwise injection (see Section 10.2 for more on embeddings). It is topologized as an open subset of the space of smooth maps $C^\infty(P, N)$ of $P$ to $N$, whose topology is given in Definition A.2.7. For the reader familiar with the terminology, the following example effectively takes Example 6.2.11 and applies the functor $\mathrm{Emb}(P, -)$ to it.

**Example 6.2.14** ([Goo], Proposition A.1) Let $M$ and $P_1, \ldots, P_n$ be as in Example 6.2.12. Let $Q$ be a smooth closed manifold of dimension $q$, and consider the $n$-cube $\mathcal{E} = S \mapsto \mathrm{Emb}(Q, M - \bigcup_{i \in \underline{n}-S} P_i)$, $S \subset \underline{n}$.[1] Then $\mathcal{E}$ is $(1 + \sum_{i \in \underline{n}} m - p_i - q - 2)$-cartesian.

We will induct on $n$ using Theorem 6.2.3. We claim that for each non-empty subset $S \subset \underline{n}$, that the $S$-cube

$$\mathcal{E}_S = \left( R \longmapsto \mathrm{Emb}(Q, (M - \bigcup_{i \in \underline{n}-S} P_i) - \bigcup_{j \in S-R} P_j) \right)$$

is $(-1 + \sum_{i \in S} m - q - p_i)$-cocartesian. The case $n = 1$ asserts that the (inclusion) map

$$\mathrm{Emb}(Q, M - P_1) \longrightarrow \mathrm{Emb}(Q, M)$$

is $(m - p_1 - q - 1)$-connected. This follows from transversality as follows. We remark here that we are going to give arguments slightly less rigorous than given, for instance, in the analogous Example 4.2.20. In particular, we will be a little looser about our definition of $k$-connected map by ignoring basepoints: $X \to Y$ is $k$-connected if for every map $S^j \to Y$, $j \le k$, there is a lift to a map $S^j \to X$, and a lift for every homotopy $S^j \times I \to Y$, $j < k$, with lifts $S^j \times \{i\} \to X$ for $i = 0, 1$.

Given a map $S^j \to \mathrm{Emb}(Q, M)$, consider the problem of lifting it to a map $S^j \to \mathrm{Emb}(Q, M - P_1)$. The map $S^j \to \mathrm{Emb}(Q, M)$ corresponds via Theorem 1.2.7 and Theorem A.2.9 to a smooth map $S^j \times Q \to M$. Inside the space of smooth maps $\mathrm{Emb}(S^j \times Q, M)$ is an open dense set of maps which are transverse to $P_1$, which follows from Corollary A.2.20.[2] In particular, transverse means the intersection is empty if $j + q < m - p_1$, or $j < m - p_1 - q$. A similar argument holds for a 1-parameter family of maps, and so the map in question is $(m - p_1 - q - 1)$-connected.

A similar transversality argument works in general. Note that Proposition 5.8.26 applies to the cube $\mathcal{E}_S$, and we may reduce to studying the connectivity of the map

$$\operatorname*{colim}_{R \subsetneq S} \mathcal{E}_R \longrightarrow \mathcal{E}_S(S).$$

Consider a map $S^j \to \mathcal{E}_S(S) = \mathrm{Emb}(Q, M - \bigcup_{i \in \underline{n}-S} P_i)$, given by $s \mapsto f_s$. By Theorem A.2.23,[3] in $\mathrm{Emb}(Q, M - \bigcup_{j \in S} P_j)$ there is an open dense set of maps $f$ such that the map

---

[1]  This is actually a cube of fibers of an $(n + 1)$-cube, namely the map of $n$-cubes
    $(S \mapsto \mathrm{Emb}(Q \sqcup \sqcup_{i \in \underline{n}-S} P_i, M)) \longrightarrow (S \mapsto \mathrm{Emb}(\sqcup_{i \in \underline{n}-S} P_i, M))$.
[2]  Technically this result applies to the space of all smooth maps, but inside the space of all smooth maps is the open set of embeddings, so this is the intersection.
[3]  With $k = 0$ and $s = |S|$ in that theorem.

$$S^j \times Q^S \longrightarrow (M - \cup_{i \in \underline{n}-S} P_i)^S$$

given by

$$(s, x_1, \ldots, x_S) \mapsto (f_x(x_1), \ldots, f_s(x_S))$$

is transverse to $\prod_{j \in S} P_j \subset (M - \cup_{i \in \underline{n}-S} P_i)^S$. In particular the intersection is empty if $j + q|S| < \sum_{j \in S} m - p_j$, or $j < \sum_{j \in S} m - p_j - q$. A similar argument with 1-parameter families establishes that the map in question is in fact $(-1 + \sum_{j \in S} m - p_j - q)$-connected. Then the sum over partitions appearing in Theorem 6.2.3 is minimized when the partition of $\underline{n}$ consists of singletons, and the $n$-cube $\mathcal{E}$ is therefore $(1 + \sum_{i \in \underline{n}} m - p_i - q - 2)$-cartesian. $\qquad \square$

We finish this section with a proof of the Berstein–Hilton Theorem, which gives sufficient conditions for a space to be the suspension of another. Our proof follows that of [KSV97, Section 4], and that paper also has a generalization which is proved in the same vein as the proof sketched below.

Let $X$ be a based space with basepoint $*$, and let $c_* \colon X \to X$ be the constant map at the basepoint. A *co-H space* structure on a based space $X$ is a *comultiplication map* $\mu \colon X \to X \vee X$ such that the compositions $\langle 1_X, c_* \rangle \circ \mu, \langle c_* \rangle, 1_X \rangle \circ \mu \colon X \to X$ are homotopic to the identity $1_X$. Here $\langle 1_X, c_* \rangle$ denotes the composition of the wedge $1_X \vee c_*$ with the fold map $\nabla \colon X \vee X \to X$.

**Theorem 6.2.15** (Berstein–Hilton Theorem) *Let $n \geq 2$ and suppose $X$ is an $(n-1)$-connected CW complex of dimension at most $3n - 3$ with a co-H space structure. Then there exists a based CW complex $K$ and a weak homotopy equivalence $\Sigma K \to X$.*

*Proof* Consider the diagram

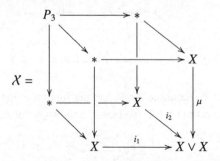

where $P_3 = \text{holim}_{\mathcal{P}_0(\underline{3})} X$ is the homotopy limit of the punctured cube, $i_j \colon X \to X \vee X$ is the inclusion of the $j$th summand, and the maps $* \to X$ are all the inclusion of the basepoint. Thus, as a 3-cube, it is by definition

homotopy cartesian. To apply Theorem 6.2.4, we need to compute how cartesian the other faces are. First, we claim the front, right, and bottom faces are
all $(2n - 3)$-cartesian. For the bottom face, this is Proposition 4.6.1. The front
face is homotopy cocartesian because the map $1_X \vee \mu \colon X \vee X \to X \vee X$ is
homotopic to $1_{X \vee X}$ since $\mu$ is the comultiplication on $X$; by Theorem 4.2.1, it
then follows that this face is $(2n - 3)$-cartesian. The argument for the right face
is the same. Since $X$ is $(n - 1)$-connected, the maps $i_1, i_2, \mu \colon X \to X \vee X$ are all
$(n-1)$-connected. It follows from Theorem 6.2.4 that $X$ is $(3n-2)$-cocartesian.
We will use this computation later.

Define $R_2 = \mathrm{holim}(X \xrightarrow{i_1} X \vee X \xleftarrow{i_2} X)$, and $P_2 = \mathrm{holim}(* \to X \leftarrow *)$. Since
the bottom face of $X$ is $(2n - 3)$-cartesian, we have that the map $* \to R_2$ is
$(2n - 3)$-connected.

By Lemma 5.3.6, we have a homeomorphism $P_3 \cong \mathrm{holim}(* \to R_2 \leftarrow P_2)$,
and hence the square

is homotopy cartesian. By 2(a) of Proposition 3.3.11, the map $P_3 \to P_2$ is
$(2n - 3)$-connected, and since $P_2 \simeq \Omega X$ (Example 3.2.10) and $n \geq 2$, $P_3$ is
itself $(n - 2)$-connected (and in particular connected since $n \geq 2$).

Let $M_3 = \mathrm{hocolim}_{\mathcal{P}_1(3)} X$. That is, $M_3$ is the homotopy colimit of the diagram

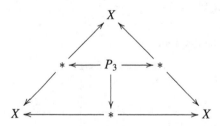

Using Example 3.6.6, Example 3.6.9, and taking homotopy colimits along the
rows above using Lemma 5.7.6, we have a homotopy cocartesian square

$$\begin{CD} \Sigma P_3 @>r_2>> X \\ @Vq_3VV @VVV \\ X \vee X @>g_3>> M_3 \end{CD}$$

We showed above that $X$ is $(3n-2)$-cocartesian, which means that the canonical map $r_3 \colon M_3 \to X \vee X$ is $(3n-2)$-connected. Since $r_3 \circ g_3 \sim 1_{X \vee X}$, it follows from statement 3 of Proposition 2.6.15 that $g_3$ is $(3n-3)$-connected. We cannot quite deduce the connectivity of $r_2$ from this because Proposition 3.7.13 does not apply. However, using Proposition 3.8.9, we have isomorphisms $H_i(r_2) \cong H_i(g_3)$ for all $i$, and since $g_3$ is $(3n-3)$-connected, $H_i(g_3) = 0$ for $0 \le i \le 3n-3$. Since both $X$ and $\Sigma P_3$ are 1-connected, it follows from Theorem 4.3.3 that $r_2$ is $(3n-3)$-connected as well.

Using Theorem 1.3.7, there exists a weak homotopy equivalence $P_3 \to K$ for some CW complex $K$. The map $r_2$ then induces a map $\Sigma K \to X$. Since $\dim(X) \le 3n-3$, $H_{3n-3}(X)$ is free abelian, generated by the $(3n-3)$-cells of $X$. The map $\Sigma K \to X$ is a homology isomorphism in dimension less than $3n-3$ and is onto in dimension $3n-3$. It follows that there is a CW complex $Y$ with the same $(3n-1)$-skeleton as $K$ and a map $Y \to K$ such that the composite $\Sigma Y \to X$ is an isomorphism in homology. Since $\Sigma Y$ and $X$ are both simply-connected, it follows from Theorem 4.3.2 that this map is a weak equivalence and hence a homotopy equivalence by Theorem 1.3.10.                    □

## 6.3 Proofs of the Blakers–Massey Theorems for $n$-cubes

We will first prove Theorem 6.2.1 and will deduce Theorem 6.2.3 from it. To prove Theorem 6.2.1, we begin by presenting two proofs of the geometric step, then present a proof of the formal step in the three-dimensional case before presenting the proof of the formal step in the $n$-dimensional case.

### 6.3.1 The geometric step

The key to the induction is the following lemma (compare to Lemma 4.4.1).

Suppose $X = (S \mapsto X_S)$ is an $n$-cube of spaces formed by attaching cells $e_i$ of dimension $d_i + 1$ for $1 \le i \le n$ to a space $X_\emptyset = X(\emptyset)$. Thus $X_S = X_\emptyset \cup \{e_i : i \in S\}$ for $S \subset \{1, \ldots, n\}$ for some choice of attaching maps $\partial e_i \to X$. The cube $X$ is strongly cocartesian (and strongly homotopy cocartesian because the maps in it are cofibrations).

**Lemma 6.3.1** ([Goo92, Lemma 2.7]) *Let $X$ be as above. Choose a base-point $x \in X_{\{n\}}$, and for $T \subset \{1, \ldots, n-1\}$ define an $(n-1)$-cube $\mathcal{F}$ by $\mathcal{F}(T) = (\mathrm{hofiber}(X(T) \to X(T \bigcup \{n\})))$. Then $\mathcal{F}$ is $(-1 + \sum_{i=1}^{n} d_i)$-cocartesian. Equivalently, the pair $\left( \mathcal{F}(n-1), \bigcup_{i \in n-1} \mathcal{F}(n-1-i) \right)$ is $(-1 + \sum_{i=1}^{n} d_i)$-connected.*

*Proof*   The equivalence of the two connectivity statements is the content of Proposition 2.6.9. Choose points $p_s \in e_s$ in the interior of each cell $e_i$ with $p_n \neq x$. Define a new $n$-cube $X^*$ by $X_T^* = X_T - \{p_i : i \notin T\}$. The inclusion $X_T \to X_T^*$ is natural in $T$ and is a homotopy equivalence for each $T$, and so it suffices to prove the result for the cube $X^*$, where we denote by $\mathcal{F}^*$ the $(n-1)$-cube corresponding to $\mathcal{F}$. Let $C = \mathrm{hofiber}_x(X_{n-1}^* \to X_{n-1}^*)$. Both $C$ and $C \cap \mathcal{F}^*(T) = \mathrm{hofiber}_x(X_T^* \to X_T^*)$ are contractible. Consider the square

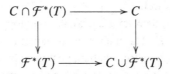

We claim that it is homotopy cocartesian for each $T$, and hence, by 1(a) of Proposition 3.7.13, the inclusion $\mathcal{F}^*(T) \to C \cup \mathcal{F}^*(T)$ is a weak equivalence for each $T$. That the square above is homotopy cocartesian follows from Example 3.7.5 – this is a square of maps which are inclusions of open sets. To verify this carefully is straightforward enough, but it would distract us from our immediate goal so we offer an outline and reference to the necessary results. First, if $X \subset Y$ is the inclusion of an open set, then the induced inclusion $\mathrm{Map}(I, X) \to \mathrm{Map}(I, Y)$ is open by definition of the compact-open topology. Next, if $(X \to Z \leftarrow Y) \to (X' \to Z' \leftarrow Y')$ is a map of diagrams such that $X \to X', Y \to Y'$, and $Z \to Z'$ are inclusions of open sets, then the induced map of limits is the inclusion of an open set. Finally we use Examples 3.1.6 and 3.1.7 to express the mapping path space and homotopy fiber respectively as limits of punctured squares.

Continuing, we observe that the map $\mathrm{hocolim}_{T \subsetneq S} C \cup \mathcal{F}^*(T) \to \mathrm{colim}_{T \subsetneq S} C \cup \mathcal{F}^*(T)$ is a weak equivalence by Example 5.8.3 (noting that the open sets $\mathcal{F}^*(\underline{n-1} - \{i\})$ have the property that $\mathcal{F}^*(\underline{n-1} - R) = \cap_{i \in R} \mathcal{F}^*(\underline{n-1} - \{i\})$). Hence it suffices to prove that the pair

$$(A, B) = \left( C \bigcup \mathcal{F}^*(\underline{n-1}), C \bigcup_i \mathcal{F}^*(\underline{n-1} - i) \right)$$

is $(-1 + \sum_{i=1}^n d_i)$-cocartesian.

For the remainder of the proof we switch the roles of 0 and 1 in the homotopy fiber. That is, if $f: X \to Y$ is a map and $y \in Y$ a point, write $\mathrm{hofiber}_y(f) = \{(x, \gamma): \gamma(0) = y, \gamma(1) = f(x)\}$ (compare to Definition 2.2.3). This convention is to make things as comparable with Goodwillie's original work as possible, should the reader want to consult it. Let $\Phi: (I^k, \partial I^k) \to (A, B)$ be a map of pairs. The map $\Phi$ corresponds via Theorem 1.2.7 to a map $\Psi: I^k \times I \to X_{\underline{n}}$ such that

- $\Psi(z, 0) = x$ for all $z \in I^k$;
- $\Psi(z, 1) \neq p_n$ for all $z \in I^k$;
- for each $z \in \partial I^k$, there exists $s$ such that $\Psi(z, t) \neq p_s$ for all $t \in I$.

When $k \leq -1 + \sum_i d_i$, we will show that there is a homotopy of $\Psi$ which preserves these conditions and deforms $\Psi$ into a map such that the last condition above holds for all $z \in I^k$ (not just on the boundary). To avoid manifolds with corners and boundaries, extend $\Psi$ to a slightly larger cube $I^k_\delta = (-\delta, 1 + \delta)^k$ for $\delta > 0$ and a slightly longer interval $[0, 1 + \epsilon)$ for $\epsilon > 0$. Consider the map

$$\overline{\Psi} : I^k_\delta \times [0, 1 + \epsilon)^n \longrightarrow (X_{\underline{n}})^n$$

given by

$$\overline{\Psi}(z, t_1, \ldots, t_n) = (\Psi(z, t_1), \ldots, \Psi(z, t_n)).$$

Since $X_{\underline{n}}$ has the structure of a manifold near each of the points $p_i$ (each $p_i$ lies in the interior of a cell of dimension $d_i + 1$), $(X_{\underline{n}})^n$ has the structure of a manifold near the point $(p_1, \ldots, p_n)$. By a small homotopy near the preimage of $(p_1, \ldots, p_n)$, we may assume $\overline{\Psi}$ is transverse to $(p_1, \ldots, p_n)$ using Theorem A.2.10.[4] By Theorem A.2.18, the space $W = \overline{\Psi}^{-1}(p_1, \ldots, p_n)$ is a submanifold of $I^k_\delta \times [0, 1 + \epsilon)^n$ of dimension $k + n - (n + \sum_{i=1}^k d_i) = k - \sum_{i=1}^k d_i$, since $(p_1, \ldots, p_n) \in (X_{\underline{n}})^n$ has codimension $n + \sum_{i=1}^k d_i$. The manifold $W$ is clearly empty if $k \leq -1 + \sum_{i=1}^n d_i$. This means that, for each $z \in I^k$, there exists $i$ such that $\Psi(z, t) \neq p_i$ for all $t$, which is what we set out to prove. $\square$

## 6.3.2 Alternative proof of the key lemma

Here we give an alternative proof of Lemma 6.3.1 by replacing the "dimension counting" (transversality) argument above with an analogous "coodinate counting" argument. It is a reproduction of an argument given by the first author in [Mun14]. We first observe that, in the hypotheses of Lemma 6.3.1, if $d_i = -1$ for all $i$, then the conclusion of Lemma 6.3.1 is vacuously true. Without loss of generality we may assume $d_n \geq 0$. The basepoint in $x \in X_{\{n\}}$ can therefore be joined by a path to some point in $X_\emptyset$, so we may also assume the basepoint lies in $X$ by the invariance of homotopy fibers over path components (Corollary 3.2.18). If $d_i = -1$ for any other value of $i$, then $X_\emptyset \to X_{\{i\}}$ is the inclusion of $X_\emptyset$ with a disjoint point added. This point plays no role in any of the homotopy fibers appearing in the cube $\mathcal{F}$, and we may ignore it altogether. That is, for this value of $i$ and a basepoint $x \in X_\emptyset$, we have $\mathrm{hofiber}_x(X_T \to X_{T \cup \{n\}}) = \mathrm{hofiber}_x(X_{T \setminus \{i\}} \to X_{T \setminus \{i\} \cup \{n\}})$ for all $T \subset \underline{n-1}$. Thus

---

[4] This can be done more carefully as in our first proof of Lemma 4.4.1.

we may assume $d_j \geq 0$ for all $1 \leq i \leq n$. The remainder of this section is a generalization of material from Section 4.4.2.

**Definition 6.3.2**   For a cube $W = W(a, \delta, L)$ in $\mathbb{R}^k$ (as in Definition 4.4.3), for each $j = 1, \ldots, n$ define

$$K_p^{j,n}(W)$$

$$= \left\{ x \in W : \frac{\delta(j-1)}{n} + a_i < x_i < \frac{\delta j}{n} + a_i \text{ for at least } p \text{ values of } i \in L \right\}.$$

If $p \leq q$, then $K_q^{j,n}(W) \subset K_p^{j,n}(W)$. The following lemma gives the basic technical deformation result, analogous to [tD08, 6.9.1] and appearing in this work as Lemma 4.4.5.

**Lemma 6.3.3**   *Let $Y$ be a space with a subspace $A \subset Y$, $W$ a cube, $j, n$ positive integers with $j \leq n$, and $f : W \to Y$ a map. Suppose that for $p \leq \dim(W)$ we have*

$$f^{-1}(A) \cap W' \subset K_p^{j,n}(W')$$

*for all cubes $W' \subset \partial W$. Then there exists a map $g : W \to Y$ homotopic to $f$ relative to $\partial W$ such that*

$$g^{-1}(A) \subset K_p^{j,n}(W).$$

*Proof*   Without loss of generality $W = I^k$, $k \geq 1$. We will construct a map $h : I^k \to I^k$ homotopic to the identify and define $g$ to be the composition of $f$ with $h$. Let $x = ((2j-1)/2n, \ldots, (2j-1)/2n)$ be the center of the cube $[(j-1)/n, j/n]^k$. For a ray $y$ emanating from $x$, let $P(y)$ be its intersection with $\partial[j-1/n, j/n]^k$ and $Q(y)$ its intersection with $\partial I^k$. Let $h$ map the segment from $P(y)$ to $Q(y)$ onto the point $Q(y)$ and the segment from $x$ to $P(y)$ affinely onto the segment from $x$ to $Q(y)$. Clearly $h$ is homotopic to the identity of $I^k$ relative to $\partial I^k$, and so $g = f \circ h$ is homotopic to $f$ relative to $\partial I^k$. It remains to check that $g$ satisfies the property in the conclusion of the lemma.

Suppose $g(z) \in A$. Write $z = (z_1, \ldots, z_k)$. If $z \in ((j-1)/n, j/n)^k$, then $z \in K_k^{j,n}(W) \subset K_p^{j,n}(W)$ and we are done. Suppose then that there exists $i$ so that either $z_i \geq j/n$ or $z_i \leq (j-1)/n$. Then, by definition of $h$, we have $h(z) \in \partial I^k$, so $h(z) \in W'$ for some face $W'$ of dimension $k - 1$. Since $g(z) = f(h(z)) \in A$, $h(z) \in f^{-1}(A)$, and by hypothesis $h(z) \in K_p^{j,n}(W')$. Thus, for at least $p$ values of $i$, we have $(j-1)/n < h(z)_i < j/n$, where $h(z)_i$ denotes the $i$th coordinate of $h(z)$. By definition of $h$,

$$h(z)_i = \frac{2j-1}{2n} + t\left(z_i - \frac{2j-1}{2n}\right) \qquad \text{for } t \geq 1.$$

Inserting this expression into the previous inequalities for $h(z)_i$ and solving for $z_i$ yields

$$-\frac{1}{t2n} + \frac{2j-1}{2n} < z_i < \frac{1}{t2n} + \frac{2j-1}{2n}.$$

Since the lower bound increases with $t$ and the upper bound decreases with $t$, substituting $t = 1$ into each gives the desired inequalities

$$\frac{j-1}{n} < z_i < \frac{j}{n}$$

so that $z \in K_p^{j,n}(W)$. □

Suppose $Y$ is a space with open subsets $Y_0, Y_1, \ldots, Y_n$ such that $Y$ is the union of $Y_1, \ldots, Y_n$ along $Y_0$. Let $f : I^k \to Y$ be a map. By the Lebesgue covering lemma, Lemma 1.4.11, we can decompose $I^k$ into cubes $W$ such that $f(W) \subset Y_j$ for some $j$ depending on $W$. We may also assume that if $W = (a, \delta, L)$ is such a cube, then $\delta$ is independent of $W$. We will use this near the end of our alternative proof of Lemma 6.3.1. The following is a straightforward generalization of our alternative proof of Lemma 4.4.1, which appears in Section 4.4.2, as is its proof.

**Theorem 6.3.4** *With the $Y_j$ and $f$ as above, assume that, for each $j$, $(Y_j, Y_0)$ is $d_j$-connected, with $d_j \geq 0$ (i.e. the inclusion map $Y_0 \to Y_j$ is $d_j$-connected). Then there is a homotopy $f_t$ of $f$ such that*

1. *if $f(W) \subset Y_j$, then $f_t(W) \subset Y_j$ for all $t$;*
2. *if $f(W) \subset Y_0$, then $f_t(W) = f(W)$ for all $t$;*
3. *if $f(W) \subset Y_j$, $f_1^{-1}(Y_j \setminus Y_0) \cap W \subset K_{d_j+1}^{j,n}(W)$.*

*Proof* Let $C^l$ be the union of cubes $W$ with $\dim(W) \leq l$. The homotopy $f_t$ is constructed inductively over $C^l \times I$. If $\dim(W) = 0$, then if $f(W) \subset Y_0$ we let $f_t = f$, which achieves the second condition. Note that $f(W) \subset Y_i \cap Y_j$ for $i \neq j$ implies $f(W) \subset Y_0$. Hence we only need to deal with the case where $f(W) \subset Y_j$ and $f(W) \not\subset Y_i$ for all $i \neq j$. In this case, since $(Y_j, Y_0)$ is $d_j$-connected and $d_j \geq 0$, choose a path from $f(W)$ to some point in $Y_0$ and use this as the homotopy, so that $f_1(W) \subset Y_0$. Then the first condition holds and so does the third (in this case, the third condition is vacuously satisfied since the inverse image in question is empty). This proves the base case.

Since the inclusion $\partial W \subset W$ is a cofibration (the proof is similar to Example 2.3.5) for any cube $W$, we may extend over all cubes $W$ so that the first and second conditions hold. By induction, suppose that $f$ has been changed by a homotopy satisfying all three conditions for cubes of dimension less than $l$, and let $W$ be a cube with $\dim(W) = l$. If $f(W) \subset Y_0$, we let $f_t = f$ as above. If $f(W) \subset Y_j$ and $f(W) \not\subset Y_i$ for all $i \neq j$, then we have the following.

- If $\dim(W) = l \leq d_j$, then, as $(Y_j, Y_0)$ is $d_j$-connected, there is a homotopy $f_t$ of $f$ relative to $\partial W$ such that $f_1(W) \subset Y_0$, and clearly the first and third conditions hold.
- If $\dim(W) = l > d_j$, we employ Lemma 6.3.3. Let $A = Y_j \setminus Y_0 \subset Y_j$. By induction we have that, for all $W' \subset \partial W$,

$$f^{-1}(Y_j \setminus Y_0) \cap W' \subset K_l^{j,n}(W') \subset K_{d_j+1}^{j,n}(W'),$$

and, by Lemma 6.3.3, there is a homotopy $f_t$ of $f$ relative to $\partial W$ such that $f_1^{-1}(Y_j \setminus Y_0) \cap W \subset K_{d_j+1}^{j,n}(W)$. $\qquad\square$

We need to convert the strongly (homotopy) cocartesian cube $X$ in the statement of Lemma 6.3.1 into one where the maps are inclusions of open sets in order to apply the previous results. For each $1 \leq j \leq n$ and corresponding cell $e_j$ with attaching map $f_j : \partial e_j \to X_0$, assume $e_j = D^{d_j+1}$, put $N_j = D^{d_j+1} - \{0\}$ and let $V_j$ be the interior of $D^{d_j+1}$. Define an $n$-cube $\mathcal{Y} = (S \mapsto Y_S)$ for $S \subset \underline{n}$ as follows. Let $U = X_0 \cup_j N_j$; the inclusion $X_0 \to U$ is a homotopy equivalence, and $U$ is open in $X_{\underline{n}}$. For $S \subset \underline{n}$ let $Y_S = U \cup_{j \in S} V_j$. Then $Y_S$ is open in $Y_{\underline{n}} = X_{\underline{n}}$ for each $S$ and the inclusion map $X_S \to Y_S$ is a homotopy equivalence which therefore gives rise to a map of cubes $X \to \mathcal{Y}$.

*Alternative proof of Lemma 6.3.1*   With $\mathcal{Y} = (S \mapsto Y_S)$ as above, choose a basepoint $y \in Y_0$, put $\mathcal{F}'(T) = \mathrm{hofiber}_y(Y_T \to Y_{T \cup \{k\}})$ for $T \subset \underline{n-1}$, and let $C$ be the contractible space $C = \mathrm{hofiber}_y(Y_{\underline{n-1}} \to Y_{\underline{n-1}})$. As indicated in the proof of Lemma 6.3.1, $\mathcal{F}'(T) \simeq \mathcal{F}(T)$. It is therefore enough to show that the cube $T \mapsto \mathcal{F}^*(T) = \mathcal{F}'(T) \cup C$ is $(-1 + \sum_j d_j)$-cocartesian; that is, that the pair

$$(A, B) = \left( \mathcal{F}^*(\underline{n-1}), \cup_{j \in \underline{n-1}} \mathcal{F}^*(\underline{n-1-j}) \right)$$

is $(-1 + \sum_j d_j)$-connected. Let $\phi : (I^k, \partial I^k) \to (A, B)$ be a map. We will show that $\phi$ is homotopic relative to $\partial I^k$ to a map $I^k \to B$ for all $-1 \leq k \leq -1 + \sum_j d_j$ (see Definition 2.6.4 to recall the definition). In the case $k = -1$ there is nothing to show, and the case $k = 0$ is trivial.

The map $\phi$ is adjoint to a map $\Phi : I^k \times I \to Y_{\underline{n}}$ with the following boundary conditions.

(B0)  $\Phi(z, 0) = y \in Y_0$ is the basepoint for all $z \in I^k$.

(B1)  $\Phi(z, 1) \in \cup_{j \in \underline{n-1}} Y_j = Y_{n-1}$ for all $z \in I^k$.

(B2)  For each $z \in \partial I^k$ there exists $i(z) \in \underline{n}$ such that $\Phi(z, t) \in \cup_{j \neq i(z)} Y_j$ for all $t \in I$.

We will make a homotopy of $\Phi$ preserving (B0)–(B2) such that (B2) holds for each $z \in I^k$, not just on the boundary. To achieve this we apply Theorem 6.3.4 to $\Phi \colon I^k \times I \to Y_{\underline{n}}$ and obtain a decomposition of $I^k \times I$ into cubes $W$ such that for each $W$ there is some $j$ such that $\Phi(W) \subset Y_j$, and a homotopy $\Phi_r$, $0 \leq r \leq 1$, of $\Phi = \Phi_0$ such that

(1)  $\Phi(W) \subset Y_j$ implies $\Phi_r(W) \subset Y_j$ for all $r$;

(2)  $\Phi(W) \subset Y_0$ implies $\Phi_r(W) = \Phi(W)$ for all $r$;

(3)  $\Phi(W) \subset Y_j$ implies $\Phi_1^{-1}(Y_j \setminus Y_0) \cap W \subset K_{d_j+1}^{j,n}(W)$.

First we prove that $\Phi_r$ satisfies (B0)–(B2) for all $r$.

(B0)  Since $\Phi(z, 0) = y \in Y_0$ is the basepoint for all $z \in I^k$, for all cubes $W \subset I^k \times \{0\}$ we have that $\Phi(W) = y$, and (2) above implies $\Phi_r(W) = \Phi(W)$ for all $r$, so that $\Phi_r(z, 0) = y$ for all $r$.

(B1)  Since $\Phi(z, 1) \in \bigcup_{j \in \underline{n-1}} Y_j = Y_{n-1}$ for all $z \in I^k$, then for all cubes $W \subset I^k \times \{1\}$, $\Phi(W) \subset Y_j$ for some $1 \leq j \leq n - 1$. Hence $\Phi_r(W) \subset Y_j$ by (1) above, and thus $\Phi_r(z, 1) \subset Y_{n-1}$ for all $r$.

(B2)  We know that for each $z \in \partial I^n$ there exists $i(z) \in \underline{n}$ such that $\Phi(\{z\} \times I) \subset \bigcup_{j \neq i(z)} Y_j$. Let $W_1, \ldots, W_h$ be cubes such that $\{z\} \times I \subset W_1 \cup \cdots \cup W_h$ and so that each $W_a$ contains a point of the form $(z, t)$ for some $t$. Since $\Phi(\{z\} \times I) \subset \bigcup_{j \neq i(z)} Y_j$, for each $a \in \{1, \ldots, h\}$ we must have $\Phi(W_a) \subset Y_{j(a)}$ for some $j(a) \neq i(z)$. This implies $\Phi_r(W_1 \cup \cdots \cup W_h) \subset Y_{j(1)} \cup \cdots \cup Y_{j(h)} \subset \bigcup_{j \neq i(z)} Y_j$ for all $r$.

Now we show that $\Phi_1$ actually satisfies the stronger condition that for each $z \in I^k$ there exists $i(z) \in \underline{n}$ so that $\Phi_1(z, t) \in \bigcup_{j \neq i(z)} Y_j$ for all $t \in I$. Let $\pi \colon I^k \times I \to I^k$ be the projection. We claim that

$$\bigcap_{j=1}^{n} \pi \left( \Phi_1^{-1}(Y_j \setminus Y_0) \right) = \emptyset$$

if $k < \sum_j (d_j - 1)$. Let $y \in \pi \left( \Phi_1^{-1}(Y_j \setminus Y_0) \right)$ for all $j$. For each $j$, choose $t_j$ so $(y, t_j) \in \Phi_1^{-1}(Y_j \setminus Y_0)$, so that $y = \pi(y, t_j)$ and $w(j) = (y, t_j) \in W_j$ for some cube $W_j \subset I^n \times I$. Thus, for each $j$, $w(j) \in W_j \cap \Phi_1^{-1}(Y_j \setminus Y_0) \subset K_{d_j+1}^{j,n}$ by condition 3 of Theorem 6.3.4. This means that $w(j)$ has at least $d_j + 1$ coordinates $w(j)_i$ such that $a_i + \delta(j-1)/n < w(j)_i < a_i + \delta j/n$, where $W_j = W(a, \delta, L)$. This implies that $y$ has at least $d_j$ coordinates $y_i$ with the same bounds (note here that the index $i$

only ranges from 1 to $n$). For each $j$, the projection $\pi(W_j)$ is a cube containing $y$, and since we have assumed that the parameter $\delta$ is independent of the cube, this means $\pi(W_j) = W$ is independent of $j$. Thus $y$ has at least $d_j$ coordinates $y_i$ such that $a_i + \delta(j-1)/n < w(j)_i < a_i + \delta j/n$ for all $j$ simultaneously, which is impossible if $n < \sum_j d_j$. Hence the intersection of sets above is indeed empty. Hence there is some $i(y) \in \underline{n}$ such that $y \notin \pi(\Phi_1^{-1}(Y_{i(y)} \setminus Y_\emptyset)$; that is, for every $t$, $(y, t) \notin \Phi_1^{-1}(Y_{i(y)} \setminus Y_\emptyset)$. When $k = 0$, the argument above requires $d_j \geq 1$ for at least one $j$, but the case $d_j = 0$ for all $j$ is trivial. $\qquad\square$

### 6.3.3 The formal step for $n = 3$

As an illustration, we will first prove the three-dimensional case of Theorem 6.2.3 by reducing to Theorem 6.2.1, Lemma 6.3.1, and Theorem 4.2.1. It does not present every nuance of the general case, but it more clearly communicates the spirit of the proof. The philosophy for this argument is that we imagine a set of cells, each of which is labeled by a non-empty subset of $T$, and each space $X_T$ in the cube $\mathcal{X} = (T \mapsto X_T)$ is built from $X_\emptyset$ by attaching cells with labels in all the subsets of $T$. More generally we imagine for $R \subset T$ that $X_T$ is built from $X_R$ by attaching cells with labels in the set of subsets $U$ with $R \subsetneq U \subset T$. The first case to consider is when the labeling sets for the cells are just the singleton subsets of $\{1, 2, 3\}$, so that $X_T$ is the union of the $X_i$ for $i \in T$, and then move on to the more general case. We begin by proving Theorem 6.2.1 for 3-cubes.

*Proof of Theorem 6.2.1 in the three-dimensional case*    Suppose

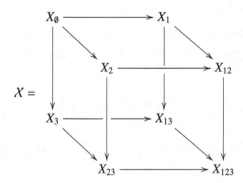

is strongly (homotopy) cocartesian and that the maps $X_\emptyset \to X_i$ are $k_i$-connected for $i = 1, 2, 3$. We first reduce to the case $k_i \geq 1$ for all $i$. Without loss of generality suppose $k_1 \leq 0$. In this case consider $\mathcal{X}$ as a map of squares $\mathcal{Y} \to \mathcal{Z}$ where

$$
\mathcal{Y} = \quad
\begin{array}{ccc}
X_\emptyset & \longrightarrow & X_2 \\
\downarrow & & \downarrow \\
X_3 & \longrightarrow & X_{23}
\end{array}
$$

and

$$
\mathcal{Z} = \quad
\begin{array}{ccc}
X_1 & \longrightarrow & X_{12} \\
\downarrow & & \downarrow \\
X_{13} & \longrightarrow & X_{123}
\end{array}
$$

By Theorem 4.2.1, $\mathcal{Y}$ is $(k_2 + k_3 - 1)$-cartesian. Since the maps $X_\emptyset \to X_i$ are $k_i$-connected for $k = 1, 2$ and the faces $\partial_\emptyset^S X$ are homotopy cocartesian for $S = \{1,2\}, \{1,3\}$, 2(a) of Proposition 3.7.13 implies that $X_1 \to X_{12}$ is $k_2$-connected and $X_1 \to X_{13}$ is $k_3$-connected. Hence, by Theorem 4.2.1, $\mathcal{Z}$ is also $(k_2 + k_3 - 1)$-cartesian. By 2(b) of Proposition 5.8.13, it follows that the 3-cube $X$ is $(k_2 + k_3 - 2)$-cartesian. Since $k_1 \leq 0$ and $k_2 + k_3 - 2 \geq k_1 + k_2 + k_3 - 2$, we obtain the desired result. Thus we may assume that $k_1 \geq 1$ and, by the same argument, that $k_i \geq 1$ for all $i$.

We may further assume that $X$ is a cofibrant pushout cube by Theorem 5.8.23, and that $(X_i, X)$ is a relative CW complex with cells of dimension $\geq k_i + 1$ by using Theorem 1.3.7 and then Theorem 2.6.26. By a similar argument to that given in Theorem 4.2.1 (see Section 4.4.3), this time using Proposition 5.4.14, we may reduce to the cube considered in Lemma 6.3.1. That is, we may assume $X$ is the cube

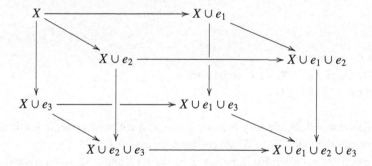

Let $x \in X \cup e_3$ be any basepoint and consider the square $\mathcal{F} =$

$$
\begin{array}{ccc}
\mathrm{hofiber}_x(X \to X \cup e_3) & \longrightarrow & \mathrm{hofiber}_x(X \cup e_1 \to X \cup e_1 \cup e_3) \\
\downarrow & & \downarrow \\
\mathrm{hofiber}_x(X \cup e_2 \to X \cup e_2 \cup e_3) & \longrightarrow & \mathrm{hofiber}_x(X \cup e_1 \cup e_2 \to X \cup e_1 \cup e_2 \cup e_3)
\end{array}
$$

By Lemma 6.3.1, $\mathcal{F}$ is $(k_1 + k_2 + k_3 - 1)$-cocartesian. By Theorem 4.2.1, the left vertical map is $(k_2 + k_3 - 1)$-connected and the upper horizontal map is $(k_1 + k_3 - 1)$-connected. It follows from Theorem 4.2.3 that $\mathcal{F}$, and hence $X$, is $(k_1 + k_2 + k_3 - 2)$-cartesian.                    □

*Proof of Theorem 6.2.3 in the three-dimensional case*    Now suppose the cube

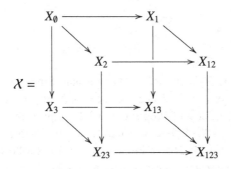

satisfies the following conditions:

1. It is is $k_{123}$-cocartesian.
2. For each $T = \{i, j\} \subset \{1, 2, 3\}$ of cardinality 2, the square

    is $k_T$-cocartesian.
3. The maps $X_\emptyset \to X_i$ are $k_i$-connected.
4. For $U \subset T$, $k_U \leq k_T$.

By Theorem 5.8.23, we may also assume $X$ is a cofibrant cube. In particular, this means the maps $X_\emptyset \to X_i$ are cofibrations for $i = 1, 2, 3$.

   *Temporary notation*: for $R, S \subset \{1, 2, 3\}$, we let $R$ and $S$ stand for $X_R$ and $X_S$ respectively, and let $R + S$ stand for the union of $X_R$ with $X_S$ along $X_{R \cap S}$. That is, $R + S$ will stand for $\mathrm{colim}(X_R \leftarrow X_{R \cap S} \to X_S)$. The maps in this diagram are cofibrations, so this colimit is weakly equivalent to the homotopy colimit of this punctured square by Proposition 3.6.17. Furthermore, we will write $ij$ in place of $\{i, j\}$, and so on.

Let $\mathcal{X}(\{1, 2, 3\})$ denote the strongly homotopy cocartesian cube

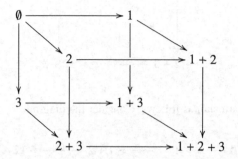

It is $(k_1 + k_2 + k_3 - 2)$-cartesian by Theorem 6.2.1; this was the content of the previous proof. Now consider the 3-cube

$$\mathcal{X}(\{12, 3\}) =$$

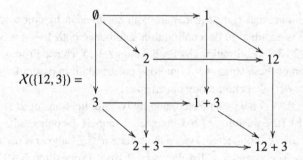

This can be factored as a "product" $\mathcal{X}(\{12, 3\}) = \mathcal{X}(\{1, 2, 3\})\mathcal{Z}_1$,

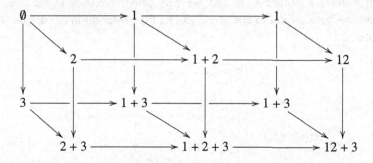

The back face $\partial_\emptyset^{13}\mathcal{Z}_1$ of $\mathcal{Z}_1$ is homotopy cartesian by 1(b) of Proposition 3.3.11, as the horizontal arrows are both weak equivalences. The front face $\partial_2^{123}\mathcal{Z}_1$, namely

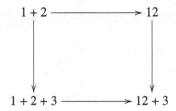

is homotopy cocartesian as follows. Consider the diagram

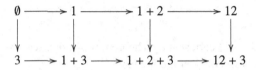

The left, middle, and right squares are both cocartesian by construction. The map $\emptyset \to 3$ is assumed to be a cofibration, and hence both $1 \to 1 + 3$ and then $1+2 \to 1+2+3$ are cofibrations by Proposition 2.3.15. Hence Proposition 3.7.6 implies each of these squares is homotopy cocartesian; in particular, the right-hand square $\partial_2^{123} \mathcal{Z}_1$ is homotopy cocartesian.

The map $\emptyset \to 3$ is $k_3$-connected, and by two applications of 2(a) of Proposition 3.7.13 (the pushout of a $k$-connected map is $k$-connected), $1 + 2 \to 1 + 2 + 3$ is also $k_3$-connected. Hence the square $\partial_2^{123} \mathcal{Z}_1$ above is $(k_3 + k_{12} - 1)$-cartesian by Theorem 4.2.1. It follows by 2(b) of Proposition 5.4.13 that $\mathcal{Z}_1$ is $(k_3 + k_{12} - 2)$-cartesian, and by 2(b) of Proposition 5.4.14 that $\mathcal{X}(\{12, 3\})$ is $(\min\{k_1 + k_2 + k_3, k_3 + k_{12}\} - 2)$-cartesian. The remainder of the argument follows a similar vein, and we will omit references to the specific results we were careful to cite above and let the reader fill in the details. The 3-cube

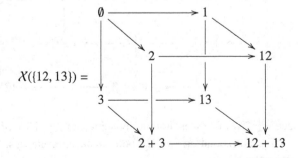

can be factored as a product $\mathcal{X}(\{12, 3\})\mathcal{Z}_2$, that is,

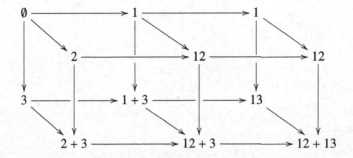

The top face $\partial_\emptyset^{12}\mathcal{Z}_2$ of $\mathcal{Z}_2$ is homotopy cartesian and its bottom face $\partial_3^{123}\mathcal{Z}_2$, namely

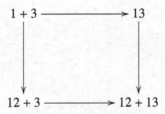

is homotopy cocartesian. Once we have a number for how cartesian $\partial_3^{123}\mathcal{Z}_2$ is, we will have one for $\mathcal{Z}_2$. The bottom face $\partial_3^{123}\mathcal{Z}_2$ itself factors as

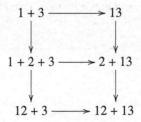

The top square is homotopy cocartesian and $(k_2 + k_{13} - 1)$-cartesian by Theorem 4.2.1, and the bottom square is also homotopy cocartesian and $(k_{12} + k_{13} - 1)$-cartesian by Theorem 4.2.1. Since $k_2 \leq k_{12}$, the product (the outer square) is $(k_2 + k_{13} - 1)$-cartesian, and so $\mathcal{Z}_2$ is $(k_2 + k_{13} - 2)$-cartesian. Hence $\mathcal{X}(\{12, 13\})$ is $(\min\{k_1 + k_2 + k_3, k_3 + k_{12}, k_2 + k_{13}\} - 2)$-cartesian. Continuing, we factor

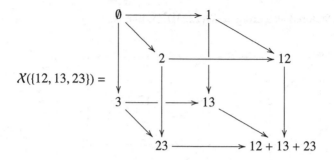

$\mathcal{X}(\{12, 13, 23\}) =$

as a product

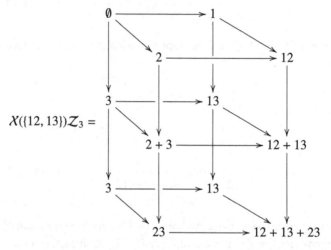

$\mathcal{X}(\{12, 13\})\mathcal{Z}_3 =$

(where $\mathcal{Z}_3$ is the bottom cube in the factorization). The back face $\partial_\emptyset^{13} \mathcal{Z}_3$ of $\mathcal{Z}_3$ is homotopy cartesian, and hence its front face $\partial_2^{123} \mathcal{Z}_3$ will determine how cartesian $\mathcal{Z}_3$ is. We have

$$\partial_2^{123} \mathcal{Z}_3 = \begin{array}{ccc} 2+3 & \longrightarrow & 12+13 \\ \downarrow & & \downarrow \\ 23 & \longrightarrow & 12+13+23 \end{array}$$

which is homotopy cocartesian. It factors as a product

$$\begin{array}{ccccccc} 2+3 & \longrightarrow & 1+2+3 & \longrightarrow & 12+3 & \longrightarrow & 12+13 \\ \downarrow & & \downarrow & & \downarrow & & \downarrow \\ 23 & \longrightarrow & 1+23 & \longrightarrow & 12+23 & \longrightarrow & 12+13+23 \end{array}$$

All three subsquares, left, middle, and right, are homotopy cocartesian. Theorem 4.2.1 implies they are, respectively, $(k_1 + k_{23} - 1)$-, $(k_{12} + k_{23} - 1)$-, and $(k_{13} + k_{23} - 1)$-cartesian. Since $k_1 \leq k_{12}, k_{13}$, the outer square $\partial_2^{123} \mathcal{Z}_3$ is $(k_1 + k_{23} - 1)$-cartesian, and hence $\mathcal{Z}_3$ is $(k_1 + k_{23} - 2)$-cartesian. Thus $\mathcal{X}(12, 13, 23)$ is $(\min\{k_1 + k_2 + k_3, k_3 + k_{12}, k_2 + k_{13}, k_1 + k_{23}\} - 2)$-cartesian. Finally, the cube

$$\mathcal{X} = \mathcal{X}(\{123\}) =$$

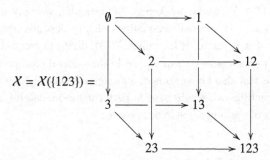

factors as a product

$$\mathcal{X}(\{12, 13, 23\})\mathcal{Z}_4 =$$

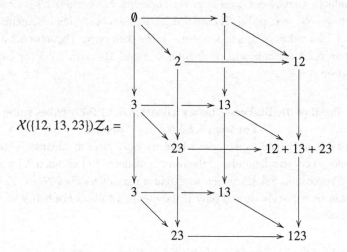

(where $\mathcal{Z}_4$ is once again the bottom cube in the factorization). The face $\partial_\emptyset^{13} \mathcal{Z}_4$ is homotopy cartesian, and hence the number for $\mathcal{Z}_4$ comes from a number for

$$\partial_2^{123} \mathcal{Z}_4 =$$

| 23 | ⟶ | 12 + 13 + 23 |
|----|----|----|
| ↓ | | ↓ |
| 23 | ⟶ | 123 |

In this case the left face, the 1-cube $23 \to 23$, is homotopy cartesian and since $12 + 13 + 23 \to 123$ is $k_{123}$-connected, the square is $(k_{123} - 1)$-cartesian. Hence $\mathcal{Z}_4$ is $(k_{123} - 2)$-cartesian. Then $X(\{123\}) = X$ is $(\min\{k_1 + k_2 + k_3, k_3 + k_{12}, k_2 + k_{13}, k_1 + k_{23}, k_{123}\} - 2)$-cartesian.                                          □

**Remark 6.3.5**    Following the proof above it is easy to see that the hypothesis $k_U \leq k_T$ for $U \subset T$ can be weakened. In particular, we only used $k_2 \leq k_{12}$, and $k_1 \leq k_{12}, k_{13}$. This set of inequalities clearly depends upon our chosen factorization of $X$ in terms of the cubes $X(\mathcal{A})$; different factorizations lead to a different set of necessary inequalities. In the general case presented below, the hypothesis can also be weakened, although to precisely determine a set of minimal inequalities we would have to perform a more careful analysis of the connectivities of various maps in our proofs.                                          □

### 6.3.4 The formal step for $n$-cubes

The induction argument used to prove Theorems 6.2.1 and 6.2.3 is organized as follows. We first prove Theorem 6.2.1 for $n$-cubes using Theorems 6.2.1 and 6.2.3 for cubes of dimension $\leq n - 1$. We then prove Theorem 6.2.3 using Theorem 6.2.1 for cubes of dimension $\leq n$ and Theorem 6.2.3 for cubes of dimension $\leq n - 1$.

#### Proof of the Blakers–Massey Theorem 6.2.1 for $n$-cubes using Theorem 6.2.3 for $(n-1)$-cubes

The goal is to reduce to the case where we only have to consider a strongly (homotopy) cocartesian cube of the form considered in Lemma 6.3.1 and then apply Proposition 5.4.12, which says that a map of $n$-cubes $\mathcal{Y} \to \mathcal{Z}$ is a $k$-cartesian $(n + 1)$-cube if and only if the $n$-cube of fibers hofiber$(\mathcal{Y} \to \mathcal{Z})$ is $k$-cartesian.

**Lemma 6.3.6**    *Theorem 6.2.1 holds if $k_j \leq 0$ for some $j \in S$.*

*Proof*    If $k_j \leq 0$ for some $j$, write $X = (\mathcal{Y} \to \mathcal{Z})$ for $(S - \{j\})$-cubes $\mathcal{Y} = (T \mapsto X_T)$ and $\mathcal{Z} = T \mapsto X_{T \cup \{j\}}$. Both $\mathcal{Y}$ and $\mathcal{Z}$ satisfy the hypotheses of Theorem 6.2.1 with the same numbers $k_i, i \in S - \{j\}$, as follows. This is trivial for $\mathcal{Y}$. For $\mathcal{Z}$, note that 1(a) of Proposition 5.8.13 says that $\mathcal{Z}$ is homotopy cocartesian since $X$ and $\mathcal{Y}$ are; this also implies that $\mathcal{Z}$ is strongly homotopy cocartesian for the same reason. The maps $Z_\emptyset \to Z_i$ are $k_i$-connected since we have for each $i$ a homotopy cocartesian square and since $Y_\emptyset \to Y_i$ is $k_i$-connected for each $i$, by 2(a) of Proposition 5.8.13 (which is a generalization of 2(a) of Proposition 3.7.13, which also applies), $Z_\emptyset \to Z_i$ is also $k_i$-connected for each $i$. Write $k = 1 - n + \sum_{i \in S} k_i$. By induction using Theorem 6.2.1, both $\mathcal{Y}$ and $\mathcal{Z}$ are $(k + 1)$-cartesian since $1 - (n - 1) + \sum_{i \neq j} k_i \geq 1 - n + 1 + \sum_{i \in S} k_i = k + 1$. By 2(b) of Proposition 5.8.13, $\mathcal{X}$ is $k$-cartesian. $\qquad\square$

Thus it suffices to prove Theorem 6.2.1 in the case $k_i \geq 1$ for all $i \in S$. We may assume $\mathcal{X}$ is a cofibrant cube by Theorem 5.8.23; in partcular $X_\emptyset \to X_i$ is a cofibration for each $i$. Using Theorem 1.3.8 and Theorem 2.6.26 we may assume $(X_i, X_\emptyset)$ is a relative CW complex where $X_i$ is obtained from $X_\emptyset$ by attaching cells of dimension $\geq k_i + 1$, and we may assume the number of cells is finite by compactness of spheres, simplices, and intervals as follows. The question of whether $\mathcal{X}$ is highly cartesian involves the connectivity of $X_\emptyset \to \mathrm{holim}_{T \neq \emptyset} X_T$, which is, roughly speaking, a problem concerned with the existence of lifts of maps $S^i \to \mathrm{holim}_{T \neq \emptyset} X_T$ and $S^i \times I \to \mathrm{holim}_{T \neq \emptyset} X_T$. These give rise by Theorem 1.2.7 to a collection of maps $S^i \times \Delta^{|T|} \to X_T$ and $S^i \times \Delta^{|T|} \times I \to X_T$ satisfying certain conditions. By compactness of the domains, the images of these maps intersect finitely many cells in $X_T$ for each $T$, so it suffices to consider this case for each such map.

Now by induction using Proposition 5.8.13, we may reduce to the case considered in Lemma 6.3.1. This argument is essentially the same one presented in the proof of Theorem 4.2.1 in Section 4.4.3, and we let the reader fill in the details.

**Lemma 6.3.7** *Theorem 6.2.1 is true for a cube of the form appearing in Lemma 6.3.1.*

*Proof* Let $\mathcal{X}$ be as in Lemma 6.3.1. Choose any basepoint in $X_n$, and let $\mathcal{F}$ be the associated $(n - 1)$-cube of homotopy fibers, which is $(-1 + \sum_{i=1}^{n} k_i)$-cocartesian. In fact, Lemma 6.3.1 implies that each face $\partial_\emptyset^T \mathcal{F}$ is $k_T$-cocartesian, where $k_T = -1 + k_n + \sum_{i \in T} k_i$, since $\partial_\emptyset^T \mathcal{F}$ is obtained from the $(|T|+1)$-cube $R \mapsto X_R$, $R \subset T \cup \{n\}$ in precisely the same way $\mathcal{F}$ is obtained from $\mathcal{X}$. Moreover, these numbers satisfy the inequalities in the statement of Theorem 6.2.1, since the $k_i$ are all positive. The sum $\sum_\alpha k_{T_\alpha}$ is minimized by the trivial partition of $\underline{n-1}$ consisting of one set because $k_n \geq 1$ and $k_T = -1 + k_n + \sum_{i \in T} k_i$. Hence

$\mathcal{F}$ is $1 - (n-1) + -1 + k_n + \sum_{i \in \underline{n-1}} k_i = 1 - n + \sum_i k_i = k$-cartesian. Thus, by Proposition 5.4.12, $X$ is $k$-cartesian.                    □

### Proof of Theorem 6.2.3 using Theorem 6.2.1 for $n$-cubes

We may assume that $S = \underline{n}$. By Theorem 5.8.23, we may assume that $X$ is a cofibrant cube, which means that, for all $T \subset \underline{n}$, the map

$$\operatorname*{colim}_{U \in \mathcal{P}_1(T)} X_U \longrightarrow \operatorname*{colim}_{U \subset T} X_U = X_T$$

is a cofibration, so the first hypothesis in Theorem 6.2.3 says that for every $\emptyset \neq T \subset S$ the map

$$\operatorname*{colim}_{U \in \mathcal{P}_1(T)} X_U \longrightarrow X_T \tag{6.3.1}$$

is $k_T$-connected.

We will be forming colimits over "convex" subsets of $\underline{n}$ other than $\mathcal{P}_1(\underline{n})$, although we can avoid using the machinery of general limits and colimits developed in the second half of this book. Since we are assuming our cube is a cofibrant cube, we may define $\operatorname{colim}_{\mathcal{A}} X$, for any collection $\mathcal{A} \subset \mathcal{P}(\underline{n})$ such that $\emptyset \in \mathcal{A}$, to be the union along $X_\emptyset$ of $X_S$ where $S \in \mathcal{A}$. Note that in the case where $\mathcal{A} = \mathcal{P}_1(S)$ for some $S \subset \underline{n}$, this colimit agrees with the already defined colimit $\operatorname{colim}_{\mathcal{P}_1(S)} X$. Call a subset $\mathcal{A} \subset \mathcal{P}(\underline{n})$ *convex* if $A \in \mathcal{A}$ and $S \leq A$ implies $S \in \mathcal{A}$. We say $A \in \mathcal{A}$ is *maximal* if whenever $S \in \mathcal{A}$ and $A \subset S$ then $S = A$.

**Lemma 6.3.8** ([Goo92, Claim 2.8])  *Suppose $\mathcal{B}$ and $C$ are convex subsets of $\mathcal{P}(\underline{n})$, and suppose $X$ is a cofibrant $n$-cube. Then the square*

$$\begin{array}{ccc}
\operatorname*{colim}_{\mathcal{B} \cap C} X & \longrightarrow & \operatorname*{colim}_{\mathcal{B}} X \\
\downarrow & & \downarrow \\
\operatorname*{colim}_{C} X & \longrightarrow & \operatorname*{colim}_{\mathcal{B} \cup C} X
\end{array}$$

*is a pushout square of cofibrations, and hence a homotopy pushout.*

A generalization of this appears as the second part of Proposition 8.6.6.

*Proof*  Both $C = \operatorname{colim}(\operatorname{colim}_C X \leftarrow \operatorname{colim}_{\mathcal{B} \cap C} X \rightarrow \operatorname{colim}_{\mathcal{B}} X)$ and $C' = \operatorname{colim}_{\mathcal{B} \cup C} X$ are quotients of $\bigsqcup_{S \in \mathcal{B} \cup C} X_S$ by an equivalence relation. We only need to show that the equivalence relations $\sim_C$ and $\sim_{C'}$ used to generate the quotients $C$ and $C'$ respectively are the same. Let $f_{S \subset T} : X_S \rightarrow X_T$ denote the maps in the cube $X$. It is clear from the canonical map $C' \rightarrow C$ that $x \sim_C y$

implies $x \sim_{C'} y$. The relation $x \sim_{C'} y$ means that $x \in X_S$ and $y \in X_T$ for some $S, T \in \mathcal{B} \cup C$ and that there exists $z \in X_R$, $R \subset S, T$, $R \in \mathcal{B} \cup C$ such that $x = f_{RCS}(z)$ and $y = f_{RCT}(z)$. By convexity of $\mathcal{B}$ and $C$, $R \in \mathcal{B} \cap C$, and so $x \sim_C y$ as well.

If $\mathcal{B} \subset C$ is an inclusion of convex sets, then the induced map $\text{colim}_{\mathcal{B}} X \to \text{colim}_C X$ is a cofibration because it is composition of cofibrations. Choose a sequence

$$\mathcal{B} = \mathcal{B}_1 \subset \mathcal{B}_2 \subset \cdots \subset \mathcal{B}_k = C$$

of convex sets $\mathcal{B}_1$ such that $\mathcal{B}_i - \mathcal{B}_{i-1}$ consists of a single set $B_i$. This can be done, for example, by choosing a minimal element $B_2$ of $C - \mathcal{B}$ and letting $\mathcal{B}_2 = \mathcal{B}_1 \cup \{B_2\}$ – convexity of $\mathcal{B}$ and $C$ ensures $\mathcal{B}_2$ is convex. Then $B_i$ is necessarily a maximal element of $\mathcal{B}_i$, and we have by the above a pushout square

$$
\begin{array}{ccc}
\underset{\mathcal{P}_1(B)}{\text{colim}\,X} & \longrightarrow & \underset{\mathcal{P}(B)}{\text{colim}\,X} \\
\downarrow & & \downarrow \\
\underset{\mathcal{P}_{i-1}(B)}{\text{colim}\,X} & \longrightarrow & \underset{\mathcal{P}_i(B)}{\text{colim}\,X}
\end{array}
$$

Since $X$ is a cofibrant cube, the top horizontal arrow is a cofibration, and since the square is a pushout, so is the lower horizontal arrow by Proposition 2.3.15.                    □

The above generalizes.

**Lemma 6.3.9** *Suppose that $\mathcal{A}_1, \ldots, \mathcal{A}_k$ are convex subsets of $\mathcal{P}(\underline{n})$, and that $X$ is a cofibrant n-cube. Then the k-cube $\mathcal{Y} = S \mapsto Y_S$ given by*

$$
Y_S = \begin{cases}
\underset{\cap_{i \in \underline{k} - S}\,\mathcal{A}_i}{\text{colim}\,X}, & S \neq \underline{k}; \\
\underset{\cup_{i \in \underline{k}}\,\mathcal{A}_i}{\text{colim}\,X}, & S = \underline{k}
\end{cases}
$$

*is cocartesian and homotopy cocartesian.*

*Proof* We use induction on $\underline{k}$ and use Lemma 6.3.8. The base case $k = 2$ is Lemma 6.3.8. By Lemma 5.6.7, we have a homeomorphism

$$\underset{S \in \mathcal{P}_1(\underline{k})}{\text{colim}\,Y_S} \cong \text{colim}\left( Y_{\underline{k}} \leftarrow \underset{R \in \mathcal{P}_1(\underline{k-1})}{\text{colim}\,Y_R} \to \underset{R \in \mathcal{P}_1(\underline{k-1})}{\text{colim}\,Y_{R \cup \{k\}}} \right).$$

By induction, $\operatorname*{colim}_{R \in \mathcal{P}_1(\underline{k-1})} Y_R \cong Y_{\underline{k-1}} = \operatorname*{colim}_{\mathcal{A}_1 \cup \cdots \cup \mathcal{A}_{k-1}} X$, and by definition

$$\operatorname*{colim}_{R \in \mathcal{P}_1(\underline{k-1})} Y_{R \cup \{k\}} = \operatorname*{colim}_{R \in \mathcal{P}_1(\underline{k-1})} \operatorname*{colim}_{i \in \cap_{\underline{k}-R}(\mathcal{A}_i \cap \mathcal{A}_k)} X,$$

which, again using induction (this time with the $\mathcal{A}_i$ replaced with $\mathcal{A}_i \cap \mathcal{A}_k$), is homeomorphic to

$$\operatorname*{colim}_{\mathcal{A}_k \cap (\mathcal{A}_1 \cup \cdots \cup \mathcal{A}_{k-1})} X.$$

Thus we have a pushout square

$$
\begin{array}{ccc}
\operatorname*{colim}_{\mathcal{A}_k \cap (\mathcal{A}_1 \cup \cdots \cup \mathcal{A}_{k-1})} X & \longrightarrow & \operatorname*{colim}_{\mathcal{A}_k} X \\
\downarrow & & \downarrow \\
\operatorname*{colim}_{\mathcal{A}_1 \cup \cdots \cup \mathcal{A}_{k-1}} X & \longrightarrow & \operatorname*{colim}_{S \in \mathcal{P}_1(\underline{k})} Y_S
\end{array}
$$

and, by Lemma 6.3.8, $\operatorname*{colim}_{\mathcal{P}_1(\underline{k})} Y_S \simeq \operatorname*{colim}_{\mathcal{A}_1 \cup \cdots \cup \mathcal{A}_k} X$. That the original cube is homotopy cocartesian follows from the fact that $X$ is cofibrant. $\quad\square$

**Lemma 6.3.10** *Let $X = S \mapsto X_S$ be a cofibrant n-cube such that $\operatorname*{colim}_{U \in \mathcal{P}_1(T)} X_U \to X_T$ is $k_T$-connected for all $T \subset \underline{n}$. If $\mathcal{A}$ is a convex subset of $\mathcal{P}(\underline{n})$ and $A \in \mathcal{A}$ is a maximal element, then the map*

$$\operatorname*{colim}_{\mathcal{A}-\{A\}} X \longrightarrow \operatorname*{colim}_{\mathcal{A}} X$$

*is $k_A$-connected.*

*Proof* We proceed as in the proof of Lemma 6.3.8. Let $\mathcal{B} = \mathcal{A} - A$ and $C = \mathcal{P}(A)$. These are clearly convex. Then $\mathcal{A} - A \cap \mathcal{P}(A) = \mathcal{P}_1(A)$ and by Lemma 6.3.8 we have a pushout square of cofibrations

$$
\begin{array}{ccc}
\operatorname*{colim}_{\mathcal{P}_1(A)} X & \longrightarrow & \operatorname*{colim}_{\mathcal{P}(A)} X \\
\downarrow & & \downarrow \\
\operatorname*{colim}_{\mathcal{A}-\{A\}} X & \longrightarrow & \operatorname*{colim}_{\mathcal{A}} X
\end{array}
$$

Since $\operatorname*{colim}_{\mathcal{P}(A)} X = X(A)$, 2(a) of Proposition 3.7.13 implies that $\operatorname*{colim}_{\mathcal{A}-\{A\}} X \to \operatorname*{colim}_{\mathcal{A}} X$ is $k_A$-connected since the top horizontal map is assumed to be $k_A$-connected by hypothesis. $\quad\square$

Recall that $\partial_U^T X$ is the $(T - U)$-cube $R \mapsto X_R$ for $U \subset R \subset T$ (Definition 5.1.1); this is the face of $X$ indexed by the pair $(U, T)$. As in Section 6.3.3, we will factor $X$ by cubes $X_{\mathcal{A}}$ indexed by various convex subsets $\mathcal{A}$.

**Definition 6.3.11** For a convex subset $\mathcal{A} \subset \mathcal{P}(\underline{n})$, define $X_{\mathcal{A}}$ to be the $n$-cube

$$X_{\mathcal{A}}(T) = \operatorname*{colim}_{U \in \mathcal{A},\ U \subset T} X_U.$$

Since $\mathcal{A}$ is convex, the indexing category for this colimit is equivalent to the convex subcategory $\mathcal{A} \cap \mathcal{P}(T)$. Thus

$$X_{\mathcal{A}}(T) = \operatorname*{colim}_{U \in \mathcal{A} \cap \mathcal{P}(T)} X_U.$$

*Proof of Theorem 6.2.3* Choose an increasing sequence $\mathcal{A}_1 \subset \mathcal{A}_2 \subset \cdots \subset \mathcal{A}_m$ of convex subsets of $\mathcal{P}(\underline{n})$ such that

- $\mathcal{A}_1 = \{S \in \mathcal{P}(\underline{n}) : |S| \leq 1\}$;
- $\mathcal{A}_m = \mathcal{P}(\underline{n})$;
- $\mathcal{A}_j - \mathcal{A}_{j-1} = \{S_j\}$ is a single subset $S_j \in \mathcal{P}(\underline{n})$;
- the sequence $|S_j|$ is non-decreasing in $j$.

For $j \geq 2$, we will show below how to factor $X_{\mathcal{A}_j}$ as $X_{\mathcal{A}_{j-1}} \mathcal{Z}_j$ for some $n$-cube $\mathcal{Z}_j$ (recall the product notation $\mathcal{Z} = X\mathcal{Y}$ for cubes from Remark 5.4.15). Lemmas 6.3.12 and 6.3.13 will show that $\mathcal{Z}_j$ is $(1 - n + \min_{\{T_\beta\}}\{\sum_\beta k_{T_\beta}\})$-cartesian where the minimum is taken over all partitions $\{T_\beta\}$ of $\underline{n}$ which contain $S_j$ as one block, and for which all other blocks are elements of $\mathcal{A}_{j-1}$ (terminology for partitions was discussed just before the statement of Theorem 6.2.3). Applying 2(a) of Proposition 5.4.14, together with the fact that $X_{\mathcal{A}_1}$ is $(1 - n + \sum_i k_i)$-cartesian by Theorem 6.2.1, proves the result. $\square$

The choice of increasing sequence above is arbitrary, and the condition that $|S_j|$ be non-decreasing in $j$ is not strictly necessary but it streamlines the organization. We now define the factorizations of the $X_{\mathcal{A}_j}$ for $j \geq 2$. With $\mathcal{A}_j$ and $S_j$ as above, write $S_j = \{a_1, \ldots, a_k\}$, and note that by hypothesis $k \geq 2$ for all $j$.

Define $\mathcal{Z}_j$ to be the cube

$$\mathcal{Z}_j(T) = \begin{cases} X_{\mathcal{A}_{j-1}}(T \cup \{a_k\}), & \text{if } a_k \notin T; \\ X_{\mathcal{A}_j}(T), & \text{if } a_k \in T. \end{cases} \tag{6.3.2}$$

The factorization $X_{\mathcal{A}_j} = X_{\mathcal{A}_{j-1}} \mathcal{Z}_j$ is immediate, since if we write

$$\mathcal{Z}_j = \partial_{\emptyset}^{n - \{a_k\}} \mathcal{Z}_j \longrightarrow \partial_{\{a_k\}}^{n} \mathcal{Z}_j,$$

we have by definition $\partial_{\emptyset}^{n - \{a_k\}} \mathcal{Z}_j = \partial_{\{a_k\}}^{n} X_{\mathcal{A}_{j-1}}$ and $\partial_{\{a_k\}}^{n} \mathcal{Z}_j = \partial_{\{a_k\}}^{n} X_{\mathcal{A}_j}$. Note that if $R \subset \underline{n}$ is a set which does not contain $S_j$, then

$$X_{\mathcal{A}_j}(R) = X_{\mathcal{A}_{j-1}}(R) \tag{6.3.3}$$

because $\mathcal{A}_j \cap \mathcal{P}(R) = \mathcal{A}_{j-1} \cap \mathcal{P}(R)$ for such $R$. Equation (6.3.3) implies that many of the faces of $\mathcal{Z}_j$ are cartesian, and in particular means that we can find a face of $\mathcal{Z}_j$ which will tell us how cartesian $\mathcal{Z}_j$ itself is.

**Lemma 6.3.12**  *With $S_j = \{a_1, \ldots, a_k\}$ as above, if $\partial^n_{S_j-\{a_k\}}\mathcal{Z}_j$ is $(K + |S_j| - 1)$-cartesian, then $\mathcal{Z}_j$ is $K$-cartesian.*

*Proof*  We need the following fact: if $U \subset R \subset \underline{n}$ satisfies $a_k \notin U$, $a_k \in R$, and $S_j \not\subset R$, then the face $\partial^R_U \mathcal{Z}_j$ is homotopy cartesian. This is true by Example 5.4.4 since

$$\partial^R_U \mathcal{Z}_j = \partial^{R-\{a_k\}}_U \mathcal{Z}_j \longrightarrow \partial^R_{U\cup\{a_k\}}\mathcal{Z}_j$$

is a map of identical $(|R| - |U| - 1)$-cubes by definition of $\mathcal{Z}_j$ and using Equation (6.3.3). Now write

$$\mathcal{Z}_j = \left(\partial^{n-\{a_1\}}_\emptyset \mathcal{Z}_j \longrightarrow \partial^n_{\{a_1\}}\mathcal{Z}_j\right).$$

Since $\partial^{n-\{a_1\}}_\emptyset \mathcal{Z}_j$ is homotopy cartesian by the fact above with $U = \emptyset$ and $R = \underline{n} - \{a_1\}$, 2(b) of Proposition 5.4.13 implies that $\mathcal{Z}_j$ is $K$-cartesian if $\partial^n_{\{a_1\}}\mathcal{Z}_j$ is $(K + 1)$-cartesian. More generally, for $l < k$ write

$$\partial^n_{\{a_1,\ldots,a_{l-1}\}}\mathcal{Z}_j = \left(\partial^{n-\{a_l\}}_{\{a_1,\ldots,a_{l-1}\}} \mathcal{Z}_j \longrightarrow \partial^n_{\{a_1,\ldots,a_{l-1},a_l\}}\mathcal{Z}_j\right).$$

Then $\partial^{n-\{a_l\}}_{\{a_1,\ldots,a_{l-1}\}}\mathcal{Z}_j$ is homotopy cartesian by the fact above with $U = \{a_1,\ldots,a_{l-1}\}$ and $R = \underline{n}-\{a_l\}$. Again by 2(b) of Proposition 5.4.13, $\partial^n_{\{a_1,\ldots,a_{l-1}\}}\mathcal{Z}_j$ is $(K + l - 1)$-cartesian if $\partial^n_{\{a_1,\ldots,a_{l-1},a_l\}}\mathcal{Z}_j$ is $(K + l)$-cartesian. Hence $\mathcal{Z}_j$ is $K$-cartesian if $\partial^n_{S_j-\{a_k\}}\mathcal{Z}_j$ is $(K + |S_j| - 1)$-cartesian.    □

Our goal now is to use Theorem 6.2.3 for cubes of dimension less than $n$ on the $(n - |S_j| + 1)$-cube $\partial^n_{S_j-\{a_k\}}\mathcal{Z}_j$, noting that $|S_j| \geq 2$ implies this is in fact a cube of dimension strictly less than $n$.

**Lemma 6.3.13**  *The cube $\partial^n_{S_j-\{a_k\}}\mathcal{Z}_j$ is $(|S_j| - n + \min_{\{T_\beta\}}\{\sum_\beta k_{T_\beta}\})$-cartesian, where $\{T_\beta\}$ ranges over all partitions of $\underline{n}$ which contain $S_j$ as one block, and all other blocks are elements of $\mathcal{A}_{j-1}$. Hence $\mathcal{Z}_j$ is $(1 - n + \min_{\{T_\beta\}}\{\sum_\beta k_{T_\beta}\})$-cartesian, where $\{T_\beta\}$ is as above.*

*Proof*  We will use Theorem 6.2.3 for cubes of dimension less than $n$ and induction on $j$. In order to do this we need to establish how cocartesian the various faces $\partial^R_{S_j-\{a_k\}}\mathcal{Z}_j$, $S_j - \{a_k\} \subset R \subset \underline{n}$ are. We will prove the following.

1. If $|R - (S_j - \{a_k\})| = 1$, then $\partial^R_{S_j - \{a_k\}} \mathcal{Z}_j$ is

$$
\begin{cases}
k_{S_j}\text{-cocartesian} & \text{if } a_k \in R, \text{ so } R = S_j, \\
k_a\text{-cocartesian} & \text{if } a_k \notin R, \text{ and } R = S_j - \{a_k\} \cup \{a\}.
\end{cases}
$$

2. If $|R - (S_j - \{a_k\})| \geq 2$, then $\partial^R_{S_j - \{a_k\}} \mathcal{Z}_j$ is

$$
\begin{cases}
\infty\text{-cocartesian} & \text{if } a_k \in R, \\
\infty\text{-cocartesian} & \text{if } a_k \notin R, R - (S_j - \{a_k\}) \notin \mathcal{A}_{j-1}, \\
k_{R-(S_j - \{a_k\})}\text{-cocartesian} & \text{if } a_k \notin R, R - (S_j - \{a_k\}) \in \mathcal{A}_{j-1}.
\end{cases}
$$

Since the cube in question is indexed by those sets $R$ such that $S_j - \{a_k\} \subset R \subset \underline{n}$, the minimum over partitions of $\underline{n}$ appearing in Theorem 6.2.3 is over those partitions of $\underline{n}$ which have a block containing $S_j - \{a_k\}$ as a proper subset. Clearly the sum over such partitions is minimized when one of the partition blocks is the set $S_j$ itself and all others are elements of $\mathcal{A}_{j-1}$, for any other partition would yield $\infty$ for the overall estimate by the estimates claimed above.

For the first item, if $R = S_j$, then

$$
\begin{aligned}
\partial^R_{S_j - \{a_k\}} \mathcal{Z}_j &= \left( \mathcal{Z}_j(S_j - \{a_k\}) \to \mathcal{Z}_j(S_j) \right) \\
&= \left( X_{\mathcal{A}_{j-1}}(S_j) \to X_{\mathcal{A}_j}(S_j) \right) \\
&= \left( \mathrm{colim}_{U \in \mathcal{A}_{j-1} \cap \mathcal{P}(S_j)} X_U \to \mathrm{colim}_{V \in \mathcal{A}_j \cap \mathcal{P}(S_j)} X_V \right) \\
&\simeq \left( \mathrm{colim}_{U \in \mathcal{P}_1(S_j)} X_U \to \mathrm{colim}_{V \in \mathcal{P}(S_j)} X_V = X_{S_j} \right)
\end{aligned}
$$

is $k_{S_j}$-connected by hypothesis. The last weak equivalence follows from the fact that $\mathcal{A}_{j-1}$ contains all proper subsets of $S_j$, and using the fact that $S_j$ is the maximal element of $\mathcal{P}(S_j)$.

If $R = S_j - \{a_k\} \cup \{a\}$, then of course $a \notin S_j$ and, since $a_k \notin R$, we have

$$
\begin{aligned}
\partial^R_{S_j - \{a_k\}} \mathcal{Z}_j &= \left( \mathcal{Z}_j(S_j - \{a_k\}) \to \mathcal{Z}_j((S_j - \{a_k\}) \cup \{a\}) \right) \\
&= \left( X_{\mathcal{A}_{j-1}}(S_j) \to X_{\mathcal{A}_{j-1}}(S_j \cup \{a\}) \right) \\
&= \left( \mathrm{colim}_{U \in \mathcal{A}_{j-1} \cap \mathcal{P}(S_j)} X_U \to \mathrm{colim}_{V \in \mathcal{A}_{j-1} \cap \mathcal{P}(S_j \cup \{a\})} X_V \right) \\
&\simeq \left( \mathrm{colim}_{U \in \mathcal{P}_1(S_j)} X_U \to \mathrm{colim}_{V \in \mathcal{A}_{j-1} \cap \mathcal{P}(S_j \cup \{a\})} X_V \right).
\end{aligned}
$$

We have to show that this map of spaces is $k_a$-connected. Choose an increasing sequence of convex sets

$$
\mathcal{P}_1(S_j) = \mathcal{B}_1 \subset \cdots \subset \mathcal{B}_l = \mathcal{A}_{j-1} \cap \mathcal{P}(S_j \cup \{a\})
$$

such that $\mathcal{B}_i - \mathcal{B}_{i-1}$ consists of a single set $R_i$. The map

$$\operatorname*{colim}_{\mathcal{B}_{i-1}} X \longrightarrow \operatorname*{colim}_{\mathcal{B}_i} X$$

is $k_{R_i}$-connected by Lemma 6.3.10. Note that $a \in R_i$ for all $i$ since $\mathcal{A}_{j-1}$ contains every proper subset of $S_j$ but does not contain $S_j$ itself. Moreover, for some $i_0$, we have $R_{i_0} = \{a\}$. Since we assume $U \subset T$ implies $k_U \leq k_T$, it follows that the map

$$\mathcal{Z}_j(S_j - \{a_k\}) = \operatorname*{colim}_{\mathcal{B}_1} X \rightarrow \operatorname*{colim}_{\mathcal{B}_{i_1}} X = \mathcal{Z}_j((S_j - \{a_k\}) \cup \{a\})$$

is $k_a$-connected.

The second item, when $|R - (S_j - \{a_k\})| \geq 2$, proceeds in a similar fashion. We have to determine the connectivity of the map

$$\operatorname*{colim}_{S_j - \{a_k\} \subset U \subsetneq R} \mathcal{Z}_j(U) \longrightarrow \mathcal{Z}_j(R). \tag{6.3.4}$$

We will consider two cases: $a_k \in R$ and $a_k \notin R$. First suppose $a_k \in R$. The map in Equation (6.3.4) can then be written as

$$\operatorname*{colim}_{S_j - \{a_k\} \subset U \subsetneq R} \mathcal{Z}_j(U) \cong \operatorname*{colim} \begin{pmatrix} \mathcal{Z}_j(R - \{a_k\}) \\ \uparrow \\ \operatorname*{colim}_{S_j - \{a_k\} \subset V \subsetneq R - \{a_k\}} \mathcal{Z}_j(V) \\ \downarrow \\ \operatorname*{colim}_{S_j - \{a_k\} \subset U \subsetneq R - \{a_k\}} \mathcal{Z}_j(V \cup \{a_k\}) \end{pmatrix} \longrightarrow \mathcal{Z}_j(R).$$

It is enough to show that the square

$$\begin{array}{ccc} \operatorname*{colim}_{S_j - \{a_k\} \subset U \subsetneq R - \{a_k\}} \mathcal{Z}_j(U) & \longrightarrow & \mathcal{Z}_j(R - \{a_k\}) \\ \downarrow & & \downarrow \\ \operatorname*{colim}_{S_j - \{a_k\} \subset U \subsetneq R - \{a_k\}} \mathcal{Z}_j(U \cup \{a_k\}) & \longrightarrow & \mathcal{Z}_j(R) \end{array}$$

is homotopy cocartesian. Reindexing, the square above is equivalent to the square

$$
\begin{array}{ccc}
\operatorname*{colim}_{U \subsetneq R - S_j} \mathcal{Z}_j(U \cup (S_j - \{a_k\})) & \longrightarrow & \mathcal{Z}_j(R - \{a_k\}) \\
\downarrow & & \downarrow \\
\operatorname*{colim}_{U \subsetneq R - S_j} \mathcal{Z}_j(U \cup S_j) & \longrightarrow & \mathcal{Z}_j(R)
\end{array}
$$

Finally, using the definitions of $\mathcal{Z}$ and $X_{\mathcal{A}}$, we obtain the equivalent square

$$
\begin{array}{ccc}
\operatorname*{colim}_{U \subsetneq R - S_j} \operatorname*{colim}_{V \in \mathcal{A}_{j-1} \cap \mathcal{P}(U \cup S_j)} X_V & \longrightarrow & \operatorname*{colim}_{V \in \mathcal{A}_{j-1} \cap \mathcal{P}(R)} X_V \\
\downarrow & & \downarrow \\
\operatorname*{colim}_{U \subsetneq R - S_j} \operatorname*{colim}_{V \in \mathcal{A}_{j} \cap \mathcal{P}(U \cup S_j)} X_V & \longrightarrow & \operatorname*{colim}_{V \in \mathcal{A}_j \cap \mathcal{P}(R)} X_V
\end{array}
$$

Note that $\mathcal{A}_j = \mathcal{A}_{j-1} \cup \mathcal{P}(S_j)$, so $\mathcal{A}_j \cap \mathcal{P}(R) = (\mathcal{A}_{j-1} \cap \mathcal{P}(R)) \cup \mathcal{P}(S_j)$. Clearly $\mathcal{P}(S_j) \subset \mathcal{A}_j \cap \mathcal{P}(U \cup S_j)$ (in fact for every $U$). It follows that the union of the convex set $\mathcal{A}_{j-1} \cap \mathcal{P}(R)$ with the convex set $\{\mathcal{A}_j \cap \mathcal{P}(U \cup S_j) \colon U \subsetneq R - S_j\}$ is equal to $\mathcal{A}_j \cap \mathcal{P}(R)$, and Lemma 6.3.8 implies that the square above is homotopy cocartesian.

Next suppose $a_k \notin R$. In this case the map in Equation (6.3.4) can be written as

$$
\operatorname*{colim}_{U \subsetneq R - (S_j - \{a_k\})} \operatorname*{colim}_{V \in \mathcal{A}_{j-1} \cap \mathcal{P}(U \cup S_j - \{a_k\})} X_V \longrightarrow \operatorname*{colim}_{V \in \mathcal{A}_{j-1} \cap \mathcal{P}(R)} X_V.
$$

Observe that any set $V \in \mathcal{A}_{j-1} \cap \mathcal{P}(R)$ which is not in $\{\mathcal{A}_{j-1} \cap \mathcal{P}(U \cup S_j) \colon U \subsetneq R - (S_j - \{a_k\})\}$ must contain $R - (S_j - \{a_k\})$. If $R - (S_j - \{a_k\}) \notin \mathcal{A}_{j-1}$, then the map is a weak equivalence by Lemma 6.3.9 since the collection $\{\mathcal{A}_{j-1} \cap \mathcal{P}(U \cup S_j) \colon U \subsetneq R - (S_j - \{a_k\})\}$ is a covering of $\mathcal{A}_{j-1} \cap \mathcal{P}(R)$ by convex sets. If $R - (S_j - \{a_k\}) \in \mathcal{A}_{j-1}$, then choose an increasing sequence

$$
\{\mathcal{A}_{j-1} \mathcal{P}(U \cup S_j) \colon U \subsetneq R - S_j\} = \mathcal{B}_1 \subset \cdots \subset \mathcal{B}_l = \mathcal{A}_{j-1} \cap \mathcal{P}(R)
$$

such that $\mathcal{B}_i - \mathcal{B}_{i-1} = R_i$ consists of a single set $R_i$ for $i \geq 2$. Since $k_U \leq k_T$ holds for all $U \subset T$ and each $R_i$ must contain $R - (S_j - \{a_k\})$, and in fact be equal to $R - (S_j - \{a_k\})$ for some $i$, the map is therefore $k_{R - (S_j - \{a_k\})}$-connected. This establishes the connectivity estimates claimed above. It follows from Lemma 6.3.12 that $\mathcal{Z}_j$ is $(1 - n + \min_{\{T_\beta\}} \{\sum_\beta k_{T_\beta}\})$-cartesian, where $\{T_\beta\}$ ranges over all partitions of $\underline{n}$ which contain $S_j$ as one block and for which all other blocks are elements of $\mathcal{A}_{j-1}$. $\qquad \square$

## 6.4  Proofs of the dual Blakers–Massey Theorems for *n*-cubes

The proof of the dual of the generalized Blakers–Massey Theorem, Theorem 6.2.2, is considerably easier than the proof of Theorem 6.2.1; the analog of the geometric step is simpler and more straightforward as it immediately follows from Proposition 5.10.4.

*Proof of Theorem 6.2.2*    Since $X$ is strongly (homotopy) cocartesian we apply Proposition 5.10.4 to see that

$$\text{hofiber}(b(X)) = \text{hofiber}(\text{hocolim}_{S \in \mathcal{P}_1(\underline{n})} X_S \to X_{\underline{n}})$$
$$\cong \underset{i \in \underline{n}}{*} \, \text{hofiber}(X_{\underline{n}-\{i\}} \to X_{\underline{n}})$$

for any choice of basepoint in $X_{\underline{n}}$. Since the maps $X_{\underline{n}-\{i\}} \to X_{\underline{n}}$ are assumed to be $k_i$-connected for each $i$, the spaces $\text{hofiber}(X_{\underline{n}-\{i\}} \to X_{\underline{n}})$ are $(k_i - 1)$-connected for each $i$. Repeated application of Proposition 3.7.23 then shows that the $n$-fold join above is $2(n-1)+\sum_{i=1}^{n} k_i - 1 = (n-2+\sum_{i=1}^{n} k_i)$-connected, so that $b(X)$ is $(-1+n+\sum_{i=1}^{n} k_i)$-connected and $X$ is $(-1+n+\sum_{i=1}^{n} k_i)$-cocartesian.     □

We now present the proof of Theorem 6.2.4, the proof of which is precisely dual to the proof of Theorem 6.2.2, and so we omit a discussion of the three-dimensional result and give an overall much sketchier treatment.

*Proof of Theorem 6.2.4*    Assume $S = \underline{n}$. By Theorem 5.4.23 we may assume, that, for all $T \subset S$, the map

$$X_{\underline{n}-T} = \lim_{U \in \mathcal{P}(T)} X_{U \cup \underline{n}-T} \longrightarrow \lim_{U \in \mathcal{P}_0(T)} X_{U \cup \underline{n}-T}$$

is a fibration. Hence the first hypothesis in Theorem 6.2.4 says that, for all $T \subsetneq \underline{n}$, the map

$$X_{\underline{n}-T} \longrightarrow \lim_{U \in \mathcal{P}_0(T)} X_{U \cup \underline{n}-T} \tag{6.4.1}$$

is $k_T$-connected.

For a collection $\mathcal{A} \subset \mathcal{P}(\underline{n})$ such that $\underline{n} \in \mathcal{A}$, we define $\lim_{\mathcal{A}} X$ to be the fiber product of $X_S$, over $X_{\underline{n}}$, where $S \in \mathcal{A}$. Evidently if $\mathcal{A} = \mathcal{P}_0(S)$ for some $S \subset \underline{n}$, this agrees with the already defined quantity $\lim_{\mathcal{P}_0(S)} X$. We call a subset $\mathcal{A} \subset \mathcal{P}(\underline{n})$ *concave* if $A \in \mathcal{A}$ and $A \subset S$ implies $S \in \mathcal{A}$, and we call an element $A \in \mathcal{A}$ *minimal* if $S \in \mathcal{A}$ and $S \subset A$ implies $S = A$. We have the following generalizations of Lemmas 6.3.8, 6.3.9, and 6.3.10.

**Lemma 6.4.1** *Suppose $X$ is a fibrant cube and $\mathcal{B}, \mathcal{C}$ are concave. Then the square*

$$
\begin{array}{ccc}
\lim_{\mathcal{B} \cup \mathcal{C}} X & \longrightarrow & \lim_{\mathcal{B}} X \\
\downarrow & & \downarrow \\
\lim_{\mathcal{C}} X & \longrightarrow & \lim_{\mathcal{B} \cap \mathcal{C}} X
\end{array}
$$

*is a pullback square of fibrations, and hence a homotopy pullback.*

*Proof*   Dual to the proof of Lemma 6.3.8.                            □

**Lemma 6.4.2** *Suppose that $\mathcal{A}_1, \ldots, \mathcal{A}_k$ are concave subsets of $\mathcal{P}(\underline{n})$, and that $X$ is a fibrant $n$-cube. Then the $k$-cube $\mathcal{Y} = S \mapsto Y_S$ given by*

$$
Y_S = \begin{cases} \lim_{\cap_{i \in S} \mathcal{A}_i} X, & S \neq \emptyset; \\ \lim_{\cup_{i \in \underline{k}} \mathcal{A}_i} X, & S = \emptyset \end{cases}
$$

*is cartesian and homotopy cartesian.*

*Proof*   Dual to the proof of Lemma 6.3.9.                            □

**Lemma 6.4.3** *Suppose $X = S \mapsto X_S$ is a fibrant $n$-cube and the map $X_{\underline{n}-T} \to \lim_{U \in \mathcal{P}_0(T)} X_{U \cup \underline{n} - T}$ is $k_T$-connected for each $T$. If $\mathcal{A}$ is a concave subset of $\mathcal{P}(\underline{n})$ and $A \in \mathcal{A}$ is a minimal element, then the map*

$$
\lim_{\mathcal{A}} X \longrightarrow \lim_{\mathcal{A} - \{A\}} X
$$

*is $k_{\underline{n}-A}$-connected.*

*Proof*   Dual to the proof of Lemma 6.3.10.                           □

**Definition 6.4.4**   For a concave subset $\mathcal{A} \subset \mathcal{P}(\underline{n})$, define $X^{\mathcal{A}}$ to be the $n$-cube

$$
X^{\mathcal{A}}(T) = \lim_{U \in \mathcal{A}, T \subset U} X_U. \tag{6.4.2}
$$

*Proof of Theorem 6.2.4*   Choose an increasing sequence of concave subsets $\mathcal{A}_1 \subset \mathcal{A}_2 \subset \cdots \subset \mathcal{A}_m$ of $\mathcal{P}(\underline{n})$ such that

- $\mathcal{A}_1 = \{S \in \mathcal{P}(\underline{n}) : |S| \geq n - 1\}$;
- $\mathcal{A}_m = \mathcal{P}(\underline{n})$;
- $\mathcal{A}_j - \mathcal{A}_{j-1} = \{S_j\}$ is a single subset $S_j \in \mathcal{P}(\underline{n})$;
- the sequence $|S_j|$ is non-increasing in $j$.

For $j \geq 2$ we will factor $X^{\mathcal{A}_j} = Z^j X^{\mathcal{A}_{j-1}}$ for some $n$-cube $Z^j$ to be defined below. Lemmas 6.4.5 and 6.4.6 will show that $Z^j$ is $(-1 + n + \min_{\{T_\beta\}}\{\sum_\beta k_{T_\beta}\})$-cocartesian where $\{T_\beta\}$ ranges over all partitions of $\underline{n}$ which contain $S_j$ as one block, and all other blocks are elements of $\mathcal{A}_{j-1}$. Applying 2(a) of Proposition 5.8.14 and using that $X^{\mathcal{A}_1}$ is $(-1 + n + \sum_i k_i)$-cocartesian by Theorem 6.2.2 then proves the desired result.                                                    □

With $\mathcal{A}_j$ and $S_j$ as above, write $\underline{n} - S_j = \{a_1, \ldots, a_k\}$. Define $Z^j$ to be the cube

$$Z^j(T) = \begin{cases} X^{\mathcal{A}_j}(T), & \text{if } a_k \notin T; \\ X^{\mathcal{A}_{j-1}}(T - \{a_k\}), & \text{if } a_k \in T. \end{cases} \qquad (6.4.3)$$

The factorization $X^{\mathcal{A}_j} = Z^j X^{\mathcal{A}_{j-1}}$ is immediate. We claim now that the degree to which $Z^j$ is cocartesian is determined by one of its subcubes.

**Lemma 6.4.5** *With $\underline{n} - S_j = \{a_1, \ldots, a_k\}$ as above, if $\partial_\emptyset^{S_j \cup \{a_k\}} Z^j$ is $(K - n + |S_j| + 1)$-cocartesian, then $Z^j$ is $K$-cocartesian.*

*Proof* Suppose $U \subset R \subset \underline{n}$ are chosen so that $a_k \notin U, a_k \in R$, and $U \not\subset S_j$. The face $\partial_U^R Z^j$ is homotopy cocartesian since

$$\partial_U^R Z^j = \left( \partial_U^{R - \{a_k\}} Z^j \longrightarrow \partial_{U \cup \{a_k\}}^R Z^j \right)$$

is a map of identical $(|R| - |U| - 1)$-cubes by definition of $Z^j$. Indeed, the condition that $U \not\subset S_j$ implies $T \not\subset S_j$ for $U \subset T$, thus for such $T$ we have $X^{\mathcal{A}_j}(T) = X^{\mathcal{A}_{j-1}}(T)$.

Now write

$$Z^j = \left( \partial_\emptyset^{\underline{n}} Z^j = \partial_\emptyset^{\underline{n} - \{a_1\}} Z^j \longrightarrow \partial_{\{a_1\}}^{\underline{n}} Z^j \right).$$

The cube $\partial_{\{a_1\}}^{\underline{n}} Z^j$ is homotopy cocartesian because the pair $(U, R) = (\{a_1\}, \underline{n})$ satisfies the properties described above. Hence by 2(b) of Proposition 5.8.13, if $\partial_\emptyset^{\underline{n} - \{a_1\}} Z^j$ is $(K - 1)$-cocartesian then $\partial_\emptyset^{\underline{n}} Z^j$ is $K$-cocartesian. More generally, for $l < k$ we have

$$\partial_\emptyset^{\underline{n} - \{a_1, \ldots, a_{l-1}\}} Z^j = \left( \partial_\emptyset^{\underline{n} - \{a_1, \ldots, a_{l-1}, a_l\}} Z^j \longrightarrow \partial_{\{a_l\}}^{\underline{n} - \{a_1, \ldots, a_{l-1}\}} Z^j \right)$$

and $(U, R) = (\{a_l\}, \underline{n} - \{a_1, \ldots, a_{l-1}\})$ satisfies the above properties, which implies that $\partial_{\{a_l\}}^{\underline{n} - \{a_1, \ldots, a_{l-1}\}} Z^j$ is homotopy cocartesian and hence again by 2(b) of Proposition 5.8.13, $\partial_\emptyset^{\underline{n} - \{a_1, \ldots, a_{l-1}\}} Z^j$ is $(K - l + 1)$-cocartesian if $\partial_\emptyset^{\underline{n} - \{a_1, \ldots, a_{l-1}, a_l\}} Z^j$ is $(K - l)$-cocartesian. It follows that $Z^j$ is $K$-cocartesian if $\partial_\emptyset^{\underline{n} - \{a_1, \ldots, a_{k-1}\}} Z^j = \partial_\emptyset^{S_j \cup \{a_k\}} Z^j$ is $(K - n + |S_j| + 1)$-cocartesian.                                    □

**Lemma 6.4.6** *The cube* $\partial_{\emptyset}^{S_j \cup \{a_k\}} \mathcal{Z}^j$ *is* $(\min_{\{T_\beta\}} \{\sum_\beta k_{T_\beta}\} + |S_j|)$-*cocartesian, where* $\{T_\beta\}$ *ranges over all partitions of* $\underline{n}$ *which contain* $S_j$ *as one block and for which all other blocks are elements of* $\mathcal{A}_{j-1}$. *Hence* $\mathcal{Z}^j$ *is* $(-1 + n + (\min_{\{T_\beta\}} \{\sum_\beta k_{T_\beta}\}))$-*cocartesian, where* $\{T_\beta\}$ *is as above.*

*Proof* Once we establish how cartesian the faces $\partial_R^{S_j \cup \{a_k\}} \mathcal{Z}^j$, $R \subset \underline{n}$ are, this will follow from Theorem 6.2.4 for cubes of dimension less than $n$ since $|S_j| \leq n - 2$ for all $j$ using the estimates below. We must show the following.

1. If $|S_j \cup \{a_k\} - R| = 1$, then $\partial_R^{S_j \cup \{a_k\}} \mathcal{Z}^j$ is

$$
\begin{cases}
k_{\underline{n}-S_j}\text{-cartesian} & \text{if } R = S_j, \\
k_a\text{-cartesian} & \text{if } S_j \cup \{a_k\} - R = \{a\}, a \neq a_k.
\end{cases}
$$

2. If $|S_j \cup \{a_k\} - R| \geq 2$, then $\partial_R^{S_j \cup \{a_k\}} \mathcal{Z}^j$ is

$$
\begin{cases}
\infty\text{-cartesian} & \text{if } a_k \notin R, \\
\infty\text{-cartesian} & \text{if } a_k \in R, \underline{n} - (S_j \cup \{a_k\} - R) \notin \mathcal{A}_{j-1}, \\
k_{S_j \cup \{a_k\}-R}\text{-cartesian} & \text{if } a_k \in R, \underline{n} - (S_j \cup \{a_k\} - R) \in \mathcal{A}_{j-1}.
\end{cases}
$$

For the first item, if $R = S_j$, then

$$
\begin{aligned}
\partial_{S_j}^{S_j \cup \{a_k\}} \mathcal{Z}^j &= \left( \mathcal{Z}^j(S_j) \longrightarrow \mathcal{Z}^j(S_j \cup \{a_k\}) \right) \\
&= \left( X^{\mathcal{A}_j}(S_j) \longrightarrow X^{\mathcal{A}_{j-1}}(S_j) \right) \\
&= \left( \lim_{U \in \mathcal{A}_j, S_j \subset U} X_U \longrightarrow \lim_{U \in \mathcal{A}_{j-1}, S_j \subset U} X_U \right) \\
&\simeq \left( \lim_{V \in \mathcal{P}(\underline{n}-S_j)} X_{V \cup S_j} \longrightarrow \lim_{V \in \mathcal{P}_0(\underline{n}-S_j)} X_{V \cup S_j} \right) \\
&\simeq \left( X_{S_j} \longrightarrow \lim_{V \in \mathcal{P}_0(\underline{n}-S_j)} X_{V \cup S_j} \right).
\end{aligned}
$$

The penultimate equivalence follows from the concavity of the $\mathcal{A}_i$. The last map is $k_{\underline{n}-S_j}$-connected by hypothesis.

If $S_j \cup \{a_k\} - R = \{a\}$ for some $a \neq a_k$, then

$$
\begin{aligned}
\partial_R^{S_j \cup \{a_k\}} \mathcal{Z}^j &= \left( \mathcal{Z}^j(R) \longrightarrow \mathcal{Z}^j(S_j \cup \{a_k\}) \right) \\
&= \left( X^{\mathcal{A}_{j-1}}(S_j - \{a\}) \longrightarrow X^{\mathcal{A}_{j-1}}(S_j) \right)
\end{aligned}
$$

$$= \left( \lim_{U \in \mathcal{A}_{j-1}, S_j - \{a\} \subset U} X_U \longrightarrow \lim_{U \in \mathcal{A}_{j-1}, S_j \subset U} X_U \right)$$

$$\simeq \left( \lim_{U \in \mathcal{A}_{j-1}, S_j - \{a\} \subset U} X_U \longrightarrow \lim_{V \in \mathcal{P}_0(\underline{n} - S_j)} X_{V \cup S_j} \right).$$

The last equivalence follows from the concavity of $\mathcal{A}_{j-1}$. We must show that this last map is $k_a$-connected. Choose an increasing sequence of concave sets

$$\{V \in \mathcal{A}_{j-1} : V \supset S_j\} = \mathcal{B}_1 \subset \cdots \subset \mathcal{B}_l = \{V \in \mathcal{A}_{j-1} : V \supset S_j - \{a\}\}$$

such that $\mathcal{B}_i - \mathcal{B}_{i-1}$ consists of a single set $R_i$. By Lemma 6.4.3, the map $\lim_{\mathcal{B}_i} X \to \lim_{\mathcal{B}_{i-1}} X$ is $k_{\underline{n} - R_i}$-connected. For some $i_0$, we have $R_{i_0} = \underline{n} - \{a\}$. In this case, the map $\lim_{\mathcal{B}_{i_0}} X \to \lim_{\mathcal{B}_{i_0 - 1}} X$ is $k_a$-connected. Moreover, $\{a\} \not\subset R_i$ for all $i$, and thus $R_i \subset \underline{n} - \{a\}$ for all $i$. Using that $k_a \leq k_{\underline{n} - R_i}$ for all $i$, we see that the composed $\lim_{\mathcal{B}_l} X \to \lim_{\mathcal{B}_1} X$ is $k_a$-connected.

For the second item, when $|S_j \cup \{a_k\} - R| \geq 2$, there are two cases, both of which claim something about the connectivity of the map

$$\mathcal{Z}^j(R) \longrightarrow \lim_{R \subsetneq U \subset S_j \cup \{a_k\}} \mathcal{Z}^j(U). \tag{6.4.4}$$

First suppose that $a_k \notin R$. In this case the connectivity of the map in Equation (6.4.4) is equal to how cartesian the square

$$
\begin{array}{ccc}
\mathcal{Z}^j(R) & \longrightarrow & \lim\limits_{R \subsetneq U \subset S_j} \mathcal{Z}^j(U) \\[1em]
\downarrow & & \downarrow \\[1em]
\mathcal{Z}^j(R \cup \{a_k\}) & \longrightarrow & \lim\limits_{R \subsetneq U \subset S_j} \mathcal{Z}^j(U \cup \{a_k\})
\end{array}
$$

is. Using the definitions of $\mathcal{Z}^j$ and $X^{\mathcal{A}}$, this square is equal to the square

$$
\begin{array}{ccc}
\lim\limits_{V \in \mathcal{A}_j, V \supset R} X_V & \longrightarrow & \lim\limits_{R \subsetneq U \subset S_j} \lim\limits_{V \in \mathcal{A}_j, V \supset R} X_V \\[1em]
\downarrow & & \downarrow \\[1em]
\lim\limits_{V \in \mathcal{A}_{j-1}, V \supset R} X_V & \longrightarrow & \lim\limits_{R \subsetneq U \subset S_j} \lim\limits_{V \in \mathcal{A}_{j-1}, V \supset R} X_V
\end{array}
$$

Since $\mathcal{A}_j = \mathcal{A}_{j-1} \cup \{S_j\}$, for $V \in \mathcal{A}_j$ such that $V$ contains $R$, then either $V \in \mathcal{A}_{j-1}$ or $V = S_j$ (since $\mathcal{A}_{j-1}$ contains all subsets which contain $S_j$ as a subset). But in this case, since $a_k \notin R$, $R$ is a subset of $S_j$, and since $|S_j \cup \{a_k\} - R| \geq 2$, $S_j$ contains $R$ as a proper subset. Hence $R \subsetneq U \subset V$ and $V \in \mathcal{A}_j$. It follows that $\{V \in \mathcal{A}_j : V \supset R\}$ is covered by the concave subcategories $\{V \in \mathcal{A}_{j-1} : V \supset R\}$

and $\{V \in \mathcal{A}_j : V \supsetneq R\}$, and by Lemma 6.4.2, the square is homotopy cartesian. Hence the map in Equation (6.4.4) is a weak equivalence.

Now suppose $a_k \in R$. In this case, using the definition of $\mathcal{Z}^j$ and $X^{\mathcal{A}_{j-1}}$, Equation (6.4.4) can be written as

$$\lim_{V \in \mathcal{A}_{j-1}, V \supset R - \{a_k\}} X_V \longrightarrow \lim_{R \subseteq U \subset S_j \cup \{a_k\}} \lim_{V \in \mathcal{A}_{j-1}, V \supset U - \{a_k\}} X_V.$$

Observe that if $V \in \mathcal{A}_{j-1}$ contains $R - \{a_k\}$ but does not contain $R - \{a_k\} \cup U$ for all $\emptyset \ne U \subset S_j \cup \{a_k\} - R$, then $\underline{n} - V$ contains $S_j \cup \{a_k\} - R$. That is, for such $V$, we have $V \subset \underline{n} - (S_j \cup \{a_k\} - R)$. Hence if $\underline{n} - (S_j \cup \{a_k\} - R) \notin \mathcal{A}_{j-1}$, then $V \notin \mathcal{A}_{j-1}$ by concavity, so there can be no such $V$ and the map in Equation (6.4.4) must be a weak equivalence. If $\underline{n} - (S_j \cup \{a_k\} - R) \in \mathcal{A}_{j-1}$, then choose an increasing sequence of concave sets

$$\mathcal{B}_1 \subset \cdots \subset \mathcal{B}_l$$

such that

$$\mathcal{B}_1 = \{V \in \mathcal{A}_{j-1} : U \supset R - \{a_k\}\},$$

$$\mathcal{B}_l = \{V \in \mathcal{A}_{j-1} : V \supset R - \{a_k\} \cup U \text{ for } \emptyset \ne U \subset S_j \cup \{a_k\} - R\},$$

and such that $\mathcal{B}_i - \mathcal{B}_{i-1}$ consists of a single set $R_i$. By the above, $\underline{n} - R_i$ must contain $S_j \cup \{a_k\} - R$ for all $i$, and moreover it must equal $\underline{n} - (S_j \cup \{a_k\} - R)$ for some $i$. Using $k_U \le k_T$ for $U \subset T$, we see that the composed map in the sequence above is $k_{S_j \cup \{a_k\} - R}$-connected. $\quad\square$

## 6.5 Homotopy groups of cubes

As we remarked in Section 4.6, the Blakers–Massey Theorem gives us a vanishing range for homotopy groups of the total fiber of a space, but does not tell us anything about the first non-trivial homotopy group. The focus in this section will be to say something about the first non-trivial homotopy groups; better yet, we will describe a space-level version of the homotopy type of the total homotopy fiber in some special cases.

We begin by finishing what we started in Section 4.6 and first complete the discussion of the first non-trivial homotopy group of an arbitrary homotopy cocartesian square. Recall that the special case of this was considered in Proposition 4.6.1. We will now use the three-dimensional version of Theorem 6.2.1 to prove the following.

**Proposition 6.5.1**   *Suppose*

$$\mathcal{X} = \begin{array}{ccc} X_\emptyset & \longrightarrow & X_1 \\ \downarrow & & \downarrow \\ X_2 & \longrightarrow & X_{12} \end{array}$$

*is a pushout square of inclusions of CW complexes where* $(X_i, X_\emptyset)$ *is* $k_i$-*connected for* $i = 1, 2$, *and additionally assume* $X_\emptyset$ *is* $k$-*connected, where* $k \geq 2$. *Then the square is* $(k_1 + k_2 - 1)$-*cartesian and*

$$\pi_{k_1+k_2-1}(\mathrm{tfiber}(\mathcal{X})) \cong \pi_{k_1+1}(X_1/X_\emptyset) \otimes \pi_{k_2+1}(X_2/X_\emptyset).$$

*Proof*   The strategy is to reduce to Proposition 4.6.1. Consider the 3-cube

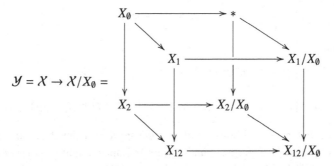

where $\mathcal{X}$ and $\mathcal{X}/X_\emptyset$ denote the left and right faces of this cube. The 3-cube $\mathcal{Y}$ is homotopy cocartesian by 1(b) of Proposition 5.8.13 because it is a map $\mathcal{Y} = \mathcal{X} \to \mathcal{X}/X_\emptyset$ of homotopy cocartesian squares, the latter of which is homotopy cocartesian using Proposition 5.8.12. Evidently the top and back faces are homotopy cocartesian as well, and it then follows from 1(a) of Proposition 5.8.13 that the 3-cube is strongly homotopy cocartesian. Since $X_\emptyset \to *$ is $(k + 1)$-connected and $X_\emptyset \to X_i$ are $k_i$-connected for $i = 1, 2$, the cube is $(k_1 + k_2 + k - 2)$-cartesian by Theorem 6.2.1. In particular, using Proposition 5.5.4 we see that the map $\mathrm{tfiber}(\mathcal{X}) \to \mathrm{tfiber}(\mathcal{X}/X_\emptyset)$ is $(k_1 + k_2 + k - 2)$-connected, and so $\pi_{k_1+k_2-1}(\mathrm{tfiber}(\mathcal{X})) \cong \pi_{k_1+k_2-1}(\mathrm{tfiber}(\mathcal{X}/X_\emptyset))$ provided $k \geq 2$. The result now follows from Proposition 4.6.1.                                                                    □

Note that the right face in the 3-cube appearing in the last proof is homotopy cocartesian and hence $X_{12}/X_\emptyset \simeq (X_1/X_\emptyset) \vee (X_2/X_\emptyset)$. We can similarly reduce the case of a general strongly homotopy cocartesian cube to the case of a cube of wedges, again under a mild hypothesis.

**Proposition 6.5.2** *Suppose* $X = (S \mapsto X_S)$ *is a strongly homotopy cocartesian n-cube of inclusions of CW complexes, that* $(X_i, X_\emptyset)$ *is* $k_i$-connected for each $i \in \underline{n}$, and that $X_\emptyset$ is $k$-connected with $k \geq 2$. Then the cube $X$ is $(-1 + n + \sum_{i \in \underline{n}} k_i)$-cartesian, and*

$$\pi_{1-n+\sum_{i \in \underline{n}} k_i}(\mathrm{tfiber}(X)) \cong \pi_{1-n+\sum_{i \in \underline{n}} k_i}(\mathrm{tfiber}(\mathcal{Y})),$$

*where* $\mathcal{Y} = (S \mapsto Y_S)$ *is a strongly homotopy cocartesian n-cube with* $Y_\emptyset = *$, $Y_i = X_i/X_\emptyset$, *and* $Y_S = \vee_{i \in S} Y_i$ *for* $|S| \geq 2$.

*Proof* We sketch the proof, which is similar to the previous one. By Theorem 6.2.1 the $n$-cube $X$ is $(1 - n + \sum_{i \in \underline{n}} k_i)$-cartesian, and so its first possibly non-trivial homotopy group occurs in dimension $-1 + n + \sum_{i \in \underline{n}} k_i$.

The cube $X/X_\emptyset = S \mapsto X_S/X_\emptyset$ is homotopy cocartesian by Proposition 5.8.12. In fact it is strongly homotopy cocartesian by the same result, because $X$ is itself strongly homotopy cocartesian. Define $\mathcal{Y} = (S \mapsto Y_S)$ by $Y_\emptyset = *$, $Y_i = X_i/X_\emptyset$, and $Y_S = \vee_{i \in S} Y_i$ for $|S| \geq 2$. Evidently $X/X_\emptyset \simeq \mathcal{Y}$ since $X_S \simeq Y_S$ for all $S \geq 2$ because $X$ and $\mathcal{Y}$ are strongly homotopy cocartesian.

We have an evident map $X \to X/X_\emptyset$ which makes a homotopy cocartesian $(n + 1)$-cube by 1(b) of Proposition 5.8.13. We claim that it is in fact strongly homotopy cocartesian. This follows from 1(a) of Proposition 5.8.13, the fact that $X$ is strongly homotopy cocartesian, and the fact that squares of the form

are homotopy cocartesian for all $i \in \underline{n}$. By Theorem 6.2.1, the $(n+1)$-cube $X \to X/X_\emptyset$ is $(-n + k + \sum_{i \in \underline{n}} k_i)$-cartesian. By Proposition 5.5.4 the map $\mathrm{tfiber}(X) \to \mathrm{tfiber}(X/X_\emptyset) \simeq \mathrm{tfiber}(\mathcal{Y})$ is then $(-n + k + \sum_{i \in \underline{n}} k_i)$-connected, and in particular is an isomorphism on $\pi_{1-n+\sum_i k_i}$ since $k \geq 2$. $\qquad\square$

Note that the statement of Proposition 6.5.2 does not include a computation of the first non-trivial homotopy group. The answer is a bit more complicated, but we can give a glimpse of it here. The above result suggests studying cubes which are comprised of wedges. We have the following stable-range description of the homotopy type of $\mathrm{tfiber}(X)$ when $X$ is made up of a wedge of suspensions, a generalization of Theorem 4.6.3.

**Theorem 6.5.3**   *Let $X_1 \ldots, X_n$ be based spaces, where $X_i$ is $k_i$-connected. Consider the strongly (homotopy) cocartesian n-cube*

$$X(S) = \bigvee_{i \in \underline{n}-S} \Sigma X_i.$$

*There exists a $(\min_i\{k_i\} + 1 + n + \sum_{i=1}^{n} k_i)$-connected map*

$$P: \bigvee_{(n-1)!} \Sigma(X_1 \wedge \cdots \wedge X_n) \longrightarrow \mathrm{tfiber}(X).$$

This follows from Theorem 6.5.7, though we do not prove that result here. The $n$-cube $X$ in question is $(1 + n + \sum_{i=1}^{n} k_i)$-cartesian by Theorem 6.2.1, and hence $\mathrm{tfiber}(X)$ is $(n + \sum_{i=1}^{n} k_i)$-connected, and so we may, as in Theorem 4.6.3, view $P$ as giving a stable-range description of $\mathrm{tfiber}(X)$ in homotopy. The map $P$ in the statement is the "total generalized Whitehead product", to be discussed below. It is built from iterating the commutator construction (see Definition 4.6.4). We will spend most of our time outlining the definition of the map $P$ and only briefly discuss its connectivity, as establishing this would require more work than we can allot this topic.

Recall the map $W: \Sigma(X_1 \wedge X_2) \to \Sigma(X_1 \vee X_2)$ from Definition 4.6.5. We want to iterate these generalized Whitehead product maps. Also recall the map $\ell: X \to \Omega\Sigma X$ for a based space $X$, which sends $x$ to the path $t \mapsto (t \wedge x)$, as well as the commutator map of Definition 4.6.4: for based spaces $X$ and $Y$, we have a map

$$C: \Omega X \times \Omega Y \longrightarrow \Omega(X \vee Y)$$
$$(\alpha, \beta) \longmapsto \alpha * \beta * \alpha^{-1} * \beta^{-1}.$$

The commutator map can be iterated as follows.

**Definition 6.5.4**   Let $X_1, \ldots, X_n$ be a based spaces, and let $\sigma \in \Sigma_n$ be a permutation of $\underline{n}$. Define

$$\widehat{C}_\sigma: \prod_{i \in \underline{n}} X_i \longrightarrow \Omega\Sigma \bigvee_{i \in \underline{n}} X_i$$

by

$$\widehat{C}_\sigma(x_1, \ldots, x_n) = C\left(\ell(x_{\sigma^{-1}(1)}), C(\ell(x_{\sigma^{-1}(2)}),\right.$$
$$\left. C(\cdots, C(\ell(x_{\sigma^{-1}(n-1)}), \ell(x_{\sigma^{-1}(n)})) \cdots ))\right).$$

Johnson [Joh95, Definition 6.6] defines a map

$$C_\sigma : \prod_{i \in \underline{n}} X_i \longrightarrow \Omega \, \text{tfiber}\left(S \mapsto \bigvee_{i \in \underline{n}-S} \Sigma X_i\right)$$

such that the composition

$$\prod_{i \ni \underline{n}} X_i \xrightarrow{C_\sigma} \Omega \, \text{tfiber}\left(S \mapsto \bigvee_{\underline{n}-S} \Sigma X_i\right) \xrightarrow{p_i} \Omega \Sigma \bigvee_{i \in \underline{n}} X_i$$

with the canonical projection $p_i$ is equal to the map $\widehat{C}_\sigma$ in Definition 6.5.4. We omit the details.

**Proposition 6.5.5** $C_\sigma$ *induces a map* $D_\sigma : \bigwedge_{i=1}^n X_i \rightarrow \Omega \, \text{tfiber}\left(S \mapsto \bigvee_{\underline{n}-S} \Sigma X_i\right)$.

*Proof* By symmetry it suffices to consider the case $\sigma = \iota$, the identity permutation. Define an $n$-cube $R \mapsto \prod_{i \in R} X_i$, where we regard $\prod_{i \in R} X_i \subset \prod_{i=1}^n X_i$ as the subspace of tuples $(x_1, \ldots, x_n)$ such that $x_j$ is the basepoint if $j \notin R$. The restriction of $C_\iota : \prod_{i=1}^n X_i$ to $\prod_{i \in R} X_i$ maps to $\Omega \, \text{tfiber}\left(S \mapsto \bigvee_{\underline{n} \cap R - S \cap R} \Sigma X_i\right)$. Thus we have a map of $n$-cubes

$$\left(R \mapsto \prod_{i \in R} X_i\right) \longrightarrow \left(R \mapsto \Omega \, \text{tfiber}\left(S \mapsto \bigvee_{\underline{n} \cap R - S \cap R} \Sigma X_i\right)\right),$$

which induces a map of total homotopy cofibers (recall Definition 5.9.1)

$$\text{tcofiber}\left(R \mapsto \prod_{i \in R} X_i\right) \longrightarrow \text{tcofiber}\left(R \mapsto \Omega \, \text{tfiber}\left(S \mapsto \bigvee_{\underline{n} \cap R - S \cap R} \Sigma X_i\right)\right).$$

From Example 5.9.4 we have that $\text{tcofiber}\,(R \mapsto \prod_{i \in R} X_i)$ is weakly equivalent to $\bigwedge_{i=1}^n X_i$. If $R$ is a proper subset of $\underline{n}$, the space $\Omega \, \text{tfiber}\left(S \mapsto \bigvee_{\underline{n} \cap R - S \cap R} \Sigma X_i\right)$ is contractible by Example 5.4.4 because two faces of the cube in question are identical. Therefore

$$\text{tcofiber}\left(R \mapsto \Omega \, \text{tfiber}\left(S \mapsto \bigvee_{\underline{n} \cap R - S \cap R} \Sigma X_i\right)\right) \simeq \Omega \, \text{tfiber}\left(S \mapsto \bigvee_{\underline{n}-S} \Sigma X_i\right),$$

and so we have an induced map

$$D_\iota : \bigwedge_{i=1}^n X_i \longrightarrow \Omega \, \text{tfiber}\left(S \mapsto \bigvee_{\underline{n}-S} \Sigma X_i\right).$$

$\square$

The map $D_\sigma$ described above is associated via Theorem 1.2.7 with a map

$$\widetilde{D}_\sigma : \Sigma \bigwedge_{i=1}^{n} X_i \longrightarrow \text{tfiber}\left(S \mapsto \bigvee_{\underline{n}-S} \Sigma X_i\right)$$

which we will call a *generalized Whitehead product*. Finally, we define the map $P$ as follows. Let $\Sigma_{n-1} \leq \Sigma_n$ denote the subgroup of all permutations which fix $1 \in \underline{n}$.

**Definition 6.5.6**   Define

$$P: \bigvee_{\sigma \in \Sigma_{n-1}} \Sigma \bigwedge_{i=1}^{n} X_i \longrightarrow \text{tfiber}\left(S \mapsto \bigvee_{i \in \underline{n}-S} \Sigma X_i\right)$$

by $P = \vee_{\sigma \in \Sigma_{n-1}} \widetilde{D}_\sigma$. We call $P$ the *total generalized Whitehead product*.

One approach to proving that $P$ has a high connectivity similar to that outlined in Section 4.6 is to once again show that it fits into a commutative diagram of highly connected maps as below. The following result is due to the first author, who owes a debt to Koschorke's work [Kos97, Theorem 3.1] as well as that of Johnson [Joh95].

**Theorem 6.5.7** ([Mun11, Theorem 4.3])   *Let $X_1 \ldots, X_n$ be based spaces, and suppose $X_i$ is $k_i$-connected. There is a commutative (up to homotopy) diagram*

$$\text{tfiber}(S \mapsto \vee_{i \in \underline{n}-S} \Sigma X_i)$$

$$\vee_{(n-1)!} \Sigma \wedge_{i=1}^{n} X_i \xrightarrow{\quad E \quad} \prod_{(n-1)!} \Omega^{n-1}\Sigma^{n-1}(\Sigma(X_1 \wedge \cdots \wedge X_n))$$

*in which $E$ is $(1 + 2\sum_{i \in \underline{n}} k_i)$-connected, $H_n$ is $(2 + n + \sum_{i=1}^{n} k_i + \min_i\{k_i\})$-connected, and hence $P$ is $(1 + n + \sum_{i=1}^{n} k_i + \min_i\{k_i\})$-connected.*

The construction of the map $H_n$ is significantly more complicated than in Theorem 4.6.6, and is due to Johnson [Joh95].

Another more direct approach is to use the Hilton–Milnor Theorem below to describe the first non-trivial group of a cube, which gives a description of the weak homotopy type of a wedge of suspensions as a weak product of smash products of the various summands. We will provide only scant details for this. The indexing set for the weak product is given by a basis for a free Lie algebra.

We start by introducing the terminology necessary for bookkeeping, following Neisendorfer [Nei10, Section 4.3].

**Definition 6.5.8** A *Hall basis* $B$ of the free Lie algebra on the symbols $x_1, \ldots, x_n$ is a countable union

$$B = \bigcup_{i=1}^{\infty} B_i$$

where $B_i \subset B_{i+1}$, $|B_{i+1} - B_i| = 1$, and the sets $B_i$ are defined by the following inductive procedure.

Define $L_1 = \{x_1, \ldots, x_n\}$, and $B_1 = \{x_1\}$, and put $B_0 = \emptyset$. For $n \geq 2$, inductively define

$$L_n = \{\mathrm{ad}(z_{n-1})^i(x) : i \geq 0, x \in L_{n-1}, x \neq z_{n-1}\}$$

where $z_{n-1}$ is the unique element in $B_{n-1} - B_{n-2}$, and $\mathrm{ad}(z) = [z, -]$ is the Lie bracket. Choose an ordering for $L_n$ in which elements of shortest length appear first (the length of an element is equal to the number of brackets appearing in that element's expression as an iterated bracket). Let $z_n \in L_n$ be the smallest element, and set $B_n = L_n \cup \{z_n\}$.

For instance, if $L = \{x_1, x_2, x_3\}$, it is easy to compute that $B_3 = \{x_1, x_2, x_3\}$ and if we assume $x_3 \in B_3 - B_2$ is the unique element, then

$$L_4 = \{[x_1, x_2], [x_2, x_3], [x_2, x_3], \text{ terms of length greater than 1}\},$$

and depending on the chosen ordering we will add one of the first three terms to the basis. The number of basis elements in which $x_i$ appears exactly $m_i$ times is equal to

$$\sum_{d \mid \gcd(m_1, \ldots, m_n)} \mu(d) \frac{\left(\frac{\sum_{i \in n} m_i}{d}\right)!}{\left(\frac{m_1}{d}\right)! \cdots \left(\frac{m_n}{d}\right)!},$$

where $\mu$ is the Möbius function. In particular, the number of basis elements in which each $x_i$ appears exactly once is equal to $(n - 1)!$.

Suppose now that $X_1, \ldots, X_n$ are based spaces. If $\omega(x_1, \ldots, x_k)$ is an element of $B$ as above, we write $\omega(X_1, \ldots, X_n)$ for the corresponding space under the correspondence $x_i \leftrightarrow X_i$ and $[x_i, x_j] \leftrightarrow X_i \wedge X_j$.

**Theorem 6.5.9** (Hilton–Milnor Theorem)   *Let $X_1, \ldots, X_n$ be based spaces. There is a weak equivalence*

$$\prod_{\omega \in B}^{w} \Omega \Sigma \omega(X_1, \ldots, X_k) \longrightarrow \Omega \Sigma \left( \bigvee_{i \in \underline{n}} X_i \right).$$

We will not discuss the definition of the map in this theorem. See [Whi78, Section XI.6] for details. The superscript w on the product over $\omega \in B$ above is meant to indicate that this product is not topologized using the product topology, but rather the weak topology. That is, there are maps

$$\prod_{\omega \in B_n} \Omega \Sigma \omega(X_1, \ldots, X_n) \longrightarrow \prod_{\omega \in B_{n+1}} \Omega \Sigma \omega(X_1, \ldots, X_n)$$

given by the inclusion, using the basepoint on the factor which is omitted. The union over $n$ of these spaces is the infinite product above, and it is topologized as such an infinite union.

The weak equivalence in Theorem 6.5.9 is natural in the variables $X_1, \ldots, X_n$ in the sense that, for each $j \in \underline{n}$, the projection map $\Omega \Sigma (\bigvee_{i \in \underline{n}} X_i) \to \Omega \Sigma (\bigvee_{i \in \underline{n} - \{j\}} X_i)$ corresponds to the projection of weak products away from those factors indexed by elements $\omega(X_1, \ldots, X_k)$ in which the space $X_j$ appears in the corresponding smash product. Thus, using Example 5.5.5, we have a weak equivalence

$$\prod_{\omega \in B'} \Omega \Sigma \omega(X_1, \ldots, X_n) \longrightarrow \mathrm{tfiber}\left( S \mapsto \bigvee_{i \in \underline{n} - S} \Omega \Sigma X_i \right), \qquad (6.5.1)$$

where $B' \subset B$ consists of those expressions $\omega$ in which each variable $x_i$ appears at least once. It is easy to see from Equation (6.5.1) that $\mathrm{tfiber}(S \mapsto \bigvee_{i \in \underline{n} - S} \Omega \Sigma X_i) \simeq \Omega \mathrm{tfiber}(S \mapsto \bigvee_{i \in \underline{n} - S} \Sigma X_i)$ is $(-1 + n + \sum_{i=1}^{n} k_i)$-connected if $X_i$ is $k_i$-connected, exactly as predicted by Theorem 6.2.1 for this cube. Furthermore, one can additionally see from Equation (6.5.1) that the inclusion

$$\prod_{(n-1)!} \Omega \Sigma \bigwedge_{i \in \underline{n}} X_i \longrightarrow \prod_{\omega \in B'} \Omega \Sigma \omega(X_1, \ldots, X_n) \simeq \mathrm{tfiber}\left( S \mapsto \bigvee_{i \in \underline{n} - S} \Omega \Sigma X_i \right)$$

is $(\min\{k_i\} + -1 + n + \sum_{i=1}^{n} k_i)$-connected, as the least connectivity of the remaining factors in the weak product is equal to $\min\{k_i\} + n + \sum_{i=1}^{n} k_i$.

# PART II

Generalizations, Related Topics, and Applications

# 7

# Some category theory

In this chapter we give the basic definitions and examples from category theory that will be needed for the development of general homotopy (co)limits in Chapter 8. Most of this material is standard, and can be found for example in [Bor94, Hir03, ML98]. Our treatment focuses on categories with (co)products, as we believe these constructions are the most familiar. Of most interest to us are the definitions of the limit and colimit of a diagram (Section 7.3). We give the standard definitions of these objects using universal properties, but also offer alternative ways to think about them (Section 7.4) that are more suitable for our development of homotopy limits and colimits and match more closely with the way pullbacks and pushouts were defined in Chapter 3. We will give special attention to examples that have already appeared in Chapters 3 and 5.

## 7.1 Categories, functors, and natural transformations

**Definition 7.1.1**  A *category* $C$ consists of

1. a class of *objects* $\mathrm{Ob}(C)$;
2. for each pair of objects $X, Y \in \mathrm{Ob}(C)$ a set of *morphisms* $\mathrm{Hom}_C(X, Y)$;
3. for each $X \in \mathrm{Ob}(C)$ an *identity morphism* $1_X \in \mathrm{Hom}_C(X, X)$;
4. a *composition* function

$$\circ \colon \mathrm{Hom}_C(X, Y) \times \mathrm{Hom}_C(Y, Z) \longrightarrow \mathrm{Hom}_C(X, Z)$$

satisfying
(a) $f \circ 1_X = 1_Y \circ f = f$;
(b) $f \circ (g \circ h) = (f \circ g) \circ h$.

We will sometimes abuse notation and write $X \in C$ for $X \in \mathrm{Ob}(C)$. A morphism $f$ from $X$ to $Y$ is typically depicted as an arrow $f \colon X \to Y$, and we call $X$ the *source* and $Y$ the *target* of $f$.

Here is some further terminology:

- A *subcategory* $C'$ of a category $C$ is a category consisting of subclasses of the classes of objects and morphisms of a category $C$.
- A subcategory $C'$ is *full* if, for all $X, Y \in \mathrm{Ob}(C')$, $\mathrm{Hom}_{C'}(X, Y) = \mathrm{Hom}_{C}(X, Y)$.
- A *discrete category* is a category with no non-identity morphisms. Any set naturally gives rise to a discrete category with the elements of that set as the objects.
- A category is *finite* if both its class of objects and its class of morphisms are finite.
- A category is *small* if the classes of objects and morphisms are both sets. We will remind the reader when the categories we are working with are required to be small, but one standing assumption is that, whenever a category is used to index a (co)product or a diagram, it is assumed to be small.

**Example 7.1.2** (Poset as a category)    A poset (partially ordered set) $(P, \leq)$ is naturally a category. Its objects are elements of the underlying set $P$, and there is a unique morphism $a \to b$ for each relation $a \leq b$. For instance, let $(\mathcal{P}(\{1, 2\}), \subset)$ denote the poset of subsets of $\{1, 2\}$, where $\subset$ denotes the usual set containment. By omitting non-identity morphisms and those that factor as non-trivial compositions (i.e. a factorization without identity morphisms), we may depict this category as a square

Note that we did not draw in the identity morphisms. This is the standard convention that we will follow throughout.

More generally, the poset $(\mathcal{P}(\underline{n}), \subset)$ is a category which was of course central in Chapter 5 (we will drop "$\subset$" from the notation from now on). Its shape is that of an $n$-dimensional cube. The two familiar full subcategories (subposets) associated with $\mathcal{P}(\underline{n})$ are $\mathcal{P}_0(\underline{n})$, the category of non-empty subsets, and the category $\mathcal{P}_1(\underline{n})$ of proper subsets. The "simplex-like" depictions of these categories should also be familiar from Chapter 5.                    □

Here is a chart of some more basic examples. The ones we will mostly use are Top and Top$_*$, the categories of topological spaces and based topological spaces, respectively, with (based) maps as morphisms. Unless we need to, we will not make a distinction between Top and Top$_*$, although we will try to remind the reader in key places that most of our constructions and results work the same way in Top as in Top$_*$.

**Example 7.1.3**

| Category | Objects | Morphisms |
|---|---|---|
| Set | sets | functions |
| poset $(P, \leq)$ | elements of $P$ | $a \to b$ means $a \leq b$ |
| Vect$_F$ | vector spaces over a field $F$ | $F$-linear maps |
| Top | topological spaces | continuous maps |
| Top$_*$ | based topological spaces | based continuous maps |
| (Ab) Grp | (abelian) groups | homomorphisms |
| Ring | commutative rings | ring homomorphisms |
| Ch$_R$ | chain complexes over a ring $R$ | chain maps |
| a group $G$ | a single object | elements of $G$, composition according to group law |
| G-Set (for fixed group $G$) | sets with (left) $G$-action | equivariant $G$-functions |
| R-Mod (for fixed ring $R$) | $R$-modules | $R$-module maps |

□

**Example 7.1.4** The category Set is a full subcategory of Top (where sets are regarded as spaces with discrete topology). Because of this, the constructions we perform in Top, such as pushouts and pullbacks (as examples of (co)limits), will automatically hold for Set. □

**Remark 7.1.5** The categories Set and Top are not small, whereas the category $\mathcal{P}(\underline{n})$ is. Another small category is the category associated to a group $G$ where there is one object and a morphism for each group element. The important property that categories like Set and Top fortunately possess is that the limits of small diagrams in those categories always exist; see Section 7.3 for more details. □

**Definition 7.1.6** Given a category $C$, define the *opposite category of $C$* to be the category $C^{\text{op}}$ whose objects are those of $C$ and whose morphism sets are defined by

$$\text{Hom}_{C^{\text{op}}}(X, Y) = \text{Hom}_C(Y, X).$$

Heuristically, $C^{\mathrm{op}}$ is $C$ with the arrows "reversed". Thus if $C$ is the category $X \to Z \leftarrow Y$ (again remember that we do not draw in the identity morphisms) then $C^{\mathrm{op}}$ can be identified with the category $X \leftarrow Z \to Y$ in an evident way.

**Definition 7.1.7** Given categories $C$ and $\mathcal{D}$, define the *product category* $C \times \mathcal{D}$ to be the category whose objects are pairs $(X, Y)$ where $X \in \mathrm{Ob}(C)$ and $Y \in \mathrm{Ob}(\mathcal{D})$ and morphisms are pairs $(f, g)$ where $f$ is a morphism in $C$ and $g$ a morphism in $\mathcal{D}$.

**Example 7.1.8** We have already (secretly) considered the product of the category $\mathcal{P}_0(\{1, 2\})$ with itself in Theorem 3.3.15. We depict it as follows

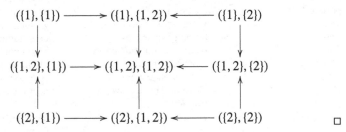

**Definition 7.1.9** A morphism $f \colon X \to Y$ in $C$ is an *isomorphism* if it has an *inverse*, that is, if there exists a morphism $g \colon Y \to X$ such that $g \circ f = 1_X$ and $f \circ g = 1_Y$. In this case we write $X \cong Y$.

**Example 7.1.10** The notion of an isomorphism of objects coincides with the usual notion of an isomorphism in the familiar categories. For example, in Set, an isomorphism is a bijection, in Top (resp. Top$_*$), it is a homoemorphism (resp. based homeomorphism), and in Grp and Ring, it is a bijective homomorphism. □

Here is a definition we will need later.

**Definition 7.1.11** A small category $C$ is said to be *acyclic* if only identity morphisms have inverses and $\mathrm{Hom}_C(X, X) = \{1_X\}$ for all objects $X \in C$.

It is easy to see that, for a finite category, acyclic means that there exists some $N \geq 1$ such that, for all compositions of morphisms $f_n \circ f_{n-1} \circ \cdots \circ f_1$ with $n > N$, at least one morphism is the identity. For example, a poset is acyclic if it is finite. We will have use for finite and acyclic categories in Chapter 9.[1]

---

[1] In the literature, a finite and acyclic category is also sometimes referred to as "very small".

**Definition 7.1.12** An object $X_i \in C$ is an *initial object* if there exists a unique morphism $X_i \to X$ for every object $X$ in $C$. Dually, an object $X_f \in C$ is a *terminal*, or *final*, *object* if there exists a unique morphism $X \to X_f$ for every object $X$ in $C$.

Initial and final objects, when they exist, are unique up to isomorphism. This follows directly from the definitions, but it is important in showing that objects defined using a universal property (such as limits and colimits) are unique up to isomorphism.

**Example 7.1.13** The empty set $\emptyset$ is an initial object in $\mathcal{P}_1(\underline{n})$ (and in $\mathcal{P}(\underline{n})$) and $\underline{n}$ is a final object in $\mathcal{P}_0(\underline{n})$ (and in $\mathcal{P}(\underline{n})$). The category $\mathcal{P}_0(\underline{n})$ has no initial object and $\mathcal{P}_1(\underline{n})$ has no final object. □

**Example 7.1.14** In Set and Top, the empty set $\emptyset$ is an initial object, and the one-point set $*$ is a terminal object. The one-point set $*$ is both initial and terminal in the pointed category Top$_*$. In Grp and Ring, the trivial group/ring is both an initial and final object. An initial/terminal object in $C$ is a terminal/initial object in $C^{\mathrm{op}}$. □

We spent Chapters 3 and 5 studying diagrams of shape $\mathcal{P}_0(\underline{n})$ and $\mathcal{P}_1(\underline{n})$ in Top and Top$_*$. This is encoded in the notion of a functor.

**Definition 7.1.15** A *(covariant) functor* $F: C \to \mathcal{D}$ between categories $C$ and $\mathcal{D}$ is a function which

1. associates to each object $X \in C$ an object $F(X) \in \mathcal{D}$;
2. associates to each morphism $f \in \mathrm{Hom}_C(X, Y)$ a morphism $F(f) \in \mathrm{Hom}_{\mathcal{D}}(F(X), F(Y))$ such that
   (a) $F(1_X) = 1_{F(X)}$ for all $X \in \mathrm{Ob}(C)$,
   (b) $F(g \circ f) = F(g) \circ F(f)$ for all $f \in \mathrm{Hom}_C(X, Y)$ and $g \in \mathrm{Hom}_C(Y, Z)$.

A *contravariant functor* is a covariant functor whose domain is $C^{\mathrm{op}}$. We often think of contravariant functors as those that "reverse arrows" (i.e. satisfy the obvious alterations to the above axioms).

**Remark 7.1.16** A *diagram* in a category $C$ is a functor $F$ from a small category $\mathcal{I}$ to $C$. The point of introducing such terminology is that it is both common in this text and the literature, and moreover it emphasizes the "picture" of the functor, and hence the shape of the indexing category, which

we shall later see plays an important role in defining homotopy (co)limits. See
Section 8.1 and Definition 8.1.8 in particular.                                    □

Functors can be thought of as functions from one category to another that
respect the categorical structure. When we speak in general of a functor without mentioning its variance or some other context, it is understood to be
covariant.

**Example 7.1.17**  Here are some standard examples of functors. Unless
otherwise indicated, they are covariant.

- $X: \mathcal{P}(\underline{n}) \to$ Top, $X(S) = X_S$, is a cubical diagram, or $n$-cubical diagram,
  or simply an $n$-cube, of topological spaces, and functors whose domain is
  $\mathcal{P}_0(\underline{n})$ or $\mathcal{P}_1(\underline{n})$ are called punctured $n$-cubical diagrams, or just punctured
  $n$-cubes.
- $I_C: C \to C$ taking objects and morphisms to themselves (identity functor).
- If $C$ is a discrete category then a functor $F: C \to \mathcal{D}$ is a collection of objects
  of $\mathcal{D}$ indexed by the objects of $C$.
- $F:$ Set $\to$ Set, taking a set to its power set.
- $F:$ Vect$_F \to$ Vect$_F$ taking $V$ to $V^*$, the dual of $V$ (contravariant).
- The inclusion functor Set $\to$ Top which regards a set as a space with the
  discrete topology.
- The forgetful functor $F:$ Top $\to$ Set, which regards a space as its underlying
  point-set.
- $\pi_k:$ Top$_* \to$ Grp taking a based space to its $k$th homotopy group.
- $H_n:$ Top $\to$ Ab, taking a space to its $n$th homology group; dually
  $H^n:$ Top $\to$ Ab, taking a space to its $n$th cohomology group (contravariant).
- For a fixed space $Z$, Map$(-, Z):$ Top $\to$ Top taking a space $X$ to the space
  of maps Map$(X, Z)$ (contravariant), and Map$(Z, -):$ Top $\to$ Top taking a
  space $X$ to the space of maps Map$(Z, X)$ (covariant). Similarly for based
  versions.                                                                         □

**Definition 7.1.18**  Functors $F: C \to \mathcal{D}$ and $G: \mathcal{D} \to C$ are *adjoint* if there is
a natural bijection

$$\text{Hom}_{\mathcal{D}}(G(X), Y) \cong \text{Hom}_C(X, F(Y))$$

for every object $X \in C$ and every object $Y \in \mathcal{D}$. Functor $F$ is called the *right
adjoint of* $G$ and $G$ is called the *left adjoint of* $F$.

**Example 7.1.19** The functor that associates a free abelian group to a set and the forgetful functor that forgets the group structure in a group are adjoint. □

**Example 7.1.20** (Loop and suspension are adjoint) The functor $\Omega$ that takes a based space to its loop space and the functor $\Sigma$ that takes a based space to its suspension are adjoint functors on $\text{Top}_*$. This is the content of Theorem 1.2.9. (These two are functors because a map can be looped and suspended; see (1.1.2) and (1.2.1).) More generally, by Theorem 1.2.7 we have that, for a fixed space $X$, functors $X \times -$ and $\text{Map}(X, -)$ are adjoint. □

In order to talk about the homotopy invariance of the homotopy limit of a punctured square, we required the notion of a "map of diagrams". Here is the formal definition.

**Definition 7.1.21** A *natural transformation* (or *map of diagrams*) $N : F \to G$ between functors $F, G : C \to \mathcal{D}$ associates to each $X \in \text{Ob}(C)$ a morphism $N_X : F(X) \to G(X)$ such that for every morphism $f : X \to Y$ the diagram

$$
\begin{array}{ccc}
F(X) & \xrightarrow{F(f)} & F(Y) \\
\Big\downarrow{\scriptstyle N_X} & & \Big\downarrow{\scriptstyle N_Y} \\
G(X) & \xrightarrow[G(f)]{} & G(Y)
\end{array}
$$

commutes.

If $F$ and $G$ are both contravariant, the horizontal arrows in the above square are reversed. We will denote the class of natural transformations between functors $F$ and $G$ by $\text{Nat}(F, G)$.

We will at times need to speak of a natural transformation $N : F \to G$ between functors $F, G : C \to \text{Top}$ (or $\text{Top}_*$) which is a "fibration", "cofibration", "weak equivalence", or some other property a map of spaces may possess. All this means is that, for each $c \in C$, the map $F(c) \to G(c)$ induced by $N$ is a fibration, cofibration, weak equivalence, etc. Other terminology we use which further emphasizes the point is to call a natural transformation a "pointwise", or "objectwise fibration, cofibration, weak equivalence, etc.".

**Example 7.1.22** Consider the category $\mathcal{P}(\emptyset)$ of subsets of the empty set. It has a single object and no non-identity morphisms. Functors $F, G : \mathcal{P}(\emptyset) \to \text{Top}$ are simply topological spaces, and if we put $F(\emptyset) = X_\emptyset$ and $G(\emptyset) = Y_\emptyset$, then a natural transformation from $F$ to $G$ amounts to a map $X_\emptyset \to Y_\emptyset$ of spaces.

Next consider the category $\mathcal{P}(\underline{1})$ of subsets of $\underline{1}$, which we may depict as $\emptyset \to \{1\}$. Let $F, G : \mathcal{P}(\underline{1}) \to \text{Top}$ be functors and put $F(\emptyset) = X_\emptyset$, $F(\{1\}) = X_1$, $G(\emptyset) = Y_\emptyset$, and $G(\{1\}) = Y_1$. Then a natural transformation from $F$ to $G$ is a commutative square of spaces

**Example 7.1.23** (Constant diagram)  If a functor $F$ sends every object of a small category $\mathcal{I}$ to an object $X$ of $C$ and every morphism to the identity morphism, we will refer to it as the *constant diagram at $X$* and will denote it by $C_X$. For example, if $C$ is Top and $X$ the one-point space $*$, the constant functor at a point will correspondingly be denoted by $C_*$. We have of course seen the constant punctured cubes already (see Example 5.1.4). Every functor $F : \mathcal{I} \to \text{Top}$ admits a unique natural transformation to $C_*$. □

**Definition 7.1.24**

- A *natural isomorphism* $I : F \to G$ of functors $F, G : C \to \mathcal{D}$ is a natural transformation for which there exists an inverse, namely a natural transformation $J : G \to F$ such that $J \circ I = 1_F$ and $I \circ J = 1_G$. Here $1_F$ and $1_G$ are the identity natural transformations from $F$ to $F$ and $G$ to $G$, respectively,
- We say categories $C$ and $\mathcal{D}$ are *equivalent* and write $C \cong \mathcal{D}$ if there exist functors $F : C \to \mathcal{D}$ and $G : \mathcal{D} \to C$ such that compositions $G \circ F : C \to C$ and $F \circ G : \mathcal{D} \to \mathcal{D}$ are naturally isomorphic to the identity funtors $I_C$ and $I_\mathcal{D}$, respectively. We say $C$ and $\mathcal{D}$ are *isomorphic* if the composites above are equal to the respective identity functors.

**Remark 7.1.25**  In general, $C$ and $C^{\text{op}}$ are not necessarily isomorphic; the obvious candidate for an isomorphism, which is the identity on objects, is not a covariant functor and thus cannot be an isomorphism by definition. □

**Example 7.1.26**  The linear transformation taking a vector space $V$ to its double dual, $(V^*)^*$, is a natural transformation between the identity functor and the double dual functor. This natural transformation is an isomorphism if $V$ if finite-dimensional. □

**Example 7.1.27** There is an isomorphism of categories $\mathcal{P}(\underline{n}) \cong \mathcal{P}(\underline{n})^{\mathrm{op}}$ given by $S \mapsto \underline{n} - S$. This functor is its own inverse. □

**Example 7.1.28** The categories $\mathcal{P}(\underline{n}) \times \mathcal{P}(\underline{1})$ and $\mathcal{P}(\underline{n+1})$ are isomorphic. We leave it to the reader to fill in the straightforward argument. □

We finish with two "meta" examples of categories that we will use.

**Example 7.1.29** (Category of small categories) The collection of small categories, where morphisms are functors between them is itself a category, called the *category of small categories*, and denoted by Cat. (We require small categories to avoid set-theoretic problems.) □

**Example 7.1.30** (Functor category) The collection of all functors between categories $C$ and $\mathcal{D}$ can itself be thought of as a category, called *functor category* and denoted by $\mathcal{D}^C$. The objects of this category are functors $F$ from $C$ to $\mathcal{D}$ and the morphisms are natural transformations of functors. The class of natural tranformation from a functor $F$ to a functor $G$ is denoted $\mathrm{Nat}_C(F, G)$ or just $\mathrm{Nat}(F, G)$ when $C$ is understood. □

## 7.2 Products and coproducts

In this section, we discuss products and coproducts from the categorical viewpoint, since they are the building blocks for limits and colimits. We have already seen how products and coproducts in Top play a role in building limits and colimits of punctured squares and cubes (see Definitions 5.2.1 and 5.6.1).

When discussing limits and colimits in a general category we will always assume that products and coproducts exist. In specific instances, such as in Top, we must prove (co)limits exist. This will be easy to do once we provide models for (co)limits built from (co)products. As our eventual goal is to discuss homotopy (co)limits, the reader may wonder why we do not just axiomatize the properties necessary to do "abstract" homotopy theory and work in a general codomain category that has all the properties of Top that we need (i.e. so that we can do "abstract" homotopy theory). There are many compelling reasons to do this, but one reason not to is to avoid more machinery. Our focus is to introduce (homotopy) (co)limits in as elementary and familiar a setting as possible. In addition, there are already many references on abstract homotopy theory [BK72a, DHKS04, Hir03, Hov99].

We start our discussion of (co)products in the relatively simple setting of the category of sets. One first notices that the product and the disjoint union of sets enjoy what is called a *universal property*. For instance, for sets $S$ and $T$, the disjoint union $S \sqcup T$ has the property that, for every set $R$ and every pair of maps $f: S \to R$, $g: T \to R$, there exists a unique map $H: S \sqcup T \to R$ compatible with the inclusions of $S$ and $T$ in the disjoint union. It is defined by the formula $h(s) = f(s)$ if $s \in S$ and $h(t) = g(t)$ if $t \in T$. Second, this map $H$ is also compatible with the maps $f$ and $g$ in the sense that $H \circ i_S = f$ and $H \circ i_T = g$, where $i_S$ and $i_T$ are the inclusions of $S$ and $T$ in the disjoint union. That is, $S \sqcup T$ is universal, in this case "initial", among all sets which admit maps from $S$ and $T$ in that every such pair of maps factors through $S \sqcup T$. Dually, the product $S \times T$ enjoys the property that, for every set $R$ and pair of maps $f: R \to S$ and $g: R \to T$, there exists a unique map $H: R \to S \times T$ compatible with the projections of $S \times T$ onto each of the factors. Thus the product is "final" among all sets admitting maps to $S$ and $T$ in the sense that every such pair of maps factors through the product.

The idea now is to look for constructions like product and disjoint union (coproduct) in general categories. They often turn out to be familiar, and thinking about them in terms of universal properties clarifies their importance. The wedge $X \vee Y$ is the coproduct in the category of based spaces in a sense similar to the disjoint union of sets above. The product $X \times Y$ of a pair of based spaces is the ordinary product, much like the product of two groups $G$ and $H$ (as sets) gives rise to a group with the appropriate universal property. But (co)products are not always easy to construct or identify. For example, the coproduct of groups is less obvious essentially because all the maps have to be homomorphisms, and so using the familiar notion of coproduct of sets does not work. The disjoint union does not satisfy the desired universal property, but it turns out that the free product $G * H$ does, and we leave it to the reader to verify this after reading the definition of the coproduct, Definition 7.2.2.

Universal properties abound and have great importance. Some of the examples that satisfy universal properties are quite familiar: The supremum and the infimum of a set of real numbers (where we think of the set of real numbers as a poset in the evident way), or the greatest common divisor and the least common multiple of a set of natural numbers (where we think of the natural numbers as a category where a morphism $a \to b$ means that $a$ divides $b$) are all examples which appear below. It is important to note that the existence of a universal object is a theorem – such objects are not guaranteed to exist, as in the case of the supremum and infimum of a bounded set of rational numbers. This very shortcoming can be used to motivate the construction of the real numbers from the rationals, at least from the perspective of real analysis.

We will revisit the infimum and the supremum in Examples 7.3.6 and 7.3.24 using the language of limits and colimits.

We begin to formalize the above observations with the following.

**Definition 7.2.1** Let $X_1, X_2$ be objects of a category $C$. The *product* of $X_1$ with $X_2$, denoted by $X_1 \times X_2$, is an object of $C$ together with two morphisms, called *(canonical) projections*,

$$\pi_i \colon X_1 \times X_2 \longrightarrow X_i$$

for $i = 1, 2$ such that the following holds: if $Y \in \mathrm{Ob}(C)$ and $f_i \colon Y \to X_i$ are morphisms for $i = 1, 2$, then there exists a unique morphism $f \colon Y \to X_1 \times X_2$ such that the diagram

$$
\begin{array}{ccc}
 & & X_1 \times X_2 \\
 & \nearrow f & \big\downarrow \pi_i \\
Y & \xrightarrow{\ f_i\ } & X_i
\end{array}
$$

commutes for each $i = 1, 2$.

That is, the product is an object which is final among all other objects which admit maps to $X_1$ and $X_2$. When the product exists, it is unique up to isomorphism. To see this is a straightforward exercise using the universal property and uniqueness of the map to the product.

Products need not exist. The notation is chosen because, in the category of sets, the product is the cartesian product. If there is ever any doubt, we assume the existence of products in any instance where we require them. The same comment applies to coproducts.

**Definition 7.2.2** Let $X_1, X_2$ be objects of a category $C$. The *coproduct* of $X_1$ with $X_2$, denoted by $X_1 \sqcup X_2$, is an object of $C$ together with two morphisms, called *(canonical) injections*,

$$j_i \colon X_i \longrightarrow X_1 \sqcup X_2$$

for $i = 1, 2$ such that the following holds: if $Y \in \mathrm{Ob}(C)$ and $f_i \colon X_i \to Y$ are morphisms for $i = 1, 2$, then there exists a unique morphism $f \colon X_1 \sqcup X_2 \to Y$ such that the diagram

commutes for each $i = 1, 2$.

The coproduct is thus initial among all objects admitting maps from $X_1$ and $X_2$. The product and the coproduct are dual in the sense that they satisfy the same universal properties with the "arrows reversed". The coproduct of two objects, when it exists, is unique up to isomorphism. More generally we can define (co)products indexed by an arbitrary indexing set.

Let $I$ be an indexing set and let $\{X_i : i \in I\}$ be a set of objects of a category $C$.

**Definition 7.2.3** The *product* of the set $\{X_i\}_{i \in I}$, denoted by $\prod_{i \in I} X_i$, is an object of $C$ and a collection of morphisms, called *(canonical) projections*,

$$\pi_i \colon \prod_{i \in I} X_i \longrightarrow X_i$$

such that the following holds: if $Y \in \mathrm{Ob}(C)$ and $f_i \colon Y \to X_i$ are a collection of morphisms, one for each $i \in I$, then there exists a unique morphism $f \colon Y \to \prod_{i \in I} X_i$ such that the diagram

$$
\begin{array}{ccc}
 & & \prod_{i \in I} X_i \\
 & \overset{f}{\nearrow} & \downarrow {\scriptstyle \pi_i} \\
Y & \underset{f_i}{\longrightarrow} & X_i
\end{array}
$$

commutes for each $i$.

In the category $C = \mathrm{Top}$ or $\mathrm{Top}_*$, we must also discuss the topology: we topologize $\prod_{i \in I} X_i$ to have the smallest topology such that the canonical projections $\prod_{i \in I} X_i \to X_i$ are continuous for all $i \in I$, and then take the compactly generated topology of the resulting topological space.

Dually, we have the following.

**Definition 7.2.4** The *coproduct* of the set $\{X_i\}_{i \in I}$, denoted by $\coprod_{i \in I} X_i$, is an object of $C$ and a collection of morphisms, called *(canonical) injections*,

$$j_i \colon X_i \longrightarrow \coprod_{i \in I} X_i$$

such that the following holds: If $Y \in \mathrm{Ob}(C)$ and $f_i \colon X_i \to Y$ are a collection of morphisms, one for each $i \in I$, then there exists a unique morphism $f \colon \coprod_{i \in I} X_i \to Y$ such that the diagram

$$
\begin{array}{ccc}
\coprod_{i \in I} X_i & & \\
\uparrow {\scriptstyle j_i} & \overset{f}{\searrow} & \\
X_i & \underset{f_i}{\longrightarrow} & Y
\end{array}
$$

commutes for each $i$.

Again if $C$ = Top or Top$_*$ we topologize the coproduct $\coprod_{i\in I} X_i$ by choosing the smallest topology such that the canonical injections $X_i \to \coprod_{i\in I} X_i$ are continuous.

Products and coproducts are already familiar in many categories. As we have mentioned, the terminology is derived from the usual product and coproduct of sets.

**Example 7.2.5**

| Category | Product | Coproduct |
|----------|---------|-----------|
| Set | Cartesian product | disjoint union |
| Vect$_F$ | Cartesian product | direct sum |
| Top | Cartesian product | disjoint union |
| Top$_*$ | Cartesian product | wedge sum |
| Grp | direct product | free product |
| Ab Grp | direct product | direct sum |
| Ring | Cartesian product | tensor product |
| Ch$_R$ | degree-wise product | degree-wise coproduct |
| R-Mod | Cartesian product | direct sum |

$\square$

**Remark 7.2.6** Suppose $C$ is a category where all products exist. For simplicity, to start let us assume we have objects $X_i, Y_i$ and morphisms $f_i\colon X_i \to Y_i$ for $i = 1, 2$. Composing the canonical projections $X_1 \times X_2 \to X_i$ with $f_i$ we see by the universal property of $X_1 \times X_2$ that the $f_i$ induce a unique morphism $X_1 \times X_2 \to Y_1 \times Y_2$, and we will denote this morphism by $f_1 \times f_2$. This is more generally true of arbitrary products indexed by an arbitrary set. That is, suppose $I$ is a set and we have for each $i \in I$ objects $X_i, Y_i$ of $C$ and a morphism $f_i\colon X_i \to Y_i$. Then there is a naturally induced morphism

$$\prod_{i\in I} f_i\colon \prod_{i\in I} X_i \to \prod_{i\in I} Y_i.$$

A dual remark applies to coproducts. Such induced morphisms will be important when we build models for limits and colimits from products and coproducts. $\square$

In the next two sections, we generalize the notion of a product and a coproduct.

## 7.3 Limits and colimits

We begin this section by discussing limits, which are one of the most important "universal" constructions. They generalize products, pullbacks, inverse limits, and other familiar ideas.

**Definition 7.3.1**   Suppose $F\colon \mathcal{I} \to C$ is a diagram in $C$ indexed on a small category $\mathcal{I}$. A *cone* on $F$ is an object $C \in \mathrm{Ob}(C)$ and, for each $X \in \mathrm{Ob}(\mathcal{I})$, a morphism $\gamma_X\colon C \to F(X)$ such that the following diagram commutes for all $X, Y \in \mathrm{Ob}(\mathcal{I})$ and $f \in \mathrm{Hom}_{\mathcal{I}}(X, Y)$:

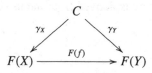

The limit of a diagram is a universal cone; every cone factors through the limit. More precisely, we have the following.

**Definition 7.3.2**   Suppose $F\colon \mathcal{I} \to C$ is a diagram in $C$ indexed on a small category $\mathcal{I}$. The *limit* of $F$, denoted by $\lim_C F$ is a cone on $F$ such that, for any other cone $C$ on $F$ and for any $X, Y \in \mathrm{Ob}(\mathcal{I})$ and $f \in \mathrm{Hom}_{\mathcal{I}}(X, Y)$, there exists a unique morphism $u\colon C \to \lim_C F$ such that the following diagram commutes (i.e. each triangle in the diagram commutes):

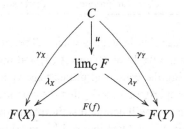

Here $\lambda_X$ and $\lambda_Y$ are morphisms associated to the cone $\lim_C F$.

If it exists at all, $\lim_C F$ is unique up to isomorphism (we leave it to the reader to think this through). When there is no danger of confusion, we may suppress the category and denote the limit by $\lim F$.

The limit of a diagram may not exist in an arbitrary category, even if products exist. A category that contains limits of all small diagrams is called *complete*. Examples of complete categories are Set, Top, Top$_*$, Grp, and R-Mod. From now on, we will assume that we are working in a complete (and small) category.

The notion of a limit reduces to some familiar constructions in many cases. Here are some simple examples.

**Example 7.3.3**   If $\mathcal{I}$ is the empty category (no objects), then the limit of any functor $F\colon \mathcal{I} \to$ Top is the one-point space $*$. In the commutative diagram from Definition 7.3.2, all we need to find is a space which admits a map from any other space. This is just a roundabout way to say that the one-point space is the final object in the category of topological spaces. □

**Example 7.3.4** (Product as limit)   Suppose $I = \{\bullet, \bullet\}$, the discrete category consisting of two objects and no non-identity morphisms, and let $F\colon I \to C$ be a functor, so $F(I) = \{X, Y\}$ for some $X, Y \in \mathrm{Ob}(C)$. The limit of this diagram is precisely the product $X \times Y$ by definition, if it exists. More generally, if $I$ is any discrete category (again with no non-identity morphisms), one recovers Definition 7.2.3. □

**Example 7.3.5** (glb in a poset as limit)   Let $(P, \leq)$ be a poset. If $a, b$ are elements of $P$, then the product of $a$ and $b$ is their greatest lower bound (glb) in $P$, since we are asking for the universal example of an element $g$ of $P$ which satisfies $g \leq a$ and $g \leq b$. If $F\colon I \to P$ is a functor, then $\lim_I F$ is the greatest lower bound of the collection $\{F(i)\colon i \in I\}$, if it exists. Here is a diagrammatic example of a poset in which greatest lower bounds do not all exist. Elements are labeled by lower-case latin letters and $x \to y$ means $x \leq y$ in the poset structure.

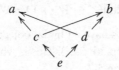

Here the greatest lower bound of $\{a, b\}$ does not exist despite the existence of common lower bounds for both $a$ and $b$ (the elements $c$ and $d$). The trouble is that $c$ and $d$ aren't directly comparable, so neither can satisfy the universal property. And $e$, despite being a common lower bound for $a$ and $b$, fails to be greater than $c$ and $d$. □

**Example 7.3.6** (Infimum as limit)   Consider the real numbers $\mathbb{R}$ as a category, with a morphism $x \to y$ if $x \leq y$. The product of real numbers $x$ and $y$ in this category is the real number $\min\{x, y\}$. In general the product of an arbitrary set of real numbers may not exist. Let $F\colon C \to \mathbb{R}$ be any functor. The limit $\lim F$ is the infimum of the set $\{F(c)\colon c \in C\}$ (which may or may not exist). This follows from the universal property: if it exists, $\lim F$ is a real number such that $\lim F \leq F(c)$ for all $c \in C$, and whenever $z \leq F(c)$ for all $c \in C$, $z \leq \lim F$. Note that if $C$ is the empty category (so we are considering the infimum of the empty set), the the most reasonable candidate is $+\infty$, though this is of course not a real number. □

**Example 7.3.7** (gcd as limit)  Consider the natural numbers $\mathbb{N}$ as a category, with a morphism $a \to b$ if $a$ divides $b$. In this category the product of natural numbers $a$ and $b$ is their greatest common divisor (gcd). More generally let $F: C \to \mathbb{N}$ be any functor where $C$ is a small category. Then $\lim F$ is the greatest common divisor of the set $\{F(c): c \in C\}$. As in Example 7.3.6, this follows from the universal property; $\lim F$ is a natural number which divides $F(c)$ for all $c \in C$, and whenever $z \in \mathbb{N}$ divides $F(c)$ for all $c$, $z$ divides $\lim F$. This limit does not exist if $C$ is empty, since $\mathbb{N}$ has no final object in this poset structure. □

**Example 7.3.8** (Intersection as limit)  Let $S$ be a set, and consider the poset $\mathcal{P}(S)$ of all subsets of $S$. The product of sets $R$ and $T$ is their intersection $R \cap T$. Let $C$ be any category and $F: C \to \mathcal{P}(S)$ any functor. Then $\lim F = \cap_{c \in C} F(c)$. □

**Example 7.3.9** (Categories with an initial object)  If $I$ has an initial object $i_0$, then if $F: I \to C$ is a functor, $\lim_I F = F(i_0)$. This is due to the fact that $F(i_0)$ satisfies the universal property required of the limit. For a more concrete example, if $I = \{\bullet \leftarrow \bullet \to \bullet\}$, then

$$\lim(X \leftarrow W \to Y) \cong W.$$

Notice that cubical diagrams have an initial object, so taking their limit is not interesting, as we saw in Remark 5.2.3. This is why we only considered limits of punctured cubes in Section 5.2. □

**Example 7.3.10** (Equalizer/kernel)  Since this example is of special importance, we will write it out in some detail for easy referencing.

If $I = \{\bullet \rightrightarrows \bullet\}$, then $\lim(X \overset{f}{\underset{g}{\rightrightarrows}} Y)$ is called the *equalizer of X and Y* (with respect to the maps $f$ and $g$), and is denoted by $\mathrm{Eq}(X \overset{f}{\underset{g}{\rightrightarrows}} Y)$ or $\mathrm{Eq}(f, g)$. According to Definition 7.3.2, if it exists, this is an object of $C$ along with a morphism $\lambda_X : \mathrm{Eq}(f, g) \to X$ such that $f \circ \lambda_X = g \circ \lambda_X$ and such that, for any morphism $\gamma: Z \to X$ satisfying $f \circ \gamma = g \circ \gamma$, there is a unique morphism $u: Z \to \mathrm{Eq}(f, g)$ making the diagram

$$\mathrm{Eq}(f, g) \overset{\lambda_X}{\longrightarrow} X \overset{f}{\underset{g}{\rightrightarrows}} Y$$

with $u: Z \to \mathrm{Eq}(f,g)$ and $\gamma: Z \to X$

(7.3.1)

commute.

In Top (and Top$_*$), Eq$(f, g)$ is given as

$$\text{Eq}(f, g) = \{x \in X \mid f(x) = g(x)\}$$

with $\lambda_X$ the inclusion map. It inherits the subspace topology. It is easy to see that this set satisfies the universal property since any $\gamma \colon Z \to X$ such that $f \circ \gamma = g \circ \gamma$ clearly factors through this subset of $X$.

In the category of abelian groups, the equalizer can be thought of as the kernel of the difference $f - g$. Thus if one of the morphisms is the zero homomorphism, we obtain the usual notion of a kernel. □

**Remark 7.3.11** The category $I = \{\bullet \rightrightarrows \bullet\}$ intuitively appears to have an initial object, but this is not the case because there are two distinct morphisms from the obvious candidate for an initial object to the other. Consequently, $\lim(X \underset{g}{\overset{f}{\rightrightarrows}} Y)$ is not equivalent to $X$ as it might visually appear, thinking of Example 7.3.9, but is rather a subset of $X$. □

**Example 7.3.12** (Inverse limit) If $I = \{\bullet \leftarrow \bullet \leftarrow \bullet \leftarrow \cdots\}$, then $F(I)$ is called an *inverse system*. The limit $\lim F$ is called the *inverse limit* and sometimes denoted by $\varprojlim F$. It is also clear from the definitions that this coincides with the usual notion of the inverse limit of, say, a tower of spaces or groups. In particular, the $p$-adic integers $\mathbb{Z}_p$ are an example of such an inverse limit.

The inverse limit is easily expressed as an equalizer as follows. It is convenient to let the natural numbers index the category $I$, so that $I = \{1 \leftarrow 2 \leftarrow \cdots\}$. Let $F(i) = X_i$, and let $p_i \colon X_i \to X_{i-1}$ denote the morphism $F(i \to i - 1)$. Let $P \colon \prod_i X_i \to \prod_i X_i$ be the map $P(x_1, x_2, \ldots) = (p_2(x_2), p_3(x_3), \ldots)$. Then by inspection there is a natural isomorphism

$$\lim_i X_i \cong \text{Eq}\left(\prod_i X_i \underset{P}{\overset{1_{\prod_i X_i}}{\rightrightarrows}} \prod_i X_i\right).$$

Thus a point in $\lim_I F$ is an infinite sequence $(x_1, x_2, \ldots)$ such that $p_i(x_i) = x_{i-1}$.

One could take this further and use Example 7.4.2 to write the equalizer as a limit of a punctured 3-cube.

Perhaps the simplest way to write $\lim_i X_i$ as a the limit of a punctured square is

$$\lim_i X_i \cong \left(\prod_i X_i \overset{\Delta}{\to} \prod_i X_i \times \prod_i X_i \overset{(1_{\prod_i X_i}, P)}{\leftarrow} \prod_i X_i\right).$$

This is straightforward to verify. □

**Example 7.3.13** (Pullback as limit)    We can now finally see why pullbacks, one of the central objects of study in Chapter 3, are instances of limits; this will justify the terminology and the notation "lim" that we used in Chapter 3.

If $\mathcal{I} = \{\bullet \to \bullet \leftarrow \bullet\}$, $\lim(X \xrightarrow{f} Z \xleftarrow{g} Y)$ is called the *pullback*, or *fiber product*, and denoted by $X \times_Z Y$. It consists of an object $X \times_Z Y$ of $C$ and morphisms $\lambda_X \colon X \times_Z Y \to X$ and $\lambda_Y \colon X \times_Z Y \to Y$ such that the diagram

$$
\begin{array}{ccc}
X \times_Z Y & \xrightarrow{\lambda_Y} & Y \\
{\scriptstyle \lambda_X} \downarrow & & \downarrow {\scriptstyle g} \\
X & \xrightarrow{f} & Z
\end{array}
$$

commutes. Further, if there is another object $C$ and morphisms $\gamma_X \colon C \to X$ and $\gamma_Y \colon C \to Y$ which make the same square commute, then there exists a unique morphism $u \colon C \to X \times_Z Y$ such that the following diagram commutes:

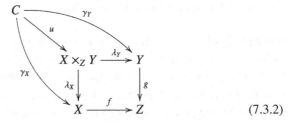

$$(7.3.2)$$

In Top (or Top$_*$, as well as in Set), the pullback is given by

$$X \times_Z Y = \{(x, y) \in X \times Y \mid f(x) = g(y)\} \tag{7.3.3}$$

with projection maps $p_1 \colon X \times_Z Y \to X$ and $p_2 \colon X \times_Z Y \to Y$ sending $(x, y)$ to $x$ and $y$ respectively. We leave it to the reader to check this. This is precisely the pullback from Definition 3.1.1. In Top, the pullback inherits the subspace topology. More specific examples of pullbacks in Top were given in Section 3.1. □

**Remark 7.3.14**    In Top, we can identify the equalizer in Example 7.3.10 as an iterated pullback using the diagonal map. Let $f, g \colon X \to Y$ be as above. We leave it to the reader to check that

$$\mathrm{Eq}(f, g) = \lim(X \xrightarrow{\Delta} X \times X \leftarrow X \times_Y X),$$

where $\Delta \colon X \to X \times X$ is the diagonal map, and the map

$$\lim(X \xrightarrow{f} Y \xleftarrow{g} X) = X \times_Y X \to X \times X$$

is the inclusion.

It might be tempting to think of $\mathrm{Eq}(X \overset{f}{\underset{g}{\rightrightarrows}} Y)$ and $\lim(X \overset{f}{\to} Y \overset{g}{\leftarrow} X)$ as the same objects. However, the former is the set of all $x \in X$ such that $f(x) = g(x)$, whereas the latter is the set of all $(x_1, x_2) \in X \times X$ such that $f(x_1) = g(x_2)$. Using the diagonal inclusion $\Delta \colon X \to X \times X$ we can realize $\mathrm{Eq}(X \overset{f}{\underset{g}{\rightrightarrows}} Y)$ as a subset of $\lim(X \overset{f}{\to} Z \overset{g}{\leftarrow} Y)$. But this inclusion need not be an isomorphism. If, for example, $X = Y = \{1, 2\}$, $f = 1_X$, and $g(i) = 1$ for $i = 1, 2$, then the equalizer consists of a single point whereas the pullback consists of two points. One case when the equalizer and pullback are isomorphic is when there exists a map

$$h \colon Y \longrightarrow X$$

such that $h \circ f = h \circ g = 1_X$ (i.e. $h$ is a common section for $f$ and $g$). In this case, applying $h$ to a point $(x_1, x_2)$ satisfying $f(x_1) = g(x_2)$ in $\lim(X \overset{f}{\to} Y \overset{g}{\leftarrow} X)$, we see that $x_1 = x_2$.[2] $\qquad\qquad\square$

**Example 7.3.15** (Homotopy pullback as limit)  We saw in Proposition 3.2.5 that the homotopy pullback $\mathrm{holim}(X \overset{f}{\to} Y \overset{g}{\leftarrow} Z)$ can be realized as a pullback since it is homeomorphic to $\lim(P_f \to Y \overset{g}{\leftarrow} Z)$. By Example 7.3.13, the homotopy pullback is an example of a limit.

As a special case, when $Y = \{z\}$ is a point in $Z$ and $g$ is the inclusion, the homotopy limit above is the homotopy fiber of $f$.

The space $P_f$ itself is the pullback of the diagram

$$X \overset{f}{\longrightarrow} Z \overset{ev_0}{\longleftarrow} Z^I$$

and is thus also an example of a limit. $\qquad\qquad\square$

We provide one more example that is peripheral to our purposes but is very important in topology. Namely, recall that a topological group $G$ *acts continuously on a space $X$* if there is a map

$$G \times X \longrightarrow X$$
$$(g, x) \longmapsto gx$$

such that $(gh)x = g(hx)$ and $ex = x$, where $e$ is the identity in $G$. We then have the set of *fixed points* of the action (set of points of $X$ fixed by all elements of

---

[2]  This situation is what happens in Section 9.4 where the existence of codegeneracies allows us to turn truncated cosimplicial digrams into punctured cubical diagrams that have the same homotopy limit as the truncations.

$G$), denoted by $X^G$, and the *orbit space* of the action (quotient of $X$ by orbits of the action), denoted by $X_G$.

Now think of a discrete group $G$ as a category $\bullet_G$ with one object $\bullet$ and one morphism for each group element which compose according to the group law. We then have the following result which follows from unravelling the definitions.

**Proposition 7.3.16** *A functor* $F: \bullet_G \to \mathrm{Top}$ *such that* $F(\bullet) = X$, *determines and is determined by a continuous group action on* $X$.

Note that $F(e) = 1_X$, where $e$ is the identity element of $G$, is forced by the axioms for a functor (see Definition 7.1.15).

We then have the following characterization of the $G$ fixed points of $X$, the space $X^G$, as a limit. The corresponding statement for $X_G$ will appear as Proposition 7.3.36.

**Proposition 7.3.17** (Fixed points as limit) *For a functor* $F: \bullet_G \to \mathrm{Top}$ *where* $F(\bullet) = X$ *and* $F(e) = 1_X$,

$$X^G = \lim_{\bullet_G} F.$$

*Proof* We will show that $X^G$ has the universal properties which defines the limit. For each $g \in G$, group action gives a map $g: X \to X$, and, as is customary, we denote its value at $x \in X$ by $gx$. Suppose $Y$ is a space with a map $f: Y \to X$ such that the following diagram commutes for all $g \in G$:

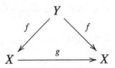

For each $y \in Y$, we have $f(y) = gx$ for all $g \in G$, which implies $f(y)$ is a fixed point of the action of $G$. Hence $f$ factors through a map $u$ to the fixed point set $X^G$. That is, we have a map $u: Y \to X^G$ which fits into the commutative diagram

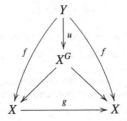

$\square$

Limits can be regarded as functors. Recall from Example 7.1.30 the category $C^{\mathcal{I}}$, which can be thought of as the category of $\mathcal{I}$-shaped diagrams in $C$. The limit is a functor[3]

$$\lim_{\mathcal{I}} : C^{\mathcal{I}} \longrightarrow C \tag{7.3.4}$$

because it associates an object in $C$ to each diagram and, if

$$N : F \longrightarrow G$$

is a natural transformation of functors, there is a naturally induced morphism

$$\lim_{\mathcal{I}} F \longrightarrow \lim_{\mathcal{I}} G, \tag{7.3.5}$$

which comes from the universal property of the limit. Observe that $\lim_{\mathcal{I}} F$ is a cone on the functor $G$ via the composition of the canonical projection to $F$ and the natural transformation $N : F \to G$. As such it is uniquely determined because of the universality of $\lim_{\mathcal{I}} F$ and $\lim_{\mathcal{I}} G$, and we have a commutative diagram

$$
\begin{array}{ccc}
\lim_{\mathcal{I}} F & \longrightarrow & \lim_{\mathcal{I}} G \\
{\scriptstyle p_{F,i}} \downarrow & & \downarrow {\scriptstyle p_{G,i}} \\
F(i) & \xrightarrow{\ N_i\ } & G(i)
\end{array}
$$

for each object $i \in \mathcal{I}$. Here $p_{F,i}$ and $p_{G,i}$ are the projections as in Definition 7.3.2, but for two different functors $F$ and $G$.

We can also consider functoriality in the $\mathcal{I}$ component of the limit. Suppose $G : \mathcal{J} \to \mathcal{I}$ is a functor. We have an induced functor $G^* : C^{\mathcal{I}} \to C^{\mathcal{J}}$ given by $F \mapsto F \circ G$. This induces a map

$$\lim_{\mathcal{I}} F \longrightarrow \lim_{\mathcal{J}} (F \circ G) \tag{7.3.6}$$

as follows. If $j \to j'$ is a morphism, then we have a commutative diagram

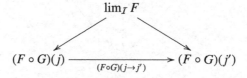

where the vertical arrows are the canonical projections, which says that $\lim_{\mathcal{I}} F$ is a cone on the functor $F \circ G$, and hence there exists a unique map $\lim_{\mathcal{I}} F \to \lim_{\mathcal{J}} (F \circ H)$.

---

[3] This is right adjoint to the constant diagram functor.

**Remark 7.3.18** An easy way to remember which way the arrow from (7.3.6) goes is to consider any example of a functor $G\colon \mathcal{J} \to \mathcal{I}$ which is the inclusion of a subcategory and any functor $F\colon \mathcal{I} \to \text{Top}$. Since $\lim_{\mathcal{I}} F$ is a subspace of $\prod_{i \in \mathcal{I}} F(i)$ and $\lim_{\mathcal{J}}(F \circ G)$ is a subspace of $\prod_{j \in \mathcal{J}} F(j)$, the natural projection $\prod_{i \in \mathcal{I}} F(i) \to \prod_{j \in \mathcal{I}} F(j)$ induces the desired map of limits.                     □

The remainder of this section essentially consists of dual versions of the definitions and various examples from the first part of this section in the sense that the arrows are reversed. Colimits are defined using a universal property and generalize direct sums, pushouts, direct limits, and other familiar constructions. Again suppose throughout that $F\colon \mathcal{I} \to C$ is a diagram in $C$ indexed on a small category $\mathcal{I}$.

**Definition 7.3.19** A *co-cone* on $F$ is an object $C \in \text{Ob}(C)$ and, for each $X \in \text{Ob}(\mathcal{I})$, a morphism $\gamma_X \colon F(X) \to C$ such that the following diagram commutes for all $X, Y \in \text{Ob}(\mathcal{I})$ and $f \in \text{Hom}_{\mathcal{I}}(X, Y)$:

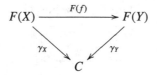

**Definition 7.3.20** The *colimit* of $F$, denoted by $\text{colim}_C F$, is a co-cone on $F$ such that, for any other co-cone $C$ on $F$ and for any $X, Y \in \text{Ob}(\mathcal{I})$ and $f \in \text{Hom}_{\mathcal{I}}(X, Y)$, there exists a unique morphism $u\colon \text{colim}_C F \to C$ such that the following diagram commutes:

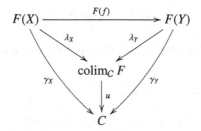

Here again $\lambda_X$ and $\lambda_Y$ are morphisms associated to the co-cone $\text{colim} F$.

As was the case with limits, $\text{colim}_C F$ is unique up to isomorphism. We will often denote the colimit simply by $\text{colim} F$. Colimits in general may not exist. A category that contains colimits of all small diagrams is called *cocomplete* and we will always assume that we are working in such a category. Some cocomplete categories are Set, Top, Top$_*$, Grp, and R-Mod.

**Example 7.3.21** If $I$ is the empty category, then the colimit of any functor $F$ from $I$ to Top is the empty set, because, according to the commutative diagram from Definition 7.3.20, all we need to find is a space which admits a map to any other space. That is, the empty set is the initial object in the category of topological spaces. In the category Top$_*$ of based spaces, however, the colimit is the one-point space. □

**Example 7.3.22** (Coproduct as colimit) Suppose $I = \{\bullet, \bullet\}$, the discrete category consisting of two objects and no non-trivial morphisms, and $F : I \to C$ a functor. Then $F(I) = \{X, Y\}$ for some $X, Y \in \mathrm{Ob}(C)$. The colimit of this diagram is precisely the coproduct $X \sqcup Y$ by definition. More generally, if $I$ is any discrete category, one recovers Definition 7.2.4. □

**Example 7.3.23** (lub in a poset as colimit) Let $(P, \leq)$ be a poset. If $a, b$ are elements of $P$, then the coproduct of $a$ and $b$ is their least upper bound (lub) in $P$, which follows from the universal property in a way dual to the discussion in Example 7.3.5. Reversing the arrows in the poset given in Example 7.3.5 gives an example of a poset where the elements $a$ and $b$ have common upper bounds but no least upper bound.

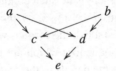

□

**Example 7.3.24** (Supremum as colimit) Consider the real numbers $\mathbb{R}$ as a category, with a morphism $x \to y$ if $x \leq y$. The coproduct of two real numbers is their maximum. Let $F : C \to \mathbb{R}$ be any functor. Then colim $F$, if it exists, is the supremum of the set $\{F(c) : c \in C\}$. □

**Example 7.3.25** (lcm as colimit) Consider the natural numbers $\mathbb{N}$ as a category, with a morphism $a \to b$ if $a$ divides $b$. Let $F : C \to \mathbb{R}$ be any functor. Then colim $F$ is the least common multiple of the set $\{F(c) : c \in C\}$. If $C$ is empty then the colimit is 1, since this is an initial object in this poset structure. □

**Example 7.3.26** (Union as colimit) For $S$ a set and for $F : C \to \mathcal{P}(S)$ a functor, the union $\bigcup_{c \in C} F(c)$ is the colimit of $F$. □

**Example 7.3.27** (Categories with a final object) If $I$ has a final object $i_1$, then for a functor $F: I \to C$, $\mathrm{colim}_I F = F(i_1)$, again because of the fact that it satisfies the required universal property.

More concretely, if for example $I = \{\bullet \to \bullet \leftarrow \bullet\}$, then

$$\mathrm{colim}(X \to Z \leftarrow Y) = Z.$$

Cubical diagrams have a final object (see Remark 5.6.3), so considering colimits of punctured cubes is more interesting; this is why we considered colimits of punctured cubes only, and not cubes, in Section 5.6. □

**Example 7.3.28** (Direct sum/product of vector spaces) Consider $\mathbb{R}$ as a 1-dimensional vector space over itself. The direct sum $\mathbb{R} \oplus \mathbb{R}$ and direct product $\mathbb{R} \times \mathbb{R}$ are isomorphic. Viewing these constructions categorically, think of $\underline{2} = \{1, 2\}$ as a discrete category, and let $F: \underline{2} \to \mathrm{Vect}_{\mathbb{R}}$ be the constant functor $F(i) = \mathbb{R}$ for all $i$. Then $\lim_{\underline{2}} F = \mathbb{R} \times \mathbb{R}$ and $\mathrm{colim}_{\underline{2}} F = \mathbb{R} \oplus \mathbb{R}$, as can be readily checked using the universal properties.

However, if the indexing category is infinite, the infinite direct sum and infinite direct product are different. Think of $\mathbb{N}$ as a discrete category and define $F: \mathbb{N} \to \mathrm{Vect}_{\mathbb{R}}$ as above. Then $\lim_{\mathbb{N}} F$ is the countable product of copies of $\mathbb{R}$, and $\mathrm{colim}_{\mathbb{N}} F$ is the countable direct sum of copies of $\mathbb{R}$. It is natural to think of a point in each as a sequence $\{x_i\}_{i \in \mathbb{N}}$, but in $\mathrm{colim}_{\mathbb{N}} F$ such a sequence must eventually be zero, whereas there is no such condition on sequences in $\lim_{\mathbb{N}} F$. It should be evident that $\lim_{\mathbb{N}} F = \prod_{\mathbb{N}} \mathbb{R}$ by the universal properties, establishing the latter claim. To establish the former, we need only see that $\bigoplus_{\mathbb{N}} \mathbb{R}$, defined to be the set of all sequences which are eventually zero, satisfies the universal property. We leave the details to the reader. □

The following two examples are dual to Examples 7.3.10 and 7.3.13.

**Example 7.3.29** (Coequalizer/cokernel) If $I = \{\bullet \rightrightarrows \bullet\}$, then $\mathrm{colim}(X \underset{g}{\overset{f}{\rightrightarrows}} Y)$ is called the *coequalizer of X and Y* (with respect to the maps $f$ and $g$), and is denoted by $\mathrm{Coeq}(X \underset{g}{\overset{f}{\rightrightarrows}} Y)$ or $\mathrm{Coeq}(f, g)$. The pertinent diagram, dual to the one in (7.3.1), is

$$
\begin{array}{ccc}
X \underset{g}{\overset{f}{\rightrightarrows}} Y & \overset{\lambda_Y}{\longrightarrow} & \mathrm{Coeq}(f, g) \\
& \searrow{\gamma} & \downarrow{u} \\
& & Z
\end{array}
\tag{7.3.7}
$$

In Set and Top, elements of $\mathrm{Coeq}(f, g)$ are points in $Y/\sim$ where $\sim$ is generated by relations $f(x) = g(x)$ for all $x \in X$. In Top, $\mathrm{Coeq}(f, g)$ inherits the quotient topology. It is again easy to verify that this quotient satisfies the required universal property.

In the category of abelian groups, the coequalizer is the quotient of $Y$ by the image of the difference $f - g$. Thus if one of the homomorphisms is zero, this reduces to the usual notion of a cokernel. □

**Remark 7.3.30** We leave it to the reader to formulate the dual of Remark 7.3.11. □

**Example 7.3.31** (Direct limit) If $\mathcal{I}$ is the opposite category of the directed poset in Example 7.3.12, namely $\bullet \to \bullet \to \bullet \to \cdots$, then for a functor $F: \mathcal{I} \to C$, $F(\mathcal{I})$ is called a *directed system*. The colimit $\mathrm{colim}_{\mathcal{I}} F$ is called the *direct limit* and sometimes denoted by $\varinjlim F$ in the literature. We prefer and will always use colim. This coincides with the usual notion of the direct limit in other contexts.

Let $C = \mathrm{Top}$ and $F(i) = X_i$, so that $F(\mathcal{I}) = (X_1 \to X_2 \to \cdots)$, and let $a_i: X_i \to X_{i+1}$ denote the maps in the diagram. The direct limit $\mathrm{colim}_{\mathcal{I}} F = \mathrm{colim}_i X_i$ can be expressed as a coequalizer very simply. There is a natural homeomorphism

$$\mathrm{colim}_i X_i \cong \mathrm{Coeq}\left( \coprod_i X_i \underset{A}{\overset{1_{\coprod_i X_i}}{\rightrightarrows}} \coprod_i X_i \right),$$

where $A: \coprod_i X_i \to \coprod_i X_i$ is the map defined by $A(x_i) = a_i(x_i)$ if $x_i \in X_i \subset \coprod_i X_i$. Thus a point in $\mathrm{colim}_{\mathcal{I}} F$ is an equivalence class of points in $\coprod_i X_i$, where we identify $x \in X_i$ with $a_i(x) \in X_{i+1}$ and let this generate the evident equivalence relation.

We can also express $\mathrm{colim}_i X_i$ as the colimit of a punctured square. There is a natural homeomorphism

$$\mathrm{colim}_i X_i \cong \left( \coprod_i X_i \overset{\nabla}{\leftarrow} \coprod_i X_i \coprod \coprod_i X_i \overset{1_{\coprod_i X_i} \coprod A}{\longrightarrow} \coprod_i X_i \right),$$

where the left arrow is the "fold" map $\nabla$ which is the identity on each copy of $\coprod_i X_i$, and the right arrow is the identity on the first copy of $\coprod_i X_i$ and $A$ on the second copy.

We have seen an example of a direct limit already just after Definition 2.7.22. There we defined the infinite symmetric product of a based space $(X, x_0)$ as a union $\mathrm{SP}(X) = \bigcup_n \mathrm{SP}_n(X)$. In categorical language, write $\mathcal{I} = 1 \to 2 \to$

$3 \to \cdots$, let $F(i) = \mathrm{SP}_i(X)$, and let $\mathrm{SP}_i(X) \to \mathrm{SP}_{i+1}(X)$ be the map sending $[x_1, \dots, x_i]$ to $[x_1, \dots, x_i, x_0]$. Then $\mathrm{colim}_I F = \mathrm{SP}(X)$. $\qquad\square$

**Example 7.3.32** (Pushout as colimit)    Dually to Example 7.3.13, we can now see why pushouts from Chapter 3 are examples of colimits.

Namely, if $I = \{\bullet \leftarrow \bullet \to \bullet\}$, then $\mathrm{colim}(X \xleftarrow{f} W \xrightarrow{g} Y)$ is called the *pushout* and denoted by $X \amalg_W Y$. The relevant commutative diagram is

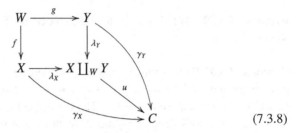

$$\tag{7.3.8}$$

Here $C$ is any object that makes the outer square commute, and $u$ is a unique map.

In Top, the pushout is given by

$$X \coprod_W Y = X \coprod Y / \sim, \tag{7.3.9}$$

where the equivalence relation $\sim$ is generated by $f(w) = g(w)$ for all $w \in W$. It inherits the quotient topology. To see that this satisfies the universal property, suppose $C$ is another set making the diagram (7.3.8) commute. Hence $\gamma_X \circ f = \gamma_Y \circ g$ and letting $u = \gamma_X \coprod \gamma_Y$ defines it uniquely. This, of course, coincides with the definition of the pushout from Definition 3.5.1.

If $W$ is the intersection of $X$ and $Y$, and $f$ and $g$ are inclusions of subsets, then pushout is precisely the union of $X$ and $Y$ along their common intersection. This is succinctly stated in a square which we have encountered repeatedly:

$$\begin{array}{ccc} X \cap Y & \longrightarrow & Y \\ \downarrow & & \downarrow \\ X & \longrightarrow & X \cup Y \end{array} \tag{7.3.10}$$

is pushout.

Furthermore, if one of the maps, say $g$, is the inclusion of a subspace, then pushout is the *union of $X$ and $Y$ along $f$*, denoted by $X \cup_W Y$, that is, it is the space obtained from $Y$ by gluing $X$ to it along $W$ using the attaching map $f$. More examples of pushouts in Top were given in Section 3.5. $\qquad\square$

**Remark 7.3.33** In analogy with Remark 7.3.14, in Top we can think of the coequalizer in Example 7.3.29 as an iterated pushout. We have that

$$\mathrm{Coeq}(f, g) = \mathrm{colim}(X \xleftarrow{\nabla} X \sqcup X \xrightarrow{f \sqcup g} Y)$$

where $\nabla$ is the codiagonal map (the map which is the identity on each summand $X$ in the coproduct). This is an iterated pushout since $X \sqcup X$ is itself a pushout. □

**Example 7.3.34** (Homotopy pushout as colimit)  We saw in Proposition 3.6.5 that the homotopy pushout $\mathrm{hocolim}(X \xleftarrow{f} W \xrightarrow{g} Y)$ can be thought of as a pushout: it is homeomorphic to $\mathrm{colim}(M_f \leftarrow W \xrightarrow{g} Y)$, and so by Example 7.3.32 the homotopy pushout is an example of a colimit. If $Y$ is a one-point space, then the pushout of the diagram

$$X \xleftarrow{\ f\ } W \xrightarrow{\ g\ } * \tag{7.3.11}$$

is the cofiber of $f$, namely $X/f(W)$; see Example 3.5.6. According to Example 3.6.8 (and Proposition 3.6.5), the homotopy cofiber of $f$ is the pushout of the diagram

$$M_f \xleftarrow{\ f\ } W \xrightarrow{\ g\ } *$$

where $W \to M_f$ is the inclusion in the mapping cylinder. The mapping cylinder itself is a pushout of the diagram

$$X \xleftarrow{\ f\ } W \longrightarrow W \times I$$

where the right map sends $w$ to $(w, 1)$. □

**Example 7.3.35** (Pushout in groups)  Finally, in the category of groups, the pushout of the diagrams of groups $H \leftarrow K \to G$ is the *amalgamated product* $G *_K H$. This is the precisely the construction one sees, for example, in the Seifert–van Kampen Theorem (Theorem 1.4.12). □

Our last example is the dual of Proposition 7.3.17, so recall the notion of the orbit space $X_G$ from the discussion prior to that result. We then have the following.

**Proposition 7.3.36** (Orbit space as colimit)  *For a functor $F \colon \bullet_G \to \mathrm{Top}$ where $F(\bullet) = X$,*

$$X_G = \mathrm{colim}_{\bullet_G} F.$$

*Proof*  Dual to the proof of Proposition 7.3.17, suppose $Z$ is a space together with a map $h: X \to Z$ such that the following diagram commutes for all $g \in G$:

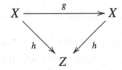

Then, for each $x \in X$, we have $h(x) = h(gx)$ for all $g \in G$, which implies $h$ factors through a map from the orbit space $X_G$ to $Z$. That is, there exists a map $u: X_G \to Z$ and a commutative diagram

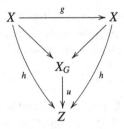

□

The colimit can also be regarded as a functor[4]

$$\operatorname*{colim}_{I} : C^I \longrightarrow C$$

since, given a natural transformation

$$N: F \longrightarrow G,$$

there is a unique morphism

$$\operatorname*{colim}_{I} F \longrightarrow \operatorname*{colim}_{I} G \tag{7.3.12}$$

induced by the universal property of the colimit, in analogy with the one in (7.3.5).

Moreover, if $G: \mathcal{J} \to \mathcal{I}$ and $F: \mathcal{I} \to C$ are functors, then we have a unique induced morphism

$$\operatorname*{colim}_{\mathcal{J}} (F \circ G) \longrightarrow \operatorname*{colim}_{I} F. \tag{7.3.13}$$

**Remark 7.3.37**  A remark similar to Remark 7.3.18 applies here. For functors to Top, and $\mathcal{J} \to \mathcal{I}$ the inclusion of a subcategory, we have an induced map of coproducts $\coprod_{j \in \mathcal{J}} F(j) \to \coprod_{i \in \mathcal{I}} F(i)$ which induces the map of colimits above; this is one way to remember which way the arrow goes.                    □

---

[4] This is left adjoint to the constant diagram functor.

## 7.4 Models for limits and colimits

In this section we present models for (co)limits in categories with (co)products. In general of course one can still discuss (co)limits in a category without reference to (co)products, since they are defined by universal properties. However, defining (co)limits by universal properties can make the object which *is* the (co)limit seem a little elusive and abstract, while using products makes the discussion more concrete and hands-on. The latter also translates well when discussing (co)limits in Top, which is what we are ultimately interested in anyway. The models presented here will serve as a basis for our models of homotopy (co)limits in the next chapter.

We will use (co)products as encountered in Section 7.2 to give models for (co)limits using (co)equalizers. At the end of the section, we will also give further models for limits and colimits in Top using natural transformations and quotients of disjoint unions; these will follow easily from the (co)equalizer model we will have already established.

To start, we formally define equalizers as objects that satisfy a universal property with respect to a certain diagram – rather then think of them as examples of limits as in Example 7.3.10.

**Definition 7.4.1** Let $c_1, c_2$ be objects of $C$ and $f, g$ morphisms from $c_1$ to $c_2$. The *equalizer of f and g*, denoted by $\text{Eq}(c_1 \overset{f}{\underset{g}{\rightrightarrows}} c_2)$ or $\text{Eq}(f, g)$, is an object of $C$ together with a morphism $e\colon \text{Eq}(f, g) \to c_1$ such that

- $f \circ e = g \circ e$;
- if $c$ is any other object and $\gamma\colon c \to c_1$ is a morphism satisfying $f \circ \gamma = g \circ \gamma$, then there exists a unique morphism $u\colon c \to \text{Eq}(f, g)$ making the following diagram commute:

$$\text{Eq}(f, g) \overset{e}{\longrightarrow} c_1 \overset{f}{\underset{g}{\rightrightarrows}} c_2$$

$$\begin{array}{c} u \uparrow \quad \nearrow \gamma \\ c \end{array}$$

(7.4.1)

Equalizers need not exist in $C$, even in a category with all products, but we henceforth assume they do. One standard result that we will appeal to in the proof of Proposition 7.4.3 is that, under the assumption that products and equalizers exist, $C$ is then in fact complete (all limits exist).

**Example 7.4.2** (Equalizer as limit of a punctured 3-cube) Let $X, Y$ be topological spaces and let $f, g\colon X \to Y$ be maps. The equalizer of $X \rightrightarrows Y$ can be expressed as the limit of a punctured 3-cube. Note that

$$\lim(X \rightrightarrows Y) = \lim(X \xrightarrow{\Delta} X \times X \xleftarrow{proj} \lim(X \xrightarrow{f} Y \xleftarrow{g} X)),$$

and the latter limit is

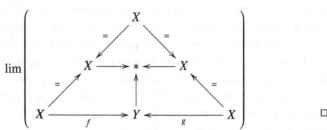

Now suppose as usual that $I$ is a small category and $F \colon I \to C$ is a diagram in $C$. For the sake of brevity, we write a product indexed by the objects $i$ of $I$ as $\prod_i$ and a product indexed by the morphisms $i \to i'$ of $I$ as $\prod_{i \to i'}$. Recall from Remark 7.2.6 that, in a category with products, a morphism $X_i \to Y_i$ for each $i \in I$ induces a morphism $\prod_i X_i \to \prod_i Y_i$. On the one hand, the definition of $F$ gives a morphism $F(i \to i') \colon F(i) \to F(i')$ for each morphism $i \to i'$, and so induces a morphism

$$a \colon \prod_i F(i) \longrightarrow \prod_{i \to i'} F(i'),$$

where the factor indexed by $i$ maps to all factors indexed by the morphisms with source $i$. On the other hand we have the identity $1_{F(i')} \colon F(i') \to F(i')$, and this induces a morphism

$$b \colon \prod_{i'} F(i') \longrightarrow \prod_{i \to i'} F(i'),$$

where the factor indexed by $i'$ maps to all factors indexed by the morphisms with target $i'$.

The following is standard, and we will not reproduce its proof here. See [ML98, V.2 Theorem 2].

**Proposition 7.4.3** (Limit as equalizer) *Let $F \colon I \to C$ be a functor, where $I$ is small and $C$ contains all products and equalizers. The equalizer*

$$\mathrm{Eq}(a, b) = \mathrm{Eq}\left( \prod_i F(i) \underset{b}{\overset{a}{\rightrightarrows}} \prod_{i \to i'} F(i') \right) \tag{7.4.2}$$

*is canonically isomorphic to $\lim_I F$.*

Dually, we make a definition of a coequalizer endowed with the universal property as follows.

**Definition 7.4.4** Let $c_1, c_2$ be objects of $C$ and $f, g$ morphisms from $c_1$ to $c_2$. The *coequalizer of $f$ and $g$*, denoted by $\mathrm{Coeq}(c_1 \overset{f}{\underset{g}{\rightrightarrows}} c_2)$ or $\mathrm{Coeq}(f, g)$, is an object of $C$ together with a morphism $e \colon c_2 \to \mathrm{Coeq}(f, g)$ such that

- $e \circ f = e \circ g$;
- if $c$ is any other object and $\gamma \colon c_2 \to c$ a morphism satisfying $\gamma \circ f = \gamma \circ g$, then there exists a unique morphism $u \colon \mathrm{Coeq}(f, g) \to c$ making the following diagram commute:

$$
c_1 \overset{f}{\underset{g}{\rightrightarrows}} c_2 \overset{e}{\longrightarrow} \mathrm{Coeq}(f, g)
$$

$$\gamma \searrow \quad \downarrow u$$

$$c \tag{7.4.3}$$

As before, we assume all coproducts and coequalizers exist, which implies that that $C$ is cocomplete (all colimits exist). Now let $I$ be an indexing category and $F \colon I \to C$ a diagram in $C$. Again for brevity, write a coproduct indexed by the objects $i \in I$ as $\coprod_i$ and a coproduct indexed by the morphisms $i \to i'$ as $\coprod_{i \to i'}$. We have two maps

$$
a, b \colon \coprod_{i \to i'} F(i) \longrightarrow \coprod_i F(i)
$$

where $a$ is induced by the morphism $F(i \to i')$ and $b$ is induced by the identity morphism.

Dual to Proposition 7.4.3, we then have the following.

**Proposition 7.4.5** (Colimit as coequalizer)  *For a functor $F \colon I \to C$ with $I$ small and $C$ containing all coproducts and coequalizers, the coequalizer*

$$
\mathrm{Coeq}(a, b) = \mathrm{Coeq}\left( \coprod_{i \to i'} F(i) \overset{a}{\underset{b}{\rightrightarrows}} \coprod_i F(i) \right) \tag{7.4.4}
$$

*is canonically isomorphic to $\mathrm{colim}_I F$.*

We now give another way to think of (co)limits, namely as (co)ends. Underlying this model are again (co)equalizers, but now of a bifunctor; this initially makes the (co)end perspective more complicated, but it will be useful to have it to hand when discussing homotopy limits and colimits in the next chapter. For simplicity we again assume that all categories we encounter here are complete and cocomplete, although (co)ends can be discussed from the perspective of universal properties.

Recall the notion of a product category from Definition 7.1.7 and let $F \colon I^{\mathrm{op}} \times I \to C$ be a functor (such a functor is called a *bifunctor*).

**Definition 7.4.6**  The *end* of $F$, denoted by $\int_I F$, is defined as

$$\int_I F = \mathrm{Eq}\left( \prod_i F(i,i) \overset{a}{\underset{b}{\rightrightarrows}} \prod_{i \to i'} F(i,i') \right).$$

The maps are induced as follows. Let $m \colon i \to i'$ be a morphism and let $1_i$ and $1_{i'}$ denote the identity morphisms. The morphism $(1_i, m) \colon (i,i) \to (i,i')$ induces a morphism $F(1_i, m) \colon F(i,i) \to F(i,i')$ and this in turn induces the morphism $a$. A morphism $(m, 1_{i'}) \colon (i,i') \to (i',i')$ induces a morphism $F(m, 1_{i'}) \colon F(i',i') \to F(i,i')$ and that in turn induces the morphism $b$.

**Example 7.4.7**  Suppose $I$ is a small category, and let $F, G \colon I \to \mathrm{Top}$ be functors. We have a bifunctor $\mathrm{Map}(F,G) \colon I^{\mathrm{op}} \times I \to \mathrm{Top}$ given by $\mathrm{Map}(F,G)(i,j) = \mathrm{Map}(F(i), G(j))$. Moreover, it is straightforward to verify from the definition that

$$\int_I \mathrm{Map}(F,G) \cong \underset{I}{\mathrm{Nat}}\,(F,G)$$

is the space of natural transformations from $F$ to $G$.  □

**Example 7.4.8**  Consider the bifunctor $F \colon \mathcal{P}_0(\underline{2})^{\mathrm{op}} \times \mathcal{P}_0(\underline{2}) \to \mathrm{Top}$ given by $F(S,T) = X_T$. Unraveling the definition we see that $\int_{\mathcal{P}_0(\underline{2})} F$ is

$$\mathrm{Eq}\left( X_1 \times X_2 \times X_{12} \overset{a}{\underset{b}{\rightrightarrows}} X_1 \times X_2 \times X_{12} \times X_{12} \times X_{12} \right),$$

where $a$ and $b$ are precisely as given before Proposition 7.4.3, and hence $\int_{\mathcal{P}_0(\underline{2})} F \cong \lim(X_1 \to X_{12} \leftarrow X_2)$. We leave it to the reader to check the details.  □

We did not use the full bifunctoriality in the last example. Any functor $F \colon I \to C$ can be trivially viewed as a bifunctor $F \colon I^{\mathrm{op}} \times I \to C$ via composition with the canonical projection $I^{\mathrm{op}} \times I \to I$. Such a composition yields a functor which is independent of the first argument.

**Proposition 7.4.9** (Limit as end)  *Let $F \colon I \to C$ be a functor with $C$ complete and $I$ small, and let $P \colon I^{\mathrm{op}} \times I \to I$ be the canonical projection. Then there is a natural isomorphism*

$$\lim_{I} F \cong \int_{I} F \circ P.$$

*Proof*   Clear using the coequalizer model (7.4.2) and Proposition 7.4.3.     □

Building on Example 7.4.8, and finally making use of bifunctoriality, we can realize the homotopy pullback of spaces as an end of a bifunctor.

**Example 7.4.10**   Recall the covariant functor $\Delta(\bullet) \colon \mathcal{P}(\underline{n}) \to$ Top from Example 5.1.5. Let $F \colon \mathcal{P}_0(\underline{2})^{\mathrm{op}} \times \mathcal{P}_0(\underline{2}) \to$ Top be given by $F(S,T) = \mathrm{Map}(\Delta(S), X_T)$. Then

$$\int_{\mathcal{P}_0(\underline{2})} F \cong \mathrm{holim}(X_1 \to X_{12} \leftarrow X_2).$$

To see this, we simply unravel the definitions. For brevity write $\mathrm{Map}(X, Y) = Y^X$, and let $\gamma \in \mathrm{Map}(\Delta(12), Y)$ in coordinates be given by $(t_1, t_2) \mapsto \gamma(t_1, t_2)$, where of course $t_1 + t_2 = 1$. We have

$$\int_{\mathcal{P}_0(\underline{2})} F = \mathrm{Eq}\left( X_1^{\Delta(1)} \times X_2^{\Delta(2)} \times X_{12}^{\Delta(12)} \underset{b}{\overset{a}{\rightrightarrows}} X_1^{\Delta(1)} \times X_2^{\Delta(2)} \times X_{12}^{\Delta(12)} \times X_{12}^{\Delta(1)} \times X_{12}^{\Delta(2)} \right)$$

where

$$a(x_1, x_2, \gamma_{12}) = (x_1, x_2, \gamma_{12}, f_1(x_1), f_2(x_2))$$

and

$$b(x_1, x_2, \gamma_{12}) = (x_1, x_2, \gamma_{12}, \gamma_{12}(1, 0), \gamma_{12}(0, 1)).$$

Thus the equalizer is homeomorphic to the subspace of $X_1 \times X_{12}^I \times X_2$ of all $(x_1, \gamma_{12}, x_2)$ such that $\gamma_{12}(0) = f_1(x_1)$ and $\gamma_{12}(1) = f_2(x_2)$, which is by definition the homotopy limit of $X_1 \to X_{12} \leftarrow X_2$, Definition 3.2.4.     □

Dual to the notion of an end, Definition 7.4.6, we have the following.

**Definition 7.4.11**   The *coend* of a bifunctor $F \colon I^{\mathrm{op}} \times I \to C$, denoted by $\int^I F$, is defined as

$$\int^I F = \mathrm{Coeq}\left( \coprod_{i \to i'} F(i', i) \underset{b}{\overset{a}{\rightrightarrows}} \coprod_i F(i, i) \right),$$

where the maps are induced as follows: let $m \colon i \to i'$ be a morphism and let $1_i$ and $1_{i'}$ denote the identity morphisms. Then $a$ is induced by $F(1_i, m)$ and $b$ is induced by $F(m, 1_{i'})$.

Dual to Example 7.4.8, we have the following.

**Example 7.4.12**  Consider the bifunctor $F \colon \mathcal{P}_1(\underline{2})^{\mathrm{op}} \times \mathcal{P}_1(\underline{2}) \to$ Top given by $F(S, T) = X_T$. Then $\int^{\mathcal{P}_1(\underline{2})} F \cong \operatorname{colim}(X_1 \leftarrow X_{\emptyset} \to X_2)$. We leave this as an exercise for the reader.                                                     □

Then we have the analog of Proposition 7.4.9.

**Proposition 7.4.13** (Colimit as coend)  *Let $F \colon I \to C$ be a functor with $C$ cocomplete and $I$ small, and let $P \colon I^{\mathrm{op}} \times I \to I$ be the canonical projection. Then there is a natural isomorphism*

$$\operatorname*{colim}_{I} F \cong \int^{I} F \circ P.$$

*Proof*  Again clear using the coequalizer model (7.4.4) and Proposition 7.4.5.
                                                                        □

**Example 7.4.14**  Recall the contravariant functor $\Delta(\underline{n} - \bullet) \colon \mathcal{P}(\underline{n}) \to$ Top from Example 5.1.5. Consider the bifunctor $F \colon \mathcal{P}_1(\underline{2})^{\mathrm{op}} \times \mathcal{P}_1(\underline{2}) \to$ Top given by $F(S, T) = \Delta(\underline{2} - S) \times X_T$. Then

$$\int^{\mathcal{P}_1(\underline{2})} F \cong \operatorname{hocolim}(X_1 \xleftarrow{f_1} X_{\emptyset} \xrightarrow{f_2} X_2).$$

The proof is dual to the explanation given in Example 7.4.10 and we leave it to the reader.                                                                □

We now briefly discuss Kan extensions. These are indexed by over- and under-categories which appear in Definition 8.1.1. The reader who does not to wish to jump ahead to that definition at the moment may safely skip the next example; Kan extensions will not appear again until Definition 8.4.2.

**Example 7.4.15** (Kan extensions)  Let $F \colon I \to C$ be functor with $C$ complete and cocomplete, and let $G \colon I \to \mathcal{J}$ be a functor, with both $I$ and $\mathcal{J}$ small. There are two natural notions of extensions of $F$ to $\mathcal{J}$. Define $L_G F, R_G F \colon \mathcal{J} \to C$ by the formulas

$$L_G F(j) = \operatorname*{colim}_{(i,m) \in G \downarrow j} F(i)$$

and

$$R_G F(j) = \operatorname*{lim}_{(i,m) \in j \downarrow G} F(i),$$

called the *left (resp. right) Kan extension of $F$ along $G$*.

One important special case to consider is where $G \colon \mathcal{I} \to \mathcal{J}$ is the inclusion of a subcategory. Note that if $j$ is an object of $\mathcal{I}$, then $I \downarrow j$ has a final object, namely $j \downarrow j$, and hence $L_G F(j) = F(j)$ by Example 7.3.27, and similarly $j \downarrow G$ has an initial object $j \downarrow j$, and hence $R_G F(j) = F(j)$ by Example 7.3.9.

Limits and colimits are Kan extensions. If we let $\mathcal{J} = *$ be the category with a single object, and $G \colon \mathcal{I} \to *$ the canonical functor, then $L_G F = \mathrm{colim}_I F$ and $R_G F = \lim_I F$. □

We give one last model for limits and colimits, but now we will restrict our attention to the category of spaces Top (everything works the same way in Top$_*$).

Recall from Example 7.1.23 the constant functor at a point, $C_*$. Then if $X \colon \mathcal{I} \to$ Top is a diagram of spaces, one can consider the set of natural transformations between the constant diagram at a point in the shape of $\mathcal{I}$ and $X$,

$$\mathrm{Nat}_{\mathcal{I}} (C_*, X).$$

This has the structure of a topological space,, since

$$\mathrm{Nat}_{\mathcal{I}} (C_*, X) \subset \prod_{i \in I} \mathrm{Map}(*, X_i) \cong \prod_{i \in I} X_i.$$

This is analogous to the situation described in Definition 5.1.3.

**Proposition 7.4.16** (Limit as natural transformations)   *Suppose $X$ and $C_*$ are as in above. There is a canonical homeomorphism*

$$\lim_{\mathcal{I}} X \cong \mathrm{Nat}_{\mathcal{I}} (C_*, X).$$

*Proof*   Using the identification $\mathrm{Map}(*, X_i) \cong X_i$ and the canonical projections $p_i \colon \prod_i \mathrm{Map}(*, X_i) \to \mathrm{Map}(*, X_i)$, note that the following diagram commutes for all morphisms $i \to i'$:

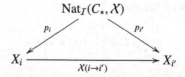

But these are the same canonical projections that are used if we replace $\mathrm{Nat}(C_*, X)$ with $\lim_I X$ (both are, after all, subspaces of $\prod_i X_i$), and so the unique canonical map $\mathrm{Nat}(C_*, X) \to \lim_I X$ must be equal to the identity.   □

The result is of course a generalization of our definition of the limit of a punctured cube, Definition 5.2.1. It says that the limit consists of those tuples in the product of the spaces in the diagram that "map compatibly through the diagram" in the sense that, whenever two points map to the same space, their images should agree; this is precisely what is dictated by naturality. We will see the analog of this model for the limit in Proposition 8.2.2 where we will describe *homotopy* limits of spaces using natural transformations.

**Remark 7.4.17** We can view Proposition 7.4.16 through the lens of ends. For concreteness, rewriting the bifunctor in Example 7.4.8 as $F(S, T) = \mathrm{Map}(*_S, X_T)$, the end of $F$ is then a subspace of $\mathrm{Map}(*, X_1) \times \mathrm{Map}(*, X_2) \times \mathrm{Map}(*, X_{12})$, and we saw it satisfied the appropriate universal property. □

Dually, instead of maps of a one-point space giving the limit of a diagram of spaces, the colimit can be realized in terms of products with the one-point space. This may seem redundant since this model will not add anything substantial to the ones we have already encountered, but it will be the one that will generalize nicely to homotopy colimit of a diagram in Section 8.2.2. It is an easy exercise to deduce the following from either of the coequalizer or the pushout models for the colimit (Proposition 7.4.5):

**Proposition 7.4.18** *For* $F \colon I \to \mathrm{Top}$ *a diagram, there is a homeomorphism*

$$\operatorname*{colim}_{I} F \cong \left( \coprod_{i \in I} X_i \right)\!/\!\sim \tag{7.4.5}$$

*where the equivalence relation is generated by: for* $x \in X_i$ *and* $x' \in X_j$, $x \sim x'$ *if there exists a map* $f$ *in the diagram such that* $f(x) = x'$.

It should be clear that this proposition is a direct generalization of the definition of the colimit of a punctured cube, Definition 5.6.1.

To really dualize the description of the limit from Proposition 7.4.16, we can simply take products with the one-point space:

$$\operatorname*{colim}_{I} F \cong \left( \coprod_{i \in I} X_i \times * \right)\!/\!\sim \tag{7.4.6}$$

where the maps in the diagram are the same as in the original one with the identity on the one-point space, and the equivalence relation is the obvious extension of the one above. It may appear that taking products with points unnecessarily complicates the pictures, but the point is that we will obtain an analogous model for the *homotopy* colimit of a diagram in Remark 8.2.13 by essentially "blowing up" the points in the above expression.

## 7.5 Algebra of limits and colimits

We studied the algebra of homotopy pullbacks and pushouts in Section 3.9, and it should be apparent that these results depended on interactions between limits and colimits since pullbacks and pushouts are examples of those. This section describes some more general results about how (co)limits interact with themselves, each other, and other functors. As usual, we will assume all the indexing categories in sight are small, and all the categories in which our functors take values are complete and cocomplete.

We first now examine the interaction of limits and colimits with the functors $\text{Map}(-, Z)$ and $\text{Map}(Z, -)$ (see Example 7.1.17). Given $F \colon I \to \text{Top}$, we can compose it with these functors to get diagrams $\text{Map}(F, Z)$ and $\text{Map}(Z, F)$, by which we mean diagrams $F$ but with each space $F(i)$ replaced by $\text{Map}(F(i), Z)$ or $\text{Map}(Z, F(i))$ (the former also has arrows reversed since $\text{Map}(-, Z)$ is contravariant). These can also be thought of as functors in the $Z$ variable. That is, if $f \colon Z \to W$ is a map, we have induced natural transformations $\text{Map}(F, Z) \to \text{Map}(F, W)$ and $\text{Map}(W, F) \to \text{Map}(Z, F)$.

For each $i \in I$ there are canonical maps $\lim_I F \to F(i)$ and $F(i) \to \text{colim}_I F$ which are compatible as $i$ varies, that is, there is a map from the space $\lim_I F$ to the $I$-diagram $F$. This induces a map $\text{Map}(Z, \lim_I F) \to \text{Map}(Z, F)$. By the universal property of limits, anything that maps to the diagram $\text{Map}(Z, F)$ factors through $\lim_I \text{Map}(Z, F)$, and hence there is a canonical map $\text{Map}(Z, \lim_I F) \to \lim_I \text{Map}(Z, F)$. Similarly, we have a map $\text{Map}(\text{colim}_I F, Z) \to \lim_{I^{op}} \text{Map}(F, Z)$.

**Proposition 7.5.1** (Maps and (co)limits) *Let* $F \colon I \to \text{Top}$ *(or* $\text{Top}_*$*) be a functor with* $I$ *small. The canonical maps*

$$\text{Map}\left(Z, \lim_I F\right) \xrightarrow{\cong} \lim_I \text{Map}(Z, F)$$

$$\text{Map}\left(\text{colim}_I F, Z\right) \xrightarrow{\cong} \lim_{I^{op}} \text{Map}(F, Z)$$

*are homeomorphisms that are natural in the Z variable.*

The proof of this statement can be found, for example, in [ML98, Theorem 1, p. 116] (for the proof of the second homeomorphism, see the discussion following the proof of that theorem).

**Remark 7.5.2**    A functor such as $\mathrm{Map}(Z, -)$ for which the first homeomorphism in the above proposition holds is said to *preserve limits*. Even though there is a map $\mathrm{colim}_I \mathrm{Map}(Z, F) \to \mathrm{Map}(Z, \mathrm{colim}_I F)$ (not hard to see), this is not necesarily a homeomorphism, so the functor $\mathrm{Map}(Z, -)$ does not preserve colimits. However, in general if a functor has a left adjoint, then it preserves limits and this adjoint preserves colimits (see e.g. [Bor94, Proposition 3.2.2] and the discussion at the end of Section 5, p. 119 of [ML98]). We could have thus used this general fact to observe that $\mathrm{Map}(Z, -)$ preserves limits since it has a left adjoint, $Z \times -$, which in turn preserves colimits (this fact is not hard to establish).                                                                        □

For the next observation, recall the notion of a product category from Definition 7.1.7 and consider the product of two small indexing categories, $I \times J$. Given a functor $F \colon I \times J \to C$, it can be regarded as a diagram of diagrams in two ways: as a diagram $J \to C^I$ of $I$-diagrams indexed by $J$ or a diagram $I \to C^J$ of $J$-diagrams indexed by $I$. Then fixing $i \in I$ or $j \in J$ gives diagrams $I \to C$ and $J \to C$. In other words, $F$ can be restricted to subcategories $i \times I$ or $J \times j$ for any $i \in I$ or $j \in J$, which are naturally isomorphic to $I$ and $J$, respectively. It thus makes sense to consider (co)lim over one of the categories and then over the other, as well as over $I \times J$, regarded as a single category. We then have the following.

**Theorem 7.5.3** ((Co)limits commute with (co)limits)    *Suppose* $F \colon I \times J \to C$ *is any functor with* $I$ *and* $J$ *small and* $C$ *complete and cocomplete. Then there are canonical isomorphisms*

$$\lim_{I \times J} F \cong \lim_{J} \lim_{I} F \cong \lim_{I} \lim_{J} F,$$

$$\mathrm{colim}_{I \times J} F \cong \mathrm{colim}_{J} \mathrm{colim}_{I} F \cong \mathrm{colim}_{I} \mathrm{colim}_{J} F$$

*Proof*    We outline the proof in the case of limits and leave the other to the reader to dualize. For $i \in I$ fixed, consider the functor $F(i, -) \colon J \to C$ given by $j \mapsto F(i, j)$, and $(j \to j') \mapsto F(1_i, j \to j')$. Let $F_J = \lim_J F$ be the limit of this functor, where $F_J(i) = \lim_{j \in J} F(i, j)$. The universal property of limits tells us that $F_J \colon I \to C$ is also a functor. Let $L = \lim_{i \in I} F_J(i)$. Now consider the diagram

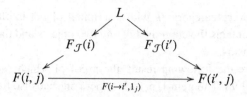

where the unmarked morphisms are the canonical projections. This tells us that $L$ is a cone on the functor $F(-, j) \colon \mathcal{J} \to C$, and hence there exists a unique morphism $L \to \lim_I F = F_I$, where $F_I(j) = \lim_{i \in I} F(i, j)$. Now we consider the diagram

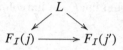

to see that $L$ is also a cone on the functor $F_I \colon \mathcal{J} \to C$, and hence there exists a unique morphism $L \to \lim_{\mathcal{J}} F_I$. What the above establishes is the existence of a unique morphism $L = \lim_I \lim_{\mathcal{J}} F \to \lim_{\mathcal{J}} \lim_I F$. By symmetry we also have a unique morphism $\lim_{\mathcal{J}} \lim_I F \to \lim_I \lim_{\mathcal{J}} F$. The composites are forced to be the identity by the uniqueness and the universal property. $\square$

We have already seen instances of (co)limits commuting. In particular, the reader should look back at Examples 2.2.12 and 2.4.15. One can also consider mixing limits and colimits, namely one can consider

$$\operatorname*{colim}_{\mathcal{J}} \lim_I F \quad \text{and} \quad \lim_I \operatorname*{colim}_{\mathcal{J}} F.$$

The two are not even weakly equivalent in general; see, for example, [Bor94, Counterexample 2.13.9] (that example shows the two are not homeomorphic but the same example works to show they cannot be weakly equivalent). However, if $F$ is a functor to sets, then something more can be said for special examples of $I$ and $\mathcal{J}$. We give some detail for this situation because it gives us a chance to define a filtered category, which is a generalization of the idea of a directed set.

**Definition 7.5.4** A (non-empty) category $\mathcal{J}$ is *filtered* if

- for any two objects $j$ and $j'$ in $\mathcal{J}$, there exists an object $k$ in $\mathcal{J}$ and morphisms $f \colon j \to k$ and $f' \colon j' \to k$;
- for any two morphisms $f, g \colon j \to j'$ in $\mathcal{J}$, there exists an object $k$ and a morphism $h \colon j' \to k$ such that $h \circ f = h \circ g$.

For example, every category $\mathcal{J}$ with a terminal object is filtered, as is any category that contains the coproduct of any two objects and the coequalizer of any two morphisms.

We then have the following result, the proof of which can be found, for example, in [ML98, Theorem 1, p. 215] (where one can also find the details on how the canonical map below is defined).

**Proposition 7.5.5** *Suppose* $F\colon \mathcal{I}\times\mathcal{J}\;\rightarrow\;$ Set *is a functor,* $\mathcal{I}$ *is a finite category (finite number of objects and morphisms) and* $\mathcal{J}$ *is a small filtered category. Then there is a canonical map*

$$\operatorname*{colim}_{\mathcal{J}}\;\operatorname*{lim}_{\mathcal{I}} F \;\longrightarrow\; \operatorname*{lim}_{\mathcal{I}}\operatorname*{colim}_{\mathcal{J}} F$$

*which is a bijection.*

We finish this section by reminding the reader that the main problem with limits and colimits is that they are not homotopy invariant. If $C = \mathrm{Top}$, even if the natural transformation $N\colon F \rightarrow G$ is a objectwise homotopy equivalence, the induced maps (7.3.5) and (7.3.12) need not be. This was in fact already illustrated by Examples 2.2 and 2.3 since fibers and cofibers are examples of limits and colimits (Examples 7.3.15 and 7.3.34). This is why it is necessary to develop the notion of *homotopy* limits and colimits, which we do in Chapter 8.

# 8

# Homotopy limits and colimits
# of diagrams of spaces

In this chapter, we generalize homotopy limits and colimits from cubical to arbitrary diagrams of spaces. It should be clear by now that the motivation for this is that limits and colimits are not homotopy invariant (see Examples 2.2 and 2.3) and that it is the job of *homotopy* limits and colimits to mend this defect. Just as homotopy limits and colimits of punctured cubes are "fattened up" versions of limits and colimits of such diagrams, so too are homotopy limits and colimits over arbitrary indexing categories. The bookkeeping for such fattenings is influenced by the "shape" of the diagram indexing the (co)limit, and involves a certain amount of machinery from the theory of simplicial sets. As we develop this machinery, we shall see how the construction of homotopy limits and colimits of punctured squares and cubes that we have already encountered in Chapters 3 and 5 is indeed a special case of this more general procedure, although this chapter can be read independently of those earlier ones.

The first detailed study of homotopy limits and colimits (in the category of simplicial sets) was undertaken by Bousfield and Kan [BK72a], who in turn followed Segal's definition of a homotopy colimit [Seg68]. Several other special cases were already considered by Puppe [Pup58], Milnor [Mil63], and others. Vogt [Vog73, Vog77] had also independently studied homotopy (co)limits in detail, although much of it in the setting of homotopy commutative diagrams. Special cases of Vogt's work were also considered by Mather [Mat76] (see Theorem 5.10.7 and Theorem 5.10.8).

Bousfield and Kan's preferred model for the homotopy limit of a diagram (of spaces or simplicial sets) is the totalization of the cosimplicial replacement of the diagram. One of the advantages of this point of view is that it leads to spectral sequences computing the homotopy and homology groups of the homotopy limit. We will devote considerable time to discussing this useful feature of the cosimplicial approach in Chapter 9.

One could also discuss homotopy (co)limits as objects that satisfy a homotopical version of the universal properties for (co)limits (see [BK72a], or [Hir03] for a more modern treatment), but the currently prevalent point of view is driven by the modern category-theoretic language of *model categories*. Homotopy (co)limits are *derived functors* of the (co)limit functors since the latter do not necessarily exist in the homotopy category and the former are thus their natural derived homotopy invariant replacements. The definition of homotopy (co)limits as derived functors also endows them with a universal property (which is "global" in the sense that it is universal with respect to all possible homotopy invariant replacements of the (co)limit). Some standard references for the derived point of view, including deeper study of the homotopy theory of diagrams are [CS02, DS95, DHKS04, DK84, DFZ86, DF87, Hir03].

While all of these approaches and models for homotopy (co)lims are useful, none of them is quite concrete enough for the scope of this book and some require a large investment for the reader in terms of learning the background material. We do not claim to improve upon this much, since we still rely upon the machinery of simplicial sets and their realizations (the necessary details of which have been relegated to the Appendix, Section A.1) and upon the familiarity of the reader with CW/simplicial complexes. However, we will in this chapter remain grounded in the category of topological spaces. In contrast, limits and colimits were relatively easy to consider in any category, as they are defined by universal properties. Such an approach, at least naively, will fail because homotopy equivalence is a weaker notion of isomorphism than the natural one in a given category. Hence there is no counterpart to Section 7.3 in this chapter. As we are driven by examples, specifically in the category of spaces, we will mainly use the approach via (co)equalizers and natural transformations since they are the most hands-on and, because they employ the definition of the geometric realization of a category, are well suited to Top.

## 8.1 The classifying space of a category

Limits (in Top) can be thought of as collections of points in the product of all the spaces in the diagram which map compatibly through the diagram. That is, if $X_i$ and $X_j$ are spaces in the diagram $X: I \to$ Top, where $I$ is small, with a map $X_i \to X_j$ arising from a morphism $i \to j$, then if $x_i \in X_i$ and $x_j \in X_j$ are part of a tuple of points in the limit, the image of $x_i$ in $X_j$ is equal to $x_j$. This is what is captured in the interpretation of the limit as the space of natural transformations of the constant diagram into the diagram $X$ (Proposition 7.4.16).

One way to think about the homotopy limit is in a similar manner, except that the points match "up to coherent homotopy". That is, with $x_i, x_j$ as above, instead of insisting that $x_i$ map to $x_j$, we insist that there is a path in $X_j$ between the image of $x_i$ and $x_j$, and this path is part of the data we collect. Moreover, we keep track of higher homotopies as well: if, for instance, part of the diagram consists of maps $X_i \to X_j \to X_k$, and $x_i, x_j, x_k$ lie in the respective spaces, then part of our data will be a path in $X_j$ between the image of $x_i$ and $x_j$, a path in $X_k$ from the image of $x_j$ to $x_k$, a path in $X_k$ between the image of $x_i$ and $x_k$ in $X_k$ via the composed map, and a "path of paths" (the image of a 2-simplex) in $X_k$ "filling in" the image in $X_k$ of the path in $X_j$ with the other two paths in $X_k$. These paths and higher homotopies are all a part of the data that comprise a point in the homotopy limit. Thus for every string of maps $X_{i_0} \to \cdots \to X_{i_k}$ we will have a map of $\Delta^k$ into $X_{i_k}$ satisfying certain conditions.

Similarly, colimits are essentially quotients of the union of the spaces in the diagram by subspaces determined by where points map. Then homotopy colimits are constructed from coherent mapping cylinders of the spaces in the diagram with face identifications according to the maps in the diagram.

The "shape" of the diagram indexing the (co)limit is therefore of interest. In the situation above we have associated to the sequence $i \to j \to k$ of morphisms in $I$ a 2-simplex which records coherence of a 2-parameter family of homotopies which takes place in $X_k$. In general we will associate to a chain $i_0 \to \cdots \to i_k$ in $I$ a $k$-simplex with a map to $X_{i_k}$ recording homotopical coherence. This shape of a diagram is encoded by the nerve of the associated category (Definition 8.1.8), which associates to a category a simplicial set whose simplices are strings of morphisms, and then, by geometric realization (Definition A.1.2), a topological space. This realization may be familiar from elsewhere; one model for the classifying space of a group regards the group as a category and then takes its realization (Example 8.1.18).

In trying to understand the system of coherent homotopies at a space, we will be interested in the local shape of the indexing category near the object indexing that space. This is captured by the notions of overcategory and undercategory.

**Definition 8.1.1** Suppose $I$ is a small category and $i$ is an object of $I$.

- The *category of objects over $i$*, denoted by $I \downarrow i$, is the category whose objects are morphisms $i' \to i$ and whose morphisms $(i' \to i) \to (i'' \to i)$ are commutative diagrams

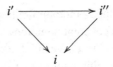

- The *category of objects under i*, denoted by $i \downarrow \mathcal{I}$, is the category whose objects are morphisms $i \to i'$ and whose morphisms $(i \to i') \to (i \to i'')$ are commutative diagrams

The identity morphism $i \to i$ is a final object in $\mathcal{I} \downarrow i$ and an initial object in $i \downarrow \mathcal{I}$. We refer to $\mathcal{I} \downarrow i$ as an *overcategory* and $i \downarrow \mathcal{I}$ as an *undercategory*. Note that the overcategory and undercategory are dual in the sense that there is a natural isomorphism of categories $(i \downarrow \mathcal{I})^{\mathrm{op}} \cong \mathcal{I}^{\mathrm{op}} \downarrow i$. Moreover, if $\mathcal{I}$ has a final object $i_f$, then there is a natural isomorphism $\mathcal{I} \downarrow i_f \cong \mathcal{I}$ given by forgetting the final object, and similarly if it has an initial object.

These definitions can be extended to functors between categories in the following way:

**Definition 8.1.2** If $F: \mathcal{I} \to \mathcal{J}$ is a functor, and $j$ is an object of $\mathcal{J}$, we define the *category of objects of $\mathcal{I}$ over $j$*, denoted $F \downarrow j$, as consisting of pairs $(i, m)$ where $m: F(i) \to j$ is a morphism. A morphism $(i, m) \to (i', m')$ in this category is a morphism $i \to i'$ such that $m' \circ F(i \to i') = m$. That is, the diagram

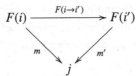

commutes. In a similar way we define $j \downarrow F$, the *category of objects of $\mathcal{I}$ under j*. We may refer to $F \downarrow j$ as an *overcategory of F* and $j \downarrow F$ as an *undercategory of F*.

In the special case where $F = 1_{\mathcal{I}}$ is the identity functor in the above definition we recover the notions of overcategory and undercategory.

**Example 8.1.3** For the category $\mathcal{P}(\underline{1}) = (\emptyset \to \{1\})$, $\mathcal{P}(\underline{1}) \downarrow \{1\}$ consists of two objects: $\emptyset \to \{1\}$ and $\{1\} \to \{1\}$ and one non-identity morphism induced by the unique morphism $\emptyset \to \{1\}$; of course $\mathcal{P}(\underline{1}) \downarrow \{1\} \cong \mathcal{P}(\underline{1})$. The category $\{1\} \downarrow \mathcal{P}(\underline{1})$ contains only the object $\{1\} \to \{1\}$ and its identity morphism. □

**Example 8.1.4** Consider $\mathcal{P}_0(\underline{2}) = (\{1\} \to \{1,2\} \leftarrow \{2\})$. Then $\mathcal{P}_0(\underline{2}) \downarrow \{1,2\}$ consists of three objects, $\{1,2\} \to \{1,2\}$, $\{1\} \to \{1,2\}$, and $\{2\} \to \{1,2\}$, and two morphisms induced by the unique morphisms $\{1\} \to \{1,2\}$ and $\{2\} \to \{1,2\}$. This can be drawn as

$$(\{1\} \to \{1,2\}) \longrightarrow (\{1,2\} \to \{1,2\}) \longleftarrow (\{2\} \to \{1,2\}),$$

and $\mathcal{P}_0(\underline{2}) \cong \mathcal{P}_0(\underline{2}) \downarrow \{1,2\}$. The category $\{1,2\} \downarrow \mathcal{P}_0(\underline{2})$ consists of a single object and morphism. □

**Example 8.1.5** Let $\mathcal{P}(\underline{n})$ be the usual cubical indexing category (see beginning of Section 5.1), and let $S \subset \underline{n}$ be fixed. Then $(\mathcal{P}(\underline{n}) \downarrow S)$ consists of an object for each subset of $S$ and a morphism for each inclusion $R_1 \to R_2$ of subsets of $S$. There is a natural isomorphism $\mathcal{P}(\underline{n} \downarrow S) \cong \mathcal{P}(S)$ given on objects by $(R \to S) \mapsto R$, as $S$ is a final object in $\mathcal{P}(\underline{n}) \downarrow S$. It restricts to an isomorphism $\mathcal{P}_0(\underline{n} \downarrow S) \cong \mathcal{P}_0(S)$

The story is similar for $S \downarrow \mathcal{P}(\underline{n})$: There is one object for each subset of $\underline{n}$ containing $S$, and one morphism for each inclusion of such subsets, and there is a natural isomorphims $S \downarrow \mathcal{P}(\underline{n}) \cong \mathcal{P}(\underline{n} - S)$ given by sending $R$ to $R - S$, and it restricts to an isomorphism $S \downarrow \mathcal{P}_1(\underline{n}) \cong \mathcal{P}_1(\underline{n} - S)$. □

**Example 8.1.6** If $I = (i_1 \overset{a}{\underset{b}{\rightrightarrows}} i_2)$, then $I \downarrow i_2$ consists of three objects and two non-identity morphisms. In fact this overcategory is isomorphic to $\mathcal{P}_0(\underline{2})$. Here is a picture of $I \downarrow i_2$:

$$(i_1 \overset{a}{\to} i_2) \overset{a}{\longrightarrow} (i_2 \overset{1_{i_2}}{\to} i_2) \overset{b}{\longleftarrow} (i_1 \overset{b}{\to} i_2)$$

Note that while $i_2$ is not a final object in $I$, the identity morphism $i_2 \to i_2$ represents a final object in $I \downarrow i_2$. □

**Example 8.1.7** If $I = (1 \leftarrow 2 \leftarrow 3 \leftarrow 4 \leftarrow \cdots)$ indexes an inverse system (as in Example 7.3.12), then $I \downarrow i$ has an object $j \to i$ for each $j > i$ and one morphism for each $j > k > i$ induced by the unique morphism $j \to k$. The category $i \downarrow I$ is finite. □

If $f : i \to i'$ is a morphism in $I$, then there are induced functors

$$I \downarrow i \longrightarrow I \downarrow i' \quad \text{and} \quad i' \downarrow I \longrightarrow i \downarrow I \tag{8.1.1}$$

given by composition with $f$. Every object over $i$ is an object over $i'$ and every object under $i'$ is an object under $i$ by virtue of the morphism $f \colon i \to i'$. The over/under categories may be thought of as values of functors

$$\mathcal{I} \downarrow - \colon \mathcal{I} \longrightarrow \text{Cat}$$

and

$$- \downarrow \mathcal{I} \colon \mathcal{I}^{\text{op}} \longrightarrow \text{Cat}$$

given on objects by $i \mapsto \mathcal{I} \downarrow i$ and $i \mapsto i \downarrow \mathcal{I}$ respectively, and on morphisms in the evident way.

We now describe a way to associate a simplicial set to a category, known as the *nerve*. The reader may wish to refer to Section A.1 for background on simplicial sets and in particular for the definition of the geometric realization (Definition A.1.2).

**Definition 8.1.8**   Given a small category $\mathcal{I}$, define a simplicial set $B_\bullet \mathcal{I}$, called the *nerve of* $\mathcal{I}$, where the set of $n$-simplices $B_n \mathcal{I}$, $n \geq 0$, is the set of all chains of morphisms

$$i_0 \to i_1 \to i_2 \to \cdots \to i_n$$

in $\mathcal{I}$. The face maps $d_j \colon B_n \mathcal{I} \to B_{n-1} \mathcal{I}$ are given by removing $i_j$ and using the composition $i_{j-1} \to i_{j+1}$. That is, $d_j$ is "compose at $i_j$"; in the case of $d_0$ (resp. $d_n$) we simply remove $i_0$ (resp. $i_n$). The degeneracy maps $s_j \colon B_n \mathcal{I} \to B_{n+1} \mathcal{I}$ are given by inserting the identity morphism at $i_j$. A chain/simplex $i_0 \to \cdots \to i_n$ is called *non-degenerate* if it is not in the image of $s_j$ for any $j$. Equivalently, no morphism in the chain is the identity map. We will denote by $\mathrm{nd}B_n(\mathcal{I})$ the set of non-degenerate $n$-simplices in the nerve of $\mathcal{I}$.

The geometric realization of $B_\bullet \mathcal{I}$ is called the *classifying space* of the category or the *(geometric) realization* of the category and is denoted by $|\mathcal{I}|$ (instead of the more cumbersome but more correct $|B_\bullet \mathcal{I}|$).

**Remark 8.1.9**   The realization $|\mathcal{I}|$ of the nerve of $\mathcal{I}$ is the realization of a simplicial complex with one simplex $\Delta^k$ for each non-degenerate string of morphisms $i_0 \to \cdots \to i_k$ (see discussion following Definition A.1.2 for why we can disregard the degenerate strings), which are glued together via the face and degeneracy maps described above. Each point in $|\mathcal{I}|$ is indexed uniquely by a non-degenerate chain $i_0 \to \cdots \to i_k$ for some $k$. We shall see that it is quite often useful to decompose the realizations of various categories into their simplices in this way.                                                                $\square$

This process that associates to a category $\mathcal{I}$ its realization $|\mathcal{I}|$ is functorial in the sense that a functor $F\colon \mathcal{I} \to \mathcal{J}$ gives rise to a continuous map $|\mathcal{I}| \to |\mathcal{J}|$, as the assignment $\mathcal{I} \to B_\bullet\mathcal{I}$ is itself a functor from categories to simplicial sets, and the realization of a simplicial set is a functor from simplicial sets to topological spaces (see (A.2)).

**Example 8.1.10**  If $\mathcal{I} = \{i_0, i_1\}$ is the discrete category with two objects, then $|\mathcal{I}|$ is a discrete set consisting of two points. The reason is that there are no non-degenerate simplices in degree greater than zero in $B_\bullet\mathcal{I}$, and there are precisely two simplices of degree zero, each associated with one of the two distinct objects of $\mathcal{I}$. More precisely, there are two simplices in $B_\bullet\mathcal{I}$ of degree $n$ for each $n$, each represented by a string of identity maps. Then, for any point $t \in \Delta^n$, $(i_j \to \cdots \to i_j, t) = (s_0^n(i_j), t) \sim (i_j, s^0(t)) = (i_j, 1)$ (recalling from Definition 1.1.1 that 1 is the unique point in the 0-simplex $\Delta^0$). Since there are obviously no strings of morphisms containing both $i_0$ and $i_1$, the realization is evidently two points, naturally labeled by the objects of $\mathcal{I}$. □

**Example 8.1.11**  For $\mathcal{P}(\underline{1}) = (\emptyset \to \{1\})$, we have $|\mathcal{P}(\underline{1})| \cong I$. To see this, the simplicial set $B_\bullet\mathcal{P}(\underline{1})$ consists of two 0-simplices, one for each of the objects, and a single non-degenerate 1-simplex given by the morphism $\emptyset \to \{1\}$. The other two 1-simplices are degenerate. For $n \geq 2$, $B_n(\mathcal{P}(\underline{1}))$ consists of $n + 2$ elements (two of which are strings of only identity morphisms, and the other $n$ of which consist of the morphism $\emptyset \to \{1\}$ in one place and identity morphisms elsewhere). Recalling the coordinates on simplices from Definition 1.1.1 and the definition of the realization, Definition A.1.2, we have

$$(\emptyset \to \{1\}, (0, 1)) = (\emptyset \to \{1\}, d^0(1)) \sim (d_0(\emptyset \to \{1\}), \{1\}) = (\{1\}, 1)$$

and

$$(\emptyset \to \{1\}, (1, 0)) = (\emptyset \to \{1\}, d^1(1)) \sim (d_1(\emptyset \to \{1\}), 1) = (\emptyset, 1).$$

Thus the non-degenerate 1-simplex associated with $\emptyset \to \{1\}$ is glued along its boundary to the two 0-simplices associated with $\emptyset$ and $\{1\}$. All higher simplices are either identified with the non-degenerate 1-simplex or its endpoints. A picture of the realization is given in Figure 8.1.

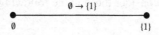

Figure 8.1 Realization of $\mathcal{P}(\underline{1})$.

□

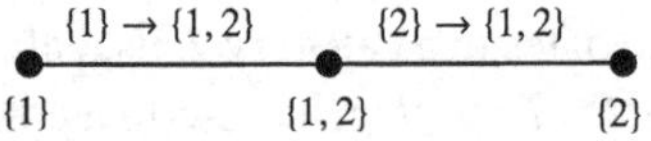

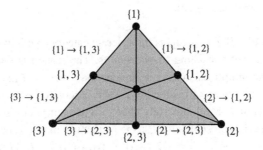

Figure 8.2  Realization of $\mathcal{P}_0(\underline{2})$.

Figure 8.3  Realization of $\mathcal{P}_0(\underline{3})$. The inside vertex is $\{1, 2, 3\}$ and the inner edges correspond to the morphisms $\{i\} \to \{1, 2, 3\}$ and $\{i, j\} \to \{1, 2, 3\}$ in the obvious way.

**Example 8.1.12**   For $\mathcal{P}_0(\underline{2}) = (\{1\} \to \{1, 2\} \leftarrow \{2\})$, we have $|\mathcal{P}_0(\underline{2})| \cong I$ as follows. The simplicial set $B_\bullet \mathcal{P}_0(\underline{2})$ consists of three 0-simplices (one for each object), two non-degenerate 1-simplices associated with the arrows $\{1\} \to \{1, 2\}$, and $\{2\} \to \{1, 2\}$, and the rest of the simplices are all degenerate. These two non-degenerate 1-simplices are glued together along the part of their boundary indexed by $\{1, 2\}$. This works in much the same way as in Example 8.1.11. A picture of the realization is given in Figure 8.2.                                                                                    □

**Example 8.1.13**   For the category $\mathcal{P}(\underline{n})$, we have $|\mathcal{P}(\underline{n})| \cong I^n$, where $I^n$ is subdivided into $n$-simplices associated with strictly increasing chains of subsets $\emptyset = S_0 \subsetneq S_1 \subsetneq \cdots \subsetneq S_n = \underline{n}$.

Also, $|\mathcal{P}_0(\underline{n})| \cong \Delta^{n-1}$, and in fact the geometric realization is a barycentrically subdivided $(n-1)$-simplex. Each non-degenerate simplex of dimension $n$ corresponds to a string $S_1 \subset S_2 \subset \cdots \subset S_n = \underline{n}$ where $|S_i| = i$. A picture for the case $n = 3$ is given in Figure 8.3. Similarly, $|\mathcal{P}_1(\underline{n})| \cong \Delta^{n-1}$. We leave the details to the reader.                                                                                    □

We will mostly be interested in the realizations of various overcategories and undercategories, so we consider those in the following examples.

**Example 8.1.14**   For $\mathcal{P}(\underline{1}) = (\emptyset \to \underline{1})$, we have $|\mathcal{P}(\underline{1}) \downarrow \emptyset| = *$. Also $\mathcal{P}(\underline{1}) \downarrow \{1\} \cong \mathcal{P}(\underline{1})$, and so as in Example 8.1.11, $|\mathcal{P}(\underline{1}) \downarrow \{1\}| \cong I$.                                                □

**Example 8.1.15** For $\mathcal{P}_0(\underline{2}) = (\{1\} \to \{1,2\} \leftarrow \{2\})$, Example 8.1.12 shows that $|\mathcal{P}_0(\underline{2}) \downarrow \{1,2\}| \cong I$ since $\mathcal{P}_0(\underline{2}) \downarrow \{1,2\} \cong \mathcal{P}_0(\underline{2})$. Also $|\mathcal{P}_0(\underline{2}) \downarrow \{i\}| \cong *$ for $i = 1, 2$ since, for such $i$, the overcategory consists of a single object and a single morphism.

Since the category $\mathcal{P}_1(\underline{2}) = (\{1\} \leftarrow \emptyset \to \{2\})$ is isomorphic to $\mathcal{P}_0(\underline{2})^{\text{op}}$, we also have $|\emptyset \downarrow \mathcal{P}_1(\underline{2})| \cong I$, and $|\{i\} \downarrow \mathcal{P}_2(\underline{2})| \simeq *$ for the same reason that $|\mathcal{P}_0(\underline{2}) \downarrow \{i\}| \cong *$. □

**Example 8.1.16** Let $\mathcal{I} = (1 \leftarrow 2 \leftarrow \cdots)$ index an inverse system (we already considered it in Example 8.1.7). Then $|\mathcal{I}|$ is an infinite-dimensional simplex whose $n$-dimensional faces are indexed by strings $i_0 \leftarrow i_1 \leftarrow \cdots \leftarrow i_n$ with $i_0 < i_1 < \cdots < i_n$. The space $|\mathcal{I} \downarrow i|$ is also an infinite-dimensional simplex for any $i$. Moreover, for each $i \to j$ in $\mathcal{I}$, the induced map $|\mathcal{I} \downarrow j| \to |\mathcal{I} \downarrow i|$ is the inclusion of one of the infinite-dimensional faces in the infinite-dimensional simplex $|\mathcal{I} \downarrow i|$. In fact, it is not hard to see that this face is of codimension $j - i$.

Also consider $\mathcal{I}^{\text{op}} = (1 \to 2 \to \cdots)$, the category indexing direct limits. The natural isomorphism $(i \downarrow \mathcal{I}^{\text{op}})^{\text{op}} \cong \mathcal{I} \downarrow i$ immediately gives the same interpretation as above for these categories and their realizations. □

As we have mentioned above, a functor $F \colon \mathcal{I} \to \mathcal{J}$ induces a map of nerves $B_{\bullet}\mathcal{I} \to B_{\bullet}\mathcal{J}$ and hence a continuous map of realizations:

$$|F| \colon |\mathcal{I}| \longrightarrow |\mathcal{J}|.$$

Furthermore, a natural transformation $N \colon F \to G$ between functors $F, G \colon \mathcal{I} \to \mathcal{J}$ gives a homotopy between the maps $|F|$ and $|G|$. To see why, note that $N$ can itself be regarded as a functor $N \colon \mathcal{I} \times (\emptyset \to \{1\}) \to \mathcal{J}$, where $N(c, \emptyset) = F(c)$ and $N(c, \{1\}) = G(c)$. Moreover, $|\mathcal{I} \times (\emptyset \to \{1\})| \cong |\mathcal{I}| \times I$ by Theorem A.1.4, and $|N| \colon |\mathcal{I}| \times I \to |\mathcal{J}|$ is in fact a homotopy between $|F|$ and $|G|$ (we omit a careful proof). We will use this last fact in the proof of the following proposition.

**Proposition 8.1.17** *If $\mathcal{I}$ has an initial or a final object, then $|\mathcal{I}|$ is contractible.*

*Proof* If $1_{\mathcal{I}} \colon \mathcal{I} \to \mathcal{I}$ is the identity functor and $F_f \colon \mathcal{I} \to \mathcal{I}$ is the constant functor which sends every object to the final object $i_f \in C$, then there is a natural transformation $1_{\mathcal{I}} \to F_f$, since there is always a unique morphism to the final object. This natural transformation gives rise to a homotopy between the identity map $|1_{\mathcal{I}}|$ and the constant map $|F_f|$, and so $|\mathcal{I}|$ is homotopy equivalent to a point. The argument is the same for the initial object, but the natural transformation of functors is from the constant functor at the initial object to the identity functor.                                                        □

The reader may wish to identify the categories with an initial or a final object in the above examples and verify Proposition 8.1.17. We close this section with one more interesting example.

**Example 8.1.18** (Classifying space of a group)    If $C = G$ is a discrete group (regarded in the usual way as a category with a single object and a morphism for each group element), then $|G|$ is precisely one model for the *classifying space of G*, most often denoted $BG$.

   It is a good exercise to work out the case $G = \mathbb{Z}/2$. In this case there is a single non-degenerate simplex of degree $n$ for each $n$, and each one is attached along its boundary to the simplex of one degree lower by a map of degree 2. The $k$-skeleton of this space is $\mathbb{RP}^k$ with its usual CW structure, and the classifying space $B\mathbb{Z}/2 = |\mathbb{Z}/2|$ itself is $\mathbb{RP}^\infty$.                    □

## 8.2 Homotopy limits and colimits

We are now ready to define homotopy limits and colimits in Top (and Top$_*$). We do the former in Section 8.2.1 and the latter in Section 8.2.2, in terms of (co)equalizers. This choice mirrors the concrete descriptions of (co)limits via (co)equalizers from Section 7.4, makes the proofs of homotopy invariance of homotopy (co)limits relatively easy to organize, and through these proofs draws special attention to the importance of punctured squares and towers.

   We defined homotopy (co)limits for punctured cubes in terms of natural transformations and products (Definitions 5.3.1 and 5.7.1). The difference between the natural transformations/products and the (co)equalizer definitions, when both are available, is minimal and is simply a matter of reorganizing the information. We will exhibit the equivalence of the definitions for homotopy limits in Proposition 8.2.2 and for homotopy colimits in Remark 8.2.13.

### 8.2.1 Homotopy limits

Let $\mathcal{I}$ be a small category and $F \colon \mathcal{I} \to$ Top a functor. At the start of Section 8.1 we vaguely described the homotopy limit as an "up to homotopy" version of the limit, and used this to conclude that the overcategories $\mathcal{I} \downarrow i$ have the necessary combinatorics to keep track of homotopies (and higher homotopies) between various points in $F(i)$. The idea is to replace $F(i)$ with the homotopy equivalent space $\text{Map}(|\mathcal{I} \downarrow i|, F(i))$ (it is homotopy equivalent because $|\mathcal{I} \downarrow i|$ is contractible by Proposition 8.1.17). This is no longer a functor of $i$ in the way that $F$ was, but rather a bifunctor. Observe that there are two maps

$$a, b \colon \prod_i \text{Map}\,(|\mathcal{I} \downarrow i|, F(i)) \longrightarrow \prod_{i \to i'} \text{Map}\,(|\mathcal{I} \downarrow i|, F(i')),$$

where $a$ is induced by the map

$$\text{Map}(|\mathcal{I} \downarrow i|, F(i)) \longrightarrow \text{Map}(|\mathcal{I} \downarrow i|, F(i'))$$

given by applying $\text{Map}(|\mathcal{I} \downarrow i|, -)$ to the map $F(i) \to F(i')$. It sends the factor indexed by $i$ to those factors indexed by $i \to i'$ for all such $i'$. The map $b$ is induced by the map

$$\mathcal{I} \downarrow i \longrightarrow \mathcal{I} \downarrow i'$$

by applying $\text{Map}(-, F(i'))$ to the induced map of realizations. It sends the factor indexed by $i'$ to the factors indexed by $i \to i'$ for all such $i$. Together, $a$ and $b$ capture the coherence in $i \in \mathcal{I}$ of the homotopies given by the maps $|\mathcal{I} \downarrow i| \to F(i)$.

**Definition 8.2.1**   Given a functor $F \colon \mathcal{I} \to \text{Top}$ (or $\text{Top}_*$), the *homotopy limit of $F$* is the equalizer of the maps $a$ and $b$ above, that is,

$$\underset{\mathcal{I}}{\text{holim}}\, F = \text{Eq}\left( \prod_i \text{Map}\,(|\mathcal{I} \downarrow i|, F(i)) \underset{b}{\overset{a}{\rightrightarrows}} \prod_{i \to i'} \text{Map}\,(|\mathcal{I} \downarrow i|, F(i')) \right).$$

To rephrase Definition 8.2.1 in terms of natural transformations (to match the definition of the homotopy limit of a punctured cube, Definition 5.3.1), we know that the realization of the nerve of the overcategory is a functor:

$$|\mathcal{I} \downarrow -| \colon \mathcal{I} \longrightarrow \text{Top}$$
$$i \longmapsto |\mathcal{I} \downarrow i|.$$

Now let $F \colon \mathcal{I} \to \text{Top}$ be a functor. The set $\text{Nat}_{\mathcal{I}}(|\mathcal{I} \downarrow -|, F)$ of natural transformations is naturally a subset of $\prod_i \text{Map}(|\mathcal{I} \downarrow i|, F(i))$, and hence itself can be thought of as a topological space by giving it the subspace topology.

Then we have the following characterization of homotopy limits in Top. It is analogous to the model for the limit given in Proposition 7.4.16.

**Proposition 8.2.2** (Homotopy limit as natural transformations)  *Suppose $F \colon \mathcal{I} \to \text{Top}$ (or $\text{Top}_*$) is a diagram. Then there is a natural homeomorphism*

$$\underset{\mathcal{I}}{\text{holim}}\, F \cong \underset{\mathcal{I}}{\text{Nat}}\,(|\mathcal{I} \downarrow -|, F).$$

*Proof* The homeomorphism is in fact the identity map. This is simply the observation that by definition the homotopy limit of $F$ is the equalizer

$$\operatorname*{holim}_{\mathcal{I}} F = \operatorname{Eq}\left( \prod_i \operatorname{Map}(|\mathcal{I} \downarrow i|, F(i)) \underset{b}{\overset{a}{\rightrightarrows}} \prod_{i \to i'} \operatorname{Map}(|\mathcal{I} \downarrow i|, F(i')) \right).$$

That is, $\operatorname{holim}_{\mathcal{I}} F \subset \prod_i \operatorname{Map}(|\mathcal{I} \downarrow i|, F(i))$ consists of those maps $f = \{f_i\}_{i \in \mathcal{I}}$ such that, for all morphisms $i \to i'$, the diagram

$$
\begin{array}{ccc}
|\mathcal{I} \downarrow i| & \longrightarrow & |\mathcal{I} \downarrow i'| \\
{\scriptstyle f_i} \downarrow & & \downarrow {\scriptstyle f_{i'}} \\
F(i) & \longrightarrow & F(i')
\end{array}
$$

commutes, where the horizontal maps are those naturally induced by functoriality. This precisely describes the space of natural transformations.    $\square$

The expression in Definition 8.2.1 should be compared with the description of the limit as an equalizer from (7.4.2). In particular, we can recover the ordinary limit from the homotopy limit by replacing the realization of the nerve of the overcategory with a point and using the canonical homeomorphism $\operatorname{Map}(*, X) \cong X$. It is clear from the comparison with (7.4.2) that there is a map

$$\lim_{\mathcal{I}} F \longrightarrow \operatorname*{holim}_{\mathcal{I}} F \tag{8.2.1}$$

induced by the natural transformation $|\mathcal{I} \downarrow -| \to C_*$ of functors from $\mathcal{I}$ to Top, where $C_*$ is the constant functor at the one-point space. More explicitly, this map is induced on equalizers by the map of diagrams

$$
\begin{array}{ccc}
\prod_i \operatorname{Map}(*, F(i)) & \overset{a}{\underset{b}{\rightrightarrows}} \prod_{i \to i'} \operatorname{Map}(*, F(i')) \\
\downarrow & & \downarrow \\
\prod_i \operatorname{Map}(|\mathcal{I} \downarrow i|, F(i)) & \overset{a}{\underset{b}{\rightrightarrows}} \prod_{i \to i'} \operatorname{Map}(|\mathcal{I} \downarrow i|, F(i'))
\end{array}
\tag{8.2.2}
$$

The maps (2.2.2) and (3.2.1) are special cases of the map (8.2.1).

It is also clear that the map of spaces $\operatorname{Map}(*, F(i')) \to \operatorname{Map}(|\mathcal{I} \downarrow i|, F(i'))$ is a homotopy equivalence for each $i$, because $|\mathcal{I} \downarrow i|$ is contractible for each $i$, but this does *not* in general induce a weak equivalence of equalizers, for the same reason that limits are not homotopy invariant. This is, however, the sense in which homotopy limit is a "fattened-up limit". This fattening will afford it the homotopy invariance property which the limit does not in general possess.

Looking back at Definition 5.3.1, the definition of the homotopy limit of a punctured cube, one easily confirms that it matches what appears in Proposition 8.2.2. Indeed, the punctured cubical simplex $\Delta(-)$ (or, as denoted there, $\Delta(\bullet)$) is precisely the diagram $|\mathcal{I} \downarrow -|$ when $\mathcal{I} = \mathcal{P}_0(\underline{n})$ (though without the barycentric subdivision which comes with realizing the overcategory; see Example 8.1.13). We will work out several special cases of punctured cubes below, but will do so straight from Definition 8.2.1.

**Example 8.2.3** If $\mathcal{I}$ is the empty category, then the homotopy limit of $F \colon \mathcal{I} \to$ Top is a point. □

**Example 8.2.4** If $\mathcal{I} = \{0, 1\}$ and $F(i) = X_i$ then $\mathrm{holim}_{\mathcal{I}} F$ is homeomorphic to $X_0 \times X_1$. □

**Example 8.2.5** Suppose a small category $\mathcal{I}$ has an initial object $i_0$. Then for a functor $F \colon \mathcal{I} \to$ Top, the projection

$$\underset{\mathcal{I}}{\mathrm{holim}}\, F \longrightarrow \mathrm{Map}(|\mathcal{I} \downarrow i_0|, F(i_0)) \simeq F(i_0)$$

is a homotopy equivalence. We will prove this in Proposition 9.3.4. This should not be too surprising in light of the corresponding fact for ordinary limits, Example 7.3.9. □

**Example 8.2.6** Let $C_X \colon \mathcal{I} \to$ Top be the constant functor at the space $X$. Then

$$\underset{\mathcal{I}}{\mathrm{holim}}\, C_X \cong \mathrm{Map}(|\mathcal{I}|, X).$$

In particular, if $C_*$ is the constant functor at the one-point space, then

$$\underset{\mathcal{I}}{\mathrm{holim}}\, C_* \cong *.$$

To see this, let $f \in \mathrm{holim}_{\mathcal{I}} C_X$. We define $F \in \mathrm{Map}(|\mathcal{I}|, X)$ by describing its values on the non-degenerate simplices of $|\mathcal{I}|$. Let $i_0 \to \cdots \to i_k$ be a non-degenerate $k$-simplex of $|\mathcal{I}|$. Let $F_{i_0 \to \cdots \to i_k}$ denote the restriction of $F$ to the $k$-simplex of $|\mathcal{I}|$ indexed by this simplex. Let $F_{i_0 \to \cdots \to i_k}$ be equal to the restriction of $f_{i_k}$ to the corresponding non-degenerate simplex $(i_0 \to \cdots \to i_k) \to i_k$ where $i_k \to i_k$ is the identity map. Similarly, the $f_i$ may be defined from a map $F \in \mathrm{Map}(|\mathcal{I}|, X)$ by letting the value of $f_i$ on the non-degenerate $k$-simplex $(i_0 \to \cdots \to i_k) \to i$ be equal to the restriction of $F$ to the $k$-simplex indexed by $i_0 \to \cdots \to i_k$.

That these correspondences determine elements of the spaces we claim they do follows from Proposition 8.2.2, as a point $f \in \mathrm{holim}_{\mathcal{I}} C_X$ amounts to maps

$f_i \colon |I \downarrow i| \to X$ for each object $i$ of $I$ such that, for every morphism $i \to i'$ in $I$, the diagram

$$
\begin{array}{ccc}
|I \downarrow i| & \longrightarrow & |I \downarrow i'| \\
\downarrow {\scriptstyle f_i} & & \downarrow {\scriptstyle f_{i'}} \\
X & \xrightarrow{\ 1_X\ } & X
\end{array}
$$

commutes. In particular, the value of $f_i$ on the non-degenerate simplex $(i_0 \to \cdots \to i_k) \to i$ is determined by the value of $f_{i_k}$ on the non-degenerate simplex $(i_0 \to \cdots \to i_k) \to i_k$, where the map $i_k \to i_k$ is the identity. $\qquad \square$

**Example 8.2.7** (Mapping path space as homotopy limit)  Consider $\mathcal{P}(\underline{1}) = (\emptyset \to \{1\})$ and $X \colon \mathcal{P}(\underline{1}) \to \mathrm{Top}$ given by $X(S) = X_S$. Let $f = X(\emptyset \to \{1\}) \colon X_\emptyset \to X_1$. Then $\mathrm{holim}_{\mathcal{P}(\underline{1})} X$ is precisely the mapping path space of $f$, $P_f$ (see Definition 2.2.1). To see why, recall from Example 8.1.14 that $|\mathcal{P}(\underline{1}) \downarrow \emptyset| = *$ and $|\mathcal{P}(\underline{1}) \downarrow \{1\}| \cong I$. It is natural to identify the map $|\mathcal{P}(\underline{1}) \downarrow \emptyset| \cong * \to |\mathcal{P}(\underline{1}) \downarrow \{1\}| \cong I$ as the inclusion of $0$ in the interval. By definition, $\mathrm{holim}_{\mathcal{P}(\underline{1})} X$ is equal to

$$
\mathrm{Eq}\left( X_\emptyset \times X_1^I \overset{a}{\underset{b}{\rightrightarrows}} X_\emptyset \times X_1^I \times X_1 \right),
$$

where $a(x,\gamma) = (x,\gamma,f(x))$ and $b(x,\gamma) = (x,\gamma,\gamma(0))$. Hence we obtain precisely the set of all $(x,\gamma) \in X_\emptyset \times X_1^I$ such that $\gamma(0) = f(x)$. $\qquad \square$

**Example 8.2.8** (Homotopy pullback as homotopy limit)  As usual, let $\mathcal{P}_0(\underline{2}) = (\{1\} \to \{1,2\} \leftarrow \{2\})$, and let $X \colon \mathcal{P}_0(\underline{2}) \to \mathrm{Top}$ be represented by the diagram $X_1 \overset{f_1}{\to} X_{12} \overset{f_2}{\leftarrow} X_2$. Let us use Definition 8.2.1 to confirm that $\mathrm{holim}_{\mathcal{P}_0(\underline{2})} X$ is homeomorphic to the homotopy pullback as defined in Definition 3.2.4. To see this, recall Example 8.1.15. We have $|\mathcal{P}_0(\underline{2}) \downarrow \{i\}| = *$ for $i = 1,2$, $|\mathcal{P}_0(\underline{2}) \downarrow \{1,2\}| \cong I$, and we may identify the endpoints $0$ and $1$ as the images of $|\mathcal{P}_0(\underline{2}) \downarrow \{1\}|$ and $|\mathcal{P}_0(\underline{2}) \downarrow \{2\}|$ in $|\mathcal{P}_0(\underline{2}) \downarrow \{1,2\}|$ respectively.

By definition, the homotopy limit of $X$ is

$$
\mathrm{Eq}\left( X_1 \times X_2 \times X_{12}^I \overset{a}{\underset{b}{\rightrightarrows}} X_1 \times X_2 \times X_{12}^I \times X_{12} \times X_{12} \right),
$$

where

$$
a(x_1, x_2, \gamma) = (x_1, x_2, \gamma, f_1(x_1), f_2(x_2))
$$

and

$$
b(x_1, x_2, \gamma) = (x_1, x_2, \gamma, \gamma(0), \gamma(1)),
$$

so that the homotopy limit is the subspace of $X_1 \times X_2 \times X_{12}^I$ of all $(x_1, x_2, \gamma)$ such that $\gamma(0) = f_1(x_1)$ and $\gamma(1) = f_2(x_2)$. But this is precisely the description of the homotopy pullback given in Definition 3.2.4. □

**Example 8.2.9** (Homotopy limit of a punctured 3-cube)  Let $I = \mathcal{P}_0(\underline{3})$ and let $X \colon \mathcal{P}_0(\underline{3}) \to \text{Top}$ be a punctured 3-cubical diagram of spaces. Let us recover the description of $\text{holim}\,\mathcal{P}_0(\underline{3})$ from Example 5.3.3 using Definition 8.2.1.

As usual let $X(S) = X_S$ for $\emptyset \neq S \subset \underline{3}$. The overcategories $\mathcal{P}_0(\underline{3}) \downarrow -$ were described in Example 8.1.5 and their realizations in Example 8.1.13.

The homotopy limit of $X$ is thus

$$\underset{\mathcal{P}_0(\underline{3})}{\text{holim}}\, X = \text{Eq}\left( \prod_S \text{Map}\left(|\mathcal{P}_0(\underline{3}) \downarrow S|, X_S\right) \underset{b}{\overset{a}{\rightrightarrows}} \prod_{S \to T} \text{Map}\left(|\mathcal{P}_0(\underline{2}) \downarrow S|, X_T\right) \right).$$

Unravelling this gives that a point in the homotopy limit is precisely the family of spaces and homotopies as described in Example 5.3.3. □

**Example 8.2.10** (Inverse homotopy limit)  Let $I = (1 \leftarrow 2 \leftarrow \cdots)$ be the category indexing inverse systems, and let $F \colon I \to \text{Top}$ be given by $F(i) = X_i$, let $p_i$ denote the maps $X_i \to X_{i-1}$. By Proposition 8.2.2, $\text{holim}_I F$ is the subspace of $\prod_i \text{Map}(|I \downarrow i|, X_i)$, called the *inverse homotopy limit of F*, such that the diagram

$$
\begin{array}{ccc}
|I \downarrow i| & \longrightarrow & |I \downarrow j| \\
\downarrow & & \downarrow \\
X_i & \longrightarrow & X_j
\end{array}
$$

commutes for each $j \leq i$. As in Example 8.1.16, the map $|I \downarrow i| \to |I \downarrow j|$ is the inclusion of an infinite-dimensional face of an infinite-dimensional simplex (in fact the inclusion of a face of codimension $i - j$), and so a point in the limit is a sequence of maps $f_i \colon |I \downarrow i| \to X_i$ such that the restriction of $f_i$ to the (codimension 1) face $|I \downarrow i + 1|$ is equal to $p_{i+1}(f_{i+1})$.

This is a rather unwieldy description of an important standard example, basically due to the fact that we are mapping in infinite-dimensional simplices. For a better (and perhaps more familiar) description, see Example 8.4.11. □

**Example 8.2.11** (Homotopy fixed points as homotopy limit)  Let $G$ be a discrete group and $\bullet_G$ the category associated with $G$ (one object $\bullet$, one morphism for each group element which composes according to the group law). Let $F \colon \bullet_G \to \text{Top}$ be a functor, with $F(\bullet) = X$.

Then one can define *homotopy fixed points* $X^{hG}$ as

$$X^{hG} = \underset{\bullet_G}{\operatorname{holim}} F.$$

Although homotopy fixed points can be defined in other ways, this definition is natural in light of Proposition 7.3.17. If the action of $G$ on $X$ is free, then the canonical map $X^G \to X^{hG}$ from the fixed points to the homotopy fixed points is a weak equivalence.                                                               □

We pause to briefly discuss functoriality. The homotopy limit has the same functorial properties as the limit does, as discussed in Section 7.3. That is,

$$\underset{I}{\operatorname{holim}}\,(-)\colon\ \operatorname{Top}^I \longrightarrow \operatorname{Top}$$

is a functor on the category of functors from $I$ to Top; it is clear from the definition that a natural transformation of functors $F \to G$ gives rise to a map of homotopy limits

$$\underset{I}{\operatorname{holim}} F \to \underset{I}{\operatorname{holim}} G. \tag{8.2.3}$$

Varying the indexing category also induces a map of homotopy limits. Suppose $G\colon \mathcal{J} \to I$ is a functor, and suppose $F\colon I \to \operatorname{Top}$ is a functor. Then $G$ gives rise to a map

$$\underset{I}{\operatorname{holim}} F \longrightarrow \underset{\mathcal{J}}{\operatorname{holim}}\,(F \circ G), \tag{8.2.4}$$

induced by $G$ in the following way. The functor $G$ induces a functor $(\mathcal{J} \downarrow j) \to (I \downarrow G(j))$ and hence a continuous map $|\mathcal{J} \downarrow j| \to |I \downarrow G(j)|$. As $\operatorname{holim}_I F \subset \prod_i \operatorname{Map}(|I \downarrow i|, F(i))$, on the level of mapping spaces the induced map is given by composition of the projection $\prod_i \operatorname{Map}(|I \downarrow i|, F(i)) \to \prod_j \operatorname{Map}(|I \downarrow G(j)|, F \circ G(j))$ away from those objects $i$ not in the image of $G$, followed by the map induced by $|\mathcal{J} \downarrow j| \to |I \downarrow G(j)|$. An important special case occurs when $G\colon \mathcal{J} \to I$ is the inclusion of a subcategory of $I$ (see e.g. Theorem 8.6.1). In this case we usually suppress the functor $G$ from the notation.

## 8.2.2 Homotopy colimits

To dualize Section 8.2.1, we start by giving the definition of the homotopy colimit using coequalizers. After giving this definition we will immediately see that it is equivalent to the definition of the homotopy colimit of a punctured cube from Definition 5.7.1.

First, there are maps

$$a, b \colon \coprod_{i \to i'} F(i) \times |(i' \downarrow I)^{\mathrm{op}}| \longrightarrow \coprod_i F(i) \times |(i \downarrow I)^{\mathrm{op}}|$$

again induced by the morphisms $i \to i'$. The map $a$ is induced by the map

$$F(i \to i') \times \mathrm{id} \colon F(i) \times |(i' \downarrow I)^{\mathrm{op}}| \longrightarrow F(i') \times |(i' \downarrow I)^{\mathrm{op}}|$$

and the map $b$ is induced by the map

$$\mathrm{id} \times ((i' \downarrow I)^{\mathrm{op}} \to (i \downarrow I)^{\mathrm{op}}) \colon F(i) \times |(i' \downarrow I)^{\mathrm{op}}| \longrightarrow F(i) \times |(i \downarrow I)^{\mathrm{op}}|.$$

**Definition 8.2.12** Given a diagram $F \colon I \to \mathrm{Top}$, the *homotopy colimit of $F$* is defined as the coequalizer of the maps $a$ and $b$ above, that is,

$$\operatorname*{hocolim}_{I} F = \mathrm{Coeq}\left( \coprod_{i \to i'} F(i) \times |(i' \downarrow I)^{\mathrm{op}}| \overset{a}{\underset{b}{\rightrightarrows}} \coprod_i F(i) \times |(i \downarrow I)^{\mathrm{op}}| \right).[1]$$

**Remark 8.2.13** It follows immediately from the definition, and dually to Proposition 8.2.2, that $\operatorname{hocolim}_I F$ is a quotient of $\coprod_i F(i) \times |(i \downarrow I)^{\mathrm{op}}|$ by an equivalence relation generated by the maps $a$ and $b$, in analogy with ordinary colimits (see Proposition 7.4.18) and the cubical homotopy colimit (see Definition 5.7.1). □

**Remark 8.2.14** (Pointed homotopy colimit) The homotopy colimit construction we have described does not make a based space from a diagram of based spaces. We fix that defect here. We can make a *pointed* homotopy colimit, defined by

$$\operatorname*{hocolim}_{I} F / \operatorname*{hocolim}_{I} C_*,$$

where $C_*$ is the constant $I$-diagram of one-point spaces (see Example 7.1.23). As we shall see in Example 8.2.18, $\operatorname{hocolim}_I C_* \cong |I^{\mathrm{op}}|$, and so the pointed homotopy colimit is not necessarily equivalent to the one defined above. We will not use special notation to denote the homotopy colimit in the category of based spaces, and the reader should assume if a functor takes values in based spaces, then when we construct the homotopy colimit we mean the based homotopy colimit. □

The expression in Definition 8.2.12 should be compared with (7.4.4) which gives a coequalizer description of an ordinary colimit. Therefore, analogous to

---

[1] The reason we use $(i \downarrow I)^{\mathrm{op}}$ (following [Hir03]) is that it makes the proof of Proposition 8.5.4 essentially trivial. It would be equally correct to use $i \downarrow I$ instead of its opposite.

the discussion leading to (8.2.1), if we replace the realizations $|(i \downarrow \mathcal{I})^{\mathrm{op}}|$ with one-point spaces in Definition 8.2.12, we obtain the ordinary colimit and we hence have a canonical map

$$\operatorname*{hocolim}_{\mathcal{I}} F \longrightarrow \operatorname*{colim}_{\mathcal{I}} F. \qquad (8.2.5)$$

induced by the map $|(i \downarrow \mathcal{I})^{\mathrm{op}}| \to *$. It "collapses" the homotopy colimit into the colimit. The map (2.4.2) is a special case of this map.

**Example 8.2.15** If $\mathcal{I}$ is the empty category, then the homotopy colimit of $F : \mathcal{I} \to \mathrm{Top}$ is the empty set. □

**Example 8.2.16** If $\mathcal{I} = \{0, 1\}$ and $F(\mathcal{I}) = \{X_0, X_1\}$ then $\operatorname{hocolim}_{\mathcal{I}} F$ is homeomorphic to $X_0 \sqcup X_1$ if the target of $F$ is Top and $X_0 \vee X_1$ if the target of $F$ is $\mathrm{Top}_*$. □

**Example 8.2.17** Suppose a small category $\mathcal{I}$ has a final object $i_1$. Then, for a functor $F : \mathcal{I} \to \mathrm{Top}$, the canonical inclusion

$$F(i_1) \cong F(i_1) \times |(i_1 \downarrow \mathcal{I})^{\mathrm{op}}| \longrightarrow \operatorname*{hocolim}_{\mathcal{I}} F$$

is a homotopy equivalence. We demonstrate this in Proposition A.1.16. The corresponding fact for ordinary colimits is Example 7.3.27. □

Dual to Example 8.2.6, we have the following.

**Example 8.2.18** Let $C_X : \mathcal{I} \to \mathrm{Top}$ be the constant functor at the space $X$. Then

$$\operatorname*{hocolim}_{\mathcal{I}} C_X \cong X \times |\mathcal{I}^{\mathrm{op}}|.$$

In particular, if $C_*$ is the constant functor at the one-point space, then, as we have mentioned in Remark 8.2.14,

$$\operatorname*{hocolim}_{\mathcal{I}} C_* \cong * \times |\mathcal{I}^{\mathrm{op}}| \cong |\mathcal{I}^{\mathrm{op}}|.$$

Though the proof is dual to that given in Example 8.2.6, we will demonstrate the second fact. By definition,

$$\operatorname*{hocolim}_{\mathcal{I}} C_* = \operatorname{Coeq}(a, b) = \operatorname{Coeq}\left( \coprod_{i \to i'} |(i \downarrow \mathcal{I})^{\mathrm{op}}| \underset{b}{\overset{a}{\rightrightarrows}} \coprod_{i} |(i \downarrow \mathcal{I})^{\mathrm{op}}| \right),$$

where $a$ takes a summand indexed by $i \to i'$ to the summand indexed by $i$ via the identity map, and $b$ takes the summand indexed by $i \to i'$ to the summand indexed by $i'$ using the induced map $|(i \downarrow I)^{\mathrm{op}}| \to |(i' \downarrow I)^{\mathrm{op}}|$. Send the point $x$ in the $n$-cell of $|(i \downarrow I)^{\mathrm{op}}|$ indexed by the non-degenerate chain $i' \to (i_0 \to \cdots \to i_n)$ in $B_*(i' \downarrow I)^{\mathrm{op}}$ to itself via the identity map (there is a canonical homeomorphism between the $n$-simplex indexed by $i' \to (i_0 \to \cdots \to i_n)$ in $|(i' \downarrow I)^{\mathrm{op}}|$ and the one indexed by $i_0 \to \cdots \to i_n$ in $|I^{\mathrm{op}}|$). This defines a surjective map $\mathrm{hocolim}_I C_* \to |I|$, and it is injective as well, since points $x$ and $y$ in $\mathrm{hocolim}_I C_*$ are mapped to the same point if and only if they are identified by the maps $a, b$. It is easy to see this has a continuous inverse. □

**Example 8.2.19** (Mapping cylinder as homotopy colimit)  Consider the category $\mathcal{P}(\underline{1}) = (\emptyset \to \{1\})$ and $X: \mathcal{P}(\underline{1}) \to$ Top given by $X(S) = X_S$. Let $f = X(\emptyset \to \{1\}): X_0 \to X_1$. Then $\mathrm{hocolim}_{\mathcal{P}(\underline{1})} X$ is precisely the mapping cylinder of $f$, as in Definition 2.4.1. In more detail, to compute the homotopy limit, we look at maps

$$a, b: X_0 \times I \sqcup X_0 \sqcup X_1 \longrightarrow X_0 \times I \sqcup X_1.$$

The map $a$ is induced by the functor, so in this case that means that it is induced by the map $f$ or the identity maps. So

$$a((x_0, t)) = (x_0, t), \quad a(x_0') = f(x_0'), \quad a(x_1) = x_1.$$

The map $b$ is induced by the maps of undercategories, so

$$b((x_0, t)) = (x_0, t), \quad b(x_0') = (x_0', 1), \quad b(x_1) = x_1.$$

Then the only identifications we obtain are $(x_0', 1) \sim f(x_0')$, which gives the mapping cylinder. □

**Example 8.2.20** (Mapping torus as homotopy colimit)  Let $F$ be a functor from the (co)equalizer category $\bullet \rightrightarrows \bullet$ that sends both objects to the space $X$, one morphism to the identity map and the other morphism to a self-map $f$. Then the homotopy colimit of $F$ is the mapping torus of $f$, the quotient of $X \times I$ by $(x, 0) \sim (f(x), 1)$. □

**Example 8.2.21** (Homotopy pushout)  Let $\mathcal{P}_1(\underline{2}) = (\{1\} \leftarrow \emptyset \to \{2\})$ and let $X: \mathcal{P}_1(\underline{2})$ be represented by the diagram $X_1 \xleftarrow{f_1} X_0 \xrightarrow{f_2} X_2$. Then $\mathrm{hocolim}_{\mathcal{P}_1(\underline{2})} X$ is precisely the homotopy pushout from Definition 3.6.3. We leave the details to the reader. □

Dual to Example 8.2.10, we have the following.

**Example 8.2.22** (Direct homotopy colimit)    Let $I = (1 \leftarrow 2 \leftarrow \cdots)$ be the category indexing inverse systems, and $I^{op} = (1 \rightarrow 2 \rightarrow \cdots)$ its opposite, the category indexing directed systems (introduced in Example 7.3.31). Let $F \colon I^{op} \rightarrow$ Top be given by $F(i) = X_i$, and let $a_i$ denote the maps $X_i \rightarrow X_{i+1}$. Using Definition 8.2.12, hocolim$_{I^{op}}$ $F$ is a quotient space of $\coprod_i |(i \downarrow I^{op})^{op}| \times X_i$ called the *direct homotopy colimit of F*.

We have already observed in Example 8.1.16 that $|(i \downarrow I^{op})^{op}|$ is an infinite-dimensional simplex for each $i$ and that a morphism $j \rightarrow i$ in $I^{op}$ induces an inclusion $|(j \downarrow I^{op})^{op}| \rightarrow |(i \downarrow I^{op})^{op}|$ of an infinite-dimensional face. Hence the homotopy colimit consists of equivalence classes of points $(t_i, x_i) \in |(i \downarrow I^{op})^{op}| \times X_i$ where we identify $(t_i, x_i)$ with $(t_i, a_i(x_i))$ (here $t_i$ is regarded as a point in $|(i + 1 \downarrow I^{op})^{op}|$ via the inclusion described above).

As with Example 8.2.10, this is a rather unwieldy space due to the presence of infinite-dimensional simplices. See Example 8.4.11 for a more a homotopy equivalent and more concrete description of this homotopy colimit as a "mapping telescope".                                                                    □

**Example 8.2.23** (Homotopy orbits as homotopy colimit)    Dual to Example 8.2.11, let $\bullet_G$ be the category associated to a discrete group $G$ and let $F \colon \bullet_G \rightarrow$ Top be a functor, with $F(\bullet) = X$. Then the *homotopy orbit space* $X_{hG}$ is defined as

$$X_{hG} = \operatorname*{hocolim}_{\bullet_G} F.$$

Again, it makes sense to define the homotopy orbit space this way since this is the natural extension of orbit spaces as described in Proposition 7.3.36. If the action of $G$ on $X$ is free, then the canonical map $X_{hG} \rightarrow X_G$ from the homotopy orbits to the orbits is a weak equivalence.                                                        □

The homotopy colimit is a functor. We may regard

$$\operatorname*{hocolim}_{I} (-) \colon \text{Top}^I \longrightarrow \text{Top}$$

as functors on the category of functors from $I$ to Top. It should be clear from the definition that a natural transformation of functors gives rise to an induced map of homotopy (co)limits, namely, given a natural transformation $F \rightarrow G$, we have a map

$$\operatorname*{hocolim}_{I} F \longrightarrow \operatorname*{hocolim}_{I} G. \qquad\qquad (8.2.6)$$

Varying the indexing category induces a map of homotopy colimits. Suppose $G \colon \mathcal{J} \rightarrow I$ is a functor, and suppose $F \colon I \rightarrow$ Top is a functor. Then $G$ gives rise to a map

$$\operatorname*{hocolim}_{\mathcal{J}} (F \circ G) \longrightarrow \operatorname*{hocolim}_{I} F. \tag{8.2.7}$$

An important special case occurs when $G\colon \mathcal{J} \to I$ is the inclusion of a subcategory of $I$, which will be discussed in Theorem 8.6.1.

## 8.3 Homotopy invariance of homotopy limits and colimits

Here we discuss homotopy invariance of homotopy (co)limits. This was done for cubical diagrams in Theorems 5.3.9 and 5.7.8. We will tackle the homotopy invariance of the homotopy limit first. It will require a little bit of background on inverse limit towers, since we filter the spaces $\operatorname{Map}(|I \downarrow i|, F(i))$ by filtering the simplicial complexes $|I \downarrow i|$ by their skeleta.

**Theorem 8.3.1** (Homotopy invariance of homotopy limits)  *Let $I$ be a small category and $F, G\colon I \to \mathrm{Top}$ (or $\mathrm{Top}_*$) functors. Suppose we have a natural transformation $N\colon F \to G$ such that $F(i) \to G(i)$ is a homotopy (resp. weak) equivalence for all $i$. Then the natural map $\operatorname{holim}_I F \to \operatorname{holim}_I G$ from (8.2.3) is a homotopy (resp. weak) equivalence.*

The key ingredients in the proof of this theorem are the homotopy invariance of homotopy pullbacks of punctured squares (Theorem 3.2.12), as well as the homotopy invariance of limits of towers of fibrations (Theorem 8.3.2), which we will establish below. The rest is a matter of using the simplicial machinery for bookkeeping.

**Theorem 8.3.2**  *Suppose that in the diagram*

$$
\begin{array}{ccc}
X_1 & \longleftarrow X_2 \longleftarrow \cdots \\
{\scriptstyle f_1}\big\downarrow & {\scriptstyle f_2}\big\downarrow \\
Y_1 & \longleftarrow Y_2 \longleftarrow \cdots
\end{array}
\tag{8.3.1}
$$

*all of the horizontal maps are fibrations and all of the vertical maps $f_i$ are homotopy (resp. weak) equivalences for all $i \in \mathbb{N}$. Then the induced map $f\colon \lim_i X_i \longrightarrow \lim_i Y_i$ is a homotopy (resp. weak) equivalence.*

We split the proofs for the two notions of equivalence into separate proofs. Our presentation in the case of homotopy equivalences below is based on [Geo79]. To the best of our knowledge the result in the case of homotopy equivalence first appeared as [EH76, 3.4.1].

*Proof of Theorem 8.3.2 for homotopy equivalences*   The idea is to construct
a sequence of compatible homotopy inverses for the $f_n$. Let $M_{f_n}$ denote the
mapping cylinder of $f_n$ for each $n$. For each $n$ let $j_n : Y_n \to M_{f_n}$ be the inclusion.
Let $h_1 : M_{f_1} \to X_1$ be the homotopy inverse of the canonical inclusion $X_1 \to$
$M_{f_1}$ (the inclusion is a homotopy equivalence because $f_1$ is), and define $g_1 =$
$h_1 \circ j_1 : Y_1 \to X_1$. Let $H_1 : X_1 \times I \to X_1$ be the composition of the inclusion
of $X_1 \times I$ in $M_{f_1}$ followed by the map $h_1$. Then $H_1(x_1, 0) = x_1$ and $H_1(x_1, 1) =$
$(g_1 \circ f_1)(x_1)$, and so $H_1$ is a homotopy from $1_{X_1}$ to $g_1 \circ f_1$.

Let $q_n : M_{f_{n+1}} \to M_{f_n}$ be the induced map of mapping cylinders. Suppose
by induction that for $i = 1$ to $n$ there exist maps $h_i : M_{f_i} \to X_i$ such that the
following diagrams

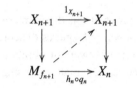

commute for $i = 2$ to $n$, where by definition $g_i = h_i \circ j_i$. Consider the lifting
problem

$$
\begin{array}{ccc}
X_{n+1} & \xrightarrow{1_{X_{n+1}}} & X_{n+1} \\
\downarrow & \nearrow & \downarrow \\
M_{f_{n+1}} & \xrightarrow[h_n \circ q_n]{} & X_n
\end{array}
$$

A solution $h_{n+1}$ to this lifting problem exists by Theorem 2.5.4, since the left
vertical map is a cofibration which is also a homotopy equivalence, and the
right vertical map is a fibration by hypothesis. Then we define $g_{n+1} = h_{n+1} \circ$
$j_{n+1}$. Note that by construction the map $h_{n+1}$ is compatible with the maps $q_i$ in
the sense that the first square diagram appearing in this proof is commutative
for $i = n + 1$, and again by construction the second square diagram commutes
when $i = n + 1$.

With the $h_i$ as above, let $H_i : X_i \times I \to X_i$ be the composition of the inclusion
$X_i \times I \to M_{f_i}$ with the map $h_i$. Then

$$
\begin{array}{ccc}
X_i \times I & \xrightarrow{H_i} & X_i \\
\downarrow & & \downarrow \\
X_{i-1} \times I & \xrightarrow[H_{i-1}]{} & X_{i-1}
\end{array}
$$

commutes for all $i$, where the vertical maps are the evident ones induced by $X_i \to X_{i-1}$, and as with the case $i = 1$, $H_i$ is a homotopy from $1_{M_{f_i}}$ to $g_i \circ f_i$. Then the map $g: \lim_i Y_i \to \lim_i X_i$ induced by the $g_i$ is homotopy inverse to the map $f: \lim_i X_i \to \lim_i Y_i$ induced by the $f_i$ because of the commutativity of the previous square. To see that $f \circ g$ is homotopic to the identity of $\lim_i Y_i$, we proceed similarly. This time, however, we find maps $f'_i: X_i \to Y_i$ which induce a map $f': \lim_i X_i \to \lim_i Y_i$, and a homotopy $1_Y \sim f' \circ g$. This implies $f \sim f' \circ g \circ f \sim f'$, so that $f$ is homotopy inverse to $g$. □

We need some algebraic results before we can proceed with the proof in the case of weak equivalences. For a tower $Z_1 \leftarrow Z_2 \leftarrow \cdots$ of based spaces with $Z = \lim_i Z_i$, there is a canonical map

$$\pi_k(Z) \longrightarrow \lim_i \pi_k(Z_i).$$

This is because a map $S^k \to Z$ gives rise to a sequence of maps $S^k \to Z_i$ via the canonical projections $\lim_i(Z_i) \to Z_i$; it is not hard to see that this induces a map to the inverse limit of the homotopy groups by the universal property. Let $K_k(Z) = \ker(\pi_k(Z) \to \lim_i \pi_k(Z_i))$ be the kernel of this canonical map. When $n = 0$ by "kernel" we mean the set of components of $Z$ which map to the inverse limit of the components of the $Z_i$ which contain the basepoints of the $Z_i$. An important observation is the following.

**Lemma 8.3.3** *Let $Z_1 \leftarrow Z_2 \leftarrow \cdots$ and $Z = \lim_i Z_i$ be as above. For all $k \geq 0$, the canonical map $\pi_k(Z) \to \lim_i \pi_k(Z_i)$ is surjective. That is, we have a short exact sequence (of groups if $k \geq 1$, pointed sets if $k = 0$)*

$$\{e\} \longrightarrow K_n(Z) \longrightarrow \pi_k(Z) \longrightarrow \lim_i \pi_k(Z_i) \longrightarrow \{e\}.$$

*Proof* For each $i$, let $p_i: Z_i \to Z_{i-1}$ denote the given fibration, and let $(\alpha_1, \alpha_2, \ldots)$ be an element of $\lim_i \pi_k(Z_i)$. Thus $(p_i)_*(\alpha_i) = \alpha_{i-1}$. Let $f_i: S^k \to Z_i$ represent the class $\alpha_i$. Then we also have homotopies $H_i: S^k \times I \to Z_i$ from $p_{i+1}(f_{i+1})$ to $f_i$. Consider the lifting problem

$$
\begin{array}{ccc}
S^k \times \{0\} & \xrightarrow{f_2} & Z_2 \\
\downarrow & \nearrow & \downarrow{\scriptstyle p_2} \\
S^k \times I & \xrightarrow{H_1} & Z_1
\end{array}
$$

Since $p_2$ is a fibration, this has a solution $\widehat{H}_1: S^k \times I \to X_2$ which represents a homotopy from $f_2$ to a map $\widehat{f}_2$ such that $p_2(\widehat{f}_2) = f_1$. Let $\widehat{f}_1 = f_1$, and by induction assume we have constructed $\widehat{f}_2, \ldots, \widehat{f}_{n-1}$ so that $p_i(\widehat{f}_i) = \widehat{f}_{i-1}$ and

$\widetilde{f_i} \sim f_i$ for $i = 2$ to $n - 1$. Let $\widetilde{H}_{n-1} : S^k \times I \to X_{n-1}$ be a homotopy from $p_n(f_n)$ to $\widetilde{f}_{n-1}$. Then, as in the previous lifting problem, there is a lift of this homotopy which represents a homotopy from $f_n$ to a map $\widehat{f_n}$ such that $p_n(\widehat{f_n}) = \widehat{f}_{n-1}$. Thus the sequence $(\widehat{f_1}, \widehat{f_2}, \dots,)$ represents a map $f : S^k \to Z$, so its homotopy class $[f] \in \pi_k(Z)$ maps to the given sequence $(\alpha_1, \alpha_2, \dots)$. □

We also need the following lemma to complete the proof of Theorem 8.3.2 in the case of weak equivalences. The proof can be found in [Hir15b] and we omit it.

**Lemma 8.3.4** *Suppose that in the diagram*

$$
\begin{array}{ccc}
X_1 & \longleftarrow X_2 \longleftarrow \cdots \\
\phantom{X_1} f_1 \downarrow \phantom{X_2} f_2 \downarrow \phantom{X_2} \\
Y_1 & \longleftarrow Y_2 \longleftarrow \cdots
\end{array}
$$

*all of the horizontal maps are fibrations and all of the vertical maps $f_i$ are weak homotopy equivalences. Then, for any choice of basepoint of $X$, the induced map $K_n(X) \to K_n(Y)$ is surjective for all $n \ge 0$.*

*Proof of Theorem 8.3.2 for weak equivalences* Let $X = \lim_i X_i$ and $Y = \lim_i Y_i$, and let $f : X \to Y$ be the map induced by the $f_i$. According to Proposition 2.6.17 we must show that for all $n \ge 0$ and for all commutative diagrams

$$
\begin{array}{ccc}
\partial D^n & \xrightarrow{\ g\ } & X \\
i \downarrow & \nearrow & \downarrow f \\
D^n & \xrightarrow{\ h\ } & Y
\end{array}
$$

a solution to the lifting problem $\widehat{h} : D^n \to X$ exists satisfying $\widehat{h} \circ i = g$ and $f \circ \widehat{h} \sim h$ relative to $\partial D^n$. Let $g_j : \partial D^n \to X_j$ and $h_j : D^n \to Y_j$ be the compositions of $g$ and $h$ with the canonical projections $X \to X_j$ and $Y \to Y_j$. Write $a_j$ for the map $X_j \to X_{j-1}$ and $b_j$ for the map $Y_j \to Y_{j-1}$.

Consider the lifting problem

$$
\begin{array}{ccc}
\partial D^n & \xrightarrow{\ g_1\ } & X_1 \\
i \downarrow & \nearrow & \downarrow f_1 \\
D^n & \xrightarrow{\ h_1\ } & Y_1
\end{array}
$$

Since $f_1$ is a weak equivalence, Proposition 2.6.17 implies that there exists a lift $\widehat{h}_1 : D^n \to X_1$ such that $\widehat{h}_1 \circ i = g_1$ and $f_1 \circ \widehat{h}_1 \sim h_1$ relative to $\partial D^n$. Suppose by induction that lifts $\widehat{h}_1, \ldots, \widehat{h}_{k-1}$ have been constructed so that $\widehat{h}_j \colon D^n \to X_j$ satisfies

(1) $\widehat{h}_j \circ i = g_j$;
(2) $f_j \circ \widehat{h}_j \sim h_j$ relative to $\partial D^n$;
(3) $a_j \circ \widehat{h}_j = \widehat{h}_{j-1}$

for all $j = 1$ to $k - 1$. We seek to construct $\widehat{h}_k \colon D^n \to X_k$ satisfying (1)–(3).

As was the case with $f_1$, since $f_k$ is a weak equivalence there exists a map $\widetilde{h}_k \colon D^n \to X_k$ such that (1) and (2) hold for $j = k$. We will make a homotopy of $\widetilde{h}_k$ relative to $\partial D^n$ to a map satisfying (3). The following diagram will make the argument easier to follow:

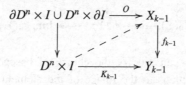

Since $f_k \circ \widetilde{h}_k \sim h_k$ relative to $\partial D^n$, $b_k \circ f_k \circ \widetilde{h}_k \sim b_k \circ h_k = h_{k-1}$ relative to $\partial D^n$. Using commutativity of the right square above we have $f_{k-1} \circ a_k \circ \widetilde{h}_k \sim f_{k-1} \circ \widehat{h}_{k-1}$ relative to $\partial D^n$. Let $K_{k-1}$ denote a homotopy from $f_{k-1} \circ a_k \circ \widetilde{h}_k$ to $f_{k-1} \circ \widehat{h}_{k-1}$ relative to $\partial D^n$. Consider the diagram

$$
\begin{array}{ccc}
\partial D^n \times I \cup D^n \times \partial I & \xrightarrow{\ O\ } & X_{k-1} \\
\downarrow & \nearrow & \downarrow{\scriptstyle f_{k-1}} \\
D^n \times I & \xrightarrow[K_{k-1}]{} & Y_{k-1}
\end{array}
$$

where the left vertical map is the inclusion, and $O \colon \partial D^n \times I \cup D^n \times \partial I \to Y$ is the map defined by $O(s,t) = g_{k-1}(s)$ for all $(s,t) \in \partial D^n \times I$, $O(s,0) = a_k \circ \widetilde{h}_k(s)$ and $O(s,1) = \widehat{h}_{k-1}(s)$ for $s \in D^n$. By Proposition 2.6.17 the dotted arrow exists, call it $\widehat{K}_{k-1}$, and represents a homotopy from $a_k \circ \widetilde{h}_k$ to $\widehat{h}_{k-1}$ relative to $\partial D^n$. Now consider the lifting problem

$$
\begin{array}{ccc}
\partial D^n \times I \cup D^n \times \{0\} & \xrightarrow{\ C\ } & X_k \\
\downarrow & \nearrow & \downarrow{\scriptstyle a_k} \\
D^n \times I & \xrightarrow[\widehat{K}_{k-1}]{} & X_{k-1}
\end{array}
$$

where the left vertical map is the inclusion and $C \colon \partial D^n \times I \cup D^n \times \{0\} \to X_k$ is the map $C(s,t) = g_k(s)$ for $(s,t) \in \partial D^n \times I$ and $C(s,0) = \widetilde{h}_k(s)$ for $s \in D^n$. Since $a_k$ is a fibration, a solution to this lifting problem exists by Proposition 2.1.9, and if we let $\widehat{h}_k$ denote the restriction of this solution to $D^n \times \{1\}$, the map $\widehat{h}_k$ is homotopic to $\widetilde{h}_k$ relative to $\partial D^n$ and $a_{k-1} \circ \widetilde{h}_k = \widetilde{h}_{k-1}$. Thus $\widetilde{h}_k$ satisfies (1)–(3) above.

The maps $\{\widehat{h}_j\}_{j=1}^{\infty}$ now determine a map $\widehat{h} \colon D^n \to X$ such that $\widehat{h} \circ i = g$. Thus we have maps $h, f \circ \widehat{h} \colon D^n \to Y$ which agree on $\partial D^n$, and hence determine an element $[(f \circ \widehat{h}) \cup h] = \beta \in \pi_n(Y)$ (we assume some basepoint in $\partial D^n$ has been chosen, and the image of this basepoint by the map $g$ bases $X$ and hence $Y$). The canonical projections $h_j$ and $f_j \circ \widehat{h}_j$ agree on $\partial D^n$ and also determine elements of $\pi_n(Y_j)$ for all $j$. Since $h_j \sim f_j \circ \widehat{h}_j$, the elements of $\pi_n(Y_j)$ are all zero (when $n = 0$ we mean these elements are all the components containing the basepoints of the $Y_j$). Thus $\beta \in K_n(Y) = \ker(\pi_n(Y) \to \lim_i \pi_n(Y_i))$, and by Lemma 8.3.4 there exists $\alpha \in K_n(X) \subset \pi_n(X)$ such that $f_*(\alpha) = \beta$. Assume for the moment $n \geq 1$. Choose $l \colon D^n \to X$ such that $l|_{\partial D^n} = \widehat{h}|_{\partial D^n}$ and such that $l \cup \widehat{h} \colon S^n = D^n \cup_{\partial D^n} D^n \to X$ represents the homotopy class $\alpha^{-1}$. Hence $e = \beta^{-1} * \beta = [(f \circ l) \cup (f \circ \widehat{h})] * [(f \circ \widehat{h}) \cup h] = [(f \circ \widehat{h}) \cup h]$ is the trivial class, and so $f \circ l \sim h$, as desired.

When $n = 0$, we need a separate argument because of the lack of a group structure on $\pi_0$. Choose a path component of $Y$, and choose a basepoint of $X$. We have the diagram of pointed sets

$$
\begin{array}{ccccc}
K_0(X) & \longrightarrow & \pi_0(X) & \longrightarrow & \lim_i \pi_0(X_i) \\
\downarrow & & \downarrow & & \downarrow \\
K_0(Y) & \longrightarrow & \pi_0(Y) & \longrightarrow & \lim_i \pi_0(Y_i)
\end{array}
$$

By Lemma 8.3.3 we can modify the choice of basepoint of $X$ so that its image in $\lim_i \pi_0(X_i)$ maps to the image of the element of $\pi_0(Y)$ we started with because the right vertical map is an isomorphism since the $f_i$ are weak equivalences (i.e. the induced maps $(f_i)_* \colon \pi_0(X_i) \to \pi_0(Y_i)$ are isomorphisms). But now the element of $\pi_0(Y)$ we started with is an element of $K_0(Y)$, and the map $K_0(X) \to K_0(Y)$ is surjective by Lemma 8.3.4, so we are done. $\square$

In the proof of Theorem 8.3.1, we will study homotopy limits by breaking them into pieces using the skeletal decomposition of the realization of the overcategories used to define them. This decomposes the homotopy limit into a tower of "truncated" homotopy limits. We formalize this as follows.

**Definition 8.3.5** Let $\mathcal{I}$ be a small category and $F \colon \mathcal{I} \to \mathrm{Top}$ a functor. For each $n \geq 0$ define

$$\text{holim}_{I}^{\leq n} F = \text{Eq}\left( \prod_{i} \text{Map}\left(|\text{sk}_n(I \downarrow i)|, F(i)\right) \underset{b}{\overset{a}{\rightrightarrows}} \prod_{i \to i'} \text{Map}\left(|\text{sk}_n(I \downarrow i)|, F(i')\right) \right).$$

(8.3.2)

Here $a$ and $b$ are the restrictions of the maps defining $\text{holim}_I F$.

One should note that this definition relies on the fact that the map $b$, which is induced by the map $I \downarrow i \to I \downarrow i'$ for morphisms $i \to i'$, takes $n$-simplices in the nerve of $I \downarrow i$ to $n$-simplices in the nerve of $I \downarrow i'$.

The following lemma constitutes most of the proof of Theorem 8.3.1, but is also a useful result in its own right. For a category $I$ and an integer $n \geq 0$, recall that $\text{nd}B_n(I)$ denotes the set of non-degenerate $n$-simplices in the nerve of $I$ (see Definition 8.1.8).

**Lemma 8.3.6** *Let $I$ be a small category and $F \colon I \to \text{Top}$ a functor. Then for each $n \geq 0$ we have a pullback square*

$$\begin{array}{ccc}
\text{holim}_{I}^{\leq n} F & \longrightarrow & \displaystyle\prod_{(i_0 \to \cdots \to i_n) \in \text{nd}B_n(I)} \text{Map}(\Delta^n, F(i_n)) \\
\downarrow & & \downarrow \\
\text{holim}_{I}^{\leq n-1} F & \longrightarrow & \displaystyle\prod_{(i_0 \to \cdots \to i_n) \in \text{nd}B_n(I)} \text{Map}(\partial \Delta^n, F(i_n))
\end{array}$$

*where the vertical maps are fibrations, and the horizontal maps are restrictions.*

*Proof* We begin by more explicitly describing the horizontal maps in the square above. By definition, $\text{holim}_{I}^{\leq n} F$ is a subspace of $\prod_i \text{Map}(|\text{sk}_n(I \downarrow i)|, F(i))$. Furthermore, for each $i$ $|\text{sk}_n(I \downarrow i)|$ is a simplicial complex, and in particular a quotient space of the coproduct over all simplices of dimension $\leq n$. Thus

$$\text{holim}_{I}^{\leq n} F \subset \prod_{i} \prod_{k \leq n} \prod_{(i_0 \to \cdots \to i_k) \to i} \text{Map}(\Delta^k, F(i)).$$

The top horizontal map above is this inclusion followed by projection onto those factors indexed by non-degenerate chains $(i_0 \to \cdots \to i_n) \to i$ where $i_n \to i$ is the identity morphism. The lower horizontal arrow is similar, but a little trickier. In this case we treat the boundary $\partial \Delta^n$ as a union of simplices indexed by the faces $d_i(i_0 \to \cdots \to i_n)$; the restriction $\text{holim}_{I}^{\leq n} F \to \text{holim}_{I}^{\leq n-1} F$ keeps track of the value of an element on the faces of each $n$-simplex.

By Proposition A.1.13, we have for each $n \geq 0$ and for each $i \in \mathcal{I}$ a pushout square

$$\coprod_{ndB_n(\mathcal{I}\downarrow i)} \partial\Delta^n \longrightarrow |\text{sk}_{n-1}(\mathcal{I} \downarrow i)|$$
$$\downarrow \qquad\qquad\qquad\qquad \downarrow$$
$$\coprod_{ndB_n(\mathcal{I}\downarrow i)} \Delta^n \longrightarrow |\text{sk}_n(\mathcal{I} \downarrow i)|$$

in which the vertical maps are cofibrations. Apply $\prod_{i\in\mathcal{I}} \text{Map}(-, F(i))$ everywhere to obtain the square

$$\prod_{i\in\mathcal{I}} \text{Map}(|\text{sk}_n(\mathcal{I} \downarrow i)|, F(i)) \longrightarrow \prod_{i\in\mathcal{I}} \prod_{ndB_n(\mathcal{I}\downarrow i)} \text{Map}(\Delta^n, F(i))$$
$$\downarrow \qquad\qquad\qquad\qquad\qquad\qquad\qquad \downarrow$$
$$\prod_{i\in\mathcal{I}} \text{Map}(|\text{sk}_{n-1}(\mathcal{I} \downarrow i)|, F(i)) \longrightarrow \prod_{i\in\mathcal{I}} \prod_{ndB_n(\mathcal{I}\downarrow i)} \text{Map}(\partial\Delta^n, F(i))$$

Since the functor $\text{Map}(-, Z)$ takes cofibrations to fibrations by Proposition 2.5.1, pushouts to pullbacks by Proposition 3.5.14, pointwise products of pullbacks are pullbacks (by Proposition 3.3.6), and products of fibrations are fibrations by Example 2.1.4, the square above is a pullback square in which the vertical maps are fibrations. For brevity we will write the pullback square above as

$$M_n(F) \longrightarrow M_{\Delta^n}(F)$$
$$\downarrow \qquad\qquad\qquad \downarrow$$
$$M_{n-1}(F) \longrightarrow M_{\partial\Delta^n}(F)$$

Recall also that, by definition, $\text{holim}_{\mathcal{I}}^{\leq j}$ is a subspace of $M_j(F)$.

Now let $f^n \in \text{holim}_{\mathcal{I}}^{\leq n} F$, and let $f^{n-1} \in \text{holim}_{\mathcal{I}}^{\leq n-1} F$ be its restriction. The element $f^n$ also determines a point in $M_{\Delta^n}(F) = \prod_{i\in\mathcal{I}} \prod_{ndB_n(\mathcal{I}\downarrow i)} \text{Map}(\Delta^n, F(i))$. Let $x$ (resp. $x_i$) stand for the non-degenerate chain $(i_0 \to \cdots i_n) \to i_n$ in $B_n(\mathcal{I} \downarrow i_n)$ where $i_n \to i_n$ is the identity (resp. the non-degenerate chain $(i_0 \to \cdots \to i_n) \to i$ in $B_n(\mathcal{I} \downarrow i)$), and let $f_x$ (resp. $f_{x_i}$) denote the restriction of $f$ to these simplices. By definition of $\text{holim}_{\mathcal{I}}^{\leq n} F$ as an equalizer, we have a commutative diagram

$$\Delta^n \xrightarrow{f_x} F(i_n)$$
$$f_{x_i} \searrow \quad \downarrow F(i_n \to i)$$
$$F(i)$$

Hence the value of $f^n \in \text{holim}_{\mathcal{I}}^{\leq n} F$ on the $n$-simplex $x_i$ is completely determined by its value on the $n$-simplex labeled by the non-degenerate chain

$(i_0 \to \cdots \to i_n) \to i_n$, where the map $i_n \to i_n$ is the identity. We can encode this in a diagram as follows. First, define a map

$$\prod_{(i_0 \to \cdots \to i_n) \in \mathrm{nd} B_n(I)} \mathrm{Map}(\Delta^n, F(i_n)) \longrightarrow \prod_{i \in I} \prod_{\mathrm{nd} B_n(I \downarrow i)} \mathrm{Map}(\Delta^n, F(i)),$$

whose projection onto the factor indexed by $(i_0 \to \cdots \to i_n) \to i$ is the composition of the map $\mathrm{Map}(\Delta^n, F(i_n)) \to \mathrm{Map}(\Delta^n, F(i))$ induced by the map $F(i_n \to i)$ with the projection onto the factor $\mathrm{Map}(\Delta^n, F(i_n))$ indexed by $i_0 \to \cdots \to i_n$.

Letting $\Phi_n(F) = \prod_{(i_0 \to \cdots \to i_n) \in \mathrm{nd} B_n(I)} \mathrm{Map}(\Delta^n, F(i_n))$, we have just observed that, by restricting attention to the $n$-simplices, $f^n$ determines a point in

$$\lim (M_n(F) \longrightarrow M_{\Delta^n}(F) \longleftarrow \Phi_n(F)).$$

Moreover, the behavior of $f^n$ on the $n$-simplices is determined by an element of the limit above which agrees with $f^{n-1}$ along the boundary of the $n$-simplices. We can capture this coherence as follows. Let

$$\partial \Phi_n(F) = \prod_{(i_0 \to \cdots \to i_n) \in \mathrm{nd} B_n(I)} \mathrm{Map}(\partial \Delta^n, F(i_n)).$$

Then $\mathrm{holim}_I^{\leq n} F$ is the limit of the punctured 3-cube

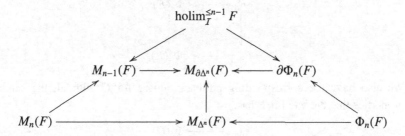

The middle row captures the coherence between $f^{n-1}$ and the point in the limit of the bottom row determined by $f^n$. We can compute the limit of this punctured 3-cube using our observation that the second displayed square in this proof, appearing in the above diagram as the lower-left square, is a pullback as follows. Using Lemma 5.2.8, the limit of this diagram is homeomorphic to

$$\lim \left( \begin{matrix} M_n(F) \\ \downarrow \\ \lim \left( M_{\Delta^n}(F) \to M_{\partial\Delta^n}(F) \leftarrow M_{n-1}(F) \right) \\ \uparrow \\ \lim \left( \Phi_n(F) \to \partial\Phi_n(F) \leftarrow \operatorname{holim}_I^{\leq n-1} F \right) \end{matrix} \right)$$

Since the second displayed square in this proof is a pullback, the top vertical map in this diagram is a homeomorphism, and by Example 3.1.4 the limit appearing above is therefore homeomorphic to the bottom-most space in this diagram. That is, we have the desired pullback diagram

$$
\begin{array}{ccc}
\operatorname{holim}_I^{\leq n} F & \longrightarrow & \Phi_n(F) \\
\downarrow & & \downarrow \\
\operatorname{holim}_I^{\leq n-1} F & \longrightarrow & \partial\Phi_n(F)
\end{array}
$$

Unraveling the notation, it is easy to see that the right vertical map in this square is a fibration. This follows from Proposition 2.5.1 and Example 2.1.4 since the inclusion $\partial\Delta^n \to \Delta^n$ is a cofibration (using Example 2.3.5 and the homeomorphism of pairs $(\Delta^n, \partial\Delta^n) \cong (D^n, \partial D^n)$). $\qquad\square$

*Proof of Theorem 8.3.1* For brevity write the square in the statement of Lemma 8.3.6 as

$$
\begin{array}{ccc}
F_n & \longrightarrow & \Phi_n(F) \\
\downarrow & & \downarrow \\
F_{n-1} & \longrightarrow & \partial\Phi_n(F)
\end{array}
$$

We also have the corresponding pullback square for $G$ with all the same properties that the one for $F$ has:

$$
\begin{array}{ccc}
G_n & \longrightarrow & \Phi_n(G) \\
\downarrow & & \downarrow \\
G_{n-1} & \longrightarrow & \partial\Phi_n(G)
\end{array}
$$

It is enough by Theorem 8.3.2 to prove that the induced map

$$F_n = \operatorname{holim}_I^{\leq n} F \longrightarrow \operatorname{holim}_I^{\leq n} G = G_n$$

is a homotopy (resp. weak) equivalence for each $n$. By induction using Corollary 3.2.15 (an immediate corollary of Theorem 3.2.12), the vertical maps in the diagram

$$
\begin{array}{ccccc}
F_{n-1} & \longrightarrow & \partial\Phi_n(F) & \longleftarrow & \Phi_n(F) \\
\downarrow & & \downarrow & & \downarrow \\
G_{n-1} & \longrightarrow & \partial\Phi_n(G) & \longleftarrow & \Phi_n(G)
\end{array}
$$

are homotopy (resp. weak) equivalences, and hence induce a homotopy (resp. weak) equivalence of homotopy limits and hence of limits $F_n \to G_n$. To get the induction started we need to verify that the map $F_0 \to G_0$ is a homotopy (resp. weak) equivalence. It is not difficult to verify that $F_0 \cong \prod_i F(i)$ and that the map $F_0 \to G_0$ is the product, over $i$, of the maps $F(i) \to G(i)$. This is evidently a homotopy (resp. weak) equivalence, since it is a product of homotopy (resp. weak) equivalences.                               □

We now present the dual to Theorem 8.3.1 establishing the homotopy invariance of the homotopy colimit.

**Theorem 8.3.7** (Homotopy invariance of homotopy colimits)  *Let $I$ be a small category and $F, G \colon I \to$ Top functors. Suppose we have a natural transformation $N \colon F \to G$ such that $F(i) \to G(i)$ is a homotopy (resp. weak) equivalence for all $i$. Then the natural map (8.2.6) is a homotopy (resp. weak) equivalence.*

Again the key ingredients in the proof will be the homotopy invariance of homotopy pushout (Theorem 3.6.13) and the homotopy invariance of colimits of towers of cofibrations, which we will establish below as Theorem 8.3.8.

**Theorem 8.3.8**  *Suppose that in the diagram*

$$
\begin{array}{ccccc}
X_1 & \longrightarrow & X_2 & \longrightarrow & \cdots \\
f_1 \downarrow & & f_2 \downarrow & & \\
Y_1 & \longrightarrow & Y_2 & \longrightarrow & \cdots
\end{array}
$$

*all of the horizontal maps are cofibrations and all of the vertical maps $f_i$ are homotopy (resp. weak) equivalences for all $i \in \mathbb{N}$. Then the induced map $f \colon \mathrm{colim}_i X_i \longrightarrow \mathrm{colim}_i Y_i$ is a homotopy (resp. weak) equivalence.*

As we mentioned, the proof for homotopy equivalences is dual to the proof presented above. However, since this is a fundamental result, let us outline how to use Theorem 8.3.2 to prove this.

*Proof of Theorem 8.3.8 for homotopy equivalences using Theorem 8.3.2*    We use Proposition 1.3.2, which says that a map $X \to Y$ is a homotopy equivalence if and only if the induced map $\mathrm{Map}(Y, Z) \to \mathrm{Map}(X, Z)$ is a homotopy equivalence for all $Z$. Let $Z$ be arbitrary, and apply $\mathrm{Map}(-, Z)$ to the diagram in the statement of the lemma. We then obtain the following diagram

$$
\begin{array}{ccc}
\mathrm{Map}(Y_1, Z) \longleftarrow \mathrm{Map}(Y_2, Z) \longleftarrow \cdots \\
\downarrow \qquad\qquad \downarrow \\
\mathrm{Map}(X_1, Z) \longleftarrow \mathrm{Map}(X_2, Z) \longleftarrow \cdots
\end{array}
$$

By Proposition 1.3.2, the vertical maps are all homotopy equivalences since the maps $X_i \to Y_i$ are. By Proposition 2.5.1, the horizontal maps are all fibrations. Hence by Theorem 8.3.2, the induced map

$$
\lim_i \mathrm{Map}(Y_i, Z) \longrightarrow \lim_i \mathrm{Map}(X_i, Z)
$$

is a homotopy equivalence (note that the limit is taken over the opposite of the indexing category we started with). By Proposition 7.5.1, $\lim_i \mathrm{Map}(Y_i, Z) \cong \mathrm{Map}(\mathrm{colim}_i Y_i, Z)$ and $\lim_i \mathrm{Map}(X_i, Z) \cong \mathrm{Map}(\mathrm{colim}_i X_i, Z)$, and so the natural map $\mathrm{Map}(\mathrm{colim}_i Y, Z) \to \mathrm{Map}(\mathrm{colim}_i X_i, Z)$ is a homotopy equivalence for all $Z$, and hence, by Proposition 1.3.2 once again, the natural map $\mathrm{colim}_i X_i \to \mathrm{colim}_i Y_i$ is a homotopy equivalence.    $\square$

To prove Theorem 8.3.8 for weak equivalences we follow [DI04], and we thank Phil Hirshhorn for helping us incorporate this work. The proof is considerably simpler than its dual, basically because the homotopy groups of a direct limit of inclusions of Hausdorff spaces are the direct limits of the homotopy groups.

*Proof of Theorem 8.3.8 for weak equivalences*    Let $X = \mathrm{colim}_i X_i$ and $Y = \mathrm{colim}_i Y_i$. Consider the diagram of groups

$$
\begin{array}{ccc}
\mathrm{colim}_i \pi_n(X_i) & \longrightarrow & \pi_n(X) \\
\downarrow & & \downarrow \\
\mathrm{colim}_i \pi_n(Y_i) & \longrightarrow & \pi_n(Y)
\end{array}
$$

The top horizonal map is the map induced on homotopy groups by the canonical maps $X_i \to \mathrm{colim}_i X_i = X$; the same is the case for the lower horizontal map. The vertical maps are induced by the map of towers. Observe that every map of a compact space $K$ to $\mathrm{colim}_i X_i$ factors through some $X_n$. This follows

from the fact that cofibrations are closed inclusions (Proposition 2.3.2) and
that our spaces are Hausdorff. See, for example, Lemma A.3 of [DI04], where
a more general statement is proved. This applies in particular to $K = S^n$ for all
$n$. It follows that the horizontal arrows are isomorphisms. It is enough to prove
that the left vertical map is an isomorphism for each $n$. For $n = 0$ it is plainly
true since $\pi_0(X_i) \cong \pi_0(Y_i)$ as sets and the diagram

$$
\begin{array}{ccc}
X_i & \longrightarrow & X_{i+1} \\
\downarrow & & \downarrow \\
Y_i & \longrightarrow & Y_{i+1}
\end{array}
$$

commutes. For $n \geq 1$, the proof is really no different. The inverses $g_i$ to the
isomorphisms $(f_i)_* : \pi_n(X_i) \to \pi_n(Y_i)$ make the diagram

$$
\begin{array}{ccc}
\pi_n(Y_i) & \longrightarrow & \pi_n(Y_{i+1}) \\
{\scriptstyle g_i}\downarrow & & \downarrow{\scriptstyle g_{i+1}} \\
\pi_n(X_i) & \longrightarrow & \pi_n(X_{i+1})
\end{array}
$$

commute, and hence they induce an inverse $g$ to the canonical map
$f_* : X \to Y$.                                                                            □

A proof dual to that given for Theorem 8.3.1 can also be given to prove
Theorem 8.3.7. Here is a definition and lemma dual to Definition 8.3.5 and
Lemma 8.3.6 that may be useful in such a proof.

**Definition 8.3.9** Let $\mathcal{I}$ be a small category and $F : \mathcal{I} \to$ Top a functor. For
each $n \geq 0$ define

$$
\operatorname*{hocolim}_{\mathcal{I}}^{\leq n} F = \operatorname{Coeq}\left( \coprod_{i \to i'} F(i) \times |\mathrm{sk}_n(i' \downarrow \mathcal{I})^{\mathrm{op}}| \underset{b}{\overset{a}{\rightrightarrows}} \coprod_i F(i) \times |\mathrm{sk}_n(i \downarrow \mathcal{I})^{\mathrm{op}}| \right).
$$

$$(8.3.3)$$

Here $a$ and $b$ are the restrictions of the maps defining $\operatorname{hocolim}_{\mathcal{I}} F$.

**Lemma 8.3.10** *Let $\mathcal{I}$ be a small category and $F : \mathcal{I} \to$ Top a functor. Then
for each $n \geq 0$ we have a pushout square*

$$
\begin{array}{ccc}
\displaystyle\coprod_{i_0 \to \cdots \to i_n \in \mathrm{ndB}_n(\mathcal{I})} \partial\Delta^n \times F(i_0) & \longrightarrow & \operatorname*{hocolim}_{\mathcal{I}}^{\leq n-1} F \\
\downarrow & & \downarrow \\
\displaystyle\coprod_{i_0 \to \cdots \to i_n \in \mathrm{ndB}_n(\mathcal{I})} \Delta^n \times F(i_0) & \longrightarrow & \operatorname*{hocolim}_{\mathcal{I}}^{\leq n} F
\end{array}
$$

*where the vertical maps are cofibrations.*

## 8.4 Models for homotopy limits and colimits

In this section we give some other ways to view the homotopy (co)limit, and give a few more examples. Other useful models we will encounter later are cosimplicial replacements for homotopy limits and simplicial replacements for homotopy colimits; we shall see those in Section 9.3 and Section A.1, respectively.

### 8.4.1 Homotopy (co)limits as (co)ends and as Kan extensions

To start, homotopy (co)limits can be expressed as (co)ends of certain bifunctors. In fact, the entire theory of homotopy (co)limits can be neatly organized using this notion. Given a functor $F \colon I \to$ Top, we have naturally induced bifunctors $F_0, F_1 \colon I^{\mathrm{op}} \times I$ given by

$$F_0(i, j) = \mathrm{Map}(|I \downarrow i|, F(j)) \tag{8.4.1}$$

and

$$F_1(i, j) = |(i \downarrow I)^{\mathrm{op}}| \times F(j). \tag{8.4.2}$$

Now recall the definitions of an end and coend, Definitions 7.4.6 and 7.4.11. Then we have the following statement, which follows immediately from the definitions. The reader should compare them with Propositions 7.4.9 and 7.4.13.

**Proposition 8.4.1** (Homotopy (co)limits as (co)ends)  *The end of $F_0$ is the homotopy limit of $F$, and coend of $F_1$ is the homotopy colimit of $F$. That is, there are natural homeomorphisms*

$$\operatorname*{holim}_{I} F \cong \int_{I} F_0$$

*and*

$$\operatorname*{hocolim}_{I} F \cong \int^{I} F_1.$$

Recall the left and right Kan extensions, $L_G F(j)$ and $R_G F(j)$, from Example 7.4.15. We have the following homotopy invariant analog.

**Definition 8.4.2**    Let $I, J$ be small categories, $F \colon I \to$ Top a functor, and $G \colon I \to J$ a functor. There are two natural notions of extensions of $F$ to $J$. We define the *homotopy right (resp. left) Kan extensions of $F$ along $G$* by

$$hR_G F(j) = \underset{(i,m)\in j\downarrow G}{\text{holim }} F(i)$$

and

$$hL_G F(j) = \underset{(i,m)\in G\downarrow j}{\text{hocolim }} F(i).^2$$

To justify calling the above "extensions" of $F$, suppose $G\colon \mathcal{I} \to \mathcal{J}$ is the inclusion of a subcategory. If $i \in \mathcal{I} \subset \mathcal{J}$, then $(i, 1_i)$ is an initial object in the undercategory $i \downarrow G$, and hence $hR_G F(i) \simeq F(i)$ by Example 8.2.5. Similarly $(i, 1_i)$ is a final object in the overcategory $G \downarrow i$ and so $hL_G F(i) \simeq F(i)$ by Example 8.2.17. The values at objects $j \in \mathcal{J}$ which are not objects of $\mathcal{I}$ are an "average" of the values of the functor on the objects under/over $j$.

The homotopy limit and colimit of $F$ can be realized as homotopy Kan extensions:

**Proposition 8.4.3** (Homotopy (co)limits as homotopy Kan extensions)    *Suppose $\mathcal{I}$ and $F$ are as in Definition 8.4.2, $\mathcal{J} = *$ is the category with a single object $*$ and a single morphism, and $G\colon \mathcal{I} \to \mathcal{J}$ is the unique functor. We have canonical homeomorphisms*

$$hR_G F(*) \cong \underset{\mathcal{I}}{\text{holim }} F$$

*and*

$$hL_G F(*) \cong \underset{\mathcal{I}}{\text{hocolim }} F.$$

*Proof*    The categories $* \downarrow G$ and $G \downarrow *$ are canonically isomorphic to $\mathcal{I}$.    □

## 8.4.2 Homotopy (co)limits of towers and (co)fibrant replacements

A classical topic we have mostly neglected so far is the homotopy (co)limit of a tower of spaces. We considered towers briefly in Examples 8.2.10 and 8.2.22, but the treatment there was not particularly satisfactory because our definitions in this case do not give rise to nice models and instead produce realizations of indexing categories that are infinite-dimensional. The "familiar" models (if they are familiar to the reader at all), are the mapping telescope and its dual, but our definitions do not yield these explicitly. The familiar models are nonetheless valid and useful, and discussing them gives us an excuse to talk briefly about (co)fibrant diagrams and (co)fibrant replacements. We give a

---

[2]   We think "homotopy initial Kan extension" (resp. "homotopy terminal Kan extension") are better names for homotopy right Kan extension (resp. homotopy left Kan extension). Dan Dugger has suggested "relative homotopy (co)limit".

sketchy treatment of these topics; for a more thorough analysis (in the context of model categories), see [Hir03].

The notion of (co)fibrant diagrams and replacements is useful in answering the following natural question: for a diagram $F: \mathcal{I} \to \mathrm{Top}$, when are the canonical maps

$$\lim_{\mathcal{I}} F \longrightarrow \operatorname*{holim}_{\mathcal{I}} F$$

and

$$\operatorname*{hocolim}_{\mathcal{I}} F \longrightarrow \operatorname*{colim}_{\mathcal{I}} F$$

(weak) homotopy equivalences? This would mean that, for such $F$, its limit or colimit would itself be homotopy invariant. We explored this in the case of punctured cubes in Propositions 5.4.26 and 5.8.25. Finding a reasonably general set of hypotheses which assures that the (co)limit and homotopy (co)limit are weakly equivalent is more difficult, but we can supply the answer for a class of diagrams of which the punctured cubes are a member.

**Definition 8.4.4**  We call a small category $\mathcal{I}$ a *Reedy category* if there exist subcategories $\overrightarrow{\mathcal{I}}$ (the *direct subcategory*) and $\overleftarrow{\mathcal{I}}$ (the *inverse subcategory*) and a function deg: $\mathrm{Ob}(\mathcal{I}) \to \mathbb{N}$ called the *degree* such that the following hold:

- If $m: i \to i'$ is a non-identity morphism in $\overrightarrow{\mathcal{I}}$, then $\deg(i) < \deg(i')$.
- If $m: i \to i'$ is a non-identity morphism in $\overleftarrow{\mathcal{I}}$, then $\deg(i) > \deg(i')$.
- Every morphism $m$ in $\mathcal{I}$ has a unique factorization $m = r \circ l$, where $r$ is a morphism in $\overrightarrow{\mathcal{I}}$ and $l$ is a morphism in $\overleftarrow{\mathcal{I}}$.

For example, $\mathcal{P}(\underline{n})$ is a Reedy category, and $\mathcal{P}_0(\underline{n})$ and $\mathcal{P}_1(\underline{n})$ inherit their Reedy category structure from this one. In this case the degree is given by $\deg(S) = |S| + 1$, and $\overrightarrow{\mathcal{P}(\underline{n})}$ can be taken to be $\mathcal{P}(\underline{n})$ itself, and the inverse subcategory can be taken to contain only the identity maps.

Other examples of Reedy categories include the categories indexing the (co)equalizer (two parallel arrows), the inverse system (see Example 7.3.12), and the directed system (see Example 7.3.31). The notion of a degree and of the direct and inverse subcategories in each case is straightforward and left to the reader. In all of these examples, $\overrightarrow{\mathcal{I}}$ or $\overleftarrow{\mathcal{I}}$ is trivial, in which case the category itself is called "direct" or "inverse", respectively. Another Reedy category of importance to us is the cosimplicial indexing category $\Delta$, which we will study in depth in Chapter 9 (see in particular the discussion preceeding Example 9.1.7).

One family of examples of categories which are not Reedy are those associated with non-trivial groups: the category with one object and one morphism for each group element which compose according to the group law cannot be Reedy. Furthermore, any category which has objects with non-trivial automorphisms cannot be Reedy.

**Definition 8.4.5** Let $(\overrightarrow{I} \downarrow i)_1$ and $(i \downarrow \overleftarrow{I})_0$ denote the full subcategories of $\overrightarrow{I} \downarrow i$ and $i \downarrow \overleftarrow{I}$ with the identity morphism $1_i \colon i \to i$ removed. We call $(\overrightarrow{I} \downarrow i)_1$ the *latching category of* $I$ *at* $i$, and $(i \downarrow \overleftarrow{I})_0$ the *matching category of* $I$ *at* $i$.

**Definition 8.4.6** Given a functor $F \colon I \to \text{Top}$ with $I$ Reedy and an object $i \in I$, define the *matching space of* $F$ *at* $i$ to be

$$M_i(F) = \varprojlim_{(i \downarrow \overleftarrow{I})_0} F.$$

There is an evident canonical map

$$F(i) \longrightarrow M_i(F)$$

called the *matching map of* $F$ *at* $i$. If, for every $i \in I$, the matching map of $F$ at $i$ is a fibration, then the functor $F$ is said to be *fibrant*.

**Definition 8.4.7** Dually, we can define the *latching space of* $F$ *at* $i$ to be

$$L_i(F) = \underset{(\overrightarrow{I} \downarrow i)_1}{\text{colim}} F.$$

There is again a canonical map

$$L_i(F) \longrightarrow F(i)$$

called the *latching map of* $F$ *at* $i$. If, for every $i \in I$, the latching map of $F$ at $i$ is a cofibration, then the functor $F$ is said to be *cofibrant*.

**Remark 8.4.8** We may regard the category Top as Top$^*$, where $*$ is the category with a single object and a single morphism, and we can thus say what it means for a space to be (co)fibrant. Thinking of a space $X$ in this way, i.e. as a functor $X \colon * \to \text{Top}$, the above conditions say that $X$ is fibrant (resp. cofibrant) if the map $X \to *$ (resp. $\emptyset \to X$) is a fibration (resp. cofibration). This is always true for topological spaces and is trivial to check from our definition of a (co)fibration. It is also true for Top$_*$; the key point here is that we assume our based spaces are well-pointed. However, we want to reiterate that we are not specifically using any model structure in this text, and "fibration" and "cofibration" as we know them here are merely examples of a fibration/cofibration in some model structure on Top.                    □

We then have the following result, the proof of which we omit. Before we state it, we need to define a *connected category* as a category where every pair of objects is connected by a finite zig-zag of morphisms.

**Theorem 8.4.9** ([Hir03, Theorem 19.9.1 and Proposition 15.10.2])   *Let $I$ be a Reedy category.*

1. *If every latching category is empty or connected and $F: I \to$ Top is a fibrant diagram, then the canonical map $\lim_I F \to \operatorname{holim}_I F$ is a weak equivalence.*
2. *If every matching category is empty or connected and $F: I \to$ Top is a cofibrant diagram, then the canonical map $\operatorname{hocolim}_I F \to \operatorname{colim}_I F$ is a weak equivalence.*

Since the inverse and direct categories, as well as categories indexing (punctured) cubes, have empty latching and matching categories, we then immediately have

**Corollary 8.4.10**

1. *If $X: \mathcal{P}_0(\underline{n}) \to$ Top is a fibrant punctured n-cube, then the canonical map $\lim_{\mathcal{P}_0(\underline{n})} X \to \operatorname{holim}_{\mathcal{P}_0(\underline{n})} X$ is a weak equivalence.*
2. *If $X: \mathcal{P}_1(\underline{n}) \to$ Top is a cofibrant punctured n-cube, then the canonical map $\operatorname{hocolim}_{\mathcal{P}_1(\underline{n})} X \to \operatorname{colim}_{\mathcal{P}_1(\underline{n})} X$ is a weak equivalence.*
3. *If $I = (\bullet \leftarrow \bullet \leftarrow \cdots)$ is the category which indexes the inverse limit, and $F: I \to$ Top is a fibrant diagram, then the canonical map $\lim_I F \to \operatorname{holim}_I F$ is a weak equivalence.*
4. *If $I = (\bullet \to \bullet \to \cdots)$ is the category which indexes the direct limit, and $F: I \to$ Top is a cofibrant diagram, then the canonical map $\operatorname{hocolim}_I F \to \operatorname{colim}_I F$ is a weak equivalence.*

Notice that the first two parts of the above corollary are restatements of Propositions 5.4.26 and 5.8.25. (Those two results are in terms of homotopy equivalences but the statements above have weak equivalences because they use Theorem 8.4.9.) In fact, Corollary 8.4.10 in a sense brings us back to the beginning of the story of homotopy (co)limits, which began for us with homotopy pullbacks and pushouts. An important observation that we made early on, Proposition 3.2.13 (or Example 7.3.15), states that if in the diagram

$X \to Z \leftarrow Y$ either of the maps were fibrations, then the pullback of this diagram was also its homotopy pullback; that is, it enjoyed the homotopy invariance property. In light of the previous corollary it is enough if this diagram is fibrant, which means in this case that both maps $X \to Z$ and $Y \to Z$ are fibrations (this is stricly more than is necessary, but Corollary 8.4.10 is a very general statement). Similarly for the diagram $X \leftarrow W \to Y$ and the equivalence of its pushout and homotopy pushout; the diagram being cofibrant means each of the maps $W \to X$ and $W \to Y$ are cofibrations (see Proposition 3.6.17 or Example 7.3.34). Propositions 5.4.26 and 5.8.25 are extensions of these important observations from squares to cubes.

**Example 8.4.11** (Inverse homotopy limit and direct homotopy colimit) Using Corollary 8.4.10, we can now give a alternative way of constructing the inverse homotopy limit (as it appears in Example 8.2.10) and the direct homotopy colimit (as it appears in Example 8.2.22) of towers of spaces. Suppose $I = \{0 \leftarrow 1 \leftarrow 2 \leftarrow \cdots\}$ and let $X \colon I \to$ Top be a functor with $X(i) = X_i$ and $X(i \to i - 1) = f_i$. Thus the diagram is an inverse system of spaces

$$X_0 \xleftarrow{f_1} X_1 \xleftarrow{f_2} X_2 \xleftarrow{f_3} \cdots$$

For any $X_i$, the matching object $M_i(X)$ is equal to $X_{i-1}$ by Example 7.3.9, as the object $i \to i-1$ is initial in the category indexing the limit which defines $M_i(X)$. For the above tower to be a fibrant diagram, it therefore suffices that the maps $f_i \colon X_i \to X_{i-1}$ are fibrations for each $i$. Hence, according to Corollary 8.4.10, a model for the homotopy inverse limit of a tower can be achieved by replacing all of the maps systematically by fibrations.

To do this, we will define a new tower $PX$ and a map of towers $X \to PX$ that will be a homotopy equivalence. More explicitly, we will construct a commutative diagram

$$
\begin{array}{ccccccc}
X_0 & \xleftarrow{f_1} & X_1 & \xleftarrow{f_2} & X_2 & \longleftarrow & \cdots \\
\downarrow & & \downarrow & & \downarrow & & \\
PX(0) & \xleftarrow{p_1} & PX(1) & \xleftarrow{p_2} & PX(2) & \longleftarrow & \cdots
\end{array}
$$

where all the vertical maps are equivalences.

The tower $PX$ is defined inductively: let $PX(0) = X_0$, with the vertical map being the identity. Then let $PX(1) = P_{f_1}$, with the maps $X_1 \to PX(1)$ and $PX(1) \to PX(0)$ being the natural maps. The vertical maps are homotopy equivalences, and the map $PX(1) \to PX(0)$ is a fibration.

If $PX(k-1)$ has been constructed, then define $PX(k)$ via the pullback square

$$\begin{array}{ccc} PX(k) & \longrightarrow & P_{f_k} \\ {\scriptstyle p_k}\downarrow & & \downarrow \\ PX(k-1) & \longrightarrow & X_{k-1} \end{array}$$

where $PX(k-1) \to X_{k-1}$ is the composition of the projection

$$PX(k-1) = \lim\left(PX(k-2) \to X_{k-2} \leftarrow P_{f_{k-1}}\right) \longrightarrow P_{f_{k-1}}$$

with the canonical projection $P_{f_{k-1}} \to X_{k-1}$.

Since the right vertical map in the above square is a fibration, so is the map $p_k$. Since the bottom horizontal map is a homotopy equivalence, so is the map $PX(k) \to P_{f_k}$.

To define the map $X_k \to PX(k)$: we have a natural map $X_k \to P_{f_k}$ and the composition

$$X_k \longrightarrow X_{k-1} \longrightarrow PX(k-1),$$

and if we compose both of those with maps to $X_{k-1}$, they agree. Since $X_k \to P_{f_k}$ and $PX(k) \to P_{f_k}$ are both homotopy equivalences, the map $X_k \to PX(k)$ is also a homotopy equivalence.

Because the maps $p_i$ are fibrations, by the homotopy invariance of the homotopy limit (Theorem 8.3.1), the induced map $\operatorname{holim}_I PX \to \operatorname{holim}_I X$ is a homotopy equivalence. Since $PX$ is fibrant, Corollary 8.4.10 implies that the natural map $\lim_I PX \to \operatorname{holim}_I PX$ is a weak equivalence.

It remains to describe $\lim_I PX$ explicitly. The map $p_k\colon PX(k) \to PX(k-1)$ sends a pair $(z_{k-1}, (x_k, \gamma_{k-1}))$ to $z_{k-1}$. It is then straightforward to verify that a point in $\lim_I PX$ consists of a sequence of points $(x_0, x_1, \ldots)$ and paths $(\gamma_0, \gamma_1, \ldots)$ where $x_i \in X_i$ and $\gamma_i\colon I \to X_i$ has the property that $\gamma_i(0) = f_{i+1}(x_{i+1})$ and $\gamma_i(1) = x_i$. A rough picture, the *mapping microscope* (the terminology parallels that of the mapping telescope, to be encountered shortly), is given below.

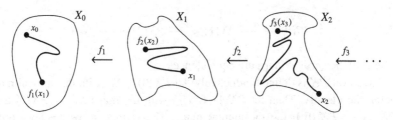

Figure 8.4 The mapping microscope of an inverse system of spaces.

Another way to achieve the same model is to use the description of $\lim_I X$ given in Example 7.3.12. There we showed that there is a homeomorphism

$$\lim_I X \cong \lim\left(\prod_i X_i \xrightarrow{\Delta} \prod_i X_i \times \prod_i X_i \overset{(1_{\prod_i X_i}, F)}{\longleftarrow} \prod_i X_i\right)$$

where $F = (f_1, f_2, \ldots)$, and $\Delta$ is the diagonal. We leave it to the reader to check that

$$\lim\left(P_\Delta \longrightarrow \prod_i X_i \times \prod_i X_i \overset{(1_{\prod_i X_i}, F)}{\longleftarrow} \prod_i X_i\right)$$

yields the same model as given above, where $P_\Delta$ is the mapping path space. The motivation for this model is simple: to create a homotopy invariant limit, we replace one map by a fibration.

Dually, suppose $I = \{0 \to 1 \to 2 \to \cdots\}$ indexes a directed system of spaces $\mathcal{Y}\colon I \to \text{Top}$, namely

$$Y_0 \xrightarrow{g_0} Y_1 \xrightarrow{g_1} Y_2 \xrightarrow{g_2} \cdots$$

In this case, the latching object $L_i(\mathcal{Y})$ of $Y_i$ is $Y_{i-1}$ and it thus suffices for all the maps in this system to be cofibrations in order for it to be cofibrant. Define a tower $M\mathcal{Y}$ inductively as follows: $M\mathcal{Y}(0) = Y_0$, $M\mathcal{Y}(1) = M_{g_0}$, and

$$M\mathcal{Y}(k) = \text{colim}(M\mathcal{Y}(k-1) \longleftarrow Y_{k-1} \longrightarrow M_{f_{k-1}}).$$

The maps $m_k\colon M\mathcal{Y}(k) \to M\mathcal{Y}(k+1)$ are the canonical inclusions $M\mathcal{Y}(k) \to \text{colim}(M\mathcal{Y}(k) \leftarrow X_k \to M_{f_k})$. Then, by an argument dual to the above, this gives a tower of cofibrations with a map $\mathcal{Y} \to M\mathcal{Y}$ which is a pointwise homotopy equivalence, and so $\text{hocolim}_I \mathcal{Y} \simeq \text{colim}_I M\mathcal{Y}$ by Corollary 8.4.10.

This colimit $\text{colim}_I M\mathcal{Y}$ is called the *mapping telescope* of the diagram $\mathcal{Y}$ and can be pictured as in Figure 8.5.

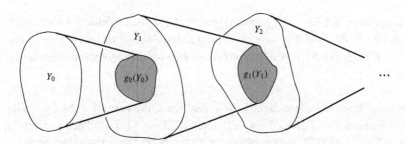

Figure 8.5 The mapping telescope of a directed system of spaces.

Alternatively, we could take the description of the direct limit given in Example 7.3.31, where we showed that there is a homeomorphism

$$\mathrm{colim}_I \; \mathcal{Y} \cong \left( \coprod_i \xleftarrow{\nabla} \coprod_i Y_i \coprod \coprod_i Y_i \xrightarrow{1_{\coprod_i Y_i} \coprod G} \coprod_i G_i \right),$$

where $\nabla$ is the "fold" map and $G(y_i) = g_i(y_i)$ for $y_i \in Y_i$. In this case we turn $\nabla$ into a cofibration, and the space

$$\left( M_\nabla \longleftarrow \coprod_i Y_i \coprod \coprod_i Y_i \xrightarrow{1_{\coprod_i Y_i} \coprod G} \coprod_i G_i \right)$$

gives precisely the mapping telescope described above.                         □

We now return to general diagrams. The following says that every diagram can be replaced by a (co)fibrant one, by a process called *(co)fibrant replacement*.

**Definition 8.4.12**   Let $F : \mathcal{I} \to$ Top be a functor. Define $\check{F}, \hat{F} : \mathcal{I} \to$ Top by

$$\check{F}(i) = \operatorname*{holim}_{(i',m) \in (i \downarrow \mathcal{I})} F(i')$$

and

$$\hat{F}(i) = \operatorname*{hocolim}_{(i',m) \in (\mathcal{I} \downarrow i)} F(i').$$

$\check{F}$ is called the *fibrant replacement of F* and $\hat{F}$ is called the *cofibrant replacement of F*.

The following result says that the formulas defining $\check{F}$ and $\hat{F}$ are functors which are pointwise homotopy equivalent to $F$, so it is in this sense that $\check{F}$ and $\hat{F}$ are "replacements" of $F$. The justification of the "(co)fibrant" part of the terminology is beyond the scope of this book; for more details, see [DHKS04, Section 23].

**Proposition 8.4.13**   *The formulas for $\check{F}$ and $\hat{F}$ in Definition 8.4.12 define functors $\check{F}, \hat{F} : \mathcal{I} \to$ Top, and there exist natural transformations $F \to \check{F}$ and $\hat{F} \to F$ such that $F(i) \to \check{F}(i)$ and $\hat{F}(i) \to F(i)$ are homotopy equivalences for each i.*

*Proof*   We verify these claims for $\check{F}$ and leave the case of $\hat{F}$ to the reader. A morphism $i \to j$ gives rise to a functor $j \downarrow \mathcal{I} \to i \downarrow \mathcal{I}$, and hence defines a map $\check{F}(i) \to \check{F}(j)$. It is straightforward to verify that the identity morphism is sent to the identity map and that the composition of two morphisms is sent to the evident composite from the definitions. Hence the formula defining $\check{F}(i)$ is a functor of $i$.

Since $i \downarrow I$ has initial object $(i, 1_i)$, using Example 7.3.9, we have a canonical map

$$F(i) \cong \lim_{(i',m)\in(i\downarrow I)} F(i') \longrightarrow \underset{(i',m)\in(i\downarrow I)}{\text{holim}} F(i') = \check{F}(i),$$

which is a functor of $i$. These maps are homotopy equivalences, since by Example 8.2.5 the projection $\text{holim}_{i\downarrow I} F \to \text{Map}(|(i \downarrow I) \downarrow (i, 1_i)|, F(i)) \simeq F(i)$ is a homotopy equivalence, and the map $F(i) \cong \text{Map}(*, F(i)) \to \text{Map}(|(i \downarrow I) \downarrow (i, 1_i)|, F(i))$ induced by the canonical map above is clearly a homotopy equivalence, as it is induced by the canonical map $|(i \downarrow I) \downarrow (i, 1_i)| \to *$. □

By the homotopy invariance of homotopy (co)limits, Theorems 8.3.1 and 8.3.7, the natural transformations $F \to \check{F}$ and $\hat{F} \to F$ induce homotopy equivalences

$$\underset{I}{\text{holim}}\, F \longrightarrow \underset{I}{\text{holim}}\, \check{F}$$

and

$$\underset{I}{\text{hocolim}}\, \hat{F} \longrightarrow \underset{I}{\text{hocolim}}\, F.$$

The following result says that the (co)limit and homotopy (co)limit of (co)fibrant replacement agree.

**Theorem 8.4.14** *The canonical maps*

$$\underset{I}{\lim}\, \check{F} \longrightarrow \underset{I}{\text{holim}}\, \check{F}$$

*and*

$$\underset{I}{\text{hocolim}}\, \hat{F} \longrightarrow \underset{I}{\text{colim}}\, \hat{F}.$$

*are homotopy equivalences.*

*Proof* We will sketch the proof of the statement for homotopy limits and leave the rest of the proof and its dual to the reader. We will show how to construct a homeomorphism $h \colon \lim_I \check{F} \to \text{holim}_I F$ such that

$$\begin{array}{ccc}
 & & \text{holim}_I\, F \\
 & \overset{h}{\nearrow} & \downarrow {\scriptstyle \simeq} \\
\lim_I \check{F} & \longrightarrow & \text{holim}_I\, \check{F}
\end{array}$$

commutes, where the vertical arrow is the homotopy equivalence given by Proposition 8.4.13 and Theorem 8.3.1, and the diagonal arrow is the map in question.

For each object $i$, by definition, $\check{F}(i) = \operatorname{holim}_{i \downarrow I} F$, and we define, for each non-negative integer $n$,

$$\check{F}^{\leq n}(i) = \operatorname*{holim}_{i \downarrow I}{}^{\leq n} F.$$

Thus,

$$\lim_I \check{F}^{\leq n} = \operatorname{Eq}\left( \prod_i \check{F}^{\leq n}(i) \overset{a}{\underset{b}{\rightrightarrows}} \prod_{i \to i'} \check{F}^{\leq n}(i') \right),$$

and clearly $\lim_I \check{F} \cong \lim_n \lim_I \check{F}^{\leq n}$. We will argue by induction on $n$ that there are homeomorphisms $\lim_I \check{F}^{\leq n} \cong \operatorname{holim}_I^{\leq n} F$ for each $n$, and that these are compatible in the sense that they give rise to a homeomorphism $\lim_I \check{F} \cong \operatorname{holim}_I F$. We leave it to the reader to check that there is a natural homeomorphism $\lim_I \check{F}^{\leq 0} \cong \operatorname{holim}_I^{\leq 0} F$.

We claim that the square

$$
\begin{array}{ccc}
\lim_I \check{F}^{\leq n} & \longrightarrow & \displaystyle\prod_{(i_0 \to \cdots \to i_n) \in \mathrm{nd}B_n(I)} \operatorname{Map}(\Delta^n, F(i_n)) \\
\downarrow & & \downarrow \\
\lim_I \check{F}^{\leq n-1} & \longrightarrow & \displaystyle\prod_{(i_0 \to \cdots \to i_n) \in \mathrm{nd}B_n(I)} \operatorname{Map}(\partial\Delta^n, F(i_n))
\end{array}
$$

is a pullback square (similar to the square appearing in Lemma 8.3.6). The main point is that the value of a point in $\lim_I \check{F}^{\leq n}$ on a non-degenerate simplex indexed by $i \to (i_0 \to \cdots \to i_n)$ is determined by its value on the non-degenerate simplex indexed by $i_0 \to (i_0 \to \cdots \to i_n)$, where the map $i_0 \to i_0$ is the identity map. This follows from the description of $\lim_I \check{F}^{\leq n}$ above as an equalizer. We know from Lemma 8.3.6 that we have a pullback square

$$
\begin{array}{ccc}
\operatorname{holim}_I^{\leq n} F & \longrightarrow & \displaystyle\prod_{(i_0 \to \cdots \to i_n) \in \mathrm{nd}B_n(I)} \operatorname{Map}(\Delta^n, F(i_n)) \\
\downarrow & & \downarrow \\
\operatorname{holim}_I^{\leq n-1} F & \longrightarrow & \displaystyle\prod_{(i_0 \to \cdots \to i_n) \in \mathrm{nd}B_n(I)} \operatorname{Map}(\partial\Delta^n, F(i_n))
\end{array}
$$

and by induction a homeomorphism $h_{n-1} \colon \lim_I \check{F}^{\leq n-1} \to \operatorname{holim}_I^{\leq n-1} F$ induces a homeomorphism $h_n \colon \lim_I \check{F}^{\leq n} \to \operatorname{holim}_I^{\leq n} F$ by inducing an evident home-omorphism of pullbacks of the above two squares. The homeomorphisms $h_n$ collectively define the desired homeomorphism $h \colon \lim_I \check{F} \to \operatorname{holim}_I F$. ☐

## 8.5 Algebra of homotopy limits and colimits

In this section we explore some properties and interactions of homotopy (co)limits with themselves and with other functors. This section parallels and extends various results we have already established for punctured squares and cubes, as well as for ordinary limits and colimits. We will thus in this section see generalizations of results from Sections 3.3, 3.7, 3.9, 5.4, 5.8, 5.10, and 7.5. In Section 8.6, we also extend some of those results in a different direc-tion, as we are now not limited by a particular shape of the diagram (punctured cubical) and can also consider functorial properties of homotopy (co)limits in the indexing variable.

From Theorems 8.3.1 and 8.3.7 and from the discussion preceding them, we have that $\operatorname{holim}_I$ and $\operatorname{hocolim}_I$ preserve (weak) homotopy equivalences as functors on the category of functors from $I$ to Top. The next result says they also preserve (co)fibrations. Recall that when we speak of a natural transfor-mation of diagrams that is a (co)fibration, we mean that it is an objectwise (co)fibration.

**Theorem 8.5.1** (Ho(co)lims preserve (co)fibrations)  *Suppose $F, G \colon I \to$ Top are diagrams and $N \colon F \to G$ is a natural transformation.*

1. If $N$ is a fibration, then the map $\operatorname{holim}_I F \to \operatorname{holim}_I G$ is a fibration.
2. If $N$ is a cofibration, then the map $\operatorname{hocolim}_I F \to \operatorname{hocolim}_I G$ is a cofibration.

The argument is very similar to the one used to prove Theorem 8.3.1. So simi-lar, in fact, that we will use parts of that argument below. We first need a lemma concerning towers of fibrations.

**Lemma 8.5.2**  *Let $\mathcal{X}_0 = (X_0 \leftarrow X_1 \leftarrow \cdots)$ and $\mathcal{Y}_0 = (Y_0 \leftarrow Y_1 \leftarrow \cdots)$ be towers of fibrations, and let $\mathcal{X}_0 \to \mathcal{Y}_0$ be a map of diagrams such that $X_0 \to Y_0$ is a fibration and the canonical map $X_{i+1} \to \lim(X_i \to Y_i \leftarrow Y_{i+1})$ is a fibration for all $i \geq 0$. Then the induced map $\lim \mathcal{X}_0 \to \lim \mathcal{Y}_0$ is a fibration.*

*Dually, let $X_1 = (X_0 \to X_1 \to \cdots)$ and $\mathcal{Y}_1 = (Y_0 \to Y_1 \to \cdots)$ be towers of cofibrations, and let $X_1 \to \mathcal{Y}_1$ be a map of diagrams such that $X_0 \to Y_0$ is a cofibration and the canonical map $\mathrm{colim}(Y_i \leftarrow X_i \to X_{i+1}) \to Y_{i+1}$ is a cofibration for all $i \geq 0$. Then the induced map $\mathrm{colim}\, X_1 \to \mathrm{colim}\, \mathcal{Y}_1$ is a cofibration.*

*Proof* We prove the first statement; the second is dual and we omit it (or it can be proved using the first statement after applying the functor $\mathrm{Map}(-, Z)$).

Let $p_i \colon X_i \to X_{i-1}$ be the maps in the diagram $X_0$. A solution to the lifting problem

$$
\begin{array}{ccc}
W \times \{0\} & \xrightarrow{\,g\,} & \lim X_0 \\
{\scriptstyle i_0}\big\downarrow & \nearrow & \big\downarrow \\
W \times I & \xrightarrow[\ h\ ]{} & \lim \mathcal{Y}_0
\end{array}
$$

amounts to a sequence of solutions

$$
\begin{array}{ccc}
W \times \{0\} & \xrightarrow{\,g_i\,} & X_i \\
{\scriptstyle i_0}\big\downarrow & {\scriptstyle \hat{h}_i}\nearrow & \big\downarrow \\
W \times I & \xrightarrow[\ h_i\ ]{} & Y_i
\end{array}
$$

such that $p_i \circ \hat{h}_i = \hat{h}_{i-1}$. When $i = 0$, there is a solution $\hat{h}_0$ since $X_0 \to Y_0$ is a fibration. Now suppose by induction that we have constructed lifts $\hat{h}_0, \ldots, \hat{h}_{i-1}$. Define $\hat{h}_i$ as follows. Let $P_i = \lim(X_{i-1} \to Y_{i-1} \leftarrow Y_i)$, and let $u \colon X_i \to P_i$ be the canonical map. Then in the diagram

$$
\begin{array}{ccccc}
W \times \{0\} & \xrightarrow{\,u \circ g_i\,} & P_i & \longrightarrow & X_{i-1} \\
{\scriptstyle i_0}\big\downarrow & & \big\downarrow & & \big\downarrow \\
W \times I & \xrightarrow[\ h_i\ ]{} & Y_i & \longrightarrow & Y_{i-1}
\end{array}
$$

the right-most square is a pullback, and hence the given lift $\hat{h}_{i-1} \colon W \times I \to X_{i-1}$ gives rise to a lift $\tilde{h}_i \colon W \times I \to P_i$ by the universal property of $P_i$. This gives rise to a lifting problem

$$
\begin{array}{ccc}
W \times \{0\} & \xrightarrow{\,g_i\,} & X_i \\
{\scriptstyle i_0}\big\downarrow & \nearrow & \big\downarrow{\scriptstyle u} \\
W \times I & \xrightarrow[\ \tilde{h}_i\ ]{} & P_i
\end{array}
$$

The right vertical map is a fibration by hypothesis, and the dotted arrow is the desired lift. □

**Remark 8.5.3** The hypothesis that $X_{i+1} \to \lim(X_i \to Y_i \leftarrow Y_{i+1})$ is a fibration implies that $X_i \to Y_i$ is a fibration for all $i$ by induction using the fact that $X_0 \to Y_0$ is a fibration. However, it is not enough to assume that $X_i \to Y_i$ is a fibration for all $i$ to obtain a fibration of limits. We leave it to the reader to dualize this remark for the second statement above. □

*Proof of Theorem 8.5.1* We prove the first statement and leave the dualization to the reader. By Lemma 8.3.6, the square

$$
\begin{array}{ccc}
\operatorname{holim}_I^{\leq n} F & \longrightarrow & \prod_{(i_0 \to \cdots \to i_n) \in \mathrm{nd}B_n(I)} \mathrm{Map}(\Delta^n, F(i_n)) \\
\downarrow & & \downarrow \\
\operatorname{holim}_I^{\leq n-1} F & \longrightarrow & \prod_{(i_0 \to \cdots \to i_n) \in \mathrm{nd}B_n(I)} \mathrm{Map}(\partial\Delta^n, F(i_n))
\end{array}
$$

is a pullback and the vertical maps are fibrations. Recall that the products on the right side of the square are indexed by non-degenerate chains $i_0 \to \cdots \to i_n$ in the nerve of the indexing category. For brevity write

$$F_n = \operatorname{holim}_I^{\leq n} F,$$

$$\partial\Phi_n(F) = \prod_{(i_0 \to \cdots \to i_n) \in \mathrm{nd}B_n(I)} \mathrm{Map}(\partial\Delta^n, F(i_n)),$$

$$\Phi_n(F) = \prod_{(i_0 \to \cdots \to i_n) \in \mathrm{nd}B_n(I)} \mathrm{Map}(\Delta^n, F(i_n)),$$

and similarly for $G$. Then the above becomes a pullback square

$$
\begin{array}{ccc}
F_n & \longrightarrow & \Phi_n(F) \\
\downarrow & & \downarrow \\
F_{n-1} & \longrightarrow & \partial\Phi_n(F)
\end{array}
$$

in which the vertical maps are fibrations. We are going to apply Lemma 8.5.2, so we begin by noting that $F_0 \cong \prod_i F(i)$ and $G_0 \cong \prod_i G(i)$, and since $F(i) \to G(i)$ is a fibration for all $i$, $F_0 \to G_0$ is a fibration (the map $F_0 \to G_0$, when composed with the homeomorphisms $F_0 \cong \prod_i F(i)$ and $G_0 \cong \prod_i G(i)$, becomes the obvious map). We then wish to prove that $F_n \to \lim(F_{n-1} \to G_{n-1} \leftarrow G_n)$ is a fibration. By induction we may assume $F_{n-1} \to G_{n-1}$ is a fibration. Using the pullback square above, we are led to the following lifting problem:

$$W \times \{0\} \xrightarrow{g} \lim(F_{n-1} \to \partial\Phi_n(F) \leftarrow \Phi_n(F))$$

$$\downarrow{i_0} \qquad\qquad\qquad \downarrow$$

$$W \times I \xrightarrow{h} \lim(F_{n-1} \to G_{n-1} \leftarrow G_n)$$

The composition of $h$ with the canonical projection gives a map $h_{F_{n-1}} : W \times I \to F_{n-1}$. We now seek a lift $h_{\Phi_n(F)} : W \times I \to \Phi_n(F)$ of $h$ such that $h_{F_{n-1}}$ and $h_{\Phi_n(F)}$ equalize when projected to $\partial\Phi_n(F)$. That is, in the following diagram, we seek a solution to a lifting problem making the diagram commute.

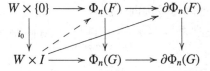

Unraveling the definitions, it is straightforward to see that $\Phi_n(F) \to \lim(\partial\Phi_n(F) \to \partial\Phi_n(G) \leftarrow \Phi_n(G))$ is a fibration by Proposition 3.1.11. Hence the desired lift exists; we first pull the given lift back to $\lim(\partial\Phi_n(F) \to \partial\Phi_n(G) \leftarrow \Phi_n(G))$ using the universal property of the limit, and then pull this lift back to $\Phi_n(F)$ using that the aforementioned map is a fibration. □

Next we have an analog of Proposition 7.5.1 and a generalization of Proposition 5.10.1 (and of Proposition 3.9.1).

**Proposition 8.5.4** (Maps and homotopy (co)limits) *Let* $F : \mathcal{I} \to$ Top *(or* Top$_*$*) be a functor. The canonical maps*

$$\mathrm{Map}\left(Z, \operatorname*{holim}_{\mathcal{I}} F\right) \xrightarrow{\cong} \operatorname*{holim}_{\mathcal{I}} \mathrm{Map}(Z, F),$$

$$\mathrm{Map}\left(\operatorname*{hocolim}_{\mathcal{I}} F, Z\right) \xrightarrow{\cong} \operatorname*{holim}_{\mathcal{I}^{op}} \mathrm{Map}(F, Z)$$

*are homeomorphisms that are natural in the $Z$ variable.*

*Proof* For the first statement, by definition $\operatorname*{holim}_{\mathcal{I}} \mathrm{Map}(Z, F) \subset \prod_i \mathrm{Map}(|\mathcal{I} \downarrow i|, \mathrm{Map}(Z, F(i)))$ consists of those maps $f = \{f_i\}_{i \in \mathcal{I}}$ such that for all morphisms $i \to i'$, the diagram

$$|\mathcal{I} \downarrow i| \longrightarrow |\mathcal{I} \downarrow i'|$$

$$\downarrow{f_i} \qquad\qquad \downarrow{f_{i'}}$$

$$\mathrm{Map}(Z, F(i)) \longrightarrow \mathrm{Map}(Z, F(i'))$$

commutes. Here the horizontal arrows are induced by the morphism $i \to i'$. By Theorem 1.2.7 we may regard the above square as a commutative diagram

$$
\begin{array}{ccc}
|\mathcal{I} \downarrow i| \times Z & \longrightarrow & |\mathcal{I} \downarrow i'| \times Z \\
\widetilde{f_i} \downarrow & & \downarrow \widetilde{f_{i'}} \\
F(i) & \longrightarrow & F(i')
\end{array}
$$

where the top vertical map is the identity on the $Z$ component for all $i \to i'$, and the $\{\widetilde{f_i}\}_{i \in I}$ are the adjoints of $\{f_i\}_{i \in I}$. This precisely describes a $Z$-parameter family of elements of $\mathrm{holim}_I \, F$; in other words, a point in $\mathrm{Map}(Z, \mathrm{holim}_I \, F)$.

For the second statement, by definition

$$
\operatorname*{hocolim}_I F = \mathrm{Coeq}\left( \coprod_{i \to i'} F(i) \times |(i' \downarrow \mathcal{I})^{\mathrm{op}}| \underset{b}{\overset{a}{\rightrightarrows}} \coprod_{i} F(i) \times |(i \downarrow \mathcal{I})^{\mathrm{op}}| \right).
$$

Now apply $\mathrm{Map}(-, Z)$ to this diagram. The isomorphism of categories $(i \downarrow \mathcal{I})^{\mathrm{op}} \cong (\mathcal{I}^{\mathrm{op}} \downarrow i)$,[3] the fact that $\mathrm{Map}(-, Z)$ takes coproducts to products (Proposition 3.5.14), and the exponential law (Theorem 1.2.7) transform the above diagram to obtain

$$
\mathrm{Map}\left(\operatorname*{hocolim}_I F, Z\right) = \mathrm{Eq}\left( \begin{array}{c} \prod_i \mathrm{Map}\left(|\mathcal{I}^{\mathrm{op}} \downarrow i|, \mathrm{Map}(F(i), Z)\right) \\ \mathrm{Map}(b,Z) \Big\downarrow\Big\downarrow \mathrm{Map}(a,Z) \\ \prod_{i \to i'} \mathrm{Map}\left(|\mathcal{I}^{\mathrm{op}} \downarrow i|, \mathrm{Map}(F(i'), Z)\right) \end{array} \right)
$$

Finally we need to note that $\mathrm{Map}(-, Z)$ takes coequalizers to equalizers to obtain the result. $\square$

Next we have analogs of Theorem 7.5.3 and Proposition 7.5.5 (note that the latter is a statement for functors to sets only, while its analog below is for spaces). They are also generalizations of Theorem 3.3.15 and Theorem 3.7.18. In addition, a part of the result below is a generalization of Theorems 5.4.9 and 5.8.9. A nice exposition of the first two statements in the result below can be found in [Vog77], which is where they first appear in this generality.

Recall the setup leading to Theorem 7.5.3, the notion of a filtered category from Definition 7.5.4, and the definition of an acyclic category from Definition 7.1.11.

---

[3] Following [Hir03], it is for this reason we have defined homotopy colimits using $(i \downarrow \mathcal{I})^{\mathrm{op}}$ rather than $i \downarrow \mathcal{I}^{\mathrm{op}}$.

**Proposition 8.5.5** (Commuting ho(co)lims)    *Let $I$ and $J$ be small categories, and $F: I \times J \to$ Top (or Top$_*$) a functor. Then*

$$\operatorname*{holim}_{I \times J} F \cong \operatorname*{holim}_{J} \operatorname*{holim}_{I} F \cong \operatorname*{holim}_{I} \operatorname*{holim}_{J} F,$$

$$\operatorname*{hocolim}_{I \times J} F \cong \operatorname*{hocolim}_{J} \operatorname*{hocolim}_{I} F \cong \operatorname*{hocolim}_{I} \operatorname*{hocolim}_{J} F.$$

*If in addition $I$ is finite and acyclic and $J$ is filtered, then there is a zig-zag of weak equivalences which produce a weak equivalence*

$$\operatorname*{hocolim}_{J} \operatorname*{holim}_{I} F \simeq \operatorname*{holim}_{I} \operatorname*{hocolim}_{J} F.$$

*Proof*   We prove the first two statements and give a sketchy treatment of the (highly non-trivial) proof of the last, along with a reference.

That homotopy limits commute follows from the following facts about the functor $\operatorname{Map}(-, -)$:

- There is a natural homeomorphism

$$\operatorname{Map}(Z, \operatorname{Eq}(X \underset{b}{\overset{a}{\rightrightarrows}} Y)) \cong \operatorname{Eq}(\operatorname{Map}(Z, X) \underset{\operatorname{Map}(Z,b)}{\overset{\operatorname{Map}(Z,a)}{\rightrightarrows}} \operatorname{Map}(Z, Y))$$

  (straightforward from the definition).
- $\operatorname{Map}(Z, \prod_i X_i) \cong \prod_i \operatorname{Map}(Z, X_i)$ (by the universal property of the product; see Definition 7.2.3).
- The exponential law, Theorem 1.2.7: $\operatorname{Map}(Z, \operatorname{Map}(X, Y)) \cong \operatorname{Map}(Z \times X, Y)$.
- For small categories $C$ and $\mathcal{D}$, $|C \times \mathcal{D}| \cong |C| \times |\mathcal{D}|$ (Theorem A.1.4).

The first two facts are easily checked, and we leave the rest of the proof, and its dual, to the reader.

For the last statement, here is an outline due to Phil Hirschhorn. A more complete proof can be found in [Hir15c]. In the category of sets, finite limits commute with filtered colimits (Proposition 7.5.5). Since limits and colimits in the category of simplicial sets are constructed dimensionwise, finite limits also commute with filtered colimits for diagrams of simplicial sets.

Filtered colimits of simplicial sets have two other useful properties:

1. For filtered diagrams of simplicial sets, the homotopy colimit is weakly equivalent to the colimit.
2. If $K$ is a simplicial set with finitely many non-degenerate simplices, then taking the space of maps from $K$ commutes with filtered colimits.

If $I$ is a finite and acyclic category and $J$ is a filtered category, then given an $I \times J$-diagram of topological spaces we can take the total singular complex (see Remark A.1.5) of each space $F(i, j)$ to obtain an $I \times J$-diagram of fibrant simplicial sets. If $\mathcal{E}$ is the indexing category for equalizer diagrams (Definition 7.4.1), we can then construct a $\mathcal{E} \times J$-diagram $\overline{F}$ from this such that $\lim_{\mathcal{E}} \overline{F}$ is isomorphic to $\operatorname{holim}_I \overline{F}$, and for this diagram we have that $\operatorname{colim}_J \lim_{\mathcal{E}} \overline{F} \cong \lim_{\mathcal{E}} \operatorname{colim}_J \overline{F}$. Since $I$ is finite and acyclic, all the nerves of the overcategories will have finitely many non-degenerate simplices, and so taking $\operatorname{colim}_J \overline{F}$ leaves us with a diagram whose limit over $\mathcal{E}$ is isomorphic to its homotopy limit over $I$. Thus, up to weak equivalences, $\lim_{\mathcal{E}} \operatorname{colim}_J \overline{F}$ is weakly equivalent to $\operatorname{holim}_I \operatorname{hocolim}_J F$ and $\operatorname{colim}_J \lim_{\mathcal{E}} \overline{F}$ is weakly equivalent to $\operatorname{hocolim}_J \operatorname{holim}_I F$. We omit the discussion of how this result for simplicial sets implies the result for topological spaces. □

Next we have a generalization of Proposition 5.4.5 (and its consequence Corollary 5.4.6).

**Corollary 8.5.6** *Let $F, G \colon I \to$ Top be functors with $I$ small, and let $F \times G \colon I \to$ Top be the functor given by $(F \times G)(i) = F(i) \times G(i)$ with the obvious maps given on factors by $F$ and $G$. Then*

$$\operatorname{holim}_I (F \times G) \cong \operatorname{holim}_I F \times \operatorname{holim}_I G.$$

*In particular, for $C_V \colon I \to$ Top the constant functor at the space $V$, we have, by Example 8.2.6,*

$$\operatorname{holim}_I (F \times C_V) \cong \operatorname{holim}_I F \times \operatorname{Map}(|I|, V).$$

*Proof* We can write the product as a homotopy limit: $F \times G = \operatorname{holim}(F \to C_* \leftarrow G)$. Alternatively, this result follows from the homeomorphism $\operatorname{Map}(X, Y \times Z) \cong \operatorname{Map}(X, Y) \times \operatorname{Map}(X, Z)$ using the definition of homotopy limit. □

Note that the second statement above is consistent with Corollary 5.4.6 since the punctured cubical indexing category $\mathcal{P}_0(\underline{n})$ is in that case contractible.

Dually, we have a generalization of parts 1 and 3 of Proposition 5.8.5 (and its consequence Corollary 5.8.6) as well as of Proposition 5.8.7.

**Corollary 8.5.7** *If $F, G \colon I \to$ Top are diagrams, let $F \sqcup G \colon I \to$ Top be the diagram given by $(F \sqcup G)(i) = F(i) \sqcup G(i)$. Then*

$$\operatorname{hocolim}_I (F \sqcup G) \cong \operatorname{hocolim}_I F \sqcup \operatorname{hocolim}_I G.$$

*If $C_V: \mathcal{I} \to \mathrm{Top}$ is the constant functor at $V$, by Example 8.2.18, we have*

$$\mathrm{hocolim}_{\mathcal{I}} (F \sqcup C_V) \cong \mathrm{hocolim}_{\mathcal{I}} F \sqcup (V \times |\mathcal{I}^{op}|).$$

*If $F$ and $G$ are diagrams of based spaces, then*

$$\mathrm{hocolim}_{\mathcal{I}} (F \vee G) \cong \mathrm{hocolim}_{\mathcal{I}} F \vee \mathrm{hocolim}_{\mathcal{I}} G,$$

*with the understanding that the homotopy colimit in $\mathrm{Top}_*$ is the ordinary homotopy colimit quotiented by the homotopy colimit of the constant functor at the one-point space (see Remark 8.2.14).*

*If $C_X$ is the constant functor at the space $X$, then*

$$\mathrm{hocolim}_{\mathcal{I}} (C_X \times F) \cong X \times \mathrm{hocolim}_{\mathcal{I}} F.$$

*Proof* The first three statements follow from the fact that disjoint unions (and wedges) are themselves homotopy colimits. For the last statement, we have a homeomorphism $X \times (Y \sqcup Z) \cong X \times Y \sqcup X \times Z$. $\qquad\square$

Recall that the loop space can be expressed naturally and functorially as a homotopy limit (Example 3.2.10), and the suspension as a homotopy colimit (Example 3.6.9). Proposition 8.5.5 thus provides the following useful consequence, which is in turn a generalization of Corollaries 5.4.10 and 5.8.10. We note that the homotopy colimits appearing in the next result are in the category of based spaces (see Remark 8.2.14).

**Corollary 8.5.8** *For $F: \mathcal{I} \to \mathrm{Top}_*$ a functor and $V$ a based space, there are natural homeomorphisms*

$$\mathrm{holim}_{\mathcal{I}} \Omega F \cong \Omega \, \mathrm{holim}_{\mathcal{I}} F,$$

$$\mathrm{hocolim}_{\mathcal{I}} \Sigma F \cong \Sigma \, \mathrm{hocolim}_{\mathcal{I}} F,$$

$$\mathrm{hocolim}_{\mathcal{I}} (C_V \vee F) \cong V \vee \mathrm{hocolim}_{\mathcal{I}} F,$$

$$\mathrm{hocolim}_{\mathcal{I}} (C_V \wedge F) \cong V \wedge \mathrm{hocolim}_{\mathcal{I}} F,$$

*where $(C_V \vee F)(i) = V \vee F(i)$ and $(C_V \wedge F)(i) = V \wedge F(i)$.*

*Proof* For the first statement, $\Omega F(i) = \mathrm{holim}(C_*(i) \to F(i) \leftarrow C_*(i))$, and $\mathrm{holim}_{\mathcal{I}} C_* \cong *$ by Example 8.2.6. The second statement is a consequence of the fourth statement when $V = S^1$. For the third statement, $F \vee G = \mathrm{hocolim}(F \leftarrow C_* \to G)$, and we need to recall that the homotopy colimit in the categoory of based spaces is as in Remark 8.2.14. For the fourth statement, $C_V \wedge F = \mathrm{hocolim}(C_V \times F \leftarrow C_V \vee F \to C_*)$, and we once again need the based version of homotopy colimit together with the last statement in Proposition 8.5.7 and the third statement here. □

The following is a generalization of the "distributive law", Corollary 3.9.4.

**Proposition 8.5.9** *Let $\mathcal{I}$ be a small category and $\mathcal{X}\colon \mathcal{I} \to$ Top a functor. Suppose $Y \in$ Top is a co-cone on $\mathcal{X}$. Then, for any $y \in Y$, there is a natural homeomorphism*

$$\text{hofiber}_y\left(\underset{\mathcal{I}}{\text{hocolim}}\, \mathcal{X} \to Y\right) \cong \underset{\mathcal{I}}{\text{hocolim}}\,\text{hofiber}_y(\mathcal{X} \to Y),$$

*where* $\text{hofiber}_y(\mathcal{X} \to Y)$ *is the functor whose value at $i$ is* $\text{hofiber}_y(\mathcal{X}(i) \to Y)$.

*Proof* The proof boils down to the fact that if $X \times Z \to X$ is the projection, and $f\colon X \to Y$ is any map, then the homotopy fiber of the composed map $X \times Z \to Y$ is homeomorphic to $Z \times \text{hofiber}_y(X \to Y)$. □

## 8.6 Algebra in the indexing variable

This section gives an overview of the behavior of homotopy (co)limits under the change of indexing category by a functor. We supply selected details and leave others to the literature. This section does not have a counterpart in earlier chapters since there we were restricted by the punctured cubical shape of the indexing category.

To begin, we have already seen in Equation (8.2.4) and Equation (8.2.7) that for $F\colon \mathcal{I} \to$ Top and $G\colon \mathcal{J} \to \mathcal{I}$, there are induced maps

$$\underset{\mathcal{I}}{\text{holim}}\, F \longrightarrow \underset{\mathcal{J}}{\text{holim}}\, F \circ G \qquad (8.6.1)$$

and

$$\underset{\mathcal{J}}{\text{hocolim}}\, F \circ G \longrightarrow \underset{\mathcal{I}}{\text{hocolim}}\, F. \qquad (8.6.2)$$

We would like to know under what circumstances these maps are weak equivalences. We would like the answer to be "when $G$ is a weak equivalence", and this turns out to be true in a certain sense. The obvious definition of $G$ being a weak equivalence would for us be to say that it is a weak equivalence if it induces a weak equivalence of realizations, and this turns out to be more or less correct. A more precise statement is Theorem 8.6.5, but we begin with a few basic facts about those $G$ which are inclusions of sub-categories as a warm-up. The following is a generalization of Propositions 5.4.29 and 5.8.28, and the technique of proof of Theorem 8.5.1 is used to prove it.

**Theorem 8.6.1**  *Given a functor $F: \mathcal{I} \to$ Top, the inclusion $\mathcal{J} \to \mathcal{I}$ of a subcategory induces a fibration*

$$\operatorname*{holim}_{\mathcal{I}} F \longrightarrow \operatorname*{holim}_{\mathcal{J}} F$$

*and a cofibration*

$$\operatorname*{hocolim}_{\mathcal{J}} F \longrightarrow \operatorname*{hocolim}_{\mathcal{I}} F.$$

*Proof*  We will prove the first statement; the second is dual and left to the reader. Recall $\operatorname{holim}_{\mathcal{I}}^{\leq n} F$ from Definition 8.3.5. We claim that the inclusion $\mathcal{J} \to \mathcal{I}$ induces a fibration

$$\operatorname*{holim}_{\mathcal{I}}^{\leq 0} F \longrightarrow \operatorname*{holim}_{\mathcal{J}}^{\leq 0} F$$

and a fibration

$$\operatorname*{holim}_{\mathcal{I}}^{\leq n} F \longrightarrow \lim\left(\operatorname*{holim}_{\mathcal{J}}^{\leq n} F \to \operatorname*{holim}_{\mathcal{J}}^{\leq n-1} F \leftarrow \operatorname*{holim}_{\mathcal{I}}^{\leq n-1} F\right)$$

for all $n \geq 1$. It then follows from Lemma 8.5.2 that the map

$$\operatorname*{holim}_{\mathcal{I}} F \longrightarrow \operatorname*{holim}_{\mathcal{J}} F$$

is a fibration. In the first case, this is the statement that $\prod_{i \in \mathcal{I}} F(i) \to \prod_{j \in \mathcal{J}} F(j)$ is a fibration, which is evident since it is a projection away from those objects not in $\mathcal{J}$. By induction assume that the map $\operatorname{holim}_{\mathcal{I}}^{\leq n-1} F \to \operatorname{holim}_{\mathcal{J}}^{\leq n-1} F$ is a fibration.

By Lemma 8.3.6, for $C = \mathcal{I}, \mathcal{J}$ the squares

$$\mathcal{S}(C) = \begin{array}{ccc} \operatorname*{holim}_{C}^{\leq n} F & \longrightarrow & \displaystyle\prod_{(i_0 \to \cdots \to i_n) \in \operatorname{nd}B_n(C)} \operatorname{Map}(\Delta^n, F(i_n)) \\ \downarrow & & \downarrow \\ \operatorname*{holim}_{C}^{\leq n-1} F & \longrightarrow & \displaystyle\prod_{(i_0 \to \cdots \to i_n) \in \operatorname{nd}B_n(C)} \operatorname{Map}(\partial\Delta^n, F(i_n)) \end{array}$$

are pullback squares in which the vertical maps are fibrations. For brevity, as in the proof of Theorem 8.5.1, write

$$F_n(C) = \operatorname*{holim}_{C}^{\leq n} F,$$

$$\partial\Phi_n(F(C)) = \prod_{(i_0 \to \cdots \to i_n) \in \operatorname{nd}B_n(C)} \operatorname{Map}(\partial\Delta^n, F(i_n)),$$

$$\Phi_n(F(C)) = \prod_{(i_0 \to \cdots \to i_n) \in \operatorname{nd}B_n(C)} \operatorname{Map}(\Delta^n, F(i_n))$$

for $C = I, J$. The fact that the above square is a pullback says in our new notation that $F_n(C) \cong \lim(F_{n-1}(C) \to \partial\Phi_n(F(C)) \leftarrow \Phi_n(F(C)))$, and the inductive hypothesis says $F_{n-1}(I) \to F_{n-1}(J)$ is a fibration.

Unraveling all of this, we have to find a solution to the lifting problem

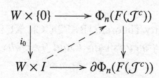

A solution $\hat{h}$ to this lifting problem amounts to lifts $\hat{h}_{F_{n-1}(I)} \colon W \times I \to F_{n-1}(I)$ and $\hat{h}_{\Phi_n(I)} \colon W \times I \to \Phi_n(F(I))$, which equalize when mapped to $\partial\Phi_n(F(I))$.

Write $h = (h_{F_{n-1}(I)}, h_{F_n(J)})$. Define $\hat{h}_{F_{n-1}(I)} = h_{F_{n-1}(I)}$, and define $\hat{h}_{\Phi_n(I)}$ as follows. There is a natural homeomorphism $\Phi_n(F(I)) \cong \Phi_n(F(J)) \times \Phi_n(F(J^c))$, where $\Phi_n(F(J^c))$ is the product of those factors of $\Phi_n(F(I))$ indexed by non-degenerate chains with at least one morphism not in $J$. On the factor $\Phi_n(F(J))$ we let $\hat{h}_{\Phi_n(I)}$ equal the composition of $h_J \colon W \times I \to F_n(J)$ with the canonical projection $F_n(J) \to \Phi_n(F(J))$. On the factor $\Phi_n(F(J^c))$, we encounter the lifting problem

$$
\begin{array}{ccc}
W \times \{0\} & \longrightarrow & \Phi_n(F(J^c)) \\
{\scriptstyle i_0} \downarrow & \nearrow & \downarrow \\
W \times I & \longrightarrow & \partial\Phi_n(F(J^c))
\end{array}
$$

where the bottom horizontal arrow is the composition of $h_{F_n(J)}$ with the projection $F_n(I) \to \partial\Phi_n(F(J^c))$. This lifting problem has a solution since the right vertical map is evidently a fibration.

It is then straightforward to see that the maps $\hat{h}_{F_{n-1}(I)}$ and $\hat{h}_{\Phi_n(I)}$ define the desired lifts. □

Continuing to try to answer the question of when the maps (8.6.1) and (8.6.2) are weak equivalences, we have the following.

**Definition 8.6.2**  A functor $G \colon J \to I$ is *homotopy initial* if, for every object $i$ of $I$, the realization $|G \downarrow i|$ is contractible, and it is *homotopy terminal* if $|i \downarrow G|$ is contractible.[4]

**Example 8.6.3**  Consider the category $\mathcal{P}(\underline{1}) = (\emptyset \to \{1\})$. The inclusion $I_\emptyset \colon \emptyset \to \mathcal{P}(\underline{1})$ of the subcategory with the single object $\emptyset$ in $\mathcal{P}(\underline{1})$ is homotopy initial, but it is not homotopy terminal, since $\{1\} \downarrow I_\emptyset$ is the empty

---

[4] These conditions are in the literature sometimes called "left cofinal" and "right cofinal".

category, whose realization is the empty set. Dually, the inclusion of the subcategory consisting only of the object $\{1\}$ is homotopy terminal but not homotopy initial. □

**Example 8.6.4** Consider the category $\mathcal{P}_0(\underline{2}) \times \mathcal{P}(\underline{1})$, depicted as

$$
\begin{array}{ccccc}
(\{1\}, \emptyset) & \longrightarrow & (\{1,2\}, \emptyset) & \longleftarrow & (\{2\}, \emptyset) \\
\downarrow & & \downarrow & & \downarrow \\
(\{1\}, \{1\}) & \longrightarrow & (\{1,2\}, \{1\}) & \longleftarrow & (\{2\}, \{1\})
\end{array}
$$

The inclusion of the subcategory $\mathcal{P}_0(\underline{2}) \times \emptyset$ (the top row) is homotopy initial, while the inclusion of the subcategory $\mathcal{P}_0(\underline{2}) \times \{1\}$ is homotopy terminal. □

The following theorem finally answers the question of when the maps (8.6.1) and (8.6.2) are weak equivalences. This is known as the cofinality theorem for homotopy (co)limits, and is due to Bousfield and Kan. We omit the proof.

**Theorem 8.6.5** (Cofinality Theorem [BK72a, Ch. XI, 9.2]) *Suppose* $F \colon \mathcal{I} \to$ Top *and* $G \colon \mathcal{J} \to \mathcal{I}$ *are functors with* $\mathcal{I}$ *and* $\mathcal{J}$ *small. If* $G$ *is homotopy initial, then the induced map*

$$
\operatorname*{holim}_{\mathcal{I}} F \longrightarrow \operatorname*{holim}_{\mathcal{J}} (F \circ G)
$$

*is a homotopy equivalence. If* $G$ *is homotopy terminal, then the induced map*

$$
\operatorname*{hocolim}_{\mathcal{J}} (F \circ G) \longrightarrow \operatorname*{hocolim}_{\mathcal{I}} F
$$

*is a homotopy equivalence.*

What follows are a couple of relatively straightforward consequences of Theorem 8.6.1. These serve as an introduction to a family of statements which includes Quillen's Theorems A and B (here Theorems 8.6.10 and 8.6.11), and Thomason's homotopy colimit theorem (here Theorem 8.6.17).

**Proposition 8.6.6** ([Goo92, Proposition 0.2]) *Suppose* $\mathcal{I}$ *is a category with subcategories* $\mathcal{I}_1$ *and* $\mathcal{I}_2$ *such that* $B_\bullet \mathcal{I} = B_\bullet \mathcal{I}_1 \cup B_\bullet \mathcal{I}_2$. *Then, for any functor* $F \colon \mathcal{I} \to$ Top, *the square of fibrations*

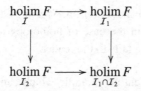

*is a pullback square, and hence a homotopy pullback square. Dually, the square of cofibrations*

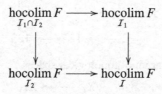

*is a pushout square, and hence a homotopy pushout square.*

*Proof* We prove the first statement and leave its dualization to the reader. The hypothesis that $B_\bullet I = B_\bullet I_1 \cup B_\bullet I_2$ implies that, for all objects $i \in I_1 \cup I_2$, $|(I_1 \cup I_2) \downarrow i| \cong |I_1 \downarrow i| \cup_{|(I_1 \cap I_2) \downarrow i|} |I_2 \downarrow i|$. Hence the square

$$
\begin{array}{ccc}
\mathrm{holim}_I F & \longrightarrow & \mathrm{holim}_{I_1} F \\
\downarrow & & \downarrow \\
\mathrm{holim}_{I_2} F & \longrightarrow & \mathrm{holim}_{I_1 \cap I_2} F
\end{array}
$$

is a pullback square by definition of the homotopy limit, since $\mathrm{Map}(A \cup_C B, Z) \cong \mathrm{Map}(A, Z) \times_{\mathrm{Map}(C,Z)} \mathrm{Map}(B, Z)$. That the square comprises fibrations is Theorem 8.6.1, and that it is a homotopy pullback square follows from Proposition 3.3.5. □

More generally, we have the following.

**Proposition 8.6.7** ([Goo92, Lemmas 1.9 and 1.10]) *Let $S$ be a finite set, and suppose $I$ is a category with a finite collection $\{I_s\}_{s \in S}$ of subcategories such that $\bigcup_{s \in S} B_\bullet I_s = B_\bullet I$ and $\bigcup_{s \in T} B_\bullet I_s = B_\bullet \bigcup_{s \in T} I_s$ for all $T \subset S$. Then, for any functor $F: I \to \mathrm{Top}$, the canonical maps*

$$
\mathrm{holim}_I F \longrightarrow \mathrm{holim}_{T \in \mathcal{P}_0(S)} \mathrm{holim}_{\cap_{s \in T} I_s} F
$$

*and*

$$
\mathrm{hocolim}_{T \in \mathcal{P}_i(S)} \mathrm{hocolim}_{\cap_{s \in S-T} I_s} F \longrightarrow \mathrm{hocolim}_I F
$$

*are weak equivalences.*

*Proof* This follows from induction on $|S|$ using Proposition 8.6.6 and 1(a) of Proposition 3.3.20 or, in the case of homotopy colimits, 1(a) of Proposition 3.7.26. We leave the details to the reader. □

**Example 8.6.8** We present an example which can be used to establish the key result of Proposition 3.3.20. Consider the category

$$\mathcal{I} = \quad \begin{array}{ccc} 1 & \longrightarrow & 13 \\ \downarrow & & \downarrow \\ 2 & \longrightarrow 12 \longrightarrow & 123 \end{array}$$

which gives rise to a diagram of topological spaces

$$\begin{array}{ccc} X_1 & \longrightarrow & X_{13} \\ \downarrow & & \downarrow \\ X_2 & \longrightarrow X_{12} \longrightarrow & X_{123} \end{array}$$

Let

$$\mathcal{I}_1 = \quad \begin{array}{c} 13 \\ \downarrow \\ 2 \longrightarrow 12 \longrightarrow 123 \end{array}$$

and

$$\mathcal{I}_2 = \quad \begin{array}{ccc} 1 & \longrightarrow & 13 \\ \downarrow & & \downarrow \\ 12 & \longrightarrow & 123 \end{array}$$

Then $\mathcal{I}_1 \cup \mathcal{I}_2 = \mathcal{I}$ (in the sense of Proposition 8.6.6) and $\mathcal{I}_1 \cap \mathcal{I}_2 = (12 \rightarrow 123 \leftarrow 13)$. The inclusions $(2 \rightarrow 123 \leftarrow 13) \rightarrow \mathcal{I}_1$ and $(2 \rightarrow 12 \leftarrow 1) \rightarrow \mathcal{I}_2$ are homotopy initial, and $\mathcal{I}_2$ has initial object 1. It follows from Theorem 8.6.5 and Example 8.2.5 that the square

$$\begin{array}{ccc} \text{holim}(X_2 \rightarrow X_{12} \leftarrow X_1) & \longrightarrow & \text{holim}(X_2 \rightarrow X_{123} \leftarrow X_{13}) \\ \downarrow & & \downarrow \\ X_1 & \longrightarrow & \text{holim}(X_{12} \rightarrow X_{123} \leftarrow X_{13}) \end{array}$$

is homotopy cartesian. □

**Remark 8.6.9** Let $\mathcal{I}$ be a small category and $F: \mathcal{I} \rightarrow$ Top a functor. Let $\mathcal{P}(\mathcal{I})$ denote the poset of subcategories of $\mathcal{I}$: objects are subcategories of $\mathcal{I}$, and morphisms are inclusions of subcategories (meaning both object and morphism sets satisfy the containment). Call a square

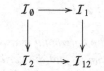

in $\mathcal{P}(\mathcal{I})$ a *pushout* if $B_\bullet \mathcal{I}_{12} = B_\bullet \mathcal{I}_1 \cup_{B_\bullet \mathcal{I}_0} B_\bullet \mathcal{I}_2$. Then we can think of Proposition 8.6.6 as saying that the functor $\mathrm{holim}_{(-)} F \colon \mathcal{P}(\mathcal{I}) \to \mathrm{Top}$ takes pushouts of categories to homotopy pullbacks, and dually that $\mathrm{hocolim}_{(-)} F \colon \mathcal{P}(\mathcal{I}) \to \mathrm{Top}$ takes pushouts of categories to homotopy pushouts. This is analogous with Proposition 3.9.1, which says that $\mathrm{Map}(-, Z)$ takes certain kinds of homotopy pushouts to homotopy pullbacks, and with Proposition 3.7.11, which says that homotopy cocartesian squares are preserved by pointwise products. Proposition 8.6.7 says more generally that $\mathrm{holim}_{(-)} F$ takes strongly homotopy cocartesian cubes to homotopy cartesian cubes, and dually that $\mathrm{hocolim}_{(-)} F$ takes strongly homotopy cocartesian cubes to homotopy cocartesian cubes. □

In light of the following two results, given a functor $G \colon \mathcal{J} \to \mathcal{I}$, one may think of the categories $G \downarrow i$ and $i \downarrow G$ as the fibers of $G$ in some sense.

**Theorem 8.6.10** (Quillen's Theorem A [Qui73]) *Suppose $G \colon \mathcal{J} \to \mathcal{I}$ is a functor such that $|i \downarrow G|$ is contractible for all $i$. Then the induced map $|G| \colon |\mathcal{J}| \to |\mathcal{I}|$ is a homotopy equivalence. The same result is true if instead we assume $|G \downarrow i|$ is contractible for all $i$.*

We will show that this is a consequence of Theorem 8.6.11. In fact, the part of the proof of Theorem 8.6.11 that we omit (which amounts to Lemma 8.6.14) is precisely what is necessary to prove Theorem 8.6.10.

　　Quillen's Theorem B, stated below, says that the spaces $|i \downarrow G|$ measure the extent to which $|G|$ fails to be a homotopy equivalence, as they are, up to homotopy, the homotopy fibers of $|G|$. Recall that $|i \downarrow \mathcal{I}|$ is contractible for all $i$.

**Theorem 8.6.11** (Quillen's Theorem B [Qui73]) *Suppose $G \colon \mathcal{J} \to \mathcal{I}$ is a functor such that, for every morphism $i \to i'$ in $\mathcal{I}$, the induced map $|i' \downarrow G| \to |i \downarrow G|$ is a homotopy equivalence. Then the square*

$$|i \downarrow G| \longrightarrow |\mathcal{J}|$$
$$\downarrow \qquad\qquad \downarrow |G|$$
$$|i \downarrow I| \longrightarrow |I|$$

*is homotopy cartesian. Here the horizontal maps are induced by the forgetful maps (forget the morphism from i), and the vertical maps are induced by G. Dually, if $|G \downarrow i| \to |G \downarrow i'|$ is a homotopy equivalence for all $i \to i'$, then the square with $|G \downarrow i|$ replacing $|i \downarrow G|$ and $|I \downarrow i|$ replacing $|i \downarrow I|$ is homotopy cartesian.*

We will need the following lemma and follow Quillen's original presentation for its proof. Recall the notion of a quasifibration from Definition 2.7.1.

**Lemma 8.6.12** ([Qui73, Lemma for Theorem B])    *Suppose $X\colon I \to$ Top, $i \mapsto X_i$ is a functor. The natural transformation $X \to C_*$ to the constant functor at the one-point space induces a map*

$$p\colon \operatorname*{hocolim}_I X \longrightarrow \operatorname*{hocolim}_I C_* \cong |I^{\mathrm{op}}|,$$

*and if, for every morphism $i \to i'$, the map $X_i \to X_{i'}$ is a homotopy equivalence, then p is a quasifibration whose fiber over a point in the interior of a simplex indexed by the non-degenerate chain $i_0 \to \cdots \to i_n$ in $I^{\mathrm{op}}$ is equal to $X_{i_n}$.*

*Proof*    We begin by noting that $p^{-1}(|\mathrm{sk}_n I^{\mathrm{op}}|) = \operatorname*{hocolim}_I^{\leq n} X$, which follows from the identification $\operatorname*{hocolim}_I^{\leq n} C_* \cong |\mathrm{sk}_n I^{\mathrm{op}}|$ indicated in the statement of the lemma. By Proposition 2.7.9 it is enough to prove that each induced map $p_n\colon \operatorname*{hocolim}_I^{\leq n} X \to |\mathrm{sk}_n I^{\mathrm{op}}|$ is a quasifibration. For $i = 0$ this is trivial; the map is in fact a fibration whose fiber over a point indexed by the object $i$ is $X_i$.

By induction assume $p_{n-1}$ is a quasifibration. By Proposition 2.7.7 it is enough to cover $|\mathrm{sk}_n I^{\mathrm{op}}|$ by open sets $U$ and $V$ such that $p_n^{-1}(U) \to U$, $p_n^{-1}(V) \to V$ and $p_n^{-1}(U \cap V) \to U \cap V$ are quasifibrations. We define $U$ and $V$ as follows. Recall from Proposition A.1.13 that we have a pushout (and homotopy pushout) square

$$\coprod_{(i_0 \to \cdots \to i_n) \in \mathrm{end} B_n(I^{op})} \partial\Delta^n \longrightarrow |\mathrm{sk}_{n-1} I^{\mathrm{op}}|$$
$$\downarrow$$
$$\coprod_{(i_0 \to \cdots \to i_n) \in \mathrm{end} B_n(I^{op})} \Delta^n \longrightarrow |\mathrm{sk}_n I^{\mathrm{op}}|$$

Let $b$ denote the barycenter of each copy of the simplex $\Delta^n$. Define

$$U = \mathrm{colim}\left( \coprod_{(i_0 \to \cdots \to i_n) \in \mathrm{end} B_n(\mathcal{I}^{\mathrm{op}})} \Delta^n - \{b\} \leftarrow \coprod_{(i_0 \to \cdots \to i_n) \in \mathrm{end} B_n(\mathcal{I}^{\mathrm{op}})} \partial \Delta^n \to |\mathrm{sk}_{n-1} \mathcal{I}^{\mathrm{op}}| \right)$$

and

$$V = |\mathrm{sk}_n \mathcal{I}^{\mathrm{op}}| \setminus |\mathrm{sk}_{n-1} \mathcal{I}^{\mathrm{op}}|.$$

Note that $U \cap V$ is the disjoint union of copies of $\mathring{\Delta}^n - \{b\}$, one for each non-degenerate chain $i_0 \to \cdots \to i_n$ in $\mathcal{I}^{\mathrm{op}}$. The map $p_n^{-1}(V) \to V$ is a quasifibration since it is a disjoint union of projection maps

$$\coprod_{(i_0 \to \cdots \to i_n) \in \mathrm{end} B_n(\mathcal{I}^{\mathrm{op}})} \mathring{\Delta}^n \times X_{i_n} \longrightarrow \coprod_{(i_0 \to \cdots \to i_n) \in \mathrm{end} B_n(\mathcal{I}^{\mathrm{op}})} \mathring{\Delta}^n.$$

In fact, this map is clearly a fibration. Hence $p_n^{-1}(U \cap V) \to U \cap V$ is a fibration, as it is the pullback of the fibration $p_n^{-1}(V) \to V$ to $U \cap V$. It remains to show $p^{-1}(U) \to U$ is a quasifibration. This follows from Lemma 2.7.11, as follows. We have

$$p^{-1}(U) = \mathrm{colim}\left( \coprod_{(i_0 \to \cdots \to i_n) \in \mathrm{end} B_n(\mathcal{I}^{\mathrm{op}})} X_{i_n} \times \Delta^n - \{b\} \right.$$

$$\left. \leftarrow \coprod_{(i_0 \to \cdots \to i_n) \in \mathrm{end} B_n(\mathcal{I}^{\mathrm{op}})} X_{i_n} \times \partial \Delta^n \to \mathrm{hocolim}_{\mathcal{I}}^{\leq n-1} X \right),$$

and it is clear that the inclusion $\mathrm{hocolim}_{\mathcal{I}}^{\leq n-1} X \to p^{-1}(U)$ is a deformation retract (retract radially from the barycenters of the $n$-simplices). Moreover this can clearly be done fiberwise over $U$, and the hypothesis that $X_i \to X_{i'}$ is a homotopy equivalence for all morphisms $i \to i'$ now guaranteess the hypotheses of Lemma 2.7.11 are satisfied. $\square$

Before we give a sketch of the proof of Theorem 8.6.11, we state a definition.

**Definition 8.6.13** Let $G: \mathcal{J} \to \mathcal{I}$ be a functor. Define the *twisted arrow category of $G$* to be the category $S(G)$ whose objects are triples $(j, i, m)$, where $j$ is an object of $\mathcal{J}$, $i$ is an object of $\mathcal{I}$, and $m: i \to G(j)$ is a morphism in $\mathcal{I}$. A morphism $(j, i, m) \to (j', i', m')$ consists of morphims $n: j \to j'$ and $m'': i' \to i$ such that the diagram

$$\begin{array}{ccc} i' & \xrightarrow{\ m''\ } & i \\ {\scriptstyle m'} \downarrow & & \downarrow {\scriptstyle m} \\ G(j') & \xleftarrow[G(n)]{} & G(j) \end{array}$$

commutes. For a category $\mathcal{I}$, define $S(\mathcal{I}) = S(1_{\mathcal{I}^{\mathrm{op}}})$.

Note that there are forgetful functors, "source" and "target", $s\colon S(G) \to \mathcal{J}$ and $t\colon S(G) \to \mathcal{I}^{\mathrm{op}}$. We will utilize the following results of Quillen about the realizations of these functors in our proof of Theorem 8.6.11 below. We omit the proof.

**Lemma 8.6.14** *Let $\mathcal{I}$ and $\mathcal{J}$ be small categories. The source and target functors $s\colon S(\mathcal{I}) \to \mathcal{I}^{\mathrm{op}}$ and $t\colon S(\mathcal{I}) \to \mathcal{I}$ induce homotopy equivalences $|s|\colon |S(\mathcal{I})| \to |\mathcal{I}^{\mathrm{op}}|$ and $|t|\colon |S(\mathcal{I})| \to |\mathcal{I}|$.*

*If $G\colon \mathcal{J} \to \mathcal{I}$ is a functor, then the source functor $s\colon S(G) \to \mathcal{J}$ induces a homotopy equivalence $|s|\colon |S(G)| \to |\mathcal{J}|$.*

*Sketch of the proof of Theorem 8.6.11* For each object $i$ of $\mathcal{I}$, let $S(G)(i)$ denote the subcategory of $S(G)$ consisting of those tuples $(j, i, m)$ and morphisms $(j, i, m) \to (j', i, m)$ where $i \to i$ is the identity morphism. This defines a functor $S(G)(-)\colon \mathcal{I}^{\mathrm{op}} \to \mathrm{Cat}$ and so a functor $|S(G)(-)|\colon \mathcal{I}^{\mathrm{op}} \to \mathrm{Top}$ by composing with geometric realization. Moreover, there is a canonical isomorphism $S(G)(i) \cong i \downarrow G$ given on objects by $(j, i, m) \mapsto (m\colon i \to G(j))$, and on morphisms in the evident way. Our hypothesis says that, for all $i \to i'$, $|i' \downarrow G| \to |i \downarrow G|$ is a homotopy equivalence, Hence the functor $|S(G)(-)|$ satisfies the hypothesis of Lemma 8.6.12 and we have a quasifibration

$$\underset{\mathcal{I}^{\mathrm{op}}}{\mathrm{hocolim}}\, |S(G)(-)| \longrightarrow |\mathcal{I}|.$$

Moreover, there is an evident homeomorphism $\mathrm{hocolim}_{\mathcal{I}^{\mathrm{op}}}\, |S(G)(-)| \cong |S(G)|$ (the same technique of proof as in Example 8.2.18 can be used). Hence we have a homotopy cartesian square

$$
\begin{array}{ccc}
|i \downarrow G| & \longrightarrow & |S(G)| \\
\downarrow & & \downarrow \\
* & \longrightarrow & |\mathcal{I}|
\end{array}
$$

The functor $S(G) \to \mathcal{I}^{\mathrm{op}}$ (which induces the map of realizations above) which on objects sends $(j, i, m)$ to $i$ and is the evident functor on morphisms factors as a composition of functors $S(G) \to S(1_{\mathcal{I}}) \to \mathcal{I}^{\mathrm{op}}$. Here $S(G) \to S(1_{\mathcal{I}})$ on objects sends $(j, i, m)$ to $(G(j), i, m)$ and is given in the obvious way on morphisms, and $S(\mathcal{I}) \to \mathcal{I}^{\mathrm{op}}$ is the source functor $s$. Moreover, this induces the functor $i \downarrow G \to i \downarrow \mathcal{I}$ in the statement of the theorem. Thus we have a diagram

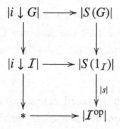

By Lemma 8.6.14, $|s|: |S(1_I)| \to |I^{op}|$ is a homotopy equivalence, and since $|i \downarrow I|$ is contractible, the left vertical map in the bottom square above is also a homotopy equivalence, and hence the bottom square is homotopy cartesian by 1(b) of Proposition 3.3.11. It follows from 1(b) of Proposition 3.3.20 that the top square is homotopy cartesian as well. Finally, the functor $i \downarrow G \to \mathcal{J}$ factors through $S(G)$ in the evident way, and similarly $i \downarrow I \to I$ factors through $S(1_I)$, and we have another diagram

$$
\begin{array}{ccc}
|i \downarrow G| \longrightarrow |S(G)| \xrightarrow{|s|} |\mathcal{J}| \\
\downarrow \qquad\qquad \downarrow \qquad\qquad \downarrow {|G|} \\
|i \downarrow I| \longrightarrow |S(1_I)| \xrightarrow{|t|} |I|
\end{array}
$$

Again using Lemma 8.6.14, the maps $|s|: |S(G)| \to |\mathcal{J}|$ and $|t|: |S(1_I)| \to |I|$ are homotopy equivalences, and again using 1(b) of Proposition 3.3.11 the right-most square above is homotopy cartesian. It follows from 1(a) of Proposition 3.3.20 that the outer square is homotopy cartesian as well. □

Note that Theorem 8.6.10 is an easy consequence of Theorem 8.6.11 since $|I \downarrow i| \simeq *$, and so if $|G \downarrow i| \simeq *$ for all $i$, then since the square appearing in Theorem 8.6.11 is homotopy cartesian, we have that $|\mathcal{J}| \to |I|$ is a weak homotopy equivalence by Proposition 3.3.18.[5]

We now return to Proposition 8.6.6, especially Remark 8.6.9. Given $F: I \to$ Top, Remark 8.6.9 says that in the category of subcategories of $I$, denoted here by $\mathrm{Cat}_{\leq I}$, the functor $\mathrm{holim}_- F: \mathrm{Cat}_{\leq I} \to \mathrm{Top}$ (resp. $\mathrm{hocolim}_- F: \mathrm{Cat}_{\leq I} \to$ Top) takes pushouts of categories to homotopy pullbacks (resp. homotopy pushouts) of spaces. From the perspective of homotopy (co)limits, our notion of pushout of categories (the nerve of the union is the union of the nerves) in

---

[5] For the spirit of the idea of the proof, think of the $I \downarrow i$ as a neighborhood of $i$, and so the collection of all such neighborhoods forms a cover of $I$. Thus $|I| \simeq \mathrm{hocolim}_I |I \downarrow i|$. The categories $G \downarrow i$ form a cover of $\mathcal{J}$ by pullback by $G$, and so $|\mathcal{J}| \simeq \mathrm{hocolim}_I |G \downarrow i|$. The hypothesis $|G \downarrow i| \simeq *$ means the natural map $|G \downarrow i| \to |I \downarrow i|$ is a weak equivalence, and so Theorem 8.3.7 implies $|\mathcal{J}| \simeq |I|$.

Cat$_{\leq I}$ is therefore the right one, and we would like to elaborate on this a little here. Recall from Example 7.1.29 the category Cat of small categories. The basic question is: given a functor $D\colon C \to$ Cat, what should we mean by the category colim$_C$ $D$?

**Definition 8.6.15** Let $C$ be a small category and $D\colon C \to$ Cat a functor. Define the *Grothendieck construction of* $D$, denoted $\Sigma_C(D)$, as the category with

- objects the pairs $(c, i) \in \mathrm{Ob}(C) \times \mathrm{Ob}(D(c))$;
- morphisms $\mathrm{Hom}_{\Sigma_C(D)}((c_1, i_1), (c_2, i_2))$ the pairs

$$(\alpha, \mu) \in \mathrm{Hom}_C(c_1, c_2) \times \mathrm{Hom}_{D(c_2)}(D(\alpha)(i_1), i_2).$$

We leave the description of the composition of morphisms to the reader, as well as the verification of the axioms for a category.

**Example 8.6.16** If $C = *$ is the category with one object and one morphism, then $\Sigma_*(D)$ is canonically isomorphic to the category $D(*)$.

If $D\colon C \to$ Cat is the constant functor $C_*$ with value $D(c) = \mathcal{D}$, then $\Sigma_C(D)$ is canonically isomorphic to $C \times \mathcal{D}$. □

We now can state Thomason's Homotopy Colimit Theorem, which describes the realization of the Grothendieck construction as a homotopy colimit.

**Theorem 8.6.17** (Thomason's Homotopy Colimit Theorem, [Tho79, Theorem 1.2]) *Let $C$ be a small category and $D\colon C \to$ Cat a functor. Then there is a natural homotopy equivalence*

$$\mathop{\mathrm{hocolim}}_{C} |D| \longrightarrow |\Sigma_C(D)|.$$

We omit the proof, but remark that the philosophy here is that $\Sigma_C(D)$ plays the role of the homotopy pushout of the categories $D(c)$.

# 9

## Cosimplicial spaces

Cosimplicial spaces were championed by Bousfield and Kan [BK72a] who, in the process of defining $R$-completions (see Example 9.2.8), initiated a careful study of their properties. The main feature of cosimplicial spaces is that one way to construct the homotopy limit of a diagram is to first perform its "cosimplicial replacement". Theorem 9.3.3 says that the "totalization" of this cosimplicial replacement is equivalent to the homotopy limit of the original diagram. In fact, because of this result, the cosimplicial approach is often taken as the definition of the homotopy limit.

For our purposes, defining homotopy limits this way requires too much extra machinery, but there are other important properties of cosimplicial spaces that we wish to explore, and they are undoubtedly a useful organizational tool. Most importantly, a feature that is important in the context of this book is that the results in Section 9.4 say that any diagram can be turned into a sequence of punctured cubical ones by passing through a cosimplicial diagram. Thus studying homotopy limits in some sense reduces to studying punctured cubical ones. We will elaborate on this at the end of Section 9.4.

Another quality of cosimplicial spaces is that they come equipped with spectral sequences converging (under favorable circumstances) to the homology and homotopy groups of their totalizations. This gives an important computational tool for studying homotopy limits of diagrams. We will describe these spectral sequences in Section 9.6 and will see some of their applications in Section 10.4.

Because of the duality of cosimplicial and simplicial spaces as well as homotopy limits and colimits, this entire chapter could have been written in terms of simplicial spaces and homotopy colimits. The choice of taking the cosimplicial/homotopy limit point of view is due to the fact that the authors initially came to the subject that way. In addition, the applications in Chapter 10 use the language of homotopy limits more than homotopy colimits. Nevertheless, we will in Section A.1 list some of the most important simplicial/homotopy colimits versions of the constructions and results from this chapter.

Some additional topics on cosimplicial spaces, such as the model category structure on the category of cosimplicial diagrams, can be found in [BK72a, GJ99]. Other references are provided throughout the chapter.

## 9.1 Cosimplicial spaces and totalization

Consider the category $\Delta$ whose objects are $[n]$ for $n \geq 0$ and whose morphisms are weakly monotone functions. Thus for a morphism $f \colon [m] \to [n]$, $f(i) \leq f(j)$ for all $0 \leq i \leq j \leq m$. The category $\Delta$ is called the *cosimplicial indexing category*. The morphisms are generated by (i.e. are compositions of) *coface maps*, denoted by $d^i$, and *codegeneracy maps*, denoted by $s^j$. The coface maps are injections that "omit $i$" and are defined by

$$d^i \colon [n-1] \longrightarrow [n], \qquad 0 \leq i \leq n,$$

$$k \longmapsto \begin{cases} k, & k < i; \\ k+1, & k \geq i; \end{cases}$$

and the codegeneracy maps are surjections that "double $i$" and are defined by

$$s^i \colon [n+1] \longrightarrow [n], \qquad 0 \leq i \leq n,$$

$$k \longmapsto \begin{cases} k, & k \leq i; \\ k-1, & k > i. \end{cases}$$

They satisfy the relations, called *cosimplicial identities*,

$$d^j d^i = d^i d^{j-1}, \quad i < j;$$

$$s^j d^i = \begin{cases} d^i s^{j-1}, & i < j; \\ \mathrm{Id}, & i = j, \ i = j+1; \\ d^{i-1} s^j, & i > j+1; \end{cases} \qquad (9.1.1)$$

$$s^j s^i = s^{i-1} s^j, \quad i > j.$$

The category $\Delta$ is usually represented as

$$[0] \rightleftarrows [1] \mathrel{\substack{\longrightarrow \\ \rightleftarrows \\ \longrightarrow}} [2] \mathrel{\substack{\Longrightarrow \\ \Longrightarrow}} \cdots$$

**Remark 9.1.1** Dually, the *simplicial indexing category* is defined as $\Delta^{\mathrm{op}}$. We already needed the simplicial language in Section 8.1 to define the nerve of a category; more can be found in Section A.1. □

**Remark 9.1.2** In Chapter 5, we used $\Delta$ as the functor from $\mathcal{P}(\underline{n})$ or its punctured versions to Top whose image was a (punctured) cubical simplex (see Example 5.1.5). Here $\Delta$ is used as notation for an indexing category. This duplication in notation should not cause confusion and, should the two notations appear near each other, the argument of the functor $\Delta$ will also be provided so as to distinguish the two. □

**Definition 9.1.3** A functor $F \colon \Delta \to C$ is called a *cosimplicial diagram in C*.

In the literature, the functor $F$ and the category $\Delta$ are often suppressed, and a bullet decoration is used instead, such as $X^\bullet$. A cosimplicial diagram is then simply described as a sequence of objects $X^{[n]}$, namely

$$X^\bullet = \{X^{[n]}\}_{n=0}^\infty \quad \text{or} \quad X^\bullet = \{X^{[n]}\},$$

along with maps between these objects satisfying the identities from (9.1.1). These maps, which are the images of cofaces and codegeneracies via the underlying functor $X^\bullet(d^i)$ and $X^\bullet(s^i)$, are by abuse of terminology also called cofaces and codegeneracies and denoted by $d^i$ and $s^i$ (so the functor is suppressed). Here we will almost exclusively be interested in cosimplicial spaces, i.e. functors $\Delta \to \text{Top}$ (or $\text{Top}_*$).

It is sometimes convenient to visualize a cosimplicial diagram as a diagram where the only arrows are the generating morphisms $d^i$ and $s^i$, as in

$$X^\bullet = \left( X^{[0]} \underset{\underset{d^1}{\longrightarrow}}{\overset{\overset{d^0}{\longrightarrow}}{\xleftarrow{s^0}}} X^{[1]} \underset{\underset{d^2}{\longrightarrow}}{\overset{\overset{d^0}{\longrightarrow}}{\underset{\xleftarrow{s^1}}{\overset{\xleftarrow{s^0}}{\overset{d^1}{\longrightarrow}}}}} X^{[2]} \cdots \right).$$

$$(9.1.2)$$

**Example 9.1.4** (Cosimplicial simplex) One important cosimplicial space is the *cosimplicial simplex* $\Delta^\bullet$, which is a cosimplicial diagram in Top given by the functor

$$\Delta^\bullet \colon \Delta \longrightarrow \text{Top}$$
$$[n] \longmapsto \Delta^n,$$

where $\Delta^n$ is the standard $n$-simplex. The morphisms $d^i, s^i \colon [n] \to [k]$ correspond to inclusions of and projections onto faces. It is convenient to have an explicit description of these maps. Write

$$\Delta^n = \left\{ (t_0, \ldots, t_n) \colon t_i \in I, \sum_i t_i = 1 \right\}.$$

For $i = 0$ to $n$, define

$$d^i \colon \Delta^{n-1} \longrightarrow \Delta^n$$

$$(t_0, \ldots, t_{n-1}) \longmapsto (t_0, \ldots, t_{i-1}, 0, t_i, \ldots, t_{n-1})$$

and

$$s^i \colon \Delta^{n+1} \longrightarrow \Delta^n$$

$$(t_0, \ldots, t_{n+1}) \longmapsto (t_0, \ldots, t_{i-1}, t_i + t_{i+1}, t_{i+2}, \ldots, t_{n+1}).$$

Then, for example, $d^1$ includes a 1-simplex into the edge of $\Delta^2 \subset \mathbb{R}^3$ with endpoints $t_0 = 1$ and $t_2 = 1$, while $s^1$ is the projection of $\Delta^2 \subset \mathbb{R}^3$ onto the 1-simplex in the $t_0 t_2$-plane.                                                    □

More examples of cosimplicial spaces are given in Section 9.2. Dualizing the notion of the realization of a simplicial set, Definition A.1.2, we have the notion of totalization. Its usefulness will become apparent in Section 9.3 when we prove that the homotopy limit of an arbitrary functor can be computed as the totalization of its cosimplicial replacement.

**Definition 9.1.5**  For a cosimplicial space $X^\bullet$, define its *totalization* $\mathrm{Tot}\, X^\bullet$ to be the space $\mathrm{Nat}(\Delta^\bullet, X^\bullet)$ of natural transformations from $\Delta^\bullet$ to $X^\bullet$, i.e. the space of sequences of maps

$$f_n \colon \Delta^n \longrightarrow X^{[n]}, \quad n \geq 0,$$

such that the diagrams

$$\begin{array}{ccc}
\Delta^n & \xrightarrow{\ d^j\ } & \Delta^{n+1} \\
{\scriptstyle f_n}\downarrow & & \downarrow{\scriptstyle f_{n+1}} \\
X^{[n]} & \xrightarrow{\ d^j\ } & X^{[n+1]}
\end{array}
\qquad \text{and} \qquad
\begin{array}{ccc}
\Delta^{n+1} & \xrightarrow{\ s^j\ } & \Delta^n \\
{\scriptstyle f_{n+1}}\downarrow & & \downarrow{\scriptstyle f_n} \\
X^{[n+1]} & \xrightarrow{\ s^j\ } & X^{[n]}
\end{array}$$

commute for all cofaces $d^j$ and codegeneracies $s^j$.

The space $\mathrm{Tot}\, X^\bullet$ is topologized as the subspace of the product $\prod_n \mathrm{Map}(\Delta^n, X^{[n]})$.

We will refer to a point $x \in X^{[n]}$ as a *cosimplex* and will call it *codegenerate* if it is in the image of a codegeneracy.

If $X^\bullet$ is a based cosimplicial space, then $\mathrm{Tot}\, X^\bullet$ has a natural basepoint given by the sequence of maps $f^n$ that send the simplices $\Delta^n$ to the basepoint of $X^{[n]}$.

**Remark 9.1.6** (Totalization as equalizer) In analogy with Remark A.1.3, totalization can be written as an equalizer. Namely, we have

$$\operatorname{Tot} X^{\bullet} = \operatorname{Eq}\left( \prod_{[n]} \operatorname{Map}(\Delta^n, X^{[n]}) \underset{b}{\overset{a}{\Longrightarrow}} \prod_{[n]\to[k]} \operatorname{Map}(\Delta^n, X^{[k]}) \right).$$

$$(9.1.3)$$

The maps $a$ and $b$ are defined as in Definition 8.2.1. That is, a map $[n] \to [k]$ induces a map $\operatorname{Map}(\Delta^n, X^{[n]}) \to \operatorname{Map}(\Delta^n, X^{[k]})$ in the evident way, and this induces the map $a$. Similarly we have an induced map $\operatorname{Map}(\Delta^k, X^{[k]}) \to \operatorname{Map}(\Delta^n, X^{[k]})$ which induces $b$. □

Given a map of cosimplicial spaces $f \colon X^{\bullet} \to Y^{\bullet}$, by which we mean a natural transformation of functors, there is an evident induced map

$$\operatorname{Tot} f \colon \operatorname{Tot} X^{\bullet} \longrightarrow \operatorname{Tot} Y^{\bullet}$$

given by composition of the maps $\Delta^{\bullet} \to X^{\bullet}$ with $f$. Totalization is thus a functor, but it is not a homotopy functor. That is, if $f$ is an objectwise weak equivalence, it does not necessarily follow that $\operatorname{Tot} f$ is a weak equivalence.

This situation is reminiscent of the problem we had with ordinary limits that prompted us to devise the homotopy invariant homotopy limit. The totalization fits between the limit and the homotopy limit in the sense that the natural map from the limit to the homotopy limit of a cosimplicial space factors through the totalization. To explain, first note that there is an obvious map of cosimplicial spaces

$$\Delta^{\bullet} \longrightarrow C_{*}^{\bullet},$$

$$(9.1.4)$$

where $C_{*}^{\bullet}$ is the cosimplicial space with $X^{[n]} = *$ for all $n$. This gives rise to a map

$$\operatorname{Nat}(C_{*}^{\bullet}, X^{\bullet}) \longrightarrow \operatorname{Nat}(\Delta^{\bullet}, X^{\bullet}).$$

The domain is precisely the limit of the diagram $X^{\bullet}$, while the codomain is its totalization.

There is also a natural transformation

$$(\Delta \downarrow \bullet) \longrightarrow \Delta^{\bullet},$$

$$(9.1.5)$$

where $(\Delta \downarrow \bullet)$ is the cosimplicial space given by $[n] \mapsto |\Delta \downarrow [n]|$ and whose cofaces and codegeneracies are induced in the obvious way from those in $\Delta$ (there is nothing special going on here; for any category $\mathcal{I}$ the overcategory $\mathcal{I} \downarrow -$ is a functor from $\mathcal{I}$ to categories). The map (9.1.5) is given as follows: We have a functor of $[n]$

$$(\Delta \downarrow [n]) \longrightarrow \mathcal{P}_0([n])$$

which associates to an object $[m] \to [n]$ its image in $[n]$ as a (necessarily) non-empty subset. This in turn induces a natural transformation of functors of $[n]$

$$|\Delta \downarrow [n]| \longrightarrow |\mathcal{P}_0([n])| \cong \Delta^n$$

(the homeomorphism on the right follows by exactly the same argument as in Example 8.1.13). We therefore have a map

$$\mathrm{Nat}(\Delta^\bullet, X^\bullet) \longrightarrow \mathrm{Nat}(|\Delta \downarrow \bullet|, X^\bullet).$$

The left side is again the totalization, but the right is precisely the homotopy limit of the diagram $X^\bullet$ by Proposition 8.2.2. To summarize, we have maps

$$\lim X^\bullet = \mathrm{Nat}(C^\bullet_*, X^\bullet)$$

$$\downarrow$$

$$\mathrm{Tot}\, X^\bullet = \mathrm{Nat}(\Delta^\bullet, X^\bullet)$$

$$\downarrow$$

$$\mathrm{holim}\, X^\bullet = \mathrm{Nat}(|\Delta \downarrow \bullet|, X^\bullet)$$

and a natural question to ask is when these maps are weak equivalences. One situation where we know that the limit and the homotopy limit are weakly equivalent is in the setting of Theorem 8.4.9 when a diagram is fibrant and the indexing category has empty or contractible latching objects. Replacing the constant cosimplicial space $C^\bullet_*$ by the thicker cosimplicial space $\Delta^\bullet$ does not quite give the totalization enough "room" for homotopy invariance, but it does get it closer.[1] It turns out that for totalization to be homotopy invariant, it is only necessary that the diagram be fibrant. We will state this more precisely below as Proposition 9.1.8, but we first want to be more explicit about what it means for a cosimplicial space to be fibrant.

Recall the notion of a Reedy category, Definition 8.4.4, and note that the cosimplicial indexing category is a Reedy category; we can take $\overrightarrow{I}$ to consist of all the objects and iterated coface maps (i.e. all maps generated by the cofaces) and $\overleftarrow{I}$ to consist of all the objects and iterated codegeneracy maps. Then, recalling Definition 8.4.6, the matching space at $[n]$, for $n \geq 0$, of a cosimplicial space $X^\bullet$, is given by

$$M_n(X^\bullet) = \lim_{([n]\downarrow \overleftarrow{I})_0} X^\bullet.$$

---

[1] In the sense that $\Delta^\bullet$ is cofibrant; see [Hir03, Corollary 15.9.11].

It is easy to see (essentially from the definition) that this limit can be rewritten as

$$M_n(X^\bullet) = \left\{ (x_0, x_1, \ldots, x_{n-1}) \in (X^{[n-1]})^n \mid s^i(x_j) = s^{j-1}(x_i), 0 \le i < j \le n-1 \right\}.$$
(9.1.6)

The $n$th matching map is then given by

$$X^{[n]} \longrightarrow M_n(X^\bullet) \qquad (9.1.7)$$

$$x \longmapsto (s^0(x), s^1(x), \ldots, s^{n-1}(x)).$$

We then say that $X^\bullet$ is *fibrant* if these matching maps are fibrations.

**Example 9.1.7** We will see many examples of fibrant cosimplicial spaces in Section 9.2. An example of a cosimplicial space that is not fibrant is the cosimplicial simplex from Example 9.1.4. Namely, the map $X^{[2]} \to M_2(\Delta^\bullet)$ is the map $X^{[2]} \to \Delta^1 \times \Delta^1$ given by two projections. But this map is not surjective (for example, $(1, 0)$ is not in the image) and hence not a fibration. $\quad\square$

**Proposition 9.1.8** ([BK72a], XI.4.4) *For $X^\bullet$ fibrant, the natural map*

$$\mathrm{Tot}\, X^\bullet \longrightarrow \mathrm{holim}\, X^\bullet$$

*is a homotopy equivalence.*

We omit the proof (the jist of which is that the map $|\Delta \downarrow \bullet| \to \Delta^\bullet$ is a transformation of cofibrant diagrams which induces an equivalence of the mapping spaces to an arbitrary fibrant diagram).

Because homotopy limits preserve weak equivalences, we then immediately have the following.

**Corollary 9.1.9** *Suppose $X^\bullet$ and $Y^\bullet$ are fibrant and $f \colon X^\bullet \to Y^\bullet$ is an objectwise weak equivalence. Then the induced map*

$$\mathrm{Tot}\, X^\bullet \longrightarrow \mathrm{Tot}\, Y^\bullet$$

*is also a weak equivalence.*

To make a totalization that respects weak equivalences, one can thus first perform a fibrant replacement of $X^\bullet$, for instance as it was done in Definition 8.4.12.

**Definition 9.1.10** The totalization of a fibrant replacement of $X^\bullet$ is called the *homotopy invariant totalization* and denoted by $\mathrm{hoTot}\, X^\bullet$ or $\widetilde{\mathrm{Tot}} X^\bullet$.

For simplicity, we will usually use $\text{Tot}\, X^\bullet$ but will be explicit about whether we have assumed $X^\bullet$ is fibrant or not.

A natural question is why we need totalization at all – why not just take the homotopy limit of a cosimplicial space? For one, totalization is the dual of the geometric realization, which is a natural construction (albeit also not homotopy invariant in general). Secondly, the homotopy limit is unwieldy since $|\Delta \downarrow [n]|$ is infinite-dimensional, so it is nice to have the more manageable construction at hand. Lastly, some cosimplicial spaces are already fibrant, so we can work with totalization right away and not worry about fibrant replacements. This is in fact the case with the central example we care about, the *cosimplicial replacement of a diagram*. We will devote Section 9.3 to cosimplicial replacements.

Another way to arrive at a homotopy invariant space from a cosimplicial space is to consider cosimplicial spaces without codegeneracies, the so-called *restricted cosimplicial spaces* or *semicosimplicial spaces*. Dropping the codegeneracies does not affect the totalization. More precisely, if $\Delta_{\text{inj}}$ is the cosimplicial indexing category but only with cofaces, that is, the injective maps, then the totalization is defined as

$$\text{Tot}_{\Delta_{\text{inj}}}\, X^\bullet = \text{Nat}(\Delta_{\text{inj}}, X^\bullet).$$

This is the same as taking the equalizer from Remark 9.1.6 but only over the injective maps $[n] \to [k]$. We then have the following.

**Proposition 9.1.11** *The inclusion $\Delta_{\text{inj}} \hookrightarrow \Delta$ is homotopy initial, and so, for a fibrant $X^\bullet$, the induced map*

$$\text{Tot}\, X^\bullet \longrightarrow \text{Tot}_{\Delta_{\text{inj}}}\, X^\bullet$$

*is a homotopy equivalence.*

The proof can be found in, for example, [DD77, Lemma 3.8] or [Lur09, Lemma 6.5.3.7], but it also follows readily from Theorem 8.6.5.

It should be noted that the fibrancy condition in Proposition 9.1.11 is important. In general, $\text{Tot}_{\Delta_{\text{inj}}}\, X^\bullet$ is not equivalent to $\text{Tot}\, X^\bullet$.

**Remark 9.1.12** The totalization $\text{Tot}_{\Delta_{\text{inj}}}\, X^\bullet$ is the dual of the "fat realization". The dual to the above proposition is Proposition A.1.8.                                    □

A nice feature of a semicosimplicial space is that it is always fibrant (we leave it to the reader to check this) and so $\text{Tot}_{\Delta_{\text{inj}}}\, X^\bullet$ is always homotopy invariant. Thus an alternative to making a homotopy invariant totalization is to drop the codegeneracies in $X^\bullet$ and then take the semicosimplicial totalization $\mathrm{Tot}_{\Delta_{\mathrm{inj}}} X^\bullet$. But, as is the case with replacing $\Delta$ by $|\Delta \downarrow \bullet|$, dropping the codegeneracies will typically produce a totalization that is much larger.

Now let $\Delta^{[\leq n]}$ be the full subcategory of $\Delta$ whose objects are the sets $[k]$ for $0 \leq k \leq n$. We then have the following.

**Definition 9.1.13** The *nth truncation of* $X^\bullet$ is the composite

$$\mathrm{tr}_n X^\bullet : \Delta^{[\leq n]} \hookrightarrow \Delta \xrightarrow{X^\bullet} \mathrm{Top} .$$

Using truncations, we can define the *n-skeleton* of a cosimplicial space $X^\bullet$, denoted by $\mathrm{sk}_n X^\bullet$, as the cosimplicial space that is in degree $k$ given by

$$\operatorname*{colim}_{(\Delta^{[\leq n]} \downarrow [k])} X^\bullet.$$

The category over which the colimit is taken consists of an integer $j$ with $j \leq n$ and a cosimplicial map $[j] \to [k]$, and the value of $X^\bullet$ on that object is $X^{[j]}$. Because we are taking the colimit, it suffices to take it over just the coface maps. As a result, in degree $\leq n$, $\mathrm{sk}_n X^\bullet$ agrees with $X^\bullet$ (the identity of $X^{[j]}$, $j \leq n$, is terminal), and in degree $> n$, consists of the image of the iterated cofaces from $\Delta^{[j]}$, $j \leq n$.

Note that, in degree $n + 1$, $\mathrm{sk}_n X^\bullet$ is precisely the latching object $L_{n+1}(X^\bullet)$. It is also clear that

$$X^\bullet \cong \mathrm{colim}(\mathrm{sk}_0 X^\bullet \to \mathrm{sk}_1 X^\bullet \to \mathrm{sk}_2 X^\bullet \to \cdots) \tag{9.1.8}$$

(remembering that we are taking the colimit of diagram indexed on $\Delta \times \mathbb{Z}_{\geq 0}$ and that taking the colimit of an $I \times J$-diagram amounts to taking it separately for each object $i \in I$).

**Remark 9.1.14** It can easily be seen that the *n*-skeleton is the left Kan extension (see Example 7.4.15) along the inclusion of the *n*th truncation of $\Delta$ into $\Delta$, that is,

$$L_{\Delta^{[\leq n]} \hookrightarrow \Delta} \mathrm{tr}_n X^\bullet \simeq \mathrm{sk}_n X^\bullet. \tag{9.1.9}$$

It is also a matter of unravelling the definitions that this is a left adjoint to the truncation functor $X^\bullet \mapsto \mathrm{tr}_n X^\bullet$. □

Now we can define the "partial totalizations" of a cosimplicial space, dual to the realizations of the *n*-skeleta of a simplicial space (Definition A.1.10).

**Definition 9.1.15**  Define the *nth partial totalization* $\text{Tot}^n X^\bullet$ of $X^\bullet$ as the space

$$\text{Tot}^n X^\bullet = \text{Nat}(\text{sk}_n \Delta^\bullet, X^\bullet).$$

Then we have the following.

**Proposition 9.1.16**  *There is an equivalence, for each $n \geq 0$,*

$$\text{Tot}^n X^\bullet \simeq \text{Nat}(\text{tr}_n \Delta^\bullet, \text{tr}_n X^\bullet).$$

On the right is the space of natural transformations between the $n$th truncations of $\Delta^\bullet$ and $X^\bullet$, that is, maps $f_i \colon \Delta^i \to X^i$, $0 \leq i \leq n$ that are compatible with cofaces and codegeneracies.

*Proof*  From (9.1.9), we have that

$$\text{Tot}^n X^\bullet = \text{Nat}(\text{sk}_n \Delta^\bullet, X^\bullet) \simeq \text{Nat}\,(L_{\Delta^{[\leq n]} \hookrightarrow \Delta} \text{tr}_n \Delta^\bullet, X^\bullet),$$

and since $L_{\Delta^{[\leq n]} \hookrightarrow \Delta}$ is adjoint to truncation, we have that

$$\text{Nat}\,(L_{\Delta^{[\leq n]} \hookrightarrow \Delta} \text{tr}_n \Delta^\bullet, X^\bullet) \simeq \text{Nat}(\text{tr}_n \Delta^\bullet, \text{tr}_n X^\bullet). \qquad \square$$

We prefer to work with truncations $X^{[\leq n]}$ since those are what will be turned into punctured cubical diagrams in Section 9.4.

The same proof as in Proposition 9.1.8 shows that, for a fibrant $X^\bullet$, the natural map

$$\text{Tot}^n X^\bullet \longrightarrow \underset{\Delta^{[\leq n]}}{\text{holim}} \, \text{tr}_n X^\bullet \qquad\qquad (9.1.10)$$

is a weak equivalence (in fact, one way to prove Proposition 9.1.8 is to prove the above and then use Proposition 9.1.17). Just as we will on occassion use $\text{hoTot}\, X^\bullet$ for the homotopy invariant totalization of $X^\bullet$, we will also sometimes use $\text{hoTot}^n X^\bullet$ for the $n$th partial totalization of a fibrant replacement of $X^\bullet$.

For $n \geq 1$, there is a natural map

$$\text{Tot}^n X^\bullet \longrightarrow \text{Tot}^{n-1} X^\bullet \qquad\qquad (9.1.11)$$

given by restricting the collection of maps $f_i \colon \Delta^i \to X^{[i]}$, $0 \leq i \leq n$, to the collection $f_i$, $0 \leq i \leq n - 1$.

We thus get a tower of partial totalizations $\text{Tot}^0 X^\bullet \leftarrow \text{Tot}^1 X^\bullet \leftarrow \cdots$ called the *total tower*, *totalization tower*, or *Tot tower*, and we have the following.

**Proposition 9.1.17**  *For any cosimplicial space $X^\bullet$,*

$$\text{Tot}\, X^\bullet \cong \lim(\text{Tot}^0 X^\bullet \longleftarrow \text{Tot}^1 X^\bullet \longleftarrow \cdots).$$

*Proof* From (9.1.8), we have

$$\Delta^\bullet \cong \operatorname*{colim}_i \operatorname{sk}_i \Delta^\bullet$$

and thus

$$\operatorname{Tot} X^\bullet = \operatorname{Nat}(\Delta^\bullet, X^\bullet) \cong \operatorname{Nat}\left(\operatorname*{colim}_i \operatorname{sk}_i \Delta^\bullet, X^\bullet\right)$$

(since Nat take homeomorphisms to homeomorphisms). But by Proposition 7.5.1, the right side is homeomorphic to

$$\lim_i \operatorname{Nat}(\operatorname{sk}_i \Delta^\bullet, X^\bullet),$$

which is, by Definition 9.1.15, precisely $\lim_i \operatorname{Tot}^i X^\bullet$. □

The totalization $\operatorname{Tot} X^\bullet$ is sometimes denoted by $\operatorname{Tot}^\infty X^\bullet$ as an indication that it is an inverse limit of a tower. The corresponding statement for simplicial spaces is Proposition A.1.14.

**Remark 9.1.18** One can also work backward from the total tower to construct a homotopy invariant totalization. Namely, an alternative way to construct a homotopy invariant totalization of $X^\bullet$ is to replace $\operatorname{Tot}^n X^\bullet$ by the homotopy limit of the restriction of $X^\bullet$ to $\Delta^{\leq n}$ (or as the homotopy limit of the pullback to the poset of non-empty subsets of $[n]$). Then $\operatorname{hoTot} X^\bullet$ is the inverse (homotopy) limit of the tower of these homotopy limits. □

For $X^\bullet$ a based cosimplicial space, we observed earlier that $\operatorname{Tot} X^\bullet$ has a natural basepoint given by the sequence of constant maps $\Delta^i \to X^{[i]}$, and hence $\operatorname{Tot}^n X^\bullet$ has a natural basepoint given by truncating the sequence at $f^n$. We will be interested in the fibers, over these basepoints, of the maps in the totalization tower for a fibrant $X^\bullet$. To get a handle on these, consider the fiber of the $n$th matching map (9.1.7) (we can talk about *the* fiber of this map since $X^\bullet$ is assumed to be fibrant), called the *$n$th normalization of $X^\bullet$*. This is given by

$$N^n X^\bullet = \operatorname{fiber}\left(X^{[n]} \longrightarrow M_n(X^\bullet)\right) = X^{[n]} \cap \bigcap_{i=0}^{n-1} \operatorname{fiber}(s^i). \tag{9.1.12}$$

The following result then says that the totalization tower for a based fibrant cosimplicial space is in fact a tower of fibrations and gives a description of its fibers.

**Proposition 9.1.19** (Fibers of the totalization tower) *For a fibrant cosimplicial space $X^\bullet$, the map $\operatorname{Tot}^n X^\bullet \to \operatorname{Tot}^{n-1} X^\bullet$ is a fibration with fiber*

$$L^n X^\bullet := \Omega^n N^n X^\bullet.$$

**Remark 9.1.20** It should be noted that the square encountered at the beginning of the proof below is reminiscent of one in the statement of Lemma 8.3.6. □

*Proof of Proposition 9.1.19* We will first argue that there exists a pullback square[2]

$$
\begin{array}{ccc}
\operatorname{Tot}^n X^\bullet & \longrightarrow & \operatorname{Map}(\Delta^n, X^{[n]}) \\
\downarrow & & \downarrow \\
\operatorname{Tot}^{n-1} X^\bullet & \longrightarrow & \operatorname{Map}(\partial\Delta^n, X^{[n]}) \times_{\operatorname{Map}(\partial\Delta^n, M_n(X^\bullet))} \operatorname{Map}(\Delta^n, M_n(X^\bullet))
\end{array}
\qquad (9.1.13)
$$

Since

$$
\operatorname{Tot}^{n-1} X^\bullet = \operatorname{Nat}(\operatorname{sk}_{n-1}\Delta, X^\bullet) \simeq \operatorname{Nat}(\Delta^{[\leq(n-1)]}, X^{[\leq(n-1)]})
$$

and

$$
\operatorname{Tot}^n X^\bullet = \operatorname{Nat}(\operatorname{sk}_n\Delta, X^\bullet) \simeq \operatorname{Nat}(\Delta^{[\leq n]}, X^{[\leq n]})
$$

we are in fact trying to extend a map of $(n-1)$-truncations to a map of $n$-truncations. To do this, we must choose a map $\Delta^n \to X^{[n]}$ that behaves well with respect to the coface and codegeneracy operators.

A point $s$ of $\operatorname{Tot}^{n-1} X^\bullet$ (i.e. a map of $(n-1)$-truncations) defines a map $L(s)\colon L_n(\Delta) \to L_n(X^\bullet)$ of latching objects and a map $M(s)\colon M_n(\Delta) \to M_n(X^\bullet)$ of matching objects. Since $L_n(\Delta) = \partial\Delta^n$, each point $s$ of $\operatorname{Tot}^{n-1} X^\bullet$ defines a solid arrow diagram

$$
\begin{array}{ccccc}
\partial\Delta^n & \xrightarrow{\ i\ } & \Delta^n & \xrightarrow{\ p\ } & M_n(\Delta) \\
{\scriptstyle L(s)}\downarrow & & \downarrow & & \downarrow{\scriptstyle M(s)} \\
L_n(X^\bullet) & \xrightarrow[\ i'\ ]{} & X^{[n]} & \xrightarrow[\ p'\ ]{} & M_n(X^\bullet)
\end{array}
\qquad (9.1.14)
$$

which in turns defines the point $(i' \circ L(s), M(s) \circ p)$ of

$$
\operatorname{Map}(\partial\Delta^n, X^{[n]}) \times_{(\partial\Delta^n, M_n(X^\bullet))} \operatorname{Map}(\Delta^n, M_n(X^\bullet)). \qquad (9.1.15)
$$

We thus have a map

$$
\operatorname{Tot}^{n-1} X^\bullet \longrightarrow \operatorname{Map}(\partial\Delta^n, X^{[n]}) \times_{(\partial\Delta^n, M_n(X^\bullet))} \operatorname{Map}(\Delta^n, M_n(X^\bullet)).
$$

Extending a map of $(n-1)$-truncations to a map of $n$-truncations is equivalent to choosing a dotted arrow in diagram (9.1.14) (i.e. a point $t$ of $\operatorname{Map}(\Delta^n, X^{[n]})$)

---

[2] This pullback square appears in [GJ99, p. 391] but the proof presented here that it is a pullback is due to Phil Hirschhorn.

that makes both squares commute (see [Hir03, Section 15.2.11]). Each dotted arrow $t\colon \Delta^n \to X^{[n]}$ in diagram (9.1.14) determines the point $(t \circ i, p' \circ t)$ of (9.1.15), and that dotted arrow $t$ is an extension of the point $s$ in $\mathrm{Tot}^{n-1} X^\bullet$ exactly when its image in (9.1.15) equals the image there of the point $s \in \mathrm{Tot}^{n-1} X^\bullet$, namely when $i' \circ L(s) = t \circ i$ and $M(s) \circ p = p' \circ t$. That is, the square (9.1.13) is a pullback.

Since the inclusion $\partial \Delta^n \to \Delta^n$ is a cofibration, and since the map $X^{[n]} \to M_n(X^\bullet)$ is a fibration as $X^\bullet$ is fibrant, the right vertical map in the square (9.1.13) is a fibration by Proposition 3.1.11. Therefore the left vertical map is also a fibration since the square is a pullback, using Proposition 2.1.16.

Proposition 2.1.16 also tells us that the fibers of the two vertical maps are homeomorphic, and clearly the fibers of the right vertical map are the iterated fibers of the square

$$\begin{array}{ccc} \mathrm{Map}(\Delta^n, X^{[n]}) & \longrightarrow & \mathrm{Map}(\Delta^n, M_n(X^\bullet)) \\ \downarrow & & \downarrow \\ \mathrm{Map}(\partial\Delta^n, X^{[n]}) & \longrightarrow & \mathrm{Map}(\partial\Delta^n, M_n(X^\bullet)) \end{array}$$

But the map of horizontal fibers is

$$\mathrm{Map}(\Delta^n, N^n X^\bullet) \longrightarrow \mathrm{Map}(\partial\Delta^n, N^n X^\bullet)$$

and the fibers of this map, by Proposition 2.1.11, are

$$\mathrm{Map}(\Delta^n/\partial\Delta^n, N^n X^\bullet) \cong \Omega^n N^n X^\bullet. \qquad \square$$

We close this section with the discussion of the augmentation of a cosimplicial space and cosimplicial contractions, which we will have use for on occassion (e.g. in Examples 9.2.8 and 9.6.19).

**Definition 9.1.21** An *augmentation* of a cosimplicial space $X^\bullet$ is a space $X^{-1}$ and a map $d^0\colon X^{-1} \to X^{[0]}$ such that

$$d^1 d^0 = d^0 d^0 \colon X^{-1} \longrightarrow X^{[1]}. \qquad (9.1.16)$$

(Thus $d^0$ equalizes the two cofaces from $X^{[0]}$ to $X^{[1]}$.)

Notice that if $X^{-1}$ augments $X^\bullet$, then there is a unique map

$$X^{-1} \longrightarrow X^{[n]}, \quad n > 0, \qquad (9.1.17)$$

given by composing $d^0\colon X^{-1} \to X^{[0]}$ with the cofaces in $X^\bullet$; the condition (9.1.16) and cosimplicial identities ensure that the choice of which coface to compose with at each stage does not matter. We can thus equivalently think of the augmentation as a cosimplicial map from $C^\bullet_{X^{-1}}$, the constant cosimplicial space at $X^{-1}$ (see Example 9.2.1), to $X^\bullet$. Consequently, there is a map

$$\overline{d^0}\colon \operatorname{Tot}(C^\bullet_{X^{-1}}) \cong X^{-1} \longrightarrow \operatorname{Tot} X^\bullet. \tag{9.1.18}$$

This is analogous to the fact we have encountered before that if a space maps into a diagram, then it canonically maps into its homotopy limit. In fact, we shall see in Remark 9.4.5 that the question of how connected the above map is is precisely the question of how cartesian certain cubical diagrams are.

**Remark 9.1.22**  An alternative definition of an augmented cosimplicial space is that this is a covariant functor from $\Delta_+$ to Top, where $\Delta_+$ is the usual cosimplicial indexing category (i.e. the category of finite totally ordered sets) but *with the empty set*, and morphisms the usual ones.                                    □

**Remark 9.1.23**  Augmentation can be used to define a *based*, or *pointed* cosimplicial space as a cosimplicial space which is augmented by a point; this provides a natural basepoint to each $X^{[n]}$ and the assignnment is compatible with the cosimplicial maps. Notice that a pointed cosimplicial space gives rise to a pointed Tot tower – the collection of constant maps from $\Delta^n$ to the basepoint in $X^{[n]}$, $n \geq 0$, serves as the basepoint for $\operatorname{Tot}^n X^\bullet$.                        □

**Definition 9.1.24**  A *left contraction* of an augmented cosimplicial space $X^{-1} \to X^\bullet$ is a collection of maps, called *extra codegeneracies*,

$$s^{-1}\colon X^{n+1} \longrightarrow X^n, \quad n \geq -1,$$

satisfying

$$\begin{aligned}
s^{-1}d^0 &= Id, \\
s^{-1}d^i &= d^{i-1}s^{-1}, \quad i \geq 1, \\
s^{-1}s^i &= s^{i-1}s^{-1}, \quad i \geq 0.
\end{aligned}$$

**Proposition 9.1.25**  *If an augmented cosimplicial space $X^{-1} \to X^\bullet$ has a left contraction, then the map $\overline{d^0}$ from (9.1.18) is a homotopy equivalence.*

*Proof*  Think of the map $\overline{d^0}$ as $d^0$ composed with itself $n$ times (by the comments above, any composition of cofaces gives the same map). Now, composing $s^{-1}$ with itself $n$ times gives a map $X^{[n]} \to X^{-1}$. Furthermore, this gives a map $X^\bullet \to C^\bullet_{X^{-1}}$ because of the identities defining the contraction $s^{-1}$, and hence a map $\overline{s^{-1}}\colon \operatorname{Tot} X^\bullet \to \operatorname{Tot}(C^\bullet_{X^{-1}}) \cong X^{-1}$. Since $s^{-1}d^0$ is the identity, it follows that the composed map $X^{-1} \to X^{[n]} \to X^{-1}$ is the identity and thus the composition $\overline{s^{-1} d^0}$ is the identity on $\mathrm{Tot}(C^{\bullet}_{X_{-1}}) \cong X^{-1}$.

It is also not hard to see that $\overline{d^0 s^{-1}}$ is homotopic to the identity on $\mathrm{Tot}\, X^{\bullet}$, and hence that $\overline{d^0}$ is a homotopy equivalence. The homotopy is given by the maps $s^{-1}$. Namely, suppose $(f_0, f_1, \ldots, f_n, \ldots)$ is a point in $\mathrm{Tot}\, X^{\bullet}$. To start inducting on $n$, we want $d^0 \circ s^{-1} \circ f_0 \sim f_0$. In other words, we want a path between points $(d^0 \circ s^{-1} \circ f_0)(\Delta^0)$ and $f_0(\Delta^0)$ in $X^0$. Call these points $x$ and $x'$, respectively. Now, we have a path between $d^0(x)$ and $d^1(x)$ given by $f_1(\Delta^1)$, which is a path in $X^1$ with endpoints $d^0(x)$ and $d^1(x)$. Because of the identities satisfied by $s^{-1}$, we then have $(s^{-1} \circ d^0)(x) = x$ and $(s^{-1} \circ d^1)(x) = (d^0 \circ s^{-1})(x')$. In other words, applying $s^{-1}$ to the path between $d^0(x)$ and $d^1(x)$ gives the desired path between $x$ and $x'$.

We can now proceed inductively: we want $(d^0)^{n+1} \circ (s^{-1})^{n+1} \circ f_n \sim f_n$. We have

$$
\begin{aligned}
(d^0)^{n+1} \circ (s^{-1})^n \circ f_n &= (d^0)^n \circ (s^{-1})^{n+1} \circ d^{n+1} \circ f_n \\
&\sim (d^0)^n \circ (s^{-1})^{n+1} \circ d^0 \circ f_n \\
&= (d^0)^n \circ (s^{-1})^n \circ f_n \\
&\sim Id \circ f_n = f_n.
\end{aligned}
$$

The homotopy between $d^{n+1} \circ f_n$ and $d^0 \circ f_n$ is given by the map $f_{n+1} \colon \Delta^{n+1} \to X^{n+1}$ (since the two compositions are images under $f_{n+1}$ of two faces of $\Delta^{n+1}$ because of cosimplicial compatibility conditions) and the second homotopy is given by induction. $\qquad\square$

The homotopy equivalence from Proposition 9.1.25 is called a *cosimplicial contraction* of the augmented cosimplicial space $X^{-1} \to X^{\bullet}$.

**Remark 9.1.26** The maps $s^{-1}$ are an example of a *cosimplicial homotopy* [Mey90, Definition 2.1] which induces a homotopy equivalence $X^{-1} \cong \mathrm{Tot}\, C^{\bullet}_{X_{-1}} \simeq \mathrm{Tot}\, X^{\bullet}$. The notion of a cosimplicial homotopy is more general – if there is a cosimplicial homotopy between maps $f, g \colon X^{\bullet} \to Y^{\bullet}$, then $\mathrm{Tot}\, X^{\bullet} \simeq \mathrm{Tot}\, Y^{\bullet}$. $\qquad\square$

There is an obvious parallel notion of a *right* contraction, where the extra codegeneracy is $s^{n+1}$, which enjoys similar properties. In particular, Proposition 9.1.25 still holds. For more details, see, for example, [CDI02, Section 3.2] or [DMN89, Section 6].

## 9.2 Examples

**Example 9.2.1** (Constant (or discrete) cosimplicial space)    Given a space $X$, consider the cosimplicial space $C_X^\bullet$ with $X^{[n]} = X$ for all $n$ and all cofaces and codegeneracies identity maps. This cosimplicial space is fibrant because $X$ is fibrant and the identity map is a fibration (recall that, according to our definitions, every space is fibrant; see Remark 8.4.8). Its totalization is $X$ because any map $f \colon \Delta^\bullet \to X^\bullet$ is determined by what it does on $\Delta^0$ in degree 0. That is, we have for example that the commutativity of the square

$$
\begin{array}{ccc}
\Delta^1 & \xrightarrow{\ s^j\ } & \Delta^0 = * \\
{\scriptstyle f_1}\big\downarrow & & \big\downarrow{\scriptstyle f_0} \\
X & \xrightarrow{\ Id\ } & X
\end{array}
$$

forces the image of $f_1$ to be that of $f_0$, namely a point in $X$. Inductively, the same is true for $f_n$, $n > 1$. Thus

$$\operatorname{Tot} X^\bullet \cong \operatorname{Map}(*, X) \cong X. \qquad \square$$

**Example 9.2.2** (Cosimplicial resolution of a space)    A *cosimplicial resolution of a space* $X$ is any cofibrant cosimplicial space that is weakly equivalent to the constant cosimplicial space at $X$ (see Example 9.2.1).[3] The canonical example is the cosimplicial space $X \times \Delta^\bullet$ which in degree $n$ has $X \times \Delta^n$. The coface and codegeneracies are the product of the identity on $X$ with the coface and codegeneracies on the cosimplicial simplex, described in Example 9.1.4.    $\square$

**Example 9.2.3** (Cosimplicial model for the loop space)    The following is an instance of a *geometric cobar construction*. Suppose $X$ is a based space with basepoint $*$. Consider the cosimplicial space $X^\bullet$ whose $n$th space is $X^n = \operatorname{Map}(\{0, 1, \ldots, n\}, X)$ ($X^0 = *$ is the 0th space). Now define the cofaces by

$$d^i \colon X^n \longrightarrow X^{n+1} \tag{9.2.1}$$

$$
(x_1, x_2, \ldots, x_n) \longmapsto
\begin{cases}
(*, x_1, x_2, \ldots, x_n), & i = 0; \\
(x_1, x_2, \ldots, x_i, x_i, \ldots, x_n), & 1 \le i \le n; \\
(x_1, x_2, \ldots, x_n, *), & i = n + 1,
\end{cases}
$$

and codegeneracies by

$$s^i \colon X^n \longrightarrow X^{n-1} \tag{9.2.2}$$

$$(x_1, x_2, \ldots, x_n) \longmapsto (x_1, x_2, \ldots, x_i, x_{i+2}, \ldots x_n).$$

---

[3]  Here $X$ should also be cofibrant, but this is always the case with our definitions; see
   Remark 8.4.8.

This is a fibrant cosimplicial space because, by (9.1.6), the matching map is given by

$$(x_1, x_2, \ldots, x_n) \longmapsto ((x_2, \ldots, x_n), (x_1, x_3, \ldots, x_n), \ldots, (x_1, x_2, \ldots, x_{n-1})),$$

which has an evident inverse, so the map $X^n \to M_n(X^n)$ is a homeomorphism (an hence a fibration).

Given a map $f \colon \Delta^\bullet \to X^\bullet$, consider, for $n > 1$, the square

$$
\begin{array}{ccc}
\Delta^n & \xrightarrow{\ s^j\ } & \Delta^{n-1} \\
{\scriptstyle f_n}\downarrow & & \downarrow{\scriptstyle f_{n-1}} \\
X^n & \xrightarrow{\ s^j\ } & X^{n-1}
\end{array}
$$

Here $f_n = (f_n^1, f_n^2, \ldots, f_n^n)$ since $f_n$ maps to a product. Then

$$s^j \circ f_n = (f_n^1, f_n^2, \ldots, f_n^j, f_n^{j+2}, \ldots, f_n^n).$$

However, by commutativity of the diagram, this must be the same as

$$f_{n-1} \circ s^j = (f_{n-1}^1 \circ s^j, f_{n-1}^2 \circ s^j, \ldots, f_{n-1}^{n-1} \circ s^j).$$

So $f_n$ is determined by $f_{n-1}$. All the relevant information to describe the totalization is thus contained in the diagram

$$
\begin{array}{ccc}
* = \Delta^0 & \underset{s^0}{\overset{d^0, d^1}{\rightleftarrows}} & \Delta^1 = I \\
{\scriptstyle f_0}\downarrow & & \downarrow{\scriptstyle f_1} \\
* & \underset{s^0}{\overset{d^0, d^1}{\rightleftarrows}} & X
\end{array}
$$

The commutativity of the squares and cosimplicial identities imply that $f$ is a map $I \to X$ which sends both endpoints to the basepoint. Thus we get a homeomorphism

$$\Omega X \longrightarrow \mathrm{Tot}\, X^\bullet$$

given by projecting $\mathrm{Tot}\, X^\bullet$ homemorphically to a smaller product, and this smaller product is identified with $\Omega X$.

Another way to think about this is that there is an evaluation map, for all $n \geq 0$,

$$\Omega X \times \Delta^n \longrightarrow X^n \tag{9.2.3}$$

$$(\alpha, x_1, \ldots, x_n) \longmapsto (\alpha(x_1), \ldots, \alpha(x_2))$$

with the obvious adjoint

$$\Omega X \longrightarrow \mathrm{Map}(\Delta^n, X^n). \tag{9.2.4}$$

It is not hard to see that the collection of maps, for all $n$, is compatible with the cofaces and codegeneracies in $\Delta^\bullet$ and $X^\bullet$, so that we get a map

$$\Omega X \longrightarrow \operatorname{Tot} X^n.$$

This map can be shown to be a homeomorphism by an argument that amounts to what was already done above.

For an operad point of view on the cobar construction described in this example, see Example A.4.14.                                              □

**Example 9.2.4** (Cosimplicial model for the free loop space)   A version of the cosimplicial space from Example 9.2.3 whose totalization gives $LX$, the free loop space on $X$, is obtained by taking the $n$th space to be $X^{n+1}$ (so the 0th space is $X$, first is $X \times X$, etc.) and only the diagonal cofaces (those that do not insert the basepoint). This is sometimes called the *cyclic cobar construction*.   □

**Example 9.2.5** (Cosimplicial model for the homotopy pullback)   Another version of the cosimplicial space from Example 9.2.3 gives a cosimplicial model for the homotopy pullback of the diagram $X = (A \xrightarrow{f} X \xleftarrow{g} B)$. This is called the *two-sided geometric cobar construction*. It is performed as follows. Consider the cosimplicial space $X^\bullet$ whose $n$th space is

$$A \times X^n \times B$$

and whose cofaces and codegeneracies are as in (9.2.1) and (9.2.2) with a slight modification that the basepoint is replaced by the images of $f$ and $g$. Define the cofaces by

$$d^i : A \times X^n \times B \longrightarrow A \times X^{n+1} \times B$$

$$(a, x_1, x_2, \ldots, x_n, b) \longmapsto \begin{cases} (a, f(a), x_1, x_2, \ldots, x_n, b), & i = 0; \\ (a, x_1, x_2, \ldots, x_i, x_i, \ldots, x_n, b), & 1 \le i \le n; \\ (a, x_1, x_2, \ldots, x_n, g(b), b), & i = n + 1, \end{cases}$$

and codegeneracies by

$$s^i : A \times X^n \times B \longrightarrow A \times X^{n-1} \times B$$

$$(a, x_1, x_2, \ldots, x_n, b) \longmapsto (a, x_1, x_2, \ldots, x_i, x_{i+2}, \ldots x_n, b).$$

It is then not hard to see, following the same argument as in Example 9.2.3, that a point in the totalization of this space is a point in $A$, a point in $B$, and a path in $X$ that begins and ends at the images of the points in $A$ and $B$ under $f$ and $g$, respectively. This is precisely the description of the homotopy limit of the diagram $X$ from Definition 3.2.4. Replacing $A$ and $B$ by one-point spaces and setting $f$ and $g$ to be inclusions of the basepoint of $X$, we obtain Example 9.2.3. □

**Example 9.2.6** (Cosimplicial model for the mapping space)   Given a simplicial model $Y_\bullet$ for a space $Y$ (by applying the singular complex functor to $Y$; see Remark A.1.5) and given a space $X$, one gets a cosimplicial model for the mapping space $\mathrm{Map}(Y, X)$ simply by applying the functor $\mathrm{Map}(-, X)$ objectwise to $Y_\bullet$. In fact, taking the standard simplicial model for $S^1$ consisting of one 0-simplex, one 1-simplex and with all other simplices degenerate, we get precisely the cosimplicial model for $LX = \mathrm{Map}(S^1, X)$ from Example 9.2.4. □

**Example 9.2.7** (Triple resolution)   A *triple* (or *monad*) $\langle F, \phi, \psi \rangle$ on the category of spaces (or simplicial sets) is a functor $F$ along with natural transformations $\phi \colon I_{\mathrm{Top}} \to F$ and $\psi \colon F^2 \to F$ (where $F^2$ means the composition of $F$ with itself) satisfying $\psi \circ (F\psi) = \psi \circ (\psi F)$ and $\psi \circ (F\phi) = \psi \circ (\phi F)$. Then, given a space $X$, there is a cosimplicial space $F^\bullet X$ with $(F^\bullet X)^{[n]} = F^{n+1} X$, called the *triple resolution of $X$*. The cofaces and codegeneracies are given by

$$d^i = F^i \circ \psi \circ F^{n-i+1} \colon (F^\bullet X)^{[n]} \longrightarrow (F^\bullet X)^{[n+1]},$$
$$s^i = F^i \circ \phi \circ F^{n-i} \colon (F^\bullet X)^{[n+1]} \longrightarrow (F^\bullet X)^{[n]}.$$

This cosimplicial space is naturally augmented by $X$, with the augmentation map $X \to (F^\bullet X)^{[0]} = F(X)$.

More on triple resolutions can be found, for example, in [Bou03, BK72a]. □

**Example 9.2.8** (*R*-completion of a simplicial set)   As an example of a triple resolution from the previous example, suppose $X_\bullet = \{X_n\}_{n=0}^\infty$ is a based simplicial set with basepoint $*$ in each $X_n$ and suppose $R$ is a commutative ring with unit. Let $R \otimes X_\bullet$ denote the free simplicial $R$-module generated by $X_\bullet$. More precisely, this is at the $n$th level given as

$$R \otimes X_n / R \otimes *$$

where $R \otimes X_n$ is the free $R$-module generated by the set $X_n$. The faces and degeneracies in $R \otimes X_\bullet$ are induced by the ones in $X_\bullet$. Now let $RX_\bullet$ be the subset of $R \otimes X_\bullet$ generated by simplices $\sum r_i x_i$ with $\sum r_i = 1$ (1 is the multiplicative identity). This subset turns out to also be a simplicial $R$-module [BK72a, I.2.2].

It is clear that $R$ is a functor from the category of simplicial sets to itself. One of its most important properties is a version of the Dold–Thom Theorem [BK72a, I.2], which we already encountered as Theorem 2.7.23:

$$\pi_*(|RX_\bullet|) \cong \widetilde{H}_*(|X_\bullet|; R). \qquad (9.2.5)$$

One can iterate this and define $R^n X_\bullet = R(R^{n-1} X_\bullet)$ with $R^0 X_\bullet = X_\bullet$. These simplicial sets were first studied in detail in [BK72a].

We then have maps of simplicial sets (that are also $R$-module homomorphisms)

$$\phi: X_\bullet \longrightarrow RX_\bullet$$
$$x \longmapsto 1 \otimes x$$

and

$$\psi: R^2 X_\bullet \longrightarrow RX_\bullet$$
$$(r \otimes (s \otimes x)) \longmapsto rs \otimes x$$

satisfying the conditions for a triple from Example 9.2.7. The collection $\{R^{n+1} X_\bullet\}_{n=0}^{\infty}$ thus forms a cosimplicial simplicial set $R^\bullet X_\bullet$, called the $R$-*resolution of* $X_\bullet$. Explicitly, the cofaces and codegeneracies are given by

$$d^i: R^n X_\bullet \longrightarrow R^{n+1} X_\bullet$$
$$r_1 \otimes r_2 \otimes \cdots \otimes r_n \otimes x \longmapsto r_1 \otimes r_2 \otimes \cdots \otimes r_i \otimes 1 \otimes r_{i+1} \cdots \otimes r_n \otimes x$$
$$s^i: R^{n+1} X_\bullet \longrightarrow R^n X_\bullet$$
$$r_1 \otimes r_2 \otimes \cdots \otimes r_{n+1} \otimes x \longmapsto r_1 \otimes r_2 \otimes \cdots \otimes r_i r_{i+1} \otimes r_{i+1} \otimes \cdots \otimes r_n \otimes x$$

A cosimplicial simplicial set is said to be *grouplike* if the cofaces (except possibly $d^0$) are group homomorphisms [BK72a, X.4.8]. Grouplike cosimplicial simplicial sets are fibrant [BK72a, X.4.10]. Since $R^\bullet X_\bullet$ is grouplike, it is thus also fibrant.

The totalization Tot $R^\bullet X_\bullet$ is a simplicial set called the $R$-*completion of* $X_\bullet$. The map $\phi$ is an augmentation (Definition 9.1.21) so that one has a natural map

$$f: |X_\bullet| \longrightarrow |\operatorname{Tot} R^\bullet X_\bullet|,$$

which has useful properties. For example, if $f$ induces isomorphisms on reduced homology (i.e. $X_\bullet$ is $R$-*good*), then a map $|X_\bullet| \to |Y_\bullet|$ induces isomorphisms on reduced homology if and only if the induced map $|\operatorname{Tot} R^\bullet X_\bullet| \to |\operatorname{Tot} R^\bullet Y_\bullet|$ is a homotopy equivalence [BK72a, I.5.5]. In addition, the homotopy type of $|\operatorname{Tot} R^\bullet X_\bullet|$ is determined by $\widetilde{H}^*(|X_\bullet|; R)$. $\qquad \square$

**Example 9.2.9** ($\mathbb{Z}$-nilpotent completion of a space)   For a based space $X$, there is a cosimplicial space where $X^{[n]} = \vee_n \Sigma X$ (and $X^{[0]} = CX$). This can be thought of as the cosimplicial space $\operatorname{sk}_0 \Delta^\bullet * X$, where $\operatorname{sk}_0 \Delta^{[n]}$ is the 0-skeleton of $\Delta^n$, that is, the set of its vertices, and $*$ is the join. Since $\operatorname{sk}_0 \Delta^n * X \simeq \vee_n \Sigma X$ (the homotopy equivalence is induced by quotienting out by one copy of the cone, the inclusion of which is a cofibration and which is a contractible subspace), the maps in this cosimplicial space are induced by the maps in $\Delta^\bullet$.

If $X$ is connected, the totalization of this cosimplicial space is $\mathbb{Z}_\infty X$, the $\mathbb{Z}$-*nilpotent completion of* $X$. This completion was first studied in detail in [BK72a], and computations using the spectral sequence associated to this cosimplicial space (called the *Barratt's desuspension spectral sequence*) were performed in [Hop, Goe93]. □

**Example 9.2.10** (Box product) Given cosimplicial spaces $X^\bullet$ and $Y^\bullet$, one can form a new cosimplicial space called the *box product*, denoted by $X^\bullet \square Y^\bullet$. The $n$th space in this product is

$$(X^\bullet \square Y^\bullet)^{[n]} = \mathrm{Coeq}\left( \coprod_{p+q=n-1} X^p \times Y^q \rightrightarrows \coprod_{p+q=n} X^p \times Y^q \right),$$

where the two maps are $(d^{p+1}, 1_{Y^q})$ and $(1_{X^p}, d^0)$. The cofaces and the codegeneracies are given by

$$d^i = \begin{cases} (d^i, 1_{Y^q}), & 0 \le i \le p+1; \\ (1_{X^p}, d^{i-p-1}), & p+2 \le i \le n+1, \end{cases}$$

and

$$s^i = \begin{cases} (s^i, 1_{Y^q}), & 0 \le i \le p-1; \\ (1_{X^p}, s^{i-p}), & p \le i \le n-1. \end{cases}$$

Batanin [Bat98] has shown that this product provides a monoidal structure on the category of cosimplicial spaces. More details about this product can also be found in [MS04a, Section 2]. The form presented here comes from [AC15, Definition 1.4]. □

**Example 9.2.11** (Hochschild homology) This example takes us outside of the realm of topological spaces, but we include it for the interested reader since it is of much importance. Let $k$ be a field, $R$ a $k$-algebra, and $M$ a bimodule over $R$. There is a cosimplicial $R$-module given by $[n] \mapsto \mathrm{Hom}_k(R^{\otimes n}, M)$ with cofaces and codegeneracies given by

$$(d^i f)(r_0, \ldots, r_n) = \begin{cases} r_0 f(r_1, \ldots, r_n), & i = 0; \\ f(r_0, \ldots, r_{i-1}r_i, \ldots, r_n), & 0 < i < n; \\ f(r_0, \ldots, r_{n-1})r_n, & i = n, \end{cases}$$

and

$$(s^i f)(r_0, \ldots, r_n) = f(r_0, \ldots, r_i, 1, r_{i+1}, \ldots, r_n).$$

Letting $d$ denote the alternating sum of cofaces, there is also a cochain complex

$$0 \xrightarrow{d} \mathrm{Hom}_k(R, M) \xrightarrow{d} \mathrm{Hom}_k(R^{\otimes 2}, M) \longrightarrow \cdots$$

Then the *Hochschild homology of R with coefficients in M* is defined as the cohomology of this cochain complex. (We shall see the alternating sum of cofaces repeatedly when we study the spectral sequences associated to cosimplicial spaces in Section 9.6.) □

For another example of an interesting cosimplicial space that is important in applications of manifold calculus of functors to knot theory, see Proposition 10.4.8.

## 9.3 Cosimplicial replacement of a diagram

In this section, we describe a way of turning any diagram of spaces into a cosimplicial one in such a way that the totalization of the cosimplicial diagram is the homotopy limit of the original diagram. This is especially useful since, as we shall see in Section 9.6, one has spectral sequences trying to compute the homotopy and the homology of the totalization, so this gives a way to get at the homotopy and the homology of the homotopy limit of an arbitrary diagram. In fact, many authors take the totalization of the cosimplicial replacement of a diagram as the definition of its homotopy limit.

Of special interest for us is that this process can be taken a step further into the setting of cubical diagrams. Namely, a cosimplicial space can be turned into a sequence of punctured cubical diagrams whose homotopy limits converge to the totalization. This will be discussed in Section 9.4.

Given a diagram $F \colon \mathcal{I} \to \mathrm{Top}$, with $\mathcal{I}$ small, consider the space

$$\Pi^n F = \prod_{i_0 \to \cdots \to i_n} F(i_n). \tag{9.3.1}$$

The product is taken over all composable morphisms of length $n$ in $\mathcal{I}$. (Composable morphisms of any length always exist in a diagram because of the existence of identity maps.) Note that

$$\Pi^0 F = \prod_{i \in \mathcal{I}} F(i).$$

We have already seen a construction like (9.3.1) in Definition 8.1.8 of the nerve $N(\mathcal{I})$ of $\mathcal{I}$. In fact, the coface and codegeneracy maps in the cosimplicial replacement are determined by faces and degeneracies in $N(\mathcal{I})$. The coface maps $d^j$ are defined as follows. The projection of

$$d^j \colon \prod_{i_0 \to \cdots \to i_{n-1}} F(i_{n-1}) \longrightarrow \prod_{i'_0 \to \cdots \to i'_n} F(i'_n), \quad 0 \le j \le n, \qquad (9.3.2)$$

onto the factor $F(i'_n)$ indexed by $i'_0 \to \cdots \to i'_n$ is the composition of the identity map of $F(i'_n)$ with the projection onto the factor indexed by $i'_0 \to i'_1 \to \cdots \to i'_{j-1} \to i'_{j+1} \to \cdots \to i'_n = d_j(i'_0 \to \cdots \to i'_n)$. The codegeneracies

$$s^j \colon \prod_{i_0 \to \cdots \to i_{n+1}} F(i_{n+1}) \longrightarrow \prod_{i'_0 \to \cdots \to i'_n} F(i'_n), \quad 0 \le j \le n, \qquad (9.3.3)$$

are defined similarly. The projection onto the factor $F(i'_n)$ indexed by $i'_0 \to \cdots \to i'_n$ is the composition of the identity map of $F(i'_n)$ with the projection onto the factor indexed by $i'_0 \to i'_1 \to \cdots \to i'_j \to i'_j \to \cdots \to i'_n = s_j(i'_0 \to \cdots \to i'_n)$.

These cofaces and codegeneracies satisfy the cosimplicial identities since the maps $d_j$ and $s_j$ in the definition of $N(\mathcal{I})$ satisfy the simplicial identities.

**Definition 9.3.1** The cosimplicial space

$$\Pi^\bullet F = \{\Pi^n F\}_{n=0}^\infty$$

with coface and codegeneracy maps as described above is called the *cosimplicial replacement of the diagram $F \colon \mathcal{I} \to$ Top.*

**Remark 9.3.2** We could have used $F(i_0)$ instead of $F(i_n)$ in (9.3.1); this would result in a cosimplicial replacement whose totalization is homeomorphic to the totalization of the replacement defined above. □

**Theorem 9.3.3** (Cosimplicial model for the homotopy limit) *Suppose $F \colon \mathcal{I} \to$ Top is a diagram of spaces with $\mathcal{I}$ small. Then there is a homeomorphism*

$$\operatorname*{holim}_{\mathcal{I}} F \cong \operatorname{Tot} \Pi^\bullet F.$$

*Proof* Recall the truncated homotopy limits $\operatorname{holim}_{\mathcal{I}}^{\le n} F$ from Definition 8.3.5. We will show that there are homeomorphisms $h_n \colon \operatorname{holim}_{\mathcal{I}}^{\le n} F \to \operatorname{Tot}^n \Pi^\bullet F$ for each $n \ge 0$ such that the diagram

$$\text{holim}_{I}^{\leq n} F \xrightarrow{h_n} \text{Tot}^n \Pi^\bullet F$$

$$\downarrow \qquad\qquad \downarrow$$

$$\text{holim}_{I}^{\leq n-1} F \xrightarrow{h_{n-1}} \text{Tot}^{n-1} \Pi^\bullet F$$

commutes. It then follows that

$$\text{holim}_{I} F = \lim_{n} \text{holim}^{\leq n}_{I} F \cong \lim_{n} \text{Tot}^n \Pi^\bullet F = \text{Tot} \Pi^\bullet F.$$

By definition,

$$\text{Tot}^n \Pi^\bullet F \subset \prod_{0 \leq k} \text{Map}\left( \text{sk}_n \Delta^k, \prod_{i_0 \to \cdots \to i_k} F(i_k) \right).$$

We claim that the projection

$$p_1 : \text{Tot}^n \Pi^\bullet F \longrightarrow \prod_{0 \leq k \leq n} \text{Map}\left( \text{sk}_n \Delta^k, \prod_{i_0 \to \cdots \to i_k} F(i_k) \right)$$

$$\cong \prod_{0 \leq k \leq n} \text{Map}\left( \Delta^k, \prod_{i_0 \to \cdots \to i_k} F(i_k) \right)$$

is a homeomorphism onto its image. To see why, let $f = \{f_k\}_{k=0}^{\infty} \in \text{Tot}^n \Pi^\bullet F$. We claim that the maps $f_k$ for $k > n$ are determined by $f_1, \ldots, f_n$. Let $k \geq n$ be fixed. For each $0 \leq j \leq k+1$, we have a commutative diagram

$$\text{sk}_n \Delta^k \xrightarrow{f_k} X^{[k]}$$

$$d^j \downarrow \qquad\qquad \downarrow d^j$$

$$\text{sk}_n \Delta^{k+1} \xrightarrow{f_{k+1}} X^{[k+1]}$$

Every point in $\text{sk}_n \Delta^k$ is in the image of some $d^j$ since $k \geq n$, and so the value of $f_{k+1}$ is determined by the value of $f_k$ for $k \geq n$.

We will now construct a homeomorphism from $\text{holim}_{I}^{\leq n} F$ to the image of $p_1$. In the proof of Lemma 8.3.6 we observed that

$$\text{holim}_{I}^{\leq n} F \subset \prod_{i} \prod_{0 \leq k \leq n} \prod_{(i_0 \to \cdots \to i_k) \to i} \text{Map}\left( \Delta^k, F(i) \right).$$

We have a projection map

$$\prod_{(i_0 \to \cdots \to i_k) \to i} \text{Map}\left( \Delta^k, F(i) \right) \longrightarrow \prod_{i_0 \to \cdots \to i_k} \text{Map}\left( \Delta^k, F(i_k) \right)$$

onto those factors indexed by simplices $(i_0 \to \cdots \to i_k) \to i_k$ in the overcategory for which the map $i_k \to i_k$ is the identity. Using the homeomorphism

$$\prod_{i_0 \to \cdots \to i_k} \operatorname{Map}\left(\Delta^k, F(i_k)\right) \cong \operatorname{Map}\left(\Delta^k, \prod_{i_0 \to \cdots \to i_k} F(i_k)\right),$$

we then have a map

$$p_2 \colon \operatorname{holim}_I^{\leq n} F \longrightarrow \prod_{0 \leq k \leq n} \operatorname{Map}\left(\Delta^k, \prod_{i_0 \to \cdots \to i_k} F(i_k)\right)$$

by composing with the above projection. The map $p_2$ is a homeomorphism onto its image as follows. An element $G \in \operatorname{holim}_I^{\leq n} F$ is a collection $G = \{G_i\}_{i \in I}$ of maps $G_i \colon \operatorname{sk}_n |I \downarrow i| \to F(i)$, and each $G_i$ is itself a collection $G_i = \{g_{\alpha \to i}\}$, where $\alpha \to i = (i_0 \to \cdots \to i_k) \to i$ is a $k$-simplex in the nerve of the overcategory $I \downarrow i$. For each $\alpha \to i = (i_0 \to \cdots \to i_k) \to i$ we have, by the definition of $\operatorname{holim}_I^{\leq n} F$ as an equalizer, a factorization

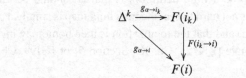

where $\alpha \to i_k = (i_0 \to \cdots \to i_k) \to i_k$ is the $k$-simplex in the overcategory $I \downarrow i_k$ for which the map $i_k \to i_k$ is the identity. The images of $p_1$ and $p_2$ are equal because the coface and codegeneracy maps in $\operatorname{Tot}^n \prod^\bullet F$ are determined by the face and degeneracy maps of the nerve of $I$, which are precisely the same instructions we use to build $\operatorname{sk}_n |I \downarrow i|$. Define $h_n \colon \operatorname{holim}_I^{\leq n} F \to \operatorname{Tot}^n \prod^\bullet F$ to be $p_2$ composed with the inverse of $p_1 \colon \operatorname{Tot}^n \prod^\bullet F \to p_1(\operatorname{Tot}^n \prod^\bullet F)$. It is clear from the constructions above that, for each $n$, $h_n$ and $h_{n-1}$ are compatible in the sense described at the beginning of this proof, and the collection of the $h_n$ give the desired homeomorphism. $\qquad\square$

We can combine Theorem 9.3.3 with Proposition 9.1.25 to prove without reference to Theorem 8.6.5 that the homotopy limit of a functor over a category with an initial object is the value of the functor at that initial object. This is the content of Example 8.2.5.

**Proposition 9.3.4** *Suppose $I$ is a small category with initial object $i_0$, and let $F \colon I \to \operatorname{Top}$ be a functor. Then the projection*

$$\operatorname*{holim}_I F \longrightarrow F(i_0)$$

*is a homotopy equivalence.*

*Proof*   Define $\prod^{[-1]} F = F(i_0)$, and define $d^0 \colon F(i_0) \to \prod^{[1]} F = \prod_i F(i)$ by using the map $F(i_0) \to F(i)$ induced by the unique morphism $i_0 \to i$ for each $i$. It is easy to see that this satisfies the conditions in Definition 9.1.21 and defines an augmentation of the cosimplicial replacement. Define $s^{-1} \colon \prod^{[n]} F \to \prod^{[n-1]} F$ by taking the factor $F(j_n)$ indexed by $i_0 \to j_1 \to \cdots \to j_n$ by the identity map to the factor $F(j_n)$ indexed by $j_1 \to \cdots \to j_n$, and projecting away from all factors not indexed by strings starting with the initial object. It is straightforward to check this satisfies the hypotheses of Proposition 9.1.25 and therefore the induced map $\overline{d^0} \colon F(i_0) \to \mathrm{Tot}\, \prod^\bullet F$ is a homotopy equivalence. Composing this with the homeomorphism given by Theorem 9.3.3 gives the desired result.                                                                                □

**Remark 9.3.5**   As mentioned earlier, because of Theorem 9.3.3, the definition of the homotopy limit of a diagram is often taken to be the totalization of the associated cosimplicial replacement. One important result that has to be established for this to be a valid definition is that $\prod^\bullet$ should be a fibrant cosimplicial space (this is not hard to see; the matching map is simply a projection and hence a fibration) and that the totalization is then homotopy invariant. For details, see for example [BK72a, Ch. XI, Section 5] or [GJ99, Ch. VIII, Section 2].                                                                         □

**Remark 9.3.6**   Note that, from Proposition 7.4.3, we have

$$\lim F = \mathrm{Eq}\left( \prod{}^0 F \mathrel{\substack{\xrightarrow{d^0} \\ \xrightarrow[d^1]{}}} \prod{}^1 F \right).$$

One has an evident map from this to $\mathrm{Tot}\, \prod^\bullet F$, and this is precisely the usual map $\lim F \to \mathrm{holim}\, F$.                                                                           □

Before we provide some examples, we observe that the identity maps in the strings of morphisms indexing the cosimplicial replacement in a sense do not matter. Namely, let $S_n$ denote the set of $n$-simplices in $B_\bullet \mathcal{I}$ (the nerve of $\mathcal{I}$; see Definition 8.1.8), and let $\mathrm{nd}(S_n)$ denote the set of non-degenerate $n$-simplices, namely those strings that do not contain any identity maps. Then we have a projection

$$p \colon \prod_{n \geq 0} \mathrm{Map}\!\left( \Delta^n, \prod_{i_0 \to \cdots \to i_n} F(i_n) \right) \longrightarrow \prod_{n \geq 0} \mathrm{Map}\!\left( \Delta^n, \prod_{i_0 \to \cdots \to i_n \in \mathrm{nd}(S_n)} F(i_n) \right).$$

Because $\mathrm{Map}(\Delta^n, -)$ commutes with products, this is homeomorphic to the projection

$$p: \prod_{n \geq 0} \prod_{i_0 \to \cdots \to i_n} \mathrm{Map}(\Delta^n, F(i_n)) \longrightarrow \prod_{n \geq 0} \prod_{i_0 \to \cdots \to i_n \in \mathrm{nd}(S_n)} \mathrm{Map}(\Delta^n, F(i_n)).$$

**Proposition 9.3.7** *Let $I$ be a small category, $F: I \to$ Top a functor, and $\Pi^\bullet F$ its cosimplicial replacement. Then the restriction of $p$ to $\mathrm{Tot}(\Pi^\bullet F)$ is a homeomorphism onto its image.*

*Proof* The main point is that an element of $F(j_k)$ in the totalization indexed by the codegenerate simplex $J_k = j_0 \to \cdots \to j_k$ is determined by the element of the factor $F(i_n)$ indexed by $I_n = i_0 \to \cdots \to i_n$ which is the unique non-degenerate simplex such that $s(I_n) = J_k$, where $s$ is a composition of degeneracy operators. More precisely, let $f_{J_k} \in \mathrm{Map}(\Delta^k, F(i_k))$ denote the projection of an element $f \in \mathrm{Tot}(\Pi^\bullet F)$ onto the factor indexed by the degenerate simplex $J_k$, and let $s$ be as above. Then $f_{I_n} \circ s = f_{J_k}$.

The inverse $q$ of the restriction of $p$ maps the factor indexed by the non-degenerate simplex $I_n$ by sending an element $f_{I_n}$ of that factor to all factors indexed by $s(I_n)$ for all $s$ which are compositions of degeneracy operators by sending $f_{I_n}$ to the map $f_{I_n} \circ s$. □

Now recall the notion of a finite and acyclic indexing category from the list of terminology following Definition 7.1.1; this is a finite category with a bound on the length of composable non-identity morphisms. For such a category, the cosimplicial replacement beyond some stage will necessarily be indexed on strings that all contain identity maps. Then an immediate consequence of Proposition 9.3.7 is the following.

**Corollary 9.3.8** *For a finite and acyclic diagram $F$ with maximum length of composable non-identity morhisms $k$,*

$$\mathrm{holim}\, F \simeq \mathrm{Tot}^k \Pi^\bullet F.$$

In particular, we can perform the replacement only up to $\Pi^k F$ since, by Proposition 9.1.16, that will determine $\mathrm{Tot}^k$.

**Example 9.3.9** Consider the diagram $X = (X_1 \xrightarrow{f} X_2)$. Then

$$\Pi^\bullet X = \left( X_1 \times X_2 \underset{\substack{\xrightarrow{d^0} \\ \xleftarrow{s^0} \\ \xrightarrow{d^1}}}{} X_1 \times X_2 \times X_2 \underset{\substack{\xrightarrow{d^0} \\ \xleftarrow{s^0} \\ \xrightarrow{d^1} \\ \xleftarrow{s^1} \\ \xrightarrow{d^2}}}{} X_1 \times X_2 \times X_2 \times X_2 \cdots \right).$$

The various copies of $X_2$ correspond to the various ways that the identity morphisms and the map $f$ can be composed with $X_2$ as the final target. By Corollary 9.3.8, we can truncate this diagram after the second stage. The cofaces and the codegeneracy are

$$d^0(x_1, x_2) = (x_1, x_2, x_2),$$
$$d^1(x_1, x_2) = (x_1, x_2, f(x_1)),$$
$$s^0(x_1, x_2, x_2') = (x_1, x_2).$$

The totalization is the space of all maps $f = (f_0, f_1)$ making the diagram

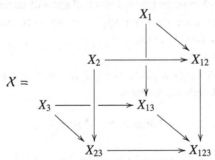

commute. A point in $\mathrm{Tot}\, \Pi^\bullet X$ then consists of a point $x_1 \in X_1$ and a path in $X_2$ that begins at $f(x_1)$ (and ends at $x_2$, but this could be any point in $X_2$). But this is precisely the description of the mapping path space from Definition 2.2.1 (which is of course the homotopy limit). $\qquad \square$

**Example 9.3.10**  Consider the punctured 3-cube

$$X = $$

commutative cube diagram with vertices $X_1$, $X_2$, $X_{12}$, $X_3$, $X_{13}$, $X_{23}$, $X_{123}$

Because of Corollary 9.3.8, the part of the cosimplicial replacement $\Pi^\bullet X$ relevant for the totalization then looks like

$$\Pi^0 X = X_1 \times X_2 \times X_3 \times X_{12} \times X_{13} \times X_{23} \times X_{123},$$
$$\Pi^1 X = X_1 \times X_2 \times X_3 \times X_{12}^3 \times X_{13}^3 \times X_{23}^3 \times X_{123}^4,$$
$$\Pi^2 X = X_1 \times X_2 \times X_3 \times X_{12}^4 \times X_{13}^4 \times X_{23}^4 \times X_{123}^{13}.$$

Following the same procedure as in the previous example, it is not hard to see that the totalization can then be efficiently represented as the space of natural transformations

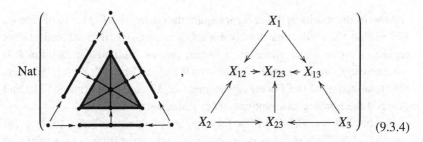

$$(9.3.4)$$

But this is homeomorphic to the space of natural transformations producing the homotopy limit of $X$ in Example 5.3.3. □

**Example 9.3.11** The cosimplicial replacement behaves as one would expect it to for cosimplicial diagrams and their truncations (see Definition 9.1.13). A simple application of (9.1.10) and Theorem 9.3.3 shows that, if $X^\bullet$ is a fibrant cosimplicial space,

$$\mathrm{Tot}(\Pi^\bullet X^{[\leq n]}) \simeq \mathrm{Tot}^n X^\bullet,$$

and hence $\mathrm{Tot}(\Pi^\bullet X^\bullet) \simeq \mathrm{Tot}\, X^\bullet$. In other words, the cosimplicial replacement of a cosimplicial space preserves the totalization. □

## 9.4 Cosimplicial spaces and cubical diagrams

In this section we explain how a truncated cosimplicial diagram $X^{[\leq n]}$ can be turned into a punctured cubical diagram $\mathcal{X}$ in such a way that the totalization of the truncation, namely the partial totalization $\mathrm{Tot}^n X^\bullet$, is equivalent to $\mathrm{holim}\,\mathcal{X}$. Thus the techniques of Chapter 5 can be used for studying partial totalizations and Tot towers. In fact, combining the notion of a cosimplicial replacement of a diagram from Section 9.3 with the results in this section, we get that *the homotopy limit of any diagram can be studied using homotopy limits of punctured cubical diagrams*. We will say more about this at the end of this section.

Similarly, if one has a tower of homotopy limits arising from a sequence of punctured cubical diagrams, one can look for a cosimplicial space that models it and use cosimplicial techniques to study it. An example of this will be given in Section 10.4.2.

Many of the definitions and results in this section first appeared in [Sin09], with some expansions, details, and proofs that did not exist in the literature provided in [MV14]. We expand futher here and provide some dual results at the end of Section A.1.

Most of the results in this section require the cosimplicial space to be fibrant. We remind the reader that this is not a big imposition since we can always replace a cosimplicial space by a fibrant one as discussed in Section 9.1. Alternatively, we could have written $\mathrm{hoTot}\, X^\bullet$ and $\mathrm{hoTot}^n X^\bullet$ for the (partial) totalization of the fibrant replacement of $X^\bullet$ (see Definition 9.1.10) and dropped the fibrancy assumptions in our statements.

In what follows, $\mathcal{P}([n])$ will be the category of subsets of $[n] = \{0, 1, \dots, n\}$ and $\mathcal{P}_0([n])$ will be the category of non-empty subsets of $[n]$ (not to be confused with $\mathcal{P}(\underline{n})$ and $\mathcal{P}_0(\underline{n})$, which were the categories of (non-empty) subsets of $\underline{n} = \{1, 2, \dots, n\}$). Thus $\mathcal{P}([n])$ and $\mathcal{P}_0([n])$ can be used as indexing categories for (punctured) $(n + 1)$-cubes. In this section, we prefer the indexing set for cubes that includes 0 since the first space in a cosimplicial space is indexed on 0 and it will thus be easier to keep track of some of the functors we are about to define.

**Definition 9.4.1** ([Sin09, Definition 6.3]) Let $c_n \colon \mathcal{P}_0([n]) \to \Delta$ be the (covariant) functor defined by $S \mapsto [|S| - 1]$, which sends $S \subset S'$ to the composite $[|S| - 1] \cong S \subset S' \cong [|S'| - 1]$, where the two isomorphisms are the unique isomorphisms of ordered sets.

**Definition 9.4.2** Given a cosimplicial space $X^\bullet$, the *punctured $(n + 1)$-cube associated to (the nth truncation of)* $X^\bullet$ is the pullback

$$c_n^*(X^\bullet) = X^\bullet \circ c_n \colon \mathcal{P}_0([n]) \longrightarrow \mathrm{Top}.$$

**Example 9.4.3** Let $n = 2$, $S = \{1, 2\}$, and $S' = \{0, 1, 2\}$. Then $S \cong [1]$ by $1 \mapsto 0$ and $2 \mapsto 1$, and $S' \cong [2]$ by $i \mapsto i$ for $i = 0, 1, 2$. Hence the map $[1] \to [2]$ induced by the inclusion $S \subset S'$ is the map $0 \mapsto 1$ and $1 \mapsto 2$. The diagram that indexes the punctured 3-cube associated to $X^\bullet$ is

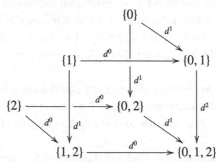

The maps are cofaces in $\Delta$ and which one goes where in the cube is determined by which element of a subset of $\{0, 1, 2\}$ was missed. Recalling that we write

$X^{[n]}$ for the value of $X^\bullet$ on $[n]$, the punctured 3-cube $X^\bullet \circ c_2$ associated to $X^\bullet$ is thus

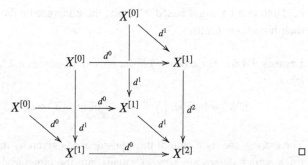

Note that the codegeneracies play no role in the construction of the punctured cube. This is fine, since, in light of Proposition 9.1.11, they do not affect the (partial) totalization of a fibrant $X^\bullet$.

The following is [Sin09, Theorem 6.6] (see also [Sin06, Lemma 2.9] or [DGM13, Lemma 7.1.1]).

**Theorem 9.4.4** *For a cosimplicial space $X^\bullet$, there is a weak equivalence*

$$\operatorname{Tot}^n X^\bullet \simeq \operatorname*{holim}_{\mathcal{P}_0([n])} (X^\bullet \circ c_n).$$

**Remark 9.4.5** If $X^\bullet$ is augmented, namely if there exists a space $X^{-1}$ and a map $d^0 \colon X^{-1} \to X^{[0]}$ satisfying $d^1 d^0 = d^0 d^0$ (Definition 9.1.21), then this means precisely that $X^{-1}$ fits as the initial space in the punctured cube $X^\bullet \circ c_n$ (the condition $d^1 d^0 = d^0 d^0$ makes the resulting cube commutative). Thus if we are interested in the connectivity of the map

$$X^{-1} \longrightarrow \operatorname{Tot}^n X^\bullet,$$

by Theorem 9.4.4, this is the same as the connectivity of the map

$$X^{-1} \longrightarrow \operatorname*{holim}_{\mathcal{P}_0([n])} (X^\bullet \circ c_n).$$

But the latter is precisely the question of how cartesian the cubical diagram resulting from the augmented cosimplicial space is. □

For the following lemma it will be convenient to write the homotopy limits in a way that indicates functoriality more explicitly for reasons that will become apparent shortly. Thus we will sometimes write $\operatorname{holim}_{c \in C} F(c)$ in place of $\operatorname{holim}_C F$.

Recall from Proposition 9.1.19 that we define $L^n X^\bullet$ as the fiber of the fibration $\mathrm{Tot}^n X^\bullet \to \mathrm{Tot}^{n-1} X^\bullet$. Here is a description of $L^n X^\bullet$ in terms of the functor $c_n$. Note that having a based $X^\bullet$ bases the punctured cubes $X^\bullet \circ c_n$ and hence their homotopy limits.

**Lemma 9.4.6** *For a based fibrant cosimplicial space $X^\bullet$, there is an equivalence*

$$L^n X^\bullet \simeq \mathrm{hofiber}\left(X^{[0]} \longrightarrow \mathop{\mathrm{holim}}_{S \in \mathcal{P}_0([n-1])} (X^\bullet \circ c_{n-1}(S \cup \{n\}))\right).$$

The map on the right side of the equivalence is given by the cosimplicial identities, which determine how $X^{[0]}$ maps into the punctured cube $X^\bullet \circ c_{n-1}$ and hence to its homotopy limit. We leave it to the reader to fill in the details.

The above lemma thus says that $L^n X^\bullet$ is the total homotopy fiber of an $n$-cube of spaces. Another description in terms of a total fiber of a different cube will be given in Proposition 9.4.10 below.

*Proof* By Theorem 9.4.4,

$$L^n X^\bullet = \mathrm{hofiber}\left(\mathop{\mathrm{holim}}_{T \in \mathcal{P}_0([n])} (X^\bullet \circ c_n(T)) \to \mathop{\mathrm{holim}}_{S \in \mathcal{P}_0([n-1])} (X^\bullet \circ c_{n-1}(S))\right).$$

By Lemma 5.3.6,

$$\mathop{\mathrm{holim}}_{T \in \mathcal{P}_0([n])} (X^\bullet \circ c_n(T)) \cong \mathrm{holim}\left(\begin{array}{c} X^{[0]} \\ \downarrow \\ \mathop{\mathrm{holim}}_{S \in \mathcal{P}_0([n-1])} (X^\bullet \circ c_{n-1}(S)) \\ \uparrow \\ \mathop{\mathrm{holim}}_{S \in \mathcal{P}_0([n-1])} (X^\bullet \circ c_{n-1}(S \cup \{n\})) \end{array}\right)$$

Fibering this over $\mathop{\mathrm{holim}}_{\mathcal{P}_0([n-1])} (X^\bullet \circ c_{n-1}(S))$ and using Theorem 3.3.15 (recalling that a homotopy fiber is a homotopy limit) yields

$$L^n X^\bullet \simeq \mathrm{hofiber}\left(X^{[0]} \to \mathop{\mathrm{holim}}_{\mathcal{P}_0([n-1])} (X^\bullet \circ c_{n-1}(S \cup \{n\}))\right). \qquad \square$$

**Example 9.4.7** When $n = 2$, Lemma 9.4.6 says that $L^2 X^\bullet$ is the total homotopy fiber of the square

$$\begin{array}{ccc} X^{[0]} & \xrightarrow{d^0} & X^{[1]} \\ {\scriptstyle d^0}\downarrow & & \downarrow{\scriptstyle d^1} \\ X^{[1]} & \xrightarrow{d^0} & X^{[2]} \end{array}$$

$\square$

**Definition 9.4.8** ([Sin09, Definition 7.2]) Let $c_n^! \colon \mathcal{P}(\underline{n}) \to \Delta$ be the functor defined by $c_n^!(S) = [n - |S|]$, and which associates to an inclusion $S \subset S'$ the map

$$[n - |S|] \cong [n] - S \longrightarrow [n] - S' \cong [n - |S'|].$$

Here the middle map sends $i \in [n] - S$ to the largest element of $[n] - S'$ less than or equal to $i$.

**Example 9.4.9** Suppose $n = 3$, $S = \{2\}$, and $S' = \{2, 3\}$. Then $[2] \cong [3] - \{2\} = \{0, 1, 3\}$ by the map $0 \mapsto 0$, $1 \mapsto 1$, and $2 \mapsto 3$. The map $\{0, 1, 3\} \to \{0, 1\} = [3] - \{2, 3\}$ sends 0 to 0, 1 to 1, and 3 to 1. Finally, $[3] - \{2, 3\} \cong [1]$ by the identity map. Putting this all together, the map $[2] \to [1]$ induced by the above inclusion sends 0 to 0, 1 to 1, and 2 to 1. □

Proposition 9.1.19 says that, for a fibrant $X^\bullet$, the fiber $L^n X^\bullet$ of the fibration $\mathrm{Tot}^n X^\bullet \longrightarrow \mathrm{Tot}^{n-1} X^\bullet$ is $\Omega^n N^n X^\bullet$, where $N^n X^\bullet$ is the $n$th normalization of $X^\bullet$ from (9.1.12). Here is another description of $L^n X^\bullet$ as a loop space in terms of cubical diagrams. Although the codegeneracies did not play a role previously, they will play an important one in the following proposition. This result is useful in that it gives a way of checking whether the homotopy spectral sequence for $X^\bullet$ converges to the totalization (see Corollary 9.6.8). Again note that having a based $X^\bullet$ bases the cubes $X^\bullet \circ c_n^!$ and hence their total fibers.

**Proposition 9.4.10** ([Sin09, Theorem 7.3]) *For a based fibrant cosimplicial space $X^\bullet$, there is an equivalence*

$$L^n X^\bullet \simeq \Omega^n \, \mathrm{tfiber}(X^\bullet \circ c_n^!).$$

*Proof* By Theorem 9.4.4, we have an equivalence

$$L^n X^\bullet \simeq \mathrm{hofiber}\!\left( X^{[0]} \longrightarrow \underset{S \in \mathcal{P}_0([n-1])}{\mathrm{holim}} (X^\bullet \circ c_{n-1}(S \cup \{j\})) \right).$$

We may write homotopy fiber on the right side of this equivalence as the total homotopy fiber of an $n$-cube $\mathrm{tfiber}(S \mapsto (X^\bullet \circ c_{n-1}(S)))$. Then by Proposition 5.5.7 we obtain an equivalence

$$\mathrm{tfiber}(S \mapsto (X^\bullet \circ c_{n-1}(S))) \simeq \Omega^n \, \mathrm{tfiber}(S \mapsto (X^\bullet \circ c_n(\underline{n} - S))).$$

We leave it to the reader to check, using codegeneracies and the cosimplicial identities, that the hypotheses of Proposition 5.5.7 are satisfied. □

The results of this and the previous section say that, to understand homotopy limits of diagrams, it suffices to understand homotopy limits of punctured cubical diagrams (and inverse limits). Namely, given a diagram $F\colon I \to \mathrm{Top}$, we have a cosimplicial diagram $\Pi^\bullet F$ such that

$$\operatorname*{holim}_{I} F \cong \operatorname{Tot} \Pi^\bullet F$$

(Theorem 9.3.3). But this totalization is the inverse limit of the tower of fibrations of partial totalizations $\operatorname{Tot}^n \Pi^\bullet F$ (Proposition 9.1.17). In turn, each partial totalization can be realized as the homotopy limit a punctured cube, that is, by Theorem 9.4.4, we have

$$\operatorname{Tot}^n \Pi^\bullet F \simeq \operatorname*{holim}_{\mathcal{P}_0([n])} (\Pi^\bullet F \circ c_n).$$

Combining all this, we thus have

$$\operatorname*{holim}_{I} F \simeq \lim_{n} \left( \operatorname*{holim}_{\mathcal{P}_0([n])} (\Pi^\bullet F \circ c_n) \right).$$

One could go even further and reduce to the case of homotopy pullbacks, since, by Lemma 5.3.6, the homotopy limit of a punctured cube is an iterated pullback. This brings us back to the main theme of this book: homotopy pullbacks are the building blocks of all homotopy limits. In summary, we thus have that

*the homotopy limit of any diagram of spaces is an inverse limit of spaces, each of which can be obtained by a sequence of homotopy pullbacks.*

## 9.5 Multi-cosimplicial spaces

One is often interested in generalizations of cosimplicial spaces to diagrams of spaces that are "cosimplicial in each direction" [DMN89, Eld13, Goe96, MV14, Shi96].

Let

$$(\Delta)^m = \Delta \times \Delta \times \cdots \times \Delta$$

be the $m$-fold product of the category $\Delta$ with itself.

**Definition 9.5.1** An *m-cosimplicial space* is a covariant functor

$$X^{\vec{\bullet}}\colon (\Delta)^m \longrightarrow \mathrm{Top}.$$

The notation for the functor $X^{\vec{\bullet}}$ makes no reference to $m$, but this will be understood from the context.

Fixing a simplex in all but one of the factors of $(\Delta)^m$ gives an ordinary cosimplicial space $X^\bullet$ with cofaces and codegeneracies which satisfy the usual identities (9.1.1). Let $\vec{n} = (n_1, n_2, \ldots, n_m)$ and denote by $X^{[\vec{n}]}$ the value of the functor $X^{\vec{\bullet}}$ on $[\vec{n}] = ([n_1], [n_2], \ldots, [n_m]) \in (\Delta)^m$.

Now let $\Delta^{\vec{\bullet}}$ be the $m$-cosimplicial version of $\Delta^\bullet$, namely the $m$-cosimplicial space whose $\vec{n}$ entry is

$$\Delta^{[\vec{n}]} = \Delta^{n_1} \times \Delta^{n_2} \times \cdots \times \Delta^{n_m}.$$

In analogy with Definition 9.1.5, we then have the following.

**Definition 9.5.2** The *totalization* $\operatorname{Tot} X^{\vec{\bullet}}$ of an $m$-cosimplicial space $X^{\vec{\bullet}}$ is the space of cosimplicial maps from $\Delta^{\vec{\bullet}}$ to $X^{\vec{\bullet}}$.

We then have a straightforward generalization of Proposition 9.1.8. Now that we know that a totalization is the inverse limit of partial totalization, one way to prove it is to use Proposition 9.5.4 below and simply take the limit as all $n_i$ go to infinity.

**Proposition 9.5.3** *For $X^{\vec{\bullet}}$ fibrant,*

$$\operatorname{Tot} X^{\vec{\bullet}} \longrightarrow \operatorname{holim} X^{\vec{\bullet}}$$

*is a homotopy equivalence.*

We again have partial totalizations of $X^{\vec{\bullet}}$. Namely, fixing $\vec{n} = (n_1, n_2, \ldots, n_m)$ we define

$$\operatorname{Tot}^{\vec{n}} X^{\vec{\bullet}} \subset \prod_{\vec{i} \leq \vec{n}} \operatorname{Map}(\operatorname{sk}_{\vec{i}} \Delta^{\vec{\bullet}}, X^{\vec{\bullet}})$$

(where $\operatorname{sk}_{\vec{i}}$ is the natural generalization of $\operatorname{sk}_i$) as the subset of such maps which are compatible with all the cofaces and codegeneracies (i.e. compatible with cosimplicial maps "in any direction" in $X^{\vec{\bullet}}$). Here $\vec{i} \leq \vec{n}$ means $i_j \leq n_j$ for all $j = 1$ to $m$. In analogy with (9.1.10), we have that, for a fibrant $X^{\vec{\bullet}}$, the map

$$\operatorname{Tot}^{\vec{n}} X^{\vec{\bullet}} \longrightarrow \operatorname*{holim}_{\Delta^{\leq \vec{n}}} X^{\vec{\bullet}} \qquad (9.5.1)$$

is a weak equivalence. Here $\Delta^{\leq \vec{n}}$ is the subcategory of $\Delta^{\vec{\bullet}}$ generalizing $\Delta^{\leq n}$ as a subcategory of $\Delta^\bullet$ in the obvious way.

To state the next result, it will be useful to have notation which says to which "direction" a partial totalization applies. Let $\operatorname{Tot}_i^k$ be the $k$th partial totalization of a multicosimplicial space $X^{\vec{\bullet}}$ in the $i$th direction (or $i$th variable).

**Proposition 9.5.4** *For a fibrant $X^{\vec{\bullet}}$, there is a homotopy equivalence*

$$\mathrm{Tot}^{\vec{n}} X^{\vec{\bullet}} \simeq \mathrm{Tot}_1^{n_1} \, \mathrm{Tot}_2^{n_2} \cdots \mathrm{Tot}_m^{n_m} \, X^{\vec{\bullet}}.$$

*Moreover, the order of the $\mathrm{Tot}_i^{n_i}$ does not matter.*

*Proof* From (9.5.1), we have

$$\mathrm{Tot}^{\vec{n}} X^{\vec{\bullet}} \simeq \underset{\Delta^{\leq \vec{n}}}{\mathrm{holim}} \, X^{\vec{\bullet}}. \tag{9.5.2}$$

We also have

$$\underset{\Delta^{\leq \vec{n}}}{\mathrm{holim}} \, X^{\vec{\bullet}} = \underset{\Delta^{\leq n_1} \times \Delta^{\leq n_2} \times \cdots \times \Delta^{\leq n_m}}{\mathrm{holim}} \, X^{\vec{\bullet}},$$

and, by Proposition 8.5.5,

$$\underset{\Delta^{\leq n_1} \times \Delta^{\leq n_2} \times \cdots \times \Delta^{\leq n_m}}{\mathrm{holim}} \, X^{\vec{\bullet}} \cong \underset{\Delta^{\leq n_1}}{\mathrm{holim}} \, \underset{\Delta^{\leq n_2}}{\mathrm{holim}} \cdots \underset{\Delta^{\leq n_m}}{\mathrm{holim}} \, X^{\vec{\bullet}}.$$

The order in which we write the homotopy limits does not matter, again by Proposition 8.5.5. Using (9.5.1) again finishes the proof. $\square$

One nice feature of an $m$-cosimplicial spaces is that, for the purposes of totalization, it can be reduced to a single cosimplicial space.

**Definition 9.5.5** Given $X^{\vec{\bullet}}$, define its *diagonal cosimplicial space* $X^{\vec{\bullet}}_{\mathrm{diag}}$ to be the composition of $X^{\vec{\bullet}}$ with the diagonal functor $\Delta \to (\Delta)^m$. More explicitly,

$$X^{\vec{\bullet}}_{\mathrm{diag}} = \{X^{([i],[i],\ldots,[i])}\}_{i=0}^{\infty}$$

with the cofaces and codegeneracies the compositions of cofaces and codegeneracies from $X^{\vec{\bullet}}$ with the same index:

$$s^j = (s^j, s^j, \ldots, s^j) \quad \text{and} \quad d^j = (d^j, d^j, \ldots, d^j).$$

It is not hard to see that the identities (9.1.1) are satisfied with this definition.

We then have the following useful result. The proof can be found in [Shi96, Proposition 8.1] (although that proof is for bicosimplicial spaces, it generalizes immediately to $m$-cosimplicial spaces).

**Proposition 9.5.6** *The diagonal functor $\Delta \to (\Delta)^m$ induces a weak equivalence*

$$\mathrm{Tot} \, X^{\vec{\bullet}} \longrightarrow \mathrm{Tot} \, X^{\vec{\bullet}}_{\mathrm{diag}}. \tag{9.5.3}$$

**Remark 9.5.7** Proposition 9.5.6 is not true for partial totalizations, namely, it is *not true* that

$$\text{Tot}^n X^{\vec{\bullet}}_{\text{diag}} \simeq \text{Tot}^n_1 \text{Tot}^n_2 \cdots \text{Tot}^n_m X^{\vec{\bullet}}.$$

However, if $\text{Tot}^{p_i}_i X^{\bullet} \simeq \text{Tot}_i X^{\bullet}$ (i.e. totalization is realized at a finite stage in each direction of $X^{\bullet}$), then we have

$$\text{Tot}^{p_1 + \cdots + p_m} X^{\vec{\bullet}}_{\text{diag}} \simeq \text{Tot}^{p_1}_1 \text{Tot}^{p_2}_2 \cdots \text{Tot}^{p_m}_m X^{\vec{\bullet}}.$$

This result is proved in [Eld13, Proposition 3.2] where the approach is the study of cosimplicial spaces and their totalization through coskeleta. More details can also be found in [Eld08]. □

We finally have the following result the proof of which can be found in [Hir15e, Corollary 2.4]. Combined with the previous proposition, it allows us to say that the homotopy totalizations of $X^{\vec{\bullet}}$ and $X^{\vec{\bullet}}_{\text{diag}}$ are weakly equivalent.

**Proposition 9.5.8** *If $X^{\vec{\bullet}}$ is fibrant, so is $X^{\vec{\bullet}}_{\text{diag}}$.*

# 9.6 Spectral sequences

We now look at some spectral sequences computing the homology and homotopy of homotopy (co)limits of cubical diagrams and totalizations of cosimplicial spaces. Most of the spectral sequences considered here arise from towers of fibrations and cofibrations, so we first make some general remarks about towers and their spectral sequences in Section 9.6.1.

In Section 9.6.2, we apply the setup from Section 9.6.1 to (punctured) cubes to set up spectral sequences that compute the homotopy and homology of homotopy (co)limits and total (co)fibers. To do this, we first construct towers of fibrations and cofibrations that in a sense filter these homotopy (co)limits. Such an approach to spectral sequences of cubical diagrams is not well known and, to the best of our knowledge, has only been considered in [Sin01, Proposition 3.3]. We will also see another, more standard, way to arrive at these spectral sequences via cosimplicial replacements (see Examples 9.6.20 and A.1.18).

By contrast, the homotopy and homology spectral sequences of a cosimplicial space, which are the subject of Sections 9.6.3 and 9.6.4, have been studied in depth. This is partly due to the fact that any diagram $F: I \to \text{Top}$ can be turned into a cosimplicial diagram $\Pi^{\bullet} F$ (Section 9.3) and that the homotopy limit of $F$ is equivalent to the totalization of $\Pi^{\bullet} F$ (Theorem 9.3.3), so that these spectral sequences provide a computational tool for understanding the homotopy limit of any diagram.

The homotopy spectral sequence was developed by Bousfield and Kan [BK72a, BK72b, BK73], and it arises from the tower of fibrations of partial totalizations. Our goal is to recall the main features of this construction (one further reading that provides more details is [GJ99]). In [BK72b], the authors consider the special case of this spectral sequence associated to their motivating construction, the $R$-completion of a space $X$ (see Example 9.2.8) which we briefly mention in Example 9.6.19. This is an example of an *unstable Adams spectral sequence* since it has homology as input and tries to compute the (unstable) homotopy groups of $X$. This spectral sequence has been further studied and generalized in [Bou89, BCM78, BT00, Bou03]. One of its important applications was in the solution of the Sullivan Conjecture [Mil84] (see also [DMN89]). Other applications and extensions of the theory include [BB10, BD03, Goe90, LL06, Smi02]. For an algorithmic approach to computations in the Bousfield–Kan unstable Adams spectral sequence, see [Rom10].

The homology spectral sequence of a cosimplicial space does not arise from a tower but rather from a double complex. It was first constructed by Anderson [And72]. A special case was considered earlier in [Rec70], where the Eilenberg–Moore spectral sequence is exhibited as the homology spectral sequence of a cosimplicial space obtained as a two-sided cobar construction on a pullback square where one of the maps is a fibration. This is why the general homology spectral sequence of a cosimplicial space is sometimes called the *generalized Eilenberg–Moore spectral sequence*. The convergence properties of this spectral sequence were studied by Bousfield in [Bou87] (based on the results of [Dwy74]) and further by Shipley in [Shi96] (see also [Goo98]). An alternative spectral sequence for computing the homology of homotopy limits is given in [Goe96]. The operations in the homology spectral sequence of a cosimplicial infinite loop space were considered in [Tur98] and more recently in [Hac10]. Other related results and applications appear in [BO05, Pod11], among others.

Both homotopy and homology spectral sequences of a cosimplicial space have in recent years also been used for the study of spaces of knots and links [ALTV08, LTV10, MV14, Sak08, Sin09]. More details about this will be given in Section 10.4.4.

For the remainder of this section, we will assume that

- the reader is familiar with the basics of spectral sequences (the standard introduction is [McC01b]);
- all cosimplicial spaces are fibrant (if not, we use their fibrant replacements).

### 9.6.1  Spectral sequences for towers of fibrations and cofibrations

What follows is a brief review of the construction of the homotopy spectral sequence for a tower of fibrations and the homology spectral sequence of a tower of cofibrations. Further details can be found in [BK72a, GJ99, KT06].

Consider the pointed tower of fibrations (fibers $F^i$ are also pictured)

$$
\begin{array}{ccccccc}
Y^0 & \xleftarrow{\ f^1\ } & Y^1 & \xleftarrow{\ f^2\ } & Y^2 & \xleftarrow{\ f^3\ } & \cdots \\
\ \big\uparrow{=} & & \big\uparrow{i^1} & & \big\uparrow{i^2} & & \\
F^0 & & F^1 & & F^2 & &
\end{array}
\qquad (9.6.1)
$$

To simplify the exposition, we will assume that each $Y^i$ is connected with abelian fundamental group. The discussion of the generalization to the case when this is not assumed can be found in [BK72a, Chapter IX, §4] or [GJ99, Section VI.2]. Applying the homotopy groups, one gets exact sequences of abelian groups

$$
\begin{array}{ccccccc}
\pi_*(Y^0) & \xleftarrow{\ f^1_*\ } & \pi_*(Y^1) & \xleftarrow{\ f^2_*\ } & \pi_*(Y^2) & \xleftarrow{\ f^3_*\ } & \cdots \\
& \searrow{\partial}\ \nwarrow{i^1_*} & & \searrow{\partial}\ \nwarrow{i^2_*} & & \searrow{\partial} & \\
& \pi_*(F^1) & & \pi_*(F^2) & & \pi_*(F^3) &
\end{array}
\qquad (9.6.2)
$$

which in turn give rise to an exact couple and hence a spectral sequence (see [McC01b, Theorem 2.8] and the discussion leading up to it for details of how this works in general). Here $\partial$ is the connecting homomorphism that reduces the degree as in Theorem 2.1.13.

Now define, for $q \geq p \geq 0$ and $r \geq 1$,

$$
\pi_i(Y^{p,r}) = \operatorname{im}(\pi_i(Y^{p+r}) \longrightarrow \pi_i(Y^p)).
$$

Here the map is the composition of the maps $f^j_*$. Then consider

$$
Z^r_{p,q} = \ker\left(\pi_{q-p}(F^p) \longrightarrow \pi_{q-p}(Y^p)/\pi_{q-p}(Y^{p,r-1})\right),
$$

where the map is given by $i^p_*$ (followed by the quotient map), and

$$
B^r_{p,q} = \ker\left(\pi_{q-p+1}(Y^{p-1}) \longrightarrow \pi_{q-p+1}(Y^{p-r})\right),
$$

where the map is again the composition of the $f^j_*$s. Finally set

$$
E^r_{p,q} = Z^r_{p,q}/\partial(B^r_{p,q}),
$$

where $\partial$ is the connecting homomorphism from $\pi_{*+1}(Y^{p-1})$ to $\pi_*(F^p)$. If $p = q$, then $B^r_{p,q}$ acts on $Z^r_{p,q}$ and the quotient above is meant to be the set of orbits;

otherwise $\partial(B^r_{p,q})$ is a normal subgroup of $Z^r_{p,q}$ so the quotient makes sense as written.[4] The differential

$$d^r : E^r_{p,q} \longrightarrow E^r_{p+r,q+r-1}$$

is defined through the composition $E^r_{p,q} \to \pi_{q-p}(Y^{p,r-1}) \to E^r_{p+r,q+r-1}$. The first map is induced by $i^p_* : \pi_{q-p}(F^p) \to \pi_{q-p}(Y^p)$ and the second by applying $\delta$ to a preimage of an element in $\pi_{q-p}(Y^p)$ in $\pi_{q-p}(Y^{p+r-1})$ (under the compositions of the maps $f^i_*$ from $\pi_{q-p}(Y^{p+r-1})$ to $\pi_{q-p}(Y^p)$). It is easy to see that this is independent of the choice of the preimage. We will say more about the particular case of $d^1$ below.

**Definition 9.6.1** ([BK72a, Ch. IX, §4]) The collection $\{E^r_{p,q}, d^r\}_{r \geq 1}$ as defined above is the *homotopy spectral sequence of the tower of fibrations* (9.6.1).

For a fixed $r$, $\{E^r_{p,q}, d^r\}$ is called the *rth page of the spectral sequence*. Because the differentials $d^r$ raise degree, this is a *second quadrant spectral sequence* and is usually depicted in the second quadrant by placing $E^*_{p,q}$ in the slot $(-p, q)$. Because of this, the notation for its pages in the literature is sometimes $E^r_{-p,q}$.

**Remark 9.6.2** Another point of view on this spectral sequence is via the *derived* homotopy long exact sequences built out of the $\pi_i(Y^{p,r})$. See [BK72a, Ch. IX, 4.1] or [GJ99, Ch. VI, Lemma 2.8] for details. □

It is worth observing what the $\{E^1_{p,q}, d^1\}$ page of this spectral sequence is. We have that

$$\pi_{q-p}(Y^{p,0}) = \pi_{q-p}(Y^p),$$

and so

$$Z^1_{p,q} = \ker\left(\pi_{q-p}(F^p) \longrightarrow \pi_{q-p}(Y^p)/\pi_{q-p}(Y^p)\right)$$
$$= \ker\left(\pi_{q-p}(F^p) \longrightarrow \{e\}\right) = \pi_{q-p}(F^p),$$
$$B^1_{p,q} = \ker\left(\pi_{q-p+1}(Y^{p-1}) \longrightarrow \pi_{q-p+1}(Y^{p-1}))\right) = \{e\}.$$

Thus

$$E^1_{p,q} = \pi_{q-p}(F^p). \tag{9.6.3}$$

---

[4] Some authors follow the notational convention where $E^{r+1} = Z^r/B^r$; we are following the conventions from [GJ99].

The first differential is given by maps in (9.6.2) as

$$d^1 : E^1_{p,q} = \pi_{q-p}(F^p) \longrightarrow E^1_{p+1,q} = \pi_{q-p-1}(F^{p+1}) \qquad (9.6.4)$$

$$\alpha \longmapsto \partial(i^p_*(\alpha))$$

To start answering the natural first question as to what

$$E^\infty_{p,q} = \lim_r E^r_{p,q}$$

might be computing (the hope is that it computes $\pi_*(\lim Y^j)$), let

$$e^\infty_{p,q} = \ker\Big( \operatorname{im}(\pi_{q-p}(\lim Y^j) \to \pi_{q-p}(Y^p)) \longrightarrow \operatorname{im}(\pi_{q-p}(\lim Y^j) \to \pi_{q-p}(Y^{p-1})) \Big).$$

There is a natural inclusion

$$e^\infty_{p,q} \hookrightarrow E^\infty_{p,q}, \qquad (9.6.5)$$

and if this inclusion is an isomorphism, then the spectral sequence *converges conditionally* to the inverse limit $\lim \pi_*(Y^j)$ (see [GJ99, Ch. VI, Lemma 2.17 and discussion following Definition 2.19] or [BK72a, Ch. IX, §5.3]).

The second issue is whether the canonical map

$$\pi_i(\lim Y^j) \longrightarrow \lim \pi_i(Y^j) \qquad (9.6.6)$$

is an isomorphism. This map was described and studied in Lemma 8.3.3 and in the discussion leading to it. To handle this, one has to consider the $\lim^1$ terms of the tower of fibrations (which were called $K_n(Z)$ in Lemma 8.3.3). These are the terms that make the sequences

$$\{e\} \longrightarrow \lim^1 \pi_{i+1}(Y^j) \longrightarrow \pi_i(\lim Y^j) \longrightarrow \lim \pi_i(Y^j) \longrightarrow \{e\} \qquad (9.6.7)$$

exact for $j, i \geq 0$. More details about how the $\lim^1$ is defined can be found in [GJ99, pp. 317–319] and [BK72a, Ch. IX, §2]. Clearly, if $\lim^1 \pi_{i+1}(Y^j) = 0$, then (9.6.6) holds. If $\lim^1 \pi_{i+1}(Y^j) \neq 0$, it is not clear what the spectral sequence converges to; see [GJ99, Ch. VI, Example 2.18].

Combining these two conditions, we get the following.

**Proposition 9.6.3** *The spectral sequence from Definition 9.6.1 converges to* $\pi_i(\lim Y^j)$ *for $i \geq 1$ if*

1. *the inclusion $e^\infty_{p,q} \hookrightarrow E^\infty_{p,q}$, is an isomorphism for all $q > p \geq 0$;*
2. $\lim^1 \pi_{i+1}(Y^j) = 0$ *for $i \geq 0$.*

When the above two conditions are satisfied, we say that the spectral sequence *converges completely*.

A condition equivalent to the two conditions above is the content of the following lemma, which is useful for computations. First recall that, for a tower of abelian groups and homomorphisms $\{G_n, f_n\}_{n \geq 0}$, $\lim_n^1$ is defined by the exact sequence

$$\{e\} \longrightarrow \lim_n G_n \longrightarrow \prod_n G_n \longrightarrow \prod_n G_n \longrightarrow \lim_n^1 G_n \longrightarrow \{e\}$$

where the middle map is given by $g_n \mapsto (g_n - f_{n+1}(g_{n+1}))$. Since $\{E_{p,q}^r\}_{r \geq 0}$ form such a tower of groups, we can consider $\lim_r^1 E_{p,q}^r$. For more details about $\lim^1$ for towers, see [GJ99, Ch. VI, Section 2].

**Lemma 9.6.4** (Complete Convergence Lemma, [BK72a, Ch. IX, 5.4]) *The two conditions from Proposition 9.6.3 are satisfied if and only if*

$$\lim_r^1 E_{p,q}^r = 0$$

*for all $q > p \geq 0$.*

For a detailed proof of this result, see [GJ99, Ch. VI, Lemma 2.20].

Another useful characterization of complete convergence that follows from Lemma 9.6.4 is given as follows.

**Definition 9.6.5** ([BK72a, Ch. IX, 5.5]) The spectral sequence from Definition 9.6.1 is *Mittag–Leffler* if, for every $p \geq 0$ and every $i \geq 1$, there exists an integer $N(p) > p$ such that

$$E_{p,p+i}^{N(p)} \cong E_{p,p+i}^\infty.$$

This condition says that, for each $(p, q)$ the spectral sequence converges after finitely many steps. From this, the following is easy to establish (details are in [GJ99, Corollary 2.22 and Lemma 2.23]).

**Lemma 9.6.6** ([BK72a, Ch. IX, Lemma 5.6]) *If the spectral sequence from Definition 9.6.1 is Mittag–Leffler, then it converges completely.*

The main reason we have introduced the Mittag–Leffler condition is because it leads to the following two simple and useful characterizations of convergence (Proposition 9.6.7 and Corollary 9.6.8).

**Proposition 9.6.7** *Suppose that the spectral sequence from Definition 9.6.1 has the property that, for each total degree $i$, there exists a page $E^r$ such that* *there are finitely many p with $E^r_{p,p+i} \neq 0$. Then the spectral sequence converges completely.*

*Proof* For the Mittag–Leffler condition to be satisfied, it is sufficient that there be only finitely many non-zero differentials going to or from bidegree $(p, p+i)$ (because once $r$ is high enough so that all those possibly non-zero differentials have been used, $E^r_{p,p+i} = E^\infty_{p,p+i}$ at that bidegree). Since every differential lowers total degree by 1, the hypothesis implies that this is true. □

Note that one way that a spectral sequence can satisfy the hypothesis from the previous result is if there exists an $r$ such that $E^r_{p,q} = 0$ for $q < kp$ where $k > 1$ ($k$ is a real number), i.e. if there is a vanishing line that is steeper than the diagonal.

**Corollary 9.6.8** *The spectral sequence from Definition 9.6.1 converges completely if the connectivity of the fibers $F^j$ of the maps $f^j : Y^j \to Y^{j-1}$ from (9.6.1) increases with $j$ (and $F^1$ is connected).*

*Proof* In this situation we have a vanishing line at $E^1$ that is steeper than the diagonal and so, by the comment above, the spectral sequence converges completely. □

**Remark 9.6.9** One can set up the above spectral sequence for a finite tower of fibrations. The spectral sequence then has a finite number of columns and converges to the initial space as the conditions of Proposition 9.6.7 are automatically satisfied. □

Analogously to what has been said so far, we can also consider a tower of cofibrations

$$Y^0 \xrightarrow{f^0} Y^1 \xrightarrow{f^1} Y^2 \xrightarrow{f^2} \cdots .$$

$$
\begin{array}{ccc}
\downarrow{=} & \downarrow{i^1} & \downarrow{i^2} \\
C^0 & C^1 & C^2
\end{array}
\tag{9.6.8}
$$

where the $C_i$ are the cofibers.

Taking homology, we get

$$H_*(Y^0) \xrightarrow{f^0_*} H_*(Y^1) \xrightarrow{f^1_*} H_*(Y^2) \xrightarrow{f^2_*} \cdots$$

$$
\begin{array}{ccccc}
& {}^{\partial}\diagdown & \downarrow{i^1_*} & {}^{\partial}\diagdown & \downarrow{i^2_*} \\
& H_*(C^1) & & H_*(C^2)
\end{array}
\tag{9.6.9}
$$

where connecting homomorphism $\partial$ shifts the degree down (see Theorem 2.3.14 for the long exact sequence of a cofibration).

We then have the following.

**Proposition 9.6.10** (Homology spectral sequence of a tower of cofibrations)
*Let $p, q \geq 0$. Given a tower of cofibrations as in (9.6.8), there is a first quadrant spectral sequence with*

$$E^1_{p,q} = H_{q+p}(C^p)$$

*(with any coefficients) and*

$$d^1 : E^1_{p,q} \longrightarrow E^1_{p-1,q}$$
$$\alpha \longmapsto \partial(i^p_*(\alpha)).$$

This is a standard homological spectral sequence of a double complex and can also be thought of as a generalization of the homology spectral sequence associated to a filtered complex (for which the cofibrations are inclusions of skeleta). Because the literature on this abounds (for spectral sequences of general double complexes, see [McC01b], and for the homology spectral sequence of a filtered complex, see e.g. [MT68, Chapter 7]), we omit the more general description of the pages, the differentials, and the conditions under which this spectral sequence converges to the (homotopy) colimit of the tower. In particular, this is true if the tower of cofibrations is finite or if some page of the spectral sequence has a vanishing line of slope greater than 1. We also leave it to the reader to set up the cohomological version of this spectral sequence.

## 9.6.2 Spectral sequences for cubical diagrams

The homotopy spectral sequence of a tower of fibrations and the homology spectral sequence of a tower of cofibrations can be used to construct homotopy and homology spectral sequences associated with punctured cubical and cubical diagrams, as we will explain below. Another, less direct (but perhaps more familiar), way to construct the same spectral sequences for the punctured cubical case as in Propositions 9.6.11 and 9.6.13 is to first perform the cosimplicial replacement as explained in Section 9.3 and then use the spectral sequences for cosimplicial spaces which are discussed in Sections 9.6.3 and 9.6.4. The advantage of the approach given in this section is that it gives spectral sequences not only for the homotopy (co)limit of a punctured cube but also for the total (co)fiber of a cube.

We first have the following.

**Proposition 9.6.11** (Homotopy spectral sequence of a punctured cube)   *Let* $X: \mathcal{P}_0(\underline{n}) \to \mathrm{Top}_*$ *be a punctured cube of connected based spaces. Then there is an n-column homotopy spectral sequence with $E^1$ term given by*

$$E^1_{p,q} = \prod_{\substack{S \subset \mathcal{P}_0(\underline{n}) \\ |S|=p+1}} \pi_q(X_S)$$

*and with the first differential*

$$d^1: \prod_{\substack{S \subset \mathcal{P}_0(\underline{n}) \\ |S|=p+1}} \pi_q(X_S) \longrightarrow \prod_{\substack{T \subset \mathcal{P}_0(\underline{n}) \\ |T|=p+2}} \pi_q(X_T),$$

*whose projection on the factor indexed by T is given by*

$$d^1 = \sum_{i \in T}(-1)^{|T| \cdot i} f(T \setminus \{i\} \to T)_*.$$

*This spectral sequence converges to $\pi_*(\mathrm{holim}\, X)$.*

*Proof*   We illustrate the argument on the case $n = 3$. The general case is completely analogous and we will make some comments about it at the end of the proof. Given a based punctured cube

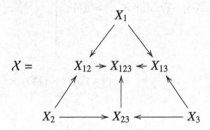

consider the punctured cubes

and

$$X_{(1)} = \begin{array}{c} X_1 \\ \swarrow \quad \searrow \\ * \to * \leftarrow * \\ \nearrow \quad \uparrow \quad \nwarrow \\ X_2 \longrightarrow * \longleftarrow X_3 \end{array}$$

The subscript $(i)$ indicates that all spaces $X_S$ with $|S| > i$ have been replaced by one-point spaces.

The obvious maps of cubes $X \to X_{(2)} \to X_{(1)}$ are fibrations, and, by Theorem 8.5.1, we thus get a tower of fibrations

$$\operatorname{holim} X \longrightarrow \operatorname{holim} X_{(2)} \longrightarrow \operatorname{holim} X_{(1)}.$$

From (9.6.3), the spectral sequence for this tower has as its $E^1$ page the homotopy groups of $F^0 = \operatorname{holim} X_{(1)}$, $F^1 = \operatorname{hofiber}(\operatorname{holim} X_{(2)} \to \operatorname{holim} X_{(1)})$, and $F^2 = \operatorname{hofiber}(\operatorname{holim} X \to \operatorname{holim} X_{(2)})$. These spaces are

$$F^0 = \operatorname{holim} X_{(1)} \simeq X_1 \times X_2 \times X_3$$

$$F^1 = \operatorname{hofiber}(\operatorname{holim} X_{(2)} \to \operatorname{holim} X_{(1)}) \simeq \operatorname{holim} \left( \begin{array}{c} * \\ \swarrow \quad \searrow \\ X_{12} \to * \leftarrow X_{13} \\ \nearrow \quad \uparrow \quad \nwarrow \\ * \longrightarrow X_{23} \longleftarrow * \end{array} \right)$$

$$\simeq \Omega X_{12} \times \Omega X_{13} \times \Omega X_{23}$$

$$F^2 = \operatorname{hofiber}(\operatorname{holim} X \to \operatorname{holim} X_{(2)}) \simeq \operatorname{holim} \left( \begin{array}{c} * \\ \swarrow \quad \searrow \\ * \to X_{123} \leftarrow * \\ \nearrow \quad \uparrow \quad \nwarrow \\ * \longrightarrow * \longleftarrow * \end{array} \right)$$

$$\simeq \Omega^2 X_{123}$$

An easy way to see this is to use Lemma 5.3.6, which expresses the homotopy limit of a puctured cube as an iterated homotopy pullback, along with Example 3.2.10 as well as the fact that homotopy limits commute with homotopy

fibers (as homotopy limits commute with themselves; see Proposition 8.5.5).
So we have

$$E^1_{0,q} = \pi_q(X_1 \times X_2 \times X_3) \cong \pi_q(X_1) \times \pi_q(X_2) \times \pi_q(X_3),$$
$$E^1_{1,q} = \pi_{q-1}(\Omega X_{12} \times \Omega X_{13} \times \Omega X_{23}) \cong \pi_q(X_{12}) \times \pi_q(X_{13}) \times \pi_q(X_{23}),$$
$$E^1_{2,q} = \pi_{q-2}(\Omega^2 X_{123}) \cong \pi_q(X_{123}).$$

In the more general case, one would "filter" the punctured cube $\mathcal{X}$ by punctured cubes $\mathcal{X}_{(i)}$ where all the spaces $X_S$ for $|S| > i$ are replaced by one-point spaces. This would result in a tower of fibrations of height $n$ with holim $\mathcal{X}$ as the initial space. The fibers of this tower are homotopy limits of cubes that have one-point spaces except for a fixed $|S|$ and this homotopy limit is the product $\prod_S \Omega^{|S|-1} X_S$.

We leave it to the reader to verify that the first differential is indeed the $d^1$ given in the statement of the proposition.

That this spectral sequence converges to holim $\mathcal{X}$ follows from the fact that this is an $n$-column spectral sequence (and is thus Mittag–Leffler; see Definition 9.6.5).                                                      $\square$

We can easily extend the previous result to total fibers of cubical diagrams.

**Proposition 9.6.12** (Homotopy spectral sequence of the total fiber)  *Let* $\mathcal{X}: \mathcal{P}(\underline{n}) \to \mathrm{Top}_*$ *be a cube of connected based spaces. Then there is an* $(n + 1)$-*column homotopy spectral sequence with* $E^1$ *term given by*

$$E^1_{p,q} = \prod_{\substack{S \subset \mathcal{P}(\underline{n}) \\ |S|=p}} \pi_q(X_S)$$

*with the first differential*

$$d^1 : \prod_{\substack{S \subset \mathcal{P}(\underline{n}) \\ |S|=p}} \pi_q(X_S) \longrightarrow \prod_{\substack{T \subset \mathcal{P}(\underline{n}) \\ |T|=p+1}} \pi_q(X_T)$$

*given in the same way as in Proposition 9.6.11. This spectral sequence converges to* tfiber($\mathcal{X}$).

*Proof*   The proof is the same as in Proposition 9.6.11, except all the punctured cubes filtering the original punctured cube encountered there are now cubes with the initial space $X_\emptyset$. The columns of the spectral sequence are obtained by taking total fibers. So we thus have, in the case $n = 3$,

$$F^0 = \text{tfiber} \begin{pmatrix} \begin{array}{c} X_\emptyset \longrightarrow * \\ \\ * \longrightarrow * \\ \\ * \longrightarrow * \end{array} \end{pmatrix} \simeq X_\emptyset$$

$$F^1 = \text{tfiber} \begin{pmatrix} \begin{array}{c} * \longrightarrow X_1 \\ X_2 \longrightarrow * \\ X_3 \longrightarrow * \\ * \longrightarrow * \end{array} \end{pmatrix} \simeq \Omega X_1 \times \Omega X_2 \times \Omega X_3$$

$$F^2 = \text{tfiber} \begin{pmatrix} \begin{array}{c} * \longrightarrow * \\ * \longrightarrow X_{12} \\ * \longrightarrow X_{13} \\ X_{23} \longrightarrow * \end{array} \end{pmatrix} \simeq \Omega^2 X_{12} \times \Omega^2 X_{13} \times \Omega^2 X_{23}$$

$$F^3 = \text{tfiber} \begin{pmatrix} \begin{array}{c} * \longrightarrow * \\ * \longrightarrow * \\ * \longrightarrow * \\ * \longrightarrow X_{123} \end{array} \end{pmatrix} \simeq \Omega^3 X_{123}$$

The last two total fibers have been computed using Examples 2.2.10 and 5.3.5, respectively. The fibers here are the loopings of the fibers from the previous proposition because total fiber is the homotopy fiber of the map from the initial space to the homotopy limit of the rest of the cube, which in these cases is the map from a point to the homotopy limits of punctured cubes from the previous proposition. Then we use that the homotopy fiber of $* \to X$ is $\Omega X$ (Example 2.2.9). □

We can also obtain analogous results for the spectral sequences computing the homology of the homotopy colimit of a punctured cube and the total cofiber of a cube using Proposition 9.6.10.

**Proposition 9.6.13** (Homology spectral sequence of a punctured cube)  *Let* $X\colon \mathcal{P}_1(\underline{n}) \to$ Top *be a punctured cube of spaces. Then there is an n-column homology spectral sequence whose* $E^1$ *term is*

$$E^1_{p,q} = \bigoplus_{\substack{S \subset \mathcal{P}_0(\underline{n}) \\ |S|=n-p-1}} \mathrm{H}_q(X_S)$$

*with the first differential*

$$d^1 \colon \bigoplus_{\substack{S \subset \mathcal{P}_0(\underline{n}) \\ |S|=n-p-1}} \mathrm{H}_q(X_S) \longrightarrow \bigoplus_{\substack{T \subset \mathcal{P}_0(\underline{n}) \\ |T|=n-p-2}} \mathrm{H}_q(X_T)$$

*whose restriction to the summand indexed by S is*

$$d^1|_{\mathrm{H}_q(X_S)} = \sum_{i \notin S} (-1)^{|S| \cdot i} f(S \to S \cup \{i\})_*.$$

*This spectral sequence converges to* $\mathrm{H}_*(\operatorname{hocolim} X)$.

*Proof*  Consider the sequence of punctured cubes $X_{(i)}$ which, for $|S| < i$, have one-point spaces and otherwise agree with $X$. Then, because homotopy colimits preserve cofibrations of diagrams by Theorem 8.5.1, there is a sequence of cofibrations

$$\operatorname{hocolim} X_{(n-1)} \longrightarrow \operatorname{hocolim} X_{(n-2)} \longrightarrow \cdots \longrightarrow \operatorname{hocolim} X_{(1)} \longrightarrow \operatorname{hocolim} X.$$

For $n = 3$, we get the cofibers

$$C^0 = \operatorname{hocolim} X_{(2)} \simeq \left( \begin{array}{c} X_{12} \\ \nearrow \quad \nwarrow \\ \emptyset \leftarrow \emptyset \rightarrow \emptyset \\ \swarrow \qquad \downarrow \qquad \searrow \\ X_{13} \leftarrow\!\!-\!\!- \emptyset -\!\!\!-\!\!\!\longrightarrow X_{23} \end{array} \right) \simeq X_{12} \amalg X_{13} \amalg X_{23}$$

$$C^1 = \operatorname{hocofiber}(\operatorname{hocolim} X_{(2)} \to \operatorname{hocolim} X_{(1)}) \simeq \operatorname{hocolim} \left( \begin{array}{c} * \\ \nearrow \quad \nwarrow \\ X_1 \leftarrow \emptyset \rightarrow X_2 \\ \swarrow \qquad \downarrow \qquad \searrow \\ * \leftarrow\!\!-\!\!- X_3 -\!\!\!-\!\!\!\longrightarrow * \end{array} \right)$$

$$\simeq \Sigma X_1 \amalg \Sigma X_2 \amalg \Sigma X_3$$

$$C^2 = \text{hocofiber}(\text{hocolim}\,\mathcal{X}_{(1)} \to \text{hocolim}\,\mathcal{X}) \simeq \text{hocolim} \left( \begin{array}{c} \ast \\ \ast \leftarrow X_\emptyset \to \ast \\ \ast \leftarrow \ast \to \ast \end{array} \right)$$

$$\simeq \Sigma^2 X_\emptyset$$

(For the based case, the disjoint unions are replaced by wedges.) An easy way to see that these are the correct homotopy colimits is to use Lemma 5.7.6, along with Examples 3.6.6 and 3.6.9 and the fact that homotopy colimits commute, Proposition 8.5.5.

In general, the cofibers are $C^i \simeq \coprod_{|S|} \Sigma^i X_S$ where $|S| = n - i - 1$. Then

$$E^1_{0,q} = H_q(X_{12} \amalg X_{13} \amalg X_{23}) \cong H_q(X_{12}) \oplus H_q(X_{13})\, H_q(X_{23}),$$

$$E^1_{1,q} = H_{q+1}(\Sigma X_1 \amalg \Sigma X_2 \amalg \Sigma X_3) \cong H_q(X_1) \oplus H_q(X_2) \oplus H_q(X_3),$$

$$E^1_{2,q} = H_{q+2}(\Sigma^2 X_\emptyset) \cong H_q(X_\emptyset).$$

We again leave it to the reader to verify that the differential $d^1$ is the one given in the statement of the proposition.

The convergence is due to the fact that this spectral sequence has finitely many columns.                                                                                    □

Finally, in analogy with Proposition 9.6.12, the spectral sequence from the previous result can be extended to total cofibers. We leave the details of the proof to the reader.

**Proposition 9.6.14** (Homology spectral sequence of the total cofiber) *Let* $X: \mathcal{P}(\underline{n}) \to \text{Top}$ *be a cube of spaces. Then there is an* $(n+1)$*-column homology spectral sequence whose* $E^1$ *term is*

$$E^1_{p,q} = \bigoplus_{\substack{S \subset \mathcal{P}(\underline{n}) \\ |S|=n-p}} H_q(X_S)$$

*with the first differential*

$$d^1: \bigoplus_{\substack{S \subset \mathcal{P}(\underline{n}) \\ |S|=n-p}} H_q(X_S) \longrightarrow \bigoplus_{\substack{T \subset \mathcal{P}(\underline{n}) \\ |T|=n-p-1}} H_q(X_T)$$

*given in the same way as in Proposition 9.6.13. This spectral sequence converges to* tcofiber($X$).

### 9.6.3 Homotopy spectral sequence of a cosimplicial space

The (Bousfield–Kan) homotopy spectral sequence for $X^\bullet$ is the spectral sequence for a pointed tower of fibrations from Section 9.6.1 as it applies to the totalization tower

$$\mathrm{Tot}^0 X^\bullet \longleftarrow \mathrm{Tot}^1 X^\bullet \longleftarrow \cdots.$$

from Proposition 9.1.17 (see Remark 9.1.23 for the pointed setup). We will examine only the first page of the homotopy spectral sequence for this tower and leave it to the reader to translate what has been developed so far for a general tower of fibrations to this special case.

Recall that $L^p X^\bullet$ stands for the (homotopy) fiber of the map from $\mathrm{Tot}^p X^\bullet$ to $\mathrm{Tot}^{p-1} X^\bullet$. Then, from (9.6.3), Proposition 9.1.19, and (9.1.12), the first page of the homotopy spectral sequence is given by

$$E^1_{p,q} = \pi_{q-p}(L^p X^\bullet) = \pi_{q-p}(\Omega^p N^p X^\bullet) = \pi_q(N^p X^\bullet) = \pi_q\left(X^{[p]} \cap \bigcap_{i=0}^{p-1} \mathrm{fiber}(s^i)\right)$$

(9.6.10)

It turns out that this can be rewritten in a way that is more amenable for computation.

**Proposition 9.6.15** ([BK72a, Ch. X, 6.2]; see also [GJ99, Ch. VIII, Lemma 1.8]) *There is an equivalence*

$$\pi_q\left(X^{[p]} \cap \bigcap_{i=0}^{p-1} \ker(s^i)\right) = \pi_q(X^{[p]}) \cap \bigcap_{i=0}^{p-1} \ker(s^i_*),$$

*where $s^i_* \colon \pi_q(X^{[p+1]}) \to \pi_q(X^{[p]})$ are the maps induced on the qth homotopy groups by the codegeneracies $s^i$ in $X^\bullet$.*

We can therefore write

$$E^1_{p,q} = \pi_q(X^{[p]}) \cap \bigcap_{i=0}^{p-1} \ker(s^i_*).$$

(9.6.11)

The expression on the right side of this equivalence is called the *pth normalization of the cosimplicial group* $\pi_* X^\bullet$, and is denoted by $N^p \pi_* X^\bullet$.[5] The statement of Proposition 9.6.15 is then simply that normalization commutes with homotopy groups:

$$\pi_*(N^p X^\bullet) \cong N^p \pi_*(X^\bullet).$$

(9.6.12)

---

[5] This is a functor from the category of cosimpicial spaces to the category of groups.

The normalization also allows us to see what the first differential $d^1$ is. In general, given a cosimplicial abelian group $G^\bullet$, taking the alternating sum of the cofaces $d^i$ gives a non-negatively graded cochain complex

$$\left(G^\bullet, \sum(-1)^i d^i\right).$$

A smaller (and usually better) complex is the *Dold–Kan normalization* $(NG^\bullet, \sum(-1)^i d^i)$, defined as above by intersecting with the kernels of codegeneracies. The *Dold–Kan correspondence* (see [Wei94, Section 8.4] or [GJ99, Ch. III, Theorem 2.5]) states that the inclusion

$$NG^\bullet \hookrightarrow G^\bullet \tag{9.6.13}$$

is a quasi-isomorphism, that is, it induces isomorphisms on cohomology.

Thus, given a cosimplicial space $X^\bullet$ we can as above form a cosimplicial abelian group $\pi_q(N^*X^\bullet)$ for each $q \geq 0$, if each $X^{[p]}$ is connected to an abelian fundamental group. Taking the alternating sum of the maps induced on homotopy by cofaces $d^i$ gives the cochain complex

$$\left(N^*\pi_q(X^\bullet), \sum(-1)^i d^i_*\right).$$

We can then take the cohomology $H^*$ of this complex (Bousfield and Kan [BK72a] call this the *cohomotopy* of the complex and denote it by $\pi^*$). If $X^\bullet$ is not connected to an abelian fundamental group, more care has to be taken (see [GJ99, p. 393]), but the following result still makes sense even in that case.

**Proposition 9.6.16** ([BK72a, Ch. X, §7]; see also [GJ99, Ch. VIII, Proposition 1.15]) *For $q \geq p \geq 0$, the $E^2$ term of the homotopy spectral sequence for $X^\bullet$ is given by*

$$E^2_{p,q} = H^q\left(N^*\pi_p(X^\bullet), d^1\right),$$

*where*

$$d^1 : E^1_{p,q} \longrightarrow E^1_{p+1,q}$$

$$\alpha \longmapsto \sum_{i=0}^{p+1} (-1)^i d^i_*(\alpha).$$

**Remark 9.6.17** One could take as the $E^1$ term the usual homotopy groups of $X^{[p]}$, in which case that page is just the cochain complex of the cosimplicial group $\pi_*(X^\bullet)$ if each $X^{[p]}$ is connected (i.e. just remove the intersections of the kernels in (9.6.10)). As mentioned above, intersecting with the kernel of the codegeneracies gives a more efficient way of constructing $E^1$, but whether $X^\bullet$ is normalized or not, the Dold–Kan correspondence (9.6.13) guarantees that the $E^2$ page will come out to be the same. A result along the same lines is Proposition 9.1.11. □

The homotopy spectral sequence for $\text{Tot}\,X^\bullet$ converges to $\pi_*(\text{Tot}\,X^\bullet)$ if any of the general conditions for complete convergence from Section 9.6.1 hold. In particular, we have from Corollary 9.6.8 that this is the case if the connectivity of the fibers $L^p X^\bullet$ grows with $p$. Note that Proposition 9.4.10 gives us a way of checking if this is the case, provided we have a handle on how cartesian the relevant cubical diagrams are.

**Remark 9.6.18** The case when $X^\bullet$ is not pointed was studied by Bousfield [Bou89], who develops obstruction theory for the problem of lifting the basepoint up the Tot tower. □

**Example 9.6.19** The homotopy spectral sequence associated to the $R$-completion of a simplicial set $X_\bullet$, $R^\bullet X$ (see Example 9.2.8), was constructed in [BK72b] and was what motivated Bousfield and Kan to define the more general homotopy spectral sequence for a cosimplicial space [BK73, BK72a]. They prove in [BK72b] that, under certain hypotheses (such as if $X$ is simply-connected), the homotopy spectral sequence for

- $\mathbb{Z}^\bullet X_\bullet$ converges to $\pi_*(|X_\bullet|)$;
- $\mathbb{Q}^\bullet X_\bullet$ converges to $\pi_*(|X_\bullet|) \otimes \mathbb{Q}$;
- $\mathbb{Z}_p^\bullet X_\bullet$ converges to $\pi_*(|X_\bullet|)/(\text{torsion prime to } p)$.

(See also [Rom10, Theorem 3].)

Because of (9.2.5), each column in the $E^1$ page can be computed as the homology of the previous one. It is in this sense that the Bousfield–Kan spectral sequence "goes from $R$-homology of $X_\bullet$ to homotopy of $X_\bullet$".

In the stable range, the Bousfield–Kan spectral sequence agrees with the Adams spectral sequence [BK72b, §5]. □

**Example 9.6.20** (Homotopy spectral sequence of a homotopy limit) Given a diagram $F \colon \mathcal{I} \to \text{Top}_*$, recall from Definition 9.3.1 that we can construct its cosimplicial replacement, $\Pi^\bullet F$, whose totalization is a model for its homotopy limit. If it converges, the homotopy spectral sequence for $\Pi^\bullet F$ then computes $\pi_*(\text{holim}_\mathcal{I} F)$. From (9.6.11), we then have

$$E_{p,q}^1 = \pi_q\left(\prod_{I_p \in \text{nd}(S_p)} F(i_p)\right) = \prod_{I_p \in \text{nd}(S_p)} \pi_q(F(i_p)),$$

where nd($S_p$) is the set of non-degenerate composable strings of length $p$ (those that do not contain identity maps). The degenerate strings have been removed since that is precisely the result of intersecting with the kernel of codegeneracies as in (9.6.11).

When $F$ is a based punctured $n$-cubical diagram, this means that the $E^1$ page is given as

$$E^1_{p,q} = \prod_{\substack{S \subset \mathcal{P}_0(\underline{n}) \\ |S|=p+1}} \pi_q(X_S).$$

This recovers the spectral sequence for the homotopy groups of the homotopy limit of a punctured cube from Proposition 9.6.11.       □

One interesting example that is outside the scope of this book is the spectral sequence for the two-sided cobar construction from Example 9.2.5. For the interested reader, a good discussion and an overview of literature on this problem is given in [KT06, Section 5.6].

Another example of a homotopy spectral sequence associated to a cosimplicial space will be given in Section 10.4.4.

### 9.6.4 Homology spectral sequence of a cosimplicial space

The homology spectral sequence for $X^\bullet$ is the usual second quadrant spectral sequence associated to a double complex.

The double complex, that is, the $E^0$ term of the homology spectral sequence, is obtained by applying the chains (with any coefficients) to $X^\bullet$. This produces, for any $q \geq 0$, a cosimplicial abelian group $C_q(X^\bullet)$. However, as in the homotopy spectral sequence case, we can use the Dold–Kan normalization functor and the quasi-isomorphism (9.6.13) to instead use $N^* C_q(X^\bullet)$ and obtain a cochain complex

$$\left(N^* C_q(X^\bullet), \sum(-1)^i d^i_*\right).$$

Now the maps $d^i_*$ are those induced on chains by the cofaces $d^i$ in $X^\bullet$. It is immediate from the cosimplicial identities that the composition of the alternating sum of the $d^i_*$ with itself is indeed zero. This differential serves as the horizontal one in the $E^0$ page.

One of course also has, for each $p$, the usual chain complex $(C_*(X^{[p]}), \partial)$, where $\partial$ is the usual differential on the chains. This differential restricts to the subcomplex $N^p C_*(X^\bullet) \hookrightarrow C_*(X^{[p]})$ and we thus get the vertical complexes in $E^0$ for each $p$,

$$(N^p \, C_*(X^\bullet), \partial).$$

Thus

$$E^0_{*,*} = N^* \, C_*(X^\bullet)$$

with the differentials as described above.

To construct the rest of the spectral sequence we proceed in the usual way when the starting point is a double complex by filtering vertically (in the $q$ direction) or horizontally (in the $p$ direction). That is, we form the *total complex* $T(E^0_{*,*})$ of the double complex $E^0_{*,*}$ which in degree $n$ is given by

$$T(E^0_{*,*})_n = \prod_{i \geq 0} N^{n+i} \, C_i(X^\bullet).$$

Notice that that $n$ need not be non-negative so this is a $\mathbb{Z}$-graded complex (and we naturally assume $N^{<0} \, C_i(X^\bullet) = 0$). The differential $\partial_T$ in $T(E^0_{*,*})$ is the sum of the vertical and the horizontal differentials in $E^0$:

$$\partial_T = \partial + (-1)^i d^i_*.$$

This total complex is filtered by subcomplexes $F^p T(E^0_{*,*})$, $p \geq 0$, which in degree $n$ are given by

$$F^p T(E^0_{*,*})_n = \prod_{i \geq p} N^{n+i} \, C_i(X^\bullet).$$

Each quotient

$$T_p(E^0_{*,*}) = T(E^0_{*,*})/F^p T(E^0_{*,*})$$

is thus a complex which can be thought of as the total complex of the double complex obtained from $E^0_{*,*}$ by replacing everything to the left of the $p$th column by zero. In particular, $T_p(E^0_{*,*})$ is zero in degrees $< -p$.

There is an evident surjection

$$T_p(E^0_{*,*}) \longrightarrow T_{p-1}(E^0_{*,*})$$

since $T_p(E^0_{*,*})$ is built out of a double complex that has one more non-zero column than the double complex for $T_{p-1}(E^0_{*,*})$ and the two are otherwise identical, so we get a tower of chain complexes

$$\left\{ T_p(E^0_{*,*}) \right\}_{p \geq 0}, \tag{9.6.14}$$

and this gives rise to the homology spectral sequence of $X^\bullet$.

The $E^1$ page is obtained by taking the homology with respect to the vertical differential, i.e.

$$E^1_{p,q} = \mathrm{H}_{q-p}\left(F^p T(E^0_{*,*})/F^{p+1}T(E^0_{*,*})\right) = N^p \, \mathrm{H}_q(X^\bullet) = \mathrm{H}_q(X^{[p]}) \cap \bigcap_{i=0}^{p-1} \ker(s^i_*).$$
$$(9.6.15)$$

This is of course completely analogous to the homotopy spectral sequence $E^1$ term from (9.6.11).

The $d^1$ differential is the horizontal differential in $E^0$, but now the alternating sum is taken over maps induced by the $d^i$ on homology (rather than chains). We thus get the following analog of Proposition 9.6.16.

**Proposition 9.6.21** ([Bou87, Section 2.1]) *For $q, p \geq 0$, The $E^2$ term of the homology spectral sequence for $X^\bullet$ is given by*

$$E^2_{p,q} = \mathrm{H}^q\left(N^* \, \mathrm{H}_p(X^\bullet), d^1\right)$$

*where*

$$d^1 : E^1_{p,q} \longrightarrow E^1_{p+1,q}$$

$$\beta \longmapsto \sum_{i=0}^{p+1}(-1)^i d^i_*(\beta).$$

The remainder of the spectral sequence is given in the usual way from the filtration $F^p T(E^0_{*,*})$, $p \geq 0$, with differentials

$$d^r : E^r_{p,q} \longrightarrow E^r_{p+r,q+r-1}.$$

**Remark 9.6.22** All of the above can be done by filtering horizontally rather than vertically, that is, truncating $N^* \, \mathrm{C}_*(X^\bullet)$ with horizontal lines. See [Goo98, Section 4] for more details about how these two ways are related. □

**Remark 9.6.23** Since we do not require $q \geq p$, it is possible that the $E^\infty$ page could have terms in negative total degree. Goodwillie [Goo98] has shown that, over $\mathbb{Z}_p$, the spectral sequence vanishes in negative dimensions. He also gives examples to demonstrate that this may not be true over $\mathbb{Z}$ or $\mathbb{Q}$. □

The issue of the convergence of the homology spectral sequence for $X^\bullet$ is more sensitive than that of the homotopy spectral sequence. Philosophically, this is because cosimplicial spaces and totalizations go with homotopy limits, and homotopy limits go with homotopy groups (essentially since there is a long exact sequence of homotopy groups for homotopy fibers). Bousfield [Bou87] and Shipley [Shi96] have used the tower from (9.6.14), as well as an exact couple that can be extracted from it, as a starting point for analyzing convergence (for any coefficients, but have obtained most results for $\mathbb{Z}/p$ and related coefficients). It would take us too far afield to discuss this in detail, and we will instead just state one of the most useful sufficient conditions for convergence.

**Theorem 9.6.24** ([Bou87, part of Theorem 3.2]) *Suppose the homology spectral sequence for $X^\bullet$ satisfies the following:*

1. *Each $X^{[p]}$ is simple (connected, has abelian fundamental group, and fundamental group acts trivially on higher homotopy groups; in particular, if $X^{[p]}$ has trivial fundamental group, it is simple).*
2. $E^r_{p,q} = 0$ *for $q \leq p$.*
3. *For each $i \geq 1$, there exist finitely many $p$ such that $E^r_{p,p+i} = 0$.*

*Then the spectral sequence converges completely, that is, it converges to $H_*(\operatorname{Tot} X^\bullet)$ (for coefficients in any ring).*

Otherwise, the best that can be said is that, if the spectral sequence converges, it converges to $H_*(\operatorname{Tot} C_* X^\bullet)$, that is, the homology of the totalization (in the category of bicomplexes; see e.g. [Wei94, Section 1.2]) of the cosimplicial chain complex obtained by applying chains to $X^\bullet$. One then has a canonical map

$$H_*(\operatorname{Tot} X^\bullet) \longrightarrow H_*(\operatorname{Tot} C_* X^\bullet), \tag{9.6.16}$$

which may or may not be an isomorphism. More about the issue of what the spectral sequence converges to (i.e. about the *exotic convergence*) can be found in [Shi96, Section 7].

**Remark 9.6.25** Everything discussed here can be dualized in order to obtain the (second quadrant) *co*homology spectral sequence of $X^\bullet$. For example, its $E_1$ page is

$$E^{p,q}_1 = H^q(X^{[p]}) \Big/ \left( \sum_{i=0}^{p} \operatorname{im}((s^i)^*) \right), \tag{9.6.17}$$

which can also be written as

$$E^{p,q}_1 = \operatorname{coker}\left( \sum_{i=0}^{p-1} \operatorname{im}((s^i)^* \colon H^q(X^{[p-1]}) \longrightarrow H^q(X^{[p]}) \right). \tag{9.6.18}$$

The first differential is again induced by the sum of cofaces (restricted to this cokernel):

$$d^1 = \sum_{i=0}^{p+1} (-1)^i (d^i)^* : E_1^{p,q} \longrightarrow E_1^{p+1,q}.$$

We leave it to the reader to fill in the details. □

**Example 9.6.26** (Homology spectral sequence of a homotopy limit) Just as in Example 9.6.20, given a diagram $F : I \to \text{Top}$ we can apply the homology spectral sequence to the cosimplicial model for its homotopy limit, $\Pi^\bullet F$. If it converges, the homology spectral sequence thus computes $H_*(\text{holim}_I F)$. □

**Example 9.6.27** Recall Example 9.2.3 and set $X = S^n$, $n \geq 1$. Since the generators (integrally) of the products of spheres are pulled back from a single sphere, most of the terms in the $E^0$ page of the spectral sequence for the cosimplicial model of $\Omega S^n$ are degenerate, and the spectral sequence has a particularly simple form, from which it is immediate, using Theorem 9.6.24, that the spectral sequence collapses. It is a simple exercise in chasing the differentials to see that we ultimately get

$$H_i(\Omega S^n) = \begin{cases} \mathbb{Z}, & (n-1)|i; \\ 0, & \text{otherwise.} \end{cases}$$

□

Another example of a homology spectral sequence associated to a cosimplicial space will be given in Section 10.4.4.

We close this section with some comments that apply to both the homotopy and the homology spectral sequences.

In the spirit of Example 9.3.11, we first have an unsurprising result about spectral sequences for truncations of cosimplicial spaces.

Recall the definition of a truncation $X^{[\leq n]}$ of $X^\bullet$ (Definition 9.1.13). For a chain complex $K$, let $K_{\leq n}$ be the *nth truncation of $K$*, the complex which agrees with $K$ up to degree $n$ but is zero after that.

**Proposition 9.6.28** ([LTV10, Lemma 4.2 and Proposition 4.3]) *Suppose the homotopy and homology spectral sequences for $X^\bullet$ converge strongly (so they converge to $\pi_*(\text{Tot}\, X^\bullet)$ and $H_*(\text{Tot}\, X^\bullet)$). Then (compare with Propositions 9.6.16 and 9.6.21) the following hold.*

1. *The homotopy and homology spectral sequences for $\Pi^\bullet X^{[\leq n]}$ have as their $E^2$ terms*

$$E_{p,q}^2 = H^q \left( N_{\leq n}^* \pi_p(X^\bullet) \right) \text{ and } E_{p,q}^2 = H^q \left( N_{\leq n}^* H_p(X^\bullet) \right).$$

2. *These spectral sequences converge strongly to $\pi_*(\text{Tot}^n X^\bullet)$ and $H_*(\text{Tot}^n X^\bullet)$.*

**Remark 9.6.29**   Lemma 4.2 in [LTV10] proves statement 1 for the homotopy spectral sequence, but it is not hard to also deduce the same for the homology spectral sequence. In addition, the statement of Proposition 4.3 in [LTV10] demonstrates part 2 for spectral sequences that are "above the diagonal", but the generalization to the hypothesis where any kind of strong convergence is assumed is also not difficult to see.                                       □

Recall the notion of a homotopy initial functor from Definition 8.6.2. Another result concerning spectral sequences of cosimplicial replacements is the following natural analog of Theorem 8.6.5.

**Proposition 9.6.30** ([LTV10, Proposition 4.5])   *Suppose* $G \colon I' \to I$ *is a functor between finite categories and* $F \colon I \to \mathrm{Top}_*$ *is a diagram of based spaces. If $G$ is homotopy initial then the homotopy and rational homology spectral sequences associated to $\Pi^\bullet F$ and $\Pi^\bullet(F \circ G)$ have isomorphic $E^2$ (and subsequent) pages.*

The last comment is that the homotopy and homology spectral sequences can be compared via a Hurewicz map. Namely, the Hurewicz maps

$$\pi_q(X^{[p]}) \longrightarrow \mathrm{H}_q(X^{[p]}), \quad q \geq 1, \; p \geq 0$$

induce a map from the homotopy spectral sequence for $X^\bullet$ to the homology spectral sequence for $X^\bullet$. In particular, on $E^1$ it will simply be the Hurewicz map restricted to the normalization

$$N^p \pi_q(X^\bullet) \longrightarrow N^p \mathrm{H}_q(X^\bullet).$$

The sequence of these maps will converge to the Hurewicz map

$$\pi_{q-p}(\mathrm{Tot}\, X^\bullet) \longrightarrow \mathrm{H}_{q-p}(\mathrm{Tot}\, X^\bullet).$$

For more details, see [Bou89, Sections 2.7 and 10.8].

# 10

# Applications

This chapter is meant to be a brief account of some of the recent developments and results that utilize some of the techniques developed in this book. As this is meant to be an overview, many details and proofs have been omitted, but ample references for further reading have been supplied. The central application is to introduce the calculus of functors, and we present two of its flavors – homotopy and manifold calculus – in Sections 10.1 and 10.2. Section 10.3 is an application of manifold calculus to spaces of embeddings, and Section 10.4 is an account of how manifold calculus, in combination with cosimplicial spaces and their spectral sequences, provides information about spaces of knots.

One important application that we did not have space to include is the *Lusternik–Schnirelmann category*. Some of the main results in that theory use Ganea's Fiber-Cofiber Construction (Proposition 4.2.14), Mather's Cube Theorems (Theorems 5.10.7 and 5.10.8) and other cubical techniques developed in this book. For more details, the reader should consult [CLOT03].

## 10.1 Homotopy calculus of functors

This and the next section are devoted to a brief outline of the calculus of functors, an organizing principle in topology which takes some inspiration from Taylor series in ordinary calculus. Our focus will be narrow, only briefly describing two flavors, known as "homotopy calculus" and "manifold calculus", but we will pay special attention to how cubical diagrams play an important role in each of these theories.

We will not attempt to answer the very general question of what a calculus of functors is, but a few philosophical remarks are in order. Given a functor $F: C \to \mathcal{D}$, the general idea is to approximate $F$ by a sequence of functors $T_k F: C \to \mathcal{D}$ which are "polynomial of degree $k$", and with natural transformations $F \to T_k F$ and $T_k F \to T_{k-1} F$ compatible in the obvious way. These functors and natural transformations form a "Taylor tower" for $F$, the analog of the Taylor series of a function $f \colon \mathbb{R} \to \mathbb{R}$. We are typically interested in the homotopy type of the values of the functor $F$, so the category $\mathcal{D}$ should be one in which we have a reasonable notion of homotopy theory. Model categories are the usual setting for that, but, as we have done throughout this book, we will stick to the case where $\mathcal{D}$ is the category of topological spaces or spectra for concreteness (all the necessary background material on spectra can be found in Section A.3).

In order for the polynomial approximations $T_k F$ to be useful, their homotopy type should be easier to compute than that of $F$. Moreover, they should be classifiable in some way. That is, the theory should compute something for free. In both of our examples, this means having a classification of "homogeneous" functors of degree $k$. We also hope that the homotopy type of the approximations $T_k F$ is related to that of $F$. If we are lucky, the map $F \to T_k F$ will be highly connected, and this connectedness will increase with $k$, that is, the Taylor tower will converge. However, this is not strictly necessary for the functors $T_k F$ to be useful, as will be illustrated in the applications of manifold calculus to knots in Section 10.4. In the next two sections, we will thus focus on three topics: the polynomial functors and polynomial approximations, the classification of homogeneous functors, and convergence.

For parallels with ordinary calculus, we will find that the difference $f(x + h) - f(x)$ appearing in the definition of the derivative of a function $f \colon \mathbb{R} \to \mathbb{R}$ has an analog – the homotopy fiber of a map between values of the functor $F$. We will also see that the degree $k$ part of the Taylor series for $f$, $f^{(k)}(0) x^k / k!$, has an analog in the description of the homogeneous layers of the Taylor tower for $F$, and whether some input for $f$ is within the radius of convergence of its Taylor series has an analog in terms of the connectivity of the input space (homotopy calculus) or handle index of the input (manifold calculus).

There are a few glaring omissions in our treatment of calculus of functors. For example, applications of homotopy calculus are largely omitted. To name a few, homotopy calculus has been used to study Waldhausen's functor $A$ [CCGH87, Goo92, Ogl13], its application to the identity functor has yielded information about the unstable homotopy groups from the stable ones [AD01, AM99, Joh95], the Taylor tower of the mapping space $\mathrm{Map}(K, X)$, where $K$ is a complex, has provided stable homotopy information about that space [AK02, Aro99], [Kuh04] exhibits a close relation between the Taylor tower and the chromatic filtration, and [Beh12] does the same for the EHP sequence. The reason a detailed treatment of these topics is omitted (we will only mention some of them in passing) is simply because to include them would require a significant investment into the machinery of stable homotopy theory and model categories, both of which go beyond the intended scope of this book. Instead, we have opted to present two applications of manifold calculus (Sections 10.3 and 10.4) which are closer in spirit to the techniques employed throughout this book.

Another topic we have not covered is the third brand of functor calculus currently in existence, namely *orthogonal calculus*, due to Weiss [Wei95]. While this theory has also resulted in interesting applications [Aro02, ALV07, Mac07, MW09], it is not based on cubical diagrams to the extent that the other two are, and hence does not fit thematically here.

We now turn to discussing the homotopy calculus of functors. The main references we draw from are Goodwillie's foundational papers [Goo92, Goo03] (see also [Kuh07] for a nice overview of the theory). Further developments of the theory can be found in [AC11, AC14, AC15, BCR07, Chi05, Chi10, Cho, Joh95, JM99, JM04, JM08, KM02, KR02, LM12, McC01a].

Homotopy calculus is concerned with functors $F: C \to D$, where $C$ is $\mathrm{Top}_*$ or $\mathrm{Top}_Y$, and $D$ is $\mathrm{Top}_*$ or Spectra. Here $\mathrm{Top}_Y$ is the category of spaces over a fixed space $Y$ where all the objects come with a map to $Y$ and all the morphisms commute with these maps. The source category will throughout this section mostly be $\mathrm{Top}_*$ and we will make some comments about $\mathrm{Top}_Y$ at the end of Section 10.1.3.

We assume our functors satisfy the following axioms.

**Definition 10.1.1**

1. A functor $F$ is a *homotopy functor* if whenever $X \to Y$ is a weak equivalence, the induced map $F(X) \to F(Y)$ is a weak equivalence.
2. A functor $F$ is *finitary* if, given a diagram $I \to C$, $i \mapsto X_i$, where $I$ is filtered (see Definition 7.5.4), the canonical map

$$\operatorname*{hocolim}_{i \in I} F(X_i) \longrightarrow F\left(\operatorname*{hocolim}_{i \in I} X_i\right)$$

   is a weak equivalence.

The condition that the functor is finitary means that it is determined by its values on finite complexes. If we do without this axiom, we could choose to define the value of our functor on infinite complexes via the homotopy colimit above, since every space can be written as a filtered homotopy colimit of finite complexes. We will generally ignore the finitary axiom, although it is crucial to the classification of homogeneous functors; see the discussion leading to Theorem 10.1.48.

We will also sometimes assume $F$ is *reduced*, which means that if $X$ is weakly equivalent to a point, then $F(X)$ is weakly equivalent to a point. If the codomain of $F$ is based spaces, then the functor $X \mapsto \text{hofiber}(F(X) \to F(*))$ is reduced for any $F$.

**Example 10.1.2** For $X$ a (based) space and $\underline{C}$ a spectrum, the following are homotopy functors:

- $X \longmapsto X^n, n \geq 1$;
- $X \longmapsto \Sigma^\infty X$;
- $X \longmapsto \Omega^\infty \Sigma^\infty X$ (see Example 10.1.10);
- $X \longmapsto \text{Map}(K, X)$, where $K$ is a finite complex (see Remark 1.3.6);
- $X \longmapsto \underline{C} \wedge \Omega^\infty X_+$;
- $\underline{C} \longmapsto \Sigma^\infty \Omega^\infty \underline{C}$. $\qquad\qquad\square$

### 10.1.1 Polynomial functors

In this section we define and discuss the "polynomial approximations" of homotopy functors. Recall the notions of homotopy cartesian (Definition 5.4.1) and strongly homotopy cocartesian (Definition 5.8.18) cubes.

**Definition 10.1.3** We call a homotopy functor $F \colon C \to \mathcal{D}$ $k$-*excisive*, or *polynomial of degree* $\leq k$, if it takes strongly homotopy cocartesian $(k + 1)$-cubes $\mathcal{X}$ to homotopy cartesian $(k + 1)$-cubes $F(\mathcal{X})$.

**Proposition 10.1.4** *If $F$ is $k$-excisive, then it is $l$-excisive for $l \geq k$.*

*Proof* It suffices to argue this for $l = k + 1$. Regard a $(k + 2)$-cube $\mathcal{Z}$ as a map of $k$-cubes $\mathcal{X} \to \mathcal{Y}$. Since the original cube is strongly homotopy cocartesian, so are $\mathcal{X}$ and $\mathcal{Y}$. Since $F$ is $k$-excisive, $F(\mathcal{X})$ and $F(\mathcal{Y})$ are homotopy cartesian. But then so is $F(\mathcal{Z}) = F(\mathcal{X} \to \mathcal{Y}) = F(\mathcal{X}) \to F(\mathcal{Y})$ by 1(b) of Proposition 5.4.13. $\qquad\qquad\square$

**Remark 10.1.5** The reason the definition of a $k$-excisive functor uses strongly homotopy cocartesian cubes rather than just cocartesian ones is because the former is necessary for the previous result to be true. $\qquad\qquad\square$

**Example 10.1.6** A 0-excisive functor $F$ takes all maps to weak equivalences and is therefore, in a sense, constant. $\qquad\qquad\square$

A 1-excisive functor, sometimes also just called *excisive* or, if it is also reduced, *linear*, is one which takes homotopy cocartesian squares to homotopy cartesian squares, that is, it satisfies the excision axiom.

**Example 10.1.7**  Consider the infinite symmetric product $X \mapsto SP(X)$, where $X$ is based (see Definition 2.7.22). Certainly SP is a homotopy functor, and the Dold–Thom Theorem, Theorem 5.10.5, says that SP is 1-excisive.                                        □

**Example 10.1.8**  The identity functor from spectra to spectra is 1-excisive (and hence $k$-excisive for any $k > 1$). This is precisely the content of Proposition A.3.13. The identity functor from spaces to space is *not* 1-excisive. For example, the square

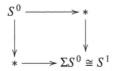

is cocartesian but not cartesian because the map $S^0 \to \mathrm{holim}(* \to S^1 \leftarrow *) \simeq \Omega \Sigma S^0$ is not an equivalence ($S^0$ has two components but $\Omega \Sigma S^0$ has countably many components).                                        □

**Example 10.1.9**  Let $\underline{C}$ be a spectrum. The functors $X \mapsto \underline{C} \wedge X$ and $X \mapsto \Omega^\infty(\underline{C} \wedge X)$ are 1-excisive (and hence $k$-excisive for any $k > 1$). In some sense these are the universal examples of 1-excisive functors from spaces to spectra or from spaces to spaces; see Theorem 10.1.48.

To see why these are 1-excisive, let

$$
\begin{array}{ccc}
X_\emptyset & \longrightarrow & X_1 \\
\downarrow & & \downarrow \\
X_2 & \longrightarrow & X_{12}
\end{array}
$$

be a homotopy pushout square. By Proposition A.3.10 we may also assume $\underline{C}$ is an $\Omega$-spectrum, so that the canonical maps $C_i \to \Omega C_{i+1}$ are weak equivalences for all $i$. Note that if $C_i$ is $n$-connected, then by statement 1 of Example 4.2.11, the canonical map $\Sigma C_i \to C_{i+1}$ is $(2n + 3)$-connected. Since $C_1$ is non-empty (our spectra are sequences of based spaces), it is $(-1)$-connected and hence $\Sigma C_1 \to C_2$ is a 1-connected map from the 0-connected space $\Sigma C_1$ to $C_2$, and it follows that $C_2$ is 0-connected. By induction, $C_n$ is $(n-1)$-connected. Thus if $X$ is non-empty, $C_n \wedge X$ is also $(n-1)$-connected by Proposition 3.7.23. Further, since smashing with a space preserves homotopy pushouts by Corollary 3.7.19, the square

$$C_n \wedge X_\emptyset \longrightarrow C_n \wedge X_1$$

$$\downarrow \qquad\qquad \downarrow$$

$$C_n \wedge X_2 \longrightarrow C_n \wedge X_{12}$$

is a homotopy pushout. The two initial maps $C_n \wedge X_\emptyset \to C_n \wedge X_i$ are $(n-1)$-connected since they are maps of $(n-1)$-connected spaces. By the Blakers–Massey Theorem, Theorem 4.2.1, this square is $(2n-3)$-cartesian. Since applying $\Omega^n$ decreases connectivity by $n$, and since $\Omega^n$ commutes with homotopy limits by iterated application of Corollary 3.3.16, the square

$$\Omega^n(C_n \wedge X_\emptyset) \longrightarrow \Omega^n(C_n \wedge X_1)$$

$$\downarrow \qquad\qquad\qquad \downarrow$$

$$\Omega^n(C_n \wedge X_2) \longrightarrow \Omega^n(C_n \wedge X_{12})$$

is $(n-3)$-cartesian, again by Blakers–Massey. Letting $n$ go to infinity in the above two squares gives the desired result. □

**Example 10.1.10** Since $\mathbb{S} \wedge X = \Sigma^\infty X$ (see Example A.3.9), where $\mathbb{S}$ is the sphere spectrum from Example A.3.4, a special case of the previous example is that the functor $X \mapsto \Sigma^\infty X$ is 1-excisive.

Now recall the functor $QX = \Omega^\infty \Sigma^\infty \colon \mathrm{Top}_* \to \mathrm{Top}_*$ from (A.7). This functor is 1-excisive. This again follows from Example 10.1.9 since $QX = \Omega^\infty(\mathbb{S} \wedge X)$. □

**Remark 10.1.11** One place where the analogy between functor calculus and ordinary calculus breaks down is that the composition of 1-excisive functors is not necessarily 1-excisive. For example, even though functors $\underline{C} \mapsto \Omega^\infty \underline{C}$ and $X \mapsto \Sigma^\infty X$ are 1-excisive, the composition $\underline{C} \mapsto \Sigma^\infty \Omega^\infty \underline{C}$ is not. □

We can generalize Example 10.1.9 to construct functors which satisfy higher-order excision, but we first need to develop a couple of tools which make this easier. This will give us a chance to use some of the material on cubical diagrams and more general homotopy limits we have developed earlier in this book.

**Proposition 10.1.12** ([Goo92, Proposition 3.3]) *Suppose $F$ is $n$-excisive and $X = S \mapsto X_S$ is a strongly homotopy cocartesian $m$-cube. Then the canonical map*

$$F(X_\emptyset) \longrightarrow \underset{|\underline{m}-S| \leq n}{\mathrm{holim}}\, F(X_S)$$

*is a weak equivalence.*

*Proof* If $m \leq n$, then the indexing category for the homotopy limit has $\emptyset$ as an initial object, and by Example 7.3.9, the homotopy limit is weakly equivalent to $F(X_\emptyset)$. If $m = n + 1$, then by definition of $n$-excisive the result is true. For $m > n + 1$ we induct on $m$. Define an $m$-cube $\mathcal{Y} = S \mapsto Y_S$ by the formula

$$Y_S = \operatorname*{holim}_{T \supset S, |\underline{m} - T| \leq n} F(X_T).$$

There is a natural map of cubes $F(\mathcal{X}) \to \mathcal{Y}$ given by composing the canonical map

$$F(X_S) \longrightarrow \operatorname*{holim}_{T \supset S} F(X_T)$$

(a weak equivalence since the indexing category has $S$ as an initial object) with the evident restriction map of homotopy limits. We are to prove that $F(X_\emptyset) \to Y_\emptyset$ is a weak equivalence. By induction, $F(X_S) \to Y_S$ is a weak equivalence for all $S \neq \emptyset$ (the inductive argument requires the case $m = n + 1$ as well). By Theorem 5.3.9 and the definition of homotopy cartesian, it is enough to show that $F(\mathcal{X})$ and $\mathcal{Y}$ are homotopy cartesian. For $F(\mathcal{X})$ this is Proposition 10.1.4. For $\mathcal{Y}$ this follows from Proposition 8.6.7 for $\mathcal{I} = \mathcal{P}(\underline{m})$ covered by the collection of subcategories $\{\mathcal{I}_j\}_{j=1}^m$, where $\mathcal{I}_j = \{T \subset \underline{m} : j \in T, |\underline{m} - T| \leq n\}$.   $\square$

The next result concerns functors of more than one variable, and gives a way to build functors satisfying higher-order excision from multivariable functors satisfying lower-order excision in each variable.

**Proposition 10.1.13** ([Goo92, Proposition 3.4])   *Suppose $F : C^k \longrightarrow \mathcal{D}$ is $d_i$-excisive in the ith variable. If $\Delta_k : C \to C^k$ denotes the diagonal inclusion, then the composition $F \circ \Delta_k$ is $\sum_i d_i$-excisive.*

*Proof* Let $\mathcal{X} = S \mapsto X_S$ be a strongly homotopy cocartesian $n$-cube, where $n > d$. Define an $n$-cube $\mathcal{Y} = S \mapsto Y_S$ by

$$Y_S = \operatorname*{holim}_{T_i \supset S, |\underline{m} - T_i| \leq d_i} F(T_1, \ldots, T_k).$$

As in the previous proof, there is an evident natural transformation of cubes $F \circ \Delta(\mathcal{X}) \to \mathcal{Y}$ which is a weak equivalence for all $S$ by Proposition 10.1.12 used $k$ times, once in each variable. It is enough now to prove that $\mathcal{Y}$ is homotopy cartesian. This follows from Proposition 8.6.7 with $\mathcal{I} = \{(T_1, \ldots, T_k) : |\underline{m} - T_i| \leq d_i \text{ for all } i\}$, which is covered by the collection $\{\mathcal{I}_j\}_{j=1}^k$, where $\mathcal{I}_j = \{(T_1, \ldots, T_k) : j \in T_i, |\underline{m} - T_i| \leq d_i \text{ for all } i\}$.   $\square$

**Example 10.1.14**   For any spectrum $\underline{C}$ the functors $X \mapsto \underline{C} \wedge (X_+)^{\wedge k}$ and $X \mapsto \Omega^\infty(\underline{C} \wedge (X_+)^{\wedge k})$ are $k$-excisive. This is because these are functors in $k$ variables that are 1-excisive in each variable by Example 10.1.9. But Proposition 10.1.13 says that if $F\colon C^k \longrightarrow \mathcal{D}$ is 1-excisive in each variable, it is $k$-excisive.

As a special case, the functors $X \mapsto \Sigma^\infty X^{\wedge k}$ and $X \mapsto \Omega^\infty \Sigma^\infty X^{\wedge k}$ are polynomial of degree $k$. □

We close with one more property of excisive functors and its useful consequence.

**Proposition 10.1.15** *Suppose $\mathcal{I}$ is a small category and we have an $\mathcal{I}$-diagram of $k$-excisive functors $F_i$, $i \in \mathcal{I}$ (so this is a functor from $\mathcal{I}$ to functors from $C$ to $\mathcal{D}$). Then $\mathrm{holim}_{i \in \mathcal{I}} F_i$ is also $k$-excisive.*

*Proof* If $\mathcal{X}$ is a strongly homotopy cocartesian $(k + 1)$-cube, then by assumption each of the functors $F_i$ has the property that $F_i(\mathcal{X})$ is homotopy cartesian. This means $F_i(\mathcal{X}(\emptyset) \to \mathrm{holim}_{S \neq \emptyset} F_i(\mathcal{X}(S))$ is a weak equivalence. Let $F = \mathrm{holim}_{\mathcal{I}} F_i$. We want to know whether the map $F(\mathcal{X}(\emptyset)) \to \mathrm{holim}_{S \neq \emptyset} F(\mathcal{X}(S))$ is a weak equivalence. Unraveling, this becomes the map

$$\mathrm{holim}_{\mathcal{I}} F_i(\mathcal{X}(\emptyset)) \longrightarrow \mathrm{holim}_{S \neq \emptyset} \mathrm{holim}_{\mathcal{I}} F_i(\mathcal{X}(S)),$$

and the last space is homeomorphic to $\mathrm{holim}_{\mathcal{I}} \mathrm{holim}_{S \neq \emptyset} F_i(\mathcal{X}(S))$ by Proposition 8.5.5. Taking the homotopy limit over $S \neq \emptyset$ gives something weakly equivalent to $F_i(\mathcal{X}(\emptyset))$ for each $i$, so by Theorem 8.3.1 we have the desired result. □

Since the fiber of a fibration can be thought of as a homotopy limit, from the previous result we immediately get the following.

**Corollary 10.1.16** *Let $G \to H$ be a natural transformation of $k$-excisive functors from $C \to \mathrm{Top}_*$. Then $F = \mathrm{hofiber}(G \to H)$ is $k$-excisive.*

*Proof* Let $\mathcal{I} = \mathcal{P}_0(\underline{2})$, and $\Phi\colon \mathcal{P}_0(\underline{2}) \to \mathrm{Top}_*^C$ be defined by $\Phi(\{1\}) = G$, $\Phi(\{1, 2\}) = H$, and $\Phi(\{2\}) = C_*$, the constant functor at a point. The natural transformations $\Phi(\{1\} \to \{1, 2\})$ and $\Phi(\{2\} \to \{1, 2\})$ are the given one and the inclusion of the basepoint, respectively. □

## 10.1.2 The construction of the Taylor tower

We now turn to the construction of polynomial approximations $P_k F$ of a homotopy functor $F$. This is done in two steps. The first is to create a new functor $T_k F$ which is in a sense closer to being $k$-excisive, and the second is to iterate this process to create a $k$-excisive functor $P_k F$. The meaning of "closer" in this context will be explained below.

**Definition 10.1.17**    A functor $F$ is *stably k-excisive* if it satisfies the condition $E_k(c, \kappa)$, given below, for some constants $c, \kappa$:

$E_k(c, \kappa)$:    If $X = S \mapsto X_S$ is any strongly homotopy cocartesian $(k + 1)$-cube such that $X_\emptyset \to X_i$ is $d_i$-connected for all $i$ and $d_i \geq \kappa$ for all $i$, then the cube $F(X)$ is $(-c + \sum_i d_i)$-cartesian.

Roughly speaking, this says that $F$ takes a special class of strongly homotopy cocartesian cubes to highly cartesian cubes. If $F$ satisfies $E_k(-\infty, -1)$, then $F$ is $k$-excisive.

**Example 10.1.18**    Let $F = I$ be the identity functor. Theorem 6.2.1 says that $I$ satisfies $E_k(k - 1, -1)$ for all $k$.                                           □

**Definition 10.1.19**    We say $F$ and $G$ *agree to order k* if there exists a natural transformation $F \to G$ that satisfies the condition $O_k(c, \kappa)$, given below, for some constants $c, \kappa$.

$O_k(c, \kappa)$:    For every $d \geq \kappa$ and every $d$-connected map $X \to *$ (i.e. for all $(d - 1)$-connected spaces $X$), the map $F(X) \to G(X)$ is $(-c + (k + 1)d)$-connected.

**Remark 10.1.20**    If $C = \text{Top}_Y$, the condition about the connectivity of the map $X \to *$ stays the same, but now for the map $X \to Y$.                           □

The construction of $T_k F$ uses the strongly homotopy cocartesian $(k + 1)$-cube $U \mapsto X * U$, $U \in \mathcal{P}(\underline{k + 1})$, studied in Examples 3.9.13 and 5.8.20 (we will set $Y = *$ in what follows).

**Definition 10.1.21**    For a functor $F$, define

$$T_k F(X) = \operatorname*{holim}_{U \in \mathcal{P}_0(\underline{k+1})} F(X * U).$$

For some intuition about this definition, see [Goo03, Remark 1.1].

**Example 10.1.22**    For $T_0 F$, note that $X * \{1\} = CX \simeq *$. Thus

$$T_0 F = \operatorname{holim} F(CX) = F(CX) \simeq F(*).$$

The last equivalence is true because $F$ is a homotopy functor. If $F$ is reduced (so $F(*) \simeq *$), $T_0 F \simeq *$. $\square$

**Example 10.1.23** To construct $T_1 F$, first note that $X * \{1\} = X * \{2\} = CX$, and $X * \{1, 2\} = \Sigma X$. Then

$$T_1 F(X) = \operatorname{holim}\left( \begin{array}{c} F(CX) \\ \downarrow \\ F(CX) \longrightarrow F(\Sigma X) \end{array} \right) \simeq \operatorname{holim}\left( \begin{array}{c} F(*) \\ \downarrow \\ F(*) \longrightarrow F(\Sigma X) \end{array} \right)$$

Thus, if $F$ is reduced, we have

$$T_1 F(X) \simeq \operatorname{holim}\left( \begin{array}{c} * \\ \downarrow \\ * \longrightarrow F(\Sigma X) \end{array} \right) \simeq \Omega F(\Sigma X)$$

by Example 3.2.7. In the further special case when $F$ is the identity functor, we have that $T_1 F(X) \simeq \Omega \Sigma X$. $\square$

Since $F$ is a homotopy functor and since homotopy limits are homotopy invariant, we have that $T_k$ is also a homotopy functor. Furthermore, there is an evident natural transformation

$$t_k(F) \colon F \longrightarrow T_k F \tag{10.1.1}$$

since $X * \emptyset = X$ is the initial space of the cube $U \mapsto X * U$. Moreover, the natural inclusion $\underline{k} \to \underline{k+1}$ gives an inclusion of categories $\mathcal{P}_0(\underline{k}) \to \mathcal{P}_0(\underline{k+1})$ and hence a natural transformation given by restriction of homotopy limits

$$T_k F \longrightarrow T_{k-1} F. \tag{10.1.2}$$

This map is a fibration by Proposition 5.4.29.

Finally, a natural transformation $F \to G$ induces a natural transformation of their homotopy limits

$$T_k F \longrightarrow T_k G. \tag{10.1.3}$$

**Proposition 10.1.24** ([Goo03, Proposition 1.4]) *Suppose $F$ is a homotopy functor that satisfies $E_k(c, \kappa)$. Then the following hold.*

1. *$T_k F$ satisfies $E_k(c - 1, \kappa - 1)$.*
2. *The transformation $F \to T_k F$ satisfies $O_k(c, \kappa)$. In particular, when $F$ is k-excisive (so $c = -\infty$), this transformation is an objectwise weak equivalence.*

*Proof* For condition 1, let $X = S \mapsto X_S$ be a strongly homotopy cocartesian $(k + 1)$-cube such that $X_\emptyset \to X_i$ is $k_i$-connected where $k_i \geq \kappa - 1$ for all $i$. We must show that the map

$$T_k F(X_\emptyset) \longrightarrow \underset{S \in \mathcal{P}_0(\underline{k+1})}{\mathrm{holim}} T_k F(X_S)$$

is $(1 - c + \sum_i k_i)$-connected (i.e. the cube $S \mapsto T_k F(X_S)$ is $(1 - c + \sum_i k_i)$-cartesian). Unraveling further, we need to show that the map

$$\underset{U \in \mathcal{P}_0(\underline{k+1})}{\mathrm{holim}} F(X_\emptyset * U) \longrightarrow \underset{S \in \mathcal{P}_0(\underline{k+1})}{\mathrm{holim}} \underset{U \in \mathcal{P}_0(\underline{k+1})}{\mathrm{holim}} F(X_S * U)$$

is $(1 - c + \sum_i k_i)$-connected. Let $U \neq \emptyset$ be fixed. The $(n + 1)$-cube $X \mapsto X_S * U$ is strongly homotopy cocartesian by Corollary 5.8.10. Moreover, the maps $X_\emptyset * U \to X_i * U$ are $(k_i + 1)$-connected by Proposition 3.7.23 since $U \neq \emptyset$, and since $F$ satisfies $E_k(c, \kappa)$, the cube $S \mapsto F(X_S * U)$ is $(-c + \sum_i(k_i + 1))$-cartesian, or $(n + 1 - c + \sum_i k_i)$-cartesian. That is, the functor $X \mapsto F(X * U)$ satisfies $E_k(c - (n + 1), \kappa - 1)$ for $U \neq \emptyset$. By Proposition 5.4.16, $T_k F$ satisfies $E_k(c - 1, \kappa - 1)$. We leave it to the reader to verify that the hypotheses of that proposition are satisfied.

For condition 2, this is also a matter of unraveling the definitions. For a $(d - 1)$-connected space $X$, the $(k + 1)$-cube $S \mapsto X * S$ is strongly homotopy cocartesian and the maps $X \to X * \{i\} = CX$ are $d$-connected, and so this satisfies the hypotheses of Definition 10.1.17. Hence the canonical map

$$F(X) \longrightarrow \underset{S \in \mathcal{P}_0(\underline{k+1})}{\mathrm{holim}} F(X * S) = T_k F(X)$$

is $(-c + (k + 1)d)$-connected, and so $F \to T_k F$ satisfies Definition 10.1.19.  □

The point of the construction so far is that the class of special strongly homotopy cocartesian cubes has been enlarged since the connectivity condition and the constant $c$ have been decreased by 1, which means that if $X$ is a strongly homotopy cocartesian cube satisfying the connectivity assumptions about the maps $X_\emptyset \to X_i$, then the cube $T_k F(X)$ is more highly cartesian than the cube $F(X)$. It is in this sense that $T_k F$ is closer to being $k$-excisive.

The construction of $T_k$ can be iterated using the map $t_k(F)$, so that we have a directed system

$$F \xrightarrow{t_k(F)} T_k F \xrightarrow{t_k(T_k F)} T_k T_k F = T_k^2 F \xrightarrow{t_k(T_k^2 F)} T_k T_k^2 F = T_k^3 F \longrightarrow \cdots$$

This is used to produce the polynomial approximations to $F$, as follows.

**Definition 10.1.25** Define the *kth polynomial approximation of F* to be

$$P_k F(X) = \underset{n}{\mathrm{hocolim}} \, T_k^n F.$$

**Example 10.1.26** Continuing Example 10.1.22, $P_0 F \simeq F(*)$ and, if $F$ is reduced, $P_0 F \simeq *$. □

**Example 10.1.27** Continuing Example 10.1.23, and again assuming $F$ is reduced, we have

$$P_1 F(X) \simeq \underset{n}{\mathrm{hocolim}} \, \Omega^n F(\Sigma^n X).$$

This can be thought of as $\Omega^\infty \underline{F}(X)$ where $\underline{F}(X)$ is the spectrum $\{F(\Sigma^n X)\}$. To see what the maps are, we start with the map $F(X) \to \Omega F(\Sigma X)$ demonstrated in Example 10.1.23. Replacing $X$ with $\Sigma X$ and looping gives that $F(\Sigma X) \to \Omega F(\Sigma \Sigma X)$ loops to $\Omega F(\Sigma X) \to \Omega^2 F(\Sigma^2 X)$. Iterating this gives the desired maps. If $F$ is the identity functor, we then have

$$P_1 F(X) \simeq \Omega^\infty \Sigma^\infty X = QX.$$ □

Again, since $F$ is a homotopy functor and homotopy colimits are homotopy invariant, $P_k F$ is a homotopy functor. Also, since $F$ maps to the directed system defining $P_k F$, there is a canonical transformation

$$p_k(F) \colon F \longrightarrow P_k F.$$

As we have mentioned above, we have a canonical natural transformation

$$q_k(F) \colon T_k F \longrightarrow T_{k-1} F \tag{10.1.4}$$

and

$$P_k F \longrightarrow P_k G. \tag{10.1.5}$$

Using the first two of the above three maps, we thus have that a homotopy functor $F$ determines a tower of functors

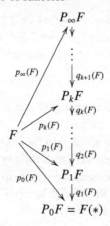

$$P_\infty F$$

$$p_\infty(F) \qquad \downarrow q_{k+1}(F)$$

$$P_k F$$

$$\downarrow q_k(F)$$

$$p_k(F)$$

$$F \qquad p_1(F) \qquad \downarrow q_2(F)$$

$$p_0(F) \qquad P_1 F$$

$$\downarrow q_1(F)$$

$$P_0 F = F(*)$$

called the *Taylor tower of F*. Here $P_\infty F = \mathrm{holim}_k P_k F$ and $p_\infty(F)$ is the map to this inverse homotopy limit of the tower induced by the maps $p_k(F)$ (for inverse limits of towers, see Example 8.4.11).

The rest of this section is devoted to the discussion of properties of $P_k F$. One important result that goes beyond the scope of this book is that the $P_k F$ are in a sense the best possible polynomial approximations to $F$. That is, they satisfy a universal property in the associated homotopy category [Goo03, Theorem 1.8].

**Proposition 10.1.28** ([Goo03, Proposition 1.5]) *If F is stably k-excisive, then*

1. *$P_k F$ is k-excisive;*
2. *F agrees to order k with $P_k F$ (in the sense of Definition 10.1.19).*

*Proof* By induction using part 1 of Proposition 10.1.24, we have that $T_k^i F$ satisfies $E_k(c-i, \kappa-i)$. Then the direct limit defining $P_k F$ satisfies $E_k(-\infty, -\infty)$ and hence $P_k F$ is $k$-excisive. By part 2 of Proposition 10.1.24, we have that $t_k(T_k^i F)$ satisfies $O_k(c-i, \kappa-i)$. But this means, by definition, that $t_k(T_k^i F)$ also satisfies $O_k(c, \kappa)$. Then the composition of all $t_k(T_k^i F)$ also satisfies $O_k(c, \kappa)$, and this is precisely the map $p_k(F) \colon F \to P_k F$. In other words, $F$ and $P_k F$ agree to order $k$. □

**Remark 10.1.29** One way to view the construction of $P_k F$ is to think of the process of iterating $T_k F$ as forcing the functor to behave well on special kinds of strongly homotopy cocartesian cubes, namely those of the form $U \mapsto X * U$. Thus one key in seeing that $P_k F$ is $k$-excisive, that is, that it takes *any* strongly homotopy cocartesian $(k+1)$-cube $X$ to a cartesian $(k+1)$-cube $P_k(X)$, is that the map $F(X) \to T_k F(X)$ factors through some homotopy cartesian $(k+1)$-cube.

**Proposition 10.1.30** ([Goo03, Proposition 1.6]) *Suppose $F \to G$ is a natural transformation between homotopy functors. If F and G agree to kth order, then the induced map $P_k F \to P_k G$ is a weak equivalence, and the converse is true if F and G are stably k-excisive.*

We omit the proof, but prove an analog of this statement in manifold calculus in the next section, Theorem 10.2.14.

We then have the following result, the proof of which we omit. The key input is [Goo03, Lemma 1.9] which says that, for any strongly homotopy cocartesian cube $X$ and a homotopy functor $F$, the map of cubes $(t_k F)(X) \colon F(X) \to (T_k F)(X)$ (see (10.1.1)) factors through a cartesian cube (for a shorter proof than the one in [Goo03], see [Rez13]).

**Theorem 10.1.31** ([Goo03, Theorem 1.8]) *For any homotopy functor $F$, $P_kF$ is $k$-excisive.*

We close with some properties that will be useful later.

**Proposition 10.1.32**

1. *If $F \to G \to H$ are natural transformations such that $F(X) \to G(X) \to H(X)$ is a fibration sequence for all $X$, then*

$$P_kF(X) \to P_kG(X) \to P_kH(X)$$

   *is a fibration sequence for all $X$ as well.*
2. *[Goo03, Corollary 1.11] For $0 \le l \le k$, the map*

$$P_lF \xrightarrow{p_l(p_k(F))} P_lP_kF$$

   *is a weak equivalence.*

*Proof* The first statement follows from the fact that homotopy limits commute with themselves and with filtered homotopy colimits (both facts appear here as Proposition 8.5.5). The second follows from formal properties of $P_kF$; see [Goo03] for details. □

### 10.1.3 Homogeneous functors

If $P_kF$ is like a degree $k$ polynomial approximation of $F$, then we should be interested in its degree $k$ term, namely the homogeneous polynomial of degree $k$ that contains information about $F$ that is not already contained in $P_{k-1}F$.

**Definition 10.1.33** A homotopy functor $F$ is $k$-reduced if $P_{k-1}F \simeq *$. It is *homogeneous of degree $k$* or $k$-homogeneous if it is $k$-excisive and $k$-reduced.

In particular, the notions of $F$ 1-homogeneous and linear (1-excisive and 1-reduced) are the same.

**Example 10.1.34** If $F \to G$ is a natural transformation with $F$ $l$-homogeneous and $G$ $k$-homogeneous, then hofiber$(F \to G)$ is $k$-excisive. This follows from Corollary 10.1.16. □

Here is the analog of Proposition 10.1.13 for homogeneous functors.

**Proposition 10.1.35** ([Goo03, Lemma 3.1]) *Suppose* $F: C^k \longrightarrow \mathcal{D}$ *is* 1-*reduced in each variable. If* $\Delta_k: C \to C^k$ *denotes the diagonal inclusion, then the composition* $F \circ \Delta_k$ *is* k-*reduced. In particular, if* $F$ *is also mulitilinear, namely linear in each variable,* $F \circ \Delta$ *is* k-*homogeneous.*

**Example 10.1.36** Functors from Example 10.1.14 are homogeneous of degree $n$ by Proposition 10.1.35 since they are multilinear.                    □

**Example 10.1.37** Continuing the previous example, recall the notion of a group acting on a spectrum and the definition of the homotopy orbits of such an action (see (A.8)). Note that $X^{\wedge k}$ has an obvious action of the symmetric group $\Sigma_k$ and suppose a spectrum $\underline{C}$ also has one. One can then consider functors

$$X \longmapsto (\underline{C} \wedge X^{\wedge k})_{h\Sigma_k} \quad \text{and} \quad X \longmapsto \Omega^\infty(\underline{C} \wedge X^{\wedge k})_{h\Sigma_k}.$$

These functors are also homogeneous of degree $k$ (because homotopy colimits commute) and are in in a way the most important ones; see Theorem 10.1.48.

                    □

**Definition 10.1.38** Define the *kth layer of the homotopy calculus Taylor tower of* $F$ to be

$$D_k F = \text{hofiber}\left(P_k F \xrightarrow{q_k(F)} P_{k-1} F\right).$$

Note that $D_k F$ is a homotopy functor as $P_k F$ and $P_{k-1} F$ are, and since $F$ is a functor to based spaces, $P_{k-1} F$ is based.

**Proposition 10.1.39** *The kth layer* $D_k F$ *is homogeneous of degree k.*

*Proof* The homotopy fiber $D_k F$ can be thought of as a homotopy limit

$$D_k F \simeq \text{holim}\left(P_k F \longrightarrow P_{k-1} F \longleftarrow *\right).$$

Then $D_k F$ is $k$-excisive by Proposition 10.1.15 because it is a homotopy limit of $k$-excisive functors.

   To see that it is $k$-reduced, consider the fibration sequence

$$D_k F \longrightarrow P_k F \longrightarrow P_{k-1} F$$

and apply the functor $P_{k-1}$ to it. By part 1 of Proposition 10.1.32, the result is again a fibration sequence. By part 2 of the same result, we have $P_{k-1} P_k F \simeq P_{k-1} F$ and $P_{k-1} P_{k-1} F \simeq P_{k-1} F$, so that the fibration sequence in fact looks like

$$P_{k-1} D_k F \longrightarrow P_{k-1} F \longrightarrow P_{k-1} F.$$

Since the right-most map is a weak equivalence, it follows that $P_{k-1}D_kF \simeq *$ and so $D_kF$ is $k$-reduced. □

Here is one of the main results about homogeneous functors.

**Theorem 10.1.40** ([Goo03, Theorem 2.1]) *Suppose $F$: Top$_*$ → Top$_*$ is a homogeneous functor of degree $k$. Then $F(X)$ is an infinite loop space for all $X \in$ Top$_*$.*

The key in proving this theorem is the following lemma.

**Lemma 10.1.41** (Delooping homogeneous functors [Goo03, Lemma 2.2]) *Suppose $F$ is a reduced homotopy functor with values in Top$_*$. Then there is a degree $k$ homogeneous functor $R_kF$ and a fibration sequence*

$$P_kF \longrightarrow P_{k-1}F \longrightarrow R_kF.$$

*Sketch of proof of Theorem 10.1.40* If $F$: Top$_*$ → Top$_*$ is homogeneous of degree $k$, then $P_kF \simeq F$ and $P_{k-1}F \simeq *$ and Lemma 10.1.41 says that there is a fibration sequence $F \to * \to R_kF$. By Example 2.2.9, we then have a weak equivalence $F \simeq \Omega R_kF$. The functor $R_kF$ is thus a delooping of $F$.

This construction can be iterated and we get a sequence of functors $R_k^nF$, $n \geq 1$, with weak equivalences $R_k^nF \simeq \Omega R_k^{n+1}F$. This defines a spectrum $\underline{R_kF}(X)$ for each $X \in$ Top$_*$ and so we therefore have a functor $\underline{R_kF}$: Top$_*$ → Spectra. That this is a homotopy functor and that it is homogeneous of degree $k$ follows from the fact that each of the functors $R_k^nF$ is.

Is it also immediate from the contruction that the composition

$$F \longmapsto \underline{R_kF} \longmapsto \Omega^\infty \underline{R_kF}$$

is a weak equivalence of functors. This shows that $F(X)$ is an infinite loop space for all $X$. □

**Remark 10.1.42** It is also not hard to show that, given a $k$-homogeneous functor $F$: Spectra → Top$_*$, the composition

$$F \longmapsto \Omega^\infty F \longmapsto \underline{R_kF}\,\Omega^\infty F$$

is a weak equivalence of functors. We therefore have a bijection

$$\{k\text{-homogeneous functors Top}_* \to \text{Top}_*\}$$

$$\updownarrow$$

$$\{k\text{-homogeneous functors Top}_* \to \text{Spectra}\}$$

□

Now remember from Proposition 10.1.35 that, if $F \colon C^k \to \mathcal{D}$ is multilinear, then $F \circ \Delta_k$ is $k$-homogeneous. If in addition $F$ is invariant under the permutations of the coordinates of $C^k$, it is said to be *symmetric*. For a symmetric multilinear functor $F$, the $k$-homogeneous functor $F \circ \Delta_k$ also has an action of $\Sigma_k$. Furthermore, since homotopy orbits are an instance of a homotopy colimit, if $F$ is a functor into spectra, then the functor

$$(F \circ \Delta_k)_{h\Sigma_k}$$

is also $k$-homogeneous. This is because, if $F$ is spectrum-valued, $T_k F$ and $P_k F$ commute with homotopy colimits.

Thus a symmetric multilinear (spectrum valued) functor gives rise to a $k$-homogeneous functor. We wish to construct an inverse, and to give some intuition for this, we recall how this works in the realm of real-valued functions (or functions over any field). If $f(x_1, x_2, \ldots, x_k)$ is multilinear, we can compose it with the diagonal map $\Delta_k$ to get a map

$$g(x) = (f \circ \Delta_k)(x) = f(x, x, \ldots, x).$$

This is a homogeneous function of degree $k$ because of multilinearity:

$$g(ax) = f(ax, ax, \ldots, ax) = a^k f(x, x, \ldots, x) = a^k g(x).$$

Thus a multilinear function gives rise to a degree $k$ homogeneous function. In this analogous example we did not need symmetry.

But we can go back as well. To do this, we use the notion of a *cross-effect*, which was classically used to study the measure of the failure of a real-valued function to be degree $n$. We give a brief review of cross-effects below; for more details, see, for example, [JM04, Section 1].

For example, given a function $g(x)$, its second cross-effect, a function of two variables, is defined to be

$$\mathrm{cr}_2 g = \mathrm{cr}_2 g(x_1, x_2) = g(x_1 + x_2) - g(x_1) - g(x_2) + g(0). \qquad (10.1.6)$$

This symmetric function is zero if and only if $g(x)$ is linear. In addition, $g(x)$ is quadratic if and only if $\mathrm{cr}_2 g(x_1, x_2)$ is linear in each variable. In this case $\mathrm{cr}_2 g(x_1, x_2) = a_2 x_1 x_2$ and so $\mathrm{cr}_2 g(x, x)/2$ recovers the quadratic term of $g(x)$ (i.e. it gives a measure of the failure of $g$ to be linear).

More generally, given a function $g(x)$, one can define the *kth cross-effect* $\mathrm{cr}_k g$, a function of $k$ variables, inductively starting with $\mathrm{cr}_1 g(x) = g(x) - g(0)$ and defining

$$\mathrm{cr}_k g(x_1, \ldots, x_k) = \mathrm{cr}_{k-1} g(x_1 + x_2, \ldots, x_k) \qquad (10.1.7)$$
$$- \mathrm{cr}_{k-1} g(x_1, x_3, \ldots, x_k) - \mathrm{cr}_{k-1} g(x_2, x_3, \ldots, x_k).$$

This is a symmetric function which can be seen to vanish if and only if $g(x)$ is a polynomial of degree $k - 1$ and it is equal to $n! a_k x_1 x_2 \cdots x_k$ if $g(x)$ is a degree $k$ polynomial (and is hence evidently multilinear). Thus again, if $g(x)$ is degree $k$, we can recover its homogeneous degree $k$ term as

$$\frac{1}{n!} f(x, x, \ldots, x) = a_k x^k.$$

The $k$th cross-effect provides an inverse to $f \circ \Delta_k$. The analog of $f \circ \Delta_k$ in our setting is $F \circ \Delta_k$, for $F$ a symmetric multilinear functor, so what remains to be done is to generalize the cross-effect to functors. This generalization is originally due to Eilenberg and MacLane [EML54] and we will review it in the language of cubical diagrams that is more suitable for our purposes. More details can be found in [BJR15, Goo03, JM04, MO06].

**Definition 10.1.43** For $F \colon \mathrm{Top}_* \to \mathrm{Top}_*$ a homotopy functor, define the *$k$th cross-effect cube*, $\mathcal{CR}_k F(X_1, \ldots, X_k)$ to be the $k$-cube given by

$$S \longmapsto \mathcal{CR}_k F(X_1, \ldots, X_k)_S = F\left( \bigvee_{i \notin S} X_i \right)$$

and with $\mathcal{CR}_k F(X_1, \ldots, X_k)_S \to \mathcal{CR}_k F(X_1, \ldots, X_k)_T$ induced by the identity on $X_i$ if $i \notin T$ and the map to the basepoint if $i \in T$.

Then define the *$k$th cross-effect* of $F$, $\mathrm{cr}_k F(X_1, \ldots, X_k)$ to be the total homotopy fiber of $\mathcal{CR}_k F(X_1, \ldots, X_k)$.

The second cross-effect of a functor $F$ is the total fiber of the square

$$
\begin{array}{ccc}
F(X_1 \vee X_2) & \longrightarrow & F(X_2) \\
\downarrow & & \downarrow \\
F(X_1) & \longrightarrow & F(*)
\end{array}
$$

Thinking of homotopy fibers as differences between spaces, the total fiber can be thought as "$(F(X_1 \vee X_2) - F(X_1)) - (F(X_2) - F(*)) = F(X_1 \vee X_2) - F(X_1) - F(X_2) + F(*)$". This is precisely the functor analog of (10.1.6). In general, the $k$th cross-effect of $F$ is in this sense precisely analogous to the expression in (10.1.7). For some examples of cross-effects, see [JM04, Example 1.4].

Some immediate properties of $\mathrm{cr}_k F$ are that it is symmetric, it is a homotopy functor (in each variable), and that it is reduced (or $(1, \ldots, 1)$-reduced). The last property is true because, if $X_i \simeq *$ for some $i$, then the cube in question is homotopy cartesian by Proposition 5.4.12, as it is a map of cubes of one lower dimension which is a pointwise weak equivalence, and hence $\mathrm{cr}_k F(X_1, \ldots, X_k) \simeq *$.

The following is analogous to the fact that $\mathrm{cr}_k g$ of a degree-$(k-1)$ function is trivial and is multilinear if $g$ is degree $k$.

**Proposition 10.1.44** ([Goo03, Proposition 3.3]) *If $F$ is $(k-1)$-excisive, then $\mathrm{cr}_k F \simeq *$. If $F$ is $k$-excisive, then $\mathrm{cr}_k F$ is multilinear (and symmetric).*

If $F$ is a functor to spectra, then the converse of the first statement is also true (see proof of [Goo03, Proposition 3.4]).

*Proof* This follows by induction by writing $\mathrm{cr}_j F(X_1, \ldots, X_{k-1}, A)$ as $\mathrm{cr}_{j-1}(\text{hofiber } F(X \vee A) \to F(X))(X_1, \ldots, X_{j-1})$ and using the fact that, if $F$ is $j$-excisive, hofiber $F(X \vee A) \to F(X)$ is $(j-1)$-excisive. $\quad\square$

Now consider any homotopy functor $F$. Since $P_{k-1}F$ is $(k-1)$-excisive, we have that $\mathrm{cr}_k P_{k-1} F \simeq *$. Thus

$$\mathrm{cr}_k P_k F \simeq \mathrm{hofiber}(\mathrm{cr}_k P_k F \to *) \simeq \mathrm{hofiber}(\mathrm{cr}_k P_k F \to \mathrm{cr}_k P_{k-1} F)$$
$$\simeq \mathrm{cr}_k(\mathrm{hofiber}(P_k F \to P_{k-1} F)) = \mathrm{cr}_k D_k F.$$

(The weak equivalence before the last equality is true because homotopy limits commute.) In particular, if $F$ is $k$-excisive, so that $F \simeq P_k F$, we have that $\mathrm{cr}_k F \simeq \mathrm{cr}_k D_k F$, that is, the $k$th cross-effect of a $k$-excisive functor detects only the $k$-homogeneous part.

We further have the following property of the cross-effect of a spectrum-valued functor.

**Proposition 10.1.45** ([Goo03, Proposition 3.4]) *Suppose $F, G \colon \mathrm{Top}_* \to$ Spectra are $k$-excisive and suppose a natural transformation $N \colon F \to G$ induces a weak equivalence $\mathrm{cr}_k F \to \mathrm{cr}_k G$. Then $N$ is a weak equivalence.*

Thus a symmetric multilinear spectrum-valued functor $F$ gives rise to a $k$-homogeneous one via the functor $F \mapsto (F \circ \Delta_k)_{h\Sigma_k}$. Conversely, a $k$-homogeneous functor gives rise to a symmetric multilinear functor via $F \mapsto \mathrm{cr}_k F$. We then have the the following important theorem.

**Theorem 10.1.46** ([Goo03, Theorem 3.5]) *Let $\mathcal{L}_k(\mathrm{Top}_*, \text{Spectra})$ denote the category of symmetric multilinear functors of degree $k$ from $\mathrm{Top}_*$ to Spectra and $\mathcal{H}_k(\mathrm{Top}_*, \text{Spectra})$ denote the category of homogeneous degree $k$ functors $\mathrm{Top}_* \to$ Spectra.*

*Up to natural weak equivalence, the functors*

$$(- \circ \Delta_k)_{h\Sigma_k} \colon \mathcal{L}_k(\mathrm{Top}_*, \text{Spectra}) \longrightarrow \mathcal{H}_k(\mathrm{Top}_*, \text{Spectra})$$

*and*

$$\mathrm{cr}_k \colon \mathcal{H}_k(\mathrm{Top}_*, \mathrm{Spectra}) \longrightarrow \mathcal{L}_k(\mathrm{Top}_*, \mathrm{Spectra})$$

*are inverses.*

This is more generally true if we replace $\mathrm{Top}_*$ with $\mathrm{Top}_Y$. In the case of the homogeneous functor $D_k F$, where $F$ is a functor to spectra, the above tells us that there is a weak equivalence

$$D_k F \simeq (\mathrm{cr}_k D_k F \circ \Delta_k)_{h\Sigma_k} = ((\mathrm{cr}_k D_k F)(X, \ldots, X))_{h\Sigma_k}.$$

**Definition 10.1.47** The spectrum-valued functor $\mathrm{cr}_k D_k F$ is denoted by $D^{(k)} F$ and called the *k-fold differential of $F$*.

If $F$ is a space-valued functor, then the above weak equivalence is given by

$$D_k F \simeq \Omega^\infty \left( (B^\infty \mathrm{cr}_k D_k F)(X, \ldots, X)_{h\Sigma_k} \right).$$

Here $B^\infty$ is the inverse to $\Omega^\infty$; these two functors give an equivalence of categories of spectrum- and space-valued multilinear symmetric functors, along the lines of the equivalence from Remark 10.1.42; see [Goo03, Proposition 3.7]. In this case the $k$-fold differential is the functor $B^\infty \mathrm{cr}_k D_k F$ and is again denoted by $D^{(k)} F$.

Now suppose $\underline{C}$ is a spectrum with an action of $\Sigma_k$. Then the functor $\underline{C} \wedge X_1 \wedge \cdots \wedge X_k$ from $\mathrm{Top}_*^k$ to Spectra is multilinear and symmetric. On the other hand, if $L$ is a multilinear functor from $\mathrm{Top}_*^k$ to Spectra, then there is an assembly map

$$L(S^0, \ldots, S^0) \longrightarrow L(X_1, \ldots, X_k)$$

(see [Goo90, page 5]) that is a weak equivalence for all finite complexes $X_j$ [Goo03, Proposition 5.8]. Since every space is a filtered homotopy colimit of finite complexes, it follows that the above map is a weak equivalence for all spaces if $L$ is finitary (see Definition 10.1.1). This means that there is a correspondence

$$\{\text{symmetric multilinear finitary functors } \mathrm{Top}_*^k \to \mathrm{Spectra}\}$$

$$\updownarrow$$

$$\{\text{spectra with } \Sigma_k\text{-action}\}$$

It follows that the differential $D^{(k)} F$ is determined (on finite complexes, or on all spaces if $F$ is finitary, as this implies $D^{(k)} F$ is finitary) by some spectrum with a $\Sigma_k$-action. This spectrum is called the *kth derivative of F at the one-point space* and denoted by $\partial^{(k)} F(*)$ (both for space-valued and spectra-valued functors). Thus, for space-valued functors $F$, we have

$$\partial^{(k)} F(*) \simeq (D^{(k)} F)(S^0, \ldots, S^0),$$

and, for spectrum-valued $F$, we have

$$\Omega^\infty \partial^{(k)} F(*) \simeq (D^{(k)} F)(S^0, \ldots, S^0).$$

We can then summarize this discussion in the following theorem that classifies homogeneous functors from based spaces.

**Theorem 10.1.48** (Classification of homogeneous functors [Goo03, Section 5]) *If $F$: $\mathrm{Top}_* \to \mathrm{Spectra}$ is a finitary homotopy functor, then the kth layer of its Taylor tower is given by*

$$D_k F(X) \simeq \left( \partial^{(k)} F(*) \wedge X^{\wedge k} \right)_{h\Sigma_k}.$$

*For a homotopy functor $F$: $\mathrm{Top}_* \to \mathrm{Top}_*$, the kth layer is given by*

$$D_k F(X) \simeq \Omega^\infty \left( \partial^{(k)} F(*) \wedge X^{\wedge k} \right)_{h\Sigma_k}.$$

**Remark 10.1.49**   One of the ways in which functor calculus resembles ordinary calculus of analytic functions is in the way the above formula for $D_k F(X)$ corresponds nicely to the term $f^{(k)}(0) \cdot x^k/k!$ of the Taylor series of $f$, where dividing by $k!$ corresponds to taking the quotient by the $\Sigma_k$ action.          □

For some model-theoretic improvements to this result, see [BCR07]. As an application of the material developed so far, see [Goo03, Section 8].

One modification that could be made in everything that has been said so far is to replace the source category by the category of spaces over $Y$, $\mathrm{Top}_Y$. This is a lot like expanding the Taylor series of $f$ at a point other than zero. In this case, if $F_Y$ denotes a functor on $\mathrm{Top}_Y$, $T_k F_Y$ is defined using the cube $F_Y(X *_Y U)$ (rather than $F(X * U)$) and the Taylor tower of $F_Y$ has $F_Y(Y)$ as its $P_0$ approximation (rather than $F(*)$). The kth layer of the Taylor tower is now $D_Y^k F$ and the kth derivative of $F$ at $(Y, y_1, \ldots, y_k)$, where $y_i$ are points in $Y$, is then denoted by

$$\partial^{(k)}_{y_1, \ldots, y_k} F(Y).$$

For $F$ a spectrum-valued functor, we then have

$$\partial^{(k)}_{y_1, \ldots, y_k} F(Y) \simeq (D_Y^k F)(Y \vee_{y_1} S^0, \cdots, Y \vee_{y_k} S^0),$$

and, for a space-valued $F$, we take $\Omega^\infty$ on the left as before.

## 10.1.4 Convergence

Recall the notion of a stably excisive functor from Definition 10.1.17.

**Definition 10.1.50** A homotopy functor $F$ is called $\rho$-*analytic* if there exists a number $q$ such that $F$ satisfies $E_k(k\rho - q, \rho + 1)$ for all $k \geq 1$.

The number $\rho$ is called the *radius of convergence of $F$* for the following reason.

**Theorem 10.1.51** (Convergence of the Taylor tower [Goo03, Theorem 1.13]) *If $F$ is $\rho$-analytic and the map $X \to *$ is $(\rho + 1)$-connected (i.e. $X$ is $\rho$-connected), the connectivity of the natural map $p_k(F)\colon F \to P_k F$ increases to infinity with $k$, so $F(X)$ is weakly equivalent to $P_\infty F$.*

**Remark 10.1.52** For functors on $\mathrm{Top}_Y$, the condition that $X \to *$ is $(\rho + 1)$-connected generalizes to the condition that the map $X \to Y$ should be $(\rho + 1)$-connected.

*Proof of Theorem 10.1.51* If $F$ is $\rho$-analytic, then it satisfies $E_k(k\rho - q, \rho + 1)$ for some $q$. By the proof of Proposition 10.1.28, $p_k(F)$ satisfies $O_k(k\rho - q, \rho + 1)$. But this means that, if $X \to *$ is $(d + 1)$-connected, the map $p_k(F)$ is $(q + d + 1 + k(d + 1 - \rho))$-connected. Thus, provided $d + 1 > \rho$, that is, as long as $X$ is at least $\rho$-connected, the connectivity of $p_k(F)$ increases to infinity with $k$. □

**Example 10.1.53** For $F\colon \mathrm{Top} \to \mathrm{Top}$ the identity functor, the Taylor tower of a space $X$ converges to $X$ if $X$ is 1-connected. This is because, by Example 10.1.18, identity is 1-analytic. In fact, the Taylor tower of the identity converges at all nilpotent spaces $X$ ($\pi_1 X$ is a nilpotent group and acts nilpotently on higher homotopy groups). For more details about the identity functor, see [Goo03, Section 8] or [Kuh07, Section 6.4]. □

**Example 10.1.54** Let $K$ be a finite CW complex. Then the functors $X \mapsto \Sigma^\infty \mathrm{Map}(K, X)$ and $X \mapsto \Omega^\infty \Sigma^\infty \mathrm{Map}(K, X)$ are $\dim(K)$-analytic [Goo92, Example 4.5]. This can be used to conclude that every Taylor coefficient spectrum for the functor from spectra to spectra given by $X \mapsto \Sigma^\infty \Omega^\infty X$ is the sphere spectrum $\mathbb{S}$. The Taylor tower of this functor was studied in detail in [Aro99, AK02]. For an exposition of these results, see [Kuh07, Sections 6.1 and 6.2]. □

The following result can be thought of as the "uniqueness of analytic continuation".

**Proposition 10.1.55** (Uniqueness of analytic continuation [Goo92, Proposition 5.1]) *Suppose $N: F \to G$ is a natural transformation where $F$ is $\rho_1$-analytic and $G$ is $\rho_2$-analytic. Suppose that, for some $k$, the map $F(X) \to G(X)$ is a weak equivalence for all $k$-connected spaces $X$. Then $F(X) \to G(X)$ is a weak equivalence for all $\rho$-connected spaces $X$, where $\rho = \max\{\rho_1, \rho_2\}$.*

*Proof* We let the reader fill in the details of the following sketch. Consider the square

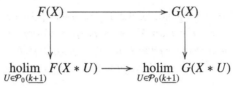

Now apply Proposition 10.1.30, Theorem 10.1.51, and part 3 of Proposition 2.6.15.                                                                          □

One way to investigate the homotopy groups of the inverse limit $P_\infty F$ of the Taylor tower is to consider its Bousfield–Kan homotopy spectral sequence from Section 9.6.3. This spectral sequence starts with the homotopy groups of the layers of the tower, namely $E_{a,b}^1 = \pi_{b-a}(D_a F)$ (see (9.6.3)). One way to ensure its convergence to $\pi_*(P_\infty F)$ is for the connectivity of the fibers $D_k F$ to increase with $k$ (see Corollary 9.6.8). This in fact happens under the conditions from Theorem 10.1.51: if $F$ is $\rho$-analytic and $X$ is $d$-connected, then we have that the map $p_k(F)$ is $(q + d + 1 + k(d + 1 - \rho))$-connected, and hence the map $q_k(F): P_k F \to P_{k-1}$ is $(q + d + 1 + (k - 1)(d + 1 - \rho))$-connected (the lower of the connectivity numbers for $p_k(F)$ and $p_{k-1}(F)$; see Proposition 2.6.15). This in turn means that $D_k F = \text{hofiber}(q_k F)$ is $(q + d + 1 + (k - 1)(d + 1 - \rho) - 1)$-connected. Again this number increases with $k$ as long as $d + 1 > \rho$, i.e. as long as $X$ is at least $\rho$-connected.

## 10.2 Manifold calculus of functors

We now turn our attention to manifold calculus of functors, developed by Weiss [Wei99] and Goodwillie–Weiss [GW99]. Another reference is the survey [Mun10]. The reader may wish to review the notion of a handlebody decomposition from Section A.2, which plays an important role in proofs throughout this section.

Let $M$ be a smooth closed manifold of dimension $m$. Let $O(M)$ denote the poset of open subsets of $M$. Thus the objects are open sets $U \subset M$ and the morphisms are inclusions $U \subset V$. The manifold calculus of functors studies contravariant functors $F: O(M) \to$ Top. It was originally built to study the space of embeddings of $M$ in a smooth manifold $N$ of dimension $n$. This space is denoted Emb$(M, N)$ and its topology is discussed below (see also Theorem A.2.12). This is a contravariant functor of $U \in O(M)$ since an inclusion $U \subset V$ gives rise to a restriction map Emb$(V, N) \to$ Emb$(U, N)$. Some authors refer to manifold calculus as "embedding calculus".

All the functors we will apply the manifold calculus to will be contravariant, and we will occasionally remind the reader that this is the case.

Before we state the axioms our functors need to satisfy, we make a few definitions. For smooth manifolds, let $C^\infty(P, N)$ denote the space of smooth maps from $P$ to $N$. This is topologized using the Whitney $C^\infty$ topology (see Definition A.2.7).

**Definition 10.2.1** Let $P$ and $N$ be smooth manifolds. The *space of embeddings*, denoted Emb$(P, N)$, is the set of all smooth maps $f: P \to N$ such that

- $f$ is a homeomorphism onto its image;
- the derivative $df: TP \to TN$ is a fiberwise injection.

The *space of immersions*, denoted Imm$(P, N)$ is the set of all smooth maps satisfying the second condition above. Both spaces are topologized as subspaces of $C^\infty(P, N)$. A path in the space of embeddings is called an *isotopy* while a path in the space of immersions is called a *regular homotopy*.

**Definition 10.2.2** An inclusion $i: U \to V$ in $O(M)$ is an *isotopy equivalence* if there exists a smooth embedding $e: V \to U$ such that the compositions $i \circ e$ and $e \circ i$ are isotopic to the identity map.

We assume all of our functors satisfy two axioms.

**Definition 10.2.3** We say a contravariant functor $F: O(M) \to$ Top is

1. an *isotopy functor* if it takes isotopy equivalences to homotopy equivalences;
2. *finitary* if for every increasing sequence $U_1 \subset U_2 \subset \cdots \subset U_k \cdots$ in $O(M)$ with $\cup_i U_i = U$, the canonical map $F(U) \to \operatorname{holim}_i F(U_i)$ is an equivalence.

The first condition says that $F$ behaves well with respect to "equivalences" (in $O(M)$ we think of equivalences as "thickenings"). The second says that $F$

is determined by its values on open sets $U$ which are the interior of a compact codimension-zero submanifold of $M$. To begin to see why, for every open set $U$ we may choose an increasing sequence $U_1 \subset U_2 \subset \cdots$ such that $\bigcup_i U_i = U$ and such that each $U_i$ is the interior of a compact codimension-zero submanifold of $U$. These conditions are analogous with those given in Definition 10.1.1.

**Remarks 10.2.4**

1. A functor that satisfies the above two axioms is sometimes referred to as *good* in the literature.
2. The second axiom is not strictly necessary in many examples, because we are often interested only in the values of the functor $F$ when the input is the interior of some manifold. We could instead restrict attention to functors defined only for such open sets and declare its values on an arbitrary open set using axiom 2. We would of course then need to show that the homotopy type of values of the functor are independent of the increasing union. We will not pursue this here.                                                  □

**Example 10.2.5** Let $N$ be a smooth manifold. The following are all contravariant finitary isotopy functors:

- $U \longmapsto \mathrm{Map}(U, X)$ (where this is the space of all maps from $U$ to $X$, with compact-open topology, $X$ an arbitrary space);
- $U \mapsto C^\infty(U, N)$ (really this is the same example as the last – see Theorem A.2.12);
- $U \longmapsto \mathrm{Imm}(U, N)$;
- $U \longmapsto \mathrm{Emb}(U, N)$.                                                      □

### 10.2.1 Polynomial functors

As in homotopy calculus of functors, the "polynomial approximations" of functors are of interest to us.

**Definition 10.2.6** A contravariant finitary isotopy functor $F \colon O(M) \to \mathrm{Top}$ is said to be *polynomial of degree* $\leq k$ if, whenever $U \in O(M)$ and $A_0, \ldots, A_k$ are pairwise disjoint closed subsets of $U$, the $(k+1)$-cube

$$\mathcal{P}(\underline{k+1}) \longrightarrow \mathrm{Top}$$
$$S \longmapsto F(U - \cup_{i \in S} A_i)$$

is homotopy cartesian.

On the face of it, this looks quite a bit different than Definition 10.1.3, but really it is not. Roughly speaking, a polynomial functor of degree $\leq k$ takes strongly homotopy cocartesian $(k + 1)$-cubes to homotopy cartesian $(k + 1)$-cubes. For the $(k + 1)$-cube $S \mapsto U - \bigcup_{i \in S} A_i$ in Definition 10.2.6, its square faces are of the form

$$
\begin{array}{ccc}
V - (A_i \cup A_j) & \longrightarrow & V - A_i \\
\downarrow & & \downarrow \\
V - A_j & \longrightarrow & V
\end{array}
$$

for some open set $V$ and $i \neq j$. This square is homotopy cocartesian by Example 3.7.5.

Note that a polynomial functor of degree $\leq 0$ is essentially constant with value $F(\emptyset)$; for $U \in O(M)$, choose $A_0 = U$, so that the map $F(U) \to F(\emptyset)$ is a weak equivalence.

One fact of note is the following, which uses 1(b) of Proposition 5.4.13. This is analogous to Proposition 10.1.4.

**Proposition 10.2.7** *If a finitary isotopy contravariant functor $F$ is polynomial of degree $\leq k$, then it is polynomial of degree $\leq l$ for all $l \geq k$.*

*Proof* It is enough to prove this for $l = k + 1$. Let $U \in O(M)$, and let $A_0, \ldots, A_{k+1}$ be pairwise disjoint closed in $U$. Let $U_S = U - \bigcup_{i \in S} A_i$. We may write the $(k + 2)$-cube $S \mapsto U_S$ as a map of $(k + 1)$-cubes $(R \mapsto U_R) \to (R \mapsto V_R)$, where $R \subset 0, \ldots, k + 1$, and $V_R = U_R - A_{k+2}$. Then the $(k + 1)$-cubes $R \mapsto F(U_R)$ and $R \mapsto F(V_R)$ are homotopy cartesian since $F$ is polynomial of degree $\leq k$. Hence the $(k+2)$-cube $(R \mapsto F(U_R)) \to (R \mapsto F(V_R))$ is homotopy cartesian by 1(b) of Proposition 5.4.13. □

**Example 10.2.8** The functor $U \mapsto \mathrm{Map}(U, X)$ is polynomial of degree $\leq 1$ (i.e. linear) for any space $X$. The idea is that, for any $U$ and $A_0, A_1$ disjoint closed in $U$, the square

$$
\begin{array}{ccc}
\mathrm{Map}(U, X) & \longrightarrow & \mathrm{Map}(U - A_0, X) \\
\downarrow & & \downarrow \\
\mathrm{Map}(U - A_1, X) & \longrightarrow & \mathrm{Map}(U - (A_0 \cup A_1), X)
\end{array}
$$

is homotopy cartesian since $\mathrm{Map}(-, X)$ takes homotopy cocartesian squares to homotopy cartesian squares by Proposition 3.9.1. This is correct in spirit, but

Proposition 3.9.1 requires the map from the homotopy colimit of the punctured square to the last space in the square to be a homotopy equivalence (not just a weak equivalence). See Remark 10.2.9. If we put $V_i = U - A_i$ above, then $U = V_0 \cup V_1$ and $U - (A_0 \cup A_0) = V_0 \cap V_1$, so the test for whether a functor $F$ is a polynomial of degree $\leq 1$ is whether $F$ takes the homotopy cocartesian square

$$
\begin{array}{ccc}
V_0 \cap V_1 & \longrightarrow & V_0 \\
\downarrow & & \downarrow \\
V_1 & \longrightarrow & V_0 \cup V_1
\end{array}
$$

to a homotopy cartesian square.                                                    □

**Remark 10.2.9**  The square

$$
\begin{array}{ccc}
V_0 \cap V_1 & \longrightarrow & V_0 \\
\downarrow & & \downarrow \\
V_1 & \longrightarrow & V_0 \cup V_1
\end{array}
$$

appearing in the previous example is homotopy cocartesian, but we establish this using Example 3.7.5 and Lemma 3.6.14, and from this we deduce that $\text{hocolim}(V_0 \leftarrow V_0 \cap V_1 \to V_1) \to V_0 \cup V_1$ is a weak equivalence. However, as mentioned in Remark 1.3.6, the functor $\text{Map}(-, X)$ does not preserve weak equivalences (it does preserve homotopy equivalences).

If $V_0$ and $V_1$ are the interiors of smooth compact codimension-zero submanifolds $L_0$ and $L_1$ of $M$ whose boundaries intersect it transversely (and hence $M$ is itself a submanifold), then $L_0 \cap L_1, L_0, L_1$, and $L_0 \cup L_1$ all have the homotopy type of (finite) CW complexes, as does $\text{hocolim}(L_0 \leftarrow L_0 \cap L_1 \to L_1)$, and the weak equivalence $\text{hocolim}(L_0 \leftarrow L_0 \cap L_1 \to L_1) \to L_0 \cup L_1$ is therefore a homotopy equivalence. Moreover, the inclusions $V_0 \cap V_1 \to L_0 \cap L_1, V_0 \to L_0$, $V_1 \to L_1$, and $V_0 \cup V_1 \to L_0 \cup L_1$ are homotopy equivalences, and the square

$$
\begin{array}{ccc}
L_0 \cap L_1 & \longrightarrow & L_0 \\
\downarrow & & \downarrow \\
L_1 & \longrightarrow & L_0 \cup L_1
\end{array}
$$

is homotopy cocartesian in the stronger sense that $\text{hocolim}(L_0 \leftarrow L_0 \cap L_1 \to L_1) \to L_0 \cup L_1$ is a homotopy equivalence. It follows that applying $\text{Map}(-, X)$ to the above square yields a homotopy cartesian square by Proposition 3.9.1. Hence the square

$$\begin{array}{ccc} \text{Map}(V_0 \cup V_1, X) & \longrightarrow & \text{Map}(V_0, X) \\ \downarrow & & \downarrow \\ \text{Map}(V_1, X) & \longrightarrow & \text{Map}(V_0 \cap V_1, X) \end{array}$$

is also homotopy cartesian, because it is pointwise homotopy equivalent to a homotopy cartesian square. We can arrange for the $V_i$ to be the interiors of such $L_i$ using the finitary axiom, and writing each as an increasing union of manifolds $K_0^i \subset K_1^i \subset \cdots$ which are smooth compact codimension-zero submanifolds such that the boundaries of $K_j^0$ and $K_j^1$ intersect transversely. See [Wei99, Example 2.3] for more details. □

**Example 10.2.10** The functor $U \mapsto \text{Imm}(U, N)$ is polynomial of degree $\le 1$, for essentially the same reasons as in the last example (immersions, like maps, are locally determined). However, the argument is a little more delicate. We refer the reader to [Wei99, Example 2.3] for the details, but we nonetheless give a sketch.

Let $U \subset M$ be open, $A_0, A_1 \subset U$ be pairwise disjoint closed sets, and put $U_S = U - \bigcup_{i \in S} A_i$. Then the square of restriction maps

$$\begin{array}{ccc} \text{Imm}(U_\emptyset, N) & \longrightarrow & \text{Imm}(U_1, N) \\ \downarrow & & \downarrow \\ \text{Imm}(U_2, N) & \longrightarrow & \text{Imm}(U_{12}, N) \end{array}$$

is by inspection categorically cartesian. The Smale–Hirsch Theorem says that if $K \subset L$ are compact codimension-zero submanifolds of $M$ whose boundaries intersect transversely, then the restriction map $\text{Imm}(L, N) \to \text{Imm}(K, N)$ is a fibration. We then use the same method as in Remark 10.2.9 together with Proposition 3.3.5 to conclude that the square above is homotopy cartesian. □

**Example 10.2.11** The functor $U \mapsto \text{Map}(U^k, X)$ is a polynomial of degree $\le k$. This follows from the pigeonhole principle as follows: let $U \in O(M)$, let $A_0, \ldots, A_k$ be pairwise disjoint closed subsets of $U$, and put $U_S = U - \bigcup_{i \in S} A_i$. Then we have a $(k + 1)$-cube $S \mapsto U_S$ (contravariant in $S$). By Proposition 5.8.26, the canonical map

$$\underset{S \neq \emptyset}{\text{hocolim}}\, (U_S)^k \longrightarrow \underset{S \neq \emptyset}{\text{colim}}\, (U_S)^k$$

is a weak equivalence. By the pigeonhole principle,

$$U^k = \underset{S \neq \emptyset}{\text{colim}}\, (U_S)^k$$

since, for each set of $k$ points in $U$, one must lie in the complement of some $A_i$. By Proposition 8.5.4, the functor $\mathrm{Map}(-, X)$ takes (homotopy) colimits to (homotopy) limits, and hence we obtain a weak equivalence

$$\mathrm{Map}(U^k, X) \longrightarrow \underset{S \neq \emptyset}{\mathrm{holim}} \, \mathrm{Map}((U_S)^k, X).$$

This implies that the cube $S \mapsto \mathrm{Map}((U_S)^k, X)$ is homotopy cartesian. As mentioned in Remark 10.2.9, this argument is technically deficient, but morally correct. □

**Example 10.2.12** Recall from Example 5.5.6 the configuration space $\mathrm{Conf}(k, U)$ of $k$ distinct ordered points in $U$. The symmetric group $\Sigma_k$ acts freely on $\mathrm{Conf}(k, U)$ by permuting the coordinates and we let $\binom{U}{k}$ denote the quotient by this action. Thus $\binom{U}{k}$ is the space of unordered configurations in $U$. The same argument as in the previous example shows that both $U \mapsto \mathrm{Map}\,(\mathrm{Conf}(k, U), X)$ and $U \mapsto \mathrm{Map}(\binom{U}{k}, X)$ are polynomial of degree $\leq k$.

More generally, let $p \colon Z \to \binom{M}{k}$ be a fibration, and let $\Gamma\left(\binom{M}{k}, Z; p\right)$ denote its space of sections. This is a contravariant functor of $U \in O(M)$, since a section defined on $\binom{U}{k}$ can be restricted to a section on $\binom{V}{k}$ for $V \subset U$. Moreover, the same argument we just cited also proves that $U \mapsto \Gamma\left(\binom{U}{k}, Z; p\right)$ is polynomial of degree $\leq k$. The reader who thinks of spaces of sections as twisted mapping spaces should find this at least plausible. The space of sections functor is almost a universal example of a polynomial functor of degree $\leq k$. We say almost since every homogeneous functor of degree $k$ is built using a space of sections functor as above. We will explore this below in more detail in Example 10.2.22 and especially in Theorem 10.2.23. □

The next result is analogous to the fact that a polynomial $p \colon \mathbb{R} \to \mathbb{R}$ of degree $k$ is completely determined by its values on a set of $k + 1$ distinct real numbers.

**Definition 10.2.13** Let $O_k(M)$ be the full subcategory of $O(M)$ consisting of those open sets $U$ which are diffeomorphic to a disjoint union of at most $k$ open balls.

**Theorem 10.2.14** *Suppose $F \to G$ is a natural transformation of contravariant finitary isotopy functors, both polynomial of degree $\leq k$, and suppose $F(U) \to G(U)$ is an equivalence for all $U \in O_k(M)$. Then $F(U) \to G(U)$ is an equivalence for all $U \in O(M)$.*

*Proof* It is enough by axiom 2 of Definition 10.2.3 to prove this in the case where $U$ is the interior of a smooth compact codimension-zero submanifold $L$ of $M$. We will proceed by induction on the handle index of $L$. We offer a slightly revised version of the argument given by Weiss in [Wei99], as the same pattern of argument will be used below in the discussion of convergence.

Suppose the handle index (see Definition A.2.5) of $L$ is equal to zero, so that $L$ is diffeomorphic to a disjoint union of finitely many closed balls, and its interior $U$ is therefore some number of open balls. If the number of balls is less than or equal to $k$, by assumption the map $F(U) \to G(U)$ is a weak equivalence and we are done. Suppose then that $U$ is diffeomorphic to $l$ open balls, where $l > k$, and write $U = B_0 \sqcup \cdots \sqcup B_{l-1}$. Let $A_i = B_i$ for all $i = 0$ to $k$. This is a collection of pairwise disjoint closed subsets of $U$ (since these are the interiors of closed pairwise disjoint balls in $M$). Put $U_S = \cup_{i \in S} A_i$ for $S \subset \{0, \ldots, k\}$. Consider the commutative diagram

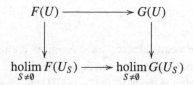

Since both $F$ and $G$ are polynomial of degree $\leq k$, the vertical arrows are weak equivalences, and by induction on $l$, the map $F(U_S) \to G(U_S)$ is a weak equivalence for all $S \neq \emptyset$, and hence the lower horizontal arrow is an equivalence. Hence the top horizontal map is a weak equivalence. The remainder of the proof is exactly the same, only we will choose the closed subsets $A_i$ differently.

Suppose the result is true for all $L$ with handle index less than $j$. Suppose $L$, again with interior $U$, has handle index $j$ and has $s$ handles of index $j$. Let $e_1, \ldots, e_s \colon D^j \times D^{m-j} \to L$ be the embeddings representing the $j$-handles of $L$. Since $j \geq 1$, we may choose pairwise disjoint closed disks $C_0, \ldots, C_k$ in $D^j$ and set $D_i = C_i \times D^{m-j}$. For each $i = 0$ to $k$, let

$$A_i = \bigcup_{h=1}^{s} e_h(C_i \times D^{m-j}) \cap U.$$

Each $A_i$ is closed in $U$ and if we set $U_S = U - \bigcup_{i \in S} A_i$ for $S \subset \{0, \ldots k\}$, for $S \neq \emptyset$, $U_S$ has a handle decomposition with handle index at most $j - 1$, so that $F(U_S) \to G(U_S)$ is an equivalence by induction for such $S$. See Figure 10.1 for a visualization.

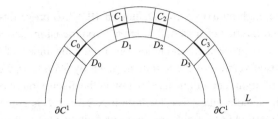

Figure 10.1 A picture of four disks $C_i$ and their thickenings $D_i$ in a 1-handle $D^1 \times D^1$ attached to $L$ along $\partial D^1 \times D^1$. Note that removing at least one of the $D_i$ leaves the manifold with one fewer 1-handles and possibly increases the number of 0-handles.

Once again consider the diagram

$$
\begin{array}{ccc}
F(U) & \longrightarrow & G(U) \\
\downarrow & & \downarrow \\
\underset{S \neq \emptyset}{\operatorname{holim}} F(U_S) & \longrightarrow & \underset{S \neq \emptyset}{\operatorname{holim}} G(U_S)
\end{array}
$$

The vertical arrows are equivalences since $F$ and $G$ are polynomial of degree $\leq k$, and the lower horizontal arrow is an equivalence since homotopy limits are homotopy invariant. Hence the map $F(U) \to G(U)$ is a weak equivalence.  □

## 10.2.2 The construction of the Taylor tower

We now define the polynomial approximations $T_k F$ to a functor $F$. The construction is simply a "restrict and extend" procedure (specifically, we are taking Kan extensions). This is the manifold calculus analog of Definition 10.1.25.

**Definition 10.2.15** For a finitary isotopy contravariant functor $F$, define, for each $U \in O(M)$, the *kth polynomial approximation of $F$* to be

$$
T_k F(U) = \underset{V \in O_k(U)}{\operatorname{holim}} F(V).
$$

In the language of Definition 8.4.2, $T_k F$ is the homotopy right Kan extension of $F$ along the inclusion $O_k(M) \to O(M)$. When $k = 0$, $O_0(M)$ consists only of the empty set, and hence $T_0 F(U) = F(\emptyset)$ for all $U$. By definition, $T_k F(U)$ is determined by its values on at most $k$ open balls in $U$, although it is somewhat delicate to prove this functor is polynomial of degree $\leq k$. The indexing category is rather unwieldy, but we will see below in Example 10.2.18 that, for any $U$ which is the interior of a smooth compact codimension zero submanifold $L$ of $M$, $T_k F(U)$ can be expressed as a homotopy limit of $F$ evaluated on at most

$k$ open balls. In this case the finite indexing category will depend, among other things, on a handle decomposition of $L$.

**Theorem 10.2.16** ([Wei99, Theorems 3.9 and 6.1])

1. $T_k F$ *is polynomial of degree* $\leq k$.
2. *There is a natural transformation of contravariant functors* $\tau_k \colon F \to T_k F$ *which is a weak equivalence if the input is diffeomorphic to at most $k$ open balls.*
3. *If $F$ is polynomial of degree* $\leq k$, *then $F \to T_k F$ is a weak equivalence.*
4. *The natural transformation* $\tau_k \colon T_k F \to T_k(T_k F)$ *is a weak equivalence.*
5. *If $F \to G$ is a natural transformation and $G$ is polynomial of degree* $\leq k$, *then up to weak equivalence the transformation factors through $T_k F$.*

*Proof*   We omit the proof of the first statement owing to its length, but we strongly encourage the reader to look at [Wei99, Theorem 3.9] and the preceding discussion. The ideas are thematically related to those used to prove homology satisfies excision, and in particular the ideas are in the same vein as our purely homotopy-theoretic proof of Theorem 6.2.1. The inclusion $O_k(U) \to O(U)$ of categories gives rise to a map of homotopy limits

$$\operatorname*{holim}_{V \in O(U)} F(V) \longrightarrow \operatorname*{holim}_{V \in O_k(U)} F(V).$$

By definition the target of this map is $T_k F(U)$. Since $U \in O(U)$ is the final object, the domain is weakly equivalent to $F(U)$ by Example 8.2.5 (remember that $F$ is contravariant, so $U$ is actually an initial object in the opposite category). The last item is a straightforward unraveling of the definitions.   □

**Example 10.2.17**   For $U \in O(M)$, let $U \mapsto \operatorname{Emb}(U, N)$ be the embedding functor. We claim that $T_1 \operatorname{Emb}(U, N) = \operatorname{Imm}(U, N)$, the space of immersions. To see this, note that if $U$ is at most one open ball, then the inclusion $\operatorname{Emb}(U, N) \to \operatorname{Imm}(U, N)$ is a homotopy equivalence (because evaluation of the derivative at some point in the ball fibers each space over the space of vector bundle monomorphisms, and the fibers of each of these fibrations are contractible). It follows from Theorem 10.2.14 that the induced map $T_1 \operatorname{Emb}(U, N) \to T_1 \operatorname{Imm}(U, N)$ is a weak equivalence for all $U$. But $\operatorname{Imm}(U, N) \simeq T_1 \operatorname{Imm}(U, N)$ since, from Example 10.2.10, we know that the immersions functor is polynomial of degree $\leq 1$.   □

**Example 10.2.18**   We will use the following in Section 10.4.1. This example details how to express $T_k F$ in a non-functorial way (i.e. by a space which is not a functor of $U$) by a homotopy limit over a finite and acyclic category.

For any functor $F$, if $U$ is the interior of a smooth compact codimension-zero submanifold $L$ of $M$, then $T_k F(U)$ can be expressed as a finite homotopy limit of values of $F$ on at most $k$ open balls in $M$. We will sketch the argument of how to do this using the proof of Theorem 10.2.14 and part 3 of Theorem 10.2.16. Suppose the handle index of $L$ is $p$. Choosing pairwise disjoint closed subsets $\{A_i\}_{i=0}^{k}$ in $U$ and letting $U_S = U - \bigcup_{i \in S} A_i$ as in the proof of Theorem 10.2.14, the canonical map

$$T_k F(U) \longrightarrow \operatorname*{holim}_{S \neq \emptyset} T_k F(U_S)$$

is an equivalence since $T_k F$ is polynomial of degree $k$ and each $U_S$ is the interior of a smooth compact codimension-zero submanifold of $M$ with handle index less than $p$. Hence $T_k F(U)$ can always be expressed by a finite homotopy limit of values of $T_k F$ on open sets which have handle index strictly smaller than that of $U$. We can continue to apply this to each $T_k F(U_S)$ for $S \neq \emptyset$, noting that each $U_S$ is the interior of a compact codimension-zero submanifold of $M$ and hence has finitely many handles of a given index. Continuing in this way, we may choose $A_i(S)$ for each $S$ so that the handle dimension of $U_S - \bigcup_{i \in R} A_i(S)$ has strictly smaller handle dimension than $U_S$. We can thus reduce to only needing the values of $T_k F$ at open sets which are themselves a finite union of open balls, noting that $F$ and $T_k F$ agree when the input is at most $k$ open balls. Furthermore, we can reduce to at most $k$ open balls using the base case of the induction in Theorem 10.2.14. Finally, part 3 of Theorem 10.2.16 tells us that we may replace $T_k F(V)$ with $F(V)$ if $V$ is a disjoint union of at most $k$ open balls. $\qquad\square$

The sequence of inclusions $O_0(M) \subset O_1(M) \subset \cdots \subset O_k(M) \subset \cdots \subset O(M)$ gives rise to a sequence of restriction maps of homotopy limits $T_k F \to T_{k-1} F$, and hence to the *Taylor tower* of $F$:

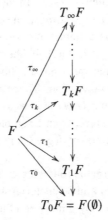

By Theorem 8.6.1, this is in fact a tower of fibrations. Here $T_\infty F$ denotes the homotopy inverse limit of this tower, $\operatorname{holim}_k T_k F$ (though we may also use the ordinary inverse limit since this is a tower of fibrations).

Our next goal is to classify homogeneous degree $k$ functors. Although their definition is independent of the Taylor tower, they arise naturally in this setting as the difference between the stages of the tower.

### 10.2.3 Homogenous functors

**Definition 10.2.19** We call a contravariant functor $F\colon O(M) \to$ Top *homogenous of degree* $k$ if it is polynomial of degree $\leq k$ and $T_{k-1}F \simeq *$. That is, $T_{k-1}F(U)$ is weakly contractible for all $U$.

Here is the analog of Definition 10.1.38.

**Definition 10.2.20** Define the *$k$th layer of the manifold calculus Taylor tower* of $F$ to be

$$L_k F = \operatorname{hofiber}(T_k F \longrightarrow T_{k-1}F).$$

**Example 10.2.21** Suppose $F\colon O(M) \to$ Top is any contravariant functor. Choose a basepoint in $F(M)$, assuming one exists. This bases $T_k F(U)$ for all $U$ and for all $k$. Then $L_k F = \operatorname{hofiber}(T_k F \to T_{k-1}F)$ is homogeneous of degree $\leq k$. It is polynomial of degree $\leq k$ since it is a homotopy limit of polynomials of degree $\leq k$ by Proposition 10.1.15, and it is homogeneous of degree $k$ since

$$T_{k-1}L_k F \simeq \operatorname{hofiber}(T_{k-1}T_k F \to T_{k-1}T_{k-1}F) \simeq *$$

as $T_k$ commutes with homotopy limits (Proposition 8.5.5) and since $T_k T_l F$ is $T_{\min\{k,l\}}F$ (this is easy to see). □

**Example 10.2.22** Recall the space of $k$ unordered configurations $\binom{M}{k}$ from Example 10.2.12, let $p\colon Z \to \binom{M}{k}$ be a fibration, and as before let $\Gamma\left(\binom{M}{k}, Z; p\right)$ denote its space of sections. As mentioned in Example 10.2.12, the functor $U \mapsto \Gamma\left(\binom{U}{k}, Z; p\right)$ is polynomial of degree $\leq k$. Let $N$ be the poset of all neighborhoods of the fat diagonal (i.e. neighborhoods containing all diagonals) in $\binom{M}{k}$. Define a contravariant functor on $O(M)$ by

$$U \longmapsto \Gamma\left(\partial\binom{U}{k}, Z; p\right) = \operatorname*{hocolim}_{Q \in N} \Gamma\left(\binom{U}{k} \cap Q, Z; p\right).$$

It can be shown that $\Gamma\left(\partial\binom{U}{k}, Z; p\right)$ is a contravariant finitary isotopy functor. It is polynomial of degree $\leq k$ basically for the same reasons $\Gamma\left(\binom{U}{k}, Z; p\right)$ is. However, it is also polynomial of degree $\leq k - 1$. The idea is to use Theorem 10.2.14 as well as its proof together with the pigeonhole principle, as follows. By Theorem 10.2.14 it is enough to prove that the map

$$\Gamma\left(\partial\binom{U}{k}, Z; p\right) \longrightarrow T_{k-1}\Gamma\left(\partial\binom{U}{k}, Z; p\right)$$

is a weak equivalence for $U \in O_k(M)$ since both domain and codomain are polynomial of degree $\leq k - 1$. If $U \in O_j(M)$ for some $j < k$ then this is automatic, as $U$ is a final object in the indexing category for $T_{k-1}$, and so we only need prove it is an equivalence when $U$ is diffeomorphic to a disjoint union of exactly $k$ open balls $B_0, \ldots, B_{k-1}$. Let $A_i = B_i$, so that the $A_i$ are pairwise disjoint closed in $U$, and let $U_S = U - \bigcup_{i \in S} A_i$. In the diagram

$$
\begin{array}{ccc}
\Gamma\left(\partial\binom{U}{k}, Z; p\right) & \longrightarrow & \underset{S \neq \emptyset}{\operatorname{holim}} \, \Gamma\left(\partial\binom{U}{k}, Z; p\right) \\
\downarrow & & \downarrow \\
T_{k-1}\Gamma\left(\partial\binom{U}{k}, Z; p\right) & \longrightarrow & \underset{S \neq \emptyset}{\operatorname{holim}} \, T_{k-1}\Gamma\left(\partial\binom{U}{k}, Z; p\right)
\end{array}
$$

the right vertical arrow is an equivalence because objectwise it is a weak equivalence and by the homotopy invariance of homotopy limits (Theorem 8.3.1). The bottom horizontal arrow is a weak equivalence since $T_{k-1}\Gamma\left(\partial\binom{U}{k}, Z; p\right)$ is polynomial of degree $\leq k-1$. The upper horizontal arrow is a weak equivalence because, by the pigeonhole principle,

$$\binom{U}{k} \cap Q = \bigcup_i \binom{U - A_i}{k} \cap Q$$

for a sufficiently small neighborhood $Q$ (the reader may wish to draw a picture of this in the case $k = 2$ with $U$ an open interval). It follows that the left vertical arrow is a weak equivalence. $\square$

It can be shown that in fact

$$T_{k-1}\Gamma\left(\binom{U}{k}, Z; p\right) \simeq T_{k-1}\Gamma\left(\partial\binom{U}{k}, Z; p\right)$$

and hence the functor

$$U \longmapsto \operatorname{hofiber}\left(\Gamma\left(\binom{U}{k}, Z; p\right) \longrightarrow \Gamma\left(\partial\binom{U}{k}, Z; p\right)\right).$$

is homogeneous of degree $k$. The next result says that this is the universal example of a homogeneous degree $k$ functor.

**Theorem 10.2.23** ([Wei99, Theorem 8.5], Classification of homogeneous functors) *Let $E: O(M) \to$ Top be homogeneous of degree $k$. Then for some fibration $p: Z \to \binom{U}{k}$ there is a weak equivalence*

$$E(U) \longrightarrow \Gamma^c \left( \binom{U}{k}, Z; p \right)$$

*natural in $U$ (i.e. a weak equivalence of functors).*

We omit the proof. For $F: O(M) \to$ Top a contravariant functor, Theorem 10.2.23 thus says that the homogeneous degree $k$ functor $L_k F$ is equivalent to the space of compactly supported sections of a fibration. That is,

$$L_k F(U) \simeq \Gamma^c \left( \binom{U}{k}, Z_k; p_k \right).$$

The fibers $p_k^{-1}(S)$ over $S \in \binom{U}{k}$ are the total homotopy fibers of a cubical diagram of values of $F$ on a tubular neighborhood of $F$. That is,

$$p_k^{-1}(S) \simeq \text{tfiber}(T \mapsto F(B_T))$$

where $T \subset S$ and $B_T$ is a tubular neighborhood of $S_T$ in $U$. In place of $p_k^{-1}(S)$ we write $F^{(k)}(\emptyset)$, and think of the fibers as the derivatives of $F$ evaluated at the empty set. We justify this as follows. By definition,

$$F^{(1)}(\emptyset) = \text{hofiber}(F(B) \to F(\emptyset))$$

is the analog of $f(h) - f(0)$, part of the difference quotient arising in the definition of the derivative of a single-variable real-valued function. The space $F^{(2)}(\emptyset)$ is the total homotopy fiber of the square

$$
\begin{array}{ccc}
F(B_1 \sqcup B_2) & \longrightarrow & F(B_1) \\
\downarrow & & \downarrow \\
F(B_2) & \longrightarrow & F(\emptyset)
\end{array}
$$

If we try to write the second derivative of $f: \mathbb{R} \to \mathbb{R}$ strictly in terms of $f$, we will encounter the expression $f(h_1 + h_2) - f(h_1) - f(h_2) + f(0)$ in our difference quotient, in analogy with thinking of the total homotopy fiber above as an iterated homotopy fiber.

Thus, through the lens of single-variable calculus, the Taylor tower of a finitary isotopy contravariant functor $F: O(M) \to$ Top is the analog of the

Maclaurin series of $f \colon \mathbb{R} \to \mathbb{R}$. The analog of $f^{(k)}(0)x^k/k!$ is precisely our section space: the space of (twisted) maps from $\binom{M}{k}$ to $F^{(k)}(\emptyset)$.

**Example 10.2.24**  We will compute the first two derivatives of the embedding functor $U \mapsto \mathrm{Emb}(U, N)$. Clearly $\mathrm{Emb}(\emptyset, N) = *$, so

$$\mathrm{Emb}^{(1)}(\emptyset) = \mathrm{hofiber}(\mathrm{Emb}(B, N) \to *) \simeq \mathrm{Imm}(B, N)$$

is the space of immersions of an open ball in $N$. For the second derivative, first note that if $B_1 \sqcup \cdots \sqcup B_k$ is a disjoint union of $k$ open balls in $M$, then $\mathrm{Emb}(B_1 \sqcup \cdots \sqcup B_k, N) \simeq \mathrm{Conf}(k, N) \times \prod_{i=1}^{k} \mathrm{Imm}(B_i, N)$, where $\mathrm{Conf}(k, N)$ is the configuration space of $k$ points in $N$, and the equivalence is given by the obvious extension of the equivalence $\mathrm{Emb}(B, N) \simeq \mathrm{Imm}(B, N)$ for an open ball $B$, described in Example 10.2.17. Therefore $\mathrm{Emb}^{(2)}(\emptyset, N)$ is homotopy equivalent to the total homotopy fiber of the square

$$\mathrm{Conf}(2, N) \times \mathrm{Imm}(B_1, N) \times \mathrm{Imm}(B_2, N) \longrightarrow \mathrm{Conf}(1, N) \times \mathrm{Imm}(B_1, N)$$
$$\downarrow \qquad\qquad\qquad\qquad\qquad\qquad\qquad\qquad\qquad \downarrow$$
$$\mathrm{Conf}(1, N) \times \mathrm{Imm}(B_2, N) \longrightarrow\qquad\qquad\qquad\qquad\qquad\qquad *$$

where all of the maps are the obvious restrictions to one or no points. Viewing this total homotopy fiber as the map from the initial space to the homotopy limit of the punctured square, it is clear that the total homotopy fiber of the square above is homotopy equivalent to the total homotopy fiber of the square

$$\mathrm{Conf}(2, N) \longrightarrow \mathrm{Conf}(1, N)$$
$$\downarrow \qquad\qquad\qquad \downarrow$$
$$\mathrm{Conf}(1, N) \longrightarrow *$$

One could be content with this answer, but it is interesting to try to pin down the homotopy type a bit better. If we think of the total homotopy fiber iteratively, by taking homotopy fibers vertically and noting that the vertical restriction maps are fibrations, we see that $\mathrm{Emb}^{(2)}(\emptyset, N) \simeq \mathrm{hofiber}(N - \{y\} \to N)$, where $(x, y) \in \mathrm{Conf}(2, N)$ is the basepoint. To analyze this space further, let $T$ be a closed tubular neighborhood of $y$ in $N$ (a closed ball surrounding $y$), and note that the inclusion $N - T \to N - \{y\}$ is a homotopy equivalence. Consider the square of inclusions

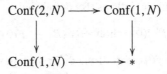

This is a homotopy pushout square, and if we fiber over $N$ everywhere, by Corollary 3.9.4 we obtain a homotopy pushout square

$$
\begin{array}{ccc}
\text{hofiber}(N - \partial T \to N) & \longrightarrow & \text{hofiber}(T \to N) \\
\downarrow & & \downarrow \\
\text{hofiber}(N - \{y\} \to N) & \longrightarrow & *
\end{array}
$$

The map of vertical cofibers is an equivalence, and the homotopy cofiber of the top horizontal map is equivalent to $\Sigma^n \Omega(N)_+$ by thinking of $T$ as a (trivial) bundle over $T$. Thus we have a weak equivalence

$$\Sigma^n \Omega(N)_+ \simeq \Sigma \, \text{hofiber}(N - \{y\} \to N) \simeq \Sigma \, \text{Emb}^{(2)}(\emptyset, N). \qquad \square$$

**Example 10.2.25** The "cancellation" of the contribution of the space of immersions in the higher derivatives persists, and the derivatives of embedding spaces are given by the total homotopy fibers of cubes of configurations as above. Although the homotopy type of $\text{Emb}^{(k)}(\emptyset, N)$ is difficult to get a handle on, one very important observation is that the connectivities of these spaces increases to infinity with $k$, provided $n > 2$. This was the subject of Examples 5.5.6 and 6.2.9. $\qquad \square$

## 10.2.4 Convergence

There are two questions to ask regarding convergence of the Taylor tower. First, does the tower converge to anything at all? More precisely, for a fixed $j \geq 0$, is $\pi_j(T_k F)$ constant for all $k \geq k_j$ for some $k_j$? The second question is whether or not the tower converges to $F$. That is, does the connectivity of the map $F \to T_k F$ increase to infinity with $k$, or, put another way, is the map $F \to T_\infty F$ a weak equivalence?

These are of course separate questions, the first analogous to the convergence of the series $\sum_{k=0}^\infty f^{(k)}(0) x^k / k!$, and the second analogous to studying for what values of $x$, if any, the difference $f(x) - \sum_{k=0}^\infty f^{(k)}(0) x^k / k!$ is equal to zero. The question of whether or not the series converges is easier to deal with, so we tackle it first. A fairly satisfactory answer can be obtained from the connectivities of the layers $L_k F = \text{hofiber}(T_k F \to T_{k-1} F)$, and these connectivities can sometimes be computed using the Blakers–Massey Theorem (Theorem 6.2.1).

The fibers of the classifying fibration give one way to answer the first question. Recall that, for spaces $X$ and $Y$ with $Y$ $k$-connected and $X$ a CW complex of dimension $d$, the space $\text{Map}(X, Y)$ is $(k - d)$-connected (see Proposition 3.3.9). The same is true of spaces of sections: the connectivity of the fibers minus the dimension of the base gives the connectivity of the space of sections. Hence we have the following general result, which follows from the classification Theorem 10.2.23.

**Proposition 10.2.26**   *For a contravariant finitary isotopy functor F, if $F^{(k)}(\emptyset)$ is $c_k$-connected, then $L_k F(M)$ is $(c_k - km)$-connected (where m is the dimension of M). More generally, if U has handle index j, then $L_k F(U)$ is $(c_k - kj)$-connected.*

In any case, the Taylor tower of $F$ converges for all $U$ of handle index $\leq j$ if $c_k - kj$ tends to infinity with $k$. This suggests the handle index should play some role in an analog of radius of convergence. To begin exploring this, consider the real-valued function case. We are often interested in the error $R_k(x) = |f(x) - T_k f(x)|$ for certain $x$, where $T_k f$ stands for the $k$th degree Taylor approximation. For $f$ smooth on $[-r, r]$ and satisfying $|f^{(k+1)}| \leq M_k$ on $(-r, r)$, we have $R_k(x) \leq M_k r^{k+1}/(k+1)!$. If $M_k r^{k+1}/(k+1)! \to 0$ as $k \to \infty$, then we would say that $f$ is analytic on $(-r, r)$; that is, its Taylor series converges to it. This will lead us to the notion of a radius of convergence, the importance of the connectivities of the derivatives of $F$ (and more general cubical diagram than just those involving balls), and some ideas about how to tackle the "error" $R_k F = \text{hofiber}(F \to T_k F)$.

The radius of convergence is a positive integer $\rho$. An open set $V$ which is the interior of a smooth compact codimension-zero submanifold $L$ of $M$ is within the radius of convergence if the handle index of $L$ is less than $\rho$. Suppose $F: O(M) \to \text{Top}$ is a contravariant functor and $\rho > 0$ is an integer. For $k > 0$, let $P$ be a smooth compact codimension-zero submanifold of $M$, and $Q_0, \ldots, Q_k$ be pairwise disjoint compact codimension-zero submanifolds of $M - \mathring{P}$. Suppose further that $Q_i$ has handle index $q_i < \rho$. Let $U_S = (P \mathbin{\mathring{\cup}} Q_S)$. We then have the following manifold calculus analog of Definition 10.1.50.

**Definition 10.2.27**   The contravariant functor $F$ is *$\rho$-analytic with excess $c$* if the $(k+1)$-cube $S \mapsto F(U_S)$ is $(c + \sum_{i=0}^{k}(\rho - q_i))$-cartesian.

This is the analog of a bound on $f^{(k)}(x)$ for $x$ close to 0. In this case, "close to zero" means having small handle index, and the $(k+1)$-cube $S \mapsto F(U_S)$ is reminiscent of our definition of $F^{(k)}(\emptyset)$, with $P$ replacing $\emptyset$, and with handles of arbitrary index added, not just handles of index zero. The number $\rho$ gives the *radius of convergence of the Taylor tower of F*; that is, it tells us the input values for which the inverse limit of the tower converges to the functor. The next theorem is an estimate for the error $R_k F = \text{hofiber}(F \to T_k F)$.

**Theorem 10.2.28** ([GW99, Theorem 2.3])  *Let F be a contravariant finitary isotopy functor. If F is $\rho$-analytic with excess c, and if $U \in O(M)$ is the interior of a smooth compact codimension 0 submanifold of M with handle index $q < \rho$, then the map $F(U) \to T_k F(U)$ is $(c + (k + 1)(\rho - q))$-connected.*

The following gives a general criterion which guarantees the Taylor tower converges to the functor being approximated.

**Corollary 10.2.29** (Convergence of the Taylor tower [GW99, Corollary 2.4]) *Suppose F is $\rho$-analytic with excess c. Then, for each $U \in O(M)$ which is the interior of a compact codimension-zero submanifold of handle index less than $\rho$, the map*

$$F(U) \longrightarrow T_\infty F = \operatorname{holim}_k T_k F(U)$$

*is a weak equivalence.*

This immediately follows since the connectivities of the maps $F(U) \to T_k F(U)$ increase to infinity with $k$ if the handle index of $U$ is less than $\rho$.

We will not give the proof of Theorem 10.2.28, but a sketch will suggest its similarity to the proof of Theorem 10.2.14. We are interested in the connectivity of the map $F(U) \to T_k F(U)$ and, as usual, it suffices to study the special case where $U$ is the interior of a smooth compact codimension-zero submanifold $L$ of $M$. Using a handle decomposition, we select pairwise disjoint closed subsets $A_0, \ldots, A_k$ such that, for $S \neq \emptyset$, $U_S = U - \bigcup_{i \in S} A_i$ is the interior of a compact smooth codimension-zero submanifold whose handle index is strictly less than the handle index of $L$. We then consider the diagram

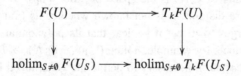

The right vertical arrow is a weak equivalence since $T_k F$ is polynomial of degree $\leq k$, and by induction we can get a connectivity estimate for the bottom horizontal arrow. We have a connectivity for the left vertical arrow by assuming $F$ is $\rho$-analytic. Together these give an estimate for the connectivity of $F(U) \to T_k F(U)$ using Proposition 2.6.15 and Proposition 5.4.17.

The next result is the analog of uniqueness of analytic continuation.

**Corollary 10.2.30** ([GW99, Corollary 2.6]) *Suppose $F_1 \to F_2$ is a natural transformation of $\rho$-analytic contravariant finitary isotopy functors, and $F_1(U) \to F_2(U)$ is a weak equivalence whenever $U \in O_k(M)$ for some $k$. Then $F_1(V) \to F_2(V)$ is a weak equivalence for each $V$ which is the interior of a smooth compact codimension-zero submanifold of handle index less than $\rho$.*

*Proof* Suppose $V \in O(M)$. Consider the following diagram.

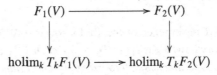

Since $F_1(U) \to F_2(U)$ is a weak equivalence whenever $U$ is in $O_k(M)$ for any $k$, it follows from Theorem 10.2.14 that $T_k F_1 \to T_k F_2$ is an equivalence for all $k$. Hence the lower horizontal arrow is an equivalence for all $V$. If the handle index of $V$ is less than $\rho$, then $F_1(V) \to \text{holim}_k T_k F_1(V)$ and $F_2(V) \to \text{holim}_k T_k F_2(V)$ are equivalences by Corollary 10.2.29, so $F_1(V) \to F_2(V)$ is an equivalence. $\qquad\qquad\qquad\qquad\qquad\qquad\qquad\qquad\qquad\qquad\qquad\qquad\square$

The question of convergence of the Taylor tower for the embedding functor $F(U) = \text{Emb}(U, N)$ is especially interesting and will be discussed separately in Section 10.3.

We end by remarking that a lack of convergence does not necessarily mean the Taylor tower does not contain anything interesting. In fact, for classical knots (where Theorem 10.3.1 does not give convergence because the codimension is 2), [Vol06b] shows that the Taylor series for the embedding functor contains finite type invariants of knots (see Section 10.4.4). On a related note, one can study multivariable contravariant functors such as $(U, V) \mapsto \text{Link}(U, V; N)$. Here $U$ and $V$ are open subsets of smooth closed manifolds $P$ and $Q$, and $\text{Link}(U, V; N)$ is the space of "link maps" $U \to N$, $V \to N$ whose images are disjoint. It is not known whether its (multivariable) Taylor series converges to it, but it is clear that its polynomial approximations are interesting, since, for example, $\text{hofiber}(\text{Link}(P, Q; N) \to T_1\text{Link}(P, Q; N))$ contains the information necessary to define the generalized linking number. See [Mun08] and [GM10].

## 10.3 Embeddings, immersions, and disjunction

Here we present an application of the manifold calculus of functors discussed in Section 10.2. The main point is to give an overview of the analyticity of the embedding functor and try to give a sense of the kind of results that go into proving such statements. In particular we want to demonstrate the importance of Theorem 6.2.3, the generalization of the Blakers–Massey Theorem. We will mostly focus our attention on the first-degree Taylor approximations to the embeddings functor, carefully obtaining a connectivity estimate for the maps $\mathrm{Emb}(M, N) \to T_1 \mathrm{Emb}(M, N) \simeq \mathrm{Imm}(M, N)$ (see Example 10.2.17) in two ways. We will also discuss the disjunction results for embeddings necessary to get estimates $\mathrm{Emb}(M, N) \to T_j \mathrm{Emb}(M, N)$ in general, although the particular results we can prove here are weaker than the best-known estimates.

### 10.3.1 Convergence of the Taylor tower for spaces of embeddings

The following theorem, due to Goodwillie and Klein, concerns the convergence of the Taylor tower for the embedding functor. A version for spaces of Poincaré embeddings has appeared in [GK08], which is itself an important step in proving the result below, which appears in [GK15].

**Theorem 10.3.1** *The functor $U \mapsto \mathrm{Emb}(U, N)$ is $(n-2)$-analytic with excess $3-n$. Hence, if $M$ is a smooth closed manifold of dimension $m$, and $N$ a smooth manifold of dimension $n$, then the map*

$$\mathrm{Emb}(M, N) \longrightarrow T_k \mathrm{Emb}(M, N)$$

*is $(k(n - m - 2) + 1 - m)$-connected. In particular, if $n - m - 2 > 0$, then the canonical map*

$$\mathrm{Emb}(M, N) \longrightarrow \underset{k}{\mathrm{holim}}\, T_k \mathrm{Emb}(M, N)$$

*is a weak equivalence.*

We will refer to a statement about the connectivity of the natural transformation $F \to T_k F$ as a *connectivity estimate*.

**Remark 10.3.2** It is shown in [Wei04] that the Taylor tower for $\mathrm{Emb}(M, N)$ also converges on homology for $n - 4m + 1 > 0$. That is, the inverse limit of the homology groups of the Taylor tower agree with those of the embedding space in this range of dimensions. This will be useful in Section 10.4.4. □

The proof of Theorem 10.3.1 is too difficult to present here, but we will illustrate a few elementary ideas that go into it.

One can obtain the connectivity of $\mathrm{Emb}(M, N) \to T_1 \mathrm{Emb}(M, N)$ "by hand" without too much work. Some of the ideas that go into one version of this computation (the second proof of Theorem 10.3.4) are important in obtaining estimates for all $k$. We will also discuss weaker estimates than predicted by Theorem 10.3.1 for the maps $\mathrm{Emb}(M, N) \to T_j \mathrm{Emb}(M, N)$ for $j \geq 2$ in Theorem 10.3.7, which relies on the material from Section 10.3.2. The techniques required for the stronger results above involve deeper relationships between embeddings, pseudoisotopies, diffeomorphisms, as well as some surgery theory.

There are two main ingredients in the proofs of the connectivity estimates we present here. One is "handle induction", as in the proof of Theorem 10.2.14, where we use the property that the polynomial approximations to a functor in the manifold calculus (a) behave well with respect to certain kinds of cubes (roughly, a polynomial of degree $\leq k$ takes strongly homotopy cocartesian $(k + 1)$-cubes to cartesian $(k + 1)$-cubes), and (b) are determined by their values on open balls to reduce to a special case where the manifold in question is built from 0-handles, that is, is a union of open balls. The other main ingredient is "multiple disjunction" for spaces of embeddings, where we use Theorem 6.2.3 to obtain certain connectivity estimates for embedding spaces necessary in the handle induction arguments. The disjunction problems are a little technical, so we have relegated that material to Section 10.3.2.

### Connectivity of the derivatives of the embedding functor

The first step in understanding some of the ideas that go into establishing the analyticity of the embedding functor is to compute the connectivity of its derivatives. This turns out to be an easy application of Theorem 6.2.3 which we have essentially already presented in Example 6.2.9. We demonstrate here how to carefully reduce to that case.

Recall from the discussion following Theorem 10.2.23 the notion of the $k$th derivative of a functor $F$ evaluated at the empty set, $F^{(k)}(\emptyset)$.

**Theorem 10.3.3** *Let $U = \coprod_i B_i \subset M$ be a disjoint union of $k$ open balls. For $S \subset \underline{k}$, let $U_S = U - \bigcup_{i \in S} B_i$. The $k$-cube $S \mapsto \mathrm{Emb}(U_S, N)$ is $((k - 1)(n - 2) + 1)$-cartesian. That is, if $E(U) = \mathrm{Emb}(U, N)$, then $\mathrm{Emb}^{(k-1)}(\emptyset, N)$ is $(k - 1)(n - 2)$-connected.*

*Proof* We will simplify to Example 6.2.9. For a subset $S$ of $\underline{k}$, the projection map $\prod_{i \notin S} B_i \times \mathrm{Emb}(U_S, N) \to \mathrm{Emb}(U_S, N)$ is a homotopy equivalence, natural in $S$, because balls and products of balls are contractible (if $S = \underline{k}$, we take $\prod_{i \notin S} B_i$ to be a point). Let $\mathrm{Conf}(j, N)$ as usual denote the configuration space of $j$ points in $N$. The map

$$\prod_{i \notin S} B_i \times \mathrm{Emb}(U_S, N) \longrightarrow \mathrm{Conf}(k - |S|, N) \times \mathrm{Imm}(U_S, N)$$

induced by the map

$$((x_1, \ldots, x_k), f) \mapsto ((f(x_1), \ldots, f(x_k)), (df_{x_1}, \ldots, df_{x_k}))$$

is a homotopy equivalence for all $S$ (where again the product of balls is taken to be a point if $S = \underline{k}$). Hence $S \mapsto \mathrm{Emb}(U_S, N)$ is $K$-cartesian if and only if $S \mapsto \mathrm{Conf}(k - |S|, N) \times \mathrm{Imm}(U_S, N)$ is $K$-cartesian. The cube $S \mapsto \mathrm{Imm}(U_S, N)$ is homotopy cartesian whenever $k \geq 2$ because $\mathrm{Imm}(-, N)$ is polynomial of degree $\leq 1$. Therefore $S \mapsto \mathrm{Emb}(U_S, N)$ is $K$-cartesian if and only if $S \mapsto \mathrm{Conf}(k - |S|, N)$ is $K$-cartesian for $k \geq 2$ by Proposition 5.4.5. But Example 6.2.9 shows that $K = (k - 1)(n - 2)$.                                    □

### Connectivity estimates for the linear stage

Theorem 10.3.3 turns out to be a very useful special case, as we shall see below, in obtaining connectivity estimates for the maps $\mathrm{Emb}(M, N) \to T_1 \mathrm{Emb}(M, N) \simeq \mathrm{Imm}(M, N)$. We begin with the map $\mathrm{Emb}(M, N) \to T_1 \mathrm{Emb}(M, N)$. Our proof requires a disjunction result from Section 10.3.2.

**Theorem 10.3.4**  *The map*

$$\mathrm{Emb}(M, N) \longrightarrow T_1 \mathrm{Emb}(M, N) \simeq \mathrm{Imm}(M, N)$$

*is $(n - 2m - 1)$-connected. In fact, if $V \subset M$ is the interior of a compact codimension-zero handlebody with handle index $k$, then the map is $(n - 2k - 1)$-connected.*

*Proof*  We induct on the handle index $k$. For the base case $k = 0$, let $l$ be the number of components of $V$. The result is trivial, and the map in question is a homotopy equivalence, when $l = 0, 1$. Suppose $l \geq 2$. Consider the sequence (from the Taylor tower of embeddings)

$$\mathrm{Emb}(V, N) \to T_l \mathrm{Emb}(V, N) \to T_{l-1} \mathrm{Emb}(V, N) \to$$
$$\cdots \to T_1 \mathrm{Emb}(V, N). \quad (10.3.1)$$

The map $\mathrm{Emb}(V, N) \to T_l \mathrm{Emb}(V, N)$ is a weak equivalence since $V$ is a final object in $O_l(V)$, using Proposition 9.3.4. By the classification Theorem 10.2.23 of homogeneous functors,

$$L_j \mathrm{Emb}(V, N) = \mathrm{hofiber}(T_j \mathrm{Emb}(V, N) \to T_{j-1} \mathrm{Emb}(V, N))$$

is weakly equivalent as a functor of $V$ to a space of sections

$$\Gamma^c\left(\binom{V}{j}, \mathrm{Emb}^{(j)}(\emptyset)\right).$$

Since $V$ has handle index 0, $\binom{V}{j}$ also has handle index 0, and so these section spaces are $(j-1)(n-2)$-connected by Theorem 10.3.3, for when $\binom{V}{j}$ has handle index zero, it has the homotopy type of a set of points. In other words, the map $T_j\,\mathrm{Emb}(V,N) \to T_{j-1}\,\mathrm{Emb}(V,N)$ is $((j-1)(n-2)+1)$-connected. This is true no matter what basepoint is chosen, provided $m < n$. It follows that the composed map $\mathrm{Emb}(V,N) \to T_1\,\mathrm{Emb}(V,N)$ is $(n-1)$-connected.

Now suppose $k > 0$. Let $V = \mathring{L}$. For $j = 1$ to $s$, let $e_j\colon D^k \times D^{n-k} \to L$ denote each of the $k$-handles. Assume $e_j^{-1}(\partial L) = \partial D^k \times D^{n-k}$ for all $j$. Since $k > 0$, as in the proof of Theorem 10.2.14 we may choose pairwise disjoint closed disks $D_0, D_1$ in the interior of $D^k$, and put $A_i^j = e_j(D_i \times D^{n-k}) \cap V$. Then each $A_i^j$ is closed in $V$, and if we set $A_i = \bigcup_{j=1}^s A_i^j$, then for each non-empty subset $S$ of $\{0,1\}$, $V_S = V - \bigcup_{i \in S} A_i$ is the interior of a smooth compact codimension-zero submanifold of $M$ of handle index strictly less than $k$.

In the diagram

$$
\begin{array}{ccc}
\mathrm{Emb}(V,N) & \longrightarrow & T_1\,\mathrm{Emb}(V,N) \\
\downarrow & & \downarrow \\
\mathrm{holim}_{S \neq \emptyset}\,\mathrm{Emb}(V_S,N) & \longrightarrow & \mathrm{holim}_{S \neq \emptyset}\,T_1\,\mathrm{Emb}(V_S,N)
\end{array}
$$

the right vertical arrow is a weak equivalence because $T_1\,\mathrm{Emb}(-,N)$ is polynomial of degree $\leq 1$. By induction for all $S \neq \emptyset$, $\mathrm{Emb}(V_S,N) \to T_1\,\mathrm{Emb}(V_S,N)$ is $(n-2(k-1)-1)$-connected, and by Proposition 5.4.17, the map of homotopy limits has connectivity equal to $n - 2(k-1) - 1 - 2 + 1 = n - 2k$. By Corollary 10.3.16, the left vertical map is $(n-2k-1)$-connected, and it follows that the top horizontal map is $(n-2k-1)$-connected. $\qquad\square$

**Remark 10.3.5** The base case of the induction on handle index above required an argument which was different than the inductive step. Although it is tempting to try to use Proposition 5.4.17 in the way we did above in the handle index zero case, it will not give the desired estimate, and genuinely appears to require a different method. Fortunately all it required was connectivity information about the derivatives, which is the content of Theorem 10.3.3. $\qquad\square$

### Connectivity estimates for the higher stages
The second proof of Theorem 10.3.4 can be adapted to prove the following restatement of Theorem 10.3.1 with very few changes.

**Theorem 10.3.6** *If M is a smooth closed manifold of dimension m, N a smooth manifold of dimension n, $n - m \geq 2$, and V is the interior of a smooth codimension-zero submanifold of handle index j, then the map*

$$\mathrm{Emb}(V, N) \longrightarrow T_k \mathrm{Emb}(V, N)$$

*is $(k(n - j - 2) + 1 - j)$-connected.*

The only changes (besides the connectivity estimates themselves) are that the pairwise disjoint closed subsets $A_i$ chosen in the proof of Theorem 10.3.4 are $k + 1$ in number, and instead of referencing Proposition 10.3.9, we reference Theorem 10.3.10. However, we do not provide a proof of Theorem 10.3.10 here, but instead a weaker version, Theorem 10.3.11, which then gives a weaker estimate than above. We leave the details to the reader to sketch a proof of the following result.

**Theorem 10.3.7** (Weak version of Theorem 10.3.6) *If M is a smooth closed manifold of dimension m, N a smooth manifold of dimension n, $n - m \geq 2$, and V is the interior of a smooth codimension-zero submanifold of handle index j, then the map*

$$\mathrm{Emb}(V, N) \longrightarrow T_k \mathrm{Emb}(V, N)$$

*is $(k(n - 2j - 2) + 1)$-connected.*

**Remark 10.3.8** The numbers in Theorem 10.3.7 and Theorem 10.3.6 agree when $k = 1$, but otherwise the estimate in Theorem 10.3.6 is better than the one in Theorem 10.3.7.

## 10.3.2 Disjunction results for embeddings

For the second proof of Theorem 10.3.4 we needed an estimate for how cartesian a square of the form

$$\mathcal{E} = \begin{array}{ccc} \mathrm{Emb}(V, N) & \longrightarrow & \mathrm{Emb}(V_0, N) \\ \downarrow & & \downarrow \\ \mathrm{Emb}(V_1, N) & \longrightarrow & \mathrm{Emb}(V_{01}, N) \end{array}$$

is. Here $V = V_\emptyset$ is the interior of some smooth compact codimension-zero submanifold of $M$ with handle index $k$, and, for $S \neq \emptyset$, the $V_S$ are the interiors of compact codimension-zero submanifolds of handle index strictly less than $k$. As in the proof of Theorem 10.3.4, let $V = \mathring{L}$. We chose each $A_i$ to be a union of products of a $k$-dimensional disk with an $(m - k)$-dimensional disk. Note that $L_S = L - \bigcup_{i \in S} A_i$ is *not* compact, but its interior is the interior of a smooth compact codimension-zero submanifold of $M$. This is important to note because below we will work not with the open sets that appear in $\mathcal{E}$, but with their closed counterparts $L$ and the $A_i$.

Let us first consider a formally similar situation.

**Proposition 10.3.9**  *Suppose $Q_0$ and $Q_1$ are smooth closed manifolds of dimensions $q_1$ and $q_2$ respectively, and $N$ is a smooth closed $n$-dimensional manifold. The square*

$$\begin{array}{ccc} \mathrm{Emb}(Q_0 \cup Q_1, N) & \longrightarrow & \mathrm{Emb}(Q_0, N) \\ \downarrow & & \downarrow \\ \mathrm{Emb}(Q_1, N) & \longrightarrow & \mathrm{Emb}(\emptyset, N) \end{array}$$

*is $(n - q_0 - q_1 - 1)$-cartesian.*

*Proof*  It is enough by Proposition 5.4.12 to choose a basepoint in $e \in \mathrm{Emb}(Q_0 \cup Q_1, N)$, take fibers vertically, and compute the connectivity of the map of homotopy fibers. By the isotopy extension theorem [Kos93, Theorem 5.2], the map $\mathrm{Emb}(Q_0 \cup Q_1, N) \to \mathrm{Emb}(Q_1, N)$ is a fibration with fiber over $e$ equal to $\mathrm{Emb}(Q_0, N - e(Q_1))$. We thus have to show that the inclusion map of vertical fibers $\mathrm{Emb}(Q_0, N - e(Q_1)) \to \mathrm{Emb}(Q_0, N)$ is $(n - q_0 - q_1 - 1)$-connected.

Let $h \colon S^k \to \mathrm{Emb}(Q_0, N)$. We may regard this as a map $\widetilde{h} \colon S^k \times Q_0 \to N$, and by a small homotopy of $h$ we may assume $\widetilde{h}$ is both smooth and transverse to $e(Q_1) \subset N$ by Theorem A.2.9 and Theorem A.2.19 with $k = 0$. By Theorem A.2.18, $\widetilde{h}^{-1}(e(Q_1))$ is a submanifold of $S^k \times Q_0$ of dimension $k + q_0 + q_1 - n$, and hence is empty if $k < n - q_0 - q_1$. A similar argument shows that any homotopy $S^k \times I \to \mathrm{Emb}(Q_0, N)$ lifts to $\mathrm{Emb}(Q_0, N - e(Q_1))$ if $k < n - q_0 - q_1 - 1$, and hence the map in question is $(n - q_0 - q_1 - 1)$-connected. □

The problem solved in Proposition 10.3.9 is known as a *disjunction problem*. The question we are answering is this: when do embeddings $Q_0 \to N$ and $Q_1 \to N$ give rise to an embedding (up to isotopy) of $Q_0 \sqcup Q_1$ in $N$? That is, we are asking when we can eliminate possible intersections between the manifolds $Q_0$ and $Q_1$ in $N$. We already discussed disjunction in Example 4.2.20, whose problem is the fiber of a disjunction problem like the one we are posing here.

The proof of the following goes beyond the scope of this book.

**Theorem 10.3.10**  (Strong disjunction [GK08, Conjecture A])  *Suppose $Q_0, Q_1, \ldots, Q_k$ are smooth closed manifolds of dimensions $q_0, \ldots, q_k$, and $N$ is a smooth closed $n$-dimensional manifold. The $k$-cube*

$$S \longmapsto \mathrm{Emb}(Q_S, N)$$

*is* $(1 - q_0 + \sum_{i=1}^{k}(n - q_i - 2))$-*cartesian.*

A more symmetric way to write the number above is $1 + (n - 2) - \sum_{i=0}^{k} q_i$. We will prove the following weaker version, which gives rise to weaker estimates for the maps $\mathrm{Emb}(M, N) \to T_j \mathrm{Emb}(M, N)$.

**Theorem 10.3.11** (Weak disjunction [Goo, Proposition A.1]) *Suppose $Q_0, Q_1, \ldots, Q_k$ are smooth closed manifolds of dimensions $q_0, \ldots, q_k$, and $N$ is a smooth closed n-dimensional manifold. Assume $q_0 \le q_i$ for $i = 1$ to $k$, and let $Q_S = \cup_{i \notin S} Q_i$. The k-cube*

$$S \longmapsto \mathrm{Emb}(Q_S, N)$$

*is* $(1 + \sum_{i=1}^{k}(n - q_0 - q_i - 2))$-*cartesian.*

The hypothesis that $q_0 \le q_i$ for all $i = 1$ to $k$ can clearly be achieved in general by choosing the minimum of $\{q_0, q_1, \ldots, q_k\}$ and relabeling so that it is $q_0$.

*Proof* The proof is very similar to the proof of Proposition 10.3.9, except that we need to invoke Theorem 6.2.3. Choose a basepoint in $e \in \mathrm{Emb}(Q_\emptyset, N)$, and consider the $k$-cube of homotopy fibers

$$R \longmapsto \mathrm{hofiber}_e \left( \mathrm{Emb}(Q_0 \bigcup_{i \notin R} Q_i, N) \to \mathrm{Emb}(\bigcup_{i \notin R} Q_i, N) \right),$$

where $R \subset \underline{k}$. By Proposition 5.4.12 the $(k + 1)$-cube $S \mapsto \mathrm{Emb}(Q_S, N)$ is $K$-cartesian if the $k$-cube

$$R \longmapsto \mathrm{hofiber}_e \left( \mathrm{Emb}(Q_0 \bigcup_{i \notin R} Q_i, N) \to \mathrm{Emb}(\bigcup_{i \notin R} Q_i, N) \right)$$

is $K$-cartesian. For each $R$ the map

$$\mathrm{Emb}\left( Q_0 \bigcup_{i \notin R} Q_i, N \right) \longrightarrow \mathrm{Emb}\left( \bigcup_{i \notin R} Q_i, N \right)$$

is a fibration with fiber over $e$ equal to $\mathrm{Emb}(Q_0, N - e(\cup_{i \notin R} Q_i))$.

Consider a face $\partial_0^R \mathrm{Emb}(Q_0, N - \cup_{i \notin R} Q_i)$. We claim this face is $(-1 + |R|(n - q_0) + \sum_{i \in R} q_i)$-cocartesian. Without loss of generality assume $R = \underline{r}$. First, note that this is a cube of inclusions of open sets, so

$$\underset{T \subsetneq R}{\mathrm{hocolim}}\, \mathrm{Emb}\left( Q_0, N - e\left( \bigcup_{i \notin R} Q_i \right) \right) \simeq \underset{T \subseteq R}{\mathrm{colim}}\, \mathrm{Emb}\left( Q_0, N - e\left( \bigcup_{i \notin R} Q_i \right) \right)$$

by Example 5.8.3. Let $S^j \to \mathrm{Emb}(Q_0, N - e(\bigcup_{i \notin R} Q_i))$ be a map, and let $h \colon S^k \times Q_0 \to N - \bigcup_{i \notin R} Q_i$ be its adjoint. We wish to deform this to a map which, for each $s \in S^j$, misses some $Q_i$ for some $i \notin R$. Consider the map

$$H \colon S^j \times \prod_{i \in R} Q_0 \longrightarrow \prod_{i \in R} N$$

given by $H(s, x_1, \ldots, x_r) = (h(s, x_1), \ldots, h(s, x_r))$. By a small homotopy of $h$ using Theorems A.2.9 and A.2.19 with $k = 0$ we may assume this map is smooth and transverse to $\prod_{i \in R} Q_i \subset \prod_{i \in R} N$, and hence by Theorem A.2.18 the inverse image $H^{-1}(\prod_{i \in R} Q_i)$ is a smooth submanifold of $S^j \times \prod_{i \in R} Q_0$ of dimension $j + |R|q_0 - \sum_{i \in R}(n - q_i)$ and is therefore empty if $j < |R|(n - q_0) - \sum_{i \in R} q_i$. A similar argument with 1-parameter families shows that a homotopy $S^j \times I \to \mathrm{Emb}(Q_0, N - \bigcup_{i \notin R} Q_i)$ lifts to the colimit in question if $j < -1 + |R|(n - q_0) - \sum_{i \in R} q_i$, and therefore $\partial_\emptyset^R \mathrm{Emb}(Q_0, N - \bigcup_{i \notin R} Q_i)$ is $(-1 + |R|(n - q_0) - \sum_{i \in R} q_i)$-cocartesian. By Theorem 6.2.3 the $k$-cube $R \mapsto \mathrm{Emb}(Q_0, N - \bigcup_{i \notin R} Q_i)$ is $(1 - k - k + k(n - q_0) - \sum_{i=1}^k q_i)$-cartesian, since the sum over partitions is clearly minimized for the partition of $\underline{k}$ consisting of blocks which are singletons. This number is equal to $1 + \sum_{i=1}^k (n - q_0 - q_i - 2)$. □

For the remainder of the details of how to turn statements like Theorem 10.3.11 into the sort of results required in the proof of Theorem 10.3.4, we will focus on Proposition 10.3.9. What we need is to "thicken up" the result in that statement so that it applies to embeddings of codimension-zero submanifolds a smooth closed manifold $M$, which is what arises in the proof of Theorem 10.3.4. First note that we can generalize the situation in Proposition 10.3.9 to a relative setting. That is, suppose $Q_0$, $Q_1$, and $N$ have boundary, and that embeddings $e_i \colon \partial Q_i \to \partial N$ have been selected to have disjoint images. Let $Q_S = \bigcup_{i \notin S} Q_i$, and let $\mathrm{Emb}_\partial(Q_S, N)$ be the space of embeddings $f \colon Q_S \to N$ such that the restriction of $f$ to $\partial Q_S$ is equal to $e_S$, and such that $f^{-1}(\partial N) = \partial Q_S$. The same argument used to prove Proposition 10.3.9 proves the following.

**Proposition 10.3.12** *With $N, Q_0, Q_1$ as described in the paragraph above, the square*

$$
\begin{array}{ccc}
\mathrm{Emb}_\partial(Q_0 \cup Q_1, N) & \longrightarrow & \mathrm{Emb}_\partial(Q_0, N) \\
\downarrow & & \downarrow \\
\mathrm{Emb}_\partial(Q_1, N) & \longrightarrow & \mathrm{Emb}_\partial(\emptyset, N)
\end{array}
$$

*is $(n - q_0 - q_1 - 1)$-cartesian.*

We can make a further generalization to the case of compact manifold triads (see the discussion after Definition A.2.5). Suppose the $Q_i$ are compact $n$-dimensional manifold triads of handle index $q_i$, where $n - q_i \geq 3$, and $N$ is an $n$-dimensional smooth manifold with boundary. In this case embeddings $e_i \colon \partial_0 Q_i \to \partial N$ have been chosen, and we let $\mathrm{Emb}_{\partial_0}(Q_S, N)$ stand for the obvious thing.

**Theorem 10.3.13** ([GW99, Theorem 1.1]) *The square*

$$\mathrm{Emb}_{\partial_0}(Q_0 \cup Q_1, N) \longrightarrow \mathrm{Emb}_{\partial_0}(Q_0, N)$$

$$\downarrow \qquad\qquad\qquad\qquad \downarrow$$

$$\mathrm{Emb}_{\partial_0}(Q_1, N) \longrightarrow \mathrm{Emb}_{\partial_0}(\emptyset, N)$$

*is* $(n - q_0 - q_1 - 1)$-*cartesian.*

This can be generalized to the case where the dimension of the $Q_i$ is $m \leq n$, essentially by a thickening of the $m$-dimensional $Q_i$ by the disk bundle of an $(n - m)$-plane bundle.

**Proposition 10.3.14** ([GW99, Observation 1.3]) *If* $\dim(Q_i) = m \leq n$ *then Theorem 10.3.13 is true.*

The rough idea of the proof is to assume that $N$ is embedded in $\mathbb{R}^{n+k}$ and let $Gr_{n-m} = \mathrm{colim}_k Gr_{n-m+k}(\mathbb{R}^{n+k})$ be a colimit of Grassmannians. Consider the map $\mathrm{Emb}(Q_S, N) \to \mathrm{Map}(Q_S, Gr_{n-m})$ given by assigning an embedding $f$ to its normal bundle $\nu_f$. The homotopy fiber of this map over some $\eta$ can be identified with the space of embeddings of the disk bundle of $\eta$ over $Q_S$. Since $S \mapsto \mathrm{Map}(Q_S, Gr_{n-m})$ is homotopy cartesian (because $\mathrm{Map}(-, X)$ is polynomial of degree $\leq 1$), by Proposition 5.4.12, the square of homotopy fibers is $(n - q_0 - q_1 - 1)$-cartesian if and only if the square $S \mapsto \mathrm{Emb}_{\partial}(Q_S, N)$ is $(n - q_0 - q_1 - 1)$-cartesian. However, this introduces more corners, since the closed disk bundle of a smooth manifold with boundary is already a compact manifold triad itself. The new corners due to the disk bundle are introduced along the corner set of the original compact manifold triad, and we will ignore the details.

We may also assume the $Q_i$ are submanifolds of an $m$-dimensional manifold $M$. Now we are in a position to describe a situation which is directly related to the square $\mathcal{E}$ from the beginning of this section, and we generalize this situation further by introducing a new manifold $P$. Suppose that $P$ is a smooth compact codimension-zero manifold triad in $M$; $Q_0, Q_1$ are smooth compact codimension-zero manifold triads in $M - \mathring{P}$; and that the handle index of $Q_i$ satisfies $n - q_i \geq 3$. Let $Q_S = \bigcup_{i \notin S} Q_i$.

**Proposition 10.3.15** *The square* $S \mapsto \mathrm{Emb}(P \cup Q_S, N)$ *is* $(n - q_0 - q_1 - 1)$-*cartesian.*

*Proof* The square $S \mapsto \mathrm{Emb}(P, N)$ is homotopy cartesian by Example 5.4.4 since all maps are the identity, and hence $S \mapsto \mathrm{Emb}(P \cup Q_S, N)$ is $(n - q_0 - q_1 - 1)$-cartesian if the square of homotopy fibers

$$S \longmapsto \mathrm{hofiber}(\mathrm{Emb}(P \cup Q_S, N) \to \mathrm{Emb}(P, N))$$

is $(n - q_0 - q_1 - 1)$-cartesian for all choices of basepoint in $\mathrm{Emb}(P, N)$ by Proposition 5.4.12. The map of squares $\mathrm{Emb}(P \cup Q_S, N) \to \mathrm{Emb}(P, N)$ is a fibration whose fiber is the square $\mathrm{Emb}_{\partial_0}(Q_S, N - P)$, which is $(n - q_0 - q_1 - 1)$-cartesian by Theorem 10.3.13.                                                                 □

We finally arrive at the technical statement which relates the open sets in the square $\mathcal{E}$ appearing at the beginning of this section with the closed sets we have been considering.

**Corollary 10.3.16** ([GW99, Corollary 1.4]) *Let* $P, Q_0, Q_1$ *be as in Proposition 10.3.15, and set* $V_S = (P \mathring{\cup} Q_S)$. *Then* $S \mapsto \mathrm{Emb}(V_S, N)$ *is* $(n - q_0 - q_1 - 1)$-*cartesian.*

To connect this explicitly with the square $\mathcal{E}$, we choose the $Q_i$ to be the $A_i$ considered in Theorem 10.3.4, and let $P$ be the closure of $L - (A_0 \cup A_1)$.

Let $Q_0, \ldots, Q_k$ be smooth closed manifolds of dimensions $q_0, \ldots, q_k$, and let $N$ be a smooth $n$-dimensional manifold. Assume we can choose a basepoint $e \in \mathrm{Emb}(Q_0 \cup \cdots \cup Q_k, N)$. Then the proof of Theorem 10.3.11 showed that the $k$-cube $R \mapsto \mathrm{Emb}(Q_0, N - \bigcup_{i \notin R} Q_i)$ is $(1 + \sum_{i=1}^{k} n - q_0 - q_i - 2)$-cartesian. Compare that with the following result.

**Proposition 10.3.17** *Let* $Q_0$ *be a smooth closed* $q_0$-*dimensional manifold, and* $Q_1, \ldots, Q_k$ *be smooth closed submanifolds of dimensions* $q_1, \ldots, q_k$ *of a smooth $n$-dimensional manifold $N$. For $S \subset \underline{k}$, let* $Q_S = \bigcup_{i \notin S} Q_i$. *Then $k$-cube*

$$S \longmapsto \mathrm{Map}(Q_0, N - Q_S)$$

*is* $(1 - q_0 + \sum_{i=1}^{k} n - q_i - 2)$-*cartesian. In fact, the same is true if we replace* $\mathrm{Map}(Q_0, N - Q_S)$ *with* $\mathrm{Imm}(Q_0, N - Q_S)$.

*Proof* The key observation is that $\text{Map}(-, N - Q_S)\colon O(Q_0) \to \text{Top}$ is polynomial of degree $\le 1$. We will prove that if $V$ is the interior of a smooth compact codimension-zero submanifold of $Q_0$ of handle index at most $j$, then $S \mapsto \text{Map}(V, N - Q_S)$ is $(1 - j + \sum_{i=1}^{k} n - q_i - 2)$-cartesian.

The base case $j = 0$ asserts that if $Q_0 = \sqcup_{i=1}^{m} B_i$ is a disjoint union of open balls, then the cube $S \mapsto \text{Map}(\sqcup_{i=1}^{m} B_i, N - Q_S)$ is $(1 + \sum_{i=1}^{k} n - q_i - 2)$-cartesian. This boils down to Example 6.2.11 as follows. The functor $\text{Map}(-; Z)\colon O(Q_0) \to \text{Top}$ is a homotopy functor, so that, given any selection of points $x_i \in B_i$, the natural map $\text{Map}(\sqcup_{i=1}^{m} B_i, Z) \to \prod_{i=1}^{m} \text{Map}(x_i, Z) \cong Z^m$ is a homotopy equivalence. By Proposition 5.4.5 it therefore suffices to prove the assertion in the case where $Q_0$ is a single point. But $\text{Map}(*, Z) \cong Z$, and so we have reduced to the case considered in Example 6.2.11.

Now suppose $V$ is the interior of a smooth compact codimension-zero submanifold of $Q_0$ with handle index $\le j$. We let the reader fill in the details of the remainder of the argument, as it is similar enough to the one presented in Theorem 10.3.4. One eventually encounters a diagram of the form

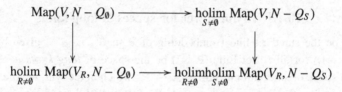

where $S \subset \underline{k}$ and $R \subset \underline{2}$, and the $V_S$ have handle index $\le j - 1$. The vertical arrows are weak equivalences since $\text{Map}(-, Z)$ is polynomial of degree $\le 1$, and the lower horizontal arrow is $(1 - j + \sum_{i=1}^{k} n - q_i - 2)$-connected by induction using Proposition 5.4.17. □

A much simpler proof can be given based on the fact that if $K$ has the homotopy type of a $p$-dimensional CW complex, then $\text{Map}(K, -)$ takes $k$-connected maps to $(k - p)$-connected maps. We also need the fact that if a smooth manifold admits a handle decomposition with handles of index at most $p$, then $M$ has the homotopy type of a CW complex of dimension $p$. The proof above, however, does not require any of this.

## 10.4 Spaces of knots

Manifold calculus of functors from Section 10.2 has in the past ten years been used effectively for the study of spaces of knots [ALTV08, LTV10, Sin09, Vol06a, Vol06b]. Recent generalizations to links [MV14] and embeddings of planes [AT14] have also demonstrated the usefulness of the functor calculus approach. Our focus here will be on spaces of knots since that is the case that has yielded the most results thus far.

Since knots are examples of embeddings and manifold calculus was designed to study spaces of embeddings, it is not surprising at first glance that this theory should have something to say about knots. However, as was first suggested in [GKW01, Example 5.1.4], one gets more mileage out of this example than the theory suggests through the use of the finite model for the Taylor tower (see Example 10.2.18). Each stage of the Taylor tower is a homotopy limit of a punctured cubical diagram (details are in Section 10.4.1) and this tower thus becomes an example of much of the theory developed in the previous chapters. The Taylor tower also has a cosimplicial model (Section 10.4.2) which demonstrates the connection between cubical diagrams and cosimplicial spaces described in Section 9.4. In addition, the spectral sequences from Section 9.6 will play an important role in the description of the rational homology and homotopy of spaces of knots in dimension $\geq 4$ (Section 10.4.4).

### 10.4.1 Taylor tower for spaces of long knots

Let $e$ be the standard linear embedding of $\mathbb{R}$ in $\mathbb{R}^n$, $n \geq 3$, given by $t \mapsto (t, 0, \ldots, 0)$ for all $t$. Let $\mathrm{Emb}_c(\mathbb{R}, \mathbb{R}^n)$ be the space of *long knots in* $\mathbb{R}^n$, that is, the space of smooth embeddings $\mathbb{R} \to \mathbb{R}^n$ which agree with $e$ (so $e$ can be thought of as the *long unknot*) outside the interval ("c" stands for "compact support"). More precisely

$$\mathrm{Emb}_c(\mathbb{R}, \mathbb{R}^n) = \{f \in \mathrm{Emb}(\mathbb{R}, \mathbb{R}^n) \mid f(t) = (t, 0, 0, \ldots, 0) \text{ for } t \notin [0, 1]\}.$$

A related space is the space of long immersions $\mathrm{Imm}_c(\mathbb{R}, \mathbb{R}^n)$ with the same prescribed linear behavior outside a compact set. As mentioned before, these spaces are equipped with the Whitney $C^\infty$ topology (and, in particular, $\mathrm{Emb}_c(\mathbb{R}, \mathbb{R}^n)$ is topologized as a subspace of $\mathrm{Emb}(\mathbb{R}, \mathbb{R}^n)$). Since there is an inclusion $\mathrm{Emb}(M, N) \hookrightarrow \mathrm{Imm}(M, N)$, we set

$$\mathcal{K}^n = \mathrm{hofiber}_e(\mathrm{Emb}_c(\mathbb{R}, \mathbb{R}^n) \hookrightarrow \mathrm{Imm}(_c\mathbb{R}, \mathbb{R}^n)).$$

From Definition 2.2.3, the definition of the homotopy fiber, a point in $\mathcal{K}^n$ is thus a long knot along with a path (isotopy) to the unknot $e$ through immersions. We will refer to the space $\mathcal{K}^n$ as the "space of knots modulo immersions" (it is also sometimes called the space of "tangentially straightened long knots"; see [DH12]).

Since, by the Smale–Hirsch Theorem [Sma59], $\mathrm{Imm}(\mathbb{R}, \mathbb{R}^n) \simeq \Omega S^{n-1}$, and since the inclusion $\mathrm{Emb}_c(\mathbb{R}, \mathbb{R}^n) \hookrightarrow \mathrm{Imm}_c(\mathbb{R}, \mathbb{R}^n)$ is null-homotopic

[Sin06, Proposition 5.17], spaces $\mathrm{Emb}_c(\mathbb{R}, \mathbb{R}^n)$ and $\mathcal{K}^n$ are related using Example 2.2.10 by

$$\mathcal{K}^n \simeq \mathrm{Emb}_c(\mathbb{R}, \mathbb{R}^n) \times \Omega^2 S^{n-1}. \tag{10.4.1}$$

**Remark 10.4.1** The reason one chooses to work with long knots rather than ordinary closed ones (embeddings of $S^1$ in $\mathbb{R}^n$ or $S^n$) in this setup is that some of the more technical arguments work out better in this case. In addition, long knots are naturally an $H$-space via the operation of stacking (or concatenation) which gives them extra structure that is useful when studying their rational homotopy type (see Section 10.4.4). In practice, however, most of what we describe here works equally well for ordinary knots. The relationship between closed knots in the sphere and long knots is also well-understood [Bud08, Theorem 2.1] and is given by

$$\mathrm{Emb}(S^1, S^n) \simeq \mathrm{Emb}_c(\mathbb{R}, \mathbb{R}^n) \times_{\mathrm{SO}_{n-1}} \mathrm{SO}_{n+1}. \qquad \square$$

Classical knot theory is mostly concerned with $\pi_0(\mathrm{Emb}_c(\mathbb{R}, \mathbb{R}^3))$, the set of *knot types* or *isotopy classes of knots*, as well as $\mathrm{H}^0(\mathrm{Emb}_c(\mathbb{R}, \mathbb{R}^3))$, the space of *knot invariants*, which are functions on isotopy classes of knots, that is, locally constant functions on $\mathrm{Emb}_c(\mathbb{R}, \mathbb{R}^3)$ (we will say more about invariants in Section 10.4.4). Working with the space of classical knots modulo immersions does not change the questions or the context very much due to the following

**Proposition 10.4.2** ([Vol06b, Proposition 5.12]) *There is a one-to-one correspondence between isotopy classes of $\mathcal{K}^3$ and isotopy classes of framed knots (not modulo immersions) whose framing number is even.*

For $\mathrm{Emb}_c(\mathbb{R}, \mathbb{R}^n)$, $n > 3$, the questions of isotopy classes or invariants are not interesting since the space of knots is in this case connected. However, higher homotopy and homology groups are very interesting, even for $n > 3$, as we shall see in Section 10.4.4.

To set up the Taylor tower for $\mathrm{Emb}_c(\mathbb{R}, \mathbb{R}^n)$ or $\mathcal{K}^n$, we can use the finite model for its stages from Example 10.2.18. We will demonstrate the construction for $\mathcal{K}^n$ since this is what we will mostly be concerned with in Section 10.4.4, but the reader should keep in mind that everything goes through the same way for $\mathrm{Emb}_c(\mathbb{R}, \mathbb{R}^n)$.

Given a $k \geq 0$, let $I_j = [5/2^{j+3}, 7/2^{j+3}]$ for $0 \leq j \leq k$. Thus $I_0, \ldots, I_k$ are pairwise disjoint subintervals of $\mathbb{R}$. (The reason for setting up the intervals like this is that we will take the stages $T_k \mathcal{K}^n$, defined below, for all $k$ at the same time when we put them together into a Taylor tower.) Let

$$\emptyset \neq S \subseteq \{0, \ldots, k\},$$

and let $e_S$ be the long unknot with (the images of) intervals indexed by $S$ removed. Then define

$$\mathcal{K}_S^n = \mathrm{hofiber}_{e_S} \left( \mathrm{Emb}_c(\mathbb{R} \setminus \cup_{i \in S} I_i, \ \mathbb{R}^n) \hookrightarrow \mathrm{Imm}_c(\mathbb{R} \setminus \cup_{i \in S} I_i, \ \mathbb{R}^n) \right). \quad (10.4.2)$$

This can be thought of as a space of "punctured knots modulo immersions". The analogous construction for ordinary knots would use $\mathrm{Emb}_c(\mathbb{R} \setminus \cup_{i \in S} I_i, \ \mathbb{R}^n)$.

Note that these are connected spaces even for $n = 3$ (and hence not interesting from the point of view of classical knot theory). However, there are maps $\mathcal{K}_S^n \to \mathcal{K}_{S \cup \{i\}}^n$ given by restricting an embedding to an embedding with one more puncture. Taken together, these spaces and maps form a punctured cubical diagram $S \mapsto \mathcal{K}_S^n$ of punctured knots. For example, when $k = 2$, we get

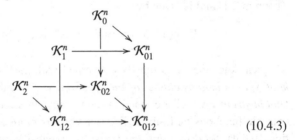

$$(10.4.3)$$

(As usual, we drop commas and parentheses from the set notation in the subscripts.) Then from Example 10.2.18, we have, for $n \geq 3$ and $k \geq 0$, the *kth (polynomial) approximation for* $\mathcal{K}^n$:

$$T_k \mathcal{K}^n = \operatorname*{holim}_{\emptyset \neq S \subseteq \{0, \ldots, k\}} \mathcal{K}_S^n.$$

(The diagram defining the stages for $\mathrm{Emb}_c(\mathbb{R}, \mathbb{R}^n)$ would consist of the spaces $\mathrm{Emb}_c(\mathbb{R} \setminus \cup_{i \in S} I_i, \ \mathbb{R}^n)$.) This kind of a homotopy limit is of course familiar from Section 5.3.

**Example 10.4.3**   We have $T_0 \mathcal{K}^n \simeq *$ (and $T_0 \mathrm{Emb}_c(\mathbb{R}, \mathbb{R}^n) \simeq *$) because the punctured cubical diagram is simply the space of long knots with one puncture, but such knots can be retracted "to infinities". (We also know from the discussion following Definition 10.2.15 that $T_0 \mathrm{Emb}(M, N) = \mathrm{Emb}(\emptyset, N) \simeq *$.) □

**Example 10.4.4**   From Example 10.2.17, we have that $T_1 \mathrm{Emb}_c(\mathbb{R}, \mathbb{R}^n) = \mathrm{Imm}(\mathbb{R}, \mathbb{R}^n)$ and this, as mentioned before, is equivalent to $\Omega S^{n-1}$ (by the Smale–Hirsch Theorem). To see this geometrically, a knot with two punctures is homotopy equivalent to a point with a vector attached – the outer arcs are retracted to infinity and the middle arc is retracted to, say, the midpoint, but since we had an embedding, the derivative information is retained in the form of the vector (this is given more precisely and more generally in (10.4.5)). Since the knot starts and ends in the direction of the long unknot, the homotopy limit defining $T_1 \, \mathrm{Emb}_c(\mathbb{R}, \mathbb{R}^n)$ gives a path of vectors that starts and ends in this standard direction. In other words, we have a loop on $S^{n-1}$. However, since $\mathcal{K}^n$ is the space of knots "modulo immersions", the derivative data is removed and so $T_1 \mathcal{K}^n \simeq *$. □

**Example 10.4.5** For $T_2 \mathcal{K}^n$, Example 5.3.3 tells us that a point in this Taylor polynomial is a point in each $\mathcal{K}_i^n$, an isotopy in each $\mathcal{K}_{ij}^n$, and a two-parameter isotopy in $\mathcal{K}_{123}^n$, where everything is compatible with the restriction maps in the diagram. □

Note that $\mathcal{K}^n$ maps into the diagram $\mathcal{K}_S^n$ by restriction (punching holes in a knot). From (5.4.1), there is therefore a canonical map

$$\mathcal{K}^n \longrightarrow T_k \mathcal{K}^n \tag{10.4.4}$$

for all $k \geq 0$. There is also a fibration

$$T_k \mathcal{K}^n \longrightarrow T_{k-1} \mathcal{K}^n$$

for all $k \geq 1$; since the diagram defining $T_{k-1} \mathcal{K}^n$ is a subdiagram of the diagram defining $T_k \mathcal{K}^n$, there is a canonical map from the homotopy limit of the bigger diagram to the homotopy limit of the subdiagram, which is a fibration by Proposition 5.4.29. Putting this together, we get the *Taylor tower for the space of knots (modulo immersions)* $\mathcal{K}^n$, $n \geq 3$:

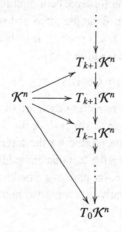

As usual, we set

$$T_\infty \mathcal{K}^n = \operatorname{holim}_k T_k \mathcal{K}^n.$$

(As in the case of the general Taylor tower in manifold calculus of functors, we could have said lim instead of holim since this is a tower of fibrations; see Example 8.4.11.)

From Theorem 10.3.1, we then have that, for $n \geq 4$,

$$T_\infty \mathcal{K}^n \simeq \mathcal{K}^n.$$

**Remark 10.4.6**  For $k \geq 2$, $\mathcal{K}^n$ is the actual pullback of the punctured cubes defining the stages $T_k \mathcal{K}^n$. One has a natural map

$$\mathcal{K}^n \longrightarrow \lim_{\emptyset \neq S \subseteq \{0,..,k\}} \mathcal{K}^n_S$$

for the same reason one has a map (10.4.4), but there is now also a map going back. This is because one can reconstruct a knot from compatible punctured knots $\mathcal{K}^n_i$; each once-punctured knot knows how to fill in $k$ holes punctured in the other knots and these filled-in pieces do not intersect. Contrast this with the case $k = 1$ where the pieces of the knot filled in by each once-punctured knot might intersect if one tries to reconstruct a knot, thereby not producing an embedding.                                                                □

### 10.4.2  Cosimplicial model for the Taylor tower for long knots

In this section we describe the cosimplicial model for the Taylor tower for $\mathcal{K}^n$ (and for $\operatorname{Emb}_c(\mathbb{R}, \mathbb{R}^n)$). That this could be done was suggested in [GKW01, Example 5.1.4] but was written down carefully in [Sin06, Sin09].

Recall the definition of the (ordered) configuration space $\operatorname{Conf}(p, M)$ of $p$ points in a manifold $M$ from Example 5.5.6 and recall that $\operatorname{Conf}(0, M) = *$, $\operatorname{Conf}(1, M) \simeq M$, and $\operatorname{Conf}(2, \mathbb{R}^n) \simeq S^{n-1}$. It is then not hard to see that there is an equivalence [Sin09, Proposition 5.15]

$$\operatorname{Emb}_c(\mathbb{R} \setminus \cup_{i \in S} I_i, \ \mathbb{R}^n) \simeq \operatorname{Conf}(|S| - 1, \mathbb{R}^n) \times S^{(|S|-1)(n-1)}. \tag{10.4.5}$$

The equivalence is given by retracting each arc of the punctured knot to, say, its midpoint. The midpoints are distinct since the punctured knot is an embedding, so the first space in the product in the target is a configuration space. The second factor comes from the fact that embeddings contain derivative data (kept track of by the immersion part of embeddings), so what is left is not just a point but also a derivative vector. Also note that, when $S = \underline{0} = \{0\}$,

$\mathrm{Conf}(|S| - 1, \mathbb{R}^n) = \mathrm{Conf}(0, \mathbb{R}^n) \simeq *$. This corresponds to the fact that, when one hole is punctured in the long knot, the resulting arcs are retracted to $\pm\infty$. The tangent vector is prescribed there – it is the vector $\vec{e} = (1, 0, \ldots, 0)$ pointing in the direction of the standard embedding $\epsilon$ or $\mathbb{R}$ in $\mathbb{R}^n$ – since long knots are linear outside a compact set.

Similarly we have

$$\mathcal{K}_S^n \simeq \mathrm{Conf}(|S| - 1, \mathbb{R}^n). \tag{10.4.6}$$

This is because modding out by immersions takes out the tangential data captured by the spheres. One of the reasons one often works with $\mathcal{K}_S^n$ rather than with $\mathrm{Emb}_c(\mathbb{R}, \mathbb{R}^n)$ is precisely because of this simplification.

The restriction maps "add a point", as pictured in Figure 10.2. To make this addition of points precise, we need to introduce a certain compactification of the configuration space $\mathrm{Conf}(k, \mathbb{R}^n)$.

Denote a configuration in $\mathrm{Conf}(p, \mathbb{R}^n)$ by $(x_1, x_2, \ldots, x_p)$. For $1 \leq i < j < k \leq p$, consider the maps

$$\phi_{ij} \colon \mathrm{Conf}(p, \mathbb{R}^n) \longrightarrow S^{n-1} \tag{10.4.7}$$

$$(x_1, x_2, \ldots, x_p) \longrightarrow \frac{x_j - x_i}{\|x_j - x_i\|}$$

and

$$\delta_{ijk} \colon \mathrm{Conf}(p, \mathbb{R}^n) \longrightarrow [0, +\infty] \tag{10.4.8}$$

$$(x_1, x_2, \ldots, x_p) \longrightarrow \frac{\|x_k - x_i\|}{\|x_j - x_i\|}.$$

(Here $[0, +\infty]$ is the one-point compactification of the positive reals.) Let

$$\phi = (\phi_{ij})_{1 \leq i < j \leq p},$$
$$\delta = (\delta_{ijk})_{1 \leq i < j < k \leq p}.$$

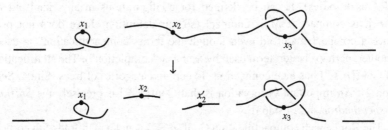

Figure 10.2 Restriction maps "add a point".

**Definition 10.4.7**

- The *Fulton–MacPherson compactification of* $\mathrm{Conf}(p, \mathbb{R}^n)$, denoted by $\mathrm{Conf}[p, \mathbb{R}^n]$, is the closure of the image of the map

$$\phi \times \delta \colon \mathrm{Conf}(p, \mathbb{R}^n) \longrightarrow \prod_{1 \leq i < j \leq p} S^{n-1} \times \prod_{1 \leq i < j < k \leq p} [0, +\infty]. \tag{10.4.9}$$

- The *Kontsevich compactification of* $\mathrm{Conf}(p, \mathbb{R}^n)$, denoted by $\mathrm{Conf}\langle p, \mathbb{R}^n \rangle$, is the closure of the image of the map

$$\phi \colon \mathrm{Conf}(p, \mathbb{R}^n) \longrightarrow \prod_{1 \leq i < j \leq p} S^{n-1}. \tag{10.4.10}$$

These definitions were independently made in [Kon99, KS00] (see also the correction by [Gai03, Section 6.2]) and in [Sin04], although Sinha was the first to explore the difference between the two spaces and to show that they are both homotopy equivalent to $\mathrm{Conf}(p, \mathbb{R}^n)$ [Sin04, Theorem 5.10] as well as to the original Fulton–MacPherson compactification [AS94, FM94] (which uses blowups along diagonals in $\mathbb{R}^{kn}$).

The main feature of the definitions is that configuration points are allowed to collide while the direction of collision is kept track of (by the various spheres $S^{n-1}$). In addition, the relative rates of approach are recorded in $\mathrm{Conf}[p, \mathbb{R}^n]$ (by the copies of $[0, +\infty]$). The space $\mathrm{Conf}\langle p, \mathbb{R}^n \rangle$ can be thought of as the quotient of $\mathrm{Conf}[p, \mathbb{R}^n]$ by subsets of three or more points colliding along a line. Since the configuration space itself does not appear in the target (some versions of the maps in Definition 10.4.7 in the literature also contain an inclusion factor), both $\mathrm{Conf}[p, \mathbb{R}^n]$ and $\mathrm{Conf}\langle p, \mathbb{R}^n \rangle$ are compactifications of $\mathrm{Conf}(p, \mathbb{R}^n)$ modulo the action of the group of translations and positive dilation of $\mathbb{R}^n$.

The Fulton–MacPherson compactification $\mathrm{Conf}[p, \mathbb{R}^n]$ is a *manifold with corners* (an $n$-dimensional manifold with corners that are modeled on $[0, \infty)^k \times \mathbb{R}^{n-k}$ as $k$ varies). It can be defined for configurations in any manifold $M$ and it is compact if $M$ is compact (so our definition above does not produce a compact manifold even though the term "compactification" is used; another, perhaps better, term used by Sinha is "completion"). The stratification of $\mathrm{Conf}[p, \mathbb{R}^n]$ has nice connections to certain categories of trees [Sin04, Section 2]. An important observation is that $\mathrm{Conf}[p, \mathbb{R}]$ is precisely the *Stasheff associahedron* $A_{p-2}$ [Sta63].

The Kontsevich compactification $\mathrm{Conf}\langle p, \mathbb{R}^n \rangle$ is not a manifold with corners (as mentioned above, it is just a quotient of one) but it turns out that this is the version that lends itself to a cosimplicial structure as follows. Recall the standard inclusion $\epsilon$ of $\mathbb{R}$ in $\mathbb{R}^n$ from Section 10.4.1. Let $\vec{e}$ be the unit vector along this inclusion, namely $\vec{e} = (1, 0, \ldots, 0)$. Then we have the coface maps

$$d^i \colon \text{Conf}\langle p, \mathbb{R}^n \rangle \longrightarrow \text{Conf}\langle p + 1, \mathbb{R}^n \rangle, \qquad 0 \le i \le p + 1, \qquad (10.4.11)$$

which are essentially given by repeating the $i$th configuration point, that is, repeating all the vectors indexed on the $i$th point. The vector between the $i$th and its double, which is now the $(i + 1)$th point in $\text{Conf}\langle p + 1, \mathbb{R}^n \rangle$, is taken to be $\vec{e}$. The first and last coface maps correspond to "doubling at $\pm\infty$". This can be made precise by considering "knots in a box", namely embeddings of $I$ in $I \times \mathbb{R}^{n-1}$ with endpoints going to prescribed points in $\{0\} \times \mathbb{R}^{n-1}$ and $\{1\} \times \mathbb{R}^{n-1}$ with prescribed derivatives. The spaces of such knots and long knots as we have defined them are homotopy equivalent. Then doubling at infinity means doubling at the endpoints of the knot in the box. See [Sin09] for details.

We also have codegeneracies

$$s^i \colon \text{Conf}\langle p + 1, \mathbb{R}^n \rangle \longrightarrow \text{Conf}\langle p, \mathbb{R}^n \rangle, \quad 0 \le i \le p, \qquad (10.4.12)$$

which simply forget the $i$th configuration point, that is, forget all the vectors indexed on the $i$th point.

**Proposition 10.4.8** ([Sin09, Corollary 4.22]) *For $n \ge 3$, the collection of spaces*

$$\{\text{Conf}\langle p, \mathbb{R}^n \rangle\}_{p \ge 0}$$

*with coface and codegeneracy maps as described above, forms a cosimplicial space denoted by $(K^n)^\bullet$.*

**Remarks 10.4.9**

1. Using the Fulton–MacPherson, rather than Kontsevich, compactification would not result in an honest cosimplicial space, but only one up to homotopy. More details about this issue (as well as a pictorial explanation of it) can be found at the end of [Sin04].
2. The above definition of $(K^n)^\bullet$ can also be made for configurations with vectors attached to them (this is in fact what Sinha's original construction has; see [Sin04, Corollary 6.8] or [Sin09, Corollary 4.22]). In that case, we get a cosimplicial space $\text{Emb}_c(\mathbb{R}, \mathbb{R}^n)^\bullet$ whose Tot tower would model ordinary long knots knots $\text{Emb}_c(\mathbb{R}, \mathbb{R}^n)$ and not knots modulo immersions.

□

The motivation for this cosimplicial space of course comes from the definition of the Taylor tower for $\mathcal{K}^n$ whose stages are homotopy limits of the punctured cubical diagram. On the other hand, from Section 9.4 (and in particular from Theorem 9.4.4), we know how to construct punctured cubes from truncations of a cosimplicial space. So, for example,

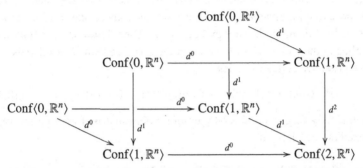

gives rise to the punctured cube

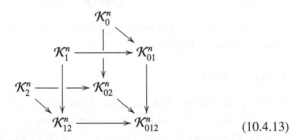

where the maps are as in Example 9.4.3. Because of the equivalence (10.4.6), the spaces in this cube are homotopy equivalent to the spaces in the cube

$$
\begin{array}{c}
\mathcal{K}_0^n \\
\mathcal{K}_1^n \longrightarrow \mathcal{K}_{01}^n \\
\mathcal{K}_2^n \longrightarrow \mathcal{K}_{02}^n \\
\mathcal{K}_{12}^n \longrightarrow \mathcal{K}_{012}^n
\end{array} \tag{10.4.13}
$$

What remains to be shown is that this spacewise equivalence is an actual equivalence of diagrams. This is essentially the argument given by Sinha in [Sin09]. In particular, we have the following.

**Theorem 10.4.10** ([Sin09, Section 6]) *There is an equivalence, for all $k \geq 0$ and $n \geq 3$,*

$$
T_k \mathcal{K}^n \simeq \mathrm{hoTot}^k (K^n)^\bullet,
$$

*where* $\mathrm{hoTot}^k (K^n)^\bullet$ *denotes the kth partial (homotopy invariant) totalization of* $(K^n)^\bullet$ *(see Definition 9.1.10 and the discussion following (9.1.10)).*

The same result is true when $\mathcal{K}^n$ is replaced by $\mathrm{Emb}_c(\mathbb{R}, \mathbb{R}^n)$, and, in fact, the statements in [Sin09] are for the latter space (the proofs are the same in both cases).

Taking the inverse limit over $k$, we then have, for $n \geq 3$,

$$T_\infty \mathcal{K}^n \simeq \mathrm{hoTot}(K^n)^\bullet.$$

From Theorem 10.3.1 we further have that, for $n \geq 4$, $\mathcal{K}^n \simeq T_\infty \mathcal{K}^n$, and so we have the following.

**Theorem 10.4.11** (Cosimplicial model for the space of knots) *For $n \geq 4$, the cosimplicial space $(K^n)^\bullet$ models the space of knots modulo immersions $\mathcal{K}^n$, that is, there is an equivalence*

$$\mathcal{K}^n \simeq \mathrm{hoTot}(K^n)^\bullet.$$

The same is true for the cosimplicial space $\mathrm{Emb}_c(\mathbb{R}, \mathbb{R}^n)^\bullet$ modeling $\mathrm{Emb}_c$ $(\mathbb{R}, \mathbb{R}^n)$ for $n \geq 4$.

The map from the equivalence in Theorem 10.4.11 is essentially given as a collection of compatible evaluation maps. Namely, given $f \in \mathcal{K}^3$, one has a map, for each $p \geq 0$,

$$ev_k(k) \colon \Delta^p \longrightarrow \mathrm{Conf}\langle p, \mathbb{R}^n \rangle \tag{10.4.14}$$

given by evaluating a knot on a point in the simplex (thought of as $p$ points on the interval). This map has been studied in detail in [Sin06, Section 5.2]. These maps are compatible with the cofaces and codegeneracies in $(K^n)^\bullet$ and so one gets a map

$$ev \colon \mathcal{K}^n \longrightarrow \mathrm{hoTot}(K^n)^\bullet. \tag{10.4.15}$$

This map is the same up to homotopy as the canonical map

$$\mathcal{K}^n \longrightarrow T_k \mathcal{K}^n \tag{10.4.16}$$

from (10.4.4) [Sin06, Theorem 5.13].

**Remark 10.4.12** In light of Example 9.2.3, it is not surprising that $(K^n)^\bullet$ models the space of knots. Namely, suppose we wanted to modify the construction described in Example 9.2.3 so that it applies to embeddings of the interval in $\mathbb{R}^n$ with fixed behavior at endpoints rather than maps (thinking of $\Omega \mathbb{R}^n$ as the space of such maps). The problem is that we would need to impose an injectivity condition on the cofaces since embeddings are injective. But this is achieved precisely by replacing $(\mathbb{R}^n)^k$ as the $k$th space in the cosimplicial model from Example 9.2.3 by $\mathrm{Conf}(k, \mathbb{R}^n)$. Another way to think about that is that replacing $\Omega X$ by the space of embeddings in the map (9.2.3) changes the codomain of that map to $\mathrm{Conf}(k, X)$. The compactificion is then necessary so that the doubling maps from (9.2.1) make sense.                        □

### 10.4.3 The Kontsevich operad and the Taylor tower for long knots

Sinha [Sin06] also noticed that another way to arrive at $(K^n)^\bullet$ is via the Gerstenhaber–Voronov/McClure–Smith framework discussed in Section A.4 (the relationship between operads and cosimplicial spaces has been relegated to the appendix since it is peripheral to the rest of this book and is only used here). Namely, the spaces $\mathrm{Conf}\langle p, \mathbb{R}^n \rangle$, $n \geq 1$, form an operad called the *Kontsevich operad*, denoted by $\mathcal{KO}^n$ (this was first noticed by Kontsevich [Kon99]) in which $\circ_i$ operations are essentially given by making a configuration "infinitesimal" and inserting it into the $k$th point of another configuration. The details of this procedure are spelled out, with pictures, in [Sin06, Section 4].

Fulton–MacPherson compactifications $\mathrm{Conf}[p, \mathbb{R}^n]$ (see Definition 10.4.7) also form an operad in the same way the Kontsevich compactifications do. For a detailed reference of this construction, see [LV14, Chapter 5]. A nice summary of the history of how compactified configurations have been used to define operads can be found at the beginning of [Sin06, Section 4]. It is also not hard to see that these two operads are homotopy equivalent (this is immediate from the fact that the two compactifications are homotopy equivalent [Sin04, Theorem 5.10] and that the operad structure maps are the same). One also has the following.

**Proposition 10.4.13** ([Sal01, Proposition 4.9])   *The little n-cubes operad $C_n$ and the Fulton–MacPherson operad (and hence the Kontsevich operad) are weakly equivalent as topological operads.*

One of the main differences between the Kontsevich and the Fulton–MacPherson operads is that Proposition 10.4.14 is true only for the former.

Recall Definition A.4.8. Then we have the following.

**Proposition 10.4.14**   *The Kontsevich operad $\mathcal{KO}^n$ is an operad with multiplication, that is, $\mathcal{KO}^n$ admits a morphism*

$$\mathcal{A}ss \longrightarrow \mathcal{KO}^n.$$

*Sketch of proof* The morphism is given as follows: spaces $\mathrm{Conf}\langle p, \mathbb{R}\rangle$ are homeomorphic to the symmetric groups $\Sigma_p$ on $p$ letters (with discrete topology). One can take the component corresponding to the linearly ordered configuration on $\mathbb{R}$. Since colinear configurations are identified in the Kontsevich compactification, we thus get an operad $\mathcal{KO}^1_{\mathrm{lin}}$ in which every space is equivalent to the one-point space. Hence

$$\mathcal{KO}^1_{\mathrm{lin}} \simeq \mathcal{A}ss.$$

On the other hand, we have the inclusion $e \colon \mathbb{R} \to \mathbb{R}^n$ from the beginning of Section 10.4.1, which induces a morphism of operads

$$e_* \colon \mathcal{KO}^1_{\mathrm{lin}} \simeq \mathcal{A}ss \longrightarrow \mathcal{KO}^n$$

sending a configuration in $\mathbb{R}$ to the configuration in the image of $e$. □

One can therefore associate a cosimplicial space $(KO^n)^\bullet$ to the operad $\mathcal{KO}^n$ as in Definition A.4.10. There is an obvious relation between the cosimplicial space $(K^n)^\bullet$ from Proposition 10.4.8 and $(KO^n)^\bullet$; the spaces comprising the two operads are the same and the cofaces and codegeneracies are given by doubling and forgetting configuration points. The only difference is in the first and the last coface map, but the two cosimplicial spaces are likely homotopy equivalent. In any case, their totalizations, at least for $n \geq 4$, are the same, as we have a result for $\mathcal{KO}^n$ that is identical to Theorem 10.4.10.

**Theorem 10.4.15** ([Sin06, Theorem 1.1]) *There is an equivalence, for all $k \geq 0$ and $n \geq 3$,*

$$T_k \mathcal{K}^n \simeq \mathrm{hoTot}^k (KO^n)^\bullet,$$

*and in particular*

$$T_\infty \mathcal{K}^n \simeq \mathrm{hoTot}(KO^n)^\bullet.$$

Again from Theorem 10.3.1 we thus further have the following.

**Theorem 10.4.16** ([Sin06, Corollary 1.2]) *For $n \geq 4$, the cosimplicial space $(KO^n)^\bullet$ models the space of knots modulo immersions $\mathcal{K}^n$, that is, there is an equivalence*

$$\mathcal{K}^n \simeq \mathrm{hoTot}(KO^n)^\bullet.$$

Combining this with Corollary A.4.12, we immediately have the following.

**Corollary 10.4.17** *For $n \geq 4$, the space of long knots $\mathcal{K}^n$ is a two-fold loop space.*

The two-fold delooping of $\mathcal{K}^n$ has recently been described as the space of derived morphisms between the little cubes operads $C_1$ and $C_n$ in [DH12, TT14].

**Remark 10.4.18**  Operad actions on knot spaces were also studied by Budney [Bud07, Bud10], although the motivation there does not come from manifold calculus or cosimplicial spaces. A variant of Corollary 10.4.17 for ordinary knots (not modulo immersions) and framed knots was proved by Salvatore in [Sal06].

## 10.4.4  Spectral sequences for spaces of long knots

One way to try to compute the homotopy and homology of $\mathcal{K}^n$ (or $\mathrm{Emb}_c(\mathbb{R}, \mathbb{R}^n)$) for $n \geq 4$ is to use the spectral sequences associated to (the fibrant replacements of) either of the cosimplicial spaces $(K^n)^\bullet$ or $(KO^n)^\bullet$ from the previous section. We choose to work with the latter cosimplicial space since that is the one mostly considered in the literature and in particular in [LTV10] where most of the results stated in this section come from. However, the reader should bear in mind that everything we say in this section can be repeated with $(K^n)^\bullet$ in place of $(KO^n)^\bullet$.

Keeping in mind that spaces $\mathrm{Conf}\langle p, \mathbb{R}^n \rangle$ are homotopy equivalent to ordinary configuration spaces $\mathrm{Conf}(p, \mathbb{R}^n)$, we have from (9.6.11) that the $E^1$ page for the homotopy spectral sequence associated to $(KO^n)^\bullet$ is

$$E^1_{p,q} = \pi_q(\mathrm{Conf}(p, \mathbb{R}^n)) \cap \bigcap_{i=0}^{p-1} \ker(s^i_*), \qquad (10.4.17)$$

where the $s^i$ are the forgetting maps from (10.4.12). Similarly, from (9.6.15) we have that the homology spectral sequence for $(KO^n)^\bullet$ starts with

$$E^1_{p,q} = \mathrm{H}_q(\mathrm{Conf}(p, \mathbb{R}^n)) \cap \bigcap_{i=0}^{p-1} \ker(s^i_*). \qquad (10.4.18)$$

These spectral sequences are defined for $n \geq 3$, but they converge to $\mathrm{hoTot}(KO^n)^\bullet$ for $n \geq 4$ because that is when they have a steep vanishing line (more on this later). By Theorem 10.4.11, they then converge to $\pi_*(\mathcal{K}^n)$ for $n \geq 4$ (because of the Goodwillie–Klein convergence of the Taylor tower; see Theorem 10.3.1) and to $\mathrm{H}_*(\mathcal{K}^n)$ for $n \geq 4$ (because of the Weiss convergence of the Taylor tower on homology; see Remark 10.3.2).

The homology spectral sequence had appeared before the approach of studying knot spaces through manifold calculus of functors was developed. Vassiliev had in [Vas90] studied the *discriminant set*, that is, the complement of the set of long knots in the space of all smooth maps from $\mathbb{R}$ to $\mathbb{R}^n$ with fixed behavior at infinity. This set consists of maps with singularities and is stratified in a way that yields a natural filtration from which a homology spectral sequence can be constructed. As is the case for the homology spectral sequences for $(K^n)^\bullet$ from (10.4.18), this *Vassiliev spectral sequence* is defined for $n \geq 3$, but it only converges to $H_*(\mathcal{K}^n)$ when $n \geq 4$ [Vas90, Section 6.6].

The relationship between the Vassiliev spectral sequence and the one from (10.4.18) was established by Turchin, who studied the combinatorics and performed many computations in the Vassiliev spectral sequence [TT04, TT06, TT07]. He proves the following.

**Proposition 10.4.19** ([TT07, Proposition 0.1])  *For $n \geq 3$, the $E^2$ page of the homology spectral sequence associated to $(KO^n)^\bullet$ is isomorphic (through a regrading) to the $E^1$ page of the Vassiliev homology spectral sequence.*

This statement was first proved for $\mathrm{Emb}_c(\mathbb{R}, \mathbb{R}^n)$ [TT04]. The result stated above is a consequence of [TT07, Theorem 8.4] and [TT06, Proposition 3.1 and Lemma 4.3].

Vassiliev has conjectured a stable splitting of the resolved discriminant which would imply that his spectral sequence collapses at the $E^1$ page [Vas99]. This collapse was proved rationally by Kontsevich in dimension $n = 3$ along the diagonal $E^1_{p,p}$. The proof uses the famous *Kontsevich Integral* which produces all *finite type invariants* of classical knots $\mathcal{K}^3$ [Kon93] (more about these will be said at the end of the section). In [CCRL02], Cattaneo, Cotta-Ramusino, and Longoni proved the collapse along the main diagonal for $n \geq 4$, but the collapse everywhere was finally established in [LTV10]:

**Theorem 10.4.20** ([LTV10, Theorems 1.1 and 1.2])  *The homology spectral sequence associated to $(KO^n)^\bullet$ collapses at the $E^2$ page rationally for $n \geq 4$. By Proposition 10.4.19, the Vassiliev homology spectral sequence thus collapses at the $E^1$ page rationally for $n \geq 4$.*

**Remark 10.4.21**  The same result for $n = 3$ was recently announced independently in [Mor12] and [ST14].  $\square$

The main ingredient in the proof of Theorem 10.4.20 is the *(stable) formality of the little n-cubes operad*, which says that singular chains on the little $n$-cubes operad is quasi-isomorphic, over the real numbers, to its homology as operads of chain complexes [Kon99, Tam03] (see also [LV14]). This applies to the Kontsevich operad and one can then use the following general result:

**Proposition 10.4.22** ([LTV10, Proposition 3.2]) *If a cosimplicial space $X^\bullet$ is stably formal over some field of characteristic zero, then the Bousfield–Kan homology spectral sequence for $X^\bullet$ collapses at the $E^2$ page over any field of characteristic zero.*

The reason this is true is essentially that the vertical differential in the spectral sequence associated to a formal cosimplicial space can be replaced by the zero differential.

The little $n$-cubes operad also turns out to be *coformal*, that is, rationally determined by its homotopy Lie algebra [NM78] (formality in turn says that the rational homotopy type is determined by the cohomology algebra). This, in conjunction with Theorem 10.4.20, leads to the collapse of the rational homotopy spectral sequence associated to $(KO^n)^\bullet$ [ALTV08] (this was reproved in [LT09] using Koszul duality).

Theorem 10.4.20 thus gives a description of any rational homology group of the space of long knots modulo immersions in codimension 3 or more. Further, the homology spectral sequence for $(KO^n)^\bullet$ has various useful combinatorial features. One can exploit the fact that the cohomology and rational homotopy of configuration spaces are well known [Arn69, CLM76, Coh73, Coh95, Koh02] and can even be expressed pictorially via *chord diagrams*.

With the combinatorics of configuration spaces in hand, it is in principle possible to compute any rational (co)homology group of a space of long knots in dimension at least 4. However, the computations are difficult, and the spectral sequence is still not very well understood. Turchin has taken the combinatorial investigation of the $E^2$ term the farthest [TT04, TT06, TT07, TT10]. Other results can be found in [CCRL02, Pel11, Sak08, Sak11, ST14]. The homotopy spectral sequence for $(KO^n)^\bullet$ was investigated, among other places, in [LT09, SS02].

We finish with some comments on that special case of much interest – that of classical long knots (modulo immersions) $\mathcal{K}^3$. Unfortunately, one no longer has the Goodwillie–Klein comparison of Theorem 10.3.1, but the Taylor tower still provides a lot of information about classical knot theory and in particular about knot invariants. Namely, it was shown in [Vol06b] that an algebraic version of the Taylor tower for $\mathcal{K}^3$ classifies real-valued *finite type knots invariants*, which is a class of invariants conjectured to separate knots. For introductory literature on finite type invariants, see for example [BN95, CDM12]. The connection between integral finite type invariants and the genuine Taylor tower (not its algebraic counterpart) was studied in [BCSS05, Con08], and there is currently evidence that the classification result also holds in this case [BCKS14].

The fact that the algebraic Taylor tower classifies finite type invariants can also be interpreted as the collapse of the homology spectral sequence associated to $(KO^3)^\bullet$ on the main diagonal, that is, in degree zero. It has recently been shown that this spectral sequence in fact collapses everywhere [Mor12, ST14], but this collapse cannot be related as easily to the homology and homotopy of $\mathcal{K}^3$ as we could for $\mathcal{K}^{>3}$. One problem is that Theorem 10.3.1 is not known to be true in this situation, so it is not known whether $\mathrm{hoTot}(KO^3)^\bullet$ models $\mathcal{K}^3$. In addition, the homology spectral sequence for $(KO^3)^\bullet$ has a vanishing line of slope $-1$, which is not steep enough to deduce the convergence of the spectral sequence. We thus only have maps

$$H_*(\mathcal{K}^3) \longrightarrow H_*(\mathrm{hoTot}(KO^3)^\bullet) \longrightarrow \bigoplus_{i=0}^{\infty} E^2_{i,*+i},$$

and it is an open question whether they are isomorphisms. The right map is in fact precisely an instance of the map (9.6.16) which is in general not known to be an isomorphism. Nevertheless, the collapse of the spectral sequence still potentially has many implications for our understanding of the topology of the space of classical knots.

# Appendix

This chapter is neither comprehensive nor thorough, but is merely in the service of other content in this book.

## A.1 Simplicial sets

The purpose of this section is both to supply the necessary machinery to define homotopy (co)limits in Chapter 8 and to highlight some facts about simplicial sets which are analogous to those in Chapter 9 and in particular in Section 9.4. For more details about simplicial sets, see for example [GJ99, May92].

Let $\Delta$ be the cosimplicial indexing category (already introduced at the beginning of Section 9.1) whose objects are the linearly ordered sets $[n] = \{0, 1, \ldots, n\}$ and the morphisms are the order-preserving functions. Here we will be interested in $\Delta^{\mathrm{op}}$ which has the same objects and whose morphisms are generated by *faces* and *degeneracies*, denoted by $d_i$ and $s_i$. These satisfy the relations

$$d_i d_j = d_{j-1} d_i, \quad i < j;$$

$$d_i s_j = \begin{cases} s_{j-1} d_i, & i < j; \\ Id, & i = j, i = j + 1; \\ s_j d_{i-1}, & i > j + 1; \end{cases} \tag{A.1}$$

$$s_i s_j = s_{j+1} s_i, \quad i \le j.$$

**Definition A.1.1** A *simplicial space* is a contravariant functor $X_\bullet \colon \Delta \to \mathrm{Top}$ (i.e. a functor $X_\bullet \colon \Delta^{\mathrm{op}} \to \mathrm{Top}$). A *based simplicial space* is the same, only we replace Top by $\mathrm{Top}_*$. The space of $n$-simplices of $X_\bullet$ is the set/space $X_n = X_{[n]}$. A *map* $X_\bullet \to Y_\bullet$ *of simplicial sets/spaces* is a natural transformation of functors.

The category Set is naturally a subcategory of Top (those spaces with the discrete topology), and so a *simplicial set* is a functor whose codomain is Set. Therefore when we speak of simplicial spaces from now on, this will implicitly include the case of simplicial sets.

As was the case with cosimplicial diagrams, we write $d_i$ and $s_i$ in place of $X_\bullet(d_i)$ and $X_\bullet(s_i)$ respectively when $X_\bullet$ is understood. We also call $d_i$ and $s_i$ the *ith face* and *degeneracy* maps respectively. For a simplex $x \in X_n$, we call $d_i x$ its *ith face* and say $x$ is *degenerate* if there is an $(n-1)$-simplex $x'$ such that $s_i(x') = x$ for some $i$.

The following construction is the dual of the totalization of a cosimplicial space (Definition 9.1.5).

**Definition A.1.2** Given a simplicial space $X_\bullet \colon \Delta^{\mathrm{op}} \to$ Top, its *geometric realization* (or *classifying space*) $|X_\bullet|$ is the space

$$|X_\bullet| = \left( \coprod_{n \geq 0} X_{[n]} \times \Delta^n \right) \Big/ \big( (d_i(x), t) \sim (x, d^i(t)), \ (s_i(x), t) \sim (x, s^i(t)) \big).$$

One observation is that the degenerate simplices in a way do not contribute to the realization, essentially because the equivalence relation says that the non-degenerate simplices determine where the degenerate ones must map. In other words, each degenerate simplex is equivalent to a non-degenerate one; in fact it is the degeneracy of a unique non-degenerate simplex (and the iterated degeneracy operator is unique as well). For more details (with pictures) about how this follows from the equivalence relations, see [Fri12, Section 4].

**Remark A.1.3** Dually to Remark 9.1.6, the realization can be described as a coequalizer (or coend of a bifunctor) as follows. For a simplicial space, we have naturally associated to it a bifunctor $X_\bullet \times \Delta^\bullet \colon \Delta^{\mathrm{op}} \times \Delta \to$ Top. Then

$$|X_\bullet| = \mathrm{Coeq}\left( \coprod_{[n] \to [m]} X_m \times \Delta^n \underset{b}{\overset{a}{\rightrightarrows}} \coprod_{[n]} X_n \times \Delta^n \right).$$

We leave it to the reader to define the maps $a$ and $b$. $\qquad\qquad\square$

One nice feature of the realization is that, for a simplicial set $X_\bullet$, $|X_\bullet|$ is a CW complex with one $n$-cell for each non-degenerate $n$-simplex of $X_\bullet$ [May92, Theorem 14.1]. Another useful result says that realization commutes with products, stated below. For a proof, see for example [May92, Theorem 14.3]. The product of two simplicial sets is defined levelwise, with the obvious faces and degeneracies.

**Theorem A.1.4** *For $X_\bullet$ and $Y_\bullet$ simplicial sets, there is a homeomorphism (in the category of compactly generated Hausdorff spaces)*

$$|X_\bullet \times Y_\bullet| \cong |X_\bullet| \times |Y_\bullet|.$$

The collection of simplicial sets with maps (natural transformations) between them forms a category, denoted by SSet. Realization is a functor from this category to Top; that is, given a map $f: X_\bullet \to Y_\bullet$, we have an induced map

$$|f|: |X_\bullet| \longrightarrow |Y_\bullet|. \tag{A.2}$$

**Remark A.1.5** The realization functor $| - |$ has an adjoint (see Definition 7.1.18) called the *total singular complex functor* $S$. This is essentially given by mapping the cosimplicial simplex into a space. That is, given a space $X$, define $X_n = \mathrm{Map}(\Delta^n, X)$. Precomposing with the cofaces $d^i$ and codegeneracies $s^i$ in the cosimplicial simplex yields the faces and degeneracies $d_i$ and $s_i$ in $SX$. We leave it to the reader to check that $S$ is the right adjoint to $|-|$. □

**Remark A.1.6** (Čech complex) Given a space $X$ and an open cover $\mathcal{U}$ of $X$, one can also construct a simplicial space called the *Čech complex*, denoted by $\check{C}(\mathcal{U})$, equipped with a canonical map $|\check{C}(\mathcal{U})| \to X$. It is a classical result of Segal [Seg68] that the canonical map $|\check{C}(\mathcal{U})| \to X$ is a homotopy equivalence if $X$ has a partition of unity subordinate to $\mathcal{U}$. This result was extended in [DI04] where it was proved that this map is a weak equivalence for any open cover $\mathcal{U}$ of $X$. □

Because realization of a simplicial space is a colimit (compare the above coequalizer to the one in (7.4.4)), it is not homotopy invariant. The most that can be said in general is that there are maps

$$\mathrm{hocolim}\, X_\bullet \longrightarrow |X_\bullet| \longrightarrow \mathrm{colim}\, X_\bullet.$$

However, for a cofibrant simplicial space, namely one for which the latching maps (see Definition 8.4.7)

$$L_n(X_\bullet) = \bigcup_{i=0}^{n-1} s_i(X_{n-1}) \longrightarrow X_n$$

are cofibrations, we have the following dual of Proposition 9.1.8.

**Proposition A.1.7** ([DI04, Appendix A]) *If $X_\bullet$ is a cofibrant simplicial space, then the natural map*

$$\mathrm{hocolim}\, X_\bullet \longrightarrow |X_\bullet|$$

*is a homotopy equivalence.*

Taking the coequalizer from Remark A.1.3 only over the injective maps $[n] \hookrightarrow [m]$ produces the *fat realization of $X_\bullet$*, denoted by $\|X_\bullet\|$. This is the same as dropping the degeneracies in a simplicial space and then taking the realization. For such a *semisimplicial space*, we have the same statement as above: given an objectwise weak equivalence of simplicial spaces $X_\bullet \to Y_\bullet$, the induced map of realizations $\|X_\bullet\| \to \|Y_\bullet\|$ is also an equivalence. In general, the simplicial and semisimplicial realization are different, but we have the following result. Recall the notion of a homotopy terminal functor from Definition 8.6.2.

**Proposition A.1.8** *The inclusion $\Delta_{\mathrm{inj}}^{\mathrm{op}} \hookrightarrow \Delta^{\mathrm{op}}$ is homotopy terminal so that, for a cofibrant simplicial space $X_\bullet$, the map*

$$\|X_\bullet\| \longrightarrow |X_\bullet|$$

*is a homotopy equivalence.*

The disadvantage of a semisimplicial realization is that it is typically very large; even when $X_\bullet$ is a constant simplicial space at a point, $|X_\bullet|$ has a simplex in each dimension.

Recall that $\Delta^{\leq n}$ is the full subcategory of $\Delta$ whose objects are the sets $[k]$ for $0 \leq k \leq n$. Here is a definition that is analogous to Definition 9.1.13.

**Definition A.1.9** The *nth truncation* of a simplicial space $X_\bullet$ is the composite functor

$$\mathrm{tr}_n X_\bullet : \Delta^{\leq n} \hookrightarrow \Delta \xrightarrow{X_\bullet} \mathrm{Set}\,/\,\mathrm{Top}\,.$$

The truncations $\mathrm{tr}_n X_\bullet$ of $X_\bullet$ themselves gives rise to simplicial spaces as follows.

**Definition A.1.10**  The *n-skeleton of a simplicial space* $X_\bullet$ is the simplicial space $\mathrm{sk}_n X_\bullet$ such that $\mathrm{sk}_n X_k = X_k$ if $k \le n$ and for $k > n$ it contains precisely the degenerate subspaces of $X_\bullet$.

**Remark A.1.11**  In analogy with Remark 9.1.14, the *n*-skeleton is really a (left) Kan extension and it is (left) adjoint to the truncation functor.    □

Note that $\mathrm{sk}_n X_\bullet \subset \mathrm{sk}_{n+1} X_\bullet$, in the sense that $\mathrm{sk}_n X_k \subset \mathrm{sk}_{n+1} X_k$ for all $k$. Dually to (9.1.8), the union of the *n*-skeleta of $X_\bullet$ is equal to $X_\bullet$. More precisely, in analogy with (9.1.8) and Proposition 9.1.17, we have the following.

**Proposition A.1.12**  *If $X_\bullet$ is a simplicial space, then $\mathrm{colim}_n \mathrm{sk}_n X_\bullet$ is naturally isomorphic to $X_\bullet$.*

The realization $|\mathrm{sk}_n X_\bullet|$ can be built out of $|\mathrm{sk}_{n-1} X_\bullet|$ using a pushout diagram that is dual to the one appearing in the proof of Proposition 9.1.19. For details, see [GJ99, Chapter VII, Proposition 1.7]. However, if $X_\bullet$ satisfies certain conditions, that pushout square simplifies. (These conditions are always satisfied in the case of the simplicial replacement diagram, defined below, which is the case we care about most.) Namely, the *n*-skeleton is roughly built out of the $(n - 1)$-skeleton by attaching the non-degenerate simplices, but this cannot be done as easily as for simplicial sets since, for a simplicial set, one would in each $X_n$ use the disjoint union over all non-degenerate simplices, but, for spaces, this would produce a discrete set of non-degenerate simplices which is usually not discrete in $X_n$. One thus defines a simplicial space $X_\bullet$ to be *degeneracy-free* if each $X_n$ contains a subspace $N_n$ as a direct summand that represent the non-degenerate part of $X_n$. This subspace is then used for building the *n*-skeleton out of the $(n - 1)$-skeleton. For a precise definition and further explanation, see [DI04, Definition A.4] (also see [GJ99, Chapter VII, Definition 1.10]).

We then have the following statement, the details of which can be found in [GJ99, Chapter 7, Equation (3.10)] (also [DI04, Equation (A.1)]).

**Proposition A.1.13**  *The commutative diagram*

$$
\begin{array}{ccc}
N_n \times \partial\Delta^n & \longrightarrow & |\mathrm{sk}_{n-1} X_\bullet| \\
\downarrow & & \downarrow \\
N_n \times \Delta^n & \longrightarrow & |\mathrm{sk}_n X_\bullet|
\end{array}
$$

*is a pushout. Therefore, since the left vertical map is a cofibration by Proposition 2.3.9 (and since $\partial \Delta^n \to \Delta^n$ is a cofibration), it follows by Proposition 2.3.15 that the right vertical map is also a cofibration.*

The following is the dual of Proposition 9.1.17 and is the analog of Proposition A.1.12 on the level of realizations.

**Proposition A.1.14** ([GJ99, Chapter 7, Equation 3.9])  *For a simplicial space* $X_\bullet$, *there is a natural homeomorphism*

$$|X_\bullet| \cong \mathrm{colim}\,(|\mathrm{sk}_0 X_\bullet| \longrightarrow |\mathrm{sk}_1 X_\bullet| \longrightarrow \cdots).$$

An *augmentation* of a simplicial space $X_\bullet$ is a space $X^{-1}$ and a map $X_0 \to X_{-1}$ that coequalizes the two faces from $X_1$ to $X_0$. For an augmented simplicial space, there is also a natural map

$$|X_\bullet| \longrightarrow X_{-1} \tag{A.3}$$

that is dual to that in (9.1.18).

For a diagram $F \colon \mathcal{I} \to \mathrm{Top}$, we can construct its *simplicial replacement* $\amalg_\bullet F$. The $n$th space of this simplicial space is given by

$$\amalg_n F = \coprod_{i_0 \xleftarrow{\alpha_0} i_1 \xleftarrow{\alpha_1} i_2 \xleftarrow{\alpha_2} \cdots \xleftarrow{\alpha_{n-1}} i_n} F(i_n).$$

We leave it to the reader to fill in what the face and degeneracy maps must be. We then have this analog of Theorem 9.3.3.

**Theorem A.1.15** (Simplicial model for the homotopy colimit)  *For $F \colon \mathcal{I} \to$ Top a small diagram of spaces, there is a homeomorphism*

$$\mathrm{hocolim}_{\mathcal{I}}\, F \cong |\amalg_\bullet F|.$$

We then also have the dual of Proposition 9.3.4 with the dual proof that uses augmentations mentioned above as well as the dual of Proposition 9.1.25 whose formulation we leave to the reader.

**Proposition A.1.16**  *Suppose $\mathcal{I}$ is a small category with final object $i_1$, and let $F \colon \mathcal{I} \to$ Top be a diagram of spaces. Then the map*

$$F(i_1) \longrightarrow \mathrm{hocolim}_{\mathcal{I}} F$$

*is a homotopy equivalence.*

We invite the reader to formulate the duals of Proposition 9.3.7 and Corollary 9.3.8.

Combining Proposition A.1.14 and Theorem A.1.15, we thus have that, given a diagram of spaces $F$, there is a tower of cofibrations

$$|\mathrm{sk}_0 \amalg_{\bullet} F| \longrightarrow |\mathrm{sk}_1 \amalg_{\bullet} F| \longrightarrow \cdots \tag{A.4}$$

whose (homotopy) colimit is hocolim $F$. One might wonder what the cofibers of this tower are, in analogy with the description of the fiber of the totalization tower from Proposition 9.1.19. While that result gives the fibers for any cosimplicial space, below we give the cofibers for the special case of simplicial replacement of a diagram and leave the general case to the reader. Recall that $B_n(\mathcal{I})$ denotes the set of $n$-simplices in $B_{\bullet}\mathcal{I}$ and $\mathrm{nd}B_n(\mathcal{I})$ denotes the set of non-degenerate $n$-simplices, namely the strings of length $n$ without the identity maps.

**Proposition A.1.17**  *For a diagram  $F\colon \mathcal{I} \to$ Top, the cofibers of the associated tower of cofibrations from (A.4) are given as*

$$\coprod_{I_n \in \mathrm{nd}B_n(\mathcal{I})} \Sigma^n F(i_n)_+.$$

*Proof*  The simplicial replacement is always degeneracy-free [Dug01, Proof of Lemma 2.7] and so we have the pushout square from Proposition A.1.13:

$$\begin{array}{ccc} N_n \times \partial\Delta^n & \longrightarrow & |\mathrm{sk}_{n-1} \amalg_{\bullet} F| \\ \downarrow & & \downarrow \\ N_n \times \Delta^n & \longrightarrow & |\mathrm{sk}_n \amalg_{\bullet} F| \end{array}$$

where $N_n$ is the non-degenerate subspace of $X_n$. The space $X_n$, in turn, is the degree $n$ part of the simplicial replacement of the diagram $F$ and it is thus the coproduct indexed by $B_n(\mathcal{I})$ of $F(i_n)$. Then $N_n$ is a subcoproduct of that; it is the coproduct, indexed by $\mathrm{nd}B_n(\mathcal{I})$, of $F(i_n)$. The above square is thus the same as

$$\begin{array}{ccc} \amalg_{I_n \in \mathrm{nd}B_n(\mathcal{I})} F(i_n) \times \partial\Delta^n & \longrightarrow & |\mathrm{sk}_{n-1} \amalg_{\bullet} F| \\ \downarrow & & \downarrow \\ \amalg_{I_n \in \mathrm{nd}B_n(\mathcal{I})} F(i_n) \times \Delta^n & \longrightarrow & |\mathrm{sk}_n \amalg_{\bullet} F| \end{array}$$

The cofiber of the right vertical map is what we are interested in. That map is a cofibration since the left vertical map is and the square is a pushout. Thus the cofibers of the two vertical maps are homeomorphic, and the left ones are those we can compute, using Example 2.4.13:

$$\coprod_{I_n \in \mathrm{end}B_n(I)} \Delta^n/\partial\Delta^n \wedge F(i_n)_+.$$

But since $\Delta^n/\partial\Delta^n = S^n$ (the sphere has a natural basepoint given by the collapsed boundary of the $n$-simplex) and smashing with the sphere is the same as suspending, the above is the same as

$$\coprod_{I_n \in \mathrm{end}B_n(I)} \Sigma^n F(i_n)_+. \qquad \square$$

**Example A.1.18** (Homology spectral sequence of a homotopy colimit)   We can use Proposition A.1.17 to set up a homology spectral sequence for the homology of the homotopy colimit of a diagram $F$. In particular, we can apply the spectral sequence associated to a tower of cofibrations from Proposition 9.6.10 to the tower from (A.4) so that we get

$$E^1_{p,q} = \mathrm{H}_{q+p}\left(\mathrm{cofiber}\left(\left|\mathrm{sk}_{p-1} \amalg_\bullet F\right| \longrightarrow \left|\mathrm{sk}_p \amalg_\bullet F\right|\right)\right).$$

But this cofiber has been computed in Proposition A.1.17 and we have

$$E^1_{p,q} = \mathrm{H}_{q+p}\left(\coprod_{I_p \in \mathrm{end}B_p(I)} \Sigma^p F(i_p)_+\right) \cong \bigoplus_{I_p \in \mathrm{end}B_p(I)} \mathrm{H}_q(F(i_p)_+).$$

The first differential $d^1$ is the alternating sum of the maps induced on homology by the face maps in $\amalg_\bullet F$.

In the special case when $F$ is a punctured cube, this spectral sequence reduces precisely to the one encountered in Proposition 9.6.14. $\qquad \square$

Recall the functor $c_n : \mathcal{P}_0([n]) \to \Delta$ from Definition 9.4.1. Here is the definition that is dual to Definition 9.4.2.

**Definition A.1.19**   Given a simplicial space $X_\bullet$, the *punctured $(n + 1)$-cube associated to (the nth truncation of) $X_\bullet$* is the pullback

$$c_n^*(X_\bullet) = X_\bullet \circ c_n : \mathcal{P}_0([n]) \longrightarrow \mathrm{Top}.$$

The contravariance of $X_\bullet$ means the puncture is in the final entry in the cube.

**Example A.1.20** Let $n = 2$. Given a simplicial set $X_\bullet$, we obtain the following punctured 3-cube:

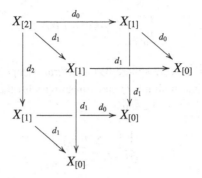

Here is the analog of Theorem 9.4.4.

**Theorem A.1.21** *For a simplicial space $X_\bullet$, there is a weak equivalence*

$$|\mathrm{sk}_n X_\bullet| \simeq \underset{\mathcal{P}_0([n])}{\mathrm{hocolim}}\,(X_\bullet \circ c_n).$$

In analogy with Remark 9.4.5, we have that, for an augmented simplicial space, the question of the connectivity of the map $|\mathrm{sk}_n X_\bullet| \to X_{-1}$ is the same as the question of how cocartesian the cubical diagram

$$X_\bullet \circ c_n \longrightarrow X_{-1}$$

is.

We conclude this section by observing that the dualization of Section 9.5 is straightforward and we leave it to the reader.

## A.2 Smooth manifolds and transversality

Although purely homotopical arguments are given in this book whenever possible, we still like to (and sometimes need to) invoke "dimension counting" arguments from the smooth category. We give proofs of the Blakers–Massey Theorem and its generalizations that utilize such arguments, though we also present a purely homotopy-theoretic proof. Several of our applications (see Examples 4.2.19, 4.2.20, 6.2.11, 6.2.12, and 6.2.14) of the Blakers–Massey theorem are for smooth manifolds, and for these examples we need some of the material from this section. We also need some of it in our discussion on manifold calculus in Section 10.2, particularly in Section 10.3.

At their heart, transversality arguments are an expression about our intuition regarding dimension. This intuition is developed from the study of affine spaces, where, for instance, generically the intersection of a $k$-dimensional affine set and an $l$-dimensional affine set in an $m$-dimensional affine set is an affine set of dimension $m - k - l$. In particular, they will generically not intersect at all when $k + l < m$. Although we claim such arguments are often intuitive, this is not to say that the theory behind them is not difficult or subtle.

As usual in this Appendix, this section is light on explanation and emphasizes definitions and results, sometimes not expressed in their most powerful incarnation, but rather in a way suited to how we use them. The reader may wish to consult [GG73], [Kos93], and [Lee03] for further details. Above all, we assume the reader is already familiar with smooth manifolds. We begin with a few words about handle decompositions of smooth manifolds, smooth approximation of continuous maps, and then talk about regular values and transversality before moving on to discuss jet bundles and the Whitney topology, which sets the stage for the Thom Transversality Theorem, Theorem A.2.19. More details for much of the material in this section can be found in [GG73, Hir94, Lee03, Mil63].

## A.2.1 Handle decompositions

Handle decompositions are not related to the rest of the material in this section, but they are used in our discussion of manifold calculus in Section 10.2. A handle decomposition of a smooth manifold is the smooth analog of a CW structure on a topological space. This gives us a refined notion of dimension in the following sense. Suppose $N$ is a smooth manifold of dimension $n$ and $Q \subset N$ is a smooth closed submanifold of dimension $q$. A tubular neighborhood $T$ of $Q$ has the structure of a manifold of dimension $n$, but for many purposes we may treat $T$ as a manifold of dimension $q$. We illustrate the uses of this principle in several places, including Theorems 10.2.14, 10.2.28, 10.3.4, and 10.3.6.

**Definition A.2.1** Let $m$ be a fixed non-negative integer. For each $0 \le j \le m$, we let $H^j = D^j \times D^{m-j}$ and refer to $H^j$ as a *j-handle*, where $j$ is called the *handle index*.

**Definition A.2.2** Let $L$ be a smooth $m$-dimensional manifold with boundary $\partial L$. Let $e \colon \partial D^j \times D^{m-j} \to \partial L$ be a smooth embedding. The quotient space $L \cup_e H^j$ is referred to as $L$ *with a j-handle attached*, and we call this process *attaching a j-handle to L*.

Attaching a $j$-handle to $L$ in this way is analogous to attaching a $j$-cell to $L$ in the CW sense.

**Definition A.2.3** A *handle decomposition* for a smooth compact $m$-dimensional manifold $M$ is a nested sequence

$$M_{-1} = \emptyset \subset M_0 \subset M_1 \subset \cdots \subset M_m = M,$$

where $M_j$ is obtained from $M_{j-1}$ by attaching handles of index $j$.

Given a handle decomposition of a smooth manifold $M$, we may refer to $M$ as a *handlebody*.

**Theorem A.2.4** *Let $M$ be a smooth closed manifold. Then $M$ admits a handle decomposition.*

This can be proved using Morse theory. The (non-degenerate) critical points of index $j$ for a smooth function $f \colon M \to \mathbb{R}$ correspond with handles of index $j$.

**Definition A.2.5** For a smooth compact $m$-dimensional manifold $M$, we say $M$ has *handle dimension* $j$ if $j$ is the least positive integer such that $M$ admits a handle decomposition with handles of index at most $j$.

The $m$-dimensional manifold $H^j = D^j \times D^{m-j}$ is a manifold with corners, since $\partial H^j = \partial D^j \times D^{m-j} \cup_{\partial D^j \times \partial D^{m-j}} D^j \times \partial D^{m-j}$. The union is taken along the "corner set" $\partial D^j \times \partial D^{m-j}$. This is the basic example of a "smooth manifold triad". Roughly speaking a *smooth manifold triad* is a triple $(M, \partial_0 M, \partial_1 M)$ consisting of a smooth manifold $M$ with boundary $\partial M$ decomposed as $\partial M = \partial_0 M \cup \partial_1 M$, where the union is taken along a "corner set" $\partial_0 M \cap \partial_1 M$. This corner set should have the structure of a smooth manifold of dimension $m - 2$. We do not need a precise definition for the purposes of this text and we refer the reader to [GW99, Section 0] for more details.

## A.2.2 Jet bundles and the topology on spaces of smooth maps

We will use jet bundles to define the topology of the space of smooth maps $C^\infty(M, N)$ between smooth manifolds $M$ and $N$. We use separate notation to distinguish it from the ordinary space of maps $\mathrm{Map}(M, N)$ (see Definition 1.2.1). We compare this topology with the compact open topology below in Theorem A.2.12. Our main reference for material in this section is [GG73], and the following is Chapter 2, Definition 2.1 there.

**Definition A.2.6** Let $M$ and $N$ be smooth manifolds and $x \in M$ a point, and suppose $f, g : M \to N$ are smooth maps.

1. $f$ has *zeroth order contact* with $g$ at $x$ if $f(x) = g(x) = y$. We write $f \sim_0 g$ at $x$.
2. $f$ has *first order contact* with $g$ at $x$ if $f$ has zeroth order contact with $g$ at $x$ and $D_x f = D_x g$ as maps $T_x M \to T_y N$. We write $f \sim_1 g$ at $x$.
3. $f$ has *kth order contact* with $g$ at $x$ if $Df : TM \to TN$ has $(k-1)$th order contact with $Dg$ at every point in $T_x M$. In this case we write $f \sim_k g$ at $x$.
4. Let $J^k(M, N)_{(x,y)}$ denote the set of equivalence classes of the relation $\sim_k$ at $x$ on the space of maps $f : M \to N$ such that $f(x) = y$.
5. Let $J^k(M, N) = \bigsqcup_{(x,y) \in M \times N} J^k(M, N)_{(x,y)}$, and call an element $\sigma \in J^k(M, N)$ a *k-jet* from $M$ to $N$.
6. Each $\sigma \in J^k(M, N)$ is an element of $J^k(M, N)_{(x,y)}$ for some pair $(x, y) \in M \times N$, and we call $x$ the *source* of $\sigma$ and $y$ the *target*, and we let $\alpha : J^k(M, N) \to M$ and $\beta : J^k(M, N) \to N$ denote the source and target maps respectively.

Given a smooth map $f : M \to N$ we have a map $j^k f : M \to J^k(M, N)$ called the *k-jet* of $f$, whose value at $x$ is the equivalence class of $f$ in $J^k(M, N)_{(x, f(x))}$. We have $J^0(M, N) = M \times N$ and $j^0 f = (x, f(x))$ is the graph of $f$. In fact there is a map $j^k : M \times C^\infty(M, N) \to J^k(M, N)$ which sends $(x, f)$ to $j^k f(x)$. A naive way of thinking about this map is that it associates the pair $(x, f)$ to the Taylor polynomial of degree $k$ at $x$. In fact, by [GG73, Chapter II, Corollary 2.3], $f \sim_k g$ at $x$ if and only if the Taylor expansions of $f$ and $g$ at $x$ of order $k$ agree.

If $m$ and $n$ are the dimensions of $M$ and $N$ respectively, then the $k$-jet space $J^k(M, N)$ is a smooth manifold of dimension $m + n + \dim(B_{n,n}^k)$, where $B_{m,n}^k$ is the direct sum of $n$ copies of the vector space of polynomials of degree at most $k$ in $m$ variables.

We now define the topology on the space $C^\infty(M, N)$ of smooth maps from $M$ to $N$.

**Definition A.2.7** ([GG73, Chapter II, Definition 3.1]) With $M$ and $N$ as above, fix a non-negative integer $k$. For a subset $U \subset J^k(M, N)$ let

$$B(U) = \{f \in C^\infty(M, N) \mid j^k f(M) \subset U\}.$$

The family $\{B(U)\}$, where $U$ ranges over all open sets of $J^k(M, N)$, forms a basis for a topology called the *Whitney $C^k$ topology*, and we let $W^k$ denote the set of open subsets of $C^\infty(M, N)$ in this topology. Note that $W_k \subset W_l$ if

$k \leq l$. The *Whitney $C^\infty$ topology* on $C^\infty(M, N)$ is the topology whose basis is $W = \cup_{k=0}^{\infty} W_k$.

One important property of the Whitney $C^\infty$ topology is that $C^\infty(M, N)$ is a Baire space in this topology: every countable intersection of open dense sets is dense.

### A.2.3 Smooth approximation of continuous maps

Every map between smooth manifolds is homotopic to a smooth map. The following can be found, for example, in [Lee03, Theorem 6.15].

**Theorem A.2.8** (Whitney Approximation Theorem) *Let $M$ be a smooth manifold and $f : M \to \mathbb{R}^k$ be continuous. Then for any positive continuous function $\delta : M \to \mathbb{R}$ there exists a smooth map $f' : M \to \mathbb{R}^k$ $\delta$-close to $f$.*

More generally we have the following (see e.g. [Lee03, Theorem 6.19]).

**Theorem A.2.9** (Whitney approximation for manifolds) *Suppose $M$ and $N$ are smooth manifolds and $f : M \to N$ is a continuous map. Then $f$ is homotopic to a smooth map $f' : M \to N$.*

Here is a relative version. It can be found in [AGP02, page xxvii]; the authors use it for precisely same reason we include it here – to get the connectivity of the inclusion map $X \to X \cup e^n$ when attaching an $n$-cell to $X$ (see Example A.2.16).

**Theorem A.2.10** *Let $f : U \to V$ be a map between bounded open sets, $U \subset \mathbb{R}^m$, $V \subset \mathbb{R}^n$. Given any open sets $W, W'$ such that $\overline{W} \subset W' \subset \overline{W'} \subset U$, there exists $g : U \to V$ such that $g|_W : W \to V$ is smooth, $f|_{U-W'} = g|_{U-W'}$, and $f \sim g$ relative to $U - W'$.*

**Remark A.2.11** This result can be applied to spaces which are only locally manifolds. For instance, suppose $X$ is a space and $X \cup e^n$ is the result of attaching an $n$-cell to $X$. Let $f : P \to X \cup e^n$ be a map, where $P$ is a smooth closed manifold. Let $x \in e^n$ be in the interior of the $n$-cell. We are going to change $f$ by a homotopy which is smooth in a neighborhood of $f^{-1}(x)$; the smooth structure on the interior of $e^n$ is the usual one inherited from $\mathbb{R}^n$. Let $V \subset \mathring{e}^n$ be an open set containing $x$, so that $U = f^{-1}(V)$ is an open set containing $f^{-1}(x)$. We can clearly choose open sets $W, W'$ as in Theorem A.2.10 such that

$f^{-1}(x) \subset W$ since $f^{-1}(x)$ is closed and therefore compact since $P$ is compact. By Theorem A.2.10, we can make a homotopy of $f$ relative to $U - W'$ to a map $g$ which is smooth on $W$. □

Finally we compare the Whitney $C^\infty$ topology with the compact open topology.

**Theorem A.2.12** *Let $M$ and $N$ be smooth manifolds with $M$ closed. The inclusion $C^\infty(M, N) \to \mathrm{Map}(M, N)$ is a weak homotopy equivalence.*

*Proof* We sketch the idea, using the characterization of weak equivalences given in Proposition 2.6.17. A map of pairs $f : (D^i, \partial D^i) \to (\mathrm{Map}(M, N), C^\infty(M, N))$ is homotopic relative to the boundary to a map $D^i \to C^\infty(M, N)$ as follows. Using Theorem 1.2.7 $f : D^i \to \mathrm{Map}(M, N)$ corresponds to a map $\widetilde{f} : D^i \times M \to N$, and since $M$ is compact and the inclusion $C^\infty(M, N) \to \mathrm{Map}(M, N)$ is open, there exists an open neighborhood $T$ of $\partial D^i$ such that $f|_{T \times M} : T \times M \to N$ is already smooth. Since $T$ necessarily contains a neighborhood of $\partial D^i$ of the form $\partial D^i \times [0, \epsilon)$, we may also assume $T = \partial D^i \times [0, \epsilon)$ for some $0 < \epsilon < 1$. Let $U \subset D^i$ be an open disk of radius $1 - \epsilon/2$ centered at $0 \in D^i$, so that $U \cup T = D^i$. We can clearly choose $W, W' \subset U$ as in Theorem A.2.10 with $W$ a slightly smaller open disk and $W \cup T = D^i$. Now apply Theorem A.2.10 to $\widetilde{f} : D^i \times M \to N$ to produce a map $\widetilde{g} : D^i \times M \to N$ which is smooth and equal to $\widetilde{f}$ on $T \times M$. This corresponds via Theorem 1.2.7 to the desired map $g : D^i \to C^\infty(M, N)$. □

## A.2.4 Transversality

Here we review some standard notions from differential topology. Our main reference for this part is [GG73]. The following is Chapter II, Definition 1.11 there.

**Definition A.2.13** Let $M$ and $N$ be smooth manifolds and $f : M \to N$ a continuously differentiable map. Let $x \in M$, and let $D_x f : T_x M \to T_{f(x)} N$ denote the derivative of $f$ at $x$. Then

- $\mathrm{corank} D_x f = \min\{\dim M, \dim N\} - \mathrm{rank} D_x f$;
- $x$ is called a *critical point* of $f$ if $\mathrm{corank} D_x f > 0$;
- a point $y \in N$ is called a *critical value* of $f$ if it is the image of some critical point of $f$;
- a point $y \in Y$ is called a *regular value* of $f$ if $f^{-1}(y)$ contains no critical points of $f$; in particular, if $y$ is not in the image of $f$ then $y$ is a regular value of $f$.

The following theorem is known as Sard's Theorem, which can be found for example in [GG73, Chapter 2, Theorem 1.12] (see also [Hir94] or [Mil63]). It states that most values of $f$ are regular values.

**Theorem A.2.14** (Sard's Theorem)  *Let $M$ and $N$ be smooth manifolds and $f\colon M \to N$ a smooth map. Then the set of critical values of $f$ has measure zero in $Y$.*

For the following consequence of Sard's Theorem, see for example [GG73, Chapter II, Corollary 1.14].

**Corollary A.2.15**  *The set of all regular values of $f$ is dense in $N$.*

As a consequence, we can deduce the connectivity of the inclusion $X \to X \cup e^n$. We first learned this proof from [BT82, Proposition 17.11].

**Example A.2.16**  Let $X$ be a space, and $f\colon \partial e^n \to X$ be a map encoding the attachment of an $n$-cell to $X$. Then the inclusion map $X \longrightarrow X \cup_f e^n$ is $(n-1)$-connected.

To see this, let $f\colon (D^i, \partial D^i) \to (X \cup e^n, X)$ be a map of pairs. Let $V \subset \mathring{e}^n$ be an open set. As in Remark A.2.11 we may also assume $f$ is smooth on $U = f^{-1}(V)$. By Corollary A.2.15, there is a regular value $x \in V$ for $f$. If $i < n$, this means $x$ is not in the image of $f$, so $f$ may be considered as a map $D^i \to X \cup e^n - \{x\}$. $X$ is a deformation retract of $X \cup e^n - \{x\}$, and so $f$ is homotopic relative to $\partial D^i$ to $X$. This implies that $X \to X \cup e^n$ is $(n-1)$-connected.  $\square$

The remainder of the material in this portion of the appendix is here so that we can showcase various applications of Theorems 4.2.3 and 6.2.3 to smooth manifold theory.

**Definition A.2.17**  Let $f\colon M \to N$ be a smooth map of smooth manifolds, $Y \subset N$ a submanifold, and $x \in M$ a point. We say $f$ is *transverse to $Y$ at $x$*, written $f \pitchfork Y$ at $x$, if either $f(x) \notin Y$, or $f(x) \in Y$ and

$$D_x f(T_x M) + T_{f(x)} Y = T_{f(x)} N.$$

We say $f$ is *transverse to $Y$*, written $f \pitchfork Y$, if $f$ is transverse to $Y$ at every point $x \in X$.

A straightforward and typical consequence of this is the following. Let $f\colon M \to N$ be smooth and $Y \subset N$ a submanifold. Suppose $\dim(M) + \dim(Y) < \dim(N)$, that is, $\dim(M) < \mathrm{codim}(Y)$. Then $f \pitchfork Y$ if and only if $f(M) \cap Y = \emptyset$.

**Theorem A.2.18** ([GG73, Chapter II, Theorem 4.4]) *Let M and N be smooth manifolds and $Y \subset N$ a submanifold. Suppose $f: M \to N$ is smooth and $f \pitchfork Y$. Then $f^{-1}(Y)$ is a submanifold of M and $\mathrm{codim}(f^{-1}(Y) \subset M) = \mathrm{codim}(Y \subset N)$. In fact, the normal bundle of $f^{-1}(Y)$ in M is the pullback of the normal bundle of Y in N via the mapping f.*

The following well-known result can be found, for example, in [GG73, Chapter II, Theorem 4.9].

**Theorem A.2.19** (Thom Transversality Theorem) *Let M and N be smooth manifolds and Y a submanifold of $J^k(M,N)$. Let*

$$T_Y = \{f \in C^\infty(M,N) \mid j^k f \pitchfork Y\}.$$

*Then $T_Y$ is the countable intersection of open dense sets in the Whitney $C^\infty$ topology, and if Y is closed, then $T_Y$ is also open.*

For the following consequence, see for example [GG73, Chapter II, Corollary 4.12]

**Corollary A.2.20** (Elementary Transversality Theorem) *Let M and N be smooth manifolds and Y a submanifold of N. The subset of $C^\infty(M,N)$ consisting of those smooth maps transverse to Y is dense in $C^\infty(M,N)$ and open if Y is closed.*

*Additionally, we have the following relative version. Suppose $U_1 \subset U_2 \subset M$ are open sets such that $U_2$ contains the closure of $U_1$. Let $f: M \to N$ be smooth and let V be any open neighborhood of f in $C^\infty(M,N)$. Then there exists g in V such that g is transverse to Y outside $U_2$ and $g = f$ on $U_1$.*

This follows from Theorem A.2.19 by choosing $k = 0$ so that $J^0(M,N) = M \times N$, by noting that the inverse image of Y in $M \times N$ via the projection is a submanifold of the product, and noting that if the 0-jet (i.e. graph) of $j$ is transverse to this inverse image, then $f$ is transverse to Y.

Revisiting Example A.2.16, if $Y = X \cup e^n$ has the structure of a smooth manifold of dimension $n$, then we can restate Example A.2.16 as saying that the inclusion $Y - \{x\} \to Y$ is $(n-1)$-connected. Here is a generalization.

**Example A.2.21** Suppose $M, N$ are manifolds with $M \subset N$ a smooth closed submanifold of codimension $k$. Then the inclusion $N - M \to N$ is $(k-1)$-connected.

To see this, let $f: (D^i, \partial D^i) \to (N, N-M)$ be a map of pairs. By compactness of $M$, there exists a neighborhood $U$ of $\partial D^i$ such that $f(U) \cap M = \emptyset$. By Theorem A.2.10 the $f: D^i \to N$ is homotopic relative to $U$ to a map which is smooth away from $U$ (the details are similar to those found in the proof of Theorem A.2.12). By abuse continue to call this map $f$. By Corollary A.2.20 $f$ is homotopic relative to $\partial D^i$ (in fact, relative to an open neighborhood of $\partial D^i$) to a map which is transverse to $M$. If $i + m < n$ this means $f$ misses $M$, so $f$ is homotopic to a map $D^i \to N - M$ if $i < n - m$; in other words, the inclusion $N - M \to N$ is $(n - m - 1)$-connected.

We end by discussing multijets and the relevant transversality theorem.

**Definition A.2.22** Let $M^s$ be the product of $s$ copies of $M$, and recall the configuration space $\mathrm{Conf}(s, M) \subset M^s$ as the set of all $(x_1, \ldots, x_s) \in M^s$ such that $x_i \neq x_j$ for all $i \neq j$. We then have the $s$-fold source map $\alpha^s: (J^k(M, N))^s \to M^s$, and we define

$$J_s^k(M, N) = (\alpha^s)^{-1}(\mathrm{Conf}(s, M)).$$

The space $J_s^k(M, N)$ obtains its manifold structure by being an open subset of the manifold $(J^k(M, N))^s$. Define $j_s^k f: \mathrm{Conf}(s, M) \longrightarrow J_s^k(M, N)$ by

$$j_s^k f(x_1, \ldots, x_s) = (j^k f(x_1), \ldots, j^k f(x_s)).$$

For more details of the following, see for example [GG73, Chapter II, Theorem 4.13].

**Theorem A.2.23** (Multijet Transversality Theorem)    *Let $M$ and $N$ be smooth manifolds and $Y$ a submanifold of $J_s^k(M, N)$. Let*

$$T_Y = \{f \in \mathrm{Map}(M, N): j_s^k f \pitchfork Y\}.$$

*Then $T_Y$ is the countable intersection of open dense sets in the Whitney $C^\infty$ topology, and if $Y$ is compact, then $T_Y$ is also open.*    □

# A.3 Spectra

In this section we briefly review the notion of a spectrum and some basic results about spectra. There are many good introductory sources on this material [Ada78, Hat02, tD08]. We will mostly use spectra in Section 10.1 and, as usual in this appendix, our goal is to present just enough basic definitions along with motivating examples to provide for a self-contained exposition. One important aspect of the theory of spectra that we will not pursue since we do not have use for it is that spectra "represent" cohomology theories. This was alluded to in the discussion following Theorem 2.4.20.

**Definition A.3.1** A *spectrum* $\underline{E}$ is a sequence of based spaces $\{E_n\}_{n\in\mathbb{N}}$ together with basepoint-preserving maps

$$\Sigma E_n \longrightarrow E_{n+1},$$

or equivalently by Theorem 1.2.9, maps

$$E_n \longrightarrow \Omega E_{n+1}.$$

If the latter maps are weak equivalences, then $\underline{E}$ is called an $\Omega$-*spectrum*.

**Example A.3.2** The spectrum $\underline{*}$ whose every space is a point is an $\Omega$-spectrum. □

**Example A.3.3** (Suspension spectrum)   Given a based space $X$, let $E_n = \Sigma^n X$. The structure maps $\Sigma E_n \to E_{n+1}$ are then identity and we have the *suspension spectrum of $X$*, denoted by $\Sigma^\infty X$. This is not necessarily an $\Omega$-spectrum since if $\Sigma E_n \to E_{n+1}$ is an equivalence, that does not necessarily imply that the corresponding map $E_n \to \Omega E_{n+1}$ is also an equivalence. If $X = *$, then $\Sigma^\infty *$ is precisely the spectrum $\underline{*}$ from the previous example since the (reduced) suspension of a point is a point. □

**Example A.3.4** (Sphere spectrum)   Setting $X = S^0$ in the previous example gives an important special case of a suspension spectrum, namely the *sphere spectrum* $\Sigma^\infty S^0$, which we will denote by $\mathbb{S}$. Since $\Sigma S^n = S^{n+1}$, this spectrum consists of spheres. □

**Example A.3.5** (Eilenberg–MacLane spectrum)   Suppose $G$ is an abelian group and set $E_n = K(G, n)$, an Eilenberg–MacLane space (so $K(G, n)$ has no homotopy other than in dimension $n$, where $\pi_n(K(G, n)) \cong G$). Using the fact that, for any space $X$, $\pi_{n+1}(X) \cong \pi_n(\Omega X)$ (see (1.4.2)), we have a weak equivalence

$$K(G, n) \simeq \Omega K(G, n + 1)$$

and we thus have an $\Omega$-spectrum called the *Eilenberg–MacLane spectrum*, denoted by $\underline{H}G$. □

**Example A.3.6** (Real and complex $K$-theory)   Let $O$ and $U$ be the infinite-dimensional orthogonal and unitary groups (direct limits of finite-dimensional orthogonal and unitary groups, which form sequences of inclusions). By Bott periodicity, there are equivalences $O \to \Omega^8 O$ and $U \to \Omega^2 U$, so this gives two periodic $\Omega$-spectra $KO$ and $KU$, called *complex and real K-theory* (since these spectra give rise to those two cohomology theories). □

A spectrum can be smashed with a space. That is, given a spectrum $\underline{E}$ and a based space $X$, we can define the spectrum $\underline{E} \wedge X$ by

$$(\underline{E} \wedge X)_n = \underline{E}_n \wedge X.$$

Recalling Example 1.1.16, we have

$$\Sigma(E_n \wedge X) = (\Sigma E_n) \wedge X \longrightarrow E_{n+1} \wedge X,$$

so the structure maps in $\underline{E} \wedge X$ are the product of stucture maps in $\underline{E}$ and the identity map.

Suppose $X$ is an $n$-connected space. By the generalization of the Freudenthal Suspension Theorem (see Theorem 4.2.10), the suspension map

$$\pi_i(X) \longrightarrow \pi_{i+1}(\Sigma X) \tag{A.1}$$

is an isomorphism for $i \leq 2n$. Iterating this map therefore gives a sequence

$$\pi_i(X) \longrightarrow \pi_{i+1}(\Sigma X) \longrightarrow \pi_{i+2}(\Sigma^2 X) \longrightarrow \cdots \tag{A.2}$$

where the maps necessarily eventually become isomorphisms even without any assumption on the connectivity of $X$. Given a sequence of groups and homomorphisms

$$G_0 \xrightarrow{h_0} G_1 \xrightarrow{h_1} \cdots,$$

the *direct limit*, or *colimit*, of this sequence is defined as

$$G_n = \coprod_n G_n / \sim,$$

where $g \sim g'$ if $(f_j \circ f_{j-1} \circ \cdots \circ f_{i+1} \circ f_i)(g) = g'$ for some $i, j$. (This is analogous to the colimit of sequence of spaces; see Example 7.3.31.) We can thus define the *ith stable homotopy group of $X$*, $\pi_i^s(X)$, as the colimit of the sequence (A.2):

$$\pi_i^s(X) = \operatorname*{colim}_n \pi_{i+n}(\Sigma^n X). \tag{A.3}$$

In particular, for $X = S^0$, we have the *stable homotopy groups of spheres*,

$$\pi_i^s(S^0) = \operatorname*{colim}_n \pi_{i+n}(S^n)$$

which are one of the most important (and difficult) objects in homotopy theory. Note that, again by Freudenthal Suspension Theorem,

$$\pi_i^s(S^0) = \pi_{i+n}(S^n)$$

for $n > i + 1$.

For a spectrum $\underline{E}$, there are maps

$$\pi_{i+n}(E_n) \longrightarrow \pi_{i+n+1}(E_{n+1}) \tag{A.4}$$

given as the composition of the suspension map (1.1.2) and the map induced on homotopy groups by the spectrum structure map:

$$\pi_{i+n}(E_n) \xrightarrow{\Sigma} \pi_{i+n+1}(\Sigma E_n) \longrightarrow \pi_{i+n+1}(E_{n+1}).$$

It thus makes sense to mimic (A.3) and define the *ith homotopy group of the spectrum $\underline{E}$* as

$$\pi_i(\underline{E}) = \operatorname*{colim}_n \pi_{i+n}(E_n). \tag{A.5}$$

Notice that a spectrum could have homotopy groups in negative dimensions. We say a spectrum is *connective* if it does not have any negative homotopy groups.

Thus for the suspension spectrum $\Sigma^\infty X$ of a space $X$, by definition we have

$$\pi_i(\Sigma^\infty X) = \pi_i^s(X).$$

In particular, stable homotopy groups of spheres can be regarded as the homotopy groups of the spectrum $\Sigma^\infty S^0 = \mathbb{S}$.

It follows that, for an $\Omega$-spectrum $\underline{E}$, we have

$$\pi_i(\underline{E}) = \pi_{i+n}(E_n), \quad n \ge 0. \tag{A.6}$$

In particular, we have the following.

**Example A.3.7**  Let $\underline{H}G$ be the Eilenberg–MacLane spectrum from Example A.3.5. Then

$$\pi_i(\underline{H}G) = \begin{cases} G, & i = 0; \\ 0, & \text{otherwise.} \end{cases} \qquad \square$$

**Definition A.3.8**  A *map* of spectra $f \colon \underline{E} \to \underline{F}$ is a collection of maps

$$f_n \colon E_n \longrightarrow F_n, \quad n \ge 0,$$

that commute with the structure maps in $\underline{E}$ and $\underline{F}$. Such a map is a *weak equivalence of spectra* if

$$f_* : \pi_i(\underline{E}) \longrightarrow \pi_i(\underline{F})$$

is an isomorphism for all $i$.

**Example A.3.9** From the definitions, we have a weak equivalence $\mathbb{S} \wedge X \simeq \Sigma^\infty X$.                                                              □

Using maps of spectra as defined above, we can define the *category of spectra* which we will denote by Spectra.

The following standard and easy result says that, up to weak equivalence, we can assume we are working with an $\Omega$-spectrum. For a proof, see, for example [Ada78, Sections 1.6 and 1.7].

**Proposition A.3.10**  *Every spectrum is weakly equivalent to an $\Omega$-spectrum.*

For an $\Omega$-spectrum $\underline{E}$, we have from the definitions that, for any $E_n$,

$$E_n \xrightarrow{\simeq} \Omega^k E_{n+k}.$$

Therefore $E_n$ is a $k$-fold loop space for any $k$, and is for this reason called an *infinite loop space*. From the examples we have had so far, we see that Eilenberg–MacLane spaces and the infinite-dimensional orthogonal and unitary groups are all infinite loop spaces.

One can create an infinite loop space out of any space $X$ as follows. Let

$$\Omega^\infty : \text{Spectra} \longrightarrow \text{Top}_*$$

be the functor which, given a spectrum $\underline{E}$, replaces it by an equivalent $\Omega$-spectrum $\underline{F}$ and then picks off the first space $F_0$ which is an infinite loop space by the above comments. Composing this with the functor

$$\Sigma^\infty : \text{Top}_* \longrightarrow \text{Spectra}$$

which assigns to a space $X$ its suspension spectrum, we get the infinite loop space functor, usually denoted by $Q$:

$$Q = \Omega^\infty \Sigma^\infty : \text{Top}_* \longrightarrow \text{Top}_* . \tag{A.7}$$

This can also be thought of as

$$QX = \Omega^\infty \Sigma^\infty X = \operatorname*{hocolim}_{n} \Omega^n \Sigma^n X.$$

The maps are given using Theorem 1.2.7. We have a commutative diagram

$$\begin{array}{ccc} \Sigma^n X & \xrightarrow{1_{\Sigma^n X}} & \Sigma^n X \\ \downarrow & & \downarrow \\ \Sigma^{n+1} X & \xrightarrow{1_{\Sigma^{n+1} X}} & \Sigma^{n+1} X \end{array}$$

where the vertical maps are the inclusion $Y \to \Sigma Y$ (induced by the inclusion $Y \times \{1/2\} \to Y \times I$). The horizontal maps correspond via Theorem 1.2.7 to maps $X \to \Omega^n \Sigma^n X$ and $X \to \Omega^{n+1} \Sigma^{n+1} X$ and the vertical maps above induce maps $\Omega^n \Sigma^n X \to \Omega^{n+1} \Sigma^{n+1} X$.

**Example A.3.11** Unravelling the definitions gives

$$\pi_i(\Sigma^\infty S^0) = \pi_i(QS^0). \qquad \square$$

Given a diagram $\underline{E} \to \underline{G} \leftarrow \underline{F}$, we can define its homotopy pullback

$$\mathrm{holim}(\underline{E} \to \underline{G} \leftarrow \underline{F})$$

levelwise. Namely, the $n$th space in this spectrum is

$$\mathrm{holim}(E_n \to G_n \leftarrow F_n)$$

and the structure maps

$$\mathrm{holim}(E_n \to G_n \leftarrow F_n) \longrightarrow \Omega\,\mathrm{holim}(E_{n+1} \to G_{n+1} \leftarrow F_{n+1})$$

are induced by the structure maps in the three spectra using the fact that loops commutes with homotopy pullbacks (Corollary 3.3.16).

Similarly one can define the homotopy pushout of a diagram $\underline{E} \leftarrow \underline{H} \to \underline{F}$ levelwise. For the structure maps, one uses the fact that suspension commutes with homotopy pushouts (Corollary 3.7.19).

Now, given a commutative square

$$\underline{S} = \begin{array}{ccc} \underline{H} & \longrightarrow & \underline{F} \\ \downarrow & & \downarrow \\ \underline{E} & \longrightarrow & \underline{G} \end{array}$$

of spectra, we have canonical maps

$$a(\underline{S})\colon \underline{H} \longrightarrow \mathrm{holim}(\underline{E} \to \underline{G} \leftarrow \underline{F})$$

and

$$b(\underline{S})\colon \mathrm{hocolim}(\underline{E} \leftarrow \underline{H} \to \underline{F}) \longrightarrow \underline{G}$$

induced by the canonical levelwise maps from (3.3.1) and (3.7.1).

**Definition A.3.12**   The commutative square diagram $\underline{S}$ is *homotopy cartesian* if $a(\underline{S})$ is a weak equivalence and it is *homotopy cocartesian* if $b(\underline{S})$ is a weak equivalence.

The following says that the notions of homotopy cartesian and homotopy cocartesian squares of spectra are the same. A proof can be found in [LRV03, Lemma 2.6].

**Proposition A.3.13**   *Suppose $\underline{S}$ is a commutative square of spectra. Then this square is homotopy cartesian if and only if it is homotopy cocartesian.*

**Remark A.3.14**   Using homotopy pullbacks and pushouts of spectra, we could define homotopy limits and colimits of punctured cubes of spectra via iterated homotopy pullbacks and pushouts (inspired by Lemma 5.3.6 and Lemma 5.7.6). Furthermore, we could then define what it means for a cube of spectra to be homotopy (co)cartesian or strongly (co)cartesian. Proposition A.3.13 then says that a cube of spectra is homotopy cartesian if and only if it is homotopy cocartesian.                                                                □

The previous result gives a special and important relationship between homotopy fibers and homotopy cofibers of maps of spectra. Given a map of spectra $f \colon \underline{E} \to \underline{F}$, we can define its homotopy fiber and cofiber as a levelwise homotopy fiber and cofiber. Or, since we know what homotopy pullbacks and pushouts of spectra are, as

$$\mathrm{hofiber}(f) = \mathrm{holim}(* \to \underline{F} \xleftarrow{f} \underline{E})$$

and

$$\mathrm{hocofiber}(f) = \mathrm{hocolim}(* \leftarrow \underline{E} \xrightarrow{f} \underline{F}),$$

where $*$ denotes the constant spectrum.

We then have the following consequence of Proposition A.3.13.

**Corollary A.3.15**   *Homotopy fibration sequences of spectra are the same as homotopy cofibration sequences of spectra. Namely, given a sequence $\underline{E} \to \underline{F} \to \underline{G}$, we have*

$$\underline{E} \simeq \mathrm{hofiber}(\underline{F} \to \underline{G}) \iff \underline{G} \simeq \mathrm{hocofiber}(\underline{E} \to \underline{F})$$

*Furthermore, for a map $f \colon \underline{E} \to \underline{F}$ of spectra, we have a weak equivalence*

$$\mathrm{hofiber}(f) \simeq \Omega\,\mathrm{hocofiber}(f)$$

*(where looping a spectrum means looping levelwise).*

*Proof*  Consider the square

If this square is homotopy cartesian, that means that there is a weak equivalence $\underline{E} \simeq \text{hofiber}(\underline{F} \to \underline{G})$ and, if it is homotopy cocartesian, that means that there is a weak equivalence $\underline{G} \simeq \text{hocofiber}(\underline{E} \to \underline{F})$. But the two notions are equivalent by Proposition A.3.13.

For the second part, the square

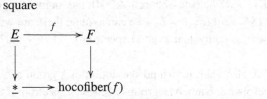

is homotopy cocartesian and so, by Proposition A.3.13, it is also homotopy cartesian. By levelwise application of Proposition 3.3.18, this means that there is a weak equivalence $\text{hofiber}(f) \simeq \text{hofiber}(* \to \text{hocofiber}(f))$. But the right side is $\Omega\,\text{hocofiber}(f)$ by levelwise application of Example 2.2.9.          □

The second part of the above result is of course false in the realm of topological spaces, but as we saw in Example 4.2.12, for a map $f \colon X \to Y$ of based spaces, there is a map $\text{hofiber}(f) \to \Omega\,\text{hocofiber}(f)$ whose connectivity is roughly the sum of the connectivity of $X$ with the connectivity of $f$.

There is another familiar result that can be deduced from Proposition A.3.13.

**Corollary A.3.16**  *For any spectrum $\underline{E}$, there are weak equivalences $\underline{E} \simeq \Sigma\Omega\underline{E}$ and $\underline{E} \simeq \Omega\Sigma\underline{E}$ (where the suspension of a spectrum is taken levelwise).*

*Proof*  The first equivalence immediately follows from considering the homotopy cartesian square

$$
\begin{array}{ccc}
\Omega E & \xrightarrow{\ f\ } & * \\
\downarrow & & \downarrow \\
* & \longrightarrow & E
\end{array}
$$

By Proposition A.3.13, this square is also homotopy cocartesian, and hence $\Sigma\Omega\underline{E} \simeq \underline{E}$. For the second statement, we start with the homotopy cocartesian square

Again by Proposition A.3.13, this square is also homotopy cartesian, and hence we have a weak equivalence $\underline{E} \simeq \Omega\Sigma\underline{E}$. □

**Remark A.3.17** One of the equivalences in the previous result can be argued directly. Namely, the map $\underline{E} \to \Omega\Sigma\underline{E}$ given levewise by the Freudenthal maps $E_n \to \Omega\Sigma E_n$ (see Theorem 4.2.10) has an inverse given levelwise by the maps $\Omega\Sigma\underline{E}_n \to \Omega E_{n+1} \simeq E_n$ (we can assume these are weak equivalences since $\underline{E}$ is weakly equivalent to an $\Omega$-spectrum). □

We also want to extend the notion of a group action to spectra. This is done levelwise. Namely, a group $G$ acts on a spectrum if there is a map $G \times E_n \to E_n$ for each $n$ satisfying the usual axioms of an action that is compatible with the spectrum structure maps. To explain this, the map $G \times E_n \to E_n$ can be regarded as a map $G \to E_n \times E_n$ and, for each $g \in G$, the suspension of the map $f_{n,g}: E_n \to E_n$ and the map $f_{n+1,g}: E_{n+1} \to E_{n+1}$ fit into the commutative diagram

$$\begin{array}{ccc} \Sigma E_n & \longrightarrow & \Sigma E_n \\ \downarrow & & \downarrow \\ E_{n+1} & \longrightarrow & E_{n+1} \end{array}$$

In terms of functors, this means that there is a functor

$$F: \bullet_G \longrightarrow \text{Spectra}$$

with $F(\bullet) = \underline{E}$ and $g \mapsto (f_g: \underline{E} \to \underline{E})$ (where $f_g$ is the collection of maps $f_{n,g}$). Then one can define the *spectra of homotopy fixed points and homotopy orbits* as

$$\underline{C}^{hG} = \operatorname*{holim}_{\bullet_G} F \quad \text{and} \quad \underline{C}_{hG} = \operatorname*{hocolim}_{\bullet_G} F, \tag{A.8}$$

where the homotopy limit and colimit are taken levelwise as in Examples 8.2.11 and 8.2.23. These levelwise homotopy (co)limits inherit structure maps from $\underline{C}$ and therefore form spectra, because homotopy limits commute with loops and homotopy colimits commute with suspensions. In Section 10.1, we are especially interested in the action of $\Sigma_k$ on certain spectra.

## A.4 Operads and cosimplicial spaces

This section contains a brief introduction to the relationship between operads and cosimplicial spaces. The details for most of what will be explained here can be found in the papers of McClure and Smith [MS02, MS04a, MS04b], who have used this relationship for a proof of the Deligne's Hochschild cohomology conjecture. An application of this operadic machinery that is useful to us is to spaces of knots, and this is the subject of Section 10.4.2. For basics on operads (in any category), including motivation for their study, see [MSS02] (see also Theorem A.4.6 below for one of the earliest and most important applications of operads).

We will only consider operads in the category of spaces. The basic definition is the following.

**Definition A.4.1** A *non-symmetric, or non-$\Sigma$, (topological) operad $O$* is a collection of spaces $\{O(k)\}_{k \geq 0}$ along with an element $1 \in O(1)$ called the *identity* and *structure maps*

$$\gamma: O(k) \times O(j_1) \times \cdots O(j_k) \longrightarrow O(j_1 + \cdots + j_k)$$

for all $k, j_1, \ldots, j_k \geq 0$, satisfying the following.

- *Identity axiom*: for $x \in O(k)$, $\gamma(1, x) = \gamma(x, 1, \ldots, 1) = x$.
- *Associativity axiom*: the diagram

$$O(k) \times \prod_{m=1}^{k}\left(O(j_m) \times \prod_{n=1}^{j_m}O(i_{mn})\right) \xrightarrow{\ Id \times \gamma\ } O(k) \times \prod_{m=1}^{k}O(i_{m1} + \cdots + i_{mj_m})$$

$$\Big\downarrow{\scriptstyle =}$$

$$\left(O(k) \times \prod_{m=1}^{k}O(j_m)\right) \times \prod_{m,n}O(i_{mn})$$

$$\Big\downarrow{\scriptstyle \gamma \times Id}$$

$$O(j_1 + \cdots + j_k) \times O(i_{11}) \times \cdots \times O(i_{kj_k}) \xrightarrow{\ \gamma\ } O(i_{11} + \cdots + i_{kj_k})$$

with right vertical map $\gamma$, commutes.

This is equivalent to the standard definition of an operad except the axioms having to do with the action of the symmetric group have been removed. Non-symmetric operads are thus also called *operads without permutations*.

A special case of the operad structure maps are the *circle-i* operations

$$\circ_i \colon O(k) \times O(j) \longrightarrow O(k + j - 1) \qquad\qquad \text{(A.1)}$$

given by

$$(x, y) \longmapsto (x, 1, \ldots, 1, y, 1, \ldots, 1),$$

where $y$ on the right appears in the slot $i + 1$. It is easy to see that all operadic structure maps are generated by these operations so that it suffices to specify the $\circ_i$ maps.

**Example A.4.2** (Associative non-symmetric operad)   The *associative non-symmetric operad $\mathcal{A}ss$* is defined by $\mathcal{A}ss(k) = *$ for all $k$. □

**Example A.4.3** (Little $n$-cubes operad)   The *little n-cubes operad* (or *little n-disks operad* or *little n-balls operad*) $C_n = \{C_n(k)\}_{k \geq 0}$, where $C_n(k)$ is the space of embeddings of $I^n \times \underline{k}$ in $I^n$ and where the embedding on each summand $I^n$ is of the form $f(\vec{x}) = D\vec{x} + \vec{c}$, for some invertible matrix $D$. We take $C_n(0)$ to be the one-point space.

   The structure maps essentially consist of "shrinking the cubes and inserting them into others". For a detailed description of these maps, see [MSS02, Section 2.2]. □

**Remark A.4.4**   One standard way to define non-$\Sigma$ operads is via a category of *rooted trees* [Boa71, MSS02, Sin06]. Roughly, these are (isomorphism classes of) rooted, planar trees with labeled leaves. The $\circ_i$ operation is given by grafting the root of one tree onto the $i$th leaf of another. A topological non-$\Sigma$ operad is then simply a functor from the category of such trees (where the morphisms are defined by contractions of edges) to Top satisfying the obvious axioms (see e.g. [Sin06, Definition 2.13]). □

**Definition A.4.5**   For $O$ a non-symmetric operad and $X$ a space, an *action of $O$ on $X$* consists of maps

$$f_k \colon O(k) \times X^k \longrightarrow X$$

satisfying

- $f_k(1, x) = x$ for all $x \in X$;

- the diagram

$$O(k) \times \left(\prod_{m=1}^{k} O(j_m)\right) \times X^{j_1+\cdots+j_k} \xrightarrow{\ =\ } O(k) \times \prod_{m=1}^{k} \left(O(j_m) \times X^{j_m}\right)$$

with vertical maps $\gamma \times Id$ on the left and $Id \times \prod f_{jm}$ on the right to $O(k) \times X^k$, then $f_k$ down to $X$, and bottom map

$$O(j_1 + \cdots + j_k) \times X^{j_1+\cdots+j_k} \xrightarrow{\ f_{j_1+\cdots+j_k}\ } X$$

commutes.

One of the main uses of operads in topology is in the following *recognition principle.*

**Theorem A.4.6** ([BV73, May72]) *If a connected space X admits an action of the little n-cubes operad, then it is homotopy equivalent to an n-fold loop space.*

**Definition A.4.7** A *morphism* $f\colon O \to P$ of operads is a collection of maps

$$f_k\colon O(k) \longrightarrow P(k), \quad k \geq 0,$$

which are compatible with the structure maps and the identity.

**Definition A.4.8** A *non-symmetric operad with multiplication* is a non-symmetric operad $O$ together with a morphism $\mathcal{A}ss \to O$.

It is not hard to see a multiplicative structure on an operad $O$ as defined above is equvalent to the existence of morphisms

$$e\colon 1 \longrightarrow O(0) \quad \text{and} \quad \mu\colon 1 \longrightarrow O(2)$$

satisfying

$$\mu \circ_1 \mu = \mu \circ_2 \mu \quad \text{and} \quad \mu \circ_1 e = \mu \circ_2 e = Id. \tag{A.2}$$

**Remark A.4.9** The motivating example for Definition A.4.8 is that of the endomorphism operad $End(A)$, where $A$ is an associative algebra. □

Using the multiplicative structure, one can then associate a cosimplicial space to an operad with multiplication as follows.

**Definition A.4.10** ([GV95])    For an operad with multiplication $O$, the associated cosimplicial space $O^\bullet$ is defined by $O^{[n]} = O^n$ with cofaces $d^i : O^n \to O^{n+1}$ defined by

$$d^i(x) = \begin{cases} \mu \circ_2 x, & i = 0; \\ x \circ_i m, & 1 \le i \le n; \\ \mu \circ_1 x, & i = n + 1, \end{cases}$$

and codegeneracies $s^i : O^{n+1} \to O^n$ defined by

$$s^i(x) = x \circ_i e, \quad 0 \le i \le n.$$

It is easy to see that, because of identities (A.2), the cosimplicial identities in $O^\bullet$ are indeed satisfied.

The following important result is due to McClure and Smith.

**Theorem A.4.11**    [MS02, Theorem 3.3] *The totalization* $\mathrm{Tot}(O^\bullet)$ *of the cosimplicial space associated to an operad with multiplication admits an action of an operad equivalent to the little 2-cubes operad* $C_2$. *The same is true for the homotopy invariant totalization* $\mathrm{hoTot}(O^\bullet)$ *[MS04a, Theorem 15.3 and Proposition 10.3] (see also [Sin06, Theorem 7.3]).*

Combining Theorem A.4.11 with Theorem A.4.6, we have

**Corollary A.4.12**    *For $O$ an operad with multiplication, there is a homotopy equivalence*

$$\mathrm{Tot}(O^\bullet) \simeq \Omega^2 X$$

*for some space $X$.*

**Remark A.4.13**    McClure and Smith prove a more general result than Theorem A.4.11 in [MS04a]: if a cosimplicial space $X^\bullet$ is an algebra over a certain *functor-operad* $\Xi^n$, then $\mathrm{Tot}\, X^\bullet$ admits an action of an operad equivalent to $C_n$ [MS04a, Theorem 9.1]. In this context, having a $\Xi^2$ structure on a cosimplicial space amounts to having an operad with multiplication that gives rise to that cosimplicial space [MS04a, Section 10], so the result reduces to Theorem A.4.11.

It should also be noted that the result which serves as the basis for this generalization is due to Batanin [Bat98] (a simplified proof is given in [MS04a]). Namely, he proves that, if $X^\bullet$ has a certain cup product (see discussion prior to Theorem 4.3 in [MS04b] or [MS02, Definition 2.1]), then $\mathrm{Tot}\, X^\bullet$ is an

$A_\infty$ space (it has an action of an $A_\infty$ operad, i.e. a non-symmetric operad whose each space is weakly equivalent to a point) and is therefore equivalent to a loop space. Batanin's proof uses the box product described in Example 9.2.10. McClure and Smith conjecture that totalization in fact gives a Quillen equivalence between cosimplicial spaces with this cup product and $A_\infty$ spaces. □

**Example A.4.14** Suppose $X$ is a topological monoid, so that there exists a continuous multiplication $X \times X \to X$, and the basepoint acts as the unit. Let $O(k) = X^k$ and define

$$\circ_i \colon O(k) \times O(j) \longrightarrow O(k + j - 1)$$

$$(x_1, \ldots, x_k) \circ_i (y_1, \ldots, y_j) \longmapsto (x_1, \ldots, x_{i-1}, x_i y_1, x_i y_2, \ldots, x_i y_j, x_{i+1}, \ldots, x_n).$$

If morphisms $e$ and $\mu$ are simply the basepoints, then $O^\bullet$ is precisely the cobar construction from Example 9.2.3. □

**Example A.4.15** The above can be generalized to produce a cosimplicial space whose totalization is the $k$-fold loop space of a based space $X$ for $k \geq 2$. Again $O(k) = X^k$ but the $\circ_i$ operations are more complicated and arise from the maps of the simplicial $k$-sphere to $X$. For details, see [MS02, Example 3.6]. □

Another example that is relevant in the study of the topology of spaces of knots is that of *compactified configuration spaces*; see Section 10.4.2.

# References

[AC11] Gregory Arone and Michael Ching. Operads and chain rules for the calculus of functors. *Astérisque*, (338):vi+158, 2011.

[AC14] Gregory Arone and Michael Ching. Cross-effects and the classification of Taylor towers. arXiv:1404.1417, 2014.

[AC15] Gregory Arone and Michael Ching. A classification of Taylor towers of functors of spaces and spectra. *Adv. Math.*, 272:471–552, 2015.

[AD01] G. Z. Arone and W. G. Dwyer. Partition complexes, Tits buildings and symmetric products. *Proc. London Math. Soc. (3)*, 82(1):229–256, 2001.

[Ada78] John Frank Adams. *Infinite loop spaces*. Annals of Mathematics Studies, Vol. 90. Princeton University Press, Princeton, NJ, 1978.

[AGP02] Marcelo Aguilar, Samuel Gitler, and Carlos Prieto. *Algebraic topology from a homotopical viewpoint*. Universitext. Springer-Verlag, New York, 2002. Translated from the Spanish by Stephen Bruce Sontz.

[AK02] Stephen T. Ahearn and Nicholas J. Kuhn. Product and other fine structure in polynomial resolutions of mapping spaces. *Algebr. Geom. Topol.*, 2:591–647 (electronic), 2002.

[ALTV08] Greg Arone, Pascal Lambrechts, Victor Turchin, and Ismar Volić. Coformality and rational homotopy groups of spaces of long knots. *Math. Res. Lett.*, 15(1):1–14, 2008.

[ALV07] Gregory Arone, Pascal Lambrechts, and Ismar Volić. Calculus of functors, operad formality, and rational homology of embedding spaces. *Acta Math.*, 199(2):153–198, 2007.

[AM99] Greg Arone and Mark Mahowald. The Goodwillie tower of the identity functor and the unstable periodic homotopy of spheres. *Invent. Math.*, 135(3):743–788, 1999.

[And72] D. W. Anderson. A generalization of the Eilenberg–Moore spectral sequence. *Bull. Amer. Math. Soc.*, 78:784–786, 1972.

[AP99] M. A. Aguilar and Carlos Prieto. Quasifibrations and Bott periodicity. *Topology Appl.*, 98(1–3):3–17, 1999. (II Iberoamerican Conference on Topology and its Applications (Morelia, 1997)).

[Ara53] Shôrô Araki. On the triad excision theorem of Blakers and Massey. *Nagoya Math. J.*, 6:129–136, 1953.

[Ark62] Martin Arkowitz. The generalized Whitehead product. *Pacific J. Math.*, 12(1):7–23, 1962.

[Ark11] Martin Arkowitz. *Introduction to homotopy theory*. Universitext. Springer, New York, 2011.

[Arn69] V. I. Arnol'd. The cohomology ring of the group of dyed braids. *Mat. Zametki*, 5:227–231, 1969.

[Aro99] Gregory Arone. A generalization of Snaith-type filtration. *Trans. Amer. Math. Soc.*, 351(3):1123–1150, 1999.

[Aro02] Gregory Arone. The Weiss derivatives of $BO(-)$ and $BU(-)$. *Topology*, 41(3):451–481, 2002.

[AS94] Scott Axelrod and I. M. Singer. Chern-Simons perturbation theory. II. *J. Differential Geom.*, 39(1):173–213, 1994.

[AT14] Gregory Arone and Victor Turchin. On the rational homology of high-dimensional analogues of spaces of long knots. *Geom. Topol.*, 18(3):1261–1322, 2014.

[Bat98] Mikhail A. Batanin. Homotopy coherent category theory and $A_\infty$-structures in monoidal categories. *J. Pure Appl. Algebra*, 123(1–3):67–103, 1998.

[BB10] Hans Joachim Baues and David Blanc. Stems and spectral sequences. *Algebr. Geom. Topol.*, 10(4):2061–2078, 2010.

[BCKS14] Ryan Budney, James Conant, Robin Koytcheff, and Dev Sinha. Embedding calculus knot invariants are of finite type. arXiv:1411.1832, 2014.

[BCM78] M. Bendersky, E. B. Curtis, and H. R. Miller. The unstable Adams spectral sequence for generalized homology. *Topology*, 17(3):229–248, 1978.

[BCR07] Georg Biedermann, Boris Chorny, and Oliver Röndigs. Calculus of functors and model categories. *Adv. Math.*, 214(1):92–115, 2007.

[BCSS05] Ryan Budney, James Conant, Kevin P. Scannell, and Dev Sinha. New perspectives on self-linking. *Adv. Math.*, 191(1):78–113, 2005.

[BD03] Martin Bendersky and Donald M. Davis. A stable approach to an unstable homotopy spectral sequence. *Topology*, 42(6):1261–1287, 2003.

[Beh02] Mark J. Behrens. A new proof of the Bott periodicity theorem. *Topology Appl.*, 119(2):167–183, 2002.

[Beh12] Mark Behrens. The Goodwillie tower and the EHP sequence. *Mem. Amer. Math. Soc.*, 218(1026):xii+90, 2012.

[BJR15] Kristine Bauer, Brenda Johnson, and McCarthy Randy. Cross effects and calculus in an unbased setting (with an appendix by Rosona Eldred). *Trans. Amer. Math. Soc.*, 367(9):6671–6678, 2015.

[BK72a] A. K. Bousfield and D. M. Kan. *Homotopy limits, completions and localizations*. Lecture Notes in Mathematics, Vol. 304. Springer-Verlag, Berlin, 1972.

[BK72b] A. K. Bousfield and D. M. Kan. The homotopy spectral sequence of a space with coefficients in a ring. *Topology*, 11:79–106, 1972.

[BK73] A. K. Bousfield and D. M. Kan. A second quadrant homotopy spectral sequence. *Trans. Amer. Math. Soc.*, 177:305–318, 1973.

[BL84] Ronald Brown and Jean-Louis Loday. Excision homotopique en basse dimension. *C. R. Acad. Sci. Paris Sér. I Math.*, 298(15):353–356, 1984.

[BL87a] Ronald Brown and Jean-Louis Loday. Homotopical excision, and Hurewicz theorems for $n$-cubes of spaces. *Proc. London Math. Soc. (3)*, 54(1):176–192, 1987.

[BL87b] Ronald Brown and Jean-Louis Loday. Van Kampen theorems for diagrams of spaces. With an appendix by M. Zisman. *Topology*, 26(3):311–335, 1987.

[BM49] A. L. Blakers and William S. Massey. The homotopy groups of a triad. *Proc. Nat. Acad. Sci. USA*, 35:322–328, 1949.

[BM51] A. L. Blakers and W. S. Massey. The homotopy groups of a triad. I. *Ann. Math. (2)*, 53:161–205, 1951.

[BM52] A. L. Blakers and W. S. Massey. The homotopy groups of a triad. II. *Ann. Math. (2)*, 55:192–201, 1952.

[BM53a] A. L. Blakers and W. S. Massey. The homotopy groups of a triad. III. *Ann. Math. (2)*, 58:409–417, 1953.

[BM53b] A. L. Blakers and W. S. Massey. Products in homotopy theory. *Ann. Math. (2)*, 58:295–324, 1953.

[BN95] Dror Bar-Natan. On the Vassiliev knot invariants. *Topology*, 34(2):423–472, 1995.

[BO05] Marcel Bökstedt and Iver Ottosen. A spectral sequence for string cohomology. *Topology*, 44(6):1181–1212, 2005.

[Boa71] J. M. Boardman. Homotopy structures and the language of trees. In *Algebraic topology (Proc. Sympos. Pure Math., Vol. XXII, Univ. Wisconsin, Madison, Wis., 1970)*, pp. 37–58. American Mathematical Society, Providence, RI, 1971.

[Bor94] Francis Borceux. *Handbook of categorical algebra*, Vol. 1. Encyclopedia of Mathematics and its Applications, Vol. 50. Cambridge University Press, Cambridge, 1994.

[Bou87] A. K. Bousfield. On the homology spectral sequence of a cosimplicial space. *Amer. J. Math.*, 109(2):361–394, 1987.

[Bou89] A. K. Bousfield. Homotopy spectral sequences and obstructions. *Israel J. Math.*, 66(1–3):54–104, 1989.

[Bou03] A. K. Bousfield. Cosimplicial resolutions and homotopy spectral sequences in model categories. *Geom. Topol.*, 7:1001–1053 (electronic), 2003.

[Bre93] Glen E. Bredon. *Topology and geometry*. Graduate Texts in Mathematics, Vol. 139. Springer-Verlag, New York, 1993.

[Bro68] Ronald Brown. *Elements of modern topology*. McGraw-Hill, New York, 1968.

[BT82] Raoul Bott and Loring W. Tu. *Differential forms in algebraic topology*. Graduate Texts in Mathematics, Vol. 82. Springer-Verlag, New York, 1982.

[BT00] Martin Bendersky and Robert D. Thompson. The Bousfield–Kan spectral sequence for periodic homology theories. *Amer. J. Math.*, 122(3):599–635, 2000.

[Bud07] Ryan Budney. Little cubes and long knots. *Topology*, 46(1):1–27, 2007.

[Bud08] Ryan Budney. A family of embedding spaces. In *Groups, homotopy and configuration spaces*. Geometry and Topology Monographs, Vol. 13, pp. 41–83. Mathematical Science Publishers, Coventry, 2008.

[Bud10]  Ryan Budney. Topology of spaces of knots in dimension 3. *Proc. London Math. Soc. (3)*, 101(2):477–496, 2010.

[BV73]  J. M. Boardman and R. M. Vogt. *Homotopy invariant algebraic structures on topological spaces.* Lecture Notes in Mathematics, Vol. 347. Springer-Verlag, Berlin, 1973.

[BW56]  M. G. Barratt and J. H. C. Whitehead. The first nonvanishing group of an $(n + 1)$-ad. *Proc. London Math. Soc. (3)*, 6:417–439, 1956.

[CCGH87]  G. E. Carlsson, R. L. Cohen, T. Goodwillie, and W. C. Hsiang. The free loop space and the algebraic $K$-theory of spaces. *K-Theory*, 1(1):53–82, 1987.

[CCRL02]  Alberto S. Cattaneo, Paolo Cotta-Ramusino, and Riccardo Longoni. Configuration spaces and Vassiliev classes in any dimension. *Algebr. Geom. Topol.*, 2:949–1000 (electronic), 2002.

[CDI02]  W. Chachólski, W. G. Dwyer, and M. Intermont. The $A$-complexity of a space. *J. London Math. Soc. (2)*, 65(1):204–222, 2002.

[CDM12]  S. Chmutov, S. Duzhin, and J. Mostovoy. *Introduction to Vassiliev knot invariants.* Cambridge University Press, Cambridge, 2012.

[Cha97]  Wojciech Chachólski. A generalization of the triad theorem of Blakers–Massey. *Topology*, 36(6):1381–1400, 1997.

[Chi05]  Michael Ching. Bar constructions for topological operads and the Goodwillie derivatives of the identity. *Geom. Topol.*, 9:833–933 (electronic), 2005.

[Chi10]  Michael Ching. A chain rule for Goodwillie derivatives of functors from spectra to spectra. *Trans. Amer. Math. Soc.*, 362(1):399–426, 2010.

[Cho]  Boris Chorny. A classification of small linear functors. arXiv:1409.8525, 2014.

[CLM76]  Frederick R. Cohen, Thomas J. Lada, and J. Peter May. *The homology of iterated loop spaces.* Lecture Notes in Mathematics, Vol. 533. Springer-Verlag, Berlin, 1976.

[CLOT03]  Octav Cornea, Gregory Lupton, John Oprea, and Daniel Tanré. *Lusternik-Schnirelmann category*, Mathematical Surveys and Monographs, Vol. 103. American Mathematical Society, Providence, RI, 2003.

[Coh73]  Fred Cohen. Cohomology of braid spaces. *Bull. Amer. Math. Soc.*, 79:763–766, 1973.

[Coh95]  F. R. Cohen. On configuration spaces, their homology, and Lie algebras. *J. Pure Appl. Algebra*, 100(1–3):19–42, 1995.

[Con08]  James Conant. Homotopy approximations to the space of knots, Feynman diagrams, and a conjecture of Scannell and Sinha. *Amer. J. Math.*, 130(2):341–357, 2008.

[CS02]  Wojciech Chachólski and Jérôme Scherer. Homotopy theory of diagrams. *Mem. Amer. Math. Soc.*, 155(736):x+90, 2002.

[DD77]  E. Dror and W. G. Dwyer. A long homology localization tower. *Comment. Math. Helv.*, 52(2):185–210, 1977.

[DF87]  E. Dror Farjoun. Homotopy theories for diagrams of spaces. *Proc. Amer. Math. Soc.*, 101(1):181–189, 1987.

[DFZ86]  E. Dror Farjoun and A. Zabrodsky. Homotopy equivalence between diagrams of spaces. *J. Pure Appl. Algebra*, 41(2–3):169–182, 1986.

[DGM13]  Bjørn Ian Dundas, Thomas G. Goodwillie, and Randy McCarthy. *The local structure of algebraic K-theory.* Algebra and Applications, Vol. 18. Springer-Verlag, London, 2013.

[DH12]  William Dwyer and Kathryn Hess. Long knots and maps between operads. *Geom. Topol.*, 16(2):919–955, 2012.

[DHKS04]  William G. Dwyer, Philip S. Hirschhorn, Daniel M. Kan, and Jeffrey H. Smith. *Homotopy limit functors on model categories and homotopical categories.* Mathematical Surveys and Monographs, Vol. 113. American Mathematical Society, Providence, RI, 2004.

[DI04]  Daniel Dugger and Daniel C. Isaksen. Topological hypercovers and $\mathbb{A}^1$-realizations. *Math. Z.*, 246(4):667–689, 2004.

[DK84]  W. G. Dwyer and D. M. Kan. A classification theorem for diagrams of simplicial sets. *Topology*, 23(2):139–155, 1984.

[DL59]  Albrecht Dold and Richard Lashof. Principal quasi-fibrations and fibre homotopy equivalence of bundles. *Illinois J. Math.*, 3:285–305, 1959.

[DMN89]  William Dwyer, Haynes Miller, and Joseph Neisendorfer. Fibrewise completion and unstable Adams spectral sequences. *Israel J. Math.*, 66(1–3):160–178, 1989.

[Dro72]  Emmanuel Dror. Acyclic spaces. *Topology*, 11:339–348, 1972.

[DS95]  W. G. Dwyer and J. Spaliński. Homotopy theories and model categories. In *Handbook of algebraic topology*, pp. 73–126. North-Holland, Amsterdam, 1995.

[DT56]  Albrecht Dold and René Thom. Une généralisation de la notion d'espace fibré. Application aux produits symétriques infinis. *C. R. Acad. Sci. Paris*, 242:1680–1682, 1956.

[DT58]  Albrecht Dold and René Thom. Quasifaserungen und unendliche symmetrische Produkte. *Ann. Math. (2)*, 67:239–281, 1958.

[Dug01]  Daniel Dugger. Universal homotopy theories. *Adv. Math.*, 164(1):144–176, 2001.

[Dwy74]  W. G. Dwyer. Strong convergence of the Eilenberg–Moore spectral sequence. *Topology*, 13:255–265, 1974.

[EH76]  David A. Edwards and Harold M. Hastings. *Čech and Steenrod homotopy theories with applications to geometric topology.* Lecture Notes in Mathematics, Vol. 542. Springer-Verlag, Berlin–New York, 1976.

[Eld08]  Rosona Eldred. Tot primer. Available at www.math.uni-hamburg.de/home/eldred/, 2008.

[Eld13]  Rosona Eldred. Cosimplicial models for the limit of the Goodwillie tower. *Algebr. Geom. Topol.*, 13(2):1161–1182, 2013.

[EML54]  Samuel Eilenberg and Saunders Mac Lane. On the groups $H(\Pi, n)$. II. Methods of computation. *Ann. Math. (2)*, 60:49–139, 1954.

[ES87]  Graham Ellis and Richard Steiner. Higher-dimensional crossed modules and the homotopy groups of $(n + 1)$-ads. *J. Pure Appl. Algebra*, 46(2-3):117–136, 1987.

[FM94]  William Fulton and Robert MacPherson. A compactification of configuration spaces. *Ann. Math. (2)*, 139(1):183–225, 1994.

[FN62]  Edward Fadell and Lee Neuwirth. Configuration spaces. *Math. Scand.*, 10:111–118, 1962.

[Fri12] Greg Friedman. Survey article: an elementary illustrated introduction to simplicial sets. *Rocky Mountain J. Math.*, 42(2):353–423, 2012.

[Gai03] Giovanni Gaiffi. Models for real subspace arrangements and stratified manifolds. *Int. Math. Res. Not.*, (12):627–656, 2003.

[Gan65] T. Ganea. A generalization of the homology and homotopy suspension. *Comment. Math. Helv.*, 39:295–322, 1965.

[Geo79] Ross Geoghegan. The inverse limit of homotopy equivalences between towers of fibrations is a homotopy equivalence – a simple proof. *Topology Proc.*, 4(1):99–101 (1980), 1979.

[GG73] M. Golubitsky and V. Guillemin. *Stable mappings and their singularities*. Graduate Texts in Mathematics, Vol. 14. Springer-Verlag, New York, 1973.

[GJ99] Paul G. Goerss and John F. Jardine. *Simplicial homotopy theory*. Progress in Mathematics, Vol. 174. Birkhäuser Verlag, Basel, 1999.

[GK08] Thomas G. Goodwillie and John R. Klein. Multiple disjunction for spaces of Poincaré embeddings. *J. Topol.*, 1(4):761–803, 2008.

[GK15] Thomas G. Goodwillie and John R. Klein. Multiple disjunction for spaces of smooth embeddings. arXiv:1407.6787, 2015.

[GKW01] Thomas G. Goodwillie, John R. Klein, and Michael S. Weiss. Spaces of smooth embeddings, disjunction and surgery. In *Surveys on surgery theory*, Vol. 2, ed. S. Cappell, A. Ranicki, and J. Rosenberg. Annals of Mathematics Studies, Vol. 149, pp. 221–284. Princeton University Press, Princeton, NJ, 2001.

[GM10] Thomas G. Goodwillie and Brian A. Munson. A stable range description of the space of link maps. *Algebr. Geom. Topol.*, 10:1305–1315, 2010.

[Goe90] Paul G. Goerss. André-Quillen cohomology and the Bousfield-Kan spectral sequence. *Astérisque*, (191):6, 109–209, 1990. International Conference on Homotopy Theory (Marseille-Luminy, 1988).

[Goe93] Paul G. Goerss. Barratt's desuspension spectral sequence and the Lie ring analyzer. *Quart. J. Math. Oxford Ser. (2)*, 44(173):43–85, 1993.

[Goe96] Paul G. Goerss. The homology of homotopy inverse limits. *J. Pure Appl. Algebra*, 111(1-3):83–122, 1996.

[Goo] Thomas G. Goodwillie. Excision estimates for spaces of diffeomorphisms. In preparation, available at http://www.math.brown.edu/~tomg/excisiondifftex.pdf

[Goo90] Thomas G. Goodwillie. Calculus. I. The first derivative of pseudoisotopy theory. *K-Theory*, 4(1):1–27, 1990.

[Goo92] Thomas G. Goodwillie. Calculus II: Analytic functors. *K-Theory*, 5(4):295–332, 1991/92.

[Goo98] Thomas G. Goodwillie. A remark on the homology of cosimplicial spaces. *J. Pure Appl. Algebra*, 127(2):167–175, 1998.

[Goo03] Thomas G. Goodwillie. Calculus. III. Taylor series. *Geom. Topol.*, 7:645–711 (electronic), 2003.

[Gra75] Brayton Gray. *Homotopy theory: An introduction to algebraic topology*, Pure and Applied Mathematics, Vol. 64. Academic Press [Harcourt Brace Jovanovich Publishers], New York, 1975.

[GV95] Murray Gerstenhaber and Alexander A. Voronov. Homotopy $G$-algebras and moduli space operad. *Internat. Math. Res. Notices*, (3):141–153 (electronic), 1995.

[GW99] Thomas G. Goodwillie and Michael Weiss. Embeddings from the point of view of immersion theory II. *Geom. Topol.*, 3:103–118 (electronic), 1999.

[Hac10] Philip J. Hackney. Homology operations in the spectral sequence of a cosimplicial space. ProQuest LLC, Ann Arbor, MI, 2010. PhD thesis, Purdue University.

[Har70] K. A. Hardie. Quasifibration and adjunction. *Pacific J. Math.*, 35:389–397, 1970.

[Hat02] Allen Hatcher. *Algebraic topology*. Cambridge University Press, Cambridge, 2002.

[Hir15a] Philip S. Hirschhorn. Functorial CW-approximation. Available at www-math.mit.edu/~psh/notes/cwapproximation.pdf, 2015.

[Hir15b] Philip S. Hirschhorn. The homotopy groups of the inverse limit of a tower of fibrations. Available at www-math.mit.edu/~psh/notes/limfibra tions.pdf, 2015.

[Hir15c] Philip S. Hirschhorn. Notes on homotopy colimits and homotopy limits. Available at http://www-math.mit.edu/~psh/notes/hocolim.pdf, 2015.

[Hir15d] Philip S. Hirschhorn. The Quillen model category of topological spaces. Available at www-math.mit.edu/~psh/notes/modcattop.pdf, 2015.

[Hir15e] Philip S. Hirschhorn. The diagonal of a multi-cosimplicial object. Available at www-math.mit.edu/ psh/diagncosimplicial.pdf, 2015.

[Hir94] Morris W. Hirsch. *Differential topology*. Graduate Texts in Mathematics, Vol. 33. Springer-Verlag, New York, 1994. Corrected reprint of the 1976 original.

[Hir03] Philip S. Hirschhorn. *Model categories and their localizations*. Mathematical Surveys and Monographs, Vol. 99. American Mathematical Society, Providence, RI, 2003.

[Hop] Michael J. Hopkins. Some problems in topology. PhD thesis, Oxford University, 1984.

[Hov99] Mark Hovey. *Model categories*. Mathematical Surveys and Monographs, Vol. 63. American Mathematical Society, Providence, RI, 1999.

[JM99] Brenda Johnson and Randy McCarthy. Taylor towers for functors of additive categories. *J. Pure Appl. Algebra*, 137(3):253–284, 1999.

[JM04] B. Johnson and R. McCarthy. Deriving calculus with cotriples. *Trans. Amer. Math. Soc.*, 356(2):757–803 (electronic), 2004.

[JM08] Brenda Johnson and Randy McCarthy. Taylor towers of symmetric and exterior powers. *Fund. Math.*, 201(3):197–216, 2008.

[Joh95] B. Johnson. The derivatives of homotopy theory. *Trans. Amer. Math. Soc.*, 347(4):1295–1321, 1995.

[KM02] Miriam Ruth Kantorovitz and Randy McCarthy. The Taylor towers for rational algebraic $K$-theory and Hochschild homology. *Homology Homotopy Appl.*, 4(1):191–212, 2002.

[Koh02] Toshitake Kohno. Loop spaces of configuration spaces and finite type invariants. In *Invariants of knots and 3-manifolds (Kyoto, 2001)*. Geometry and Topology Monographs, Vol. 4, pp. 143–160. Mathematical Science Publishers, Coventry, 2002.

[Kon93]  Maxim Kontsevich. Vassiliev's knot invariants. In *I. M. Gel'fand Seminar*. Advances in Soviet Mathematics, Vol. 16, pp. 137–150. American Mathematical Society, Providence, RI, 1993.

[Kon99]  Maxim Kontsevich. Operads and motives in deformation quantization. *Lett. Math. Phys.*, 48(1):35–72, 1999.

[Kos93]  Antoni A. Kosinski. *Differential manifolds*. Pure and Applied Mathematics, Vol. 138. Academic Press, Boston, MA, 1993.

[Kos97]  Ulrich Koschorke. A generalization of Milnor's $\mu$-invariants to higher-dimensional link maps. *Topology*, 36(2):301–324, 1997.

[KR02]  John R. Klein and John Rognes. A chain rule in the calculus of homotopy functors. *Geom. Topol.*, 6:853–887 (electronic), 2002.

[KS00]  Maxim Kontsevich and Yan Soibelman. Deformations of algebras over operads and the Deligne conjecture. In *Conférence Moshé Flato 1999*, Vol. I (Dijon), Mathematical Physics Studies, Vol. 21, pp. 255–307. Kluwer Academic, Dordrecht, 2000.

[KSV97]  J. Klein, R. Schwänzl, and R. M. Vogt. Comultiplication and suspension. *Topology Appl.*, 77(1):1–18, 1997.

[KT06]  Akira Kono and Dai Tamaki. *Generalized cohomology*. Translations of Mathematical Monographs, Vol. 230. American Mathematical Society, Providence, RI, 2006. Translated from the 2002 Japanese edition by Tamaki, Iwanami Series in Modern Mathematics.

[Kuh04]  Nicholas J. Kuhn. Tate cohomology and periodic localization of polynomial functors. *Invent. Math.*, 157(2):345–370, 2004.

[Kuh07]  Nicholas J. Kuhn. Goodwillie towers and chromatic homotopy: an overview. In *Proceedings of the Nishida Fest (Kinosaki 2003)*. Geometry and Topology Monographs, Vol. 10, pp. 245–279. Mathematical Science Publishers, Coventry, 2007.

[KW12]  John Klein and Bruce Williams. Homotopical intersection theory, iii: Multirelative intersection problems. arXiv:1212.4420, 2012.

[Lee03]  John M. Lee. *Introduction to smooth manifolds*. Graduate Texts in Mathematics, Vol. 218. Springer-Verlag, New York, 2003.

[LL06]  José La Luz. The Bousfield–Kan spectral sequence for Morava $K$-theory. *Proc. Edinb. Math. Soc. (2)*, 49(3):683–699, 2006.

[LM12]  Ayelet Lindenstrauss and Randy McCarthy. On the Taylor tower of relative $K$-theory. *Geom. Topol.*, 16(2):685–750, 2012.

[LRV03]  Wolfgang Lück, Holger Reich, and Marco Varisco. Commuting homotopy limits and smash products. *K-Theory*, 30(2):137–165, 2003. Special issue in honor of Hyman Bass on his seventieth birthday. Part II.

[LT09]  Pascal Lambrechts and Victor Turchin. Homotopy graph-complex for configuration and knot spaces. *Trans. Amer. Math. Soc.*, 361(1):207–222, 2009.

[LTV10]  Pascal Lambrechts, Victor Turchin, and Ismar Volić. The rational homology of spaces of long knots in codimension $> 2$. *Geom. Topol.*, 14:2151–2187, 2010.

[Lur09] Jacob Lurie. *Higher topos theory*. Annals of Mathematics Studies, Vol. 170. Princeton University Press, Princeton, NJ, 2009.

[LV14] Pascal Lambrechts and Ismar Volić. Formality of the little $N$-disks operad. *Mem. Amer. Math. Soc.*, 230(1079):viii+116, 2014.

[Mac07] Tibor Macko. The block structure spaces of real projective spaces and orthogonal calculus of functors. *Trans. Amer. Math. Soc.*, 359(1):349–383 (electronic), 2007.

[Mat73] Michael Mather. Hurewicz theorems for pairs and squares. *Math. Scand.*, 32:269–272 (1974), 1973.

[Mat76] Michael Mather. Pull-backs in homotopy theory. *Canad. J. Math.*, 28(2):225–263, 1976.

[May72] J. P. May. *The geometry of iterated loop spaces*. Lecture Notes in Mathematics, Vol. 271. Springer-Verlag, Berlin, 1972.

[May90] J. P. May. Weak equivalences and quasifibrations. In *Groups of self-equivalences and related topics (Montreal, PQ, 1988)*. Lecture Notes in Mathematics, Vol. 1425, pp. 91–101. Springer, Berlin, 1990.

[May92] J. Peter May. *Simplicial objects in algebraic topology*. Chicago Lectures in Mathematics. University of Chicago Press, Chicago, IL, 1992.

[May99] J. P. May. *A concise course in algebraic topology*. Chicago Lectures in Mathematics. University of Chicago Press, Chicago, IL, 1999.

[McC01a] Randy McCarthy. Dual calculus for functors to spectra. In *Homotopy methods in algebraic topology (Boulder, CO, 1999)*. Contemporary Mathematics, Vol. 271, pp. 183–215. American Mathematical Society, Providence, RI, 2001.

[McC01b] John McCleary. *A user's guide to spectral sequences*, Cambridge Studies in Advanced Mathematics, Vol. 58. Cambridge University Press, Cambridge, second edition, 2001.

[Mey90] Jean-Pierre Meyer. Cosimplicial homotopies. *Proc. Amer. Math. Soc.*, 108(1):9–17, 1990.

[Mil63] J. Milnor. *Morse theory*. Based on lecture notes by M. Spivak and R. Wells. Annals of Mathematics Studies, Vol. 51. Princeton University Press, Princeton, NJ, 1963.

[Mil84] Haynes Miller. The Sullivan conjecture on maps from classifying spaces. *Ann. Math. (2)*, 120(1):39–87, 1984.

[ML98] Saunders Mac Lane. *Categories for the working mathematician*. Graduate Texts in Mathematics, Vol. 5. Springer-Verlag, New York, second edition, 1998.

[MO06] Andrew Mauer-Oats. Algebraic Goodwillie calculus and a cotriple model for the remainder. *Trans. Amer. Math. Soc.*, 358(5):1869–1895 (electronic), 2006.

[Moo53] John C. Moore. Some applications of homology theory to homotopy problems. *Ann. Math. (2)*, 58:325–350, 1953.

[Mor12] Syunji Moriya. Sinha's spectral sequence and homotopical algebra of operads. arXiv:1210.0996, 2012.

[MS02] James E. McClure and Jeffrey H. Smith. A solution of Deligne's Hochschild cohomology conjecture. In *Recent progress in homotopy theory (Baltimore, MD, 2000)*. Contemporary Mathematics, Vol. 293, pp. 153–193. American Mathematical Society, Providence, RI, 2002.

[MS04a] James E. McClure and Jeffrey H. Smith. Cosimplicial objects and little *n*-cubes. I. *Amer. J. Math.*, 126(5):1109–1153, 2004.

[MS04b] James E. McClure and Jeffrey H. Smith. Operads and cosimplicial objects: an introduction. In *Axiomatic, enriched and motivic homotopy theory*. NATO Science Series II Mathematics, Physics and Chemistry, Vol. 131, pp. 133–171. Kluwer Academic, Dordrecht, 2004.

[MSS02] Martin Markl, Steve Shnider, and Jim Stasheff. *Operads in algebra, topology and physics*. Mathematical Surveys and Monographs, Vol. 96. American Mathematical Society, Providence, RI, 2002.

[MT68] Robert E. Mosher and Martin C. Tangora. *Cohomology operations and applications in homotopy theory*. Harper & Row, New York–London, 1968.

[Mun75] James R. Munkres. *Topology: a first course*. Prentice-Hall, Englewood Cliffs, NJ, 1975.

[Mun08] Brian A. Munson. A manifold calculus approach to link maps and the linking number. *Algebr. Geom. Topol.*, 8(4):2323–2353, 2008.

[Mun10] Brian A. Munson. Introduction to the manifold calculus of Goodwillie–Weiss. *Morfismos*, 14(1):1–50, 2010.

[Mun11] Brian A. Munson. Derivatives of the identity and generalizations of Milnor's invariants. *J. Topol.*, 4(2):383–405, 2011.

[Mun14] Brian A. Munson. A purely homotopy-theoretic proof of the Blakers–Massey theorem for *n*-cubes. *Homology Homotopy Appl.*, 16(1):333–339, 2014.

[MV14] Brian A. Munson and Ismar Volić. Cosimplicial models for spaces of links. *J. Homotopy Relat. Struct.*, 9(2):419–454, 2014.

[MW80] Michael Mather and Marshall Walker. Commuting homotopy limits and colimits. *Math. Z.*, 175(1):77–80, 1980.

[MW09] Tibor Macko and Michael Weiss. The block structure spaces of real projective spaces and orthogonal calculus of functors. II. *Forum Math.*, 21(6):1091–1136, 2009.

[Nam62] I. Namioka. Maps of pairs in homotopy theory. *Proc. London Math. Soc. (3)*, 12:725–738, 1962.

[Nei10] Joseph Neisendorfer. *Algebraic methods in unstable homotopy theory*. New Mathematical Monographs, Vol. 12. Cambridge University Press, Cambridge, 2010.

[NM78] Joseph Neisendorfer and Timothy Miller. Formal and coformal spaces. *Illinois J. Math.*, 22(4):565–580, 1978.

[Ogl13] Crichton Ogle. On the homotopy type of $A(\Sigma X)$. *J. Pure Appl. Algebra*, 217(11):2088–2107, 2013.

[Pel11] Kristine E. Pelatt. A geometric homology representative in the space of long knots. arXiv:1111.3910, 2011.

[Pod11] Semën S. Podkorytov. On the homology of mapping spaces. *Cent. Eur. J. Math.*, 9(6):1232–1241, 2011.

[Pup58] Dieter Puppe. Homotopiemengen und ihre induzierten Abbildungen. I. *Math. Z.*, 69:299–344, 1958.

[Qui73] Daniel Quillen. Higher algebraic *K*-theory. I. In *Algebraic K-theory, I: Higher K-theories (Proc. Conf., Battelle Memorial Inst., Seattle, Wash.,*

*1972)*. Lecture Notes in Mathematics, Vol. 341, pp. 85–147. Springer, Berlin, 1973.

[Rec70] David L. Rector. Steenrod operations in the Eilenberg–Moore spectral sequence. *Comment. Math. Helv.*, 45:540–552, 1970.

[Rez13] Charles Rezk. A streamlined proof of Goodwillie's *n*-excisive approximation. *Algebr. Geom. Topol.*, 13(2):1049–1051, 2013.

[Rog] John Rognes. Lecture notes on algebraic *k*-theory. Available at http://folk.uio.no/rognes/kurs/mat9570v10/akt.pdf.

[Rom10] Ana Romero. Computing the first stages of the Bousfield–Kan spectral sequence. *Appl. Algebra Engrg. Comm. Comput.*, 21(3):227–248, 2010.

[Sak08] Keiichi Sakai. Poisson structures on the homology of the space of knots. In *Groups, homotopy and configuration spaces*, Geometry and Topology Monographs, Vol. 13, pp. 463–482. Mathematical Science Publishers, Coventry, 2008.

[Sak11] Keiichi Sakai. BV-structures on the homology of the framed long knot space. arXiv:1110.2358, 2011.

[Sal01] Paolo Salvatore. Configuration spaces with summable labels. In *Cohomological methods in homotopy theory (Bellaterra, 1998)*, Progress in Mathematics, Vol. 196, pp. 375–395. Birkhäuser, Basel, 2001.

[Sal06] Paolo Salvatore. Knots, operads, and double loop spaces. *Int. Math. Res. Not.*, pages Art. ID 13628, 22, 2006.

[Seg68] Graeme Segal. Classifying spaces and spectral sequences. *Inst. Hautes Études Sci. Publ. Math.*, (34):105–112, 1968.

[Sel97] Paul Selick. *Introduction to homotopy theory*. Fields Institute Monographs, Vol. 9. American Mathematical Society, Providence, RI, 1997.

[Shi96] Brooke E. Shipley. Convergence of the homology spectral sequence of a cosimplicial space. *Amer. J. Math.*, 118(1):179–207, 1996.

[Sin01] Dev P. Sinha. The geometry of the local cohomology filtration in equivariant bordism. *Homology Homotopy Appl.*, 3(2):385–406, 2001.

[Sin04] Dev P. Sinha. Manifold-theoretic compactifications of configuration spaces. *Selecta Math. (NS)*, 10(3):391–428, 2004.

[Sin06] Dev P. Sinha. Operads and knot spaces. *J. Amer. Math. Soc.*, 19(2):461–486 (electronic), 2006.

[Sin09] Dev P. Sinha. The topology of spaces of knots: cosimplicial models. *Amer. J. Math.*, 131(4):945–980, 2009.

[Sma59] Stephen Smale. The classification of immersions of spheres in Euclidean spaces. *Ann. Math. (2)*, 69:327–344, 1959.

[Smi02] V. A. Smirnov. $A_\infty$-structures and differentials of the Adams spectral sequence. *Izv. Ross. Akad. Nauk Ser. Mat.*, 66(5):193–224, 2002.

[Spa67] E. Spanier. The homotopy excision theorem. *Michigan Math. J.*, 14:245–255, 1967.

[Spe71] C. Spencer. The Hilton–Milnor theorem and Whitehead products. *J. London Math. Soc.*, 2(4):291–303, 1971.

[SS02] Kevin P. Scannell and Dev P. Sinha. A one-dimensional embedding complex. *J. Pure Appl. Algebra*, 170(1):93–107, 2002.

[ST14] Paul Arnaud Songhafouo-Tsopméné. Formality of Sinha's cosimplicial model for long knots spaces. arXiv:1210.2561, 2014.

[Sta63] James Dillon Stasheff. Homotopy associativity of *H*-spaces. I, II. *Trans. Amer. Math. Soc.* 108, 275–292; 108:293–312, 1963.

[Sta68] James D. Stasheff. Associated fibre spaces. *Michigan Math. J.*, 15:457–470, 1968.

[Str66] Arne Strøm. Note on cofibrations. *Math. Scand.*, 19:11–14, 1966.

[Str68] Arne Strøm. Note on cofibrations. II. *Math. Scand.*, 22:130–142 (1969), 1968.

[Tam03] Dmitry E. Tamarkin. Formality of chain operad of little discs. *Lett. Math. Phys.*, 66(1–2):65–72, 2003.

[tD08] Tammo tom Dieck. *Algebraic topology*. EMS Textbooks in Mathematics. European Mathematical Society (EMS), Zürich, 2008.

[tDKP70] Tammo tom Dieck, Klaus Heiner Kamps, and Dieter Puppe. *Homotopietheorie*. Lecture Notes in Mathematics, Vol. 157. Springer-Verlag, Berlin, 1970.

[Tho79] R. W. Thomason. Homotopy colimits in the category of small categories. *Math. Proc. Cambridge Phil. Soc.*, 85(1):91–109, 1979.

[TT04] Victor Turchin (Tourtchine). On the homology of the spaces of long knots. In *Advances in topological quantum field theory*. NATO Science Series II Mathematics, Physics and Chemistry, Vol. 179, pp. 23–52. Kluwer Academic, Dordrecht, 2004.

[TT06] Victor Turchin (Tourtchine). What is one-term relation for higher homology of long knots. *Mosc. Math. J.*, 6(1):169–194, 223, 2006.

[TT07] Victor Turchin (Tourtchine). On the other side of the bialgebra of chord diagrams. *J. Knot Theory Ramifications*, 16(5):575–629, 2007.

[TT10] Victor Turchin (Tourtchine). Hodge-type decomposition in the homology of long knots. *J. Topol.*, 3(3):487–534, 2010.

[TT14] Victor Turchin (Tourtchine). Delooping totalization of a multiplicative operad. *J. Homotopy Relat. Struct.*, 9(2):349–418, 2014.

[Tur98] James M. Turner. Operations and spectral sequences. I. *Trans. Amer. Math. Soc.*, 350(9):3815–3835, 1998.

[Vas90] V. A. Vassiliev. Cohomology of knot spaces. In *Theory of singularities and its applications*. Advances in Soviet Mathematics, Vol. 1, pp. 23–69. Amer. Math. Soc., Providence, RI, 1990.

[Vas99] V. A. Vassiliev. Homology of *i*-connected graphs and invariants of knots, plane arrangements, etc. In *The Arnoldfest (Toronto, ON, 1997)*. Fields Institute Communications, Vol. 24, pp. 451–469. American Mathematical Society, Providence, RI, 1999.

[Vog73] Rainer M. Vogt. Homotopy limits and colimits. *Math. Z.*, 134:11–52, 1973.

[Vog77] Rainer M. Vogt. Commuting homotopy limits. *Math. Z.*, 153(1):59–82, 1977.

[Vol06a] Ismar Volić. Configuration space integrals and Taylor towers for spaces of knots. *Topology Appl.*, 153(15):2893–2904, 2006.

[Vol06b] Ismar Volić. Finite type knot invariants and the calculus of functors. *Compos. Math.*, 142(1):222–250, 2006.

[Wei94]  Charles A. Weibel. *An introduction to homological algebra*. Cambridge Studies in Advanced Mathematics, Vol. 38. Cambridge University Press, Cambridge, 1994.

[Wei95]  Michael Weiss. Orthogonal calculus. *Trans. Amer. Math. Soc.*, 347(10):3743–3796, 1995.

[Wei99]  Michael Weiss. Embeddings from the point of view of immersion theory I. *Geom. Topol.*, 3:67–101 (electronic), 1999.

[Wei04]  Michael S. Weiss. Homology of spaces of smooth embeddings. *Q. J. Math.*, 55(4):499–504, 2004.

[Whi78]  George W. Whitehead. *Elements of homotopy theory*. Graduate Texts in Mathematics, Vol. 61. Springer-Verlag, New York, 1978.

[Wit95]  P. J. Witbooi. Adjunction of $n$-equivalences and triad connectivity. *Publ. Mat.*, 39(2):367–377, 1995.

# Index